Kuhlenbeck: The Central Nervous System of Vertebrates

Hartwig Kuhlenbeck

# The Central Nervous System
# of Vertebrates

**A General Survey of its Comparative Anatomy
with an Introduction to the Pertinent
Fundamental Biologic and Logical Concepts**

S. Karger · Basel · München · Paris · London · New York · Sydney

Volume 3, Part II

# Overall Morphologic Pattern

With 472 figures

19 73

S. Karger · Basel · München · Paris · London · New York · Sydney

S. Karger · Basel · München · Paris · London · New York · Sydney
Arnold-Böcklin-Strasse 25, CH-4011 Basel (Switzerland)

«Die vergleichende Anatomie, Histologie, Architek-
tonik und Embryologie des Zentralnervensystems bildet
ferner einen umfangreichen Zweig und zugleich eine unent-
behrliche Methode der neurobiologischen Forschung. Sie
verrät die zahlreichen Wege, durch welche die Evolution
der Nervensysteme der verschiedenen Tiersorten im phylo-
genetischen Zusammenhang ihre heutige Verschiedenartig-
keit zustande gebracht hat. Vertieft man sich dabei genügend
in den Zusammenhang von Form und Funktion, so gelangt
man in eine wunderbare Welt der Harmonie zwischen Geist
und lebendem Nervensystem...

Wer vergleichende Anatomie des Nervensystems sagt,
sagt also auch vergleichende Physiologie – Psychologie
und – Biologie, und das ist ein Gebiet, aus welchem die
künftige Forschung mit vollen Zügen schöpfen kann...

Dass der Mensch für den Menschen sich zunächst inter-
essiert, ist verzeihlich und naheliegend. Hat er aber einmal
erkannt, dass er nur ein Glied in der Tierreihe bildet und
dass sein Hirn, das Organ seiner Seele, aus dem Tiergehirn
und somit aus der Tierseele stammt, so muss er doch zur
Erkenntnis gelangen, dass das Studium der Neurobiologie
dieser seiner Verwandten das grösste Licht auf sein eigenes
Nerven- und Seelenleben werfen muss.»

AUGUST FOREL
(«*Die Aufgaben der Neurobiologie*»)

# Preface

The lectures on the central nervous system of vertebrates, given by the author during his first sojourn in Japan, 1924–1927 (TAISHÔ 13 to SHÔWA 2), intended to foster the interest in comparative neurologic studies based upon the morphologic principles established by the *Gegenbaur* or *Jena-Heidelberg School of Comparative Anatomy*. Notwithstanding their introductory and elementary nature, these lectures, published by Gustav Fischer, Jena, in 1927, included a number of advanced as well as independent concepts, and represented, as it were, the outline of a further program.

Despite various vicissitudes, and although I found the prevailing intellectual climate in the realm of biologic sciences rather unfavorable to the pursuit of investigations related to the domain of classical morphology, I have, *tant bien que mal*, carried on with my studies as originally planned, and propose to summarize my viewpoints in the present series, designed to represent a general survey, and projected to comprise five separate volumes, of which the first two are now completed. It can easily be seen that the present series follows closely the outline of my old 'Vorlesungen', meant to stress 'die grossen Hauptlinien der Hirnarchitektur und die allgemeinen Gesetzmässigkeiten, welche in Bau und Funktion des Nervensystems erkannt werden können'.

Comparative anatomy of the vertebrate central nervous system requires a very broad and comprehensive background of biological data, evaluated by means of a rational, consistent, and appropriate logical procedure. Without the relevant unifying concepts, comparative neurology becomes no more than a trivial description of apparently unrelated miscellaneous and bewildering configurational varieties, loosely held together by a string of hazy 'functional' notions. A perusal of the multitudinous literature dealing with matters involving the morphologic aspects of neurobiology reveals, to the critical observer, considerable confusion as regards many fundamental questions.

For this reason, the present attempt at an integrated overall presentation includes a somewhat detailed scrutiny of problems concerning the significance of configuration and configurational variety with respect to evolution and to correlated reasonably 'natural' taxonomic classifications. Because comparative anatomy of the central nervous system embodies the morphological clues required to infer the presumable phylogenetic evolution of the brain, a number of general questions referring to ontogenetic evolution are critically considered: it is evident that both the inferred phylogenetic sequences and the observable ontogenetic sequences represent evolutionary processes suitable for a comparison outlining the obtaining invariants.

Moreover, the comparison of organic forms involves procedures closely related to *analysis situs*. Thus, a simplified and elementary discussion of the here relevant principles of *topology* was deemed necessary.

Finally, since vertebrate comparative anatomy and vertebrate evolution, including the origin of vertebrates, cannot be properly assessed in default of an at least moderately adequate familiarity with the vast array of invertebrate organic forms, a general and elementary survey of invertebrate comparative neurology from the vertebrate neurobiologist's viewpoint, that is as seen by an 'outsider' with a modicum of first-hand acquaintance, has been included as volume two of this series. The approximately 20 pages and 12 figures dealing with this matter in my 1927 'Vorlesungen' have thus, of necessity, become rather expanded.

US N.I.H. Grant NB 4999, which is acknowledged with due appreciation, made possible the completion of Volumes 1 and 2 of this series, and, for the time being, the continuation of these studies, by supporting a 'Research Professorship' established to that effect, following my superannuation, at the Woman's Medical College of Pennsylvania.

Concluding this preamble to the present series, I may state with CICERO (*De oratore*, III, 61, 228): '*Edidi quae potui, non ut volui sed ut me temporis angustiae coegerunt; scitum est enim causam conferre in tempus, cum afferre plura si cupias non queas.*'

H.K.

# Foreword to Volume 3, Part II

The present volume, containing chapters VI and VII, completes the whole work's *General Subdivision (Allgemeiner Teil)* which is chiefly concerned with basic topics of comparative neurology in combination with an introduction to the here significant fundamental biological and logical concepts.

Chapter VI deals with essential aspects of general organic morphogenesis, with the relevant configurational features characterizing the morphogenesis of the Vertebrate neuraxis, and with the disturbances of morphogenesis resulting in teratologic conditions. This is followed by a detailed formanalytic account of the neural tube's regional ontogenetic differentiation, including an evaluation of neuromery and of homologies. Additional sections are concerned with the comparative anatomy of meninges and blood vessels, and with the significance of studies on brain weights.

Chapter VII, on various supplementary topics involving the interdependence of peripheral nervous system and central neuraxis, deals with peripheral nerve endings for input and output, spinal nerves and morphologic problems of neuromuscular relationship, cranial nerves and the head problem, with the concept of so-called functional nerve components, with electrical nerves and organs, with Vertebrate bioluminescence, and with essential general features of the vegetative nervous system.

The subsequent volumes of this series comprising its *Special* or *Systematic Subdivision (Spezieller oder Systematischer Teil)* and planned as a critical review of the 'adult' Vertebrate central nervous system's relevant regional features, are now in preparation. As in the other volumes, numerous personal and in part previously unpublished observations shall be included.

I am obliged to the *Woman's Medical College of Pennsylvania* (since 1970: *The Medical College of Pennsylvania*) for the extended support of my work after expiration, in 1969, of the initial Federal Grants N.I.H. N.B. 4999 1–3 and 4–6. I am also thankful for the hospitality extended to me by Professor R. SHUMAN, then Acting Chairman of the Department of Pathology, during my Research Professorship of Neuropathology 1970–1971. Following my official retirement to Emeritus Professor of Anatomy in 1971, the College continued to furnish the facilities for my *Laboratory of Morphologic Brain Research*. I am especially grateful to my many former students, who, through the *Alumnae Association* of the old *Woman's Medical College*, are now providing the necessary funds for that Laboratory.

In concluding the General Subdivision of this series, I also wish to express my thanks and appreciation to those who substantially helped me in the preparation of this work: to my expert Chief Technician and trusted Secretary in the former Department of Anatomy, the late Miss VERA MENOUGH, to my present efficient Secretary Mrs. DOLORES BRENNAN, and last, but indeed not least, to my wife, who throughout long years, gave me encouragement and moral support in my endeavors.

H.K.

# Table of Contents of the Present Volume

## Volume 3, Part II: Overall Morphologic Pattern

# Table of Contents of the Complete Work

## Volume 1  Propaedeutics to Comparative Neurology

# Volume 5 Derivatives of the Prosencephalon; The Neuraxis as a Whole

# VI. Morphologic Pattern of the Vertebrate Neuraxis

## 1. Morphogenesis

### A. Some General Aspects of Vertebrate Morphogenesis

The term *morphogenesis* can be defined as referring to the series of evolutionary changes whereby an adult living organism has attained its characteristic spatial configuration. Since all definitions as well as generalizations are encumbered by considerable intrinsic ambiguities and semantic shortcomings,[1] further comments on the meanings of morphogenesis in the present context might be appropriate.

The evolution of an individual organism evidently includes phylogenetic and ontogenetic events; in the aspect here under consideration, only the latter shall be emphasized. Again, because spatial configuration, in accordance with the viewpoint expounded in volume 1 of this series, subsumes *form* and *structure*, morphogenesis *sensu latiori* refers to the evolution of both. *Form* designates here the shape, outline, or general figure ('mode of construction') of larger, compound entities, such as whole individual organisms, organs, or arbitrarily delimited major subdivisions of organs. With respect to the neuraxis, form would thus subsume the shape of differentiating and differentiated neural tube, including outlines, configuration, and topological connectedness of its constituent parts above the cytologic and *sensu stricto* histologic level.

*Structure*, on the other hand, designates here the arrangement and configuration of smaller elements in the 'substance' or 'substratum' of the larger form elements. The terms structure or 'grain' of this substratum are thus employed as referring to the cellular components and to their constituent parts as well as to their arrangement in so-called architectural patterns. Structure therefore encompasses cytology, histol-

---

[1] Relevant 'Comments on Words, Language, Thought, and Definition', pertinent to the epistemologic outlook of the present series on the central nervous sytem of vertebrates, have been elaborated by the author in his contribution to the *Helen Adolf Festschrift* (Ungar, New York 1968). A concise summary of the epistemologic problems at issue was given in the paper '*Weitere Bemerkungen zur Maschinentheorie des Gehirns*' (K., 1966).

ogy *(texture)*, and detailed aspects of microscopic anatomy pertaining to cellular arrangements *within* subsets of form elements such as '*Grundbestandteile*' and '*Formbestandteile*' (cf. vol. 1, pp. 222, 284).

In the present chapter, the term *morphogenesis* will be used mostly in the narrower sense, namely as referring to the *evolution of form*, in contradistinction to the term *histogenesis;* this latter subsumes the differentiation of structure, namely of cellular elements *qua* tissue units (cytologic and histologic aspects), and of the detailed architectural patterns *within* grisea representing morphologic units such as the just-mentioned '*Formbestandteile*' (aspects of microscopic anatomy *sensu strictiori*). The histogenesis of nervous tissue was dealt with in the preceding volume of this series (vol. 3, part I, chapter V, particularly section 1). Comments on relevant histogenetic processes pertaining to 'microscopic anatomy', in which the domains of form and of structure overlap, shall be included, where required, in the discussion of morphogenesis as well as in the subsequent volumes (4, 5) concerned with the subdivisions of the adult vertebrate neuraxis.

Although both histogenesis and morphogenesis are events doubtless depending on the activity of 'genes' (cf. chapter II, section 7, vol. 1), both these evolutionary processes may justifiably be assessed as the outcome of genetic *and* nongenetic factors. Such presumption precludes the dogmatic concept characterized by what has been called an 'autonomy of genetic factors', i.e. of factors intrinsically pertaining to the genome. With the rise of so-called 'molecular biology'[2] there origi-

---

[2] The term '*molecular biology*', as claimed to have been originally defined by W.T. ASTBURY *(Harvey Lectures 1950–1951*, Thomas, Springfield 1952), referred primarily to the configuration of biological molecules and to their structural behavior at various levels of organization. Subsequently, molecular genetics and concepts of information theory became incorporated into a domain of biology which also involves 'biophysics' and, in the wider sense, may now be called macromolecular biophysics and biochemistry. It has been objected that, since biology is the study of life, and molecules, as such, are not alive, the term 'molecular biology' is self-contradictory. Now, although I disapprove the ludicrous 'scientific' attitude displayed by many 'molecular biologists' and believe that their extravagant shenanigans deserve a full measure of ridicule, I nevertheless consider their specialty to be an important field of biology. It seems obvious that the term '*living matter*' can be applied to complex macromolecular chemical systems with definable sorts of interactions and behavior. There is no valid reason why the study of molecular behavior displayed by such systems should not be called molecular or macromolecular biology. The above-mentioned objection, by a vocal and well publicized biologist of the *Haryalton* 'establishment', is evidently highly specious. One might then also claim that the meaningful terms 'biochemistry' and 'biophysics' should be discarded, since chemicals or mass and energy, 'as such, are not alive'.

nated a highly sophisticated but neoprimitive tendency to reduce all biologic, thus including histogenetic and morphogenetic, 'explanations' upon gene regulation and related mechanisms in terms of molecular biophysics. It is hereby implied that the relevant aspect of living systems is their structure provided by arbitrarily separable small units, particularly 'macromolecules'. This has led to the view that in order to understand life we only have to know these molecules, and 'the rest will take care of itself'.[3]

With regard to morphogenesis in the wider sense, that is, to ontogenetic differentiation of form *and* structure at all levels of complexity, the occurring events are presumed to result from progressive and selective activities of an originally so-called 'totipotent' genome. It is believed, in accordance with the hypothesis of JACOB and MONOD (1961), that this differential gene action becomes mediated by specific chemical 'messengers' which 'activate' and 'repress' the genes. The repressive control of protein synthesis is assumed to occur at the locus where genetic information is transferred from DNA to messenger RNA (mRNA), namely, at the 'level of transcription'. Although all 'genes' might be considered 'at rest' until severally 'activated' by some physicochemical 'factor', it is widely believed that the continuous activity of genes is inhibited by a repressor or by multiple repressors which, in turn, might be one or more special 'gene factors'. A particular gene would then become active when the repressor effect is removed. Accordingly, genes would not become 'activated', but 'derepressed' by an 'inducer' which disengages the repressor. A 'gene' could thus consist of three 'components' which, together, constitute the '*operon*' (cf. vol. 1, p. 130 of this series). Of these components, the 'promoter' would produce the 'repressor', and the 'operator', if not inhibited by the 'repressor', could initiate the 'action' of the gene by the third or main 'structural element'. This latter controls the production of the relevant cell proteins. Another locus for control activity, differing from that stressed by the *Jacob-Monod hypothesis*, has been postulated at the so-called 'translation level', where mRNA information is converted into the production of specific proteins.

At the time of this writing, numerous investigators in the field of 'molecular biology' are attempting to isolate single 'genes' of the

---

[3] SZENT-GYÖRGI, A. (Science 161: 988, 1968) in a paper stressing the significance of 'bioelectronics', i.e. of intermolecular electron transfer at the 'submolecular' level.

above-mentioned complex type, particularly in bacteria, whose procaryote organization (cf. below, p. 17) and interactions with bacteriophages seem to provide suitable conditions for the required experimental manipulations. Preliminary results of such studies appear to indicate that such complex 'genes', although only about 20Å wide, may have a length in the order of magnitude of 10 000 Å (one μ) or somewhat more. Should or should not a molecular array of this size, or a still larger one such as an entire *'chromosome'* be considered a single *'molecule'* or a 'set' of 'molecules'? This, of course, depends on definitions which can be formulated with respect to the concept of chemical bonding. In regard to the somewhat ambiguous definition of a 'gene' one might furthermore ask whether a rather long DNA strand interpreted as a 'single' gene actually 'determines' only a 'single' somatic characteristic, or rather represents a structure with multiplex functional effects, of which only one, being easily recognizable and definable, is stressed by the designation of such linear array as a specific 'gene'. At present, it is widely assumed, in accordance with views propounded by BEADLE, that *one 'gene' corresponds to one enzyme.* The sequence of the amino acids in these particular proteins can be related to the sequence of nucleotide triplets in a given chromosomal segment of DNA (so-called principle of *colinearity* ), the DNA sequence then becoming *'transcribed'* into a complementary RNA sequence. This latter, in turn, becomes *'translated'* into the amino acid sequence.

The DNA in a mammalian chromosome appears to be sufficient for the make-up of a double helical molecule with a length of several cm, and the total DNA in a mammalian cell nucleus, if arranged as a single thread, might reach a linear dimension of about 2 m. In some bacteria, the DNA seems to occur as a single, approximately 1 200 μ long double helical molecule, whose ends are joined to form a loop. Again, it has been assumed that the actual *size* of some vertebrate chromosomes is consistent with their make-up by a single double helical DNA molecule. However, the available observations could also be interpreted to suggest that vertebrate chromosomes may contain several 'separate' DNA 'molecules'. The number of 'genes' in man, estimated some years ago at about 50 000 or less (cf. e.g. ASHBY, 1952) is now assessed at ' a few million'.

The significant events in *neurogenesis* involve the grouping of neuronal elements into specific griseal aggregates which become more or less selectively interconnected by fiber systems. The establishment of these connections suggests specific chemical affinities between the neuronal elements, respectively cell populations, which pertain to the interrelated grisea. The inference that these main connections of the nervous system are developed under genetic control appears fully justified. Yet, even the larger number of 'genes' now assumed seems insufficient to specify all details. It is therefore probable, as pointed out by ASHBY (1952), that the genes fix permanently certain *function rules*, which must be relatively few in number, but are capable to direct a sequence of further events: the function rules must be fixed, their application *flexible*, thus providing for '*plasticity*'. It seems likely that, in the relevant further events, '*superencipherment*' by additional biologic 'codes', superimposed upon the variety encoded in the genome proper, plays a substantial role.

Concerning the significance of particular 'genes', which refer to identifiable, linearly arranged chromosomal loci, as e.g. shown in the various *chromosome* (or genetical) *maps*, the following remark, supplementing a comment in volume 1, p. 129 of this series, seems perhaps pertinent. On the basis of present-day concepts it can be assumed that such genes 'control' the synthesis of specific proteins or enzymes and, moreover, may affect the activities of other genes involved in the said synthesis. The interactions of the synthetized proteins, and their amino acid composition, appear to determine the 'genetic' traits of cells and organisms.

Thus, strictly speaking, there are, in this sense, no specific genes for the observable hereditary characteristics described, e.g., as red or white flower color, 'bristles', 'hairy wing', 'thick leg', 'lysine content', etc. Loosely speaking, the code values of the chromosomal gene loci may be interpreted to represent parameters whose 'effects' are significant for the 'ultimate' determination of certain definable conditions, states, or characteristics. Evidently, numerous phenotypical differences can therefore result, as 'dependent variables', from differences in the make-up of these chromosomal loci.

Recent trends and phraseologies concerning theories of gene regulation are exemplified by BRITTEN's and DAVIDSON's paper (1969) which was prominently featured in '*Science*'. It thus bears the stamp of official approval by the strict anonymous censorship arbitrarily yielded, in order to rule and to protect their respective domains, by influential coopting groups, representing self-appointed power elites within the scientific establishment (cf. some of the comments in K., *Mind and Matter*, 1961, §79). The cited authors have elaborated a rather complex model involving several types of '*genes*'. Based on the concept of a gene as a region of the genome with definable elementary functions, an attempt is made to distinguish *producer genes, receptor genes, activator genes*, and *sensor genes*. A set of producer genes, which is activated when a particular sensor gene triggers its set of integrator genes, become a '*battery of genes*'. It is presumed that a particular cell state usually requires the operation of many 'batteries'. These latter are regulated by activator RNA molecules 'synthetized on integrator genes'. 'The effect of the integrator genes is to induce transcription of many producer genes in response to a single molecular event'. The authors state that their theory might supply a means 'of visualizing the process of evolution'. These evolutionary implications of the proposed model are said to be related to the properties of its regulatory system which suggest that 'both the rate and the direction of evolution (for example, toward greater or lesser complexity) may be subject to control by natural selection'. Be this as it may, one could at least concede that the cited model vividly illustrates the complexities of the events which it purports to describe.

So-called '*molecular hybridization*' has recently been developed as a new approach to the detection of DNA–RNA complexes and represents a powerful and intricate technical method for the detection of '*complementarity*' at the level of a single 'gene'. This procedure is based on the following sorts of interactions.

If a DNA double helix is heated, the hydrogen bonds become disrupted, and the two strands separate. If two separate strands with complementary base sequences are maintained at a lower temperature range under appropriate ionic conditions, a double helix is again formed. The degree to which such helix is reconstituted, and the 'perfection' of the new helix can be used to study the relatedness in the DNA structure of different species.

Formation of a 'hybrid' double helix can also occur by the junction of RNA and DNA strands complementary to each other. If only portions of these strands are complementary, the unpaired portion of the RNA molecule can be removed by exposure to ribonuclease, which hydrolyzes the 'backbone' of free polyribonucleotides.

HANDLER's publication (1970) summarizes some of the main conclusions reached by the use of that methodology as follows: (1) all types of cellular RNA, including messenger, ribosomal, and (transfer) t-RNA, are transcripts of the DNA; (2) in any given segment of DNA only one of the two complementary strands is used as a template to generate RNA copies; (3) invasion by a DNA virus results in cessation of transcription of the cellular DNA and the onset of transcription of viral DNA; (4) transcription of a genome is not a random process but occurs in an orderly manner, proceeding sequentially along the 'chromosome' (i.e. along the helix strand); (5) the nucleolus has been shown to contain the DNA templates for ribosomal RNA (as regards the nucleolus, cf. the comments on pp. 96–97 and on p. 112, including footnote 42 in vol. 3, part I of this series); and (6) certain tumorous cells have been shown to harbor a viral DNA which can no longer complete the viral life cycle. These problems, however, have proved to be more complex because, under certain circumstances, so-called 'reversed transcription may occur, such that RNA produces a DNA copy of itself (cf. further below p. 22).

Reverting to the overall problem of morphogenesis in the wider sense, and assuming that the protoplasm (cytoplasm and karyoplasm) of, e.g., a developing zygote is provided with the biochemical properties necessary[4] for a certain sort of differentiation, the substratum of this zygote can be said to undergo differentiation by the progressive production of specific 'protein patterns'. These latter, in turn, are correlated with additional macromolecular patterns. Such superposition and interaction of events leads to particular relatively *static*, that is enduring spatial configurations which manifest particular *dynamic*, i.e. functional properties. With respect to this configurational and functional evolution, the zygote may be considered to have 'potency'. Depending on arbitrary definition, it is '*totipotent*', '*multipotent*', or '*competent*'. Such potency, moreover, is not restricted to a zygote, but can, in particular instances, namely '*parthenogenesis*' and some experiments related to '*cloning*', be manifested by the unfertilized ovum and even by 'adult' tissue cells.

The term differentiation shall here refer to the evolution of form and structure as, e.g., recorded by visual inspection at macroscopic and microscopic levels. Determination, on the other hand, usually refers to what is called the 'irrevocable' or final imposition of a given specific differentiation upon the competent substratum by means of events de-

---

[4] *Necessary (necessity)* as a term qualifying 'condition', refers to the if-clause of implicative propositions in accordance with the principle of sufficient reason which was discussed and concisely reformulated in vol. 1, pp. 289–292. This principle, poorly understood by most contemporary philosophers and scientists, denotes the relationships of antecedent to consequent under five general aspects enumerated in the cited chapter. Taking the term *condition* as representing the above-mentioned if-clause, such condition may be merely *necessary* (required, 'contingent') or, in addition, *sufficient*. In logical notation, necessity F of A is expressed by $A(x) \sqsupset xF(x)$, and sufficiency F of A by $A(x) \equiv xF(x)$.

scribable as 'signals' or 'instructions'. 'Signals' represent here patterns encoding an orderly or ordered variety (negentropy), i.e. information.[5]

Since the differentiating metazoan zygote can be evaluated as 'totipotent' with respect to the evolving organism, the onset of specific competences manifested by embryonic cells or cell aggregates implies a reduction of potency, that is, a constraint, limiting the range of differentiation, and directing the course of events toward a particular sort of determination. The onset of determination for the various embryonic parts is inferred from the experimentally recorded loss of *'regulative plasticity'*; it seems to occur very early in some invertebrates *('mosaic eggs')*, and relatively late in other forms, which include vertebrates *('regulative eggs')*. Thus, with respect to the ontogenetic evolution of these latter, the potency of a given embryonic region remains, for a not negligible period of time, substantially greater than is shown by its actualization in the course of normal development. Its *prospective potency* is said to exceed, or to go beyond, its *prospective significance* indicated by presumptive final morphologic and histologic configuration of that region. Development reduces, as it were, by selection, a greater potential variety to a lesser actual one. In the evolution of 'mosaic eggs', prospective potency and prospective significance of embryonic parts are, to at least some extent, equivalent.

Plasticity at amphibian gastrula stages is, for instance, displayed by exchanged portions of ventral epidermis and neural plate epithelium, which will subsequently differentiate into structures corresponding to their new sites (Germ.: *'ortsgemäss'*). However, if such excision and exchange is performed at neurula stages, 'regulative plasticity' no longer obtains: an insert of epidermis in the neural tube, and an island of neural tissue (e.g. a portion of neural tube, or an eye cup) in the abdominal skin result from the experiment. The exchanged parts now differentiate in accordance with their already determined locus of origin (Germ.: *'herkunftsgemäss'*). Nevertheless, since numerous intermediate gradations between developing metazoan eggs of so-called determinate and indeterminate type exist, these two types cannot be sharply separated from each other. In some instances, substantial differences *qua* 'regulative plasticity' and cognate aspects of morphogenetic behav-

---

[5] Insofar as concepts of information, of negentropy, and problems concerning other related topics are relevant to problems of neurobiology, an elementary discussion of this subject matter has been included in the monograph *'Mind and Matter'* (K., 1961, chapt. VII).

ior seem to occur with respect to metazoan forms pertaining to rather closely related taxonomic subdivisions.

Data recorded by experimental embryology have disclosed directive effects of some embryonic parts upon others. The term *'induction'* is used for this type of control, whereby a previously determined aggregate may impose the direction of determination upon a still undetermined but competent one, in accordance with potentialities of the latter. Inducing cells are presumed to produce so-called inductors or *organizers,* which might be either actual chemical substances ('evocators') or merely represent a designation subsuming events related to biophysical interactions. In the temporal sequence of inductions, there are inductors classifiable in terms of several subordinated ranks, e.g. of first, second, and third order. As regards chemical inductors, even dead tissue[6], or again, an agar plug soaked in extract from known inductors, may be effective under certain conditions. An example of induction is illustrated by Figure 1 which shows results of well-known and often quoted experiments initiated by SPEMANN and his school (cf. SPEMANN, 1938).

Inductors may directly act on the proteins of competent cells or cell aggregates, initiating the production of the specific protein structures pertaining to the next stage of differentiation. Or, the inductors could be effective as regulators of gene activity. This latter mechanism, again, might operate upon the genes either directly, or indirectly through the mediation of a cytoplasmic factor.

Ontogenetic development of metazoa is characterized by an orderly sequence of typical configurational (morphogenetic) stages which include two different sorts of events, namely *qualitative* formative processes (differentiation *sensu stricto, 'Gestaltungsvorgänge'*), and *quantitative* growth processes *('Wachstumsvorgänge')*.[7] If both processes are closely

---

[6] AREY (1954) appropriately remarks in this respect that (in certain experiments) tissues that are ineffective when living may become effective when dead. On the other hand, it has been shown that circumscribed destruction of, or damage to, developing cell aggregates (e.g. in a gastrula) may trigger an 'inductive' effect. Evidently, if cell damage may release 'inductors' or 'inductive' activities, an additional complicating factor obtains in all experiments involving transplantation and manipulation during relevant ontogenetic stages. All interpretations of such experiments involve thus a not negligible element of doubt.

[7] These questions, concerning the evolution of morphologic pattern in ontogeny, have been dealt with from a different and more generalized viewpoint, which considers overall aspects of the form analytic approach, in vol. 1, p. 224f.

interrelated, they can be considered as merely two different aspects of ontogenetic events. Thus, the onset of some formative processes may depend on the attainment of *critical mass*. The orderly, properly spaced and timed differentiation of body parts seems to presuppose the presence of what has been called an 'adequate supply of precursors', relevant to competence and determination, at the respective loci as well as at the pertinent time.

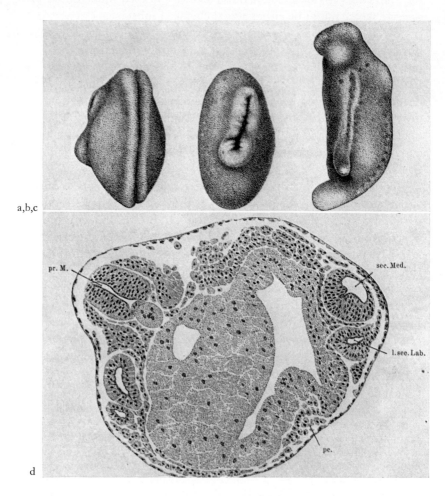

a,b,c

d

*Figure 1 A.* Effect triggered by organizer fragment from gastrula of *Triton cristatus* upon gastrula of *Triton taeniatus*. A secondary neural tube in the larva of *Triton taeniatus* is shown about 44 h (a, b) and about 5 days (c, d) following implantation (after Spemann and H. Mangold, 1924, from Dürken, 1929).

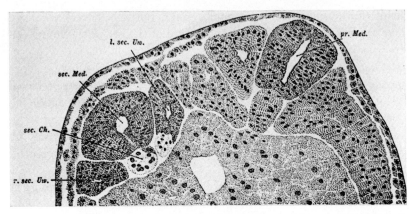

*Figure 1 B.* Cross-section through a larva of *Triton alpestris* with a secondary embryonic anlage induced by implantation, at the gastrula stage, of a portion of dorsal gastropore lip from *Triton alpestris* (after SPEMANN and H. MANGOLD, 1924, from WEISSENBERG, 1931). l sec. Uw., r. sec. Uw.: left and right secondary somite; pr. Med.: primary neural tube; sec. Med.: secondary neural tube; sec. Ch.: secondary notochord.

If this ordered temporal sequence depends essentially on the action of genes within the cells of the differentiating aggregates, the question how particular groups of genes become activated or derepressed at the required *time*, assumes considerable importance.

Thus, embryonic competence might operate by means of 'turning on and off' genes in a particular temporal order. For an understanding of such 'programming events', it would then be necessary to elucidate by what sort of interactions or mechanisms differential gene activity, corresponding to the sequence of characteristic stages, is timed. A progressive and selective 'closure' of an originally 'totipotent' genome through the action of specific chemical messengers, controlling genes in accordance with the hypothesis of JACOB and MONOD[8], has been suggested in very generalized and not very convincing nor satisfactory terms. This kind of mechanism, related to the so-called inductive component of morphogenetic events, is occasionally designated as 'genetic

---

[8] It should be added that the *Jacob and Monod hypothesis* is primarily based upon growth and multiplication phenomena manifested by *bacteria*. The biological mechanisms in these *procaryotic* 'unicellular' organisms may, in various here relevant aspects, considerably differ from many of those obtaining in, and significant for, the more complex *eucaryote* metazoa. On the other hand, it has been claimed that '*what is true for Escherichia coli is true for the elephant*'. This much applauded apophthegm by one of the *Establishment's* beatified luminaries of 'macromolecular biology' is, with respect to some highly qualified

regulation at the transcription level'. Said hypothesis, however, fails to give any indication concerning the nature of the necessary timing operations.

On the other hand, the temporal regulation of succeeding developmental stages might also occur at the above-mentioned so-called 'translation level'. Although it is taken for granted that gene action becomes differentially manifested at the unfolding stages of ontogeny, it has not been conclusively demonstrated that the postulated genes are differentially active *qua* 'transcription of information' at any given arbitrarily delimited period of time.

Data of developmental physiology suggest that DNA activity is not always synchronized with specific protein synthesis. Because differential rates of DNA synthesis and differential rates of 'transcription' do not seem to be invariably followed by corresponding stage-specific protein synthesis, at least some asynchrony in the so-called gene-protein system is presumed to occur.

If differentiating cytoplasmic aggregates elaborate their own chemical messengers, the activity of closed-loop pathways between cytoplasm and genome appears most likely. Such feedback would involve return signals to the genetic locus and might generate complex patterns of *oscillations*.

A regular *sequential induction* of embryonic parts has been conclusively demonstrated for various form elements in a number of different organisms. Nevertheless, in other organisms, particularly in some echinoderms, a progressive and rather synchronous differentiation seems to proceed without sequential induction. It may therefore be presumed that sequential induction, while doubtless representing a significant mechanism, is not the only one available or required for ontogenetic development.

The orderly succession of specific embryonic stages[9], regardless whether sequential induction, synchronous determination, differential gene activity at the transcription level, control at the translation level

---

and generalized invariants, evidently valid, but otherwise obviously false if not downright absurd; it characterizes the extreme intellectual myopia of these sophisticated gentlemen. One could also say that the many fashionable molecular biologists attempt to show how and why the elephant does not differ from *Escherichia coli*, while the few lone still surviving morphologists attempt to show how and why *Escherichia coli* and the elephant are rather different organisms. '*You, that way; we, this way.*' (SHAKESPEARE, *Love's Labour's Lost*, act V, the end).

[9] Cf. vol. 1, p. 230.

or any combination of these and other similar interactions are presumed to obtain, can evidently be conceived as the manifestation of a biological timing system roughly comparable to a clockwork. Yet, none of the various hypotheses or theories has provided a reasonably detailed or satisfactory model of temporal regulation, although the following very generalized concepts may represent a helpful approach to this problem.

If a given type of complex differentiation is to occur in accordance with a certain norm, the embryonic cells or regions must obviously be in the necessary states at the appropriate time. The duration of competence needed for subsequent determination by causal factors is believed to depend upon a so-called *biological clock* provided by a multiplicity of periodic, cyclic, or rhythmic events. Reference to the significance of circular or oscillating biological processes as 'time-measuring' arrangements was repeatedly made in the preceding volumes of this series.[10] The term '*circadian rhythms*', coined by biologists a few years ago, is frequently used to designate phenomena of this type,[11] and a symposium dealing with various aspects of biological clocks was held in 1960.[11a] Although the avowed metabolic nature of such rhythms, the circadian organization of living systems, and numerous other pertinent problems related to the subject matter under consideration were then discussed from different viewpoints, neither this symposium nor subsequent contributions to this subject have, so far, disclosed sufficiently concrete data for the formulation of convincing and useful models. In attempting such constructions, it is difficult to distinguish, by suitable definitions, biological events which are 'clock-controlled' from those biological events (e.g., biochemical changes) which are to be identified with the 'clock'. Another problem concerning the concept of 'biologic clocks' involves the question whether the periodicities of such 'clocks' depend primarily upon 'endogenous biochemical oscillations' or upon

---

[10] Vol. 1, p. 230; vol. 2, p. 180; vol. 3, part I, footnote 96.

[11] The neologism *circadian*, derived from *circa* and *dies*, refers strictly speaking to *diurnal* rhythms, namely cycles with a duration of about 24 h, and merely represents a substitution for diurnal. In the wider sense, however, circadian is frequently used as referring to biological clock phenomena in general.

[11a] Cold Spring Harbor Symposia on Quantitative Biology, vol. 25. Biological Clocks (CHOVNICK, A., ed.), Cold Spring Harbor 1960. As regards biological clock mechanisms believed to be located in the subesophageal ganglion of the cockroach *Periplaneta*, cf. the publications of HARKER (1960, etc.), listed in the treatise on the invertebrate nervous system by BULLOCK *et al.* (1965) cited in vol. 2 of this series.

still less understood 'exogenous periodic variables'. 'Clock mechanisms' are present in the absence of a nervous system, but also become significantly elaborated within neural grisea. In certain instances, single-celled neuronal pacemakers have been suggested. There is, e.g., some evidence that the subesophageal ganglion of the cockroach may control the diurnal activity rhythms, and a few nerve cells in this ganglion are believed to be concerned with this 'clock effect'. Neurosecretory factors seem likewise to play a role. In addition, the ocelli are also known to be implicated in the cockroach's diurnal rhythm. Whether the pace is here actually set by one or several nerve cells discharging at long intervals, remains, however, still an open question.

Nevertheless, the capabilities of biological clocks to act as '*sequence programmers*' for a multiplicity of events at diverse functional levels, including ontogenetic evolution, neural mechanisms, and overall behavior of animals, can reasonably be taken for granted. The behavior of extremely complex systems resulting from a patterned, that is, configurational or permutational series of superimposed mechanisms presents severe obstacles for analysis, understanding, and prediction. The analysis of its single components, taken by themselves, may entirely fail to reveal their relevant operative significance within the configurationally integrated system. Thus, experimental models constructed by 'cyberneticians' and containing randomly connected nets of even simple binary logical type can display unpredictable but statistically recordable peculiar properties of cyclic behavior restricted to regularities of periodicities far shorter than the obtaining theoretical limit on cycle length.

'*Circadian rhythms*' and '*biologic clocks*', being manifestations of the 'flow of time', are evidently related to the general *problem of aging* as discussed on p. 712f, of the preceding volume 3/I, and moreover, justify the concept of '*biologic time*'. Particularly with regard to so-called 'endogenous senescence' it seems justified to presume that such 'biologic time' is 'regulated' by the genome. On the other hand, in accordance with the apparently multifactorial aspect of the aging process, numerous additional parameters are doubtless involved. Thus, for instance, life duration in bees [12] seems to be affected by diverse factors such as various activities (e.g. 'flying hours') and nutrition.

---

[12] In the brief discussion of animal longevity on p.715 of volume 3/I, the dropping out of a line, which unfortunately remained undetected upon proof reading, resulted in the *erroneous* statement: 'The imagines of worker bees are said to attain an age of about

Computing mechanisms recording or measuring the flow of physical time [12a] involve spatial events and are based on the comparison of
physical changes displaying velocity (cm · sec $^{-1}$), with standard velocities, assumed to be constant, of selected cyclic events. In section 6 of
the preceding chapter V, the significance of periodic processes as
well as the relationship of space and time with respect to time measurement were discussed. The cited section 6, moreover, includes illustrations of relevant scales of magnitude concerning measurements of
length (distance), based on the g · cm · sec notation of physics. In the present chapter, dealing with morphogenesis characterized by the temporal succession of stages, it may be appropriate to illustrate a few comparable time scales relevant to physical, chemical and biologic events
(Figs. 2–6).

Some authors believe that most enzymes are synthesized discontin-
uously at subsequent periods in the so-called *cell cycles*, that is, in ordered temporal sequences, which are characteristic for each enzyme (e.g.
MITCHISON, 1969). A 'step' pattern, and a 'peak' pattern of enzyme
synthesis are said to be recognizable.

A hypothesis of 'oscillatory repression', which applies primarily to
'*procaryotes*', assumes that the periods of synthesis are due to oscillations in the negative feedback system of an enzyme with end product
repression. These oscillations become 'entrained' in phase with the cell
cycle.

A hypothesis of 'linear reading', primarily developed for '*eucaryotes*', assumes that the genes are transcribed in a sequence which cor-

---

5 years'. The complete passage should *correctly* read: 'The imagines of worker bees have
a short life-span varying between roughly 24 to 70 days, while the imagines of queen bees
are said to attain an age of about 5 years'. Details concerning these differences in life
durations of worker and queen bees, with appropriate references to extensive studies of
this topic, can be found in COMFORT's treatise on 'Ageing' cited in section 9 of chapter V
(COMFORT, A.: Ageing. The biology of senescence, 2nd ed. Holt, Rinehart & Winston,
New York 1964).

[12a] The distinction of experienced *perceptual time* and of imagined (fictional) conceptual
times, to which the postulated physical time pertains, and the significance of the time concept for neurologic epistemology, have been discussed in the author's monograph '*Mind
and Matter*' (1961, pp. 58–88). As far as the behavioristic aspects of natural sciences such
as physics or biology are concerned, the definition of time becomes reduced to an operational meaning in accordance with the views of BRIDGMAN (1936). The meaning of time is
thereby given by a set of operations, namely the concrete procedures of time measurement
and the hereto-related semantic rules.

responds to their 'linear' spatial order in the chromosomes. It is claimed that both hypotheses (preliminary theories, or models) 'fit some but not all of the facts'.

The pattern of synthesis for a few enzymes, which are produced continuously, is linear, with a doubling in rate at a particular point in the cycle. Points of rate doubling are believed to correspond with the functional replication of the appropriate genes.

In *bacteria*, traditionally classified as *Schizomycetes*, the functional replication is said to occur at the same time as chemical replication, while in *yeasts* (traditionally classified as *Eumycetes* or *Fungi*), such replication is reported to occur later. 'These patterns imply that the cell cycle (and cell growth) is an ordered sequence of syntheses, with continuous change in the chemical composition of the cell' (MITCHISON, 1969).

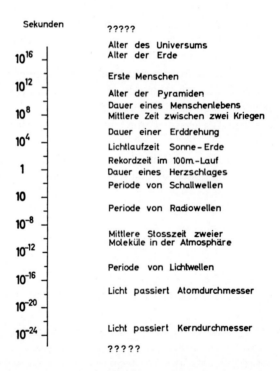

*Figure 2.* Time scale in seconds, indicating duration of a variety of events, as well as 'age' of certain states, stages, happenings, configurations, or systems which provide assumed characteristic reference data ('Fixpunkte' according to EIGEN) of our experiences (from EIGEN, 1966). Figures 2–6 should be compared with figures 238B and C of section 6 in the preceding chapter V (vol. 3, part I), which illustrate length scales in centimeters.

The present-day taxonomic terms *'procaryotes'* and *'eucaryotes'*, recently fairly often used in the literature, and employed above, represent a contemporary formulation of concepts propounded more than hundred years ago by Ernst Haeckel. This author's early and highly significant distinction of *Monera* (or *Cytoda*) as a subgroup of *Protista* without nuclei is thereby fully vindicated.

Among such Monera, at least some pleuropneumonia-like organisms (PPLO, mycoplasmas) seem to be characterized by a particularly diffuse distribution of the DNA components. A nuclear membrane is likewise absent in blue-green algae (Cyanophyceae), and in bacteria, whose loosely packed DNA material is designated as 'nuclear equivalent' or nucleoid (cf. vol. 3, part I, p. 419, of this series).

| Periode | Sekunden | Lebensdauer |
|---|---|---|
| | | $U^{238}$ |
| | $- 10^{16} -$ | |
| | - - | |
| Erdpräzession | $- 10^{12} -$ | |
| | - - | $C^{14}$ |
| Erdumlauf | $- 10^{8} -$ | |
| Mondumlauf | - - | |
| Erddrehung | $- 10^{4} -$ | |
| | - - | Neutron |
| | $- 10^{0} -$ | |
| | - - | |
| | $- 10^{-4} -$ | |
| | - - | Myon ($\mu^{-}$) |
| Kernpräzession | $- 10^{-8} -$ | Mesonen ($\pi^{+}_{,}\pi^{-}_{,}K^{+}_{,}K^{-}$) |
| (bei $10^{4}$ Gauß) | - - | Hyperonen |
| | $- 10^{-12} -$ | ($\Lambda, \Sigma^{+}_{,}\Sigma^{-}_{,}\Xi^{-}$) |
| Molekülschwingung | - - | |
| Atomschwingung | $- 10^{-16} -$ | $\pi^{\circ}$-Meson |
| | - - | |
| | $- 10^{-20} -$ | |
| Kernschwingung | - - | |
| | $- 10^{-24} -$ | ??? |

*Figure 3.* Time-constants of physics, including nuclear and astronomic events. The term *'Lebensdauer'* refers to the *'half-life'* of radioactive substances and indicates the time required for half of a population of identical radioactive atoms or particles to disintegrate. It corresponds to a reduction of measurable radioactivity by 50 percent. The orderliness characterizing 'half-life' is a statistical one based on the exponential law of decay in accordance with the expression $e^{-at}$ (from Eigen, 1966).

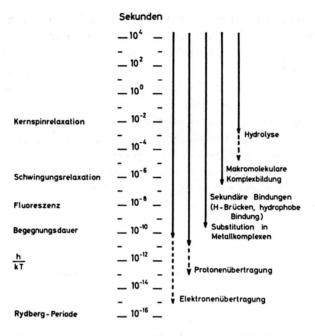

*Figure 4.* Time-constants of some chemical reactions compared with a few periods significant for small-scale physical events. The term '*relaxation*' refers to time constants based on the exponential law (from Eigen, 1966).

*Monera, Cytoda,* or *procaryocytes* (mycoplasmas, bacteria, blue-green algae) besides lacking a nuclear membrane, do not seem to manifest mitotic spindles, nor to possess endoplasmic reticulum, *Golgi apparatus,* mitochondria, and plastids.[12b] The macromolecular structure of these 'primitive' or 'simple' organisms manifests, nevertheless, a considerable degree of complexity, intrinsic to 'living matter', and already correctly pointed out by Haeckel (1893). This author stated: '*Indem ich hier nochmals die vollkommene Einfachheit des Moneren-Körpers betone,*

---

[12b] Recent studies, summarized by Raven (1970) have again revived Altmann's original hypothesis, mentioned on p. 115 (footnote 46) of the preceding volume 3, part I in this series, namely that mitochondria might be phylogenetic descendants of bacteria which associated in a symbiotic relationship with other organisms. Raven (1970) assumes a multiple origin for plastids and mitochondria, pointing out that many independent symbiotic events could have been involved in the origin of these cellular organelles. The cited author states: 'In view of much accumulating evidence, it now seems almost certain that the plastids and mitochondria in eucaryotic cells originated as free-living procaryotes

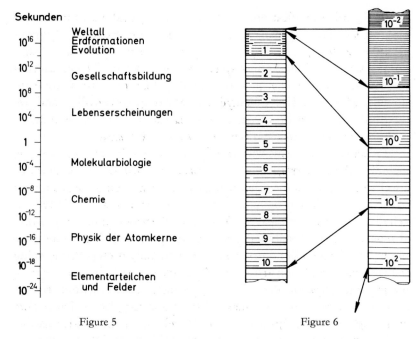

Figure 5                                    Figure 6

*Figure 5.* Time constants significant for cosmic evolution (from EIGEN, 1966).

*Figure 6.* Rough comparison between *linear* (left) and *logarithmic* (right) scales as applicable to diagrams illustrating spatial or temporal orders of magnitude displayed by physical, chemical, and biologic systems (from EIGEN, 1966).

*will ich zugleich daran erinnern, dass dadurch keineswegs eine sehr zusammengesetzte Molekular-Structur desselben ausgeschlossen ist, eine verwickelte Zusammensetzung aus organisirten Molekül-Gruppen und Molekülen ( Plastidulen oder Micellen). Im Gegentheil dürfen wir eine solche, aus allgemeinen Gründen, theoretisch mit voller Sicherheit annehmen; allein empirisch nachweisbar, durch das Mikroskop wahrnehmbar ist dieselbe nicht'* (loc. cit. pp. 427–428). At present, the data obtained by electron microscopy have begun to provide some visible evidence of this anticipated complexity.

which found shelter within primitive eucaryotic cells and eventually were established as permanent elements within them'. Although this evidence has been largely ignored in the construction of phylogenetic classifications of living organisms, RAVEN (loc. cit.) believes that the implications of such symbiotic origin of plastids and mitochondria for phylogenetic relationship should be taken into consideration. An elaborate phylogenetic theory based on concepts of this sort, and deriving ancestral Eucaryota (Protista of HEACKEL) from a 'serial symbiosis' of ancestral Procaryota (Monera of HAECKEL), has been recently propounded by MARGOLIS (1970).

As regards classification, DAVIS *et al*. (1968) deal with three major categories of bacteria, namely (I) gram-positive Eubacteria, (II) gram-negative Bacteria, excluding photosynthetic forms, (III) other major groups, which include Actinomycetales, Spirochetales, Mycoplasmata-ceae (under which the pleuropneumonia-like organisms are sub-sumed), Rickettsiaceae, Chlamydozoaceae (e.g. Chlamydia), and, on the borderline with Protozoa, Bartonellaceae. Orders are here desig-nated by the suffix -*ales*, and families by the suffix -*aceae*.

*Eucaryotes* (organisms with true nucleated cells in HAECKEL's sense, including the *Monocyta*, *Syncytia*, *Coenobia* and *Histones* of that author), are characterized, on the other hand, by a nuclear membrane, a number of distinctive chromosomes within the nucleus, and a mitotic type of cell division ('mitotic apparatus') by which the products of chromosom-al replication are distributed upon the daughter cells. Within the cyto-plasm, membrane-bound organelles are generally present. Concerning the two main cell 'compartments' of Eukaryotes, namely *nucleus* (karyo-plasm) and *cytoplasm*, the following distinction *qua* functional signifi-cance is now commonly emphasized (cf. HANDLER, 1970). The *nucleus*, enclosing the DNA chromatin material releases 'genetic information' to the cytoplasm by transcription into RNA. Both DNA and RNA synthesis, as well as replication of the chromosomes, are conspicuous aspects of *nuclear activity*.

The *cytoplasm*, in which the nucleus is imbedded, represents the main site of protein, carbohydrate and lipid syntheses. It also *provides* the precursor material for nucleic acid synthesis and constitutes an en-ergy source for nuclear events. The cytoplasm *receives* from the nucleus the messenger RNA carrying information for growth and function, moreover both the ribosomal and the t-RNA required for the perform-ance of cytoplasmic activities and the maintenance of cellular structure. There obtains thus considerable mutual interaction between nucleus and cytoplasm.

With reference to the logically and semantically important prob-lems of taxonomy, whose significance for a comprehensive neurobio-logical viewpoint was discussed in volume 1 of this series (pp. 41–84, 145–155, *et passim*) it may here be added that HAECKEL, again about hundred years ago, introduced the concept *Protista* subsuming unicel-lular organisms regardless of their further classification as 'animals' or 'plants'.

HAECKEL, who considered numerous different functional, nutri-tional, and configurational criteria of taxonomic classification, suggested, in fact, various different possible interpretations. In one of the last editions of his generalized '*Natürliche Schöpfungsgeschichte*' (9th ed., 1893), he proposed several revised versions of suitable taxonomic systems. Thus, HAECKEL (loc. cit) defined a '*kingdom*' of Protista which comprised both *Protozoa* and *Protophyta*. Both of these latter *subkingdoms*, again, included as one of their further subdivisions, *Phytomonera* and *Zoomonera*. For a simplified, but reasonably valid system of classification, as discussed in volume 1 of this series, I have preferred to distinguish plants and animals as two different taxonomic sets, representing logical classes (not, of course, taxonomic 'classes', but rather 'kingdoms'), of which none wholly includes the other, but which, nevertheless, have some members in common (loc. cit. p. 79).

As regards the phylogenetic or evolutionary implication of the classification distinguishing *Eukaryota* from *Prokaryota*, data obtained from investigations of molecular biology have been interpreted as indicating that 'the divergence of nucleated organisms and bacteria was 2.6 times more remote in evolution than the divergence of the nucleated organisms into separate kingdoms' (McLAUGHLIN and DAYHOFF, 1970). This inference, which, although not implausible, cannot be evaluated as conclusive, is based on so-called 'unit characters' in cytochrome c and in t-RNA. The development of the 'genetic code' through the differentiation of t-RNA types for different amino acids is regarded, by the cited authors, as still more remote in evolution. It is assumed that those events giving rise to greater amounts of difference have happened 'proportionally earlier in evolutionary history'.

From a viewpoint of more specialized systematics, these taxonomic problems, their semantic solutions, and their possible hypothetical phylogenetic implications, were recently reviewed by WHITTAKER (1969). *Viruses* might justifiably be included under the category of 'living organisms' (cf. vol. 1, p. 43, and vol. 3, part I, p. 413 of this series), but WHITTACKER (loc. cit.) did not deal with the relationship of these peculiar particles of 'living' substance to his revised systems for an overall classification of organisms.

*Viruses*, which are obligatory intracellular parasites, consist of individual 'entities' called virions. Each virion is made up by an array of DNA or RNA material, surrounded by a protein coat, or capsid, which, together with the nucleic acid component, forms the nucleocapsid. This latter may again be surrounded by an envelope or represent a virion of 'naked' type.

Viruses cannot replicate by themselves, but must use enzymes and precursors provided by the host cells. In accordance with these latter, animal viruses, bacterial viruses, and plant viruses can be distinguished. It is now generally accepted that all viruses contain only a single type of nucleic acid, either DNA or RNA, but not both[13], and a protein coat, while some viruses may contain, in addition, lipids and carbohydrates. The agents of psittacosis, together with those of trachoma, lymphogranuloma venereum, and others, which contain both DNA and RNA, and were formerly considered viruses, are now classified as *Chlamydia*. As mentioned above, they are believed to be more closely related to Rickettsiae and Bacteria than to viruses.

Although much progress in virology has been made since our discussion of tentative virus classification in connection with a study on neurotropic effects of vesicular stomatitis virus (H. K. and M. WIENER KIRBER, 1962), a rational taxonomy of viruses still presents many difficult and unsolved problems. The text on microbiology by DAVIS *et al.* (1968) contains pertinent fairly recent data and comments on these questions. In the preceding volume of this series (vol. 3, part I, p. 414), Figure 247 depicted high-magnification electron photomicrographs of vesicular stomatitis virions. On the basis of recent findings, this pathogen is presumed to represent a *single-stranded* RNA virus. Both *single-strandedness* and RNA constitution of viruses involve, at present, significant further questions. Although, according to the original *Watson-Crick theory*, DNA produces RNA, and RNA produces protein (amino acids), the DNA-RNA interaction does not seem to be the one-way process implied by the initial formulation of the theory. It appears now well established that the activity of RNA tumor viruses may cause replication of DNA, which, in turn, becomes a template for virus RNA synthesis (BALTIMORE, 1970; TEMIN and MIZUTANI, 1970). Commenting on these findings published in their periodical, the editors of *Nature* (226, pp. 1198, 1970) add the following questions concerning the relevant DNA polymerase:

---

[13] Although, in accordance with the original *Watson-Crick concept*, DNA produces RNA, and RNA determines the amino-acid sequences of protein, the DNA-RNA replication being a one-way process, certain difficulties arose in attempting to explain how RNA tumor viruses could transform normal cells into a 'stable line' of malignant cells. Recent findings related to this problem seem to indicate that, in certain circumstances, DNA may be synthetized from an RNA template by means of an enzyme (DNA polymerase).

'Does the enzyme make single- or double-stranded DNA or both and is the base sequence of the DNA complementary to that of the viral RNA?

Will the enzyme transcribe the whole of the viral RNA?

Does it transcribe any RNA molecule or is it specifically dependent on viral RNA templates?

Is the enzyme self-coded by a viral gene and do uninfected cells with normal karyotypes or bacteria contain similar RNA-dependent DNA polymerases?'.

Quite evidently, the subject matter here under consideration is, be-cause of its general biologic implications, of considerable import for a wider concept of neurobiology, despite its apparent remoteness. In particular, three aspects, all of them related to ontogenetic and phylogenetic morphogenesis as discussed in this section, should be pointed out, namely (1) *the definition and origin of life; (2) the evolution of precellular into cellular living systems,* and (3) *possible phylogenetic relationship based on ultrastructural 'homology'* (as distinguished from morphologic homology) *of DNA sequences.*

Since the phenomena subsumed under the abstract term 'life' display multitudinous as well as very different features, their relevant summarization within a short 'definition' must evidently remain arbitrary and unsatisfactory. It should be recalled that the concept *'life'* has a very high degree of *extension* which is inversely related to its very low degree of *intension.* Nevertheless, a pertinent general definition was discussed and proposed in the author's monograph *'Brain and Consciousness'* (1957; cf. also vol. 1, pp. 99, including footnote 28, of this series). As regards the biological characteristics of the definition significant for the present context, life may be said to represent a series of events related to various colloidal systems of proteins combined with nucleic acid and other substances. *Metabolism* (which may be 'dormant' but not irreversibly abolished during the 'life' of such systems) can be regarded as a convenient term subsuming basic vital events. Metabolism generally refers to the 'totality' of physico-chemical changes involving the 'flow' of mass and energy through the relatively 'persistent' configuration of a living system. The 'nutritional' aspect includes *assimilation* and *dissimilation,* other aspects can be designated as *intermediary, energy,* and *balance metabolism (Intermediärer Stoffwechsel, Betriebs-Stoffwechsel, Bilanz-Stoffwechsel). Anabolism* roughly denotes here the changes displayed by conversion of small molecules into larger ones, and *catabolism* refers to the conversion of large into small molecules. As a rule, anabolic processes may 'consume' and catabolic processes may 'produce' energy.

With respect to *morphogenesis*, however, *reproduction*, namely *self-re-plication*, which seems particularly related to the interactions of nucleic acid, and *growth* are of paramount importance. Metabolism, growth, and reproduction can be enumerated as the vegetative functional manifestations of life (K., 1957). With regard to *phylogenetic evolution*, the phenomena of *self-duplication* and *mutation* have been pointed out as the particularly significant properties of living systems, which, in the aspect under consideration, reproduce, mutate, and reproduce their mutations. It is, moreover, self-evident that the biologic definition of 'life' as the characteristic properties of 'open systems' implies a distinction between such '*systems*' and their 'surrounding', or '*environment*' by a '*boundary*' describable in structural terms.

Concerning the *prebiotic evolution* of chemical systems leading to 'spontaneous generation' namely to *abiogenesis* in HAECKEL's sense, it is evident that, in accordance with the adopted concept of life, a sharp demarcation between nonliving and living matter cannot be postulated or defined. Rather, a vague demarcation or boundary zone must be assumed, such that living systems did not originate at a particular given 'instant' (cf. vol. 1, pp. 99 of this series). Again, it has been claimed that the construction of proteins only from L-amino acids, but not from 'equally likely' D-amino acids suggests a monophyletic origin of life. This interpretation does not appear convincing, since numerous still unknown parameters and constraints might quite generally favor or require the participation of L-amino acids in the chemical systems likely to display lines of behavior resulting in prebiotic evolution.[13a]

If such evolution of bioblastic micellae or precellular living systems, namely *Probionta* of HAECKEL, should still take place on our planet, or if some such assumed earlier systems which did not evolve

---

[13a] A similar objection can be made concerning the validity of the argument that all present-day organisms must have evolved 'from a single, common primordial ancestor' because essentially the same triplet groups of bases provide the required variety for 'coding', 'transcribing', and 'translating' genetic information in all hitherto examined living systems. While a strictly 'monophyletic' origin of life is, of course, possible, a polyphyletic, 'staggered' (i.e. occurring at diverse succeeding periods) abiogenesis appears by no means unlikely, and this question can be regarded as still wide open (cf. vol. 1, pp. 98–99, 256–258, vol. 2, pp. 300f. of this series).

into 'higher' forms of life,[13b] should still be present, the difficulties of their positive identification are doubtless considerable. Despite recent progress in 'organic geochemistry', the available data on the terrestrial environment, particularly of the atmosphere and waters of the earth in the remote geological past, are uncertain. The question, whether 'natural' prebiotic evolution could only happen under the then obtaining conditions, and cannot any longer occur, may be considered still unanswerable at present, regardless of the now perhaps not altogether improbable possibilities of experimental 'abiogenesis'. Likewise, the question remains open whether all early living systems were necessarily based on photosynthesis, or whether some such systems could, *ab initio*, derive the necessary energy from other processes related to, and from interactions with, available prebiotic organic compounds.

It has been suggested that viruses or micellae somewhat similar to viruses might represent intermediary, ambiguous organic forms bridging the gap between prebiotic, precellular systems and *procaryotic* Cytoda or Monera. Instead of a gap, one could thus assume a blurred zone of 'demarcation' or transition, namely a domain of 'overlap', which does not separate but rather links nonliving and living matter (cf. vol. 1, pp. 43 and 93 of this series).

*Viruses*, nevertheless, as obligatory intracellular parasites, depend entirely upon the host cell for the biosynthetic processes. Some viruses may even exchange 'genes' with the host cell. It seems, therefore, quite permissible (although not necessary) to assume that viruses, as now defined, have phylogenetically evolved from host cell components rather than from early precursors of cells (DAVIS *et al.*, 1958). With regard to this hypothesis, one might, however, perhaps consider it surprising that the evolution of viruses should be related to such a wide range of host cells. It would then appear that cell components (particularly nucleic acids) from precaryote Monera (bacteria), from eucaryote Metazoa, and from eucaryote Metaphyta had a very generalized tendency to

---

[13b] It might be pointed out that present-day *Protista* could be considered essentially similar to those of Precambrian times, presumably more than 500 million years, perhaps even approximately 2 billion years ago. Unless one assumes subsequent independent origins for recent Protista at much later periods, it must be inferred that this particular organic variety (e.g. *Phylum* with its general subdivisions) has been maintained 'unchanged' throughout a very long period of time. Again, considering *Vertebrata*, present-day Urodeles are still, *qua* general organization, similar to tailed Amphibia presumed to have lived about 300 million years ago (cf. vol.1, p.102 *et passim*, of this series).

become independent, resulting in the evolution of three overall types of viruses (bacteriophages, animal viruses, plant viruses).

In accordance with the biochemical definition of life, the cited authors justly consider viruses 'alive' when they replicate in cells. Outside of cells, on the other hand, virus particles are metabolically inert, and, in the opinion of DAVIS *et al.*, 'no more alive than fragments of DNA'. One could here, nevertheless, reply that, outside of cells, their metabolism is not irreversibly abolished, but merely 'dormant'. This condition might, to some extent be compared with the seed dormancy of Metaphyta (cf. vol. 1, pp. 229 of this series). Should or should not such 'dormant' states be considered states of life subsumed under GER-LACH's (1968a, b) formulation of the *vita reducta* concept?

Concerning possible *phylogenetic relationships* based on macromolecular or ultrastructural similarity, some general comments were included in volume 1, pp. 140–141 of this series. Besides the study of protein components by means of paper chromatography, etc., and the further analysis of amino acid composition, an investigation into the base composition of DNA could provide plausible (but not necessarily compelling) evidence for evolutionary relatedness, since genetic variety (information) seems to be encoded in the nucleic acid make-up.

DAVIS *et al.* (1968) therefore speak of 'taxonomy based on DNA homology'. As organisms drift further and further apart in evolution through the accumulation of mutations, their genes not only encode different structures, but also differ increasingly in their base sequence. The cited authors point out that the infertility 'test' for a species in higher organisms is presumed to depend just upon this 'homology': 'the ability of essentially any region in any chromosome of one parent to replace the corresponding region of the other parent and to carry out its essential functions'.

The chemical study of DNA might accordingly afford a direct approach to the 'measurement' of the relatedness of different organisms. Thus, the base composition of DNA is found to be essentially the same in all vertebrates, namely about 40 mole percent guanine and cytosine (GC) and 60 mole percent adenine and thymine (AT). In bacteria, on the other hand, this composition varies considerably, from about 30 to 70 mole percent GC.

Similarity obtaining in different organisms *qua* DNA base composition, which seems to be a stable characteristic, settled over a long period of evolution, represents, however, only a minimal criterion for genetic relatedness, because organisms with a similar composition can have very different sequences. DAVIS *et al.* (1968) therefore state that in a more refined comparison, 'homology' of DNA sequence (at least in bacteria) can now be measured quantitatively 'by determining the ability of DNA strands from two different sources to form molecular hybrids with each other *in vitro*' (cf. also the experiments by HARRIS and WATKINS with 'heterokaryons' and 'heterosynkaryons', namely hybrid cells derived from mouse and man, discussed in volume 1, pp. 76–77 of this series).

In view of these developments, DAVIS *et al.* (1968) look *'hopefully'* forward 'to the growth of a firmly based, rational taxonomy of bacteria'. Because base sequences are ultimately translated into amino acid sequences of proteins, the cited authors assume that evolutionary relations may also be clarified by analysis of the amino acid sequences in 'homologous' proteins (e.g. cytochromes, various enzymes) of different microbes, 'just as the analysis of hemoglobins has yielded interesting results with vertebrates'.

An amusing controversy has recently arisen with regard to the amino acid substitutions in the cytochrome c of closely related species such as horse and donkey. Since one of two, functionally identical but structurally different, cytochromes c is characteristic for each one of these two species, it is claimed that the relevant evolutionary changes in the cytochrome c are neither adaptive nor nonadaptive, that is, are the result of *neutral mutations.*

This, of course, contradicts the totalitarian mutation-selection theory widely accepted by geneticists and other biologists according to which neutral mutations cannot occur, since all mutations must be either 'adaptive' or 'deleterious'.[13c] In volume 1, pp. 115–138 of this series, the many weaknesses intrinsic to the genetic mutation-selection

---

[13c] This evaluation of mutations evidently introduces anthropomorphic, semitheological, teleologic concepts of purpose, in addition to the obtaining inherent intrinsic tautology of terms such as adaptive, maladaptive, unfavorable, deleterious, lethal (*lethal:* nonviable *qua* individual; *maladaptive* or *deleterious:* leading to reduction or elimination of a species; *reduction*, again, implies a postulated numerical 'optimum' for a species; with respect to 'more or less', factors leading to increase and overcrowding might justly be considered 'unfavorable').

Clearly, these attributes of mutations can only be established *ex post facto*, on the basis of available data, being then also used for extrapolation in fictional thought models. Assuming that evolution is entirely due to mutations in accordance with the views of many geneticists, such mutations are then either compatible with survival, i.e. 'adaptive', even if 'neutral', or incompatible, i.e. 'deleterious'. Reduction of optic system and loss

theory of evolution have been discussed, and need not again be dealt with in this context. It will be sufficient to recall the highly ambiguous and unsatisfactory nature of terms such as '*adaptation*' and '*random*'. It is also evident that a reasonably satisfactory and comprehensive formulation or description ('explanation') of phylogenetic organic evolution ('transformism'), or, for that matter, also for ontogenetic evolution, has not been achieved up to now. The genetic mutation-selection theory embodies, of course, numerous apparently well substantiated facts. It can be extended to particular aspects of evolution by the assumption of '*genetic drift*'. This term refers to changes in gene frequency which may cause a mutant gene to be fixed in a given population, while the same mutant would not become fixed in a larger population. It is presumed that, in a small group, 'random' fluctuations such as genetic drift are not averaged out. The term '*gene flow*', which also plays a role in contemporary evolutionary speculations, refers to genetic exchange per generation for 'open populations' that are 'normal components of species'. This so-called gene flow is believed, by some biologists, to represent a 'cohesive force' holding together species as 'evolutionary units'. Other biologists question the significance of 'gene flow' and the thereto related 'evolutionary units'. On the whole, it may be presumed that the selection-mutation theory can be accepted as at least partially, and in many respects, valid, but by no means as exhaustive or exclusive.

It is, of course, entirely unjustified to designate evolutionary processes based on '*neutral mutations*' as '*non-Darwinian evolution*'. DARWIN himself postulated events, which, in present-day terminology, amount to neutral mutations (cf. volume 1, pp. 90 of this series). He stated: 'variations neither useful nor injurious would not be affected by natural selection, and would be left either a fluctuating element, as perhaps we see in some polymorphic species, or would ultimately become fixed, owing to the nature of the organism and the nature of the conditions' (*The origin of species*, *Modern Libr. ed.*, pp. 64).

---

of vision, although from an anthropomorphic viewpoint 'unfavorable', became evidently an 'adaptive' trait in Gymnophiona, other Amphibia such as *Proteus anguineus*, and in additional organic forms with reduced eyes. At most, 'adaptive' and 'maladaptive' etc. might be considered 'statistical' terms indicating, on the basis of previous data, probability of survival. Yet, these terms cannot be rigorously statistical, since all the hereby required numerical data and variables cannot be ascertained. Even a *Poisson distribution* cannot be properly calculated if the 'expectation' z for the expression $e^z$ does not remain constant for the entire domain of events under consideration.

While the behavior of cytochrome c with respect to evolution is doubtless interesting, the alluded controversy is quite meaningless, being essentially related to insufficiently rigorous semantic and logical analysis of the relevant problems.

An interesting early attempt to introduce a *genetic concept of 'homology'* was made by an anatomist (DANFORTH, 1925) who reviewed problems related to the homology and phylogeny of hairs, which are characteristic mammalian structures. DANFORTH concluded that 'homology' between two structures 'is dependent upon the similarity of genetic factors involved in their production and is consequently usually partial and not absolute'.

DANFORTH, who did not explicitly define what he understood to be 'homology', apparently referred to the phylogenetic homology concept in GEGENBAUR's formulation. The formanalytic or topologic homology concept, as adopted in this series, on the other hand, is based on purely configurational relationships within a morphologic pattern (bauplan). Mammalian hairs display here a particular case of general homology, discussed in volume 1 of this series (pp. 201–203). It may be added that, in considering the general homology of hair, the component cell aggregates are evaluated with regard to *form* rather than to *structure (texture)*, although, in this domain of anatomical space, concepts of form elements and of structural elements tend to overlap.

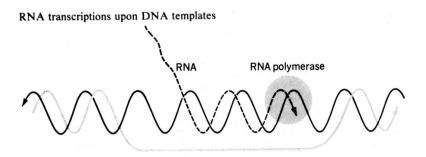

**RNA transcriptions upon DNA templates**

RNA        RNA polymerase

*Figure 6 A.* Diagram purporting to illustrate template effect of DNA (transcription) by means of a *semi-conservative* duplication mechanism. Attachment of RNA polymerase opens the DNA helix by partial separation without breakage of either strand, and bases of ribonucleotides pair complementarily with those on one selected DNA strand whose anticodons are transcribed into codons of the synthetized RNA strand which, as it grows, is said to 'peel off' by elimination of pyrophosphate (after WATSON from HANDLER, 1970).

As regards the so-called *DNA homology*, it is at present quite impossible to formulate a meaningful chain of causal events linking particular chromosomal DNA sequences to particular sorts of morphologic configurations displayed by a bauplan. It is true that DNA sequences and their complementary messenger RNA sequences appear to encode, by means of 64 redundant triplet groups, 20 different amino acids, whose sequences, again, presumably provide the specific protein molecules. But this, even if elucidated in much greater details than at the present time, does not seem to yield any useful insight into the nature of the actual morphogenetic processes. Moreover, in terms of cryptanalysis, the much vaunted and extravagantly publicized 'breaking of the genetic code' is, of course, no true code-breaking at all, but merely a preliminary general identification of the biologic coding system (based on *'superencipherment'* )[13d] and of some of its symbols. At the nucleic acid or chromosomal level, the redundant DNA code of one strand, selected by RNA polymerase (cf. Fig. 6A), is transcribed into the very similar complementary RNA code, in which U (uracil) replaces T (thymine) as one of the four basic codon symbols.[13e] The resulting RNA code is illustrated by Figure 6B. An interesting feature of the redundant nucleic acid code is the presence of 'punctuation' or 'stop' symbols which seem to indicate limits of the chain sequences.

The process of matching the *messenger RNA codons* (Fig. 6B) with the *complementary anticodons*, carried by the t-RNA polymers binding a particular amino acid, is presumed to be mediated by the activity of the cytoplasmic ribosomes (cf. Fig. 6C). These particles, discovered by means of electron microscopy, and located along the endoplasmic reti-

---

[13d] Cf. K., *Mind and Matter*, 1961, p. 393. From the viewpoint of cryptography and cryptanalysis, the transmission of information by means of symbols, i.e. by *codes sensu latiori*, can be performed by systems classified as *ciphers sensu strictiori* and as *codes sensu strictiori*. The former are transforms of clear text letter by letter, or at most by trigrams or digrams. *Codes sensu strictiori*, on the other hand, imply transforms of whole words, or even phrases, by groups of symbols. Under these premises, the genetic code would represent a *primary cipher*, related to a complex system of additional *superencipherments*. Yet, it seems, of course, permissible as well as convenient to use the term *'code'* (*sensu latiori*) for the sets of relevant transformations of variety obtaining in biologic events, genetic as well as neural.

[13e] Cf. vol. 3, part I, pp. 106–107 of this series. It should be pointed out that in footnote 35, p. 105 of that chapter, an inadvertent *lapsus calami* remained undetected upon proof reading. The statement: 'DNA contains desoxyribose, but no ribose and no cytosine' should of course read: 'and no *uracil*'.

Second position

| | | U | C | A | G | |
|---|---|---|---|---|---|---|
| | | 18 | 06 | 19 | 15 | U |
| | U | 18 | 06 | 19 | 15 | C |
| | | 05 | 06 | 31 | 33 | A |
| | | 05 | 06 | 32 | 17 | G |
| | | 05 | 08 | 20 | 12 | U |
| First position | C | 05 | 08 | 20 | 12 | C |
| | | 05 | 08 | 14 | 12 | A |
| | | 05 | 08 | 14 | 12 | G |
| | | 04 | 07 | 13 | 06 | U |
| | A | 04 | 07 | 13 | 06 | C |
| | | 04 | 07 | 11 | 12 | A |
| | | 16 | 07 | 11 | 12 | G |
| | | 03 | 02 | 09 | 01 | U |
| | G | 03 | 02 | 09 | 01 | C |
| | | 03 | 02 | 10 | 01 | A |
| | | 03 | 02 | 10 | 01 | G |

Third position

*Figure 6 B.* Messenger RNA codons represented by trinucleotide triplets specifying a particular amino acid, e.g. UUC indicating *l*-phenylalanine (18). Codons UAA, UAG, and UGA (31, 32, 33) are the so-called '*terminator triplets*' or '*punctuation marks*', not yet entirely understood, but presumed to indicate chain terminations. In the relevant literature, UAA is usually referred to as '*ochre*', and UAG is called '*amber*'. The numerals from 01 to 20 indicate amino acids: 01: glycine (gly); 02: *l*-alanine (ala); 03: *l*-valine (val); 04: *l*-isoleucine (ilu); 05: *l*-leucine (leu); 06: *l*-serine (ser); 07: *l*-theronine (thr); 08: *l*-proline (pro); 09: *l*-aspartic acid (asp); 10: *l*-glutamic acid (glu); 11: *l*-lysine (lys); 12: *l*-arginine (arg); 13: *l*-asparagine (asn); 14: *l*-glutamine (gln); 15: *l*-cysteine (cys); 16: *l*-methionine (met); 17: *l*-tryptophane (try); 18: *l*-phenylalanine (phe); 19: *l*-tyrosine (tyr); 20: *l*-histidine (his); 31, 32, 33: 'punctuations'; A: adenine; C: cytosine; G: guanine; U: uracil. C and U are pyrimidines, A and G are purines (modified and adapted from HANDLER, 1970).

culum, contain ribosomal RNA and proteins. Ribosomal RNA, representing DNA transcriptions, is apparently synthetized in the nucleolus.

Although the general features of these events seem to be reasonably well elucidated by the available data and on the basis of plausible inferences, numerous details still remain insufficiently clarified.

While there is little doubt that visible organic configuration displayed by a morphologic pattern or bauplan is intrinsically related to genetic factors including macromolecular or ultrastructural arrangement of nucleic acid sequences, it seems obvious that *morphologic homology* and *macromolecular or DNA 'homology'* (Davis *et al.*, 1968) must be subsumed under two entirely different logical categories, which should

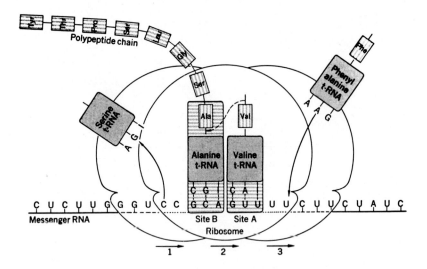

*Figure 6 C.* Diagram purporting to illustrate '*translation*' mechanism. Protein synthesis takes place on m-RNA 'threaded' through a ribosome. The protein chain grows one amino acid at a time, the sequence of amino acids being determined by the order in which m-RNA codons are matched by each amino acid t-RNA. The ribosome is supposed to move along the m-RNA from left to right. The formation of a specific polypeptide chain proceeds until reaching a terminator triplet, whereupon the completed protein is released by hydrolytic removal of the t-RNA still fixed to the last amino acid in the chain. The first amino acid representing one of the chain in process of synthesis is shown at upper left (Tyr). At the actual site of synthesis the bond between alanine and valine is just about to be effected. At left, the uncharged serine t-RNA is departing, and at right the charged phenylalanine attached to its t-RNA is shown to arrive. I (e.g. in CGI) stands for inosine, a base closely resembling guanine (after Crick, 1966, from Handler, 1970).

be clearly distinguished by an appropriate terminology. One homology refers to patterns of macromolecular structure, while the other refers to configurational patterns of whole organisms and of organs in anatomic space implying macroscopic and light microscopic orders of magnitude. Both categories, of course, have some invariants in common. Again, it might be recalled that 'homology' in rigorous algebraic topology differs entirely from morphologic 'homology' in a simplified topology applicable to the analysis of organic form.

*Experimental embryology (developmental mechanics,* including aspects of developmental genetics), despite undisputable successes[14] in providing relevant data and useful concepts, has so far failed to formulate a comprehensive and plausible theory of organic form development and thereto related overall biologic events. This failure is partly due to the complexity of the phenomena in question,[15] and partly to the fact that any satisfactory and reasonably exact detailed 'explanation' of material events requires a rigorous mathematical formulation.

In this respect, only those natural sciences which can be dealt with by means of mathematics may be considered exact sciences, such e.g. as physics and chemistry. The application of mathematics to biology, attempted in restricted fields by numerous authors for more than a hundred years, and a contemporary more generalized approach to the mathematical foundations of biology, particularly undertaken by RASHEVSKY (1960, 1962), are still encumbered with considerable difficulties. These latter may be said, on the one hand, to concern the requirement of an adequate training in advanced mathematics which, because of the nature of their own field of studies, few biologists can be expected to obtain, while few mathematicians will be capable of gaining the required insight into a wide range of biological problems. On

---

[14] Cf. vol. 1, pp. 235–239 *et passim.* The exaggeration of these successes and their overvaluation with respect to allegedly new insights into the nature of life have substantially contributed to the neglect if not disrepute of morphology in biologic sciences.

[15] Among the multitude of biochemical events relevant to ontogenetic as well as to phylogenetic morphogenesis, but, at best, only indirectly and distantly related to DNA and RNA activities, the role played by *thyroxin iodine* might be mentioned. The possible evolutionary significance of thyroxin in accordance with the theory of my former student and present colleague, Dr. MARY B. DRATMAN, was pointed out in chapter II, pp. 144–145 of volume 1 of this series. Various promising leads have already resulted from her research program in which I participated as a consultant (DRATMAN, 1967; DRATMAN *et al.* 1968, 1969).

the other hand, the very complexity of biological events, which greatly exceeds that of nonbiological physical and chemical happenings, frequently precludes a satisfactory mathematical formulation.

As regards obstacles of this latter type, there are, again, two different sorts of hindrances. The well-known embarrassment inherent in the classical 'three-body problem' illustrates the common difficulties in dealing with a system of many interacting particles. Although mathematical physicists have indeed made some progress by devising new mathematical methods suitable for the treatment of the 'many-body-problem' with respect to atomic and to solid state physics,[16] the application of these and similar advances to biology seems still rather remote.

The other sort of difficulty is metamathematical, and concerns the nature of mathematics which is a so-called syntax language, devoid of content (cf. K., 1968). Observed biological events, on the other hand, are described by means of sensory percepts (by direct inspection or through the mediation of instruments). If such biological events are then expressed in terms of mathematics, and relevant equations are set up, these latter must either be accompanied by, or at least presuppose the existence of, an additional 'text' (BRIDGMAN, 1936) in a language akin to a metalanguage which correlates object and syntax language. In other words, a referential meaning must be assigned to the symbols used in the equation.[17] In order to have an appropriate operational and predictive significance, the equations must, besides, contain numerical values, which are assigned to the relevant symbols. These values, again, must be obtainable or verifiable by measurements.

It is evident that, in comparison with physics and chemistry *qua* sciences not concerned with the behavior of living matter, the selection of relevant mathematical symbols for the multitudinous aspects of biological phenomena is beset by far greater difficulties. Moreover, as

---

[16] Cf., e.g., MARCH, N.H.; YOUNG, W.H., and SAMPANTHAR, S.: The many-body problem in quantum mechanics (Cambridge University Press, New York 1967). MATTUCK, R.D.: A Guide to *Feynman diagrams* in the many-body problem (McGraw-Hill, New York 1967).

[17] Thus, in the simple gravitational equation $\frac{dv}{dt} = g$, the '*text*' must indicate that v is a number describing a property of the moving body which can be obtained by a certain kind of measurement, and that t is the time obtained by another kind of measurement, etc. (BRIDGMAN, 1936). Comments on the significance of mathematics, relevant to the epistemologic concepts adopted as the guiding principles in preparing the present series on the CNS, have been given in section 8, chapter III of volume 1, and in §§ 64–71 of the author's monograph '*Mind and Matter*' (1961).

RASHEVSKY (1960) justly points out, it is well-known that many of these phenomena are not quantitative, but relational. Some phenomena of this sort, however, can likewise be dealt with in terms of one of the nonmetrical branches of mathematics, namely in terms of *topology* about which a few additional comments, supplementing my discussion of *analysis situs* in section 2, chapter III of volume 1 will be given further below.

*Morphogenesis* in the widest sense, including histogenesis, results from the displacement of material units, that is of mass, whereby specific and relatively permanent three-dimensional configurations are produced. This involves both growth (quantitative) and formative (qualitative) processes (cf. above, pp. 9–10). Generally speaking, the relevant events can be roughly classified (cf. e. g. AREY, 1954) in accordance with the following categories, which, however, admit considerable semantic 'overlap': (1) *localized growth*, resulting in various enlargements as well as constrictions; (2) *cell migration;* (3) *cell aggregation*, forming masses, 'cords', or 'sheets', (4) *fusion* and *splitting*, which includes *delamination* of sheets in separate *layers*, cavitation of cell masses, and bifurcation of cords, (5) *foldings*, which include evagination, invagination, eversion and inversion, (6) *bending*.

As pointed out in volume 1 (p. 231), these displacements imply changes in the motion of masses, involving acceleration, deceleration, and states of relative equilibrium. Such events, in terms of classical physics, are manifestations of 'forces', namely of interactions expressible as vectorial magnitudes. The concept of 'equilibrium' unifies the analysis of statics and dynamics by dealing with the 'force' of 'inertia'.

It can, therefore, be assumed that these biological events must be describable, that is explainable, by means of the notations of mathematical physics dealing with the motions of continuous deformable media. Presupposing the concept of *continuity*,[18] the designation 'field' is here given to the distribution of some 'condition' prevailing throughout a region of space.

If the condition is adequately described at each point by a scalar magnitude involving a single number, a *scalar field* obtains. Temperature, density, concentration, and time represent such scalar 'condi-

---

[18] *Continuity*, as manifested by a continuous medium, implies that any volume element of the medium, no matter how small, still contains points of the region or material of the medium. Local singularities manifested as material 'particles' within the *continuum* are thereby not precluded.

tions'. The position of a point in three-dimensional space, however, is determined, within a conventional coordinate system, by three numbers which indicate direction and distance from the origin to the point. 'Conditions' thus characterized by direction as well as by magnitude are *vectors*. The field defined by the 'instantaneous' velocities of material 'points' in a moving fluid, or the conditions represented by a so-called 'field of force', are two different examples of *vector fields*. In addition, there exist many complex physical conditions which require more than three numbers for their determination, such as the *stress* and *strain*[19] in an elastic medium, each of which condition is defined by nine numbers. Quantities of this sort, operated upon by an elaborate system of logical rules and notations, are called *tensors*, implying *tensor fields*.

In generalized tensor theory, scalars become included as tensors of zero order (or rank), and vectors as tensors of the first order. Tensors *sensu strictiori* are then those of second and higher orders. Operations with tensors of rank zero (scalar operations) follow rules differing from those valid for tensors of higher rank. In the n-dimensional spaces[20] of advanced mathematical formulations, a tensor of rank one (i.e. a vector) has n components, a tensor of rank two has $n^2$ compo-

---

[19] *Strain* may be loosely defined as change in relative position manifested by a medium, the changes being produced by the deformation of the medium resulting from *stress*. Accordingly, *stress* then refers to the 'force' or interaction causing the strain. In the mathematical description of a continuous medium, an interesting generalization of the vector concept by means of matrix notation presents itself. Supposing that, in an orthogonal three-dimensional coordinate system, an elementary volume element *dxdydz* is acted upon by a certain stress (force per unit area), an aggregate of nine quantities is obtained, which, in general, will be functions of x, y, z, and t. These quantities include tensile stress normal to the three orthogonal planes and tangential stresses *(shears)* along the two coordinate directions of each plane. Denoting, for each plane (1, 2, 3), the tensile stresses by $W_{11}$, $W_{22}$, $W_{33}$, and the shearing ones by $W_{12}$, $W_{13}$, $W_{21}$, $W_{23}$, $W_{31}$, $W_{32}$, the *stress tensor* S may be written in form of the following matrix, convenient for the further operations with arrays of this sort:

$$S = \left\| \begin{array}{ccc} W_{11} & W_{12} & W_{13} \\ W_{21} & W_{22} & W_{23} \\ W_{31} & W_{32} & W_{33} \end{array} \right\|$$

Additional relevant details on deformable media and tensor fields can be found in Lindsay's and Margenau's 'Foundations of Physics' (1957).

[20] It should here be added that, in four-dimensional space-time, where a vector (or tensor of rank one) has four components, one of these refers to the time dimension. Thus, although time can be dealth with as representing a scalar condition (cf. further above), it becomes here a vectorial one. In a somewhat similar manner, the time rate of

nents, and a tensor of rank three has $n^3$, etc. Thus, in three-dimensional space, stress and strain, represented by tensors of rank two, have each nine components. The complexity of the resulting notations is therefore evident.

Tensors may be said to represent mathematical or physical concepts in accordance with certain specific rules of transformation. In mathematical physics, tensor analysis provides, by means of curvilinear coordinates,[21] the fundamental equations for problems of hydrodynamics, elasticity, and electromagnetism. If generalized so-called laws of mathematical physics are formulated in terms of tensors, such 'laws' become entirely independent of any one particular coordinate system.

Moreover, comparing e.g. problems of hydrodynamics or elasticity with those of electromagnetism, it will be seen that the former concern the behavior of ponderable matter[22] (mass), while the latter concern certain conditions obtaining in space.[23] Yet, a generalized conception of 'mechanics' tends to include both sorts of problems into a single overall scheme of dynamics.

The behavior of continuous nonliving or living physical media requires, in addition to the concepts of space and time, certain mechanical categories such as velocity, acceleration, deceleration, equilibrium, force, displacement, density, strain, stress and others, as used in hydrodynamics and elastokinetics.

The media represented by *living matter* manifest *fluidity*, of which protoplasmic streaming is a conspicuous example. Fluidity precludes

---

motion, indicated by $\frac{1}{t}$, may be dealt with, depending upon circumstances, in scalar or vectorial terms, such that, in general, *speed* is a *scalar*, and *velocity* a *vector*. The inclusion of the time dimension into the space-time system of relativistic physics requires, however, the use of the operator i (that is $\sqrt{-1}$) in connection with the time dimension. This latter, namely w in the coordinate system x y z w, is assumed to be ct, where c is the velocity of light.

[21] In simplified, nonrigorous graphic interpretation, such coordinate systems in two-dimensional space are illustrated by the diagrams of THOMPSON shown in figures 11 A–E of volume 1 (chapter III, section 2, pp. 186–192).

[22] No satisfactory definition of '*mass*' exists in physics, although mass may be measured in terms of *weight* or described in terms of *inertia*. Mass is symbolized by the letter m in the well-known formulation $f = m \frac{1}{t^2}$ ($f = g \cdot cm \cdot sec^{-2}$) or $f = m \cdot a$ (force equals the product of mass and acceleration) whose circularity is quite evident.

[23] Formerly conceived as behavior of the *ether*.

the assumption of stress. However, protoplasmic aggregates also display the behavior of an *elastic solid*, in which no plastic flow can occur. In colloidal living systems undergoing sol-gel transformations, nonuniformities in the concentrations of substance presumably produce nonuniformities in expansion and contraction of a gel, thereby causing mechanical stresses.

Thus, the protoplasmic substratum can display the physical properties of an incompressible *viscous fluid* and those of an *elastic solid*. Contradictory assumptions must therefore be made about the simultaneous behavior of living systems. Consequently, in developing mathematical theories of such contradictory physico-chemical systems, somewhat contradictory procedures must be followed: for the computation of *stresses*, the system is considered an elastic solid, and subsequently, for computation of the rate of *deformation*, the system is assumed to represent a plastic fluid (RASHEVSKY, 1960).

Theories purporting to describe the behavior of matter *sensu latiori*, that is, of anorganic or organic (i.e. biologic) physical processes, are generally based on certain hypotheses whose symbolic expression takes the form of *differential equations*. In fact, the differential notation may be considered an invention for the purpose of physical description. From such equations, new equations are derived, which symbolize 'physical laws' that should be verifiable by possible laboratory operations or by reliable observations. The transition from the differential equation to the 'physical law' is obtained by the mathematical operation of *integration*. The differential equation can be considered a symbolic description based on the method of elementary abstraction and its solution by integration then represents a passage from small-scale phenomena to large-scale phenomena (cf. LINDSAY and MARGENAU, 1957; cf. also the comments in chapter III, pp. 285–287 of vol. 1).

As regards the basic differential notation, there are *ordinary differential equations* involving only one independent variable, and *partial differential equations* dealing with several independent variables. In the former equations only *ordinary derivatives*[24] appear, while the latter equations contain *partial derivatives*.

---

[24] *Derivatives* can be said to represent the limits of the quotient of the increments of the variables which describe the process over the increment of space or time in which this event takes place. A 'law of nature' of this form is the expression of the relation between one state and the neighboring in time or space, and thus implies, in terms of mathematical physics, the principle of causality (cf. HOPF, 1948; also K.: Mind and matter, 1961, §43, pp. 163–165).

In classical *Newtonian mechanics*, the motion of elementary particles is considered as the sole basis of all physical processes. The ordinary differential equation is the expression or description of laws of this sort, since there is only one independent variable, namely the time.

In the domain of field theory, on the other hand, all processes are determined by field quantities which have definable values at each point of space. These values are usually a function of (scalar) time. There obtain thus, in three-dimensional space, four independent variables, namely the three space coordinates and the time. The laws which are based on this concept are expressed by partial differential equations.[25]

Ordinary differential equations may occur, as approximations neglecting the influence of all but one of the variables, in problems not pertaining to classical *Newtonian mechanics*. As a rule, however, ordinary and partial differential equations are mathematical expressions of two separate or distinctive fundamental points of view, the synthesis of which in quantum theory is considered one of the major problems of contemporary physics (cf. HOPF, 1948).

The abstract mathematical description provided by differential equations does not, *per se*, refer to any specific physical problem. Such specific cases must be defined by initial or boundary conditions, involving the *boundary value problem*. It becomes here necessary to find a solution to a differential equation or set of equations which will meet certain specified requirements for a given set of values of the independent variables, called the boundary points in mathematical physics. In order to solve the differential equation, as many physical boundary

---

[25] Thus, the instantaneous velocity of the fluid through an orthogonal coordinate system x, y, z at a point P and at the time t is assumed to be a vector q. Through every such point in the fluid it is possible to pass a curve, the tangent to which at that point gives the direction of flow at the point. The instantaneous rate of flow at this point is called the magnitude of the instantaneous velocity at the point. Along the three axes the vector q has components u, v, w, representing the component velocities of the fluid. Denoting by $\rho$ the density or mass per unit volume of the fluid, the well-known *equation of continuity* is obtained: $\dfrac{\partial(\rho u)}{\partial x} + \dfrac{\partial(\rho v)}{\partial y} + \dfrac{\partial(\rho w)}{\partial z} = -\dfrac{\partial \rho}{\partial t}$.

It is a partial differentiation equation of the first order in which the fundamental quantities $\rho$, u, v, w are functions of both space (x, y, z) and time (t) variables. This equation simply describes, in the symbolic shorthand of mathematical language, the assumed continuity and indestructibility of the medium with which the quantities $\rho$, u, v, w are associated.

conditions[26] must be given as there are arbitrary functions or constraints in the integrated equation. It is evident that, in this respect, the
*prima facie* (particularly to the uninitiated) highly impressive equations
set up by mathematical biophysicists will, in most instances, merely
represent a platitude expressed in a redoubtable notation, and make a
very poor showing. Paraphrasing an appropriate quip aimed at BER-
TRAND RUSSELL and coined by an irate Britisher, there is much less to
them than meets the eye.

Nevertheless, despite all serious obstacles and severe drawbacks, it
may be reasonably postulated that all behavioristically observable as
well as describable phenomena of life, involving motion and equilibrium of masses or manifestations and transformations of energy, are
purely mechanistic, that is, can, in principle, be formulated ('explained') in terms of mathematical physics. The events of morphogenesis *sensu latiori*, concerning the evolution of form dealt with in this
chapter, and the processes of histogenesis, as particularly discussed in
chapter V, represent an important group of such mechanistic biologic
phenomena.

The evolution of configured living systems, in the aspect here under consideration, can be assumed to begin with the behavior of the
zygote, namely with the division of a cellular unit. The fundamental
phenomenon of self-reproduction displayed by such unit seems to be
based not only on the capacity of certain molecules to provide replicas
of themselves, but also significantly upon the *gestalt-like*[27] *properties of*
*complex open systems*. Theories derived from one of these two concepts
may differ from those derived from the other, but such dissimilar theories need not be mutually exclusive.

Although, in proliferating cells, each type of cytoplasmic organelle
is believed to double in amount during what is called the 'average cell

---

[26] A specific *boundary condition* represents a postulated event in space and time expressed
by the statement that a symbol, standing for a certain physical quantity, shall have a
definite value or set of values throughout a specified interval of time. General boundary
conditions imply fundamental restrictions on the type of activity possible for the system
considered, but do not require the specification of the exact values at the boundary.
Situations in which general boundary conditions may be introduced are, for instance,
encountered in some aspects of wave propagation and eletromagnetic radiation.

[27] '*Gestalt-like*' refers here merely to a *permutational aggregate* providing a system
whose particular properties do not pertain to its isolated elements (e.g. properties of
configurated carbohydrate chains not manifested by their constituent atoms or even
molecular subsets).

life cycle', cell division itself seems to involve primarily replication of the genetic material in the nucleus, followed by segregation of the genetic material and its distribution upon the resulting two functional nuclei.

At present, the '*cell life cycle*' (already mentioned above on p.12) is arbitrarily considered to begin after each nuclear division. It is presumed that, at the start of the cycle, a period called *gap 1* ($G_1$) corresponds to the preparation for DNA synthesis, which represents the next period (S). A period between completed synthesis and next nuclear division is called *gap 2* ($G_2$), preceding the next nuclear and cell division (D). Thus, the cell cycle $G_1$, $S_1$, $G_2$, D has been defined as alternate periods of gene replication (S) and gene segregation (D), each phase

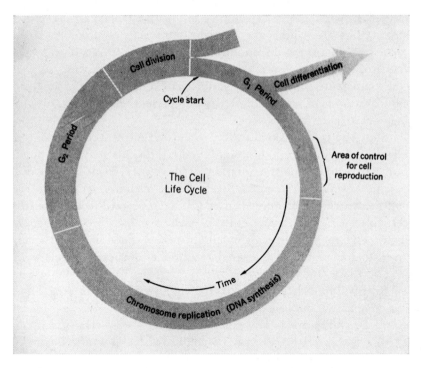

*Figure 6 D.* Diagram purporting to illustrate the so-called *cell life cycle*. Cell reproduction is said to be regulated by interruption of cycle process at a still unidentified stage in the $G_1$ phase. Depending on the cell type, the total time period of such 'cycle' evidently varies considerably. In some actively reproducing mammalian cells, the cycle is assumed to last 'typically' 10–15 h. This diagram may be compared with the 'matrix'-like tabulation of figure 43 C on p. 57 in the preceding volume (3/I) of this series (after PRESCOTT from HANDLER, 1970).

being initiated or terminated by some sort of physico-chemical event representing a 'signal'. The progress of a cell through this cycle is also considered to be 'a series of sequentially ordered gene transcriptions, with correspondingly ordered protein syntheses, which control the timing of various events. It seems obvious that this generalized so-called *cell cycle* (Figure 6 D) is, depending on numerous variables (taxonomic organic form, embryonic or adult cells, cell type *qua* perennial, stable, and labile elements) of very different length, and may not proceed (in mature perennial elements, cf. vol. 3, Part I, pp. 705–707) beyond the so-called $G_2$ period.

As regards the mechanism of *gene replication*, three possible types have been suggested, namely: (1) *conservative*, whereby the DNA helix remains intact and 'somehow' produces a new one; (2) *semi conservative*, characterized by partial separation, but without breakage, of either strand, thus maintaining the overall double helix continuity; (3) *disperse*, resulting in the disruption of the old DNA chains during synthesis of the new ones. With the exception of a few forms, such as some viruses with single stranded nucleic acid molecules, the semiconservative mode of replication seems to represent the generally obtaining normal pattern.

Since, by cell division, a given cellular element loses its original identitiy, the life span(s) of a single cell can be indicated by the notation $s \leq c$ (c denoting the duration of the cycle). Evidently, for true perennial elements, the notation $s < c$ holds. Well-known older tissue culture experiments, undertaken by CARRELL, seemed to indicate that certain explanted cellular elements, such as chick fibroblasts, could divide and multiply for an indefinite period extending over many years, thereby manifesting a very large number of successive cell cycles (cf. vol. 3/I, p. 716f, of this series).

Recent authors have contested the findings of CARRELL *et al*. Thus, in a compilation published under the auspices of the *National Academy of Science*, and entitled '*Biology and the future of man*' (HANDLER, ed., 1970), it is stated that present-day studies have dealt CARRELL's conclusion 'a body-blow'.[27a] This compilation, prepared by '*committees*' and

---

[27a] It is claimed (loc.cit.) that the earlier experiments of CARRELL *et al*. cannot be repeated: 'A series of well-controlled experiments in many different laboratories has conclusively demonstrated that normal animal and human cells grown in tissue culture divide for a finite number of generations and that cessation of division is not related to faulty culture techniques'. According to this 're-evaluation' and 'reinterpretation' of

so-called '*panels*' with numerous participants belonging to a self-appointed official 'elite', sets forth the 'ultimate' orthodox views of the 'establishment' on the significant problems in the entire domain of biology as conceived by these gentlemen.

Concerning the question of cell division *in vitro*, explanted normal cells are here said to divide for periods of time not greater than one year. Some inferences from this behavior 'may be relevant to senescence in the whole animal'. In those presumably rare instances where cells divide indefinitely, 'they are invariably found to be abnormal'. *Per contra*, 'normal human fetal cells divide approximately 50 times in tissue culture before dying, and adult human cells divide approximately 20 times before cessation of cell division' (HANDLER, ed., 1970).

Now, with regard to the just-mentioned vaguely defined 'normal human fetal cells', the following comments, based on inferences from the available data, might perhaps be appropriate in order to evaluate the quoted statements by the pandits of the *National Academy of Science*. Assuming, in a reasonably justified tentative oversimplification, that the total number of approximately $10^{13}$ ($\cong 2^{43}$) adult human body and blood cells could be reached by identical, uniform multiplication rates of the elements resulting from a division of the original single cell (zygote), about 43 successive divisions are required.

This first approximation, however, does evidently not hold, since, after 'final' morphogenesis, different cell types are present in different numbers. Accordingly, and again in a reasonably oversimplified formulation, some cell types must undergo a lesser, and others a greater number (than 43) of successive divisions, in order to reach, at an undefined early stage of morphogenetic maturity, their respective quota of the $2^{43}$ total.

---

'classical experiments purporting to show the "immortality" of cells grown outside of the animalsbody', it must be 'suspected that the nutrient fluids used periodically to feed the culture introduced new living cells into the culture'.

Because of severe limitations of our equipment, general facilities, and funds, we were unable to deal with the relevant problems in our tissue culture studies undertaken at the Medical College of Pennsylvania by my associates and myself (cf., e.g., STILWELL, 1952; K. and M.W. KIRBER, 1959). Although I cannot, therefore, express a definite opinion on the contested findings of CARRELL *et al.*, I remain rather sceptical about the allegation of faulty culture techniques in the older experiments. I am rather inclined to believe that, despite greater sophistication in the approach and the instrumentation of recent authors, these latter are perhaps somewhat less capable than the older ones. Quite evidently, in respect to the question at issue, there is something wrong somewhere, or, in German vernacular, '*da stimmt irgendwo irgend etwas nicht*' (cf. also further remarks in the text).

Moreover, this total consists of *perennial*, *stable*, and *labile* elements (cf. vol. 3, part I, pp. 705–707). Some of the labile elements (e.g. blood cells, epidermal, intestinal, uterine epithelia) have a life-span as short as only a few weeks. Therefore, in maintaining their approximately constant number during, e.g., a human life duration of about 70 years or more, such labile cell populations must be renewed by an indefinite sequence of further divisions within a numerical order of magnitude between 400 and 800 or more. This number of successive cell divisions should be added to the presumably more than 43 initial successive divisions leading to the respective quotas within the original $10^{13}$ cell population probably already approximately obtaining within the first few postnatal years. It seems evident that the claimed time limit of one year for cell division in tissue culture, and the reported restricted overall number of 50, respectively 20, successive cell divisions cannot hold with respect to the actual human (or comparable other animal) organism.

The interesting attempts of RASHEVSKY (1960) to establish physico-mathematical theories of cell division have emphasized the role of diffusion drag forces. The cited author points out the limitations of his diffusion drag force theory and concedes that there is, at present, no adequate physico-mathematical formulation of cell division, for which numerous patterns of interaction at the molecular level seem to play a role. Yet, as RASHEVSKY justly remarks, if the physical mechanisms of cell division are to be elucidated, all relevant conceivable mechanisms should first be investigated *in abstracto*. Subsequent experimental studies will then perhaps disclose which of the conceivable factors might be selected as relevant. It seems likely that a combination of several such factors actually causes cell division.

Because of the troublesome problem involving the delimitation of a diffusion field, difficulties arise in applying thermodynamics for the calculation of mechanical forces in such fields. Again, with the presently available methods and data, the equations for motions such as deformation and elongation of a dividing cell, or for dispersion and recondensation of a cell, are highly arbitrary and unsatisfactory.

Be this as it may, detailed attempts at establishing mathematical formulations 'explaining' cell division have been undertaken by RASHEVSKY (1960). External forces such as diffusion forces, electric forces, and gravity, as well as internal forces, e.g. elastic stresses due to swelling, are considered. Heed is given to laws of plastic flow, related to continuous viscous fluids in relative motion, and to the significance of the flow of water. Generalized theories of elasticity and of plasticity, using tensor notations, are discussed by the cited author. Although, because of the arbitrary nature of the assumptions, quantitative estimates of the parameters involved in relevant equations cannot be obtained as necessitated by the boundary value requirement pointed out above, certain types of equations can be suggested.

The simplest mechanical model for production of new cell material assumes that substances penetrate by diffusion into the cell, combining to form more and more complex molecules constituting the cell body, the rate of consumption of every substance being proportional to the products of the concentrations of cell substances. Temporary asymmetries of concentration, correlated with asymmetric distribution of ordinary osmotic pressure, may result in deforming forces which could even divide the cell into two. This assumption stresses a factor related to ordinary osmotic pressure, and not to diffusion drag forces. Thus, temporary periodic deformations, resulting in random-like pulsations preceding the actual division of cells, are well-known occurrences which have been often shown in time-lapse micromotion pictures of living tissues.

Another possible factor in cell division is represented by swelling stresses in the colloids which make up the cell. As regards the theory based on diffusion forces, relevant equations require the definition of a 'coefficient of diffusion', and the use of symbols (such as the *Laplacian operator* $\nabla^2$) relevant for vector and tensor analysis. The *Laplacian operator*[28] likewise plays a role in the formulation of periodicities in metabolizing systems whose activities result in a complex pattern of mechanical forces.

As regards the magnitude and behavior of such forces in diffusion fields, the diffusing substance and the solvent may be considered as two interpenetrating, continuous viscous fluids in relative motion, exerting drag forces on small particles (G. YOUNG, 1938, as quoted by RASHEVSKY, 1960). The molecules of the solute in a dilute solution behave like those of an ideal gas, confined to the volume of the solution. The small particles will therefore be in a state of thermal motion. Advanced mathematical formulations of biologic diffusion processes would require consideration of possible effects related to the shapes of

---

[28] Some general comments on this operator in connection with statistical and mathematical concepts relevant to the topics of neurological epistemology can be found in § 67 (pp. 320–343) of the monograph '*Mind and Matter*' (K., 1961).

the molecules and of the particles, and the introduction of *van der Waals' forces*. These latter subsume the relatively weak attractive forces operating between neutral atoms and molecules, but, despite their weakness, effective at comparatively great distances. Some theories of mathematical biophysics can, with a certain measure of success, be based almost exclusively on mechanical concepts pertaining to classical physics. Thus, RASHEVSKY has treated diffusion phenomena, plastic deformation, and other biological events from the point of view of a continuum. Yet, as the cited author justly remarks, an extension into domain of molecular or atomic physics will become necessary.

For a preliminary general formulation of the diffusion problem, RASHEVSKY (1960) has tabulated the following list of constants and variables, which also include electrical charges, and which must be dealt with in the relevant equations (Fig. 7 A). This tabulation will suffice to illustrate the complexity of the required mathematical operations.

If, in any volume element of a living system consisting of intracellular components, cellular units, or aggregates of cells, including intercellular material, a substance is produced at a rate proportional to the concentration in that element, this event may be physico-chemically considered to be an *autocatalytic reaction*, in which the product catalyzes its own formation. Catalyzing particles repel or attract one another,

$D =$ Diffusion coefficient of an ion in cm$^2$ sec$^{-1}$

$c$ = Concentration of an ion in g cm$^{-3}$

$B$ = Mobility of an ion in sec g$^{-1}$

$f$ = Force on an ion in dynes

$k$ = Boltzmann constant = $1.4 \times 10^{-16}$ ergs degree$^{-1}$

$T$ = Absolute temperature

$J$ = Rate of transport of an ion through the membrane in g cm$^{-2}$ sec$^{-1}$

$q$ = Rate of production of an ion in g cm$^{-3}$ sec$^{-1}$

$M =$ Molecular weight of an ion

$N =$ Avogadro's number

$\varepsilon$ = Charge on an electron in E.S.U.

$E$ = Net charge on the cell in E.S.U.

$K$ = Dielectric constant of the internal medium

$V =$ Volume of the cell in cm$^3$

$r_0$ = Radius of the cell in cm

*Figure 7 A.* Constants and variables significant for the mathematical formulation of various biologic events discussed in the text (after RASHEVSKY, 1960).

and such interactions result in microgradients or in gradients of greater intensity and extension. Accordingly, corresponding forces of repulsion or attraction must be taken into consideration.

Again, depending upon numerous hidden parameters, biochemical and biophysical processes related to morphogenesis either require the expenditure of energy or liberate, i.e. produce, energy. According to RASHEVSKY's formulation of events involved in cell division, it appears that, below a critical size, any deformation of the cell from an original spherical shape requires supply of energy, while above this critical size a complete division of the cell may liberate energy. It can also happen that for small deviations of the spherical shape such elongation results, within the limits of certain configurations, in a decrease of energy. Yet, the required equations cannot be set up with any reasonable expectation of actual validity until a general mathematical theory of stability dealing with the behavior of cells and other related protoplasmic complexes has been developed

It should here be added that 'electric phenomena', which can be detected and registered by means of microelectrodes, seem to be concomitant with cell division. Thus, mitotic chick embryonic and mouse embryonic fibroblasts in tissue culture are *'electrically coupled'* to their interphase neighbors, and the available evidence suggests that this *'coupling'* persists throughout the division cycle of the mitotic cells (O'LAGUE *et al.*, 1970). Generally speaking, so-called 'coupling' by 'low resistance junctions' between various sorts of cells *in vivo* appears to be well documented. Such cells may establish and maintain *'junctions'* of this type when grown in tissue cultures. The prevalence of so-called 'junctions' between cells *in vivo* and *in vitro* is believed to suggest some sort of fundamental significance for the type of cell contacts subsumed by the designation 'junction' with the connotation of 'electrical coupling'.

Living systems consists of orderly aggregates displaying a multiplicity of regions, many of which are separated by definable boundaries. Some of these latter are represented by ultrastructural membranes,[29] such e.g. as cellular and intracellular ones. A membrane of

---

[29] Particular reference to such membranes was made in the preceding volume (3, I, chapter V, section 6). The oversimplified but useful concept of *'unit membranes'* was there adopted with the necessary reservations. Although this concept, like the cell theory or the neuron theory, represents a convenient formulation of general principles, the details of the postulated structure of such membranes, especially as inferred from electron microscopy (which provided the first visualization of these structures), must be taken with a

this sort, which may disappear and reappear in the course of certain morphogenetic events, is exemplified by the well-known nuclear membrane. The *active* and often *selective transport* of substances through biologic membranes, for instance against the gradients of their own concentrations, differs considerably from *passive transport*, such as, e.g. by filtration or osmosis, due to physical interactions commonly manifested by nonliving systems. Active transport requires the expenditure of energy derived from specific physico-chemical biologic processes which are based on the hierarchically superimposed, composite events occurring in the living state of matter. RASHEWSKY (1960) points out that active transport may be due to the transport of one sort of molecule by another sort, which, in accordance with its own interactions, moves in a direction opposite to the decreasing concentration of the first. In case of transport through membranes, a mechanism might obtain which acts somewhat like the so-called '*Maxwell demon*'.[30] This, in other words, implies *selection* by the introduction of negentropy, pro-

---

grain of salt. It is generally maintained that these membranes are built up by parallel layers of phospholipids, but the orientation and specific composition of molecules in different sorts of 'unit membranes' may display considerable varieties. Moreover, there is some suspicion that one and the same membrane might display different configurations correlated with energetic conditions, such that the membrane exists in an energized or in a nonenergized state. These membranes obviously play a significant role in the molecular aspects of specific cell contacts.

The currently prevailing hypotheses, based on the presence of phospholipid layers, and frequently referred to as the '*Danielli-Davson theory*', assume the activity of high-energy intermediates coupling ATP and the electron transport chain. Some authors, however, have expressed dissenting views, stressing mechanical rather than chemical work. One model suggests passage of small nonelectrolytes through aqueous channels or pores. A recent publication on the movement of molecules across cell membranes was prepared by STEIN (1967). This author brings a critical survey of several aspects of membrane transport and discusses the significance of energy-yielding metabolic processes.

Again, it should be here pointed out that the prominent cell walls characterizing many sorts of plant cells, and consisting of cellulose, are additional structures external to the protoplasmic surface film of the cell ('unit membrane'). Reference to these plant cell walls was made in vol.1 (chapt.1, p.37, chapt.II, p.42) of this series. Prominent cell walls external to the surface film are also present in a number of bacteria (cf. fig.250, p.419, vol.3/I).

[30] The problem of the imaginary '*Maxwell demon*' is discussed in the author's monograph '*Mind and Matter*' (K., 1961, §19, pp.76–78). It should here again be pointed out that, in contradiction to a widely held belief, the activity of this '*demon*' does not show, for at least two valid reasons, the possibility of transferring, without expenditure of energy,

vided by the numerous '*coding systems*' now generally assumed to be characteristic for the manifestations of life by compound physico-chemical systems.

One might also say that active transport involves metabolic (vital) factors requiring energy provided by living systems, while passive transport involves purely 'physical' factors and displays phenomena that can be duplicated in artificial (model) systems.

In a simplified formulation adapted to the needs of the morphologically oriented neurologist not concerned with the complex physico-chemical aspects of biological behavior, a few appropriate generalizations may perhaps be pointed out, with particular reference to a recent summary by SCHLÖGL (1969). This author, in reviewing the pertinent concepts on selective transport through such membranes, enumerates the following main mechanisms: flow *('Strömung')*, pinocytosis (cf. vol. 3, part I, pp. 461–462 of this series), and membrane permeation, together with 'accessory' processes of diffusion in liquid or gaseous phase, and of electrolytic or ionic events.

Concerning this last-mentioned, electrical membrane behavior, it should also be recalled that (a) conductance, which may be regarded as an index of 'ion permeability', (b) capacitance, and (c) impedance, represent three relevant concepts, referred to in chapter V, section 8 (volume 3, part I) of the present treatise. In section 6 of that chapter, the

---

heat from a cooler container to a warmer one: (a) Even if the imagined microvalve operated by the *demon* involves negligible friction, and mass of small molecular magnitude, any operation implying motion with acceleration and deceleration of mass requires energy. (b) if the '*demon*' in accordance with the peculiar hypothesis of J.C.ECCLES (The neurophysiologic basis of mind, 1953, p. 283 *et passim*) consists of a soul 'substance' not pertaining to the physical space-time system, but endowed with extraordinary knowledge as well as 'perception', and operates the valve by means of PK, the operation *qua* physical motion still involves work, implying energy. If, however, the '*demon*' pertains to the physical space-time system, an intricate detection mechanism distinguishing swift from slow molecules is required. How could such recording mechanism work without energy transformations? In either case, the '*demon*' is provided with, and applies, information.

The actual significance and value of the simile is based on the fact that, by the introduction of *negentropy (information)*, a so-called irreversible process may here be made reversible by means of a control mechanism whose expenditure of energy is negligible in comparison to the energy obtained by the introduction of negentropy. The indefinite numerical relationships between expended energy and obtained energy are similar to those holding in many trigger mechanisms (e.g. muscular exertion on a trigger and resulting firing of a gun). The role of information and negentropy in the problem of *Maxwell's demon* was pointed out in a paper by SZILARD (1929).

special significance of synaptic membranes for rectification (one-way transmission) of neuronal impulses, *i.e.* of the neurokym, was also considered. In agreement with fundamental physico-chemical principles elaborated by MAX PLANCK, WALTER NERNST, and other early pioneer investigators, it is now easy, for the epigones, to set up various sorts of tentative equations expressing, by mathematical shorthand symbols, some aspects of the relevant electrical phenomena. Based upon simplified postulates, expressions of this sort may concern membrane potential, conductance and its variation with current, moreover impedance changes as functions of current and of the composition of environmental solutions (e.g. GOLDMAN, 1943–1944; HODGKIN and KATZ, 1949). Resting membrane potential might be generated by a combination of multiple mechanisms and a diversity of chemical ultrastructures characterizing the membranes. Thus, MARMOR and GORMAN (1970) report that, in a marine mollusk, the resting potential of a neuronal membrane is separable into at least two components, namely, one which shows a 'classical dependence' on ionic gradients and permeabilities, and a second which depends on temperature and on metabolic (i.e. 'vital') activity. HODGKIN and KATZ (1949) undertook their study on the giant axon of the molluskan squid, and particularly considered the physico-chemical phenomena related to the reversal of resting membrane potential during the events characterized by the action potential.

Among the diverse transport mechanisms discussed by SCHLÖGL (1969) 'membrane permeation' is characterized by considerable specificity. The permeability of a membrane for different material components can be conspicuously unequal. This 'specificity' is partly related to chemical reactions upon or within the membrane, e.g. to particular coupling of Na-ions to a carrier protein. Motility *(Beweglichkeit)* and solubility *(Löslichkeit)* are here by no means equivalent, but must be considered as representing different expressible 'magnitudes'. Numerous transport processes through membranes differ considerably from diffusion in free solutions. Thus, in membranes which are not semipermeable, so-called negative osmosis can occur, such that an *'incongruent'* transport of solvent and solved particles from the side of lower concentration to the side of higher one takes place against the concentration gradient. Membranes with *'asymmetric structure'* may mediate increased concentration differences, combined with changes of volume. Such transports are passive, i.e. not requiring metabolic energy and were, according to SCHLÖGL (loc. cit.) not sufficiently considered by biologists. In addition to electrical interactions, and to 'incongruent' transport, the cited author also refers to pressure gradients and pressure diffusion. As regards the biologically more important active transport mechanisms, catalytic processes are emphasized.

In summarizing the considerable difficulties still preventing the formulation of valid rigorous mathematical expressions for the behavior of biological membranes, SCHLÖGL (1969) concludes: '*Eine erfolgreiche Bearbeitung dieser Fragen erscheint nur in einem wissenschaftlichen Team möglich, das sich aus Vertretern verschiedener Disziplinen zusammensetzt – aus Biochemikern, Biologen, Physiologen, Physikochemikern und theoretischen Physikern.*'

With respect to cellular units, it is widely and tacitly presupposed that these elements represent, as it were, bags formed by ultrastructural membranes and containing ions as well as macromolecules in a liquid water solution. However, as already pointed out in section 6 of the preceding chapter V (volume 3, part 1 of this series), the behavior of water in biological systems is still very poorly understood. Recent evidence, based on nuclear magnetic resonance spectroscopy, seems to indicate that cell- or tissue-water has a significantly greater degree of 'crystallinity' than ordinary liquid water. Moreover, different states of 'crystallinity' may obtain, which display different sorts of behavior. A polymeric structure of water, called '*polywater*' is claimed to have been demonstrated by means of infrared and *Raman spectra* (LIPPINCOTT *et al.*, 1969). This, however, has been contested, and some investigators assume that the said anomalous viscous watery material contains high concentrations of sodium, potassium, chloride, and other compounds and does not represent a true polymerized form of water. Nevertheless, since the properties of 'polywater' have also been interpreted as resulting 'from accidental biological contamination', it is not impossible that this particular 'polywater' effect plays a significant role in the normal state of 'living matter'. Be that as it may, it seems not impossible that, if cell water should, to a relevant extent, manifest a solid rather than a liquid state, many of the current theories of ion transport across membranes, and of nerve conduction, including the postulate of an 'ion pump', would require substantial revisions.

Another recent approach to cell dynamics is based on the concept that the cell is a resonating system made up of *coupled oscillators* which interact through either *strong* or *weak coupling*. Strong couplings may comprise the direct effects of repressors, derepressors, and inhibitors on gene activity as well as the synthesis of m-RNA and proteins. Weak coupling is then assumed to comprise the indirect interaction between oscillators through the pools of metabolites which they share, and would display properties of 'random noise'. A measure of the cell's tendency to shift 'spontaneously' between various resonant modes, with oscillator energy flowing from one oscillator to another in the direction of a negative gradient has been designated as the '*talandic temperature*' by Goodwin (1969). It represents an observable quantity presumed to indicate how energy will be distributed among interaction feedback processes regulating various cell activities. This hypothesis attempts a merger of thermodynamics with parameters of cell organization. However, considerable mathematical difficulties still present serious obstacles for a detailed elaboration of this theory in quantitative terms.

Concerning the expression '*parameter*' used above as well as elsewhere in this context, objections to the misuse of this word in medical literature have been frequently voiced. It is claimed that this designation has become meaningless. Since I have adopted and defined the word 'parameter' (K., 1957, p. 109f.) in accordance with a phraseology elaborated by Ashby (1952, 1960), it is perhaps appropriate to explain that the use of this term stresses the significance, of the measure or factor so designated, as a quantity ('constant' or 'variable'), through functions of which may be expressed other variables. In other words, the at worst quite innocuous term 'parameter' justifiably emphasizes interrelationships with respect to an assumed system. Thus, e.g. so long as a parameter is constant, the related system, with changing variables, may display 'changes of state' within a stable 'field of behavior' (Ashby, 1952). A change in parameter, on the other hand, may be correlated with a major change modifying the field, i.e. the 'line of behavior'. In German, 'parameter' can, and is, appropriately rendered by '*Zustandsbedingung*' respectively '*Zustandsbedingumgen*'.

In terminating the discussion dealing with various aspects of morphogenetic behavior (*sensu latiori*) at the cellular level, including cell division, the poorly understood process of *amitosis* should perhaps be mentioned. Most contemporary authors doubt the occurrence of amitotic division 'under normal circumstances'. Yet, there are doubtless relatively rare but obvious instances in which such nuclear division, with or without correlated formation of two daughter cells, is rather strongly suggested. Findings of this sort are occasionally made upon careful (light) microscopic studies of the nervous system.

Amitosis was inferred on the basis of observations seemingly indicating a nuclear division by mere constriction of that structure without the preceding chromosomal arrangements characteristic for mitosis, and without participation of 'centrosomes', 'asters', or achromatic 'spindles' (cf. vol. 3, part I, pp. 121–129). Because various degrees of constriction provide an array of stages ending with near complete fission, supplemented by completely disconnected paired nuclei displaying direct or close contiguity, the assumption of an occasional or aberrant amitotic cleavage into separate nuclei or even cell bodies can be based on reasonably valid 'circumstantial evidence'. The relative scarcity of such findings obviously justifies an evaluation of the inferred amitosis as not 'normal'. *Mitosis,* on the other hand, although originally inferred on the basis of static pictures, is, by now, a commonly observed phenomenon in living tissues studied by means of diverse techniques. While it is possible that actual *amitosis* might have been observed and recorded *in vivo,* no convincing report has come to my attention.

Since, *prima facie,* no apparent mechanism for the equal distribution of the 'genetic material' is detectable in the assumed stages of amitosis, a 'grossly imbalanced' status of the daughter nuclei, stressed by some authors, could perhaps result. Be that as it may, the following circumstances are possibly relevant with respect to the problem of amitosis.

(1) The chromosomes of mitotic division represent, from the viewpoint of 'ultrastructure', exceedingly gross configurations presumably containing the 'genetic material' in an ordered, but still unknown arrangement providing 'equal distribution' upon chromosomal fission. This implies complex events leading, from the nondescript spatial distribution of visible chromatin particles in the 'interphase' nucleus, to the formation of specific chromosomes displayed during mitosis. Electron micrographs of early prophase stages in representative vertebrates (e.g. urodele Amphibia, cf. BLOOM, 1970) disclose sequences of progressive changes related to the condensation, into 'lumps', of 'dispersed' 'chromosomal fibrils'. These 'lumps' or 'clumps' seem to enlarge by 'continuous accretion of more and more fibrils', until the metaphase chromosomes are formed. It is doubtful whether the recorded micrographs, displaying, as it were, a coagulation picture of electron-dense, detritus-like granular filaments forming fuzzy clusters upon a dehydrated background, actually correspond to the conditions obtaining in the living state. Moreover, the effective electron microscopic resolving power has remained inadequate, so far, for an interpretation of the '*fuzz*' in terms of 'macromolecular biology'.

(2) Typical mitosis does not seem to occur in Procaryotes such as bacteria, although equal distribution of the nucleoid genetic material takes place.

(3) More or less distinct reduplication and division of Eucaryote chromosomes within an intact nuclear membrane, first observed about 1939 by GEITLER in Heteroptera (the invertebrate 'true bugs'), and likewise studied by my former associate Dr. E. F. STILWELL (1952) in chick tissue cultures, is an aberrant form of mitosis, perhaps related to 'polyploidy'.

One might thus justifiably assume that the mechanisms for the arrangement and distribution of genetic material (chromatin) can display different dynamic patterns, whose ultrastructural respectively macromolecular details remain to be elucidated. Regardless whether implying 'imbalanced', or somehow effected approximately equal distribution of significant genetic material, amitosis could represent a pattern of this sort, possibly implying undetected rearrangements preceding the apparently nondescript chromatic stages of division.

It should here be added that an equal distribution of genetic material upon somatic cells has been demonstrated by experiments with, or related to, 'cloning' ('clones' being populations of cells derived by division from a single isolated tissue cell). Thus, GURDON et al. (1966) showed a few years ago that the destroyed nucleus of a frog's ovum could be replaced by the nucleus from a somatic (intestinal) cell of another frog. Some, but not all, such manipulated eggs developed into normal adult frogs with the genetic make-up of the nucleus donor. Sensational and widely publicized comments, by journalistic writers in news-media and popular literature have predicted the *prima facie* not entirely impossible, but certainly much more difficult ultimate application of said experiment's results to the systematic cloning of human beings: '*Brave New World*', indeed.

Proceeding from *cellular morphogenesis* to that of *cell aggregates*, one may first consider a very elementary first approach pointed out by RASHEVSKY (1960). It is thereby assumed that complex nonhomogeneous structures can arise from originally homogeneous cell aggregates through interactions of relatively simple phenomena of diffusion, chemical kinetics and precipitation. Basic assumptions are here:

1. The existence of *polarity*, at least partly accountable for by diffusion forces.

2. The existence of *gradients*, which are closely connected with polarity.

3. The *mutual inhibition* of chemical reactions, a phenomenon widely observed in biological systems.

4. A simple *kinetic equation* of the type

$$\frac{dc_B}{dt} = Ac_b - ac_B,$$

where A and a are constants, B and b metabolites, and c indicates concentration.

RASHEVSKY has shown that, under rather generalized conditions, the formation of cavities will occur in cellular aggregates. Again, differences in growth rates, and processes of flow can produce invaginations and foldings characteristic for many embryologic events which may be expressed by mathematical formulations as elaborated in RASHEVSKY's attempt. At present, however, these studies, although of considerable potential interest for the embryologist, have not succeeded to give a satisfactory account of the relevant ontogenetic events. Even the relatively simple question how cell aggregates, which might otherwise separate and disperse under the action of diffusion forces, are held together, cannot be answered in a convincing manner.

In addition to RASHEVSKY's preliminary and tentative approach, *periodic propagation of 'chemical waves'* has been suggested as a relevant factor in ontogenetic pattern formation (GOODWIN and COHEN, 1969). Such events, interfering with each other, are believed to include the propagation of self-sustaining involute spirals. Recently, WINFREE (1972) has discussed the interesting geometric properties displayed by spiral waves of chemical activity observed in model experiments with non-biologic material. This author noted 'spontaneously generated pacemaker centers', moreover involutes describing waves of stably self-renewing geometry, and 'spontaneously' oscillating processes. Some of the reaction kinetics still remained incompletely analyzed for suitable mathematical formulation. It seems not unlikely that far more complex and very poorly understood processes of this type are involved in the functional structure of embryonic 'morphogenetic fields' as well as of 'biologic fields' in general. It will also be recalled that, among the factors influencing the configuration of cell aggregations, the distribution of mucopolysaccharides, including hyaluronic acid, and the responses to cell contact, depending on biochemical or electrical events affecting the cell membranes, play a significant role.

An appraisal of the problems concerning *positioning information* and the spatial pattern of cellular differentiation was also recently presented by WOLPERT (1969). This author is compelled to admit that 'one is acutely conscious of the absence of the physiological and molecular basis of positional information and of polarity'. He then continues: 'but unless the correct questions are asked one has little hope of finding out how genetic information is interpreted in terms of spatial patterns'. I would add to this, that 'asking the correct questions' implies an appropriate semantic formulation. In this respect, the fashionable and highly sophisticated representatives of the prevailing scientific establishment, which is permeated by the personal 'power politics' of 'dominant dukes', generally make, in my opinion, a very poor showing.

Populations of living cells constitute specific configurated regions conceivable as topologic neighborhoods or as morphologic sets of different orders *(Bauplan:* overall morphologic pattern; *Grundbestandteile:* fundamental pattern elements; *Formbestandteile:* form elements). Their interactions are regulated by an undefined variety of physico-chemical processes. Actually obtaining aspects of biochemical variety, representing operands in a series of transformations, can, if definable, be regarded as information, embodying negentropy, namely selected variety. *Random* or, in this respect, unselected variety may be conceived as *'disorder'*. The numerically expressible term *entropy*, introduced into physics by R. CLAUSIUS (1822–1888) as a factor indicating the unavailable energy in a thermodynamic system, derives from the Greek τροπή (turning, change, related to τρέπειν, turn about), and refers here to a change in the status of energy. Loosely speaking, a comparable change in status obtains with respect to manifestations of possible variety. Instances of such variety, about whose actual occurrence no definite knowledge (certainty) is available, evidently differ from instances whose actual occurrence is recorded, known, or otherwise certain. Thus, entropy can serve as a measure of disorder, of unselected variety, of reversibility, of temperature, of probability, of uncertainty, and of other related conditions. It may be expressed by the formulation $S = k \log_e P$, where k is a constant, and P the number of 'elementary complexions'. *Negentropy* — $S = k \log_e (\frac{1}{P})$ becomes then the reciprocal of entropy. Expressed in *bits*, negentropy can also be formulated as $-S = \log_2 (\frac{1}{P})$.

Living systems produce and consume a great number of substances. The corresponding reactions, however, are not independent of each other, but become interconnected and affect each other. Any mathematical theory of such interconnected reactions is beset with extreme difficulties. Yet, because such quantitative mathematical formulations are required, it seems obvious that they must be worked out in order to achieve a satisfactory description of morphogenesis and other fundamental life processes in terms of a desirable but still not achieved exact biologic science comparable to physics and chemistry. Such quantitatively or numerically exact biology, however, could then at best represent only one aspect of biologic science in the wider sense, since the study of life must deal with many phenomena beyond the scope of physics and chemistry (cf. K., 1966). These latter sciences, despite their exactness, their complexity, and their high degree of forbidding esoteric mathematical sophistication, may be considered elementary, simple, and primitive in comparison with biology as conceived from a comprehensive point of view.

The unsatisfactory state of theories attempting to provide an understanding of biological events leading to morphogenesis is evident not only with respect to ontogenetic but also to phylogenetic evolution. Because of the multifactorial aspects prevailing in 'phylogenetic mechanisms', a wide variety of *prima facie* plausible, but insufficiently demonstrable, perhaps partly valid, but rather unconvincing models can be elaborated (cf. e.g. the recently propounded hypothesis of OHNO, 1970, concerning 'evolution by gene duplication').

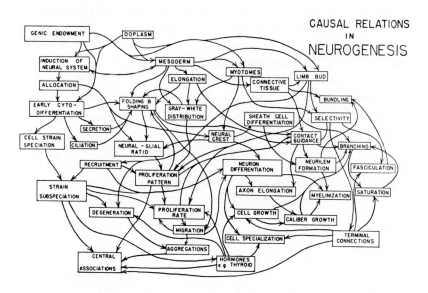

*Figure 7B.* Diagram of 'causal relations in neurogenesis' according to WEISS (1955).

WEISS (1955) justly comments on some of the intricate and complex causal events (interactions) displayed by merely one morphogenetic *(sensu latiori)* manifestation of some living systems, namely by neurogenesis. He remarks that such processes present even considerable difficulties for verbal summarization, and points out that a glibly generalized formula may end up with either meaningless platitudes or improper oversimplifications. I would, however, maintain that some fictional formulations, based on judicious oversimplifications, fully understood as such, are not only unavoidable, but definitely helpful. WEISS (loc. cit.) then elaborated a useful diagram (Fig. 7B) which graphically depicts the complexity of the assumed interrelations and interactions, expressed by single- and double-headed arrows.

It is evident that, in attempting to investigate the problems of neurogenesis in the wider sense, a variety of entirely different approaches can be chosen. Such approaches, however, should be based on appropriate semantic formulations of the problems at issue.

In this respect, P. WEISS (1950) has presented a suitable classification of the relevant aspects in the introduction to a 'Conference on Genetic Neurology'. WEISS distinguishes here the following 'separable' topics with a number of further subdivisions.

I. Differentiation of the central nervous system and of its units; (a) allocation, including 'inductions'; (b) transformations, e.g. by shift of cells and changes of mechanical configuration; (c) proliferation; (d) extensive migration, including the problems pertaining to neurobiotaxis; (3) resorption; (f) differentiation, particularly in its biochemical aspect, and (g) elaboration, particularly with regard to response patterns.

II. Outgrowth of nerve fibers; (a) ultrastructure; (b) mechanism of axon growth; (c) orientation; (d) grouping into nerves or fiber tracts, i.e. 'fasciculation'; branching and plexus formation; (f) peripheral connections, and (g) numerical saturation.

III. Neuron growth, including metabolic requirements and 'trophic effects'.

IV. Nerve regeneration, including problems of selectivity[30a].

V. Specificity and selectivity, including the problems of 'modulation' and 'resonance'.

VI. Development of behavior.

----

[30a] Problems of nerve regeneration and selectivity, discussed in the preceding chapter V (vol. 3, I, p. 701) of this series, and in the monograph *'Brain and Consciousness'* (K., 1957, pp. 122, 140–141) also involve what was there, for want of a better word,

It can easily be seen that, although WEISS (loc. cit.) justifiably regards these topics as 'relatively separable', there obtains considerable overlap of the problems under consideration. In dealing with this subject matter, I have followed a somewhat different arrangement, with particular emphasis on the distinction between structure (texture) and form. References to many of the topics enumerated by WEISS were therefore included in the preceding chapter V (volume 3, part I of this series). Additional comments on the growth of the nervous system, particularly with reference to the formulations propounded by WEISS, and based on the results of the new technical approaches, were presented by SZENTÀGOTHAI and others in a recent symposium on the growth of the nervous system (WOLSTENHOLME and O'CONNOR, eds., 1968). Yet, despite progress in techniques and a considerable amount of publications disclosing new data, it could be maintained that no significant new insights into the cited problems have been gained within the period marked by the two symposia (1950–1968).

With regard to problems of general morphogenesis, concerning *'the forces that shape the embryo'*, TRINKAUS (1969) has recently summarized some of the main points at issue in contemporary investigations of 'developmental biology'. The cell properties believed to be here of importance are: surface adhesiveness, surface deformability, the ability to form minute locomotor structures such as ruffled membranes, lobopodia, and filopodia, moreover changes in certain ultrastructures within the cytoplasm (microfilaments, microtubules), and contact inhibition (e.g. the tendency of fibroblasts to cease movements upon contacting other fibroblasts). The importance of the ultrastructural cell membrane and its chemical constitution is stressed, and the genic control of morphogenetic movements represents still another significant parameter. Cell differentiation, cell contacts, moreover displacements and distortions of cohesive sheets of cells, which involve both individual cell locomotion and the kinematics of whole aggregates, quite obviously all depend on the interplay of a multitude of physico-chemical variables.

---

termed *'lability'*. In the aspect under consideration, this term may be defined as stressing the reversion, from the adult stability of a ('fixed postmitotic') 'perennial element' to a condition compatible with extensive regenerative growth processes of 'embryonic' type (e.g. regeneration of optic nerve and tract with re-establishment of proper functional tectum opticum connections in eye transplantations performed on some urodele amphibians).

Yet, I believe that unless these diverse vaguely defined factors can (1) be expressed in terms of accurately measurable magnitudes, and (2) be dealt with in proper mathematical formulations adapted to the multifactorial interactions and to their relevant specific boundary conditions, no satisfactory or rigorous 'explanations' of morphogenetic events become possible. The required mathematical procedures for dealing with the 'forces that shape the embryo' imply problems whose extreme difficulties by far transcend those encountered in attempting to explain less complex physical, chemical or biologic events. The 'tense expectancy' said to pervade a particular and technologically sophisticated 'biological community' which looks forward to more and more, increasingly spectacular *'breakthroughs'*, might well be unduly optimistic.

The orderliness of *spatial configuration* manifested by the material aggregates providing the substratum of living organisms subsumes nonmetrical relational phenomena of primary concern for *morphology*, including its subdivision *morphogenesis*. The intuitively[30b] recognized *'sameness'* of form displayed by groups of organisms, and in different sorts of organisms, by particular stages of ontogenesis as well as by those body parts that are considered *homologous*, implies invariants which evidently fall under the purview of *topology*.

In his earlier endeavors to establish a rational theoretical basis for the controversial concepts of *morphologic pattern (bauplan)* and of *homology*, the author (K., 1929a) referred to *gestalt-theory*, which likewise implies invariants under transformation. Subsequently realizing that the relevant problems could be more conveniently formulated in terms of an elementary topology[31] (K., 1954, p. 63), a first attempt[32] at the *topologization of morphology* was undertaken in volume 1 of this series (K., 1967, pp. 165–211).

---

[30b] *'Intuition'* can here be taken in the *everyday sense*, such as directly experiencing the 'sameness' of several perceived objects (e.g. of the humerus in an array of skeletons), or in the *mathematical sense* as upheld by the 'intuitionist' POINCARÉ (cf. K., *Mind and Matter*, pp. 293–294, also, loc. cit., § 48, p. 195).

[31] The designation 'weak' topology used in vol. 1, p. 194 with reference to this simple morphologic topology was explicitly qualified as taken in a loose sense, but should be discarded in order to avoid confusion with the highly specialized term 'weak topology' which has an entirely different meaning in the rigorous topology of advanced mathematics.

[32] The cited chapter contains a generalized elaboration on the problem of *'invariance'*, involving diverse apparently heterogenous aspects of mathematics not directly related to

Although apparently very few references to topologic concepts can be found in biologic literature, a small number of authors in this field did advocate the application of topology, namely LEWIN (1936), RASH-EVSKY (1960), KRONSBEIN and STEELE (1967), and RHINES (1967). Because the topologic space of morphology represents a concrete, directly observable 'natural' space, involving 'material' neighborhoods, my own application of topology to the here pertinent problems differs significantly from that by LEWIN (1936) for psychology, and by RASHEVSKY (1960) for mathematical biophysics. Both cited authors deal with *'functional'* concepts in a highly abstract topologic space, and disregard

---

the simplified topologic space of morphology. These comments were included in order to correlate the present series with the preceding monographs (K., 1957, *Brain and Consciousness;* K., 1961, *Mind and Matter)*, and to stress the background of *neurological epistemology* upon which  the present series is based. Again, in addition to topology, several other important qualitative divisions of mathematics deal with mappings and relations, namely the theory of sets, of groups, and various subdivisions of mathematical logic.

Since the author is neither a topologist nor even an expert mathematician, but a neurobiologist aiming to familiarize himself with some aspects of mathematics highly relevant to his own field, this first attempt at the 'topologization of morphology' unavoidably suffers from a certain weakness due to relaxed rigor. This, however, does not invalidate its basic soundness. Although, because of circumstances, the author was unfortunately compelled to prepare his presentation of topology without assistance or advice by a mathematician and topologist, he subsequently had the opportunity to discuss the relevant problems with a topologist (Dr. W. BETTINGER) from the Mathematical Institute of the University of Würzburg, who kindly proffered diverse constructive criticisms which will be helpful for the future clarification of weaknesses and ambiguities.

Five accidental errata in vol. 1, chapter III, due to inadequate proof-reading, should also be pointed out: (1) In footnote 8, p. 170, line 8 from bottom, *Fermat's theorem* must read $x^n + y^n = z^n$. (2) In footnote 10, p. 172, the transformation must read $[(a_0 \, b_1) \, (b_0 \, a_1)]$. (3) On p. 174, line 12–13, it must read 'an element is called an inner element of A' instead of 'of T'. (4) On p. 177, line 8–9 from top, 'isomorphic[15]' must be substituted for 'homomorphic[15]'. (5) Likewise on p. 177, footnote 15, end of first line, 'isomorphism' must be substituted for 'homomorphism'.

I might add that, even in a fundamental innovation introduced by a mathematician of the very first rank such as B. RIEMANN (1826–1866), namely in *'Grundlagen für eine allgemeine Theorie der Funktionen einer veränderlichen complexen Grösse'* (1851), faulty reasoning concerning an important step was pointed out by WEIERSTRASS and then admitted by RIEMANN, without detriment to the significance of that dissertation. Again, in attempting to deal with psychology, metaphysics, natural philosophy, and the mechanics of audition, the valuable contributions of RIEMANN (reprinted in 'Collected Works' 1953, pp. 509–538, 338–350) are likewise not exempt of various weaknesses concerning epistemologic rigor and biology.

morphology.[33] Again, while my own topologic approach deals with problems of *form* as distinguished from *structure* (cf. above, p. 1), the topologic approach discussed by KRONSBEIN and STEELE (1967), and by RHINES (1967), who are metallurgists, concerns problems of *structure* ('grain', 'texture') in both anorganic and organic stereology.

A morphologic pattern or bauplan can be conceived as a simplified three-dimensional topologic space representing an arcwise connected set whose subsets are partially and hierarchically ordered neighborhoods corresponding to cellular aggregates and their derivatives. Regardless of their shape and size, different morphologic configurations display the same bauplan if they are topologically identical, that is *homeomorph*. The comparison of two or more morphologic configurations, indicating whether such identity does or does not obtain, is based upon an operation or 'transformation' called *mapping*, whereby the identifiable neighborhoods of a given configuration can be brought in *one-to-one* correspondence with those of other configurations such that the basic relationships of connectedness are preserved. If to each element of a space A there corresponds a unique element $f(x)$ of a space B, then there is said to be a mapping or map of the set A in the set B, and the element $f(x)$ is said to be the image of the element x.[34] A mapping is

---

[33] RASHEVSKY *(1960)*, in fact, expresses the opinion that morphologic applications of topology deal only with 'rather superficial relations'. According to this author, 'topological analogies go much deeper in the realm of the living when we observe not merely structural but functional (in a biological sense) relations'. It goes without saying that, being inclined to uphold and stress the '*gleichberechtigte*' import of morphology, I do not share RASHEVSKY's evaluation, although I readily grant the importance of the cited author's functional viewpoint. Thus, in discussing the topological properties of graphs, RASHEVSKY emphasizes the significance not only of connectedness but also of direction. Two graphs are here said to be topologically identical or homeomorph, 'if they can be obtained from each other by mere deformation, without breaking any of the lines, so as to preserve the number of points and lines, and the relations between points and lines, including the direction of the lines'. All nondirected triangles, regardless of their shape and size, are homeomorph, but in graphs with directed lines (vectors) triangles are not homeomorph, if one or more of their sides are represented by vectors running in opposite direction (e.g. from A to B and from B to A).

[34] Under certain rules, the mapping of a set A correlating points of A with points of B is said to be a mapping of A *into* B. It is mapping of A *onto* B if each point of B is the image of a least one point of A. With respect to this distinction it seems appropriate to remark that, in the simplified topologic space of morphology, in which the smallest elements are spots of finite smallness, mathematical points have no existence. Because of this and additional complications, the application of the rigorous topological concepts '*into*' and '*onto*'

said to be continuous, if neighborhoods map on neighborhoods.[35] In order to be homeomorph, a one-to-one mapping must be bicontinuous, that is, continuous both ways, whether configuration A is mapped on configuration B or *vice versa*.

For the purposes of a simplified morphologic topology, this bicontinuous one-to-one *homeomorphism* may be considered as synonymous with *isomorphism sensu strictiori*. Since morphologic patterns, however, despite fundamental invariance with respect to their basic components ('Grundbestandteile'), can display considerable differences with regard to the differentiations or further subdivisions of these components into subsets (form elements, 'Formbestandteile'), the application of one-many and many-one mappings becomes necessary. It is here convenient to use the term *homomorphism* in agreement with ASHBY's procedure (1956).[36] *Homomorphism* (as distinguished from *homeomorphism* or *isomorphism sensu stricto*) obtains, when a many-one transformation, applied to the more complex domain, can reduce it to a form that is isomorphic with the simpler. Again, it is evident that, while an isomorphic *(sensu stricto*, i.e. homeomorphic) mapping is necessarily one-to-one,

---

as referring to mappings is here beset with certain difficulties and ambiguities. It seems therefore preferable to use the noncommittal terms *on* or *upon* in connection with mappings in the *simplified topology of morphologic space*.

[35] This nonrigorous definition should be sufficient in the aspect under consideration. More rigorously, with reference to chapter III of volume 1 in this series, and amending a definition given there on p. 177, a mapping T–T′ is continuous, if the original image of every open set in T′ is open in T ('eine Abbildung T nach T′ ist stetig, wenn das Urbild jeder offenen Menge in T′ offen in T ist').

[36] Cf. also, particularly with respect to the concepts of *closed* and of *open transformations* related to *group theory*, the here-pertinent simplified definitions in K., Mind and Matter (1961), pp. 365–366, and in vol. 1, p. 166 of this series. It should, moreover, be added that the term *homomorphy* has been used by morphologists with a somewhat different denotation, implying 'similar size or shape'. Homomorphy is then meant to designate dubious instances of morphologic 'similarity' or 'sameness' whose classification as analogy or (kat-) homology remains debatable. My own terminology is based on an interpretation of JACOBSHAGEN's fundamental comparative anatomical concepts in terms of elementary topology, and has been dealt with in chapter III of volume 1. The detailed reviews of the homology concept by LUBOSCH (1925, 1931) and NAEF (1931) will disclose, to the interested reader, the multiplicity of fluctuating and ambiguous terms propounded by the numerous authors who discussed the homology problem, for which, in my opinion, simplified concepts of topology, devoid of phylogenetic implications, provide the only reasonably satisfactory formulation. This latter, of course, by no means precludes or prevents further plausible phylogenetic speculations based on the observed morphologic relationships.

a homorphic mapping is single-valued only in one direction. Thus, two different organisms, or given organs in such organisms, may each display a bauplan which is topologically identical, i.e. homeomorph with respect to 'Grundbestandteile', but involves greater or lesser complexity with respect to further subdivisions of these basic components.

It is then possible to state that all formed elements or sets of formed elements which, although occurring in different taxonomic groups of organisms, can be mapped by an isomorphic or homomorphic topologic transformation upon a given component (set or subset) of a common morphologic pattern, are the '*same*' or, in other words, *homologous*. If the mapping is isomorphic, that is one-one, *special orthohomology* obtains. If the mapping is homomorphic by one-many or many-one transformations, *special kathomology* obtains, *augmentative* in the first case, and *defective* in the second. If the homomorphism involves repetitive sequences along the main body axis (metamerism), an *allomeric* type of special kathomology obtains.

In contradistinction to special homology, *general morphologic homology* refers to grouped as well as randomly distributed repetitive elements within an extended domain for which a morphologic pattern can be formulated, and applies to multiple configurations such as bony and horny scales, teeth, feathers, hairs, claws, nails and hoofs.

A third major class of homology is represented by *promorphologic homology*, namely by isomorphism or homomorphism within promorphologic entities of a single organism such as metameres, antimeres, and parameres. Thus, in the metameric extremities of one and the same vertebrate organism, the humerus and the femur provide an example of promorphologic orthohomology (JACOBSHAGEN, 1925). Generally speaking, some authors consider homology as referring to *interindividual* morphologic relationships, while the *intraindividual* correspondences representing *promorphologic homology* are then subsumed under the term *homotypy*.

In addition to the qualifying terms as defined above, namely *ortho-* and *kat-* (*kata*, implying a downgrading), augmentative, defective, and allomeric (including 'mixed forms'), the supplemental terms *homo-* (*iso-*), and *hetero-* can be used, as shown in the following tabulation, to indicate whether homologous configurations do or do not display the same shape, structure (texture, grain), function, and anatomical (as contrasted with topological) position, or whether configurations which are homologous in their definitive (adult) connectedness, are derived from nonhomologous ontogenetic matrices (e.g. certain muscles of Mammalia in comparison with those of Anamnia and *vice versa*):

*Shape:*    *homo-* and *heteroeidetic* homology
*Structure:* *homo-* and *heterotypic* homology
*Function:* *homo-* and *heteropractic* homology
          *iso-* and *heteropractic* homology
*Position:* *homo-* and *heterotopic* homology
*Matrix:*  *homo-* and *heteromeric* homology

Different organic configurations which cannot be compared with each other by means of isomorphic or homomorphic mappings, although displaying some sorts of invariants concerning structure or function, or both, do not represent, or do not pertain to, the same bauplan, and therefore do not manifest homology. The obtaining *invariance*[37] or *similarity* is then expressed by the term (morphologic, biologic) *analogy*, of which several definitions were given in volume 1, pp. 205–210 of this series.

For practical purposes, a morphologic pattern or *bauplan*, which is an elementary three-dimensional topologic space, can conveniently be depicted by *two-dimensional diagrams* in a plane to which one of the three conventional sorts of orthogonal body axes (e.g. rostro-caudal, dorso-ventral, right-left) is perpendicular. The cutting of a configurated space in the course of a deformation, which is here provided by the mapping upon a figure or diagram, becomes a permissible procedure. It is admitted that if a figure is cut during a deformation, and the edges cut are joined after the deformation in exactly the same way as before, the process still defines a topological transformation of the original figure. In morphology, mental reconstruction by the scanning of serial sections, or actual physical reconstructions by means of graphic or plastic models provide the required joining. Such cutting-up or '*tearing*' of topologic spaces representing complex interconnected systems greatly simplifies their analysis, and methods of 'tearing' have been applied to the practical solution of certain topologic problems (cf. vol. 1, p. 178).

A well-meaning critic (*Neurology* 19, p. 517, 1969) has stated that one objection to the '*relaxed morphologic topology*' which I have introduced could be based on what he characterizes as 'a serious lack of pre-

---

[37] From a general viewpoint adapted to nonmathematicians, the rather complex and difficult concept of *invariance* (including that of so-called 'relative' and 'absolute' invariants) is discussed in a publication by BELL (1951).

ciseness which D'Arcy Thompson was able to bring to his studies on biological forms'.

To this comment, it can be replied that the preciseness displayed by Thompson's ingenious transformations is intrinsic to their metric, quantitative nature. Yet, as Thompson (1917, 1942) himself admits, 'there are many organic forms which we cannot describe, still less define, in mathematical terms'. Evidently, by 'mathematical terms', Thompson means here quantitative metric terms, and ignores the topologic approach, which is likewise 'mathematical'.

The '*relaxed morphologic topology*', however, applies just to those 'distortions' and transformations, which still completely defy quantitative metric methods. It places into the given manifold a *vaguely localized* but *combinatorially exactly* determined framework. Being thus a higher generalization of 'deformation', 'distortion', or 'transformation' than the metric methods, it is, therefore, in the relevant aspects under consideration, not only more powerful, but more accurate.

It is a common mistake to confound '*preciseness*' or '*precision*' with '*accuracy*' (cf. vol. 2, p. 310, footnote 67, of this series). Precise figures, based on unreliable data or procedures may be highly inaccurate although giving the appearance of (quite illusory) accuracy, while imprecise statements, if exactly referring to the available or obtainable data, may be completely accurate.

The considerable weaknesses inherent in Thompson's clever procedure have been justly pointed out by J. Huxley (1932) and were also discussed in volume 1 (p. 186f) of this series. Although Thompson's transformations are *quantitative* (metric) and may, therefore, give highly *precise* results, these results are intrinsically *inaccurate* and can, in Huxley's words, 'only give a general and *qualitative* picture of the mechanism at work, in place of a specific and quantitative one'.

Moreover, in order to give significant results within its limited, highly inaccurate capability, Thompson's method requires curvilinear coordinates based on tensor transformations, which necessitate the application of rather formidable advanced mathematical procedures. Thompson restricts himself to simplified graphic solutions of his complex *curvilinear* coordinate transformations. In addition, he uses as a first approximation coordinate systems of simpler type, such as *Cartesian* rectangular equidistant coordinates with subsequent rectangular deformations, oblique straight-lined coordinates, and radial or polar coordinates with circular arcs cutting the radii orthogonally. Elementary mathematical procedures, which Thompson briefly points out, are here sufficient. Needless to say, these transformations suffer from still more pronounced lack of accuracy.

THOMPSON's method of elementary coordinate transformation, in terms of rectilinear systems (equidistant *Cartesian*, rectangular, oblique, trapezoidal, hour-glass shaped rectilinear) has been recently applied to the deformations produced by the walls of a tubular viscus on a contained body, the avian egg being used as a model system (SMART, 1969). The cited paper indicates the derivation of the relevant equations and the steps involved in the transformations.

Numerous manifestations of organic *life* seem to differ fundamentally from the behavior displayed, in accordance with the known norms of physics and chemistry, by nonliving matter. The orderly events characterizing ontogenetic evolution, including the aspects of morphogenesis dealt with in this chapter, are conspicuous examples of such specifically *vital phenomena*. Nevertheless, it can be maintained that, despite the alluded apparent difficulties, all objectively[38] recordable manifestations of life may, in principle, be adequately described[38a] (i.e. explained) in terms of physics, chemistry, communication theory, and form analysis, all of which, again, involve various aspects of mathematics, including topology.

In contradiction to this view, which may be designated as '*mechanistic*' or simply as '*mechanism*', the '*vitalistic*' views deny that such mechanistic 'explanation' of life is possible, and claim that life is a *nonphysical* or *extra-physical phenomenon*, not reducible to physics in the

---

[38] ASHBY (1952) expressed the opinion that 'science' deals, and can deal, only with what one observer 'can demonstrate to another'. Although this concept of 'science' can easily be challenged, the restriction imposed by ASHBY's view may be used in defining the term 'objectively' as well as the behavioristic approach to science in general.

[38a] This includes manifestations of 'instinctive behavior' so complex and precise in their execution as to baffle even confirmed behaviorists such as the late Professor LASHLEY, who 'stood aghast' at the inadequacy of our concepts of their mechanisms (cf. vol. 2, p. 45 of this series). The behavior there referred to concerned the apparently 'purposeful' use of ingested coelenterate nematocysts by the Platyhelminth turbellarian *Microstoma*, which KEPNER interpreted in mentalistic terms. It should here be added that the 'convenient' use of ingested coelenterate nematocysts by an alien predatory organism is not restricted to the cited Platyhelminth, but is also well-known to occur in the Phylum Mollusca. Some *Nudibranch* Gastropods ('snails') pertaining to this invertebrate group store and utilize for their own defense the nematocysts of the venomous siphonophore *Physalia*. A reference to that behavior as likewise displayed by Mollusca should have been included in both section 4 and section 10 of the cited volume, but was inadvertently omitted.

wider sense. One can evidently require that, in order to rebut *vitalism*, it is first necessary to *define* vitalism. Now, in its most generalized meaning, vitalism is merely an unverified and at present still unverifiable proposition *denying* 'mechanism' as defined above. The truth of this proposition can neither be rigorously proved nor disproved, but it must be kept in mind that, for propositions of this sort, non-disprovability does not imply provability, and is, therefore, not equivalent to a proof.

As regards particular vitalistic theories, the concrete older ones, based on *'animal spirits'* and similar ill-defined 'fluids', have been generally discredited by the obtained data, and the more sophisticated newer ones, including those of a cleverly disguised 'cryptovitalistic' type, operate with insufficiently defined and therefore hazy, evasive, or slippery concepts.

Thus, recent vitalistic authors[39] (e.g. Driesch, Eccles, and others) postulate some specific 'principles' essentially differing from the manifestations of physicalisms, that is, from the behavior of matter (i.e. mass and energy) as described by physics, yet capable of acting upon matter, and supposedly required for an understanding of intrinsically biologic phenomena pertaining to the objective aspects of life. However, adherents of vitalism have failed to provide any intelligible, coherent, and reasonably satisfactory formulation concerning nature and action of such alleged 'principles'.

As regards ontogenetic events, Gurwitsch (1914) pointed out factors which, in his opinion, were of a nonmaterial nature. In his very interesting and sober paper, the cited author referred to a potentially preformed configuration *('dynamisch präformierte Morphe')*. He assumed this to be a *nonmaterial configuration* 'located' within the embryo and becoming subsequently materialized.[40] However, the potentially preformed configuration, justly postulated by Gurwitsch, can now be

---

[39] Cf. vol. 1, p. 23 of this series. It might be added that the vitalistic concept propounded by Eccles can be evaluated as more closely related to the naive older animistic 'fluid' theories than to the sophisticated recent ones (e.g. of Driesch), and fully deserves the accolade of the *Pontifical Academy*.

[40] 'Eine materiell nicht gekennzeichnete Configuration an einem bestimmten Orte des Keimes, welche erst nachträglich materialisiert wird' (Gurwitsch, 1914, p. 137). Anticipating the introduction of 'coding concepts' into biology, Gurwitsch also makes use of the term *'Zellsprache'* in this very interesting paper.

conceived as entirely '*material*', namely as the variety or information[41] encoded in the genome.

Additional views, closely related to vitalism, invoke *teleology*, that is, the concept of purpose. Since purpose represents a *phenomenon of consciousness*, such as intention ('will'), desire, like and dislike, or for that matter *qua* consciousness, the experienced color red etc., it seems obvious that this concept has no logical existence in behavioristic (objective) sciences, which exclude consciousness. It is a common error to confuse extrapolative or directed behavior, in cases where conscious activities must be precluded, with purpose, and to evaluate factors, resulting in a given effect, which is arbitrarily considered as an 'end', in terms of purposefulness. Some recent authors, including MONOD and JACOB (1961), make use of the terms '*teleonomy*' and '*teleo-*

---

[41] In an array of unselected variety, each variety can be assumed to have a certain probability, which, given the required premises, may be numerically *(logarithmically)* expressed in terms of *entropy*. Selection of variety, which then either implies certainty or at least reduction of uncertainty, can then be expressed in the *reciprocal* terms of *negentropy*.

*Information theory*, based on these concepts, can be said to use a syntax language, and completely ignores the value (in terms of referential language dealing with actual events) or the 'meaning' of transmitted and processed information. It is evident that it must be so, since such value or meaning pertains to the domain of *qualités pures*, namely of conscious interpretation which uses a truly referential object language (K., 1968).

Information has been recently defined by a behaviorist as a 'causal relation' between input and output variables. This definition may be challenged, since, in accordance with the principle of sufficient reason, the indeed existing causal relationship between input and output concerns the physical, dynamic aspects of the particular mechanisms involved, following the *principium rationis sufficientis fiendi*. The information flow, on the other hand, concerns probability and pattern aspects, which are not causal, but follow the *principium rationis sufficientis essendi* (cf. volume 1, chapter III, pp. 290–292). It is furthermore evident that the introduction of *probability*, which is incompatible with the concept of *causality*, imposes the necessity of dealing with the events under consideration by means of an entirely different logical model, as justly pointed out by MAX PLANCK (1960). In a similar manner, the application of the quantum concept likewise precludes the use of causal concepts, which, in the rigorous *Newtonian formulation* applying to classical physics, are based on *continuity*.

Finally, the superposition of logical concepts, in terms of a syntax language, upon information flow as provided by various computer mechanisms, pertains to the domain of the *principium rationis sufficientis cognoscendi*. It should therefore be stressed that information theory *sensu stricto* (as essentially founded by C.E. SHANNON about 1948) is a branch of *probability theory* concerned with the transmission of messages. Loosely speaking and *sensu latiori* information theory is, however, frequently identified with a generalized communication theory or with so-called 'cybernetics'.

*nomic'* as an adaptation of, or as a substitution for, 'teleology' and 'teleological'. There is little justification for this prodedure, which is merely an attempt at reintroducing discredited aspects of teleology in a sophisticated but thinly disguised manner.

It is evident that, because of the *brain paradox* (K., 1959, 1961, 1965, 1966), consciousness cannot be described or explained in terms of mechanism, since experienced or conceived (conceptual) 'matter' and the formulation as well as the understanding of mechanistic explanations already represent aspects of consciousness. The thereby imposed limitations and restrictions have, nevertheless, no bearing upon the significant validity of a behavioristic or 'materialistic' approach to the phenomena dealt with by natural sciences. The formulations and notations of their approach constitute a fictional 'universe of discourse' in which the concept of consciousness is not only devoid of any operational significance, but has, in fact, no logical existence. In the same manner, a rigorous behavioristic approach exludes *axiologic* valuations implying preferability or nonpreferability and based on concepts such as 'good' and 'bad', or 'like' and 'dislike'. It is merely permissible, in evaluating the significance of events, to state that 'this is compatible (accordant, consistent, consonant, congruent, in agreement) with that' ('adapted' to that), or 'this is not compatible etc. with that'. The theoretical background for an understanding of the relevant epistemologic problems has been elaborated in my previous monographs (1957, 1961) and was again summarized in two recent papers (1966, 1968).

If, transcending the behavioristic approach, biologic sciences also deal with manifestations of consciousness *qua qualités pures*, then, of course, the opinion that life is a nonphysical or extraphysical phenomenon, not reducible to physics in the wider sense, appears both justified and unjustified (rather than 'true' and 'false'). Such opinion becomes justified with respect to the *brain paradox*, and can be held unjustified because one may reasonably postulate that all biological phenomena related to the behavior of material systems (displaying mass and energy in a physical space-time system) must be described and can be thereby explained in terms of 'mechanism' in the here-adopted sense. This postulate is not invalidated by the admitted and well-known fact that all biological phenomena cannot be explained in terms of *contemporary* physics and natural sciences. Thus, the previous failures of classical physics to explain phenomena, now describable in terms of quantum physics, did not characterize said phenomena as extraphysical. Again, the recording of previously unknown effects (as e.g. radioactivity by 19th century physicists) may initiate entirely new branches of study.

In other words, the specifically vital behavior of matter, as displayed by living organisms, can be assumed to result from the *combinatorial and permutational superposition* as well as *interaction* of multitudinous physical and chemical systems, such that the behavior of complex arrays may substantially differ from that of their constituent subsets or elements at various levels of lesser complexity. Yet, despite these considerable differences, the properties of a 'whole' can still be said to remain *based* upon the *intrinsic* properties of its parts.[42] It seems therefore evident that the complex and still very poorly understood events in ontogenetic and phylogenetic morphogenesis do not require the invoking of vitalistic, cryptovitalistic, teleologic, teleonomic, and similar (e.g. parapsychologic or theological) concepts in guise of 'explanation'.

However, with respect to the aforementioned circumstances and contingencies, it is obvious that *neurobiology* assumes a *unique status*, if it be granted that consciousness represents a function of the brain. The correlation of mental phenomena (in private perceptual space-time) with the neurobiological physicalisms (in public physical space-time), which is thereby required for certain aspects of neurobiologic science, involves intricate semantic and epistemic problems. A reasonable approach to these questions must be based on rigorous definitions and distinctions concerning the phenomena pertaining to either of both *prima facie* entirely disparate domains, representing separate 'universes of discourse', namely to the domain of physicalisms or to that of mind.[43] The principles of *neurologic epistemology*, which deals with the here relevant issues, provide therefore the background for the present series on the vertebrate central nervous system. The topics considered in these volumes, although implying a strictly 'mechanistic' and rigor-

---

[42] Cf. K., Mind and Matter (1961) pp. 425–426; also vol. 1, p. 10 of this series, and K., Brain and Consciousness (1957) pp. 96–103. It may be added that 'superposition' implies an 'hierarchical' order.

[43] If the phenomena of *mind (consciousness)* are considered, then the non rational ('irrational' or extra-logical *qua* formal systems of logic) manifestations of affectivity assume a paramount significance, implying axiologic values, choice, decision, and purpose. Such decisions and similar related phenomena pertain to the domain of the *principium rationis sufficientis agendi* (cf. vol. 1, pp. 290–291, Brain and Consciousness, 1957, p. 8, Mind and Matter, 1961, pp. 102–103 *et passim*).

ously 'materialistic' approach, in which consciousness or 'mind' have no logical existence, are discussed from an advanced standpoint allowing for the use of the materialistic and behavioristic data in those formulations which necessarily require the inclusion of mental phenomena into the domain of neurobiology.

With regard to an assumed difference between *'description'* and *'explanation'*, reference may here be made to the comments, in volume 1, chapter III (p. 289f), on the significance of these terms. Depending upon arbitrary definitions, one might either say that the meanings of both words overlap, or that 'explanation' should be subsumed under 'description' as a set of special cases, namely as 'description' in accordance with the principle of sufficient reason in one of its five applications.

Thus, if description in the wider sense refers to the verbal formulation of experienced or observed data (events, 'things'), then explanation, which also implies a reduction of 'unknown' to 'known' data of experience, can be said to mean such formulation with regard to material, logical, arithmetical, geometrical, topological relationships, as the case may be, or particularly to 'cause' *sensu stricto*, or to purpose. Some persons may only consider a formulation of the 'cause' or of the 'purpose' as a 'true' explanation. Children, for instance, have the tendency to believe that everything has a purpose, as if it were made by and for man. Experts in child psychology such as PIAGET and others have justly pointed out that the well-known and often unanswerable infantile *'why'* questions are related to this mental attitude. Because children assume that everything has a purpose they may ask 'Why do the stars shine ?', or 'Why is the grass green ?'. The parents who attempt to answer such questions with a causal, physical explanation will miss the point. Accordingly, upon the question 'Why do we have snow ?' the answer 'It is for children to play in' might here provide a fully satisfactory explanation.

Again, HAECKEL, in his *'Lebenswunder'* (1906) objects to the term *'beschreibende Naturwissenschaft'* (e.g. morphology) as contrasted with *'erklärende Naturwissenschaft'* (e.g. physics). He also disputes the claim of noted scientists such as R. VIRCHOW (1821–1902) and G. R. KIRCH-HOFF (1824–1887), that complete 'description' is the ultimate aim of natural sciences. It is evident that VIRCHOW and KIRCHHOFF interpreted 'description' as including formulations in accordance with the principle of sufficient reason, while HAECKEL distinguishes description of 'facts' (*'Beschreibung der einzelnen Tatsachen'*) from explanation, which, in his opinion, must refer to 'causes' (*'Erklärung durch die bewirkenden Ur-sachen'*). The difference in opinion concerns here merely arbitrary questions of semantics.

In concluding this discussion on those aspects of morphogenesis related to the general behavioral aspects of 'living matter' which are characterized by a flow of energy and mass through particular sorts of configuration,[43a] it should again be emphasized that living organisms represent a special case of what, in communication and information theory (cf. ASHBY, 1957, 1960) is called *a very large probabilistic system*, richly cross-connected and interconnected (cf. Fig. 7B), *with stable and ultrastable properties*. The mathematical analysis of such systems requires, in addition to other procedures (cf. RASHEVSKY, 1960), a special sort of statistical approach (v. NEUMANN, 1952).

Up to now, the rigorous scientific analysis of living systems remains still at a very primitive, initial, and unsatisfactory stage. With regard to the two main aspects of their behavior, namely (a) *function* (in terms of energy flow as well as metabolic transformations), and (b) the *manifestations of configuration* (origin and evolution as well as maintenance of configurational patterns), it is evident that some special subsystems, or special properties of numerous subsystems, are mainly related to (a), and others mainly to (b). It is moreover obvious that, while a fair amount of detail has already been elucidated concerning the systems pertaining to group (a), such as the *Krebs cycle*, the pentose pathway (*Embden-Meyerhof pathway*, the cytochrome or cytochrome oxydase systems (electron transmitter systems), and the relevance as well as structure of nucleic acids, etc., nothing of any significance is known about the *modus operandi* of subsystems subsumed under group (b), that is, concerned with the displacement of mass required for the formation of specific organic configurational patterns, and with the maintenance of such patterns despite considerable interference.

Thus, in proceeding with the discussion of actual configurational aspects, no more than an account of formal stages and interrelating formal events (configurational changes) can be given. The nature, distribution, and action of the pertinent *forces* (physico-chemical *interactions*), which would require advanced mathematical procedures, if the required data were known and could be properly formulated, remain,

---

[43a] Cf. the definitions of organic life elaborated and discussed in K., Brain and Consciousness (1957), p. 12f.

except for vague and essentially meaningless conjectures,[43b] an entirely unknown domain whose exploration is beset with formidable difficulties. These difficulties, however, may be assessed as due to the complexity of events and their correlation, and not to any essential or 'transcendent' incomprehensibility within the 'materialistic' universe of discourse respectively 'semantic model', which deals with numerically expressible, in principle measurable interactions or transformations involving 'matter', i.e. 'mass' and 'energy'. On the other hand, any attempt to derive *'mind'* (synonymous with *consciousness* or *awareness*) from matter or material interactions is, *ab initio*, logically impossible, since, as already intuitively apprehended by LOCKE (1689, 1690),[43c] it involves a *petitio principii*. This latter, in neurologic epistemology, can be formulated as the *brain paradox* (K., 1959, 1961).

---

[43b] E.g. most of the conclusions based on the interesting results of experimental embryology (so-called 'developmental mechanics').

[43c] The intrinsic and logically insuperable incomprehensibility of a relationship between *'physicalisms'* (i.e. mass or energy transformations) and *consciousness (mind)* was already expressed by LOCKE (1689, 1690) as follows: 'That the size, figure, and motion of one body should cause a change in the size, figure, and motion of another body, is not beyond our conception; the separation of the parts of one body upon the intrusion of another; and the change from rest to motion upon impulses; these, and the like, seem to us to have some connexion one with another. And if we knew these *primary qualities* of bodies, we might have reason to hope we might be able to know a great deal more of these operations of them one upon another: but our minds not being able to discover any connexion betwixt these primary qualities of bodies and the *sensations* that are produced in us by them, we can never be able to establish certain and undoubted rules of the consequence or coexistence of any *secondary qualities*, though we could discover the size, figure, or motion of those invisible parts which immediately produce them. We are so far from knowing what figure, size, or motion of parts produce a yellow colour, a sweet taste, or a sharp sound, that we can by no means conceive how size, figure, or motion of any particles can possibly produce in us the idea of any colour, taste, or sound whatsoever; there is no conceivable connexion betwixt the one and the other'. LOCKE, however, failed to realize that his so-called *primary qualities*, namely extension, configuration, and motion (space and time) are no less essential attributes of consciousness than the so-called secondary ones, as subsequently shown by the epistemologic concepts of BERKELEY, HUME, KANT, and SCHOPENHAUER. The *abstraction of primary qualities*, nevertheless, permits one to formulate the operationally valid model (fiction) of a physical world representing a postulated *public space-time system*. On this basis it is indeed possible, *apparently* contradicting one aspect of LOCKE's views, to establish 'what figure, size, or motion of parts produce a yellow color, a sweet taste, or a sharp sound, both *qua* physical R-events and *encoded* physical neural events (N-events). Yet, the *'translation'* of *physical N-events* into the correlated actual *mental (psychic, conscious) P-events* by means of any intelligible 'interaction' remains intrinsically incomprehensible.

In section 1 of chapter V, the origin of the *neural tube*, and the further differentiation of its cellular constituents into neuroblastic and spongioblastic elements was discussed from the viewpoint of *histogenesis*. In contradistinction, the first section of the present chapter deals with aspects of *morphogenesis sensu strictiori* which concern the ontogenetic evolution of the vertebrate neuraxis. This latter represents a *dorsal epithelial* tube including a *lumen*, and, with few exceptions manifested by delamination or by segregation and organization of compact cell masses, is formed by the closure of the infolding medullary plate in the midline, concomitantly with separation from the integumental ectoderm.

However, preceding these particular morphogenetic events, which lead to the so-called *neurula* stage, numerous complex formative processes, characterized by displacements resulting from changes in the motion of masses as referred to further above, occur in the vertebrate embryo.[44] These events lead to the consecutive stages of ontogenesis enumerated and conveniently classified, from a practical descriptive viewpoint, in volume 1 (chapter III, p. 230) of this series. With regard to the morphogenetic aspect here under consideration, some comments on the processes of *cleavage*, of *blastulation*, and of *gastrulation* are perhaps appropriate. It is evident that the relative amount of yolk[45] is correlated with significant differences concerning these early embryonic events (cf. volume 1, p. 241 and Fig. 14).

---

[44] In some chordates, e.g. tunicates, cytoplasmic movements can already be recorded following fertilization of the zygote. Thus, in *Styela*, immediately after penetration by the spermatozoon, streaming motions of the cytoplasm of the egg result first in a stratification of gray, clear, and yellow cytoplasm, and, subsequently, but prior to the first cleavage, into a segmental or crescentic arrangement which, although obtaining in a single cell (fertilized zygote), corresponds almost exactly to the configuration of comparable 'fate areas' in the multicellular blastula of Amphioxus as shown in fig. 9 A.

[45] With regard to relative amount of yolk, ova are classified as (a) *oligolecithal*, (b) *mesolecithal*, and (c) *polylecithal*. Many invertebrates, Amphioxus, and the placental mammals, have ova of type (a). Type (b) occurs in some Cyclostomes (Petromyzon), most amphibians, Dipnoans, and, as a borderline case, in marsupials. Type (c) is found in some cyclostomes (Myxine), in Selachians, Teleosts, gymnophione amphibians, reptiles, birds, and monotreme mammals. With regard to yolk distribution, ova are classified as (1) *isolecithal* (even distribution), (2) *telolecithal* (concentration toward vegetal pole), and (3) *centrolecithal* (center of ovum). Many invertebrates and presumably all placental mammals have ova of type (1); Cyclostomes, Selachians, Teleosts, Dipnoans, amphibians, Sauropsidans, monotremes, marsupials (with yolk 'vacuole'), and cephalopod mollusks display type (2). Type (3) is found in some arthropods (insects, myriapods, and various crustaceans). A further

Thus, among chordates or vertebrates, *isolecithal ova*, as found in Amphioxus and placental mammals, show total (holoblastic) and at least approximately equal cleavage. Marsupials can be included as a borderline case. Moderately *telolecithal* ova of cyclostomes and amphibians display unequal total cleavage. In contradistinction to these *holoblastic* ova, in which the entire egg cell divides, the *meroblastic*, highly telolecithal ova of fishes, reptiles, birds, and monotreme mammals manifest discoidal partial cleavage. In other words, only the protoplasmic region about the animal pole divides, while the yolk-rich 'hemisphere' about the vegetal pole does not participate in the cleavage process and remains, in this respect, inert (Fig. 8). Among fishes, ganoids show a mode of cleavage that may be considered transitional between holoblastic and meroblastic type. Thus, Acipenser and Polypterus are essentially holoblastic, while Lepidosteus is meroblastic. Again, although the ova of teleosts are, as a rule, smaller than those of amphibians, their cleavage is of conspicuously meroblastic type. In one order of Amphibia, however, namely in Gymnophiona, whose ova contain much yolk, a transition to meroblastic cleavage obtains, resulting in early ontogenetic stages somewhat similar to an embryonic disc, followed by the formation of large cells in the yolk-rich region. Dipnoans (Ceratodus, Lepidosiren, Protopterus), on the other hand, display a cleavage similar to that of holoblastic Amphibia, but, particularly in Lepidosiren, the sluggish behavior of the vegetative hemispheres suggests a tendency toward the meroblastic mode. In meroblastic aplacental mammals (Prototheria), namely in the monotremes Echidna and Ornithorhynchus, cleavage as well as the evolution of the embryonic disc proceed in a manner closely similar to those processes in Sauropsidans, although some particular characteristics, which can be disregarded from the viewpoint of a generalized discussion, subsequently become noticeable.

---

classification of ova distinguishes *cleiodic* and *noncleiodic* eggs. The former are cut off or isolated from free material exchange with the environment because of a more or less impervious shell which is, at most, permeable to gaseous exchange only (e.g. eggs of birds). In contradistinction, noncleiodic eggs, e.g. those of most aquatic Anamnia, can absorb water and dissolved substances, particularly anorganic material, from their environment. ARISTOTLE, in '*Generation of Animals*' and '*Historia Animalium*' remarks that the eggs of various animals, particularly of oviparous fishes and of cephalopods 'grow' after deposition (cf., e.g., H.A.XVII). This observation can be regarded as an early recognition of the behavior displayed by noncleiodic ova.

With regard to holoblastic ova among Metazoa in general, three main types of *cleavage patterns* are commonly recognized, designated as *radial*, *spiral* (oblique), and *bilateral*, which may, however, include some varieties with transitional characteristics. In radial cleavage, displayed by many echinoderms, by Porifera, and by Cnidaria, the successive cleavage planes are at right angles to each other, being symmetrically arranged around a polar axis, and present a radial symmetric pattern in polar view. In spiral cleavage, a rotational displacement of cell parts around the egg axis leads to an inclination or obliqueness of the mitotic spindles, in a clockwise or anticlockwise direction which alter-

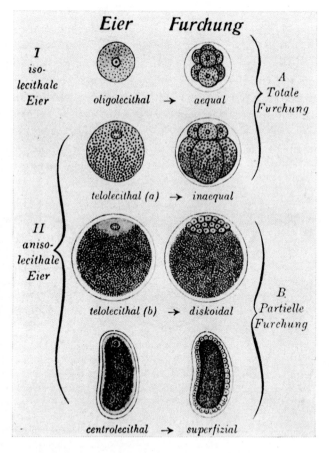

*Figure 8 A, B.* Different sorts of Metazoan ova and their corresponding cleavage types (from WEISSENBERG, 1931).

nates with regular sequences in successive cleavages, whereby a sort of 'spiral' pattern becomes apparent. At later stages of development, spiral cleavage, as a rule, changes to the bilateral type. Nemertea, Annelida, many Turbellaria, probably all Mollusca except Cephalopoda, and some other invertebrates exhibit spiral cleavage. In bilateral cleavage, the spindles and cleavage planes display a bilateral symmetry with reference to the median plane of the embryo. This type of cleavage occurs in Amphioxus, in at least some tunicates, and, among vertebrates, in amphibians and holoblastic mammals.

A peculiar type of cleavage is displayed by the meroblastic centrolecithal ova found in various arthropods (cf. above, footnote 45). Mitosis is here restricted to the peripheral cytoplasmic layer of the ovum and results in a *superficial partial cleavage* (Fig. 8 B, bottom).

As regards ontogenetic events preceding cleavage, some observations in Amphibia suggest that the plane of symmetry of the embryo may become determined in the ovum at the time of fertilization.

In frogs and various urodeles, the upper (animal) hemisphere of the egg is covered by a pigmented layer. Shortly after fertilization, a *'gray crescent'* appears approximately opposite to the locus of sperm entrance at the margin of the animal hemisphere and marks the region in which the dorsal lip of the blastopore will develop. The whole pigmented 'cap' seems to rotate toward the locus of sperm penetration (Fig. 8 C).

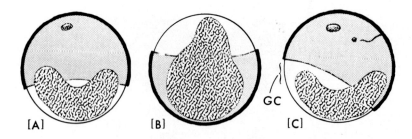

*Figure 8 C.* Formation of *gray crescent* in a frog egg (after WADDINGTON, 1966). A: frog egg before fertilization; the animal hemisphere is covered by a pigmented layer and the heaviest yolk forms a cup-shaped mass in the lower (vegetative) half of the ovum. B: when a fertilized egg is held upside-down, the heavier yolk gradually pours down to the now lower animal half; the gray crescent, and later the blastopore, will then be formed within the pigmented layer, but just above the main yolk mass. C: normal fertilization; the sperm (upper right) moves toward the egg nucleus; the whole pigmented cap is slightly rotated toward the locus of sperm penetration; the opposite region, from which the pigmented coat has moved away, becomes the gray crescent (GC).

Since the blastopore marks the caudal end of the embryo, a line bisect-
ing the crescent and connecting it with the animal pole thus indicates
the mid-dorsal line and thereby the (longitudinal) midsagittal axis of
the subsequently developing organism. As a general rule, but not invariably, the first cleavage plane corresponds to this median axis of the
embryo. The configuration observed in the fertilized ovum of the tuni-
cate Styela, and mentioned above in footnote 44, displays a somewhat
comparable arrangement. Here, a posteroventral yellow crescent indicates prospective mesoderm, and an anterior gray crescent indicates
dorsally an upper neural ectoderm, and ventrally the prospective notochord.

It should here again be mentioned that, although 'fertilization', as
a general rule, provides the initiation of ontogenetic metazoan
development, this latter can start by *parthenogenesis* without participation of a male gamete (cf. TYLER, 1941). 'Normal' parthenogenesis
is known to occur in various Invertebrates, and 'artificial parthenogenesis' has been produced, by means of diverse stimulations, in
Invertebrates as well as a few Vertebrates, including Mammals
(Rabbits). Depending on the presence or absence of the y-chromosome
in the haploid ovum, parthenogenetic individuals are always either
males (e.g. in Bees) or females (e.g. in Mammals, as demonstrated for
the Rabbit). Generally, parthenogenetic individuals are haploid, but,
in some instances, as e.g. resulting from experimental procedures,
diploidy seems to be restored during the course of early developmental
stages. Moreover, not only the haploid cell of the ovum, but also the
haploid sperm (male pronucleus) introduced into enucleated ovum
cytoplasm was shown to be sufficient for subsequent ontogenetic
development. Again, such ontogenetic events related to 'cloning', i.e.
produced by diploid nuclei of somatic cells, were briefly mentioned
above on p. 54.

The early cleavage processes of vertebrate ontogenesis are succeeded by *blastulation*, whereby, in holoblastic ova of Anamnia, the resulting cells are commonly arranged in form of a hollow sphere *(blastula)*,
containing the *blastocoel* (Fig. 9). In meroblastic ova of Anamnia and
Amniota, on the other hand, this formative event becomes modified,
and results in a *blastoderm*, i.e. a cellular plate at the animal pole. The
blastoderm may display, depending upon the developmental stage,
upon the blastodermic region, and upon the particular taxonomic
forms, one, two, or several ill-defined layers. In addition, a conspicuous cleft between blastoderm and 'inert' yolk may occur (subger-

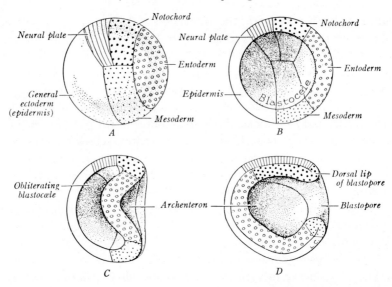

*Figure 9 A–D.* Diagram of gastrulation in Amphioxus (approx. ×350). Areas whose normal fates seem well established, are marked distinctly. A: blastula; B: hemisected blastula; C, D: early and later gastrulae, hemisected and showing invagination of entoderm (from AREY, 1954).

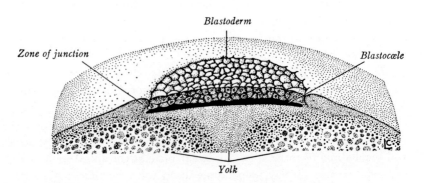

*Figure 10 A.* Diagram of avian blastoderm (pigeon), hemisected. The subgerminal cleft between blastoderm and yolk has here been interpretended by AREY as blastocoele, and this blastoderm is thus assumed to be an early blastula (from AREY, 1954).

minal cleft seen in Figure 10). Moreover, in some instances, a few apparently disconnected cells or even free nuclei appear scattered in the yolk-rich region, particularly near the blastoderm (Fig. 11 A, B). Thus, a conceptual as well as semantic difficulty arises concerning the arbitrary classification (a) of a cleft between blastoderm and yolk, or (b) of an at times well defined cleft within the blastoderm, as the 'true' morphologic equivalent homologous to the blastocoele in holoblastic Anamnia.

In the highly meroblastic Teleosts, gastrulation occurs by flattening of the blastoderm which spreads over the yolk. This morphogenetic movement of cell aggregates is termed *epiboly*, and ultimately entirely encompasses the yolk mass, closing the 'blastopore'. During these events, the margin of the blastoderm thickens to form a so-called *germ ring*. Cells of the germ ring then converge toward the middorsal line, and provide the blastoderm's *'embryonic shield'*. The rest of the blastoderm becomes the yolk sac which is subsequently vascularized. As epiboly proceeds, the germ ring thins, and the growing, elongated em-

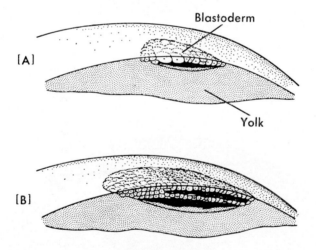

*Figure 10 B*. Simplified concept of chick cleavage (from WADDINGTON, 1966). A: early cleavage stage; B: subsequent stage, in which the lower cells are becoming separated from the upper ones to form the entoderm. Concerning the evaluation discussed in the text, one might ask whether the black cleft in A should be called blastocoele or archenteron. In B the lower cleft is evidently archenteron, and the upper one may be considered to represent the blastocoele. Because of semantic limitations as well as ambiguities in the interpretation of ill-defined motions by proliferating cell masses, it is difficult to decide whether the cleft in A is the forerunner of upper or lower cleft in B.

bryonic shield differentiates into the distinctive germ layers and primordia (TRINKAUS, 1969). Figure 11 C shows an early gastrula stage of the Teleost Fundulus. A somewhat more advanced Teleostean blastoderm ('neurula') at a stage corresponding to the differentiation of neuraxial primordium was shown in Figures 16 and 17 A (pp. 14 and 16)

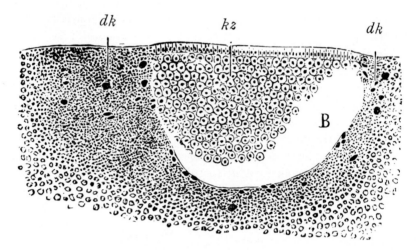

*Figure 11 A.* Blastoderm of a Selachian (Pristiurus), here interpreted by O. HERTWIG as blastula. B: blastocoele; dk: scattered cells (nuclei) in yolk (Dotterkerne); kz: embryonic cells ('Keimzellen'). The uppermost (ectodermal) layer is low columnar (from O. HERTWIG, 1915).

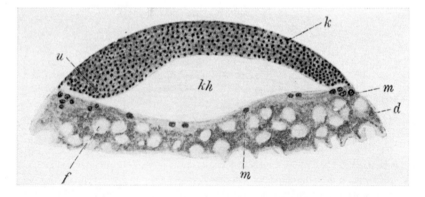

*Figure 11 B.* Blastoderm of a Teleost (trout), here interpreted by WEISSENBERG as blastula. d: yolk; f: lipid vacuoles; k: blastoderm; kh: blastocoele (Keimblasenhöhle); m: nuclei in 'yolk-syncytium; u: thickened caudal edge of blastoderm, from which entoderm later originates by a process akin to infolding (from WEISSENBERG, 1931).

of chapter V (volume 3, Part I, section 1). The topic there under discussion concerned the formation of a tubular neuraxis derived from a 'solid' neural cell cord in contradistinction to the mode of neural tube formation characterized by the folding of a neural plate into a groove, with subsequent dorsal closure.

In Selachians, however, whose ova are likewise polylecithal (megalecithal), displaying a meroblastic gastrulation and neurulation somewhat similar to these events in Teleosts (Fig. 11 D), the neural tube derives from a folded plate (Fig. 11 E) in a manner comparable to that predominantly manifested by numerous other vertebrates (cf. vol. 3/I, p. 12).

As regards holoblastic Amniota, namely marsupials and placental mammals, a *blastocyst* or blastodermic vesicle is formed. Although some such blastocystic stages look like a hollow sphere formally identical with the blastula of Anamnia, the cellular aggregate from which, in contradistinction to the embryonal membranes, the body of the embryo evolves, is located about one pole. This aggregate, either as a

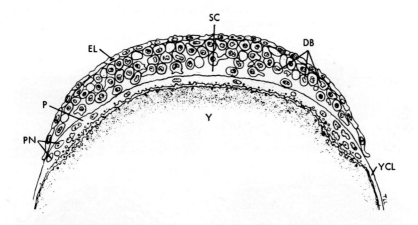

*Figure 11 C.* Diagram of an early gastrula stage of Fundulus, as interpreted by LENTZ and TRINKAUS (from TRINKAUS, 1969). 'The blastoderm is flattened and extends over one-third of the yolk as a result of epiboly. The cells of the enveloping layer (EL) are flattened and closely united. Lobopodia extend from some of the deep blastomeres (DB). Note that the periblast (P) and the yolk cytoplasmic layer (YCL) are parts of a continuous cytoplasmic layer. The periblast is the thicker nucleated (PN) part which lies between the blastoderm and the yolk (Y). SC: segmentation cavity. The clear area immediately beneath the periblast and yolk cytoplasmic layer is a region where yolk is apparently being digested.' It will be noted that in their interpretation, these contemporary authors do here not use the terms 'ectoderm' and 'entoderm'.

plate, or as an inner cell mass in which an amniotic cavity arises, finally provides a blastoderm very similar to that of meroblastic forms. Thus, a substantial difference between *blastocyst* and *blastula* obtains with re-

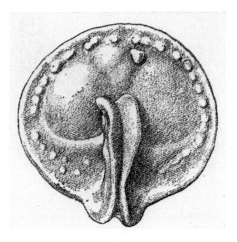

*Figure 11 D.* Blastoderm of the Selachian Torpedo at the stage of neural groove formation (after BALFOUR, from O. HERTWIG, 1915).

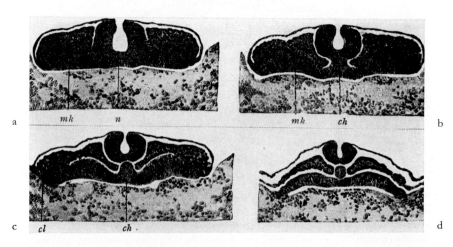

*Figure 11 E.* Cross-sections through the blastoderm of the Selachian Scyllium, showing neural groove and caudo-rostral differentiation gradient of germ-layer structures (from O. HERTWIG, 1915). a: caudal region, roughly corresponding to primitive streak neighborhood of some other vertebrates; b: somewhat more rostral section, showing notochord anlage; c and d: still more rostral levels with increasing degree of differentiation. ch: notochord anlage; cl, mk: mesoderm; n: caudal midline ectoderm.

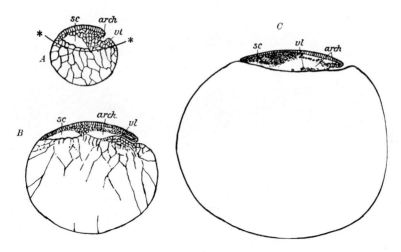

*Figure 12 A–C.* Gastrulation in Petromyzon (A, Cyclostome), Chimaera (B, Holocephalian), and true Selachian (C), showing a transition from holoblastic to meroblastic type
with blastoderm. arch: archenteron; sc: blastocoele; vl: ventral lip of blastopore. In C the
distinction between what is here interpreted as rostral part of archenteron and blastocoele
becomes arbitrary. The line marked by asterisks in A is supposed to indicate the limit of the
region corresponding to the blastoderm in the meroblastic type of development (after
DEAN, from VEIT, 1923).

gard to the prospective significance of the regions within the hollow
cell sphere's surface area. From a phylogenetic viewpoint, these early
mammalian morphogenetic events can be interpreted as an instance of
*cenogenesis* (cf. vol. 1, p. 241). Depending upon arbitrary postulates or
criteria, the early mammalian blastocyst may or may not be considered
homologous to the anamniote blastula.

    With respect to *gastrulation* (Figs. 9, 12, A–E), similar semantic and
conceptual difficulties are evident. Thus, one could define the gastrula
as a sac with a double wall, enclosing the archenteron which opens at
the blastopore, and formed by the invagination of the blastula. This
would correspond to the original concepts of *gastraea* and *gastrula* introduced about hundred years ago by ERNST HAECKEL. Or, considering gastrulation to be the process whereby the three germ layers become configurated and assume their characteristic spatial relationships, the term gastrula would subsume all stages in which first an ectoderm and an entoderm, and next a mesoderm can be clearly recog-

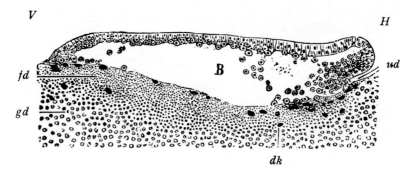

*Figure 12 D*. Details of gastrulation in a blastoderm of the Selachian Pristiurus (after RÜCKERT from O. HERTWIG, 1915). B: blastocoele; H: caudal end; V: rostral end; dk: nuclei of 'yolk-syncytium'; fd: finely granulated yolk; gk: coarsely granulated yolk; ud: primordium of archenteron (blastopore).

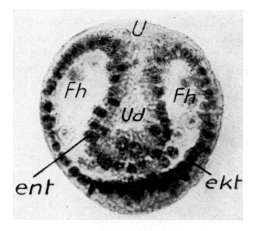

*Figure 12 E*. Gastrula of the sea-urchin Echinus (coelenterate invertebrate) for comparison with the depicted vertebrate gastrulae (from MAURER, 1928). ekt: ectoderm; ent: entoderm; Fh: blastocoel; U: blastopore; Ud: archenteron.

nized, up to the stage of *neurula*.[46] Some manifestations of vertebrate blastoderms, approximately representing a flat disc with a potential cleft between an outer and inner layer, are apparently the homeomorph

---

[46] The term *neurula* designates the embryonic stage subsequent to that of *gastrula*, and characterized by the formation of a neural tube from the neural plate. It is evident that this stage, which initiates the evolution of so-called primordial organs (cf. vol. 1, pp. 230–231) cannot be sharply or rigorously delimited from the preceding and the subsequent ones.

topologic equivalents of a gastrula, the surface facing the yolk being a combined archenteron and blastopore. Other sorts of blastoderm cannot be properly interpreted in this manner. Moreover, embryonic discs in different taxonomic groups display some differences in the formation of both entoderm and mesoderm, which defy classification in accordance with a single rigid semantic frame.[47]

As regards the motions of cell aggregates during gastrulation and neurulation, whereby prospective regions become displaced and molded into the characteristic configuration of primordial organs, it will here suffice to depict some representative as well as best known stages of Amphioxus, amphibians, and birds (Figs. 9, 13–18 A). For these stages, which illustrate the holoblastic and the meroblastic types of development, so-called fate maps (maps of prospective parts) have been worked out by VOGT (1929) and other experimental embryologists.[48]

Despite topologically irrelevant differences in shapes and dimensions, the fate maps of these different forms are almost identical or at least remarkably similar. Likewise, discounting secondary details, the morphogenetic movements seem very much alike, although the total dimension (size) of the embryo remains fairly constant throughout gastrulation and early neurulation of Amphioxus and amphibians (Fig. 18 B), while mass and dimensions of the avian blastoderm exhibit steady growth. From the viewpoint of exact sciences, as discussed above, it would be required to express these morphogenetic movements, presumed to be based on complex physico-chemical interactions, in terms of rigorous mathematical formulations. The considerable difficulties obstructing the fulfillment of this desideratum are evident. Even in the loose formulations of experimental and 'dynamic' embryology, which are unavoidably nonrigorous sciences, semantic limitations in describing the motions of folding, invagination, bending, and stretching become apparent, leading to awkward wordings as well as meaningless controversies.

At present, and taking the evolution of the primordial axial system in the holoblastic Amphibia as a significant example, the following

---

[47] Thus, as regards formation of primary entoderm in the blastoderm of birds, delamination, invagination, and various migration theories have been formulated.

[48] The techniques of marking with vital dyes (e.g. neutral red, nile blue sulfate), with carmine or with powdered carbon, as well as methods of stereophotography and of time-lapse cinematography can be used for studies of this sort.

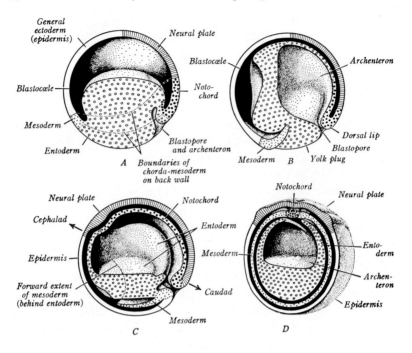

*Figure 13.* Gastrulation in tailed Amphibia, illustrating the displacements or movements of prospective neighborhoods. D represents a cross-section showing the caudal half of stage C (modified after HAMBURGER, 1942, from AREY, 1954).

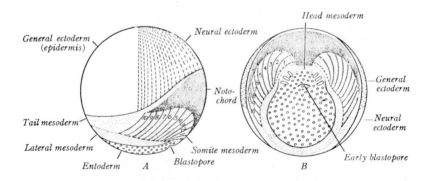

*Figure 14.* Map of prospective parts (neighborhoods) at the beginning of gastrulation in a tailed amphibian (after VOGT, from AREY, 1954). A: lateral view; B: view from posterior (vegetal) pole.

views concerning biochemical interactions have been expressed (TIE-
DEMANN, 1966). At least two loosely definable chemical factors taking
part in these events can be assumed, a '*neural factors*' and a '*mesodermal
factor*'. This latter, in highly purified form, seems to be a protein of relatively low molecular weight. Besides the inducing factors, special inhibitors appear to take part in the control of differentiation. The processes of induction and differentiation are considered as being antagonistic to repression. One of the first changes called forth by the mesodermal factor is a change of cell affinities in the reacting cells. Ribosomal
RNA and transfer RNA are synthetized in isolated ectoderm, which
does not form specialized tissues, as in whole differentiating embryos
of the same age. At least some sort of RNA shows, however, a lower activity in isolated ectoderm.

It seems perhaps permissible to consider, in accordance with
HAECKEL's *gastraea theory*, the invaginated gastrula of holoblastic Anamnia, displaying a typical blastopore, as a phylogenetically primitive

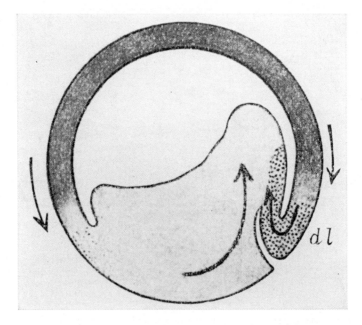

*Figure 15*. Diagram of sagittal section through an early frog gastrula, illustrating the
main directions of morphogenetic movements. Animal cells (ectoderm) dark, vegetative
cells (entoderm) lighter. Dorsal border zone in region of dorsal blastoporic lip (dl) dotted
(from WEISSENBERG, 1931).

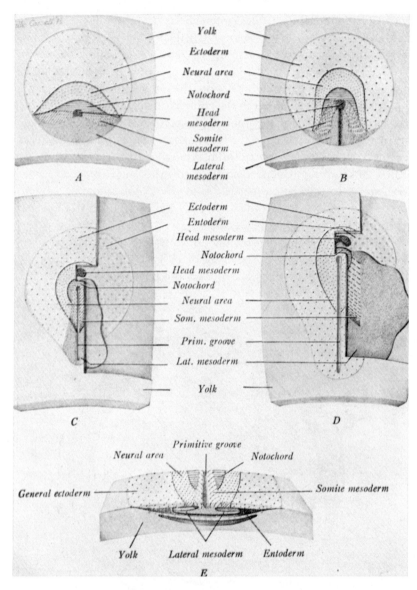

*Figure 16.* Gastrulation movements as assumed in a bird (chick). A: map of prospective parts, differentially marked on the surface of an early blastoderm. B: Formation of early primitive streak. C, D: Passage of chorda-mesoderm to a middle level. E: cross-section, looking rostrad, through middle of primitive streak in stage C (largely after Pasteels, from Arey, 1954).

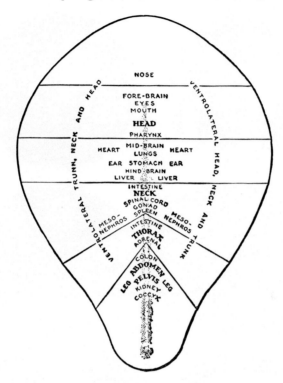

*Figure 17.* Location of prospective organs in the blastoderm of the chick at the stage of the head process, as tested by cultivating excised regions (after RAWLES, from AREY, 1954).

*Figure 18 A.* Simplified concept of the morphogenetic movement pattern during formation of the primitive streak in a chick blastoderm (from TORREY, 1967).

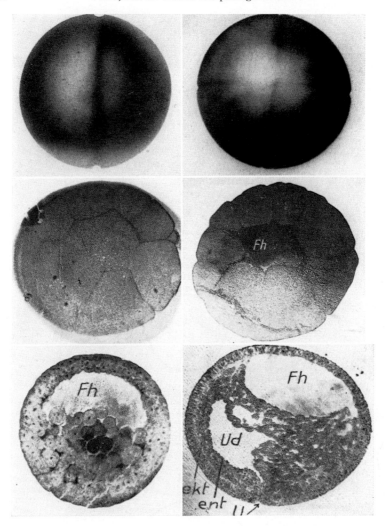

*Figure 18 B.* Six stages, from first cleavage to early blastulation, in the development of amphibian eggs, shown at approximately identical magnification, and illustrating the essentially constant overall dimension (size) maintained during these morphogenetic events. Top: two- and four-cell stages of *Rana fusca* (F. KOPSCH); middle: sections through morula and early blastula of Amblystoma (F. MAURER); bottom: sections of later morula and early blastula of Amblystoma (F. MAURER) Fh: blastocoel; ekt: ectoderm; ent: entoderm; U: blastopore; Ud: archenteron (from MAURER, 1928). The large size of the early blastomeres is related to their yolk content. It can be seen that, as cleavage and differentiation proceeds, and the yolk becomes distributed upon an increasing number of cellular units, these elements decrease in size, which latter gradually approaches the definitive orders of magnitude. *Note:* The incomplete letter at right bottom is *U.*

and primordial ontogenetic stage. The presumed cenogenetic changes
which subsequently resulted in the mode of ontogenetic evolution
from a disc-like blastoderm, are, on the other hand, correlated, in various meroblastic fishes, with thickenings and foldings at the enlarged
caudal margin of the disc. Figure 19 shows the characteristic aspect of
a blastoderm in a Selachian, where the posterior fold can be interpreted
as representing the dorsal lip of a blastopore. Figure 20 shows early
ontogenetic developmental stages in a Teleost. The neuraxis develops

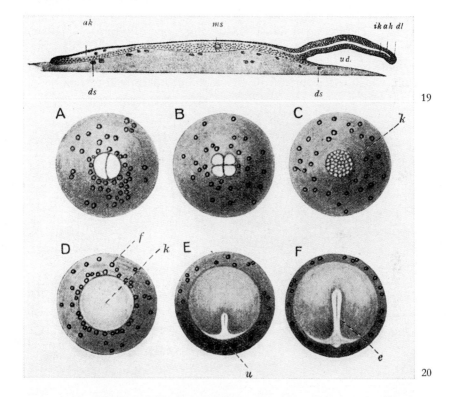

*Figure 19.* Midsagittal section through a Selachian blastoderm (Torpedo), showing
modified dorsal lip of blastopore (after ZIEGLER, from HERTWIG 1915). ak (rostral): cleft between ectoderm and mesoderm; ak (caudal): ectoderm of dorsal lip (dl) of blastopore; ds:
yolk syncytium; ik: entoderm; ms: mesoderm; ud.: 'archenteron'.

*Figure 20.* Meroblastic discoidal cleavage, blastoderm, and early embryonic development in a Teleost (trout) as seen from above (modified after PLEHN, from WEISSENBERG,
1931). A–C: early stages up to morula; D: blastula; E, F: blastoderm with embryonic anlage
arisen from margin of blastoderm. e: embryo; f: lipid droplets in yolk; k: blastoderm;
u: thickened caudal margin of blastoderm corresponding to dorsal lip of blastopore.

here in the first, rostralward protruding streak-like cell mass, displayed in Figure 17 A of volume 3, part I, and also in more caudal portions of the embryonic anlage shown in E and F of the present Figure 20. In Amniota, however, the blastoderm gives origin to a peculiar elongated cell aggregate designated as the *primitive streak*, whose knob-like rostral end is the *primitive knot* (Fig. 21), rostrally to which the neural plate develops. Again, in the region of this knot, a *primitive pit* becomes evident in numerous forms and at various stages. Moreover, a *shallow primitive* groove may be seen in the midline of the streak, rostrally ending at the primitive pit. Prospective mesodermal and notochordal tissue seems to converge toward streak and node. Presumably passing through the node, such cell aggregates form an elongated, tongue-like structure stretching rostralward under the ectoderm. This configuration, which is separated from the neural ectoderm and its non-neural

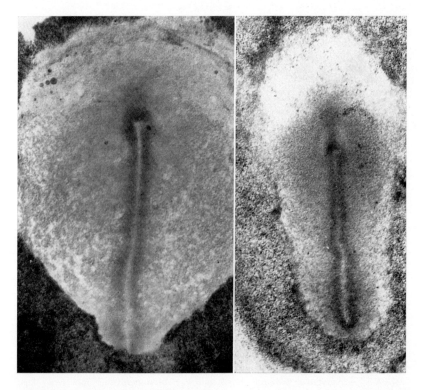

*Figure 21*. Primitive streak and primitive knot in chick blastoderms of between 13 and 19 h of incubation. These stages precede those shown in figures 4–6 of vol. 3, part I.

vicinity, but which is ventrally continuous with a lining of entoderm, represents he *head process* (notochordal plate, chorda-mesoderm plate). Along primitive knot and most of the primitive streak, however, all three germ layers are in direct continuity. Figures 3, 4, and 7 A of the preceding volume (chapter V) illustrate some of these relationships. The primitive streak and its associated structures (groove, pit, knot) may be regarded as a modified, stretched and seam-like blastopore.

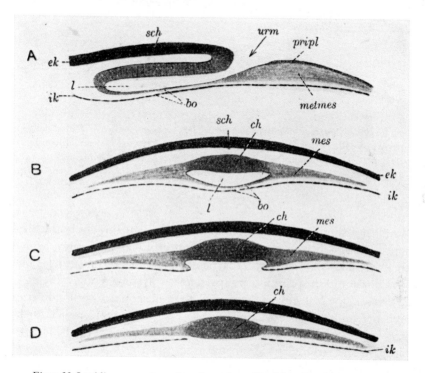

*Figure 22.* Semidiagrammatic sections through reptilian blastoderms, showing features of the primitive pit (from WEISSENBERG, 1931). A: sagittal section; B–D: cross-sections; A and B show a fully developed chorda-mesoderm sac; C: rupture of sac into subgerminal cleft; D: the interrupted lining of gut-entoderm begins to close the gap in midline. bo: bottom of sac, formed by 'sac-entoderm' above, and 'gut-entoderm' below; ch: chorda-primordium; l: lumen of chorda-mesoderm sac; ek: ectoderm; ik: gut entoderm; mes: mesoderm originated from sac; metmes: metastomal mesoderm derived from primitive streak (pripl) caudal to primitive pit (urm) considered to represent a modified blastopore. The gut-entoderm, being of greater significance, may arbitrarily be called 'primary' in contradistinction to the 'secondary' 'sac-entoderm'. A reversed designation can be justified by argumentation based on phylogenetic postulates. It is likewise evident that the so-called chorda-mesoderm sac could also be designated as chorda-entoderm sac on the basis of such argumentation.

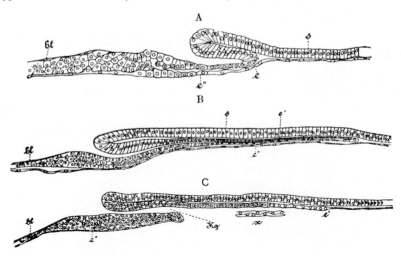

*Figure 23.* Extensive chorda-mesoderm sac in a reptilian embryo (Platydactylus), for comparison with the preceding figure (after WILL, from HERTWIG 1915). A: Early stage; B: later stage; C: final rupture. bl: ectoderm of primitive streak; c″ and e″: gut-entoderm; e and e′: chorda-mesoderm sac; Kg: neurenteric canal; s: ectoderm; x: remnant of ruptured sac wall.

Whether it should be considered homologous or analogous to the blastopore of holoblastic forms depends on arbitrary postulates, which therefore allow more than one particular answer or formulation.[49] The invagination at the primitive pit, particularly in some reptiles (Figs. 22, 23), may show features similar to those of a blastopore, leading into an archenteron-like structure, designated as *chorda-mesoderm sac* (*'Chorda-mesodermsäckchen'*). Two different entodermal layers seem to be temporarily present, arbitrarily designated as so-called *'primary'* and *'secondary'*

---

[49] Although a reasonably rigorous formanalytic homology concept based on elementary topology allows, within a wide range, the logically satisfactory unequivocal establishment of relevant homologies, it was pointed out in volume 1, pp. 208–209, 252 *et passim*, that, nevertheless, some undecidable problems remain, which require arbitrary solutions leading to different equally defensible formulations. The question at issue is somewhat analogous to that involved in the boundary value problem, discussed further above. Thus, where clearly distinctive elements in a configuration with well definable connectedness are given such as, e.g., skeletal components, many muscles, or certain clear-cut subdivisions of the neuraxis, an obviously favorable boundary condition for comparison obtains. This, however, is not always the case in ontogenetic events displaying a variety of stages with, figuratively speaking, 'blurred' transitions.

*entoderm* (Fig. 24). Subsequently, the ventral layer of the sac may break through into the cleft between embryonic disc and yolk; this is followed by the establishment of a single, definitive entoderm layer. The transitory opening results into a passage from ectoderm into definitive gut, corresponding to a *neurenteric canal* as briefly discussed further below. In some birds, the lumen of the chorda-mesoderm sac may continue rostrad into a transitory *notochordal canal* (cf. also K., 1930). Similar transitory configurational manifestations can also be observed in comparable ontogenetic stages of at least some mammals, including man (Fig. 32).

The chorda-mesoderm pocket of Amniota may be interpreted as the manifestation of an invagination process (Fig. 24) related to the gastrulation of holoblastic Anamnia, but modified in yolk-rich Anamnia and Amniota, such that this modification is also retained in oligo-

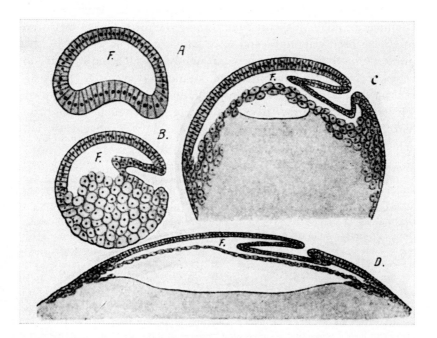

*Figure 24.* Diagrams illustrating a concept of cenogenetic entoderm formation (modified after WENKEBACH, 1897, from LUBOSCH, 1925). A: beginning oligolecithal holoblastic gastrulation in Amphioxus; B: beginning telolecithal holoblastic gastrulation in Cyclostomes and amphibians; C: hypothetical primitive meroblastic type of gastrulation; D: Formation of chorda-mesoderm sac in reptiles; F: blastocoele. This diagram should be compared with the two preceding figures and with figure 12 A–C.

lecithal mammals. One could also say that the compound events manifested by the apparently simple process of holoblastic anamniote gastrulation have become disjoined into separate sets of events (e.g. 'primary' and 'secondary' 'gastrulation'). TRIEPEL (1922), emphasizing the
events leading to the differentiation of an axial structure represented by
the notochord, distinguishes a particular *'chordula'*-stage in the ontogenesis of vertebrates, and regards the chorda-mesoderm sac as the manifestation of these events. Such axial differentiation can be disturbed by
procedures of experimental embryology (Fig. 25). From an overall
viewpoint, however, it is inconvenient to separate *chordula*[50] and *neurula*
stages. On the other hand, the difficulties of describing the relevant
complex and blurred dynamic events in terms of static formanalytic
concepts are quite evident. This however, does in no way impair the
validity, within a wide and important domain, of the formanalytic approach.

Apart from the morphologic interpretation of the primitive streak,
a number of different theories concerning the formation of this structure have been elaborated and were based on the concepts of either
proliferation *in situ* or of movements involving cell groups. Both concepts, of course, are, to some extent, not mutually exclusive. Many significant data were obtained by investigations of the chick blastoderm,
which provides a particularly suitable material for studies of this problem. Following earlier authors who emphasized cell movements, Du-

---

[50] Since the chorda-mesoderm plate may be presumed to have an inductive effect with
regard to the neural plate and neural tube, the concept of *'chordula'* and *'chordulation'* seems,
of course, quite justifiable from the viewpoint of experimental embryology. Thus, if
amphibian embryos are exposed to a hypertonic salt solution or to treatment with lithium
chloride (or magnesium chloride) before the onset of gastrulation, invagination fails to
occur. Instead, entoderm and mesoderm evaginate, and a constriction separates the
ectoderm from the two other germ layers. This peculiar developmental anomaly, which
may be partial or more rarely total, is called *exogastrulation* (HOLTFRETER, 1933). In all
instances where total exogastrulae could be raised to an advanced stage, the ectoderm
remained entirely undifferentiated, while the entomesoderm displayed a substantial degree
of differentiation, despite the complete inversion of all structures (entoderm outside,
mesoderm inside, presenting an interesting problem of topology). The supposition that
inductive stimuli from the mesoderm are necessary for neural differentiation is thereby
strongly supported (cf. also HAMBURGER, 1942). These experimental data can also be
interpreted as favoring a reasonable and flexible concept of so-called germ layer specificity
in agreement with the views adopted for the present treatise. The extensive experimental
studies with exogastrulation of echinoderms should also be mentioned but need not be
further discussed in this context which concerns vertebrate ontogenetic evolution.

VAL, about 1884, presented a concrescence theory. GRÄPER, about 1929, described a 'polonaise'-like movement of epiblast. A somewhat more complex double vortex of epiblast cells was then discovered by

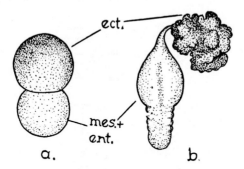

*Figure 25.* Total exogastrulation in the urodele amphibian Amblystoma as demonstrated by HOLTFRETER (1933). a: in gastrula stage; b: in tail-bud stage; ect: ectoderm; mes. +ent.: evaginated mesoderm and entoderm (after HOLTFRETER, from HAMBURGER 1942).

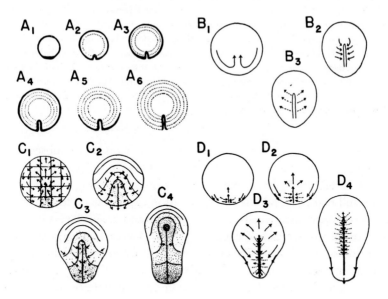

*Figure 26.* Diagrams illustrating theories of primitive streak formation in the chick blastoderm (from ROMANOFF, 1960). $A_1$–$A_6$: the concrescence theory of DUVAL; $B_1$–$B_3$: the concept of a 'polonaise' movement of epiblast cells proposed by GRÄPER; $C_1$–$C_4$: the theory of a 'double vortex' of epiblast cells as originally conceived by WETZEL; $D_1$–$D_4$: WETZEL's theory after modifications by PASTEELS.

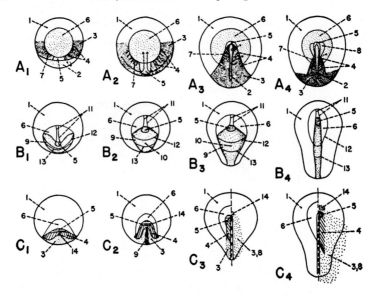

*Figure 27 A.* Diagrams illustrating three different conceptions of neighborhood configurations involving presumptive areas in the chick blastoderm before and during formation
of primitive streak (modified after GRÄPER, WERZEL, and PASTEELS, from ROMANOFF, 1960).
$A_1$–$A_4$ according to GRÄPER; $B_1$–$B_4$ according to WETZEL; $C_1$–$C_4$ according to PASTEELS; 1:
epidermal ectoderm; 2: entoderm; 3: lateral plate mesoderm; 4: somite mesoderm; 5: chorda; 6: neural ectoderm; 7: first somite; 8: invaginated mesoderm; 9: primitive streak material; 10: mesoderm arising from primitive streak; 11: lines surrounding median region of
chorda and neuraxis floor; 12: posterior boundary of neural plate (approximately corresponding to level of future 12th somite); 13: boundary between somite and lateral mesoderm; 14: prechordal (preaxial) mesoderm.

WETZEL (1929), and further sorts of movements were subsequently
pointed out by PASTEELS and others[51] (Figs. 26, 27 A). SPRATT and
HAAS (1962, 1965) have expressed doubts concerning the evidence,
from cell marking and streak explantation experiments, for the movement of upper surface cells, into the middle and lowest layers in any region of the pre-streak and early streak blastoderm, by invagination, involution or ingression. The cited authors stress an essentially entodermal cell movement, namely the radial spreading of cells from the lowest and middle layers of a 'growth center' located in the posterior region

---

[51] A detailed review of these theories, buttressed by original observations, can be
found in ROMANOFF's comprehensive work on the avian embryo (1960), which also
includes the relevant bibliographic references.

of the chick's blastoderm, while the ectodermal surface layer remains relatively immobile. These spreading cells are presumed to supply extra-embryonic entoderm and mesoderm (Fig. 27 B). The primitive streak is said to arise *in situ* as a posterior elongation of the growth center, which becomes the anterior end of the streak. Subsequent proliferative activity of the streak apparently provides the cells for the germ layers of the body of the embryo. Although the primitive streak could thus be interpreted as an elongated growth center or blastema representing the direct primordium of the principal axial organs rather than as a way station through which cell aggregates reach their particular locations, the epiblastic cell movements described by previous authors seem rather well documented and might still be considered as morphogenetic movements related to the activities of the streak. SPRATT and HAAS (1965) de-emphasize the movements of uppermost (epiblastic) cell aggregates and consider it improbable that any significant and systematic involution or invagination of upper surface areas occur. Nevertheless, these authors admit that expansion, convergence and deformation movements of the surface layer do occur. The authors consider

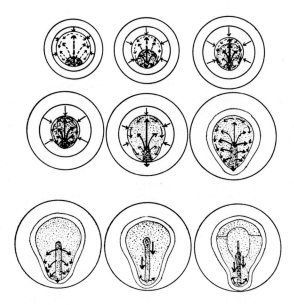

*Figure 27 B*. Diagrams illustrating the topographic pattern of lower- and middle-layer tissue movements in chick blastoderms of early stages, as determined by systematic carbon and carmine markings of the surface of the lowest layer and of selected parts of the lower middle layer (from SPRATT and HAAS, 1965).

these displacements to be associated with elongation of the pellucid area, which changes from an approximately circular to a pear-shaped outline. Subsequently, such movements are associated with elongation of the primitive streak and with the development of the primitive groove and folds. As an elongated growth center, the streak is said to have its earlier origin from the marginal zone ring, itself a circular blastema with multiple streak- and body-forming capacities.[52] However, the particular mechanics obtaining in these ontogenetic processes, regardless of their, in many respects still not entirely clarified details, can be evaluated as not affecting the morphologic, i.e. configurational interpretation of the primitive streak. For that matter, if a blastoderm, as mentioned above, is interpreted as a gastrula, the cleft between embryonic disc and yolk being a combined blastopore and archenteron, then the marginal zone ring assumes the topologic significance of a blastoporic lip (cf. the 'circular blastema' of SPRATT and HAAS mentioned in the last sentence of footnote 52).

Generally speaking, the further morphogenetic events in the evolution of an embryonic disc or blastoderm may be described as characterized by an anterior extension of head process and prechordal meso-

---

[52] SPRATT and HAAS (1965) summarize their views as follows. '(1) The cellular precursors of all three primary germ layers in the unincubated blastoderm are topographically represented by a disc-shaped, epithelial-like and complete uppermost layer; and ring-shaped, incomplete and congruent middle and lowest layers below the uppermost layer. (2) There is a gradient in cell population density decreasing from posterior to anterior. (3) Centripetal extension of a sheet of lowest and adjacent middle layer cells, primarily from the posterior portion of the ring but also from all points on its circumference, results in closing of the ring and completion of the middle and lowest layers as coherent discs. (4) No movement (invagination or involution) of uppermost, surface areas as coherent sheets into either the middle or lowest layers could be demonstrated in any region, including the primitive streak. (5) Embryonic germ layers arise *in situ* by proliferation from the streak. (6) The streak in its morphology and function is more like a blastema than a blastopore. (7) As an elongated growth center, the streak has its origin from the marginal zone ring, itself a circular blastema with multiple streak- and embryo body-forming capacities'.

With respect to the history of the germ layer theory, the following addition to my comments in chapter III (p. 159), volume 1 of this series is perhaps appropriate. Although, as there stated, H. C. PANDER (1794–1865) may perhaps be said to have 'discovered', about 1817, by more detailed examination and description, the three germ layers of the chick embryo, a proper reference to K. F. WOLFF (1733–1794) was inadvertently and unfortunately omitted. It should be stressed that, in his *Theoria generationis* (1759) WOLFF initiated a new approach to embryology based on accurate observation, and seems to be the first who recognized the significance of germ layers.

derm, accompanied by neural plate as well as neural groove formation, and by a posterior displacement of primitive node and streak. On the whole, a forward elongation of the blastoderm becomes apparent. The more massive median portion of the head process becomes the noto-chord, while the lateral wings provide para-axial or somitic mesoderm, and the entoderm separates from the mesodermal components.

At somewhat later stages of blastodermal ontogenesis, the remnant of the primitive streak forms the *end bud*, which subsequently becomes the *tail bud*. Some differences of opinion, essentially of a semantic na-ture, have been expressed about the question whether the tail bud, par-ticularly prominent in avian embryos, and investigated in detail by HOLMDAHL (1925, 1939) represents the entire streak (including node) or merely the posterior portions of the streak. At any rate, the tail bud *(Schwanzknospe)* seems to be a blastema, from which substantial por-tions of the embryonic body differentiate *in situ*.[53] The bud, like the prim-itive streak, appears to move caudalward as it decreases in size. A tail bud is likewise displayed in at least some instances of holoblastic onto-genesis, but seems, as could be expected, to play a minor role in this sort of development. In discussing the significance of the avian tail-bud, HOLMDAHL (1933) stresses the general significance of that blaste-ma as an *'indifferent cell mass'*. From this point of view he concludes that the part of the neural tube developing from said (apparently) different matrix cannot be regarded as an ectodermal organ. Quite evidently, ar-bitrary questions of classification and semantics are here involved, which are somewhat related to the definition of a so-called mesecto-derm. Thus, in certain of the various experiments by RAVEN (1933) it was shown that excised rostral marginal regions of open urodele neur-al plate, transplanted into ventral epidermis, develop into neural tissue

---

[53] The *in situ* origin of the caudal part of the neural tube within the solid blastema of the avian tail bud was pointed out and depicted in the preceding volume (chapter V, section 1, fig. 18). The peculiar mechanism of morphogenetic differentiation within an apparently uniform or homogenous blastema also occurs during the ontogenetic evolution of grisea in various regions of the neuraxis, and was pointed out since 1920 by my collabo-rators and myself (cf. chapter V, section 1). KAHLE (cf. further below in sections 5 and 6) subsequently introduced the somewhat ambiguous terms 'Matrixphasen', 'Matrixauf-brauch', and 'Matrix-Exhaustion'. Again, as regards differentiation of fine structure (histogenesis), a recent ultrastructural study on the origin of cilia in the nasal epithelium of mouse embryos is of more general interest (FRISCH and FARBMAN, 1968). It was shown that structures differentiate at random prior to and independent of the interactions that subsequently orient them in a coordinated spatial and temporal relationship.

as well as cartilage. Depending upon the chosen logical and semantic postulates, such cartilage might then be classified as ectodermal, as mesectodermal or still as mesodermal.

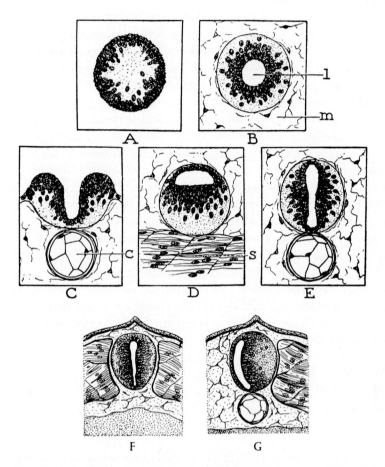

*Figure 28 A–G.* Shape of neural tube under different conditions in amphibians (combined after HOLTFRETER and HAMBURGER, 1955, and WEISS, 1955). A: solid neural mass developed in explantation; B: neural tube surrounded by mesenchyme; C: 'asyntaxia dorsalis' (failure of the tube to close, eversion with thinned floor, somewhat reminiscent of normal medulla oblongata pattern, or of telencephalic pattern in teleosts); D: neural tube with eccentric lumen, underlain by musculature; E: neural tube of normal appearance, underlain by notochord; F: neural tube of normal appearance, flanked by somite fragments; G: neural tube with eccentric lumen, underlain by notochord but with unilateral somite fragments. It is of interest to compare these configurations of the vertebrate neural tube with the patterns displayed by the vesicular cerebral ganglion of the invertebrate ectoproct Bryozoan Cristatella as depicted in figures 39, p. 66, vol. 2 of this series.

Generally speaking, the neurula stage of vertebrate ontogenesis ends with the formation of a dorsal neural tube which is produced by folding of neural plate, closure of neural groove, and separation from the integumental ectoderm. The neural tube may then be considered to have reached the stage of a so-called *primordial organ* (cf. vol. 1, p. 230). Its rostral part becomes the *brain*, and its longer caudal part provides the *spinal cord*. The neural groove closure begins as a rule in a region approximately corresponding to a posterior or middle portion of the brain, and proceeds caudalwards as well as rostralwards. Although transformation of the neural plate into the neural tube, which may take place in excised and isolated plates, is believed to occur by 'forces' (interactions) pertaining to the plate itself, events in its neighborhood seem to have an effect upon the configuration of the tube (Fig. 28).

The small terminal aperture preceding the total closure at the rostral end is the *anterior neuropore*, which thus seems to indicate the anterior end of the original neural plate. However, since this plate is not a static structure, but displays continuous growth activities, the definition of its 'end', and the identification of the locus of that end at more advanced stages of brain differentiation become ambiguous. On the basis of all available data, the location of the ultimate rostral closure or anterior end may be assumed to correspond to an undefined neighborhood of the region designated as *lamina terminalis* which is the rostral midline portion of the brain and pertains, as discussed below, first to prosencephalon, and subsequently to diencephalon *and* telencephalon[54].

The comparable terminal aperture at the caudal end of the neural tube is the *posterior neuropore*, which, in many vertebrate forms, displays a transitory communication with the gut through a *canalis neurentericus* (Figs. 29–32). It is evident that the posterior neuropore and, if present,

---

[54] The lamina terminalis can be considered as resulting from the closure of the *lateral plates* (specifically of their alar plate components) in the midline, and does thus not include the therewith continuous rostral median portion of *roof plate (lamina epithelialis)* which contains the locus of paraphysial anlage and provides the velum transversum. In some vertebrates, e.g. various Selachians, and others, a *recessus neuroporicus*, which may be related to the anterior neuropore, can be found in the lamina terminalis between anterior and pallial commissures. Sufficiently systematic and conclusive observations on the fate of anterior neuropore and the recessus neuroporicus in the vertebrate series do not seem to be available. It appears dubious whether, apart from the rather evident relationship to the lamina terminalis, an entirely 'constant' locus of anterior neuropore and recessus neuroporicus can be assumed with regard to different vertebrate forms.

the canalis neurentericus assume a different relationship in forms developing by holoblastic gastrulation, and in forms developing by means of an embryonic disc or blastoderm, usually displaying a primitive pit. The transitory opening of the chorda-mesoderm sac into the gut, as mentioned above, provides here a neurenteric canal. Caudad of

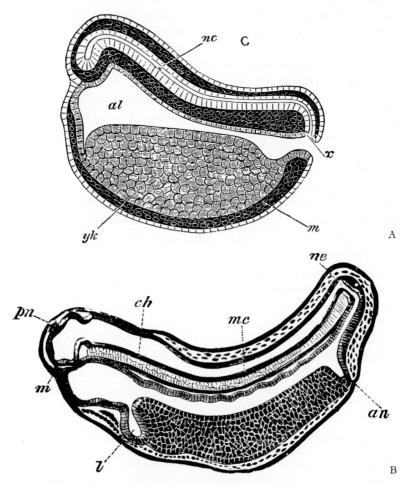

*Figure 29.* Semidiagrammatic longitudinal sections through amphibian larvae displaying a neurenteric canal. A Early frog larva (adapted after Götte by Balfour, from O. Hertwig, 1915). al: gut; C: dorso-caudal side; m: mesoderm; nc: neural canal; yk: yolk; x: canalis neurentericus. B Somewhat later larva of the anuran Bombinator (after Götte, from O. Hertwig, 1915).

an: anal region; ch: notochord; l: liver diverticulum; m: oral region; me: neural tube; ne: neurenteric canal.

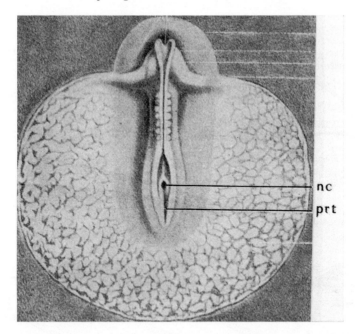

*Figure 30*. Surface view of an albatross embryo with about 7 somites. The neural groove has not yet begun to close into a neural tube by dorsal 'suture' (after SCHAUINSLAND, from O. HERTWIG, 1915). nc: pit becoming canalis neurentericus; prt: primitive groove; other leads not relevant to present context.

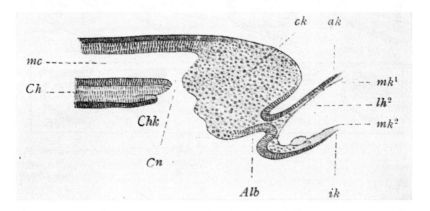

*Figure 31*. Semidiagrammatic midsagittal section through the tail-bud of an avian embryo (after SCHAUINSLAND from O. HERTWIG, 1915). Alb: allantois diverticulum; ak: ectoderm; Ch: notochord with rudiment of chorda-canal (Chk); ck: tail-bud; Cn: canalis neurentericus; ik: entoderm; lh²: extraembryonic celomic cavity; mc: medullary canal (neuraxis); mk¹, mk²: parietal and visceral layers of mesoderm.

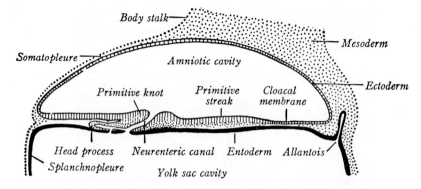

*Figure 32.* Diagrammatic sagittal section through a human embryo of 19 days, showing neurenteric canal (after SCAMMON, from AREY, 1954).

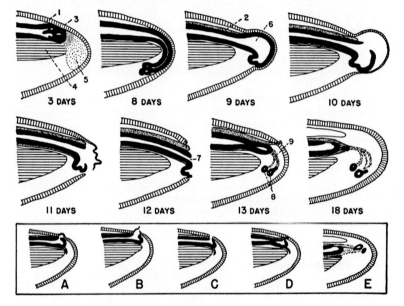

*Figure 33 A.* The secondary posterior neuropore in a chick embryo (modified after SCHUMACHER, 1928, from ROMANOFF, 1960). Formation and rupture of a terminal neural tube vesicle produces a secondary neuropore which eventually closes by a process of 'retro-gression'. The neural tube may not bend ventrally to the notochord, and the changes then occur in the manner shown by the sequence A–E. 1: neural tube; 2: peripheral layer of neur-al tube; 3: ectoderm; 4: notochord; 5: tail bud; 6: vesicle; 7: secondary posterior neuro-pore; 8: remnants of neural tube; 9: fibrous strands (unmyelinated nerve fibers ?) of neural tube. Although concerning a rather trivial matter, the variations indicated by the two se-quences provide a good illustration for the lack of precision, the 'randomness', or the 'blur-red' aspect obtaining in a wide category of biologic events.

this level, the neural tube does not develop by folding, but rather *in situ* within a formative blastema (tail bud, end bud). In birds, a secondary posterior neuropore opening through the epidermis has been described at the caudal end of the neural tube by Schumacher (1928) and others (cf. also Romanoff, 1960). One or more small detached vesicles may here be formed (Fig. 33 A). Similar secondary communications between caudal parts of neural tube and gut may likewise occur in some instances *(secondary neurenteric canals)*. Comparable manifestations of a secondary posterior neuropore and related disturbances can also be found in mammalian, including human ontogenesis. These processes involve the region of the neural tube related to the so-called sinus terminalis of the spinal cord.

In the preceding volume 3, part I (pp. 18, 52, 299) the formation of neural tube and its accompanying *neural crest* was discussed from the viewpoint of histogenesis concerning neuroectodermal elements. Nevertheless, the participation of neural crest elements in other formative events, and the concept of '*mesectoderm*' were briefly mentioned. Since the present section 1 of chapter VI refers to various different aspects and problems of organic development in general, a few additional remarks on the peculiar capacities of the neural crest seem here appropriate. Cells from that primordium are now assumed to provide not only such neuroectodermal elements as neurons of spinal, cranial, sympathetic, and parasympathetic ganglia, including borderline 'neuroectodermal' cells of adrenal medulla and chromaffin bodies, moreover lemnoblastic *Schwann cells*, but also epidermal pigment cells, as well as 'mesodermal' branchial cartilages and meningeal components.

In a recent survey of '*developmental biology*', Trinkaus (1969) reviews some of the problems involved in this migration of neural crest elements to various distant areas of the embryo. He formulates the following questions. 'Do all cells of the neural crest find their mark? Or do many of them disperse into other areas, where they either cytolyze, remain undifferentiated, or are incorporated in undetectably small numbers in the structures which form in that area? Do cells of the neural crest migrate at random from the neural tube out into the embryo, or do they move along certain pathways? If crest cells move along certain pathways, are these characteristic and well defined? If so, what features of the cells and their environment serve to direct their movements?'

Trinkaus (1969) then points out that radioautographic experiments undertaken in the chick by Weston (1963, 1967) with the tritiated thymidine method have provided some answers to the cited questions. When marked neural crest and neural tube are grafted on an unlabeled host, the movements of labeled cells can be followed.

It was thus shown that cells of the neural crest migrate in two fairly well defined streams of individual cells, a ventral one into the mesenchyme between neural tube and myotome, and a dorsal one into the epidermal ectoderm (Fig. 33 B). This migration is not considered 'random', since the cells seem to follow 'favored pathways', ceasing their movement and accumulating at certain loci such as spinal and sympathetic ganglia. The dorsal stream moves rapidly into the ectoderm and may

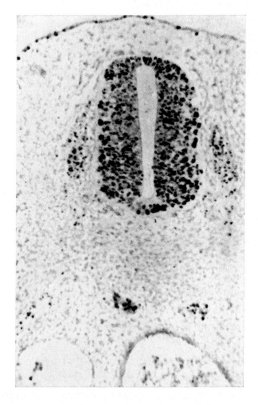

*Figure 33 B.* 'Transverse section through 17th somite of chick embryo fixed at 4 days of incubation. A piece of neural tube (with neural crest) labeled with tritiated thymidine had been grafted to this embryo 50 h previously. Radioautograph exposed 4 weeks' (after WESTON, 1963, from TRINKAUS, 1969). 'Note labeled cells in ectoderm, condensed labeled spinal ganglia, sheath cells on motor nerve in left lower corner, and labeled sympathetic ganglia adjacent to aorta (bottom).'

provide *promelanoblasts* since, when ectoderm containing these neural crest cells is grafted to a non-pigmented host, it yields pigment cells.

The direction of the ventrad migration through somitic mesenchyme appears related to the orientation of the neural tube. If this latter is grafted in an inverted orientation, the crest cells still migrate ventrad with respect to their own neuraxis, that is, now dorsad with respect to the host organism. As regards the influence of the mesodermal, somitic segmentation upon migration processes from the neural crest, it has been observed that migration between the metameric somities is 'attenuated', while migration within segmented somitic mesenchyme is 'enhanced' (TRINKAUS, 1969). With respect to the origin of *Schwann cells* (lemnoblasts), it has been shown that, after removal of the neural crest (in the chick), such tritium-labeled elements derive from the ventral part of the neural tube. This, of course, does not exclude the probability that peripheral lemnoblasts also originate from the neural crest (cf. vol. 3, part I, p. 39).

As stated above, the central nervous system of adult craniote vertebrates can be described as comprising two main subdivisions, namely brain and spinal cord. Loosely speaking, the brain is enclosed in the cranium, and the spinal cord in the vertebral canal. From a more rigorous viewpoint, however, the boundary region between cranial cavity and vertebral canal does not provide a suitable or satisfactory landmark for a conceptual delimitation of brain from spinal cord, quite apart from the obtaining circumstances implying nonlinear, gradual transitions, characterized by boundary regions instead of imaginary boundary lines.

As regards the conventional lines and planes, related to directional or positional terms, and significant for purposes of morphologic description, it is perhaps opportune to include here the following brief summary recalling appropriate definitions.

Three so-called *axial lines* which can be conceived as geodesics and therefore not necessarily 'straight', provide an essentially *nonmetric* and *non-Euclidean (Euclidoid)*, ameboid three-dimensional coordinate system *(German: 'Bezugsmollusk')* of anatomical space:

1. The *long axis (longitudinal axis, midsagittal axis)* runs rostro-caudad in most vertebrates, namely from the tip of the snout or nose, as the case may be, i.e. from the anterior *apex (rostrum*, originally the beak of a ship's prow), to the end of the tail. In human anatomy, where terms refer to an upright standing position (Fig. 34 A) such line would run from vertex to tip of coccyx. Depending on arbitrary and

not entirely standardized postulates, one could, of course, trace this axis with a curved initial segment from the tip of the nose or from the *nasion* of anthropologists. Parallel to the midsagittal axis, additional sagittal axial lines may be drawn.

2. Numerous *dorso-ventral axes*, at a right angle to the long axis, can be traced connecting a spot in the ventral surface midline with an approximately corresponding spot on the dorsal surface midline.

3. Numerous *transverse or right–left axial lines*, at a right angle to both preceding axes, can be traced to connect approximately corresponding

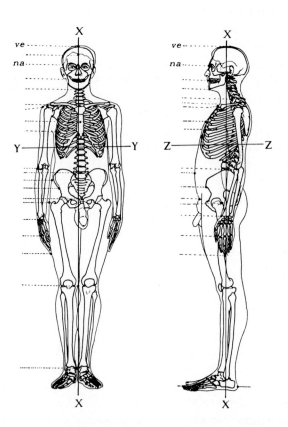

*Figure 34 A*. Axial lines of the human body in the so-called '*anatomical position*', superimposed on a diagram of SALLER (1930) indicating reference points for anthropologic measurements. X: longitudinal axis; Z: one of the numerous dorso-ventral (sagittal) axes; Y: one of the numerous transverse (or right-left) axes; na: nasion; ve: vertex. It is evident that these axes, particularly X, are not true geometrical straight lines in the *Euclidean* sense, but rather fictional sequences of loci.

lateral surface spots, which, as a rule, might be chosen either about midway between dorsal and ventral midline, or at the greatest distance from the midsagittal plane, described further below. An '*axial*' or '*paraxial*' line differs from an assumed 'true' axis, to which it must be parallel, by not being restricted to the connection of spots located in an assumed external body 'midline'.

Again, in relation to these lines, whose significance with respect to promorphology and symmetry has been discussed in volume 1 of this series (chapter III, pp. 211–222, 1967), three sorts of *planes* can be roughly defined:

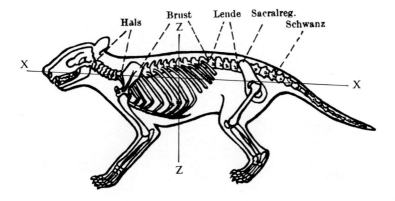

*Figure 34 B*. Longitudinal axis (X) and one of the numerous transverse (dorso-ventral) axes (Z) of a mammalian body superimposed upon a diagram of SELENKA-GOLDSCHMIDT (1923). It can be seen that X may here be called the 'sagittal' axis. However, in all vertebrates, any plane parallel to both X and Z is a sagittal plane.

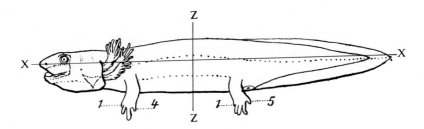

*Figure 34 C*. Longitudinal axis (X) and one of the numerous transverse (dorso-ventral) axes (Z) of a salamander larva superimposed on a diagram by WIEDERSHEIM (1893). 1–4 and 1–5: digits.

1. *Sagittal planes*, dividing the body into right and left parts, are parallel to long axis and to dorsoventral axes, but at right angle to transverse axes. Sagittal planes include dorso-ventral and longitudinal axial lines, the long axis being contained in the midsagittal plane. Some authors restrict the term 'sagittal plane' to the mediosagittal or midsagittal one, designating the other parallel planes as *parasagittal*. Since longitudinal and dorsoventral axial lines are included in sagittal planes, both sorts of lines, depending upon arbitrary rules, can be designated as 'sagittal'.

2. *Transverse or cross-sectional planes* are at right angle to the longitudinal axis, but parallel to dorso-ventral and transverse axes (Figure 35). Such planes divide the body into caudal and rostral (in man: cranial) parts. These planes, moreover, include a dorso-ventral as well as a transverse axis. In man, where the so-called anatomical position is erect, transverse planes are evidently horizontal.

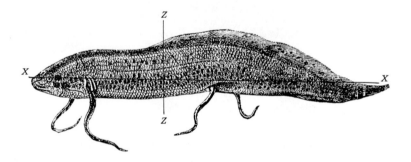

*Figure 34 D*. Body form of the lungfish, *Protopterus annectens*, for comparison with preceding figures (after Boas from R. Hertwig, 1912).

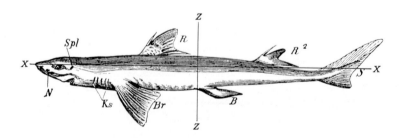

*Figure 34 E*. Body form of the Selachian *Acanthias vulgaris*, for comparison with preceding figures (from R. Hertwig, 1912). B: caudal fin; Br: rostral fin; Ks: gill slits; N: nasal pit; R: rostro-dorsal fin; R²: caudo-dorsal fin; S: tail fin; Spl: spiraculum.

3. *Frontal, coronal*, or (in most vertebrates) *horizontal planes*, dividing the body into dorsal and ventral parts, are at a right angle to the dor-so-ventral axes, but parallel to longitudinal and transverse axial lines, thereby including one of each. The terms *'frontal'* and *'coronal'* refer to human anatomy, where such planes are assumed to be roughly parallel to outer surface of frontal squama and to coronal suture. It is evident that such planes, horizontal in fishes and most tetrapods, are vertical in man.

Any plane not coinciding with one of the three fundamental planes is an *oblique plane*. Again, elongated organs, such as the central nervous system, may also be cut at right angles to their three main axes. The obtained planes can then be oriented with reference to the long axis of the organ, which may or may not correspond to that of the body. Thus

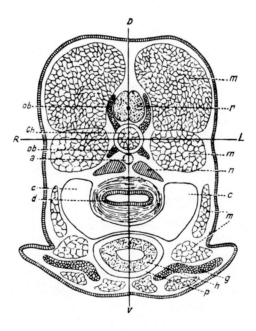

*Figure 35*. Transverse plane as displayed by a cross-section through a vertebrate body ('generalized fish'), from R. HERTWIG, 1912. D–V: dorsoventral axis; R–L: right-left (transverse) axis. The main longitudinal (here sagittal) axis may be assumed to pass, at a right angle to the plane of the picture, through the dorsal neural tube. Other longitudinal axes pass, in a similar manner, through axial skeleton (notochord), dorsal blood vessel (a), gut (d), celomic cavity (c), and heart (h).

various angles (dihedrals) with the three fundamental body planes can thereby be displayed.

It is obvious that, in addition to the intrinsic arbitrariness and ambiguity of semantic formulations, the differences between the phraseologies obtaining in applied human anatomy and in theoretical morphology, concerned with form analysis of organisms, can lead to further equivocations. Thus, anterior and posterior in man are synonymous with ventral and dorsal but may mean rostral and caudal in other vertebrates (Fig. 34B, C, D). Comparable ambiguities can involve the terms *superior* and *inferior*. *Cranial* and *cephalic*, i.e. nearer to the head, are equivalent to superior in man, and to anterior or rostral in most other vertebrates. With reference to the human head, the ventral ('anterior', rostral) position of the face, of which the nose forms an apex *(rostrum)*, while another (superior) apex is provided by the *vertex* (top of the head, approximately the *bregma* of anthropologists), must be taken into consideration. In other vertebrates, *rostral* denotes the relative position of a spot toward the anterior apex at which the midsagittal axis begins, and where dorsal and ventral surface midlines meet. In man, these two midlines can be assumed to meet at the vertex.

The expressions *on*, *over*, and *under* are often loosely used, but should be properly qualified unless the meaning is self-evident.

*Internal* and *external* denote locations relative to the center of an organ but are occasionally used as synonyms for *superficial* and *deep* which refer to the surface of the body. The terms *lateral* and *medial* are rather unambiguous, indicating positions farther from or nearer to the midsagittal plane, respectively, while *median* denotes an unpaired locus straddling the midline. In the description of extremities and appendages, *proximal* means a position nearer the trunk, and *distal* refers to a more distant position. With regard to configurations conceived as 'centers', such as the whole neuraxis, specific grisea within the neuraxis, or the perikarya (bodies) of cellular elements, *proximal* (more central) and *distal* (more peripheral) are likewise used, *mutatis mutandis*, to indicate spatial relationships concerning nerves, fiber tracts, nerve fibers, and cellular processes (outgrowths) in general.

For overall descriptive purposes, it can be assumed that the here relevant morphogenetic processes configuring the central nervous system of vertebrates begin at the *neurula stage* as defined further above (p. 86, including footnote 46). With this stage, the differentiation of primordial organs which subsequently evolve into the definitive organs, becomes apparent. Reverting to the distinction of quantitative growth processes and qualitative formative or configurating processes (cf. p. 9), a few comments on *growth*, preceding the more detailed discussions of *configuration*, might be pertinent.

*Organic growth*, i.e. increase in mass and size, depends, *inter alia*, upon the relevant processes of protoplasmic synthesis, water uptake, and deposition of intercellular substances. It can be recorded by measurements of size in one, two or three spatial dimensions, and by measurements of mass (weight).

Again, the amount of growth can be expressed in *absolute* and in *relative* terms. Thus, an absolute increase of one cm or one g from an initial value of 10 represents a relative increase of 10 percent, but of 5 percent if the initial value was 20.

The amount of growth during a period of time is the *growth rate*. The *absolute growth* rate is the amount of increase during a given period of time divided by its length in terms of the chosen time unit. The *relative growth rate* is the relative increment per unit of time, computed by dividing the absolute rate by the initial value.

If the relative growth rate stays uniform or constant, that is, if the rate of increase remains proportional to the magnitude of what is increasing, this type of growth conforms to the so-called 'natural law of growth', which can be expressed by means of the exponential factor $e^{at}$, whose reciprocal $e^{-at}$ characterizes the 'die-away'- or 'decay-curve'.

Since organic growth is not unlimited, but subject to intrinsic constraints, whereby living organisms attain sizes of a limited order of magnitude, it is evident that constant relative growth rates in accordance with the expression $e^{at}$ cannot be maintained in the course of ontogenetic evolution. Moreover, it is evident that the changes in shape, configuration, and proportion are the effects of unequal growth manifested by various body parts. This variety of form is mediated through *differential rates of growth* effective in diverse regions and directions, as well as in different organs or organ systems.

The ratios existing between the growth rates of various parts of the body in a given animal species appear to be relatively constant: 'It is these fixed relations that produce similar final form in the countless individuals of any animal-kind. And this is accomplished in spite of the fact that the constituent parts of an organism make their appearance and begin to grow at different times. The variance in starting times and growth rates among species is responsible for what may be called their growth pattern' (AREY, 1954).

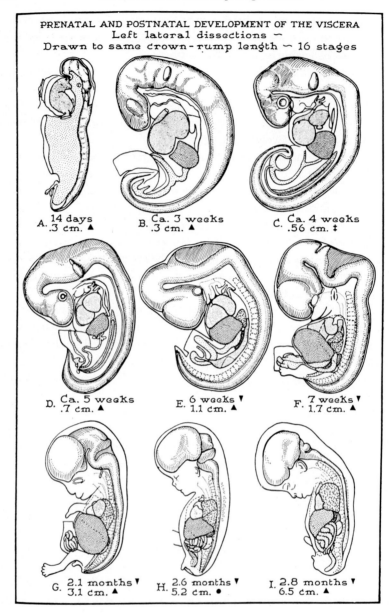

*Figure 36.* A series of semidiagrammatic left lateral views illustrating morphogenesis and growth of neuraxis and various viscera in man from early embryonic stages to maturity (after J.F.LEWIS, from SCAMMON, 1953). The relative dimensions of telencephalon in

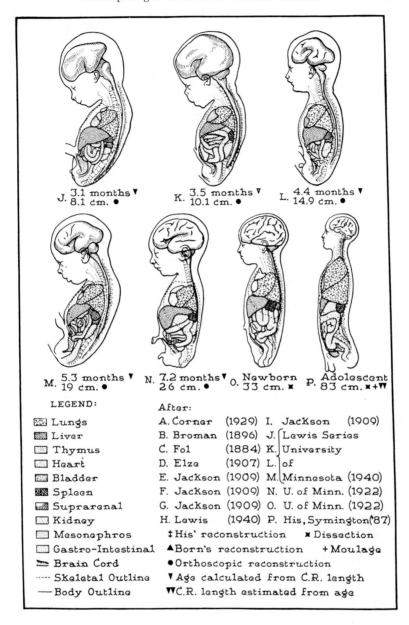

J. 3.1 months ▼
8.1 cm. ●

K. 3.5 months ▼
10.1 cm. ●

L. 4.4 months ▼
14.9 cm. ●

M. 5.3 months ▼
19 cm. ●

N. 7.2 months ▼
26 cm. ●

O. Newborn
33 cm. ✖

P. Adolescent
83 cm. ✖+▼

LEGEND:

| | | After: | | | |
|---|---|---|---|---|---|
| ▨ | Lungs | A. Corner | (1929) | I. Jackson | (1909) |
| ▩ | Liver | B. Broman | (1896) | J. ⎰Lewis Series | |
| ▢ | Thymus | C. Fol | (1884) | K. ⎱University | |
| ▢ | Heart | D. Elze | (1907) | L. ⎰of | |
| ▨ | Bladder | E. Jackson | (1909) | M. ⎱Minnesota (1940) | |
| ▨ | Spleen | F. Jackson | (1909) | N. U. of Minn. (1922) | |
| ▨ | Suprarenal | G. Jackson | (1909) | O. U. of Minn. (1922) | |
| ▨ | Kidney | H. Lewis | (1940) | P. His, Symington ('87) | |
| ▢ | Mesonephros | ‡ His' reconstruction | | ✖ Dissection | |
| ▨ | Gastro-Intestinal | ▲ Born's reconstruction | | + Moulage | |
| ➤ | Brain Cord | ● Orthoscopic reconstruction | | | |
| ---- | Skeletal Outline | ▼ Age calculated from C.R. length | | | |
| —— | Body Outline | ▼ C.R. length estimated from age | | | |

comparison with the more caudal brain subdivisions (brain stem, etc.) should be noted. It will be seen that the predominant expansion of telencephalon becomes conspicuous at about stage G.

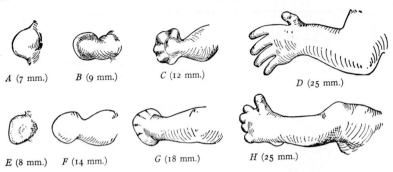

*A* (7 mm.)      *B* (9 mm.)      *C* (12 mm.)            *D* (25 mm.)

*E* (8 mm.)    *F* (14 mm.)      *G* (18 mm.)      *H* (25 mm.)

*Figure 37 A.* Stages in the development of the human limbs between the 5th and 8th week (approx. ×5). Upper row superior extremity, lower row inferior one (from Arey, 1954).

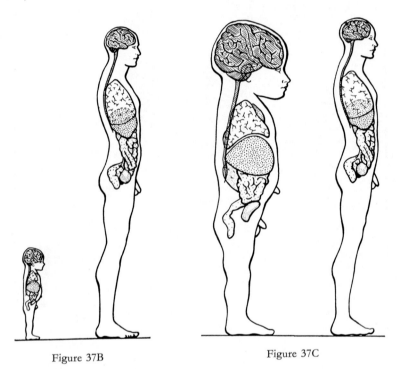

Figure 37B            Figure 37C

*Figure 37B.* Right lateral aspect of neuraxis and other viscera in human newborn and adult, at about $\frac{1}{20}$ natural size, illustrating the pertinent proportions (after H. A. Wilmer, from Scammon, 1953).

*Figure 37C.* Right lateral aspect of neuraxis etc. as shown in 37B, but drawn at an identical length (after H. A. Wilmer, from Scammon, 1953).

It seems, therefore, justified to state that, seen from this viewpoint, the qualitative formative processes (differentiation or '*Gestaltungs-vorgänge*') may be conceived in terms of *differential growth*. AREY (loc. cit.) dinstinguishes in this respect (1) local differences in the growth intensity; (2) growth gradients; (3) reduction of the early dominance of rostral over caudal levels; (4) functional demands, and (5) influence of

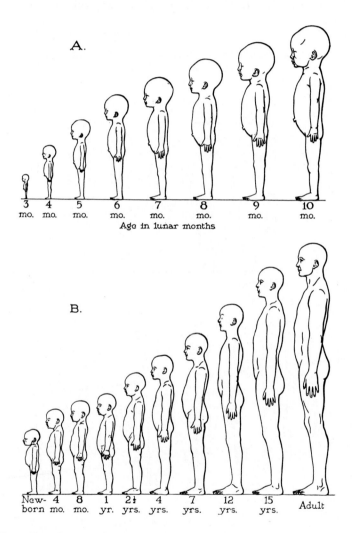

*Figure 38 A, B.* Series showing growth and development of human body form. A: eight fetal stages based on an empirical formula by CALKINS and SCAMMON. B: ten postnatal stages based on SCHADOW, 1834 (from SCAMMON, 1942, 1953).

the growth rate of a neighboring part. Thus, Figures 36 and 37 A demonstrate how the spatial distribution of diverse differential growth fields results in the progressive external modeling of the human brain and of the human limbs, which latter arrise from initially bud-like swellings. Corresponding, more complex modelings of internal configurations are displayed by the figures illustrating stages of neuraxial morphogenesis in the next subsections of this chapter. Fig-

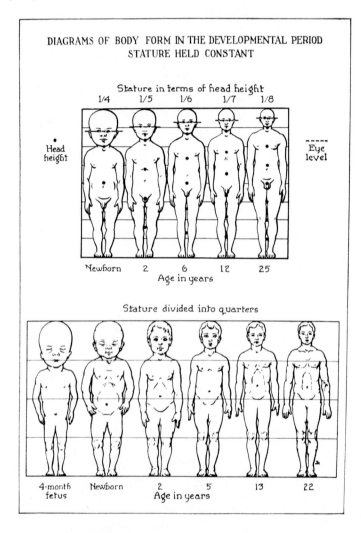

*Figure 38 C.* Diagrams illustrating changes in human body form and proportions during postnatal developmental periods (based on STRATZ, 1903, from SCAMMON, 1953).

ures 37B–38D depict changes in size and proportions of the human body and skull during prenatal as well as postnatal ontogenesis.

It has been shown that individual organs or organ systems display their characteristic growth curves. Studies concerning these questions have been undertaken especially with regard to human ontogenetic development (SCAMMON, 1953; also concisely summarized and discussed by AREY, 1954). According to the just-cited authors, human fetal organs tend to increase more or less rapidly to a maximum relative size, and then (either relatively or in some instances absolutely) to decrease throughout subsequent ontogenesis up to maturity. Moreover, various organs appear to have an *initial* period of slow growth, followed by a subsequent phase of rapid increase.

Curves illustrating human prenatal and postnatal brain as well as body growth are shown in Figure 39. Postnatal growth of human organ systems is depicted by Figure 40, which indicates that most of the organs can be roughly arranged in four main groups. Examples of postnatal neural type growth curves in rat and man are illustrated by Figure 41. From my own incidental observations concerning neural growth in the chick during *incubation* (K., 1937), some curves, shown in Figure 42, were obtained, which, despite certain irregularities, are somewhat similar to the so-called neural type of growth in man after birth, as illustrated in Figure 40. On the other hand, there occurs indeed little or even practically no initial growth during early phases fol-

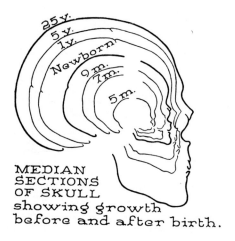

*Figure 38 D*. Diagram showing growth of human skull before and after birth (from SCAMMON, 1953).

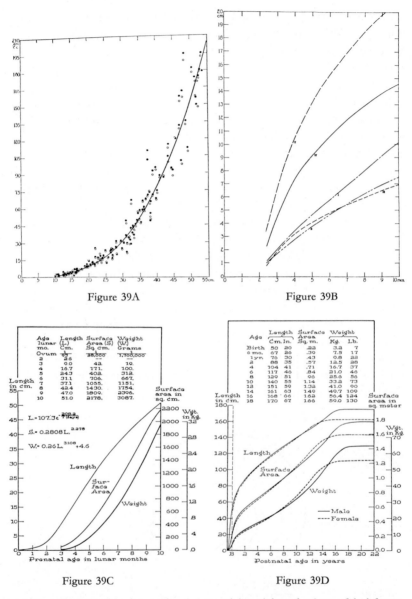

Figure 39A

Figure 39B

Figure 39C

Figure 39D

*Figure 39 A.* Graph and curve showing growth in weight and volume of the left cerebral hemisphere in human prenatal life. Abscissa: total body length in cm; ordinate: left hemisphere weight and volume in g and cc; individual cases indicated by dots for weight and circles for volume (from DUNN, 1921).

*Figure 39 B.* Curves illustrating by linear measurement the prenatal growth of human 'cerebrum', 'brain stem', and spinal cord. Abscissa: prenatal months; ordinate length of

lowing cleavage in the ontogeny of various invertebrates and vertebrates, as indicated e.g. by Figure 18B.

For additional data concerning curves, rates, and other aspects of
growths, reference may be made to DONALDSON's treatise on the rat
(1924), to THOMPSON's essays on growth and form (1942), and to the
papers by DUNN (1921), by G. J. NOBACK (1925–1926), and by
C. R. NOBACK (1944, also NOBACK and ROBERTSON, 1951).

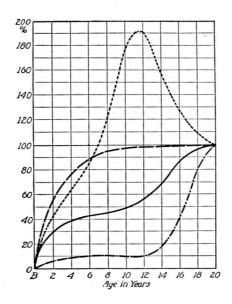

*Figure 40.* Curves illustrating postnatal growth of various types of human organs.
Growth is calculated in relation to average adult weights as 100 percent (after SCAMMON,
1953, from AREY, 1954). Lymphoid type ( - - - - ); thymus, lymph-nodes, intestinal
lymphoid masses. Neural type (– – –): brain and its parts, dura, spinal cord, optic apparatus, many head dimensions. General type (————): body as a whole, external dimensions
(with exception of head and neck), respiratory and digestive organs, kidneys, aorta and
pulmonary trunks, spleen, musculature as a whole, skeleton as a whole, blood volume.
Genital type (——·—·): testis, ovary, epididymis, uterine tube, prostate, prostatic urethra,
seminal vesicles.

---

'cerebrum', 'brain stem', and spinal cord in cm; I: fronto-occipital diameter; II: temporal
diameter; III: 'fronto-spinal length'; IV: spinal cord length; V: brain-cord length (from
DUNN, 1921).

*Figure 39C, D.* Curves, formulae and tabulations illustrating aspects of human prenatal (C) and postanatal (D) growth (after BOYD, 1941, from SCAMMON, 1953).

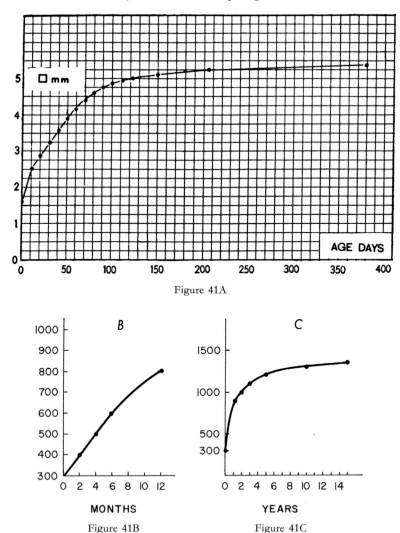

Figure 41A

Figure 41B       Figure 41C

*Figure 41 A–D*. Postnatal neural growth curves in rat and man. A: growth of midsagittal area of corpus callosum in albino rat (after SUITSU, 1920). B, C: growth curves depicting postnatal brain weight increase in man during the first postnatal year (B) and during childhood and early adolescence (C). Weight in g; the effect of time calibration upon the steepness of the curves in B and C should be noted (based on data compiled by RASMUSSEN, 1955). D: Male and female brain weight curves for the entire course of human life. Kurve der Hirngewichte in Gramm männlicher (————) und weiblicher (- - - -) Individuen, zusammengestellt nach *Quain's Anatomy*, II. Aufl., 1908, Bd. 3, p. 942, auf Grund der Angaben von R. BOYD (Phil. Trans. 1860). (After R. BOYD, 1860, and QUAIN, 1908, from R. SEMON, 1920.)

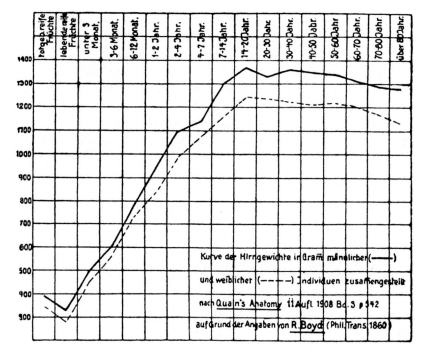

Figure 41D

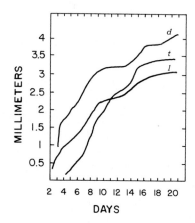

*Figure 42.* Curves illustrating the course of growth of the chick's diencephalon during incubation. Abscissa: days of incubation; ordinate: measurements in millimeters; d: dorso-basal diameter; l: greatest length; t: greatest wall thickness; the irregularities can perhaps be attributed to shrinkage effects caused by the histologic processing (after K., 1937).

Concerning time scales, discussed above in connection with '*cir-cadian rhythms*', and in the preceding volume (chapter V, section 9) with respect to life durations, aging, as well as problems of '*biologic time*', it will here be sufficient to refer very briefly to a few representative data pertaining to the ontogenetic event of *hatching* in Anamnia and Sauropsida, and of *birth* in Mammalia.

Thus, in a frog larva *( Rana pipiens )*, the first *movements* may be noticeable 70 to 84 h after cleavage, *hatching* occurs 'normally' between about 95 to 100 h, and *metamorphosis* approximately 13 to 16 days after fertilization of the egg. In a *Sauropsidan* such as the chick, *hatching* usually takes place after an incubation of about 21 days.

Concerning gestation in *Mammalia*, AREY (1954) has compiled the following tabulation:

Comparative data concerning gestation in mammals

| Animal | Gestation period | No. in litter | Animal | Gestation period | No. in litter |
|---|---|---|---|---|---|
| Opossum | 13 days | 8 | Macacus | | |
| Mouse; rat | 20; 22 days | 6; 8 | monkey | 24 weeks | 1 |
| Rabbit | 32 days | 6 | Man; | | |
| Cat; dog; | | | manlike apes | 38 weeks | 1 |
| guinea pig | 9 weeks | 4–6 | Cow | 40 weeks | 1 |
| Sow | 17 weeks | 6–12 | Mare | 48 weeks | 1 |
| Sheep; goat | 21 weeks | 1–2 | Rhinoceros | 18 months | 1 |
| | | | Elephant | 20 months | 1 |

It may be recalled that in the rat, whose metabolic rate is said to be thirty times as rapid as that of man, the life-span is about one-thirtieth, such that three years of rat life correspond to 90 years of human life (cf. vol. 3, part I, chapter V, section 9). Because sufficiently numerous and conclusive data are not yet available, it remains still questionable whether such relationships can be construed as manifesting a general 'law'.

In attempting to define the factors controlling organic growth in general, and in particular differential growth, one is faced with the various intricate problems of morphogenesis pointed out above in the initial part of this section. From a more generalized and somewhat different viewpoint, AREY (1954) has classified these factors in the following manner: (1) constitutional factors involving the genetic aspect; (2)

temperature; (3) nutritional factors; (4) growth-promoting factors such as embryonic enzymes, hormones, and vitamins (these latter also pertaining to group 3), and (5) growth-arresting factors.

### B. Some Relevant Configurational Features Related to Morphogenesis of the Vertebrate Neuraxis

If craniote vertebrates are regarded as phylogenetically derived from chordate Acrania, the neuraxis of *Amphioxus* assumes considerable morphologic significance despite evident restrictions and qualifications, which are mandatory because recent Branchiostoma can hardly be said to represent a truly ancestral form.

The neural tube of the early Amphioxus larva, at a stage in which anterior neuroporus and canalis neurentericus are well displayed, is illustrated in Figure 43. The subsequent ontogenetic development results in the formation of a definitive neuraxis whose differentiation into brain and spinal cord remains rather inconspicuous. Depending upon arbitrary criteria, several different formulations or interpretations appear permissible.

*Prima facie*, the entire central nervous system of Amphioxus seems to represent merely a spinal cord, beyond whose anterior end the noto-

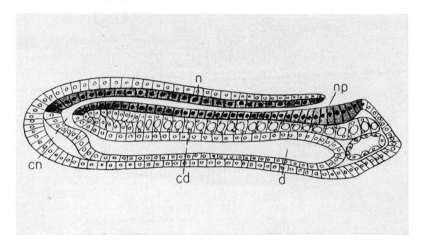

*Figure 43.* Semidiagrammatic sagittal section through an Amphioxus larva (after HATCHEK, from K., 1927). cd: notochord; cn: neurenteric canal; d: gut; n: neural tube; NP: anterior neuropore.

chord extends as far as the region of the rostrum. Nevertheless, the
most anterior part of the neuraxis shows two regions of special differentiation, an *apical* and a *postapical* one (Fig. 44, 45).

The *apical* portion can be described as a small vesicle, which is fairly
well demarcated from the rest of the neural tube, and represents the *archencephalon* of v. KUPFFER (1906). Before the anterior neuropore closes, it is rostrally connected with an unpaired ciliated ectodermal vesicular pit, subsequently somewhat displaced to the left, whose interpretation as an olfactory or chemoreceptor organ by several authors has
not been accepted by others; its functional significance may be considered controversial. In some instances it seems to remain connected
with the brain vesicle by an unpaired fibrous strand. Basal to the neuroporic region, a pigmented spot is frequently present. Still more basally,
an essentially paired apical so-called *nervus terminalis* is connected with

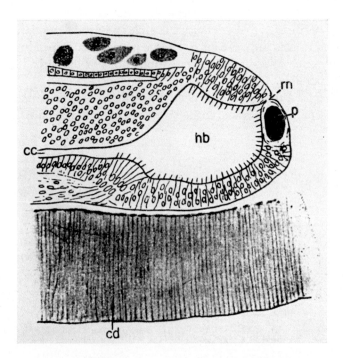

*Figure 44.* Semidiagrammatic sagittal section through the brain vesicle of Amphioxus
(modified after KUPFFER, from K., 1927). cc: central canal; cd: notochord; hb: lumen of
cerebral vesicle; p: pigmented spot; rn: recessus neuroporicus (an internal and an external
pit are indicated). The large *dorsal cells of Rhode* (not labeled) are shown in the postapical region.

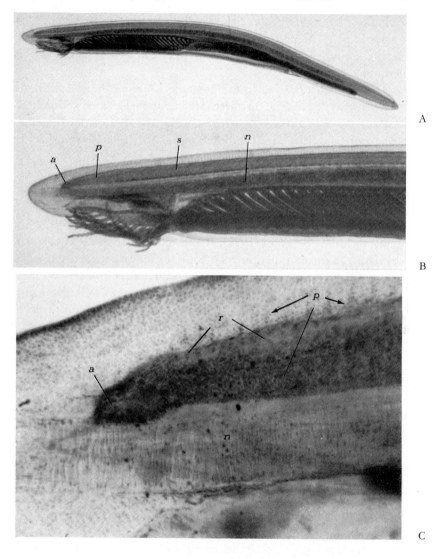

*Figure 45.* Neuraxis of a young Amphioxus as seen in a whole mount (hematoxylin-eosin stain). A General aspect of whole animal. B Rostral end of animal. C Epichordal cerebral vesicle (archenteron) and postapical region with *dorsal cells of Rhode*. A ×25; b ×57; C ×450; red ²/₃. a: archencephalon; n: notochord; p: postapical region; r: *dorsal cells of Rhode;* s: spinal cord with eye-spots.

the end of the brain vesicle. The bilateral components lie here closely together and are barely separable. This nerve supplies the rostrum and the cirri (mouth tentacles); a few thin fiber bundles reach the above-mentioned vesicular pit.

At the basal transition to the postapical region, the ciliated brain vesicle displays a patch of distinctive tall epithelial ciliated cells, the *infundibular organ* of BOEKE, which, regardless of its functional significance, can be evaluated as kathomologous to the *saccus vasculosus* of fishes.

The *postapical region* with its dorso-ventrally slit-like lumen is generally quite similar to the more caudal spinal cord, but is characterized by the presence of large dorsal cells (dorsal *cells of Rhode, cells of Joseph*). Earlier authors had interpreted these elements as true nerve cells, while others, including KAPPERS (1929), assumed that they were of neurosensory nature, since their peripheral cytoplasm shows a striped, palisaded pattern. On the basis of recent ultrastructural observations by WELSCH (1968), their significance as photoreceptor cells seems well documented. These elements are remarkably similar to the photoreceptor cells of annelids and molluscs. While the postapical region of the neuraxis lacks ventral nerve roots, two pairs of dorsal nerves are present.

The *spinal cord sensu strictiori* is distinguished by the presence of segmentally arranged, alternating dorsal and ventral roots, of ventral pigmented eye spots, consisting each of two cells, and by a peculiar arrangement of large multipolar nerve cells (median *giant cells of Rhode*) which are related to a system of intramedullary giant fibers. Relevant morphologic and structural details concerning peripheral and central nervous system of Amphioxus will be discussed in connection with the pertinent problems dealt with in chapter VII of this volume as well as in volumes 4 and 5 concerned with the standard subdivisions of the neuraxis. It should here, however, already be pointed out that the so-called ventral roots of Branchiostoma, although displaying morphological relationships formally identical with those of ventral roots in craniote vertebrates, are structurally entirely different, being essentially fibers provided by the muscular elements (FLOOD, 1966), and therefore non-nervous as well as mesodermal. A comparable type of muscular innervation obtains in some invertebrates (cf. below, chapter VII).

The apical vesicle of Amphioxus can easily be mapped, by a one-to-one reference, upon the embryonic craniote vertebrate prosencephalon. The post-apical region may then be mapped, by a one-many reference, upon the vesicles representing mesencephalon and rhombencephalon of craniote embryos. Spinal cord *sensu strictiori* of Amphioxus and that of craniote vertebrates can be easily interrelated by bicontinuous one-to-one mapping.

With respect to the overall morphologic relationships of primary relevance in the present context, the designation *archencephalon* (v. KUPFFER, 1906) for the rostral vesicle of the neuraxis in Amphioxus appears fully justified. This vesicle, orthohomologous with the early craniote prosencephalon, can be evaluated as the simplest known manifestation of a brain-like configuration significantly differing from the spinal portion of the neural tube. Ontogenetically, the homeomorph component of craniotes evolves into *diencephalon* and *telencephalon*. From a phylogenetic viewpoint, the assumption of a similar mode of evolution appears quite reasonable. Further details on the archencephalon of Amphioxus, including its so-called infundibular organ and that structure's relationship to *Reissner's fiber* shall be considered in chapter VIII of volume 4.

The post-apical neuraxis region in Amphioxus can be regarded as a slightly modified portion of spinal cord, since it differs much less significantly from this latter. The interpretation of craniote mesencephalon and rhombencephalon as kathomologous to this post-apical region is in perfect agreement with the well supported conclusion, based on additional data discussed further below, that the craniote mesencephalon and rhombencephalon, together subsumed under the term *deuterencephalon*, are promorphologically (serially) homologous (essentially orthohomologous) with the spinal cord, thus representing a portion of this latter secondarily included into brain and also mainly into the cranium. Both designations deuterencephalon and archencephalon can therefore be considered ontogenetically as well as phylogenetically appropriate. The evaluation of the deuterencephalon as a phylogenetically and ontogenetically transformed, and in the higher craniotes substantially modified rostral portion of spinal cord was accepted and maintained, on the basis of several morphologic features, more than forty years ago by the author (K., 1926, 1927, 1932).

During the ontogenesis of craniote vertebrates, at certain early stages preceding the complete closure of the neural tube, a configurational pattern of the neuraxis comparable with that in Amphioxus may be recognized. The anterior end of the neural tube displays a vesicular enlargement homologous to the apical brain vesicle in Branchiostoma, thus representing the *archencephalon*, which is here designated as *prosencephalon* in accordance with generally accepted embryologic terminolo-

gy. It is delimited from the remainder of the neural tube by a very conspicuous constriction (Fig. 46). In contradistinction to the *epichordal*
archencephalon of Amphioxus, which lies entirely upon the notochord, this latter extending beyond the neuraxis as far as the rostrum

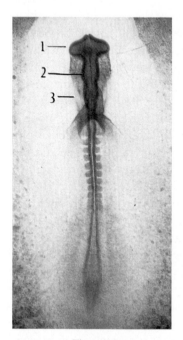

Figure 46

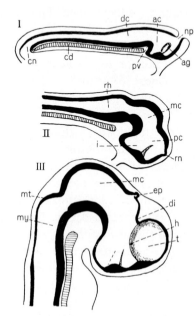

Figure 47

*Figure 46.* Closure of the brain tube in a chick embryo of 33 h. The anterior neuropore
is clearly indicated, and the spinal tube is not yet closed. The prosencephalon, whose lateral expansions are the eye diverticula, is delimited from the deuterencephalon by a conspicuous constriction. Two parts of the deuterencephalon (2, 3) can be recognized. The
rhombencephalon displays some additional relatively minor constrictions (whole mount,
hematoxylin-eosin stain, ×20; red. $^4/_5$). 1: prosencephalon; 2: mesencephalon; 3:
rhombencephalon.

*Figure 47.* Diagram of ontogenetic stages in vertebrate brain morphonenesis. I Essentially duplex stage (archencephalon and deuterencephalon). II Essentially tripartite
stage. III Definitive stage of essentially quintuple subdivision (from K., 1927). ac: archencephalon; ag: eye-stalk opening; cd: notochord; cn: neurenteric canal; dc: deuterencephalon; di: diencephalon; ep: epiphysial anlage; h: torus hemisphaericus whose basal
midline portion is the torus transversus; i: 'infundibular' region; mc: mesencephalon;
mt: metencephalon (cerebellar plate); mg: myelencephalon; np: anterior neuropore; pc:
prosencephalon; pv: ventral brain fold (plica ventralis encephali); rh: rhombencephalon;
rn: recessus neuroporicus; t: telencephalon.

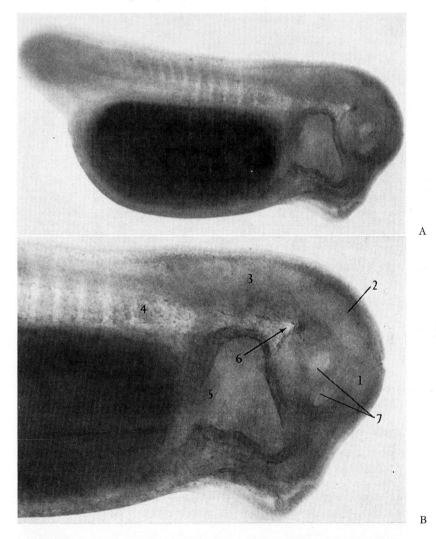

*Figure 48.* Early stage of brain differentiation as displayed in a frog larva of 4-mm length (hematoxylin-eosin stain; all magnifications red. $^2/_3$). A Total aspect of whole mount ($\times$41). B Rostral region showing further details of brain ($\times$64). C Approximately midsagittal section, showing rostral end of notochord, touching base of prosencephalon, in region of plica ventralis encephali ($\times$70). D Parasagittal section, showing tripartite subdivision of brain ($\times$54). 1: prosencephalon; 2: mesencephalon; 3: rhombencephalon; 4: notochord; 5: pharynx; 6: plica ventralis encephali; 7: eye primordia; two unlabeled dips indicate boundaries between 1 and 2, and 2 and 3; the slight protrusion seen rostral to the right dip in A and B, which is not constant at this stage, may represent the first manifestation of the parencephalic neuromere (cf. section 3), coincident with the beginning of prosencephalic differentiation into diencephalon and telencephalon.

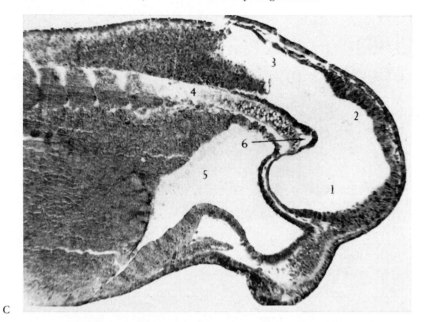

C

D

Figure 48

(Fig. 45), the craniote archencephalon is *prechordal*, lying rostrad to the end of the notochord (Figs. 47, 48, 49 A).

The basis of the archencephalon is delimited, at the rostral limb of a conspicuous fold, the *ventral brain fold ( plica ventralis encephali )*, from an epichordal enlargement of the neural tube representing the *deuterencephalon*, which displays further particular differentiations only in Craniota. This displacement in the relationship of chorda and rostral neuraxis can be evaluated as topologically nonsignificant.

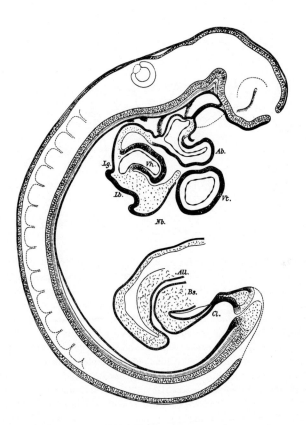

*Figure 49 A*. Midsagittal section through a human embryo of about 8.2 mm total length (after His, 1904). The neural tube, with open anterior neuropore as well as with posterior neuropore, displays a stage intermediate between those of figure 47 I and II. The entrance into the eye stalks is shown as a slit. The location of otic vesicle with respect to rhombencephalon is indicated. Ab: aortic bulb; All: allantoic duct; Bs: body stalk; Cl: cloaca; Lb: liver primordium; Lg: lung primordium; Nb: umbilical vesicle: Vh: cardiac atrium; Vt: cardiac ventricle.

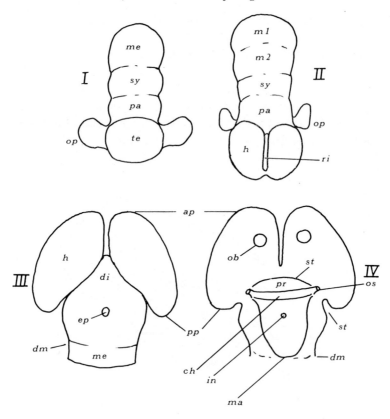

*Figure 49 B.* External aspects of early stages of ontogenetic forebrain development in man (adapted and modified from HOCHSTETTER, 1919). Since GROENBERG (1901) has depicted essentially identical stages in the insectivore *Erinaceus*, and I have myself made similar observations in various other mammalian embryos, these aspects can be regarded as significantly typical for Mammalia in general. The neuromeric segmentation (pa, sy, m1, m2, discussed in section 3) is somewhat diagrammatically exaggerated. I: Anterior (parietal) view of still unpaired evaginated telencephalon in an embryo of about 3.4 mm. II: Beginning bilateral evagination of still unpaired telencephalon in an embryo of about 7.5 mm. Between the two hemispheric diverticula, the dorsal sagittal ridge in the floor of the interhemispheric groove can be seen. III and IV: Dorsal and basal view of paired evaginated telencephalon in an embryo of about 17 mm. The interhemispheric groove has now become an interhemispheric cleft or fissure. ap: anterior (rostral) pole; ch: external chiasmatic ridge; di: diencephalon; dm: diencephalo-mesencephaic boundary; ep: epiphysial anlage; h: telencephalic hemisphere; in: infundibulum; m1, m2: suggestion of transitory mesencephalic neuromeres; ma: mammillary tubercle; me: mesencephalon; ob: anlage of olfactory bulb; op: optic vesicle resp. cup; os: optic stalk; pa: parencephalic portion of diencephalon; posterior (caudal) pole; pr: external surface of preoptic recess; ri: dorsal midsagittal ridge of telencephalon; st: sulcus telo-diencephalicus (s. hemisphaericus); sy: synencephalic portion of diencephalon; te: telencephalon.

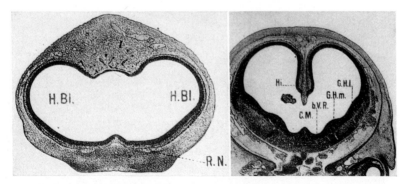

*Figure 49C.* Increasing hemispheric evagination of the still unpaired human telencephalon (from HOCHSTETTER, 1919). I Embryo of approx. 13 mm. Between the two hemispheric diverticula, the dorsal midsagittal ridge in the floor of the interhemispheric groove can be seen. II Embryo of approx. 19.5 mm. The interhemispheric groove has deepened into a fissure. b.V.R.: *'basale Vorderhornrinne'* (fold or groove between $B_2$ and $B_3$ zones of K., 1929a); C.M.: 'cavum Monroi'; G.H.l.: *'laterale Abt. des Ganglienhügels'* ($D_1$, K. 1929a); G.H.m.: *'mediale Abt. des Ganglienhügels'* ($B_1$ of K., 1929a); H.B.l.: hemispheric vesicles; Hi: hippocampal anlage ($D_3$ of K., 1929a); R.N.: fila olfactoria.

The craniote archencephalon or prosencephalon initially displays two relatively prominent ventrolateral protrusions, the primary eye diverticula. These two eye primordia soon become stalked vesicles (Fig. 50, cf. also Fig. 46), whose hollow stalks later obliterate to be transformed into the paired so-called *nervus opticus*, which is actually a brain tract.[55] The optic vesicles, by a combination of growth and invagination (infolding), are changed into the double-layered optic cups.

The subcommissural organ, whose cells commonly display secretory activity, also related to the production of *Reissner's fiber*, seems to originate in the caudal roof of the archencephalon within the neighborhood of the synencephalic neuromere.

---

[55] In one of the many so-called anatomical nomenclature commissions, the suggestion was made, about 1930, to designate the optic nerve as *'pars orbitalis tractus optici'*. Since, however, the nervous connection between retina and brain consists of three well defined portions, namely (1) from eye to optic chiasma; (2) region of optic chiasma, and (3) a rather conspicuous bundle (tractus opticus) from optic chiasma to various grisea such as lateral geniculate complex and tectum opticum, the convenient term *nervus opticus*, in contradistinction to *chiasma opticum* and *tractus opticus*, was kept, on the basis of reasonably valid practical considerations, in the conventional nomenclatures (e.g. PNA, 1955).

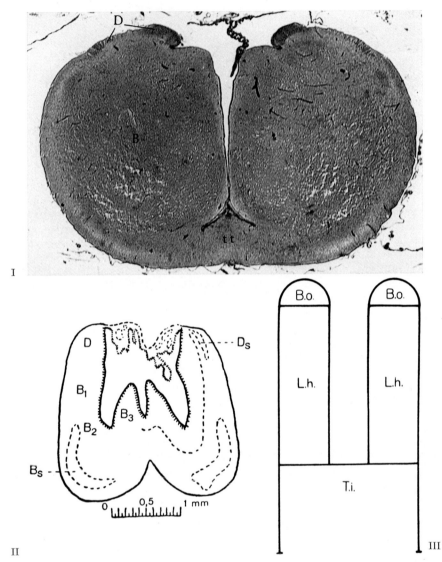

*Figure 49 D*. Persisting pars impar telencephali (telencephalon medium) in the Holocephalian Selachian Chimaera and in the Dipnoan Protopterus. I Cross-section through telencephalon impar of Chimaera (from K. and Niimi, 1969). II Cross-section through telencephalon impar of Protopterus (from K., 1929a). D, Ds: pallial region; B, B$_1$, B$_2$, B$_3$, Bs:
basal region. III Diagram of telencephalon in Protopterus, indicating main subdivisions,
including a telencephalon impar (from Gerlach, 1933). B.o.: bulbus olfactorius; L.h.:
paired evaginated part of lobus hemisphaericus; T.I.: remaining unpaired evaginated portion of lobus hemisphaericus (telencephalon impar); tt (in I): torus transversus with components of commissura anterior.

Discounting some additional secondary or at least in some stages and in many instances rather minor constrictions (cf. e.g. Fig. 46), related to the so-called *neuromeres*, which will be discussed further below, this essentially duplex stage of the brain tube (*archencephalon* and *deuterencephalon*, Fig. 47, I) is soon transformed into a tripartite configuration, whereby the rostral part of deuterencephalon, corresponding to the region of ventral brain fold, develops into a large, essentially, and particularly dorsally, prechordal vesicle, the *mesencephalon*. This, again, is separated by a prominent dorsal constricting fold from the caudal portion of deuterencephalon, which becomes the *rhombencephalon* (Fig. 47, II). The delimiting constriction, described by pioneer investigators such as HIS and v. KUPFFER, was named *plica rhombo-mesencephalica* by the latter author (Figs. 51 B, F, 52 A, C).

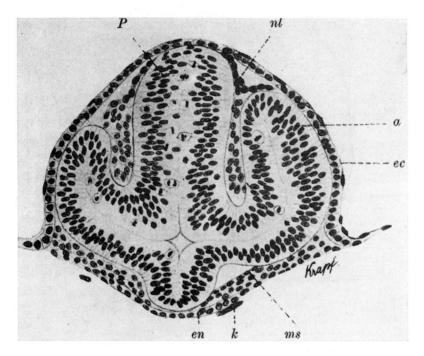

*Figure 50 A*. Prosencephalon of a trout embryo (Teleost) of 15 days, showing evagination of eye vesicles (from KUPFFER, 1906). a: eye vesicle; ec: ectoderm; en: entoderm; k: so-called 'merocyte nucleus'; nl: disintegrating neural crest, joining mesoderm on left; ms: mesoderm; P: wall of prosencephalon.

In at least some fishes such as Salmo and Esox, the ventral brain fold, forming the floor of the mesencephalon, displays a secretory ependyma, which has been designated as the '*flexural organ*' (OLSSON, 1956). This structure, like the subcommissural organ, appears related to *Reissner's fiber*. Further comments on this topic shall be included in volume 4.

The tripartite brain is subsequently transformed into a configuration which may be conveniently described as the *quintuple and definitive stage* of overall brain subdivision (Fig. 47, III). In a manner somewhat similar to the evagination of the ventrolateral eye diverticula, a more dorsal and rostral *unpaired vesicle*, involving the *lamina terminalis*, originates and becomes the *telencephalon* or *hemispheric brain* (Figs. 47, III; 49 B), enclosing a *common ventricular space*. The evagination of the telencephalon is correlated with a rostrodorsal expansion of that part of

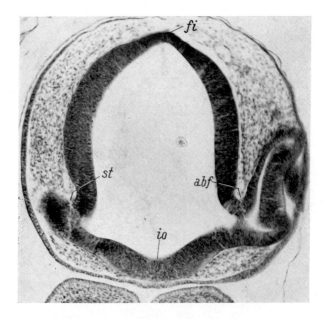

*Figure 50 B.* Prosencephalon of avian embryo (duck of 72-hours incubation), showing further differentiation with transition of eye vesicle to eye cup and ectodermal lens placode, right. The dorsal part (left) represents an early evagination of unpaired telencephalon (from K., 1936). At this stage, telencephalon and diencephalon just begin to become delimited (cf. reconstruction of this stage in fig. 131). abf: external groove of eye stalk; fi: dorsal ridge of telencephalon near telo-diencephalic boundary; io: basal interoptic groove; st: beginning sulcus telo-diencephalicus temporarily in close vicinity of external groove of eye stalk.

lamina terminalis which includes the region about and dorsal to the locus of final closure of anterior neuropore, whose remnant may or may
not still be indicated by a *neuroporic recess*. The portion of prosencephalon caudal to telencephalon becomes the *diencephalon*.

In most vertebrate groups the telencephalic ventricle becomes bi-
partite by a bilaterally symmetric expansion of its dorsal wall resulting
in the formation of two hemispheric protrusions separated by a median
sagittal depression whose bottom may, in numerous instances, display
a sagittal ridge (Fig. 49 B). At this stage, however, the two hemispheric
diverticula remain conjoint and still can be said to represent an unpaired telencephalon with common ventricular space (Fig. 49 C).

It seems therefore expedient to designate the first phase of telencephalic development in the dorsal portion of the anterior wall of the
neural tube anterior wall (lamina terminalis) as a process of *unpaired* but

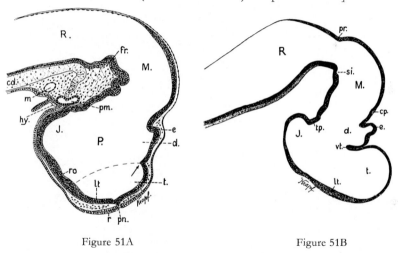

Figure 51A                                        Figure 51B

*Figure 51 A.* Approx. midsagittal section through brain of an Acanthias embryo (Selachian) of 10-mm length (from KUPFFER, 1906). cd: notochord; d: diencephalon; e: epiphysial anlage; fr: region of ventral plica rhombo-mesencephalica; hy: hypophysial anlage; J: infundibular region; lt: telencephalic lamina terminalis; M: mesencephalon; m:
'mandibular cavity'; pm: 'praemandibular cavity'; pn: recessus neuroporicus; R:rhomb-
encephalon; r: unpaired olfactory placode; ro: recessus praeopticus; t: telencephalon;
the velum transversum is indicated by an arrow; telencephalo-diencephalic boundary, according to the present treatise, is indicated by a dotted line.

*Figure 51 B.* Approx. midsagittal section through the brain of an Acanthias embryo of
about 16- to 17-mm length (from KUPFFER, 1906) cp: commissura posterior; pr: plica
rhombo-mesencephalica; si: 'sulcus intraencephalicus posterior' ('fovea isthmi' of other
authors); vt: velum transversum; other abbr. as in preceding figure.

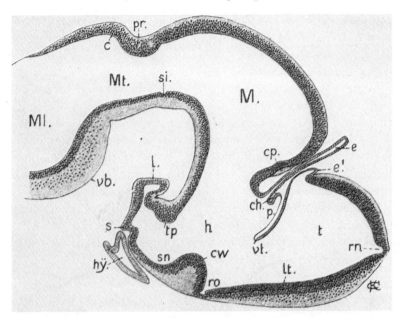

Figure 51C

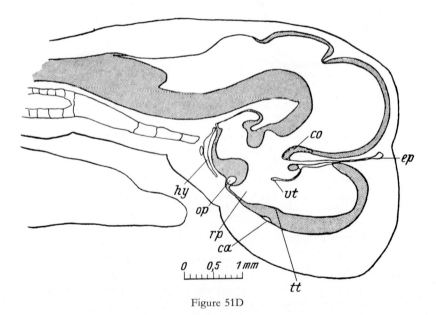

Figure 51D

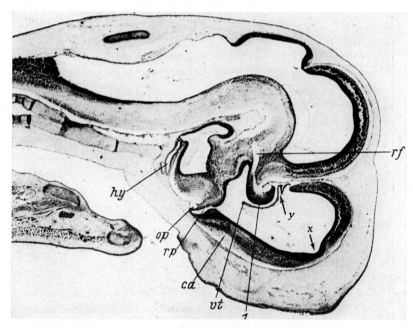

Figure 51E

*Figure 51 C.* Approx. midsagittal section through the brain of an Acanthias embryo of about 27-mm length (from KUPFFER, 1906). c: cerebellar plate; ch: commissura habenulae; cw: ridge of optic chiasma; e′: paraphysial anlage; h: diencephalic ventricle; l and s: hypothalamic evaginations; sn: postoptic recess of hypothalamus; Ml: myelencephalon; Mt: metencephalon; vb: convexity of rostral rhombencephalic flexure; other abbr. as in preceding figures.

*Figure 51 D.* Approx. midsagittal section through head and brain of an Acanthias embryo of about 45-mm length (From KUHLENBECK, 1929b). ca: commissura anterior; co: commissura posterior; ep: epiphysis; hy: hypophysial anlage (adenohypophysis); op: optic chiasma; rp: recessus praeopticus; tt: torus transversus; vt: velum transversum.

*Figure 51 E.* Sagittal section through head and brain of Acanthias embryo shown in preceding figure (From KUHLENBECK, 1929b). Because of a slight obliqueness in the sectional planes the telencephalon displays here recessus neuroporicus (arrow x) and paraphysial rudiment (arrow y), while the mesencephalo-diencephalic boundary neighborhoods are shown laterally to ventricular lumen. 1: epithalamus; rf: fasciculus retroflexus; other abbr. as in preceding figure.

*bilateral symmetric evagination.* A further phase in the process of evagination, leading to the formation of separate right and left telencephalic 'hemispheres' which are disjoined, in transverse planes, by a complete midline interhemispheric 'cleft' or 'fissure', may then be designated as *paired evagination.*

It is evident that, at the level corresponding to the original unpaired rostral evagination, the two separate hemispheres become basally united in the midline by the lamina terminalis. An extensive part of the conjoint or unpaired basal wall or floor of the hemisphere, may, in fact, be interpreted as an expanded portion of lamina terminalis. Dorsally, the two halves of the unpaired hemisphere are united in the midline by derivatives of the epithelial roof plate (such, e.g., as rostral leaf of velum transversum, and paraphysis).

In perhaps all vertebrate forms, the olfactory bulbs subsequently become separately paired evaginated by additional disjoint rostral expansions of the unpaired telencephalic vesicle. In Petromyzon (Cyclostomes) and in all Amniota (Sauropsida and Mammalia), the entire telencephalon, except for the small region corresponding to the communication between right and left ventricles at the level of interventricu-

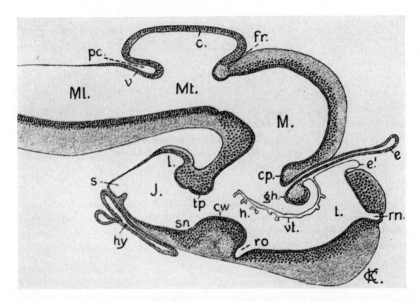

*Figure 51 F.* Approx. midsagittal section through the brain of an Acanthias embryo of 70-mm length (from KUPFFER, 1906). gh: ganglion habenulae; pc: 'plica cerebelli posterior'; v: velum medullare posterius; other abbr. as in figures 51 A–C.

lar foramina, also develops into a paired evaginated telencephalon consisting of two 'separate' hemispheres.

In some Anamnia (Selachians, Dipnoi, Amphibians) a substantial part of the originally unpaired telencephalon likewise becomes paired, with two 'separate' telencephalic hemispheres, but a conspicuous pars impar persists (Fig. 49 D).

In other Anamnia, however, particularly in Osteichthyes, almost the entire telencephalon, but generally with exception of the stalked or unstalked olfactory bulbs, may be made up by an unpaired evaginated configuration[55a] representing a massive telencephalon medium. This latter is formed, as it were, in correlation with the just-mentioned considerable expansion of lamina terminalis.

A suitable conceptual delimitation of telencephalon from diencephalon is beset with numerous difficulties and the early attempts to define a convenient as well as morphologically sound boundary have resulted in various different and controversial semantic solutions. In the course of ontogeny and phylogeny, many configurational features which pertain to the fundamental pattern common to all vertebrates become obscured. The factors involved in this 'blurring' are multiple; they include changes of shape affecting the earliest subdivisions, and the origin of primordial grisea, which then undergo various types of one-to-many differentiations into further subdivisions, moreover, the shifting of cell masses, and distortions due to the incursion of fiber pathways. Since the conceptual tracing of a morphologically well substantiated telo-diencephalic boundary must be concordant with the arrangement and confines of two fundamental longitudinal zonal systems, each one of which is characteristic for telencephalon and diencephalon, respectively, relevant details concerning the boundaries of

---

[55a] Although the distinction between *conjoint* telencephalon impar and *disjoint*, separated, paired telencephalic 'hemispheres' seems best expressed by the terms paired evagination and unpaired evagination (cf. e.g. K., 1927, p.78 *et passim*), the proper verbal formulation of these morphogenetic processes poses certain semantic problems. Thus, the term *'evagination'* is occasionally used as referring only to *'paired evagination'*, while a telencephalon impar is then referred to as *'unevaginated'*. I have myself followed this procedure in a brief report on the telencephalon of *Amia calva* (Anat. Rec. *157:* 368–369, 1967), stating that only the olfactory bulbs are evaginated (paired) and inverted, the remainder of the telencephalon being unevaginated (unpaired) and laterally everted. Adopting a more rigorous terminology it would seem preferable to say that, in *Amia*, only the olfactory bulbs are paired evaginated, the remainder of the telencephalon being unpaired evaginated and laterally everted.

these and other subdivisions of the neuraxis will be discussed further below in connection with the zonal systems.

With regard to the present context, it will suffice to describe the telo-diencephalic boundary as follows. In the dorsal midline, the conjoint hemispheric vesicles are bounded from the diencephalon by a transverse epithelial fold of roof plate, the *velum transversum* (Fig. 52–57). The extraventricular dorsal concavity of the velum represents the dorsal, transverse, midline portion of an oblique, elongated ring-like external groove, the *sulcus telo-diencephalicus*, which crosses the midline dorsally as well as basally, and can be taken as corresponding to the caudal boundary of telencephalon.

From the velum transversum, the paired (right and left) sulcus telo-diencephalicus runs in a ventrolateral or lateral direction, separating the hemispheric evaginations from the lateral surface of diencephalon, and converges toward the ventral midline. The basal transverse part of

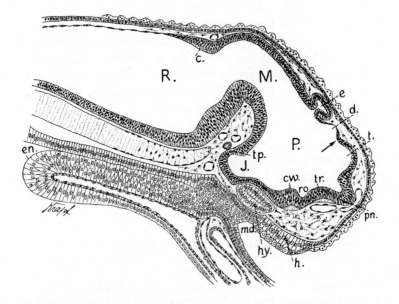

*Figure 52 A.* Approx. midsagittal section through head and brain of an embryo of Acipenser (Ganoid, Osteichthyes) at the hatching stage (from KUPFFER, 1906). c: cerebellar plate; cw: chiasmatic ridge; d: diencephalic roof; e: epiphysis; en: entoderm; h: ectodermal thickening; hy: adenohypophysial anlage; J: infundibular region; M: mesencephalon; md: oral recess; P: prosencephalon; pn: processus and recessus neuroporicus; R: rhombencephalon; ro: recessus praeopticus; t: telencephalon; tp: tuberculum posterius; tr: torus transversus; arrow indicates primordium of velum transversum.

this sulcus corresponds to the external surface of a thickened portion of lamina terminalis marking the primordial locus of commissura anterior, and designated as *torus transversus* by v. KUPFFER (1906) or *commissural plate* by HOCHSTETTER (1919). This torus transversus, a midline structure, thus separates the lamina terminalis into a dorsal or dorsorostral telencephalic, and a basal diencephalic part, which latter becomes the rostral wall of the preoptic recess.

Since some ambiguities concerning the designation lamina terminalis, also occasionally called lamina rostralis, are evident in the literature, the following additional remarks may be appropriate. At the time of its formation, which coincides with the gradual closure of the originally here wide open neuraxis (anterior neuropore), the lamina terminalis could be described as an unpaired neighborhood, comparable to a plate interconnecting right and left sides of the neuraxis by providing the anterior wall of the prosencephalon. It is generally oriented in a transverse plane approximately perpendicular to an imaginary longitudinal axis of the neural tube.

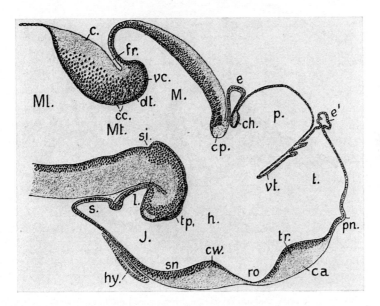

*Figure 52B.* Approx. midsagittal section through the brain of a 4-week-old developing Acipenser (from KUPFFER, 1906). ca: commissura anterior; cc: commissurae cerebellares; ch: commissura habenulae; cp: commissura posterior; dt: decussation of trochlear nerve; e: epiphysis; e': paraphysis; vc: valvula cerebelli; vt: velum transversum; other abbr. as in preceding figures from KUPFFER; fr = pr in Fig. 51 B.

During early stages of ontogenesis, the entire lamina terminalis represents a rather thin membrane which derives from a rostral midline fusion of the lateral walls of the neural tube (lateral plates). This fusion likewise implies continuity with the dorsally located roof plate, which thereafter provides the telencephalic lamina epithelialis extending rostrad from the velum transversum. Initially, a distinctive boundary between lamina terminalis and roof plate may not be recognizable in the relevant sagittal sections.

Subsequently, however, the *lamina epithelialis*, as a roof plate component, becomes distinguishable, because of its thinness, from the *lamina terminalis* pertaining to the conjoint lateral plates. The boundary between lamina terminalis and lamina epithelialis may then become indicated, at some stages, and in many instances, by a noticeable fold of transition, depicted by His (1893) and called *angulus terminalis* by this author (cf. Figs. 56 B, C, 57 D).

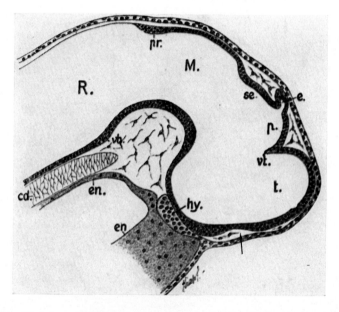

*Figure 53 A.* Approx. midsagittal section through head and brain of an 8-mm long larva of the urodele amphibian, *Salamandra atra* (from KUPFFER, 1906). cd: notochord; e: epiphysis; en: entoderm; hy: adenohypophysial anlage; M: mesencephalon; p: parencephalic roof of diencephalon; pr: plica rhombo-mesencephalica; R: rhombencephalon; se: synencephalic roof of diencephalon; vb: region of plica ventralis encephali; vt: velum transversum; arrow indicates sulcus telo-diencephalicus in accordance with present interpretation.

Although it might be argued that, at the initial stages of telence-phalic morphogenesis, the primordial prosencephalic lamina terminalis would extend as a midline structure from the floor of the diencephalic recessus praeopticus to the rostral (telencephalic) leaf of velum transversum, it seems preferable not to classify the lamina epithelialis (becoming a roof plate component) as a subdivision of the lamina terminalis (becoming a lateral plate component).

Again, while His (1893, pp. 161–163) unequivocally designated the embryonic structure derived from the closure of the anterior neuropore as lamina terminalis, he later (1904, p. 84, footnote) called it *lamina reuniens*, and claimed that the term 'lamina terminalis' should be restricted to the often relatively thin membrane which, in the adult brain, provides the rostral boundary of the third ventricle in the preoptic (or optic) recess. This lamina terminalis is, as His (1904) states, derived from the basal part of what he calls lamina reuniens, and corresponds, of course, to the diencephalic (hypothalamic) part of lamina terminalis in the nomenclature which I have adopted.

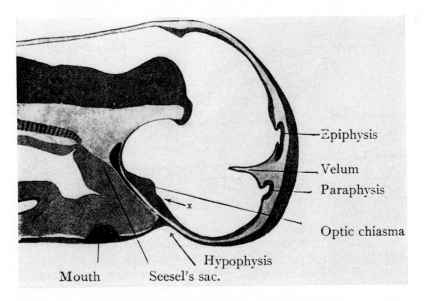

*Figure 53 B.* Approx. midsagittal section through head of an Amblystoma larva whose brain is slightly more differentiated than the stage of the preceding figure (from JOHNSTON, 1906). Arrow x points to torus transversus according to present interpretation. Between chiasmatic ridge and torus, a preoptic recess is faintly outlined. 'Mouth' indicates a thickening of oral ectoderm. *Seessel's pouch* represents a diverticulum of oral gut.

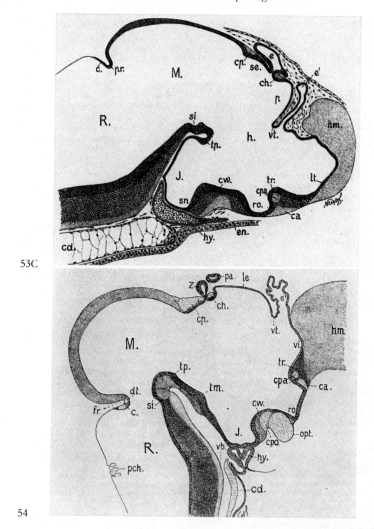

53C

54

*Figure 53 C.* Approx. midsagittal section through the brain of a 13 mm long larva of *Salamandra maculosa* (from KUPFFER, 1906). ca: commissura anterior; ch: commissura habenulae; cp: commissura posterior; cpa: commissura pallii; cw: chiasma ridge; e: epiphysis; e′: paraphysis; h: diencephalic ventricle; hm: part of obliquely cut telencephalic hemispheric wall; lt: part of telencephalic lamina terminalis; p: parencephalic roof of diencephalon; ro: recessus praeopticus; se: synencephalic roof of diencephalon; si: sulcus intraencephalicus posterior (fovea isthmi); tp: tuberculum posterius; tr: torus transversus; other abbr. as in previous figures from KUPFFER.

*Figure 54.* Approx. midsagittal section through the brain of an embryo of the reptilian *Anguis fragilis* (from KUPFFER, 1906). The developmental stage, not specifically indicated,

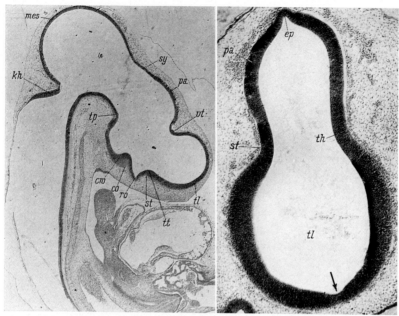

55A                                                                                                55B

*Figure 55 A.* Approx. midsagittal section through head and brain of a chick embryo of
3 days (from KUHLENBECK, 1936). co: optic chiasma; cw: chiasmatic ridge; kh: cerebellar
plate; mes: mesencephalic roof; pa: parencephalic roof of diencephalon; ro: recessus
praeopticus; st: sulcus telencephalo-diencephalicus; sy: synencephalic roof of diencephalo-
mesencephalic boundary region; tl: telencephalon (lamina terminalis); tp: tuberculum
posterius; tt: torus transversus; vt: velum transversum ($\times 20$, red. $^4/_5$).

*Figure 55 B.* Section through differentiating prosencephalon of a duck embryo of
120 h (from KUHLENBECK, 1936). The unpaired telencephalon is just beginning to become
delimited, by sulcus telo-diencephalicus (st), respectively torus hemisphaericus (th), from
the parencephalic portion of diencephalon (pa). ep: epiphysial anlage; tl: telencephalon;
arrow points to remnant of recessus neuroporicus.

---

corresponds approximately to that of *Salamandra* in the preceding figure 53 C. c: cerebellar
plate; ca: commissura anterior; cd: notochord; ch: commissura habenulae; cp: commissura posterior; cpa: commissura pallii; cpo: postoptic and supraoptic commissures; cw:
chiasmatic ridge; dt: decussation of trochlear nerve; e': paraphysis, fr: dorsal plica rhom-
bomesencephalica; hm: medial wall of telencephalic hemisphere; hy: adenophypophysial
anlage; y: postoptic hypothalamic region; le: lamina epithelialis of parencephalic diencephalon roof; M: mesencephalon; opt: optic chiasma; pa: parietal (parapineal) organ; pch:
choroid plexus of fourth ventricle; ro: recessus praeopticus; si: sulcus intraencephalicus
posterior (fovea isthmi); tm: 'tuberculum posterius inferius'; tp: tuberculum posterius; tr:
torus transversus; vb: convexity of rostral rhombencephalic flexure; vi: remnant of ventriculus impar telencephali; vt: velum transversum; z: pineal body.

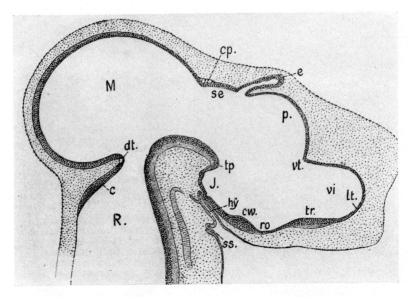

*Figure 55 C.* Approx. midsagittal section through head and brain of a chick embryo of
$4^1/_2$ days (from KUPFFER, 1906). c: cerebellar plate; cp: posterior commissure; cw: chiasmatic ridge; dt: decussation of trochlear nerve; e: epiphysial anlage; lt: telencephalic lamina terminalis; p: parencephalic roof of diencephalon; ro: recessus praeopticus; se: synencephalic roof of diencephalon; ss: *Seessel's pouch;* tp: tuberculum posterius; tr: torus transversus; vi: ventriculus impar telencephali; vt: velum transversum; other abbr. as in preceding figures from KUPFFER.

The evagination of the conjoint two telencephalic diverticula,
which is accompanied by the formation of a concavity, the sulcus telo-
diencephalicus, on the external surface of the neural tube, is correlated
with a corresponding convexity, the *torus hemisphaericus* (K., 1927), on
the internal surface and here indicates an appropriate boundary between the ventricular surfaces of telencephalon and diencephalon
(Fig. 47 C; 55 B, 57 C). The *torus transversus (commissural plate)*, and the
*velum transversum* can then be conceived as the basal and the dorsal midline portions, respectively, of torus hemisphaericus.[56]

---

[56] The term '*torus*' means here a smooth, ridge-like protuberance. In *topology*, however,
'*torus*' has a different meaning and refers to a three-dimensional solid shaped like a ring
('*doughnut*'), obtained by revolving a circle about a non intersecting and nontangential
external line in its plane, whereby a region of space outside the closed tube or solid but
inside the ring remains ('hole' in the 'doughnut'). The surface of such torus displays
topologically significant properties. Cf. the comments on '*torus*' in vol. 1, pp. 168, 170,
footnote 8.

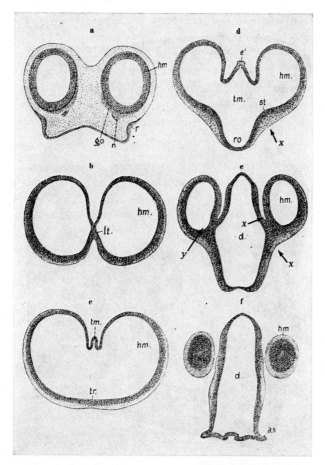

*Figure 55 D.* Cross-sections through the brain of a chick embryo of $4^{1}/_{2}$ days (after a series of G. Henrich, from Kupffer, 1906). Sequence of sections a–f as follows; a: rostral paired evaginated portion of telencephalon; b: caudal end of rostral paired evaginated portion of telencephalon, with rostral end of lamina terminalis; c: telencephalon medium s. impar (unpaired evaginated portion of telencephalon; d: section through telencephalon medium rostral to velum transversum and through diencephalic recessus praeopticus; e: section through caudal paired evaginated telencephalon and through diencephalon as well as caudal part of hemispheric stalk interconnecting telencephalon and through diencephalon in region of optic stalks. Abbreviations in sections: as: optic stalk; d: diencephalon; e': paraphysis; go: 'ganglion olfactorium' (?), probably accumulation of sheath cells for fila olfactoria; hm: telencephalic hemispheres; lt: lamina terminalis; n: olfactory nerve; r: olfactory placode; ro: recessus praeopticus; st: torus hemisphaericus (telo-diencephalic boundary) in present interpretation; tm (in section c): dorsal ridge of telencephalon medium (part of paraphysial evagination, cf. fig. 50 B) in present interpretation; tr: torus transversus; added arrows show sulcus telencephalo-diencephalicus ( x ) and sulcus terminalis (y) in present interpretation.

At a right angle to the transverse segments of sulcus telo-diencephalicus, the longitudinal (sagittal) *interhemispheric groove*, respectively *fissure*, arises in the midline. It runs dorsally rostrad from the velum transversum and then bends ventralward and caudad to the region of the commissural plate. The bottom of the dorsal interhemispheric groove is formed by lamina epithelialis of the roof plate, in which, immediately rostral to velum transversum, an unpaired epithelial dorsal

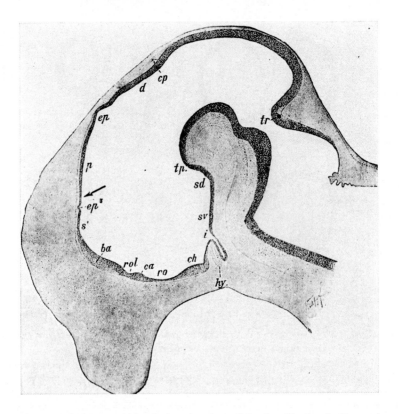

*Figure 56 A.* Approx. midsagittal section through head and brain of a 20-mm sheep embryo (after NEUMAYER, from ZIEHEN, 1906). ba: anlage of corpus callosum in lamina terminalis; ca: anlage of commissura anterior in lamina terminalis; ch: chiasmatic ridge; cp: region of commissura anterior; d: synencephalic roof of diencephalon; ep: epiphysis; ep': paraphysis; hy: hypophysial anlage; i: infundibular region; p: parencephalic roof of diencephalon; ro: recessus praeopticus; rol: recessus neuroporicus ('recessus olfactorius impar' of NEUMAYER-ZIEHEN); s': transition of lamina terminalis into telencephalic roof plate; sd: recessus mammillaris; sv: posterior hypothalamic region; tp: tuberculum posterius; tr: decussation of trochlear nerve in dorsal plica rhombo-mesencephalica; arrow points to region of barely suggested velum transversum.

diverticulum evaginates, which represents the primordium or anlage of the *paraphysis*.[56a] Rostral to, as well as adjacent to that structure, much of the lamina epithelialis, depending on the particular vertebrate form, may or may not give origin to the telencephalic choroid plexuses. The early interhemispheric groove thus very incompletely separates the two telencephalic hemispheres, which share a common, bilaterally symmetric ventricular space and a continuous common basal wall provided by the lamina terminalis. In various vertebrate forms, mentioned above, and again to be pointed out further below, this connected configuration of the hemispheres remains conspicuously permanent, representing a so-called *pars impar telencephali (telencephalon medium sive impar)*.

The further ontogenetic (and phylogenetic) morphogenesis of the originally unpaired, that is, *conjoint bilateral symmetric telencephalon* is characterized by a number of different folding processes, some of which lead to distinctive configurational patterns displayed by various groups of vertebrates. The rostrad bulging of the hemispheres beyond the expanded neighborhood of lamina terminalis results in two anterior vesicles, which, on cross-sections, appear as separate, completely disconnected structures, with their own (right and left) *lateral ventricles*. Likewise, the hemispheres may bulge caudad beyond the caudal part of torus hemisphaericus, thus resulting in a posterior portion (or *posterior recess*) of lateral ventricle which is located caudally to the pars

---

[56a] This 'ependymal circumventricular organ' (cf. vol.3/I, p.355 of this series; also BECCARI, 1943, p.723f., p.740), respectively its anlage, does not seem to be present in Cyclostomes. It can be seen in Selachians, Ganoids, Teleosts, and Dipnoans, but the details of the differentiation of this 'organ' in these forms are still insufficiently clarified. The paraphysis is particularly well developed in many amphibians and present, but perhaps generally reduced in reptiles, being, however, rather conspicuous in Chelonia. It seems to be rudimentary in birds. The paraphysial anlage is conspicuous in relevant ontogenetic stages of seemingly all examined mammals, becoming lost (i.e. 'relatively' reduced) in the course of development. In most or possibly all adult mammalian forms it is apparently not recognizable, and may be 'transformed' into a neighborhood of telencephalic choroid plexus. One must distinguish between the morphologic aspect of the paraphysis and its histologic aspect (i.e. its homotypic and its heterotypic homology, cf. tabulation on p. 65). With respect to the significance of the dorsal sagittal ridge seen in the floor of the itnerhemispheric groove at some early ontogenetic stages and in some vertebrate forms (cf. e.g. figs.49B, C, 50B), the primordium of the paraphysis seems to involve either the caudal portion of that ridge or perhaps even its entire extent.

impar telencephali. Each hemisphere thereby obtains, at the tip of its external surface, a prominent *posterior pole* (cf. Figs. 49 B, 55 D, 59).

In accordance with the terminology explained further above, this morphogenetic process may be designated as *paired evagination*. Accordingly, a *paired evaginated* and an *unpaired, conjoint (but nevertheless bilaterally-symmetric) type of telencephalic development* can be distinguished. In the

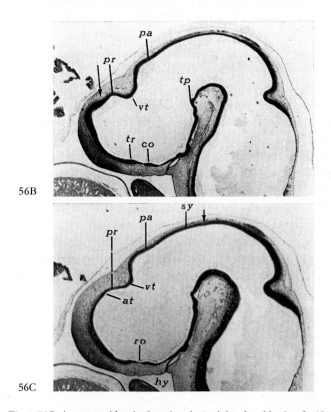

*Figure 56 B.* Approx. midsagittal section through head and brain of a 10-mm pig embryo (× 10, red. ⁷/₁₀, hematoxylin-eosin). Remnants of notochord, accompanied by vessels, can be seen in the concavity of plica ventralis encephali. co: chiasmatic ridge; pa: parencephalic roof of diencephalon; pr: paraphysial anlage; tp: region of tuberculum posterius; tr: torus transversus; vt: velum transversum; arrow indicates region of *angulus terminalis* of His (1893).

*Figure 56 C.* Sagittal section close to that of preceding figure. at: suggestion of angulus terminalis of His (1893); hy: hypophysial anlage; pa: parencephalic roof of diencephalon; pr: paraphysial neighborhood; ro: recessus praeopticus; sy: synencephalic roof of diencephalon; vt: velum transversum; arrow indicates diencephalo-mesencephalic boundary region.

aspect here under consideration, paired evagination thus refers to a separate, disjoint (but likewise bilateral symmetric) outward bending of rostral, or of rostral and caudal hemispheric wall lateral to the lamina terminalis (K., 1927).

The neural tube region including the median lamina terminalis and its paired lateral neighborhoods represents thus the zones of transition in which the telencephalon is, and remains, *ab initio*, continuous with the diencephalon, thus constituting the *primary hemispheric stalk sensu latiori*. The median portion (lamina terminalis, torus hemisphaericus) provides the passage for the commissures interconnecting the two antimeric components of the telencephalon, namely, depending on the particular vertebrate types, commissura anterior, commissura pallii, commissura hippocampi, and corpus callosum.

The lateral neighborhoods *( primary hemispheric stalk sensu strictiori )* afford a passage for the fiber bundles interconnecting telencephalon and diencephalon, namely, *medial* and *lateral forebrain bundle*. In mammals, the diverse fiber systems of capsula interna are added to the later-

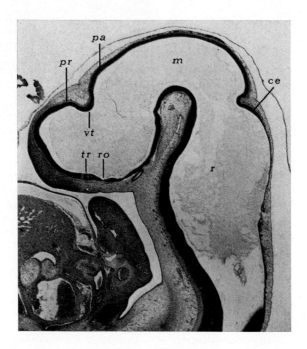

*Figure 56 D.* Sagittal section slightly lateral to those of figures 56 B and C., ce: cerebellar plate; m: mesencephalon; r: rhombencephalon; other abbr. as in preceding figures.

al forebrain bundle, while the fornix which can be considered an original component of the medial forebrain bundle, becomes a separate,
more or less distinctive tract.

In the vertebrate forms with caudal paired evagination, the sulcus
telo-diencephalicus runs, with rostral concavity, in approximately
semicircular fashion (cf. fig. 57) caudal to the telencephalic neighborhood of hemispheric stalk. This latter, concomitantly with the considerable increase of the telencephalic fiber connections, displays an especially extensive enlargement in mammals (cf. below, section 5). The
ventricular groove at the bottom (i.e. basal floor) of the posterior extension of lateral ventricle within the caudalward paired evaginated
hemisphere, and running approximately parallel to the caudodorsal
stretch of sulcus telo-diencephalicus (Fig. 55 D), corresponds to the

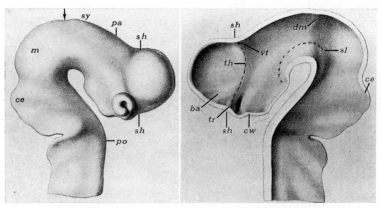

57A                                                                57B

*Figure 57A.* Brain of a 6.9 mm long human embryo in lateral view (model by His,
1904). Interpretation according to present treatise. ce: cerebellar plate; m: mesencephalon;
pa: parencephalic roof of diencephalon; po: pontine flexure; sh: sulcus telo-diencephalicus; sy: synencephalic roof of diencephalon; arrow indicates approximate region of diencephalo-mesencephalic boundary; the optic vesicle, already transformed into an optic cup,
distinctly shows the basal choroid fissure (not labeled).

*Figure 57B.* Ventricular view of brain model shown in preceding figure (from His,
1904). Interpretation according to present treatise; ba: primordia of basal ganglia; ce: cerebellar plate; cw: chiasmatic ridge (rostral to it, preoptic recess with entrance into optic
stalks, which later fuses to become sulcus intraencephalicus anterior); dm: approximate
diencephalo-mesencephalic boundary; sl: sulcus limitans; sh: sulcus telo-diencephalicus;
th: torus heimisphaericus; tr: torus transversus; vt: velum transversum; the telo-diencephalic boundary along the torus, and the assumed rostral end of sulcus limitans are indicated by broken lines; the additional ridges (ba) near basal part of torus hemisphaericus are
primordia of the basal ganglia (so-called 'ganglionic hills' or 'Ganglienhügel').

*sulcus terminalis* of informal human anatomical nomenclature, that is, to the groove located at the attachment of *lamina affixa* (BNA, PNA) along the telencephalic ventricular course of *stria terminalis* (BNA, PNA). Because of the coalescence (fusion) between epithelial (ependymal) lamina affixa with lateral surface of the thalamus, the corresponding stretch of sulcus terminalis becomes obliterated in those forms in which the medial wall of the caudally evaginated hemisphere contains a lamina epithelialis (usually in connection with a choroid plexus attached to fimbria of hippocampal formation).

The term *evagination* (Germ. *Ausstülpung*) has, of course, several different denotations (e.g. unsheathing), of which at least two are topologically significant. With regard to an open tube (either at one end or at both), it may mean: turning inside-out (Germ. *umkrempeln*), such that the inner surface now becomes the outer one. With regard to a closed tube, it may mean merely to *protrude*, or to *bulge*, the bulge being then delimited by a topologically 'circular' fold. Quite evidently, the second denotation, namely protrusion or bulging of the wall in its entire

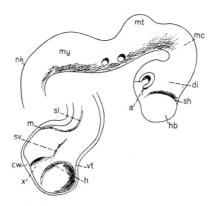

*Figure 57C.* External (lateral) and internal view of brain in an approx. 7.8 mm long human embryo (modified after HOCHSTETTER in accordance with results of my own observations, from K., 1927). a: optic stalk and cup; cw: chiasmatic ridge; di: diencephalon; h: torus hemisphaericus; hb: telencephalic vesicle; m: recessus mammillaris; mc: mesencephalon; mt: dorsal metencephalon (cerebellar anlage); my: myelencephalon; nk: cervical flexure *(Nackenkrümmung)*; sh: sulcus telo-diencephalicus (primary sulcus hemisphaericus, a secondary sulcus hemisphaericus may subsequently indicate an approximate external limit between pallial region and basal ganglia, it can be seen, but is not labeled above the lowest sh in fig. 57A); sl: sulcus limitans; sv: sulcus diencephalicus ventralis; vt: velum transversum; x: entrance into optic stalk, becoming sulcus intraencephalicus anterior.

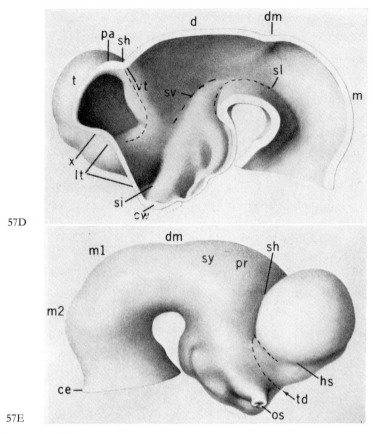

57D

57E

*Figure 57 D.* Internal view of forebrain and mesencephalon in an approx. 10.2 mm
long human embryo (from His, 1904; designations and interpretations in accordance with
the present treatise). cw: chiasmatic ridge; d: diencephalon; dm: region of diencephalo-
mesencephalic boundary; lt: lamina terminalis; m: mesencephalon; pa: rudimentary
paraphysial anlage; si: preoptic recess with entrance into optic stalk, which becomes transformed into sulcus intraencephalicus anterior; sh: sulcus hemisphaericus; sl: sulcus limitans; sv: sulcus diencephalicus ventralis; t: telencephalon; vt: velum transversum; x:
fold at transition of lamina terminalis to roof plate (distorted angulus terminalis?) for comparison with figure 57F; dotted lines indicate telo-diencephalic boundary (torus hemisphaericus), and ventricular sulci in accordance with present interpretation.

*Figure 57 E.* External (lateral) aspect of the embryonic brain depicted in preceding figure (from His, 1904; interpreted in accordance with present views). ce: cerebellar anlage;
dm: diencephalo-mesencephalic boundary zone; hs: secondary sulcus hemisphaericus,
probably a borderline shrinkage artefact (cf. figs. 57 A and G); m1 und m2: rostral and
caudal bulgings of mesencephalon; os: optic stalk; pr: parencephalic region of diencephalon; sh: sulcus hemisphaericus (telo-diencephalic boundary); sy: synencephalic region of
diencephalon; td: telo-diencephalic boundary; dotted line indicates this latter boundary in
flattened basal segment of sulcus hemisphaericus (cf. fig. 57 G).

thickness, applies in the present context. In my original discussion of these processes (K., 1927, pp. 78, 80 *et passim*), the concept of paired evagination was formulated in terms of both *paarige Ausstülpung* and *'getrennt evaginierte Endhirnblasen ( Hemisphären )'* while unpaired evagination was pointed out as *'vereinigt oralwärts' erfolgende Hemisphärenbildung*, resulting in a telencephalon *'mit einheitlich nach vorn evaginiertem Endhirn'*, provided by the *'vorgestülpten Abschnitt der Lamina*

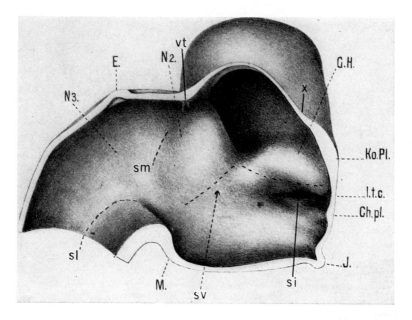

*Figure 57 F.* Internal aspect of the forebrain in a human embryo of approx. 12.8 mm (from HOCHSTETTER, 1919, interpreted in accordance with present views). This figure should be compared with figure 57 D. Ch. pl.: chiasmatic ridge; E.; epiphysial anlage; G.H.: basal ganglia primordium *('Ganglienhügel')*; J: 'Infundibulum'; Ko.Pl.: commissural plate of lamina terminalis; l.t.c.: 'lamina terminalis cinerea' (lamina terminalis of diencephalic preoptic recess); M: 'mammillary tubercle'; N2: primordium of thalamus ventralis (eminentia thalami); N3: border zone between parencephalic and synencephalic regions of diencephalon; sl: sulcus limitans; sm: early manifestation of sulcus diencephalicus medius; sv: sulcus diencephalicus ventralis; vt: velum transversum. In addition to the dotted lines indicating sl, sm, and sv, a dotted line indicating telo-diencephalic boundary along modified torus hemisphaericus has been traced from velum transversum to commissural plate; the sulcus terminalis, an intraventricular groove in the caudally evaginated hemisphere (cf. fig. 55 D) begins at the junction of sv with torus hemisphaericus. At later stages, the dorsal end of sulcus intraencephalicus anterior may likewise connect with rostral end of sulcus terminalis. Dorsal end of commissural plate at x.

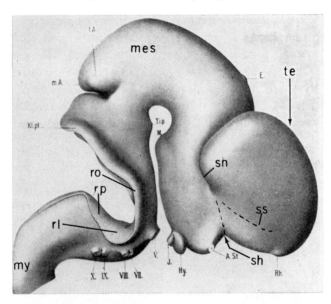

*Figure 57G.* External (lateral) aspect of the brain in a human embryo of about 19.4 mm (from HOCHSTETTER, 1919, interpreted in accordance with present views). A.st.: optic stalk; di: diencephalon; E.: epiphysial anlage; Hy.: hypophysial anlage; J.: infundibulum; Kl.pl.: cerebellar plate; l.A., m.A.: posterior bulgings of tectum mesencephali (anlage of colliculi inferiores); M.: 'mammillary tubercle'; mes: mesencephalon; my: myelencephalon; Rh.: anlage of olfactory bulb; rl: rhombic lip *sensu strictiori;* ro: upper part of rhombic lip *sensu latiori;* rp: lower (caudo-basal) part of rhombic lip *sensu latiori;* sh; sulcus hemisphaericus (s. telo-diencephalicus); ss: antero-lateral branch of sulcus hemisphaericus, approximately between pallium and basis (cf. fig. 57E, hs; it seems evident that HOCHSTETTER's material was better preserved than that of HIS); te: telencephalon; T.i.p.: 'tuberculum interpedunculare'; arrow indicates basal telo-diencephalic boundary (cf. fig. 57E); dotted lines indicate baso-lateral segment of true sulcus hemisphaericus and accessory antero-lateral branch of that sulcus.

*terminalis (dorsal bzw. rostral der Commissura anterior )'.*[57] The difference between *rostral* and *caudal paired evagination* was likewise pointed out (K., 1927, p. 80), as well as the fact that, with respect to the lamina terminalis (an unpaired, median structure), paired or separate evagination can be described as *'paarig lateralwärts'*.

---

[57] A sort of folding opposite to *evagination* is *invagination (Einstülpung)*, as e.g. manifested by typical holoblastic gastrulation. Although somewhat related to invagination, the process of *inversion* is here considered to be a different and specific type of folding as discussed further below in the text.

*Rostral paired evagination* takes place in an anterior direction beyond the region pertaining to lamina terminalis and pars impar telencephali. Therefore it should not be confused with the *unpaired evagination* as defined above, namely with the conjoint rostral bending of an expanded telencephalic lamina terminalis correlated with a large telencephalon impar. Occasionally, as mentioned above (footnote 55a), the term '*evagination*' has been restricted to what is here called *rostral paired evagination*, thus leading to an ambiguity, whereby the unpaired telencephalon medium is designated as 'nonevaginated'. This terminology, however, does by no means imply lack of evagination of lamina terminalis, but merely refers to the presence of a substantial telencephalon impar resulting from pronounced unpaired evagination of said lamina. Moreover, even in forms with a very extensive telencephalon impar, the olfactory bulbs and their immediate neighborhood are rostrally or rostro-laterally paired-evaginated. Their ventricle *(rhinocoele)* may subsequently obliterate. The telencephalon impar, on the other hand, can include the entire lobus telencephalicus as distinguished from the bulbus olfactorius, which, in turn, may or may not display an elongated stalk.

At certain ontogenetic stages of the vertebrate telencephalon, the lamina epithelialis (roof plate) ends rostrally along the antero-dorsal border of lamina terminalis. In many vertebrate forms, the configuration at adult stages displays the same relationship. Hence, the rostral paired evaginated portions of the telencephalon do not, in such representatives of Anamnia and Amniota, include the lamina epithelialis as a wall component, and are then not provided with the attachment of a

The significance of evagination, inversion and eversion with regard to the morphogenesis of the vertebrate neuraxis in general, and of the telencephalon in particular, was discussed and elaborated in my '*Vorlesungen*' (K., 1927). Much later, KÄLLEN (1951), without reference to these previous formulations, attempted to describe the processes of evagination and eversion in telencephalic ontogenesis of lower vertebrates. KÄLLEN, however, failed to apprehend the difference between paired and unpaired evagination; moreover, he included longitudinal infolding of the lamina epithelialis (actually an invagination of this lamina in the transverse plane) under his concept of evagination. *Paired evagination (mihi)* which includes components of the lamina epithelialis is called 'pseudoevagination' by KÄLLEN.

SCHOBER (1966), although adopting concepts of KÄLLEN (1951), nevertheless brings a competent and very useful survey of the different telencephalic patterns displayed by lower vertebrates (Anamnia) and related to the combinations of foldings which I prefer to designate as *unpaired evagination, paired evagination* (rostral or rostral and caudal), *inversion* and *eversion* in accordance with my views expressed in 1927 and further elaborated in the present section.

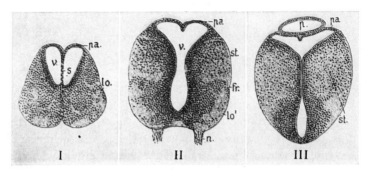

*Figure 58 A.* Cross-sections through paired evaginated (I) and unpaired evaginated portions (II, III) of everted telencephalon in a young trout shortly after hatching (from KUPFFER, 1906). fr: 'fissura lateralis'; lo: olfactory bulb; lo' (in II): caudal end-portion of olfactory bulb; n: fila olfactoria; p: protruding vesicular portion of parencephalon ( ?; paraphysis ?); pa: lamina epithelialis of telencephalic roof plate; s: medial paired evaginated portion of roof plate; st: dorsal portion of unpaired evaginated telencephalic alar plate; v: telencephalic ventricle (paired in I, ventriculus impar s. communis in II and III).

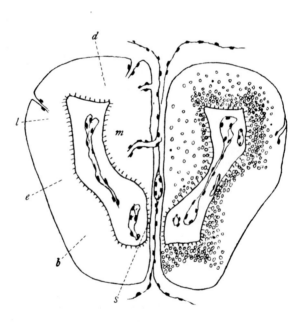

*Figure 58 B.* Cross-section through paired evaginated inverted telencephalon of a urodele amphibian *(Triton* sp.), slightly rostral to interventricular foramen. A medial paired evaginated portion of roof plate (s) extends into the paired hemispheres (from K., 1921).

choroid plexus, although, in some instances (e.g. Gymnophiona, various other Anamnia, as well as certain Amniota), tufts of choroid plexus can protrude into the anterior lateral ventricles through the interventricular communications. However, in the ontogenetic development of some Anamnia, the rostral paired evaginated parts of the telencephalon may include a medial paired evaginated component of lamina epithelialis. This latter is more dorsomedial in Osteichthyes (cf. Fig. 58 A), and more basimedial in certain Amphibia (cf. Fig. 58 B, 'septum ependymale'). In the adult dipnoan Ceratodus (Neoceratodus), such paired evaginated region of roof plate becomes very conspicuous, representing an intricately folded lamina epithelialis of perhaps borderline choroid plexus type, described by BING and BURCKHARDT (1905) as 'lingula interolfactoria' (Fig. 58 C).

Caudal paired evagination takes place in a posterior direction lateral to, and beyond, the rostral end of diencephalon, leading to the formation of a telencephalic subdivision caudal and lateral to the telencephalon impar, and ending with a *polus posterior* (Figs. 49 B, 55 D, 59). In forms with inversion, the lamina epithelialis may or may not be included. Depending on the particular sort of vertebrate organization, this

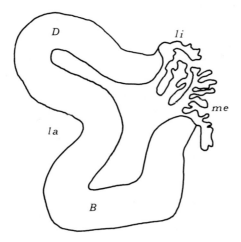

*Figure 58 C.* Cross-section through paired evaginated telencephalic hemisphere of the lung-fish, *Ceratodus*, displaying inclusion of a conspicuous medial component of folded roof plate (adapted and simplified after HOLMGREN and VAN DER HORST, 1925). Level of section is caudal to olfactory bulb but through a fairly rostral portion of lobus hemisphaericus. B: basal components of telencephalon; D: dorsal or pallial components; la: lateral side; li: lamina epithelialis forming '*lingula interolfactoria*'; me: medial side (midline).

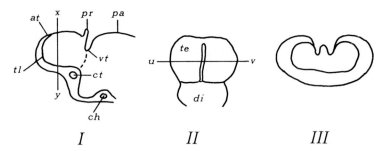

*Figure 59 A.* Diagrams illustrating an early stage of unpaired evaginated telencephalon. I Diagrammatic midsagittal section. II Diagrammatic dorsal aspect. III Diagrammatic cross-section. at: angulus terminalis (transition of lamina terminalis alar plate to roof plate); ch: chiasma ridge; ct: commissural plate of lamina terminalis with telencephalic commissures; di: diencephalon; pa: parencephalic roof of diencephalon; pr: paraphysial anlage; te: telencephalon; tl: dorsal part of telencephalic lamina terminalis; vt: velum transversum; u–v, x–y: plane of section shown in III; dotted line indicates telencephalo-diencephalic boundary.

caudal portion of telencephalon therefore may or may not be provided with a choroid plexus attached to its wall at the transition of lateral plate and epithelial roof plate.

In forms with eversion, some caudal paired evagination can likewise occur, and includes then the lateral components of lamina epithelialis (cf. Fig. 61 A). It is self-evident that, because evagination of lamina terminalis can only take place in a rostral direction, that is, away from the central lumen of the neural tube, namely in a direction opposite to invagination, caudal evagination must be paired, involving wall neighborhoods *lateral* to the lamina terminalis, which is a median (midline) neighborhood.

*Figure 59 B.* Diagrams illustrating telencephalic inversion (IIIa–Va) and eversion (IIIb–Vb). I Midsagittal section indicating morphologic landmarks (the lead to ct also indicates basal transverse segment of telo-diencephalic sulcus). II Midsagittal section showing expanded commissural plate and planes of section for figures III to V. ce: cerebellar plate; ch: chiasmatic ridge (includes optic chiasma as well as supraoptic and postoptic commissures); cp: commissura posterior; ct: torus transversus with telencephalic commissures; di: diencephalon; ep: epiphysial anlage; hy: hypothalamus; me: mesencephalon; pa: parencephalic roof of diencephalon; pr: paraphysial anlage; rp: recessus praeopticus; st: sulcus terminalis praeopticus (sulcus terminalis telencephali of MILLER, 1940); sy: synencephalic roof of diencephalon; vt: velum transversum; x, y, z: planes corresponding to figures III, IV, V: in these latter figures, sequence a indicates inverted, and sequence b everted telencephalon impar with identical degree of unpaired evagination.

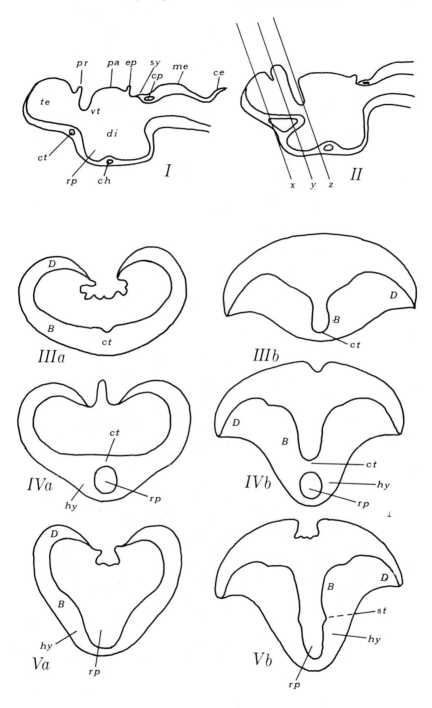

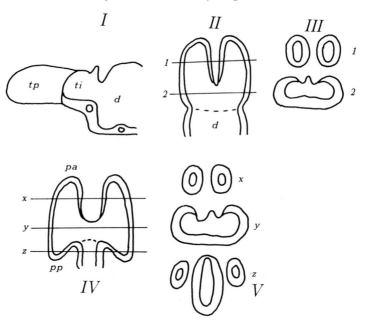

*Figure 59C.* Diagrams illustrating rostral and caudal paired telencephalic evagination. I Midsagittal section. II 'Horizontal' section through prosencephalon with telencephalon impar and rostral paired evagination. III Cross-sections in planes 1 and 2 of II, showing paired (1) and unpaired (2) evaginated portions of telencephalon. IV 'Horizontal' section through prosencephalon with rostral and caudal paired evagination. V Cross-sections in planes x, y, z of IV, showing rostral paired evaginated (x), unpaired evaginated (y), and caudal paired evaginated (z) portions of telencephalon. d: diencephalon; pa: polus anterior; pp: polus posterior; tp: rostral paired evaginated telencephalon; ti: telencephalon impar (aula); dotted line indicates telo-diencephalic boundary along torus transversus (anterior commissure).

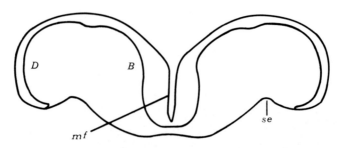

*Figure 60.* Stretched lamina epithelialis (roof plate) of everted telencephalon impar in the Ganoid Amia, showing midsagittal longitudinal fold ('median septum'). B: basal telencephalic zonal components; D: dorsal (pallial) components; mf: median fold of lamina epithelialis; se: sulcus externus.

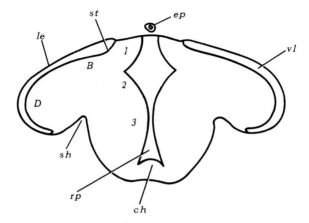

*Figure 61 A.* Caudal paired evagination of the everted telencephalon in the Teleost Corydora (based on material studied by MILLER, 1940, in my laboratory). B, D as in preceding figure; ch: rostral end of chiasmatic ridge; ep: epiphysial stalk; le: lamina epithelialis of roof plate; rp: recessus praeopticus (caudal end); sh: sulcus telo-diencephalicus (into which sulcus externus continues); st: sulcus terminalis (telo-diencephalic boundary in ventricular floor); vl: posterior recess of ventriculus lateralis telencephali; 1: epithalamus (gangl. habenulae); 2: rostral end of thalamus ventralis; 3: hypothalamus.

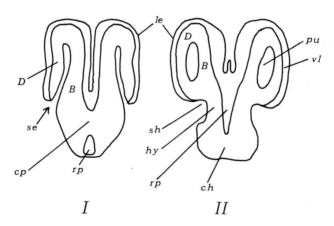

$$I \qquad\qquad II$$

*Figure 61 B.* 'Pseudoventricle' near polus posterior of telencephalon in the Ganoid Polypterus (adapted and modified after NIEUWENHUYS, 1963). I Cross-section at level of commissural plate. II Cross-section displaying '*pseudoventricle*'. cp: commissural plate of lamina terminalis; hy: hypothalamus; pu: 'pseudoventricle'; se: sulcus externus, becoming 'pseudoventricle' in II; other abbr. as in preceding figures.

It should, moreover, be understood that both rostral and caudal paired evagination merely represent an expansion or 'exaggeration' of the early embryonic process leading to the original bilateral-symmetric conjoint unpaired telencephalic evagination which nevertheless displays two rostral and two caudal '*poles*' of the hemispheres. Typical rostral or caudal paired evagination results then, as it were, from further expansion or protrusion of these 'polar' configurations (Fig. 49 B). The morphogenetic events concerning ontogenetic evolution of the vertebrate telencephalon and comprising growth as well as formative processes involving displacements of the material substratum describable in terms of unpaired or paired evagination, combined with inversion or eversion, are illustrated in the diagrams of Figure 59.

In addition to *evagination* (paired or unpaired), and to *inversion* or *eversion*, other aspects of morphogenetic displacement, folding or bulging, which may affect the neuraxis and particularly the telencephalon, can be distinguished. Thus, REMANE (1966) subsumes under the term '*internation*' several sorts of displacement whereby surface structures 'sink' into a more internal or central position. An opposite displacement might the be called '*externation*'. Particularly with respect to brain morphogenesis, SPATZ (1966) has characterized the results of such kind of processes as *introversion* and *promination*, respectively. As regards specific telencephalic configurations, which will be discussed in their appropriate context in subsequent section or chapters, the ventricular bulges related to protrusions of basal ganglia, paraterminal body, and other components, can be evaluated as internation or introversion. The grooves or recesses related to the ventricular boundaries of such prominences, although, in extreme cases (e.g. in *Chimaera* or *Latimeria*) somewhat resembling evagination, do not, as a rule, involve bulging of the external surface, and are here therefore not subsumed under true evagination as defined above. As an obvious example of externation or promination, the bulging tuberculum olfactorium in macrosmatic mammals may be cited.

The convenient designations *inversion* and *eversion*, as used in the preceding paragraphs, refer thus to a mode of folding distinctive from *telencephalic evagination*, which latter is particularly characterized by the changes of contour displayed in sagittal and 'horizontal' ('frontal') planes. Conversely, the contour features of inversion and eversion are particularly evident in the transverse plane as displayed by cross-sections. *Inversion* is here characterized, in the telencephalon impar, by the basimediad bending of the dorsal wall of the telencephalon toward, or as far as, its connection with the epithelial roof plate. This latter then lies entirely at the bottom of interhemispheric groove (except for the paraphysis, which may protrude dorsad from the midline of this groove). Inversion of the telencephalon impar can be accompanied by apparent fusion of the solid hemispheric walls with ostensible disappearance of lamina epithelialis in transverse planes (cf. Fig. 64, III). Actually, however, this configuration is caused, in the transverse plane, by a dorsal (supraneuroporic) extension of the (unpaired) evaginated massive lamina terminalis with caudal displacement of the lamina epithelialis, i.e. roof plate (cf. Fig. 62, II).

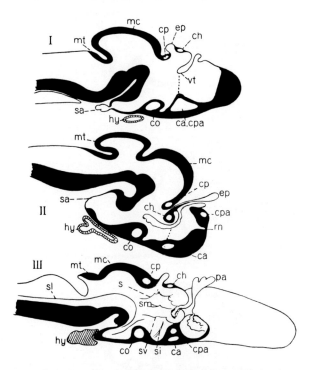

*Figure 62.* Configuration of lamina terminalis and commissural plate in representatives of Osteichthyes, Selachians, and Amphibia (from K., 1927). I Late larva of Amia calva (after BASHFORD DEAN). II Advanced embryonic stage of Acanthias (after KUPFFER). III Fully developed Amblystoma (combined after HERRICK and personal observations). ca: commissura anterior; ch: commissura habenulae; co: chiasma opticum; cp: commissura posterior; cpa: commissura pallii; ep: epiphysis; hy: adenohypophysis; mc: mesencephalon; mt: metencephalon; pa: paraphysis; rn: recessus neuroporicus; s: sulcus synencephalicus rostrally continuous with sulcus diencephalicus dorsalis; sa: saccus vasculosus; si: sulcus intraencephalicus anterior; sl: sulcus limitans; sm: sulcus diencephalicus medius; sv: sulcus diencephalicus ventralis; vt: velum transversum.

In the *paired evaginated telencephalon*, *inversion*, involving the basimediad bending of the dorsal wall, is characterized by a circular, closed
tubular configuration, whose lumen is the lateral, i.e. disjoint telencephalic ventricle. In transverse planes, the dorsal hemispheric wall becomes medially continuous with the basal wall, such that, as the result
of this connectedness, the dorsomedial neighborhood $D_3$ adjoins the
basimedial neighborhood $B_4$. The morphologic significance of these
topologic neighborhoods, pertaining to the telencephalic longitudinal

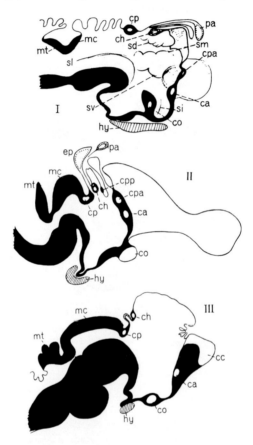

*Figure 63.* Configuration of lamina terminalis and commissural plate in late larval Petromyzon and representatives of Amniota (from K., 1927). I Immature Petromyzon (Ammocoetes). II Almost fully developed embryo of Lacerta (modified after KUPFFER in accordance with original observations). III Human embryo of about 100-mm length (after
HOCHSTETTER, in accordance with original observations). cc: corpus callosum; cpp: commissura aberrans; pa: parietal organ; sd: sulcus diencephalicus dorsalis; other abbr. as in
preceding figure.

zonal system, will be discussed further below in section 6 of this chapter.

Because, as components of the roof plate, *paraphysial anlage* and *velum transversum* are *ab initio* continuous with the dorsal or dorso-caudal end of lamina terminalis, the inverted telencephalon impar generally includes a constituent of roof plate which comprises, in many forms, a telencephalic choroid plexus, a paraphysis, and the rostral fold of velum transversum. The two first-named structures may, however, not develop, and thereby be lacking at the adult stage, although the wall neighborhoods representing their anlagen are recognizable at various embryonic stages. Likewise, in numerous vertebrate forms, the velum transversum can become greatly distorted by the subsequent ontogenetic events, or undergo considerable relative reduction to a nondescript, hardly recognizable rudiment.

*Eversion* is characterized, in the transverse plane, by a lateral or even ventrolateral bending of the original dorsal wall of the telencephalon. Simple eversion (e.g. Fig. 58A) can be distinguished from extreme eversion combined with outward folding (German: *Umstülpung, Umkrempelung*), whereby a portion of the bent dorsal neuronal wall faces, with its ventricular lining, the lateral, external aspect of the

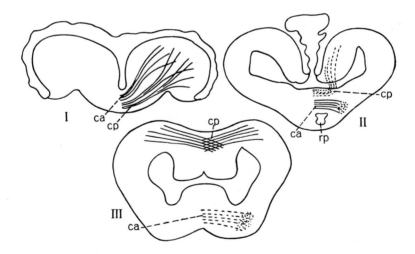

*Figure 64.* Configuration of ventriculus impar telencephali and telencephalic commissures of lamina terminalis in representatives of Anamnia (from K., 1927, modified after JOHNSTON in accordance with original observations). I Teleostean pattern. II Amphibian (urodele) pattern. III Selachian pattern. ca: commmissura anterior; cp: commissura pallii ('commissura hippocampi'); rp: recessus praeopticus.

brain, although, of course, still remaining intraventricular, the ventricle being laterally covered by lamina epithelialis (e.g. Fig. 60, 64, I). Substantial lateral eversion occurs in the pars impar telencephali of Osteichthyes. In these fishes, the bulk of the endbrain can become an everted telencephalon impar. The *lamina epithelialis* is thereby greatly stretched in the transverse plane, and the interhemispheric groove may be flattened out (cf. Fig. 64, I). In some instances, however, a longitudinal fold, appearing as a median invagination of the lamina epithelialis, may be displayed, particularly at rostral levels (cf. Fig. 60). Such invagination of lamina epithelialis can even assume the appearance of a median 'septum'.

The *olfactory bulbs*, on the other hand, which, with or without elongated stalks, represent the only rostral paired evaginated components of these strikingly everted endbrains, display inversion. Yet, at the boundary between paired rostral evaginated and unpaired evaginated telencephalic neighborhoods, various intermediate features transitional between incomplete inversion and only slight eversion, combined with inclusion of lamina epithelialis, can become evident.

*Caudal paired evagination* correlated with *eversion* is manifested in many Osteichthyes (Fig. 61 A). These forms then display a telencephalic polus posterior, differing, because of its eversion, from that of brain types with telencephalic inversion (e.g. of Petromyzon, Amphibia, and Amniota).

There are, in addition to the just-mentioned transitions involving incomplete inversion or slight eversion at the caudal boundaries of olfactory bulb or stalk, other telencephalic configurations which exhibit, as it were, limiting cases, intermediate between typical inversion and extreme eversion. The pars impar telencephali in the elasmobranch Chimaera and in extant dipnoans can be subsumed under this category of limited eversion. Caudal paired evagination with a corresponding true polus posterior is not displayed by these forms.

A very peculiar sort of *everted caudal paired evagination* occurs in the ganoid *Polypterus*. The data recorded by NIEUWENHUYS (1963) show that the extreme eversion of the telencephalon near the polus posterior is correlated with the formation of a caudally closed recessus of the external brain surface. In cross-sections (Fig. 61 B) this recess (*'pseudo-ventricle'* of NIEUWENHUYS*)* has the appearance of a real ventricular cavity although being actually a part of the brain surface. The lining of the pseudo-ventricle, provided by meninx, has topographically *inside* location, but remains topologically *outside*. The true ventricular cavity,

lined by ependyma and by lamina epithelialis of roof plate, is located
on the topographically external aspect of the polus posterior. Again,
in Polypterus, the everted dorsolateral telencephalic wall remains relatively thin, with a pronounced external concavity (sulcus externus, fissura externa) which provides the just mentioned caudal recess. In other
ganoids and particularly in teleosts, however, the everted telencephalic
wall becomes rather thick and assumes a conspicuously massive aspect
in cross-section (cf. Figs. 60, 64, I). The depth of sulcus externus
(Fig. 60), which depends on the degree of eversion as well as on the degree of wall-thickening becomes reduced in many forms. Thus, this
sulcus is e.g. quite shallow in the ganoid Acipenser and has almost entirely disappeared in the Salmonid teleost Trutta, being, however,
rather conspicuous in various Cyprinoid teleosts.

A *telencephalon medium sive impar*, which seems, in fact, to represent
the earliest stage of telencephalic ontogenetic morphogenesis in all
vertebrates, may thus be compatible with both inversion or eversion,
and, moreover, may comprehend only a relatively small part of the telencephalon, or the bulk of it. In mammals, including man, the approximately transverse (cross-sectional), but somewhat inclined planes
passing through the foramina interventricularia *(Monroi)* and the recess between columns of fornix and dorsal half of bulging shelf formed
by anterior commissure encompass the extent of an almost negligible
rudimentary telencephalon medium. The ventricular wall of this latter,
moreover, includes here a roof provided by the attachment of telencephalic lamina epithelialis (with its derived structures) to the supraforaminal portion of lamina terminalis containing subfornical organ, fornix and corpus callosum. The ventricular space (ventriculus impar) of
the adult human telencephalon medium is indicated in the ventricular
cast of Figure 78H II.

Again, a *telencephalon medium seu impar* is compatible with both rostral (or rostrolateral) and with caudal paired evagination. The former,
as far as bulbi olfactorii are concerned, seems to be an obligate feature,
combined with inversion. Caudal paired evagination, on the other
hand, combined with either inversion or eversion, and characterized
by a true polus posterior, is not an obligate morphologic feature, since
it is wanting in various fishes (e.g. in most Selachians and in Dipnoans).

Differentiation of a *paraphysis* is, *per se*, compatible with an everted
as well as an inverted telencephalon impar. The available recorded data
on paraphysial development and structure, respectively on adult pres-

ence or absence in forms with telencephalic eversion, are less numerous or conclusive than those referring to forms with inversion.

Sagittal sections (Figs. 62, 63) display the significant relationships between configuration of telencephalic lamina terminalis and presence of a telencephalon with conspicuous common ventricle in the final developmental ('adult') stages of two groups of vertebrates. In the first, which includes the *gnathostome fishes* and the *Amphibia*, the lamina terminalis retains much of its caudo-rostral orientation related to the original unpaired rostral evagination. In the second group, comprising *Cyclostomes* such as *Petromyzon* as well as *Sauropsida* and *Mammalia*, the original rostral evagination of lamina terminalis becomes modified in the course of ontogenesis, since this structure assumes here an essentially dorso-ventral orientation, whereby the *ventriculus impar s. communis telencephali* is reduced to an almost negligible remnant.

Again, it becomes evident that the *telencephalic commissures*, namely commissura anterior, commissura pallii, and in Mammalia, corpus callosum,[57a] which take their course through the torus transversus or commissural plate of lamina terminalis, display, in their arrangement, differences correlated with the particular configuration manifested by said lamina (Figs. 64, 65). Further details concerning these commissures shall be considered in the appropriate sections or chapters of the present series.

Comparing Figure 63 II, 65 II, and 77 with corresponding figures depicting the relevant configurational relationships in Anamnia, it will be seen that (in at least many, or typical) Sauropsidan brains the rather 'extreme' dorsocaudal bending of telencephalic lamina terminalis is correlated with a position of secondary foramen interventriculare above

---

[57a] The *corpus callosum*, characteristic for Eutherian mammals, consists of commissural fibers related to the so-called neopallium. It is a telencephalic commissure which can be evaluated as a 'neocortical' expansion of the commissura pallii (respectively commissura hippocampi) of 'lower' forms. In the same manner as the 'original' commissura pallii and the commissura anterior, it arises within the commissural plate of the lamina terminalis. The fibers of said commissures thus cross the midline within an *ab initio* present connection between bilaterally symmetric regions of the neuraxis. The origin and expansive growth of the human corpus callosum during ontogenesis was described in detail by HOCHSTETTER (1919), who emphasized intussusceptional growth, although some growth by direct apposition likewise seems to obtain. Despite a recent but in my opinion unconvincing attempt by RAKIC and YAKOVLEV (1968) to contest HOCHSTETTER's interpretation, this latter can be considered essentially correct (K., 1969). Further comments on this question will be found further below in section 6 of this chapter, and shall also be included in chapter XIII of this series.

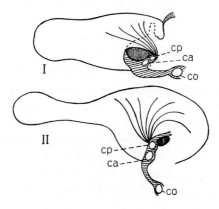

*Figure 65.* Configuration of telencephalic commissures in lamina terminalis of amphibians and reptiles (from K., 1927, modified after JOHNSTON in accordance with original
observations). I Amphibian. II Reptile. ca: commissura anterior; co: chiasma opticum;
cp: commissura pallii ('commissura pallii anterior' of Reptiles).

(i.e. essentially dorsal to) commissura pallii (anterior). This is not generally the case in Mammalia (Fig. 78 A), where, in the definitive developmental state, said foramen is located between 'pallial' commissures
(e.g. corpus callosum and commissura hippocampi of Eutheria) on the
dorsal side, and commissura anterior on the ventral side

The median portion of commissura anterior can be evaluated, in all
vertebrates *sensu strictiori*, as a morphologically significant structure
pertaining to, but delimiting the telencephalon in the midline neighborhood represented by (diencephalic and telencephalic) lamina terminalis. The telo-diencephalic boundary is here manifested as a transitional (topologically) open boundary zone in which the hypothalamic
cell aggregates of preoptic recess merge (or intermingle) with the basal
telencephalic cell aggregates forming a 'bed nucleus' of commissura
anterior. The diencephalic lamina terminalis extends thus roughly
from the chiasmatic ridge to said 'bed nucleus'. The telencephalic lamina terminalis, which includes anterior commissure and the other telencephalic commissures, extends from commissura anterior and finds its
dorsal end at the attachment of telencephalic lamina epithelialis. In
some Reptilia, an accessory pallial commissure (*comm. pallii posterior*,
Figs. 63 II, 77) becomes established within the lamina epithelialis,
caudal to paraphysis, in the region of velum transversum, within the
dorsal telo-diencephalic boundary region. Again, following this elementary discussion involving a first approximation to the complex formanalytic problems presented by the ontogenetic evolution of the ros-

tral neuraxis components, further comments on these morphologically significant boundary zones will be found in the subsequent sections (3, 5, 6) dealing with the pertinent problems.

Following this elaboration on telo-diencephalic boundary and modes of neuraxial, respectively telencephalic morphogenesis, it seems appropriate to proceed with a preliminary discussion and summary of the main subdivisions which are based on the initial ontogenetic aspects of the pattern of the brain tube, and can be suitably adopted for a characterization of the quintuple or definitive stage of overall brain configuration. The definitive main subdivisions will again be considered further below (section 3) in connection with their relationship to the neuromeres.

The *telencephalon*, consisting of *bulbus olfactorius* and *lobus hemisphaericus*, which latter may be mainly disjoint paired or mainly conjoint unpaired evaginated, and may display inversion or eversion, or both, is considered the most rostral and first main subdivision. Its caudal demarcation, i.e. the telo-diencephalic boundary was defined above.

The *diencephalon*, representing, in rostro-caudal sequence, the second main subdivision, may be delimited from the third one, namely the mesencephalon, by a conceptual boundary traced dorsally from an ill-defined region of posterior commissure (perhaps middle qua rostrocaudal sequence) to a basal rostromedian prominence, located in the ventricular floor of the neural tube, and designated as *tuberculum posterius* by KUPFFER (1906). This prominence, although rather constant and conspicuous in many embryonic and even adult stages of the vertebrate brain, may be less evident and merely vaguely indicated in some instances. The posterior commissure, like the anterior commissure a constant landmark of the vertebrate neuraxis, consists of crossing (transverse) fibers, interconnecting right and left sides, and related to grisea of diencephalon as well as mesencephalon. It should be stressed that, despite the clearly recognizable overall distinctiveness of diencephalon and mesencephalon, the boundary between the two subdivisions is not a linear (two-dimensional) plane, but a zone of transition with three-dimensional overlap of grisea. The *roof plate* of the diencephalon, which develops a *choroid plexus* in most vertebrate groups, originates, as a rule, also a single or duplex epithelial dorsal midline structure, the *pineal body (epiphysis)* and the less often differentiated *parapineal body. The floor* of the diencephalon participates in the formation of the compound *hypophysis* by providing the *neurohypophysial components.*

The *mesencephalon* is characterized as the third major subdivision of the cerebral neuraxis by a number of ontogenetic features. The dorsal part or roof *(tectum mesencephali)* displays a manifest external convexity, which, at certain embryonic stages of many vertebrates, produces the most distinctive anterior prominence of the head. This convexity corresponds to a pronounced flexure of the neural tube (Fig. 66 B), the *cephalic flexure (Kopfbeuge)*, whose external ventral concavity is the embryonic fossa interpeduncularis of His (1893). This latter is formed by the *plica ventralis encephali*, and contains the rostral end of the notochord, which may directly adjoin the caudal floor of the diencephalon (hypothalamus).

The two limbs of the cephalic flexure can be designated as *anterior* and *posterior vertex flexures (vordere und hintere Scheitelkrümmung*, this latter being the '*Isthmusbiegung*' of His, 1893). The often ill-defined neighborhood mentioned above as *tuberculum posterius* is located in the basal ventricular wall convexity of the anterior limb of the cephalic flexure and approximately corresponds to the rostral end of that convexity.

In contradistinction to telencephalon and diencephalon, the dorsal wall of mesencephalon does not, in general, display a thin epithelial roof plate, but is formed by the conjoint more massive lateral plates (alar plates), which do not provide a choroid plexus. Some cyclostomes, however, with mesencephalic choroid plexus, exhibit an exception to this rule (e.g. Petromyzon).

Caudally, the mesencephalon becomes delimited from the derivatives of rhombencephalon by a marked transverse dorsal fold with ventral convexity, the *plica encephali dorsalis* of Kupffer and of His (1893). Although a plane at right angle to a roughly drawn central axis of neural tube, passing through the bottom of this plica, would indicate the boundary between mesencephalon and rhombencephalic derivatives, the basal locus of this limit is, in general, less well demarcated. In at least some vertebrates, and at some ontogenetic stages, however, a depression or even a transverse fold in the basal ventricular wall, called *Isthmusfurche (innere Querfurche)* by His (1893), and sulcus intraencephalicus posterior by Kupffer (1906), may be interpreted as a landmark of the mesencephalon's caudal boundary. It corresponds to Hochstetter's (1919) 'Isthmusbucht', and to Kingsbury's (1920, 1922) *fovea isthmi* (Figs. 51 B, C, 52 B, 53 C, 54), whose significance will be discussed further below in section 2. Thus, the above-mentioned conceptual plane could be drawn as passing through plica encephali dorsalis and fovea isthmi. His (1893) describes an external convexity corresponding to the 'innere Querfurche', and calls its *eminentia interpe-duncularis*. HOCHSTETTER (1919) uses the terms *tuberculum interpedunculare* and *Isthmushöcker* for this slight prominence, which was also recorded by other authors (e.g. BURCKHARDT, 1895, and additional publications). Although I have frequently observed it myself, I did not

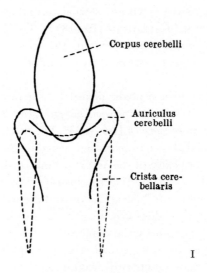

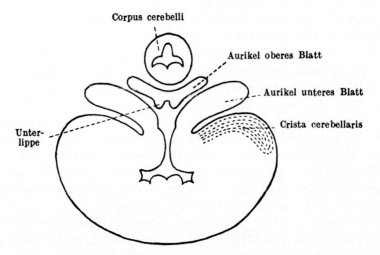

*Figure 66 A*. Diagram of cerebellar configurations in Selachians (from GERLACH, 1947). I Dorsal view. II As seen in cross-section. 'Oberes Blatt' and 'unteres Blatt' could be considered components of so-called 'rhombic lip' (cf. footnote 58). Read: Auricula.

gain the impression that it produced a generally and easily detectable landmark of adult vertebrate brains. *In toto*, the transition between mesencephalon and rhombencephalic derivatives represents a ring-like neighborhood or segment of the neural tube, designated by His and others as *isthmus rhombencephali*.

The fourth major subdivision of the brain in conventional nomenclature (BNA, PNA) is the *metencephalon*, which subsumes the rostral derivatives of the rhombencephalon, exclusive of isthmus rhombencephali. This latter, although arbitrarily assigned to the rhombencephalon as its most rostral individual component, does not, despite its significance in morphogenesis, constitute a major component of the adult vertebrate brain and might be included into the metencephalon *sensu latiori* if not into the mesencephalon.

The dorsal wall at the rostral extremity of rhombencephalon, adjacent to the bottom of plica encephali dorsalis, forms at relevant ontogenetic stages a protruding transverse plate *(cerebellar plate)*, which, like the mesencephalic roof, does not include a lamina epithelialis. This latter, however, is attached to the dorso-caudal end of cerebellar plate, and provides the roof of a large portion of rhombencephalon. In the dorsal wall of the neural tube, the cerebellar plate thus forms a prominent dorsal bulge, called genu rhombencephali by His (1893). At the same level, the floor expands as a basally directed bulge of the wall, resulting in an external ventral convexity, the *rostral rhombencephalic flexure*, called *pontine flexure* in mammals including man (Fig. 66B).

The *cerebellar plate*, to which the lamina epithelialis is attached, expands laterally into an ear-like, leaf-like, or lip-like fold of the rhombencephalic tube, bordering or surrounding a lateral recess (Fig. 57G). The everted portion of this fold, forming the floor of recessus lateralis

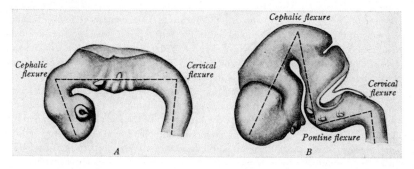

*Figure 66B.* Flexures of the brain tube during human ontogenesis (after Arey, 1954). A: at 6 mm; B: at 14 mm.

ventriculi quarti, was designated as the rhombic lip[58] by His (1891). In Anamnia, the laterally expanded leaf-like folds become the structures known as the paired *auricles*, which are especially conspicuous in some forms with extensively developed cerebellum (e.g. in Selachians, cf. Figure 66 A). Each auricle, again, consists of an anterior or superior fold, continuous with the corpus cerebelli derived from the cerebellar plate, and of an inferior or posterior (posterolateral) fold, caudally continuous with the lateral (alar) plate edge of myelencephalon. The tip of the rhombencephalon's lateral recess corresponds approximately to the dorsolateral concavity of the rostral rhombencephalic flexure.

In *Osteichthyes*, although the early anlage of corpus cerebelli does not differ from that in other Vertebrates, its rostral portion subsequently tends to be invaginated into the mesencephalic ventricle, developing the so-called *valvula cerebelli* (cf. Figs. 52 A, B, 62 I, 71 A, B).

---

[58] Since some of the figures published by His show indeed somewhat exaggerated folded outlines presumably caused by postmortal deformation, Hochstetter (1929) does not accept the term 'rhombic lip', about which some confusion has arisen in the literature (cf. the discussion by Ellenberger *et al.*, 1969). There is, however, no doubt that, in the region of recessus lateralis, the lateral part of alar plate forming the floor of this recess becomes externally everted at some ontogenetic stages of numerous, if not all mammals and various additional vertebrate forms. Accordingly, in at least partial agreement with His, the term rhombic lip *sensu strictiori* shall here be used to designate the lateral, more or less everted portion of the alar plate in the region of recessus lateralis, and providing the floor of that recess. The distinction of a primary and secondary rhombic lip, made by His on the basis of dubious data, may be discounted. Caudalward, in the region of myelencephalon proper, the dorsal portion of alar plate becomes slightly (secondarily) inverted within a pattern of overall eversion of the neural tube. This inverted portion consists essentially of glial subependymal cell plate with a few neuronal components, and provides a rather narrow shelf or ledge with a tapering medial edge (taenia ventriculi quarti), to which the lamina epithelialis of pars inferior ventriculi quarti is attached. It represents the *ponticulus* of Henle (cf. also Hikiji, 1933), not to be confused with the ponticulus or *propons* of Obersteiner (1912), who used these terms to designate transverse fiber bundles externally to the pyramids at the caudal end of the human pons. However, sufficient data are not available concerning the degree of eversion of alar plate in lateral recess, its changes of shape during ontogenetic development, and its various morphologic aspects in different classes, orders, and species of animals. It seems therefore permissible to use the term *rhombic lip* in the wider sense as a designation for the vaguely circumscribed 'lateral part' of the overall-everted rhombencephalic alar plate (amplifying the suggestion by Ellenberger *et al.*, 1969, who merely refer to 'the lateral part of the alar plate'). One might, of course, also describe the outline of a pronounced embryonic lateral recess corresponding to the pontine flexure, and displaying a lip-like pattern, as 'rhombic lips' (cf. fig. 57 G), which would comprise a lower and an upper lip (rl and ro, fig. 57 G). These two 'lips', in turn, would correspond to the cerebellar auricular folds of, e.g., Selachians (fig. 66, '*Aurikel oberes Blatt*' and '*Aurikel unteres Blatt*').

Especially in Teleosts, this valvula, particularly but not exclusively related to the lateralis system, becomes greatly expanded and can occupy a large part of the ventricular lumen, into which a recess of extraventricular i.e. extracerebral space may also thereby protrude. Details concerning this peculiar configuration shall be dealt with in chapter X of volume 4, in which the extreme spread of the *Mormyrid* cerebellum and of its enlarged valvula will likewise be taken into account.

The *cerebellum* provides the roof of the enlarged rostral portion of rhombencephalon designated as *metencephalon*, whose basis, but only in mammals, subsequently becomes an externally and grossly delimitable region called *pons*, characterized by transverse fiber systems intermingled with grisea. A well definable pons, however, suitable for the establishment of a significant gross subdivision, is only present in mammalian brains.[59] In taxonomically lower forms, particularly in Anamnia, it is expedient to distinguish merely two rhombencephalic components, namely (1) a *medulla oblongata* extending from the ill-defined boundary region of spinal cord to the isthmus (which might or might not be included in this medulla oblongata), and (2) a *cerebellum* dorsally and dorsolaterally superimposed upon the rostral part of oblongata. Nevertheless, by considering the distinctive transverse levels corresponding to the cerebellum as representing the metencephalon, regardless of presence or absence of a delimitable pons, the convenient and operationally valid useful fiction of a quintuple definitive stage characterizing the brain of all craniote vertebrates can be justifiably upheld.

In conventional terminologies based upon the configuration of the adult human brain, (BNA, PNA), the *myelencephlon*, here synonymous with medulla oblongata *sensu strictiori*, becomes then the fifth major subdivision of the vertebrate brain. The neural tube in this subdivision has rostrally a roof of thin lamina epithelialis (plexus chorioideus). Caudally, the myelencephalon's roof is entirely massive. Thus, the rostral portion of the myelencephalon is *everted*, and can be loosely called the 'open' part. The caudal portion is *unpaired inverted* (*i.e.* inverted as far as the midline), whereby a simple tube with massive roof is formed. This 'closed' part of the myelencephalon becomes a tube with caudalward gradually narrowing lumen which is directly continuous with the *central canal*, i.e. the lumen of spinal tube (spinal cord).

---

[59] Although a few grisea in some respects comparable to the nn. pontis of mammals seem to be present in birds and possibly even in reptiles, an externally and grossly delimitable pons is not found in Sauropsida.

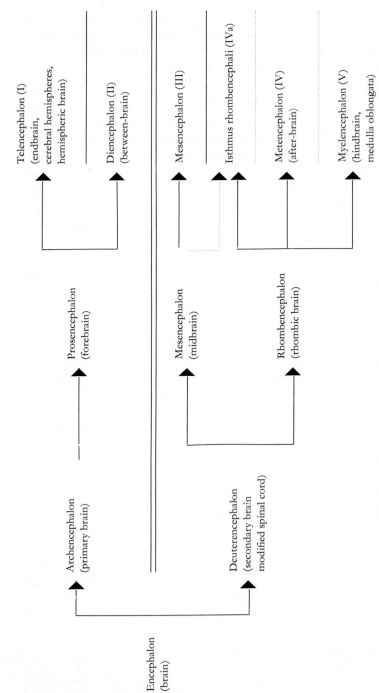

*Figure 66 C.* Simplified tabulation showing ontogenetic derivation of conventional brain subdivisions from the embryonic brain anlage (slightly modified after K., 1927).

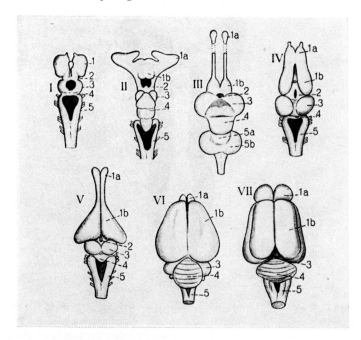

*Figure 67.* Simplified semidiagrammatic sketches depicting overall external configuration of brain in different classes of vertebrates as seen in dorsal view (from K., 1927). I Petromyzon (Cyclostome). II Scyllium (Selachian). III Cyprinus (carp, Teleost, Actinopterygian or Osteichthyan). IV Frog (Amphibian). V Lacerta (lizard, Reptilian). VI Columba (dove, Aves). VII Erinaceus (hedgehog, Mammalian). 1: telencephalon; 1a: bulbus olfactorius; 1b: lobus hemisphaericus; 2: diencephalon; 3: mesencephalon; 4: cerebellum (dorsal component of metencephalon); 5: medulla oblongata (myelencephalon); 5a: lobus facialis; 5b: lobus vagi (5a and 5b characteristic for some Teleosts, especially Cyprinoids).

The transition between caudal myelencephalon and spinal cord is gradual, lacking a distinctive and clear-cut boundary. In this zone of transition, a ventrally concave embryonic flexure of the neural tube is displayed in many vertebrates, particularly in Amniota. This *cervical flexure* (Nackenkrümmung) is especially pronounced in mammals and man (Fig. 66B).

The morphogenetic events, whereby the vertebrate neuraxis reaches a so-called quintuple stage of brain outline, upon which a convenient subdivision of the definitive configuration of that organ can be based are illustrated in the following tabulation (Fig. 66C). With respect to the term '*forebrain*' it should here be added that not only numerous older pioneer authors but also some more recent ones, e.g. HOLMGREN (1922) and myself in early publications (K., 1921), have

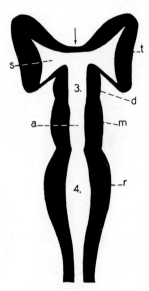

*Figure 68.* Diagram of ventricular system in the vertebrate brain (from K., 1927). a: mesencephalic ventricle *(aquaeductus Sylvii* of mammals); m: mesencephalon; r: rhombencephalon; s: lateral ventricle of telencephalon (the lead actually ends in lateral portion of ventriculus impar s. communis); 3: third ventricle; 4: fourth ventricle (rhomboid fossa); arrow points to telencephalic portion of lamina terminalis.

used said designation as synonymous with telencephalon. Although this could perhaps be justified, if the early embryonic prosencephalon is conceived as the *'primary forebrain'*, the telencephalon then being the *'secondary'* or *'definitive'*, *forebrain*, it seems preferable to avoid ambiguities by strictly adhering to the generally adopted conventional terminology indicated in the tabulation.[60]

Figure 67 depicts, in an appropriately simplified semidiagrammatic fashion, the overall external configurations displayed by these subdivisions of the vertebrate brain in representatives of different classes at the 'adult' stage. In forms (such as e.g. Mammals) with a predominant development of the telencephalon, this latter does not yet display its pre-eminence at early stages of ontogenesis (cf. also e.g. fig. 341). Even in Man, the dimensions of diencephalon and deuterencephalon ('brain stem', cerebellum) are not significantly overtaken by those of telen-

---

[60] This tabulation, which disregards the behavior of neuromeres, is supplemented in section 3 by a similar tabulation which includes the relationship of neuromeres to the definitive brain subdivisions of the here-adopted terminology.

cephalon until about the second months of intrauterine development (ca. 30 mm; cf. fig. 36). This, again, is fully consistent with an overall conformity in the succession of ontogenetic and presumed phylogenetic events.

The diverse foldings and constrictions affecting the cerebral portion of the neuraxis result in a rostro-caudal sequence of intercommunicating fluid-filled cavities, designated as *ventricles*.[61] These spaces constitute a more or less tubular channel which displays a number of recesses, and bifurcates at the rostral end (Figure 68).

The two cavities of the *paired evaginated telencephalon* are the *lateral ventricles* which open through interventricular passages into the *common ventricle of telencephalon impar*. This single cavity (Figures 49C, 59A), very conspicuous at early ontogenetic stages, has been termed *cavum Monroi* (e.g. HOCHSTETTER, 1919). With increasing inversion and paired evagination (Figures 49B, 59C), two lateral portions of the common telencephalic ventricle become, as it were, progressively segregated from a correspondingly reduced median portion. This reduction, of course, is generally not one of actual size by 'shrinkage', but rather a relative, or proportional one resulting from differential rates of further growth, concomitant with 'folding' processes, displayed by diverse telencephalic wall segments.

A description of these morphogenetic changes, and the rigorous definition of suitable anatomical terms involve some arbitrary and even awkward questions of semantics. If the designation '*cavum Monroi*' is subsequently restricted to the above-mentioned 'reduced' median portion, and the two lateral portions are regarded as representing parts of the overall lateral ventricles included in the telencephalon impar, then each one of the communications between such reduced '*cavum Monroi*' and the two lateral portions may be termed *foramen interventriculare Monroi (foramen Monroi primitivum* of HOCHSTETTER, 1919), being the direct forerunner of that passage as seen in the adult human brain.

In the brain of adult Amphibia, the reduced median ventricular space of the inverted telencephalon impar has been called *aula (viz. ventriculorum lateralium)* by early investigators of comparative neurology

---

[61] The suffix *-coele* or *-cele* (κοιλία, cavity; κοῖλος, hollow) is sometimes used in terms referring to brain ventricles. Occasionally, the following less common designations will thus be encountered in the literature: *rhinocoele (rhinocele)* for the ventricular space in the olfactory bulbs, *telocoele* for telencephalic ventricles, *diocoele* for diencephalic ventricle, *mesocoele* for mesencephalic ventricle, and *rhombocoele* for metencephalic ventricle.

(cf. also K., 1921). The amphibian aula, whose floor is provided by the lamina terminalis with its expanded commissural plate, opens laterad into the two lateral ventricles (as defined above) of telencephalon impar. Caudad, it is continuous with the diencephalic ventricle.

In the adult human brain, the ventricular space corresponding to the aula is very inconspicuous and distorted, being mainly represented by a triangular recess whose floor is the anterior commissure, and whose roof (apex) is the attachment of choroid plexus to subfornical organ. The caudal end of the septum provide the anterior wall, and the columnae fornicis the lateral boundaries (sides) of said recess. The delimitation of the human ventriculus impar telencephali from the ventriculus diencephali will be defined further below.

As regards the *everted telencephalon*, its common ventricle evidently corresponds not only to the 'reduced' *cavum Monroi*, but also to the so-called lateral ventricles within the inverted telencephalon impar. In teleosts (Figure 64, I), the aula is represented by the slit-like median space between the 'internated' bilateral basal grisea, while the space between lamina epithelialis of roof plate and ependymal lining of the more laterally everted telencephalic grisea is directly comparable to the afore-mentioned so-called lateral ventricles. It is here understood that the term '*lateral ventricles*', although primarily and rigorously referring to the spaces within the paired evaginated hemispheres, might justifiably allow for the subsumption of the direct extension of the said paired ventricles into the lateral parts of the inverted telencephalon impar. If, however, the telencephalon impar is not inverted, but displays only very moderate eversion, as for instance in the holocephalian Plagiostome *Chimaera* or in the Dipnoan *Protopterus* (Figure 49 D), then an uncomplicated ventriculus impar is present, corresponding to an ontogenetically early *cavum Monroi* or primordial *aula*.

Originally, the amphibian aula was described as the most rostal part of the third ventricle, which, by protruding into the telencephalon, would thereby include a telencephalic portion (cf. K., 1921). However, since a morphologically definable as well as form-analytically significant boundary between telencephalon and diencephalon can be traced, it appears preferable to use a terminology which clearly distinguishes telencephalic and diencephalic ventricular spaces. If one calls the communication between aula and lateral ventricles the *paired* or *secondary foramen interventriculare Monroi*, then the boundary between aula and diencephalic ventricle may be designated as the *unpaired, primary foramen interventriculare (Monroi)* as mentioned in footnote 62. This un-paired primary foramen should not be confused with HOCHSTETTER's (1919) *foramen Monroi primitivum*, which is an already (paired) early forerunner or stage of the definitive secondary or paired one.

In Anamnia, the boundaries of the primary interventricular fora-men are usually indicated by the caudal edge of the commissural plate *(torus transversus laminae terminalis)* basally, by the *velum transversum* (if recognizable) dorsally, and laterally by a mostly noticeable transition, frequently involving either a bulge (remnant of original torus telo-diencephalicus) or a sulcus resulting from secondary distortion, be-tween caudal telencephalic structures and rostral diencephalic compo-nents, such as ganglion habenulae, eminentia thalami, and hypothal-amic neighborhoods pertaining to recessus praeopticus.

In adult man and many other Mammalia, the outlines of this 'pri-mary' unpaired foramen are blurred by the considerable deformation of the primordial pattern by massive and in part disruptive ontogenetic growth processes. The edge of the median shelf formed by the exposed *anterior commissure*, the faint groove indicating the *transition of columna fornicis into pars tecta*, an undefined neighborhood at the rostral end of *sulcus terminalis*, and a nondescript tuft of choroid plexus derived from the original *velum transversum* are structures approximately related to the *rim of the 'primary foramen'*. This latter, despite its irrelevance and al-most complete defacement in the brain of adult man and perhaps of most mammals, is nevertheless a morphologically very significant structure not only in the adult brain of many Anamnia, but particularly also in the ontogenetic development of the brain in all vertebrates including man.

Again, it should be recalled that the *telencephalon impar*, although representing in ontogenesis the first anlage of the telencephalon in all vertebrates, and remaining a substantial part of this brain subdivision in many adult Anamnia, may become greatly reduced in the adult brain configuration of numerous forms. In adult Amniota, including mam-mals and man, the telencephalon impar has indeed become anatomical-ly almost negligible, unless rigorous morphologic or ontogenetic con-cepts are taken into consideration.

By comparing Figure 63 with Figure 62 it can be seen that the on-togenetic behavior of the *lamina terminalis* in Amniota (Sauropsida, Mammalia) has led to a configuration correlated with a reduction of the telencephalon impar approaching almost complete disappearance.

Among Anamnia, cyclostomes such as *Petromyzon* (Figure 63, I) likewise lack a significant telencephalon impar in late larval (Ammocoetes) and adult stages. The highly 'aberrant' myxinoids *Bdellostoma* and *Myxine* (Kupffer, 1900, 1906; Conel, 1929, 1931; Jansen, 1930; Holmgren, 1946) exhibit a very peculiar pattern deformation, of which some features are, to a lesser degree, also recognizable in Petro-

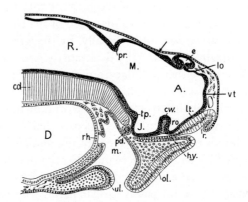

*Figure 69 A.* Approx. sagittal section through the head of an Ammocoetes (Petromyzon larva) of about 4-mm length (from Kupffer, 1906, interpreted in accordance with the present treatise). A: archencephalon; D: gut; I: infundibular region; M: mesencephalon; R. rhombencephalon; cd: notochord; cw: chiasmatic ridge; e: epiphysial anlage (including upper parietal organ; hy: adenohypophysial anlage; lt: lamina terminalis; lo: anlage of lower parietal organ; m: oral cavity; ol: 'upper lip'; pd: preoral gut structure; pr: plica rhombo-mesencephalica; r: olfactory placode; rh: pharyngeal membrane; ro: recessus praeopticus; tp: tuberculum posterius; vt: velum transversum; ul: 'lower lip'. The velum transversum, according to this interpretation, displays a rostral displacement still more pronounced in Bdellostoma and presumably Myxine (cf. fig. 69 D). Arrow indicates prosencephalo-mesencephalic boundary region.

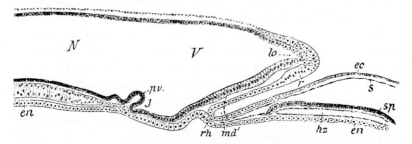

*Figure 69 B.* Approx. midsagittal section through the head of an early embryo of Bdellostoma (after Kupffer, 1900). ec: ectoderm; en: entoderm (at left underlying notochord); hz: heart anlage; J: infundibular region; lo: lobus olfactorius impar; md': primary oral cavity; N: deuterencephalon; pv: plica encephali ventralis; r: unpaired olfactory placode; rh: pharyngeal membrane; s: somatopleure; sp: splanchnopleure. A slight dip in the dorsal tube wall indicates transition between archencephalon and deuterencephalon (prosencephalo-mesencephalic boundary). V: Archencephalon.

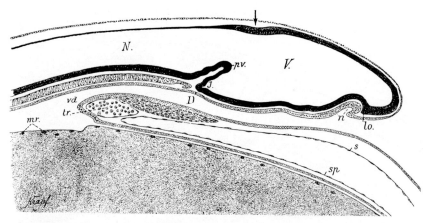

*Figure 69C.* Approx. midsagittal section through the head of an early embryo of Bdellostoma, somewhat more advanced than that of preceding figure (from KUPFFER, 1900). D: branchial gut; J: infundibular region; N: deuterencephalon; V: archencephalon; lo: lobus olfactorius impar; mr: 'merocyte' nuclei; n: nasal diverticle; pv: plica encephali ventralis; s: somatopleure; sp: splanchnopleure; tr: truncus arteriosus; vd: anterior gut opening; arrow indicates plica prosencephalo-mesencephalica.

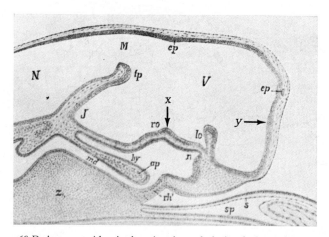

*Figure 69 D.* Approx. midsagittal section through the head of an embryo of Bdellostoma, more advanced than that of preceding figure (from KUPFFER, 1900). J: infundibular region; M: mesencephalon; N: deuterencephalon; V: archencephalon; ap: accessory fold related to hypophysial invagination; cp: commissura posterior; ep: epiphysial anlage; hy: hypophysial anlage; lo: lobus olfactorius impar; md: definitive oral gut; n: nasal diverticulum; rh': secondary pharyngeal membrane; ro: recessus praeopticus; s: somatopleure; sp: splanchnopleure; tp: tuberculum posterius; arrow y indicates rostroventrally displaced velum transversum; arrow x indicates torus transversus; z: oro-pharyngeal mesodermal blastema. The fold rostral to lo is interpreted as an accessory fold related to the peculiar lobus olfactorius impar.

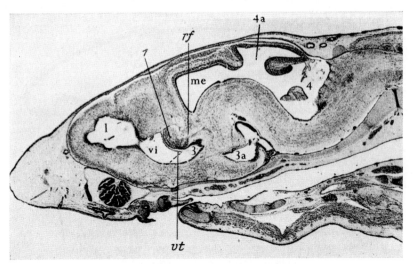

I

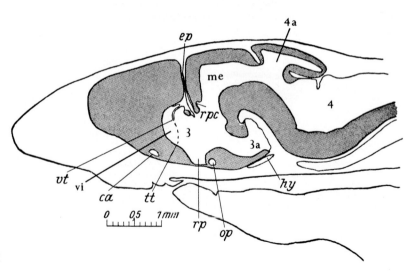

II

*Figure 70*. Sagittal sections through head and brain in a developing Scyllium canicula (Selachian) of about 45-mm length (from K., 1929a). I Paramedian section. II Approx. midsagittal section. ca: commissura anterior; ep: epiphysis; hy: adenohypophysis; l: ventriculus lateralis telencephali; me: mesencephalic ventricle; op: optic chiasma; rf: fasciculus retroflexus; rp: recessus praeopticus; rpc: recessus postcommissuralis (commissurae posterioris); tt: torus transversus; vi: ventriculus impar telencephali; vt: velum transversum; 1: epithalamus (ganglion habenulae); 3: third ventricle (diencephalon); 3a: infero-posterior (hypothalamic) recess of third ventricle; 4: fourth ventricle; 4a: cerebellar recess of fourth ventricle; dotted line indicates telo-diencephalic boundary. The relatively caudal position of nasal (olfactory) cavity should be compared with the rostral one in figure 76 I.

myzon. Figure 69 A–D shows, at early ontogenetic stages, a rostral and slightly basal displacement affecting the locus of velum transversum, correlated with a rostral extension of diencephalic neighborhoods dorsal to telencephalon impar (cf. also Figure 63, I). Further comments about the Myxinoid brain configuration, whose interpretation presents various difficulties, can here be omitted and shall be included in other sections of this series dealing with the relevant structures of the fish brain.

In accordance with the adopted demarcation of a morphologically significant telo-diencephalic boundary, the *ventricular lumen restricted to the diencephalon* is here designated as the *third ventricle*, which becomes rostrally continuous with the ventriculus impar telencephali,[62] whose

---

[62] Conventional anatomical nomenclatures of the human body (e. g. BNA and PNA), which are not concerned with the somewhat complex implications of comparative anatomy and ontogeny, seem to include the rudimentary remnant of ventriculus impar telencephali into the third ventricle. Again, the *foramen of Monro* in the adult human brain, designated as *foramen interventriculare (Monroi*, BNA) in both BNA and PNA, is usually described, e. g. in GRAY's Anatomy (27th, centennial edition, Philadelphia 1959) as the opening between lateral and third ventricles. For practical purposes, this terminology is, of course, quite adequate. From a more advanced morphological viewpoint, the ventricular boundary between telencephalon impar and diencephalon might be called the *primary, unpaired interventricular foramen (primary foramen Monroi)*. The entrance into each one of the paired or lateral ventricles could then be designated as the *secondary (paired) foramen Monroi* (cf. K., 1927). In adult Amniota as well as in Petromyzon, where the telencephalon impar becomes almost negligible, primary and secondary foramina Monroi roughly coincide. It should also be recalled that, in the embryonic human brain, HOCHSTETTER (1919) described the telencephalon medium, whose ventricle he appropriately called '*cavum Monroi*'. HOCHSTETTER, however, also distinguished a *foramen Monroi primitivum* which merely represents the entrance into lateral expansions of telencephalon impar. HOCHSTETTER (1919) made no attempt to establish a morphologically acceptable boundary between telencephalon and diencephalon.

Right and left antimeric structures (such as somites, extremities, and other paired organs) are not usually distinguished from each other, *qua* right and left, as first and second, but rather, in their proper longitudinal sequence, as e. g. first right and first left, second right and second left somites. Again, because the conventional nomenclatures (BNA, PNA) do not refer to a lateral ventricle as being the first or the second, it seems appropriate to designate the paired lateral ventricles as first right and first left. The *ventriculus impar telencephali* (aula) becomes then the *second ventricle*, and the *ventriculus diencephali* is properly called the *third*, thus avoiding any apparent inconsistency attributable to the morphologically justified distinction of three telencephalic ventricles. It should be added that both BNA and PNA retain the traditional designation *fourth ventricle (ventriculus quartus)* for the *metencephalic ventricle*, thereby ignoring the *ventriculus mesencephali* ('cerebral aqueduct') as a 'true' ventricular space, although this mesencephalic ventricle is of relatively considerable size not only during ontogenetic development but also in many adult vertebrates.

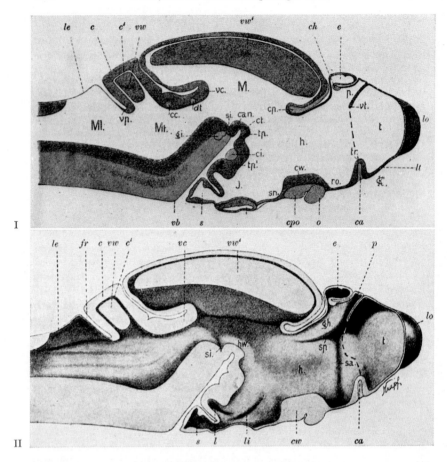

*Figure 71.* Sagittal sections through the brain of a representative of Osteichthyes. Trout (Trutta), just hatched (after KUPFFER, 1906). I Midsagittal section. II Identical section with reconstruction of inner surface of brain tube wall. c: cerebellum; c': ventricular cleft of cerebellum; ca: commissura anterior; can: 'commissura ansulata'; cc: commissura cerebellaris; ci: 'commissura infundibularis'; ch: commissura habenulae; cp: commissura posterior; cpo: postoptic (and supraoptic) commissures; ct: 'commissura tubercularis'; cw: chiasmatic ridge; dt: decussation of trochlear nerve; e: epiphysis; fr: plica rhombo-mesencephalica; gh: ganglion habenulae; gi: ganglion interpedunculare; h: diencephalon; hw: tegmental ridge (dorsally bounded by the not labelled sulcus limitans; 1: entrance into 'lobus posterior' hypothalami; le: lamina epithelialis (of fourth ventricle); lo: olfactory bulb; lt: lamina terminalis telencephali; o: optic chiasma; p: parencephalic roof of diencephalon; s: 'saccus infundibuli'; sa: sulcus intraencephalicus anterior (presumably exaggerated in KUPFFER's drawing, and dorsally connected with sulcus terminalis, which, in turn, connects with sulcus diencephalicus dorsalis (sp); si: sulcus intraencephalicus posterior; sn: 'sinus postopticus'; sp: 'sulcus parencephalicus' (actually here already transformed

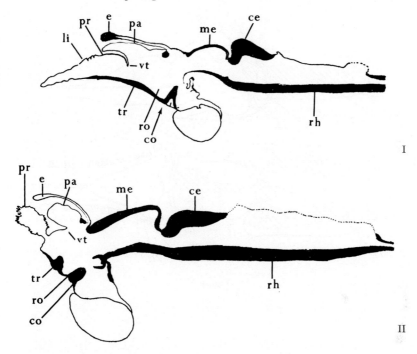

*Figure 72.* Sagittal sections through the brain of the Dipnonan Ceratodus. I After BING and BURCKHARDT (1905). II After HOLMGREN and VAN DER HORST (1925). ce: cerebellum; co: chiasma opticum; e: epiphysis; li: 'lingula interolfactoria' (cf. fig. 58 C); me: mesencephalon; pa: parencephalic roof of diencephalon; pr: paraphysis; rh: rhombencephalon; ro: preoptic recess; tr: torus transversus (including neighborhood of anterior commissure); vt: velum transversum. The designations and leads have been supplied in agreement with the viewpoints adopted in the present treatise.

---

into sulcus diencephalicus dorsalis); ro: recessus praeopticus; t: telencephalon (ventriculus impar); tp: tuberculum posterius; tp': 'tuberculum posterius inferius'; tr: torus transversus; vb: rostral thombencephalic flexure (at entrance to plica ventralis encephali); vc: valvula cerebelli; vp: 'velum medullare posterius; vt: velum transversum; vw, vw': midline secondary fusions of cerebellar and dorsal mesencephalic components, respectively; J: 'infundibulum' (posterior hypothalamus); M: mesencephalon; Ml: myelencephalon; Mt: metencephalon; dotted lines indicate telo-diencephalic boundary.

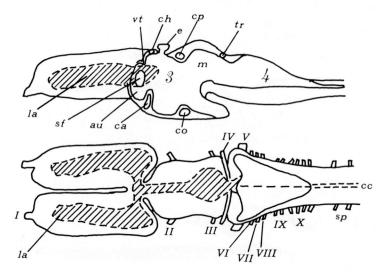

Figure 73

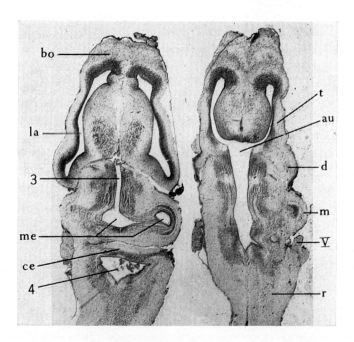

Figure 74

caudo-basal limit is usually indicated by a protruding neighborhood of lamina terminalis containing the *commissura anterior*. A rostral protrusion of third ventricle basal to commissura anterior represents the *preoptic (or optic) recess*. In the here-adopted classification and terminology, as elaborated in previous studies (K., 1927, and other publications), the preoptic recess, pertaining to the hypothalamus (cf. further below, section 5) is an intrinsic part of the diencephalon. The basal portion of the third ventricle, enclosed by a hypothalamic wall region, is thus subdivided by the transverse chiasmatic ridge, which contains the decussation of optic nerve (and supra-, respectively postoptic commissures), into a rostral preoptic and a caudal postoptic part. This latter can be designated as the *infundibulo-mammillary region*. Depending on the particular vertebrate forms, this caudal subdivision of the hypothalamus and its ventricular space display variously differentiated features, such as, in dorso-basal sequence, a mammillary recess, a saccus vasculosus, a so-called tuber cinereum, and the distinctive neurohypophysial structures. These configurations shall be dealt with in chapter XII of volume 5, but a few additional brief comments are also included further below in section 5 of the present chapter.

At the *diencephalo-mesencephalic boundary* the third ventricle is continuous with the ventricular space of mesencephalon *(mesocoele)*, which is relatively very large at ontogenetic stages, and may remain an exten-

---

*Figure 73*. Ventricular system in the brain of a diagrammatic 'generalized amphibian' (modified after Larsell, 1942). au: aula (reduced ventriculus impar telencephali); ca: commissural ridge of anterior and pallial commissures; cc: central canal of spinal cord; ch: commissura habenulae; co: chiasmatic ridge with optic chiasma, supraoptic and postoptic commissures; cp: commissura posterior; e: epiphysis; la: lateral ventricle of telencephalon; m: mesencephalic ventricle; sf: secondary foramen interventriculare between reduced ventriculus impar *(aula)* and paired lateral ventricle; sp: first spinal nerves; tr: decussation of trochlear nerve; vt: velum transversum (the paraphysis has been omitted); I: fila olfactoria; II: optic nerve; III: n. oculomotorius; IV: n. trochlearis; V: n. trigeminus; VI: n. abducens; VII: n. facialis; VIII: n. acusticus; IX: n. glossopharyngeus; X: vagus group (including n. XI). Upper sketch indicates median view, lower drawing represents dorsal view.

*Figure 74*. Two approx. horizontal sections displaying aspects of the ventricular system in the anuran amphibian brain *(Bufo vulgaris, Nissl stain*, approx. ×13.5, red. $^7/_{10}$). au: aula; bo: bulbus olfactorius; ce: cerebellum; d: diencephalon; la: lateral telencephalic ventricle; m: mesencephalon; me: mesencephalic ventricle; r: rhombencephalon; t: telencephalon. 3: third (or diencephalic) ventricle; 4: fourth ventricle; V: portions of trigeminal root; surface grooves indicate approximate boundaries of telencephalon, diencephalon, mesencephalon, and rhombencephalon. Left section more dorsal than right one; right half of each section slightly more dorsal than left half.

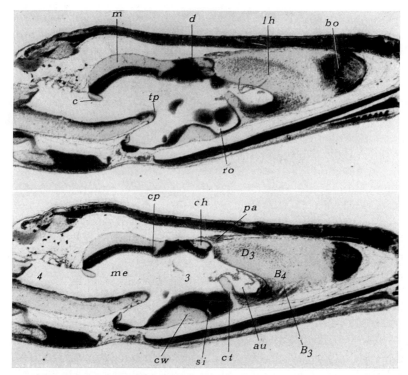

*Figure 75.* Two paramedian sagittal sections through the brain *(in situ craniale)* of a urodele amphibian *(Triton* sp., *van Gieson stain,* ×approx. 29, red. $^{7}/_{10}$). Bottom section slightly lateral of top one. au: aula (ventriculus impar telencephali); bo: bulbus olfactorius; c: cerebellum (midline commissura cerebellaris); ch: commissura habenulae; cp: commissura posterior; ct: commissural plate with telencephalic commissures (dorsal: comm. pallii; basal: comm. anterior); cw: chiasmatic ridge with optic chiasma, supra- and postoptic commissures); d: diencephalon; lh: lobus hemisphaericus telencephali; m: mesencephalon; me: mesencephalic ventricle; pa: paraphysis; ro: recessus praeopticus; si: cut through sulcus intraencephalicus anterior (with may display variable anterior and posterior branches); tp: tuberculum posterius; $D_3$, $B_3$, $B_4$: telencephalic longitudinal zones; 3: third ventricle; 4: fourth ventricle.

*Figure 76.* Three paramedian sagittal sections (I–III) through head and brain of the Gymnophione amphibian Schistomepum. Section III (from K. *et al.,* 1966) is almost midsagittal (hematoxylin-eosin, I approx. ×13.5, II and III approx. ×20, red. $^{2}/_{3}$). au: aula; bo: bulbus olfactorius; c: cerebellar rudiment; cc: region of rhombencephalic and spinal canalis centralis; d: diencephalon; la: lateral ventricle of telencephalon; m: mesencephalon; me: mesencephalic ventricle; na: nasal cavity; rf: pronounced convexity of rostral rhombencephalic flexure (typical for Gymnophiona); rl: lateral recess of fourth ventricle; si: sulcus intraencephalicus anterior in preoptic recess; 3: third ventricle; 4: fourth ventricle; Ah: adenohypophysis; Bp: approximate rostral end of deuterencephalic basal

(Continued on page 202)

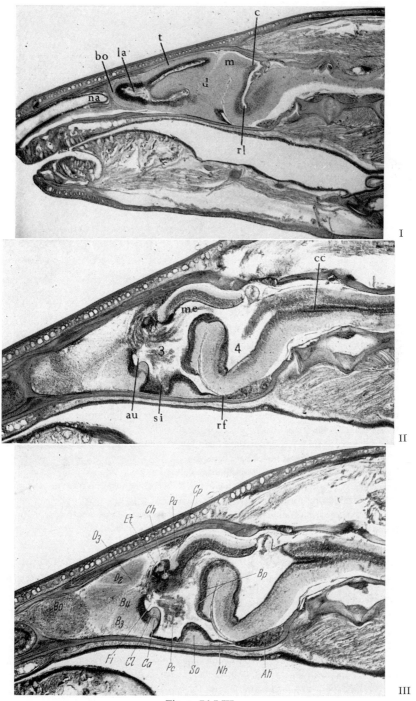

Figure 76 I-III

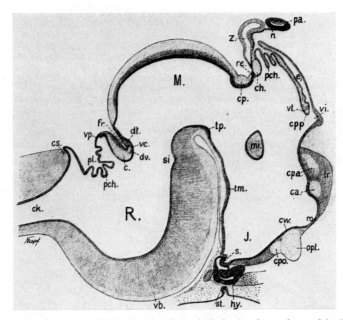

*Figure 77.* Approx. midsagittal section through the brain of an embryo of the Sauropsidan *Anguis fragilis* in advanced stage of development (from KUPFFER, 1906). ci: cerebellum; ca: commissura anterior; ch: commissura habenulae; ck: canalis centralis (bulbi); cp: commissura posterior; cpa: commissura pallii (anterior); cpp: commissura pallii posterior; cpo: postoptic commissures; cs: *commissura infima Halleri;* cw: chiasmatic ridge; dt: decussation of trochlear nerve; dv: decussatio veli (isthmi); e': paraphysis; fr: fissura rhombo-mesencephalica; hy: adenohypophysis; mi: massa intermedia (commissura mollis diencephali); n: nerve to parietal (parapineal) organ; opt: optic chiasma; pa: parietal eye (parapineal organ); pch: choroid plexus; pl: 'plica chorioidea'; re: recessus pinealis; ro: recessus praeopticus; s: infundibular evagination (neurohypophysis); si: sulcus intraencephalicus posterior; st: *Seessel's pouch;* tm: 'tuberculum posterius inferius'; tp: tuberculum posterius; tr: torus transversus; vb: basal convexity of rostral rhombencephalic flexure; vi: ventriculus impar telencephali at level of (secondary) *foramen Monroi;* vt: velum transversum; z: pineal body (epiphysis).

sive cavity in numerous submammalian forms. In the adult brain of man and other mammals, it is a rather narrow channel designated as *cerebral aqueduct (aquaeductus Sylvii)*.

At the levels of isthmus rhombencephali, the mesencephalic ventri-

plate; Ca: commissura anterior; Ch: commissura habenulae; Cl: commissura pallii; Cp: commissura posterior; Fi: aula (ventriculus impar telencephali leading to interventricular foramen); Nh: rostral part of neurohypophysis; Pa: paraphysis; Pc: choroid plexus of third ventricle reaching preoptic recess; So: supraoptic commissures in chiasmatic ridge. The telencephalic longitudinal D and B zones indicated as in figure 75.

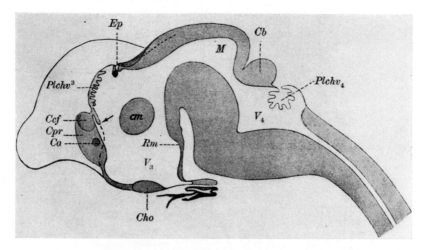

*Figure 78 A*. Approx. midsagittal section through the brain in a fairly advanced fetal developmental stage of the insectivore mammalian Erinaceus europ. (after GROENBERG, 1902, from ZIEHEN, 1906). Ca: commissura anterior; Cb: cerebellum; Ccf: joint anlage of corpus callosum and commissura hippocampi within telencephalic lamina terminalis; Cho: chiasma opticum; cm: 'commissura mollis (massa intermedia of 3rd ventricle); Cpr: 'concrescentia primitiva' (primary enlargement of median commissural plate); Ep: epiphysis; M: mesencephalon; Plchv₃, Plchv₄: choroid plexus of third respectively fourth ventricle; Rm: recessus mammillaris; $V_3$, $V_4$: third respectively fourth ventricle. Added notations: arrow: velum transversum; dotted line: telencephalo-diencephalic boundary; overlapping of diencephalic hypothalamic (preoptic) cell zone and telencephalic basal B zones in basal part of lamina terminalis.

cle is continuous with a common ventricle of metencephalon and myelencephalon, listed as the *fourth ventricle* in conventional terminologies (BNA, PNA). Because of its approximately rhombic outline in dorsal view, this cavity is also known as the *rhomboid fossa*.[63] Caudalward, the fourth ventricle, as stated above, is continuous with the *central canal* of lower medulla oblongata and of spinal cord.

---

[63] The roof of rhomboid fossa is rostrally (i.e. in the metencephalon) provided by the cerebellum, and more caudally (i.e. in the myelencephalon) by lamina epithelialis with choroid plexus. The anterior apex of the 'rhombus' corresponds to the constriction at the isthmus rhombencephali, the lateral apices are indicated by the lateral recesses, where eversion is most pronounced, and the posterior apex is represented by the transition into the central canal of the inverted or 'closed' part of medulla oblongata. The term rhombencephalon was apparently suggested by this roughly rhomboid outline of fourth ventricle in projections upon a frontal ('horizontal') plane. Openings in the epithelial roof of the rhomboid fossa of Mammals, providing a communication between ventricular and subarachnoid spaces, are briefly dealt with in section 7, p. 694, footnote 309.

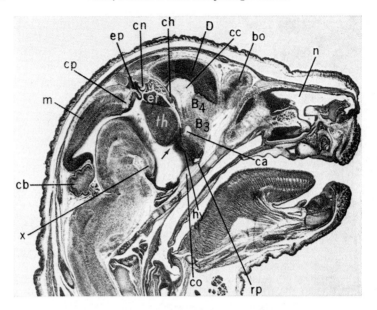

Figure 78B

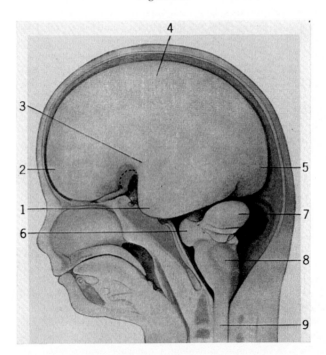

Figure 78C

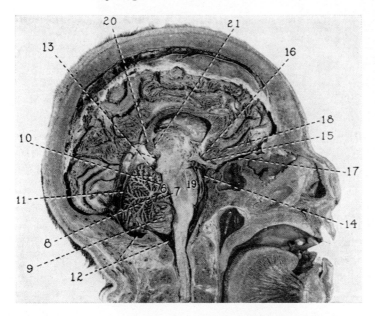

*Figure 78 D.* Approx. midsagittal section through head and brain of adult man, showing topographic relations of cerebral structures (from K., 1940). For explanation of labels, cf. figure 78F.

*Figure 78 B.* Approx. midsagittal section through head and brain of a near-term mouse fetus (from Kalter, 1968, labeled in conformity with the present treatise). bo: bulbus olfactorius; ca: commissura anterior; cb: cerebellum; cc: corpus callosum; co: chiasma opticum and supraoptic commissures; cn: commissura habenulae; cp: commissura posterior; ep: epiphysis; et: epithalamus (ganglion habenulae); hy: hypothalamus; m: mesencephalon; n: nasal cavity; rp: recessus praeopticus; th: thalamus (dorsalis); x: rostral end of deuterencephalic basal plate (tegmental cell cord); arrow: sulcus diencephalicus ventralis (the oblique dark cell bands dorsal to this sulcus represent medial portions of thalamus ventralis); $B_4$, $B_3$: basal telencephalic cell zones; D: cortex telencephali derived from pallial D-zones; ch: rostral end of choroid plexus of 3rd ventricle.

*Figure 78 C.* Lateral view of brain in a 14-week-old human embryo, as exposed *in situ* (after His, from Arey, 1954). 1: temporal lobe; 2: frontal lobe; 3: primordium of fissura cerebri lateralis (*fossa Sylvii*); 4: parietal lobe; 5: occipital lobe (1–5: telencephalon); 6: pons; 7: cerebellum; 8: myelencephalon; 9: spinal cord; dotted lined, added at bottom of *fossa Sylvii* indicates limen insulae (lateral border of 'rhinencephalon').

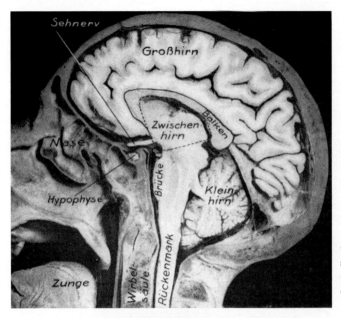

Figure 78 E. Sagittal section through head and brain of adult man, slightly lateral to the midline as shown in preceding figure (from MAURER, 1928).

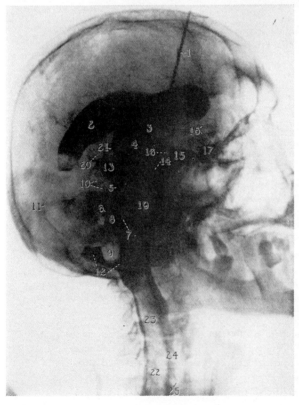

Figure 78 F. Lateral (dextro-sinistral) roentgenologic view of adult human head (cadaver) with barium sulfate injection into lateral ventricle overflowing into subarachnoid spaces (from K., 1940). 1: injection needle in left lateral ventricle; 2: right lateral ventricle with obliterated posterior and temporal horns; 3: interventricular foramen; 4: third ventricle; 5: *aqueductus Sylvii;* 6: fourth ventricle (fastigium); 7: region of choroid plexus of fourth ventricle; 8: nodulus cerebelli; 9: tonsilla cerebelli; 10: moniticulus and lobus centralis; 11: outline of tentorium cerebelli; 12: cisterna magna; 13: cisterna ambiens; 14: cisterna interpeduncularis; 15: cisterna chiasmatis; 16: arteria communicans ant. in cisterna laminae terminalis; 17: cisterna olfactoria; 18: sulci rostrales; 19: pons; 20: splenium corporis callosi; 21: recessus suprapinealis; 22: intumescentia cervicalis medullae spinalis; 23: dorsal roots of fourth cervical nerve; 24: roots of fifth cervical nerve; 25: roots of sixth cervical nerve.

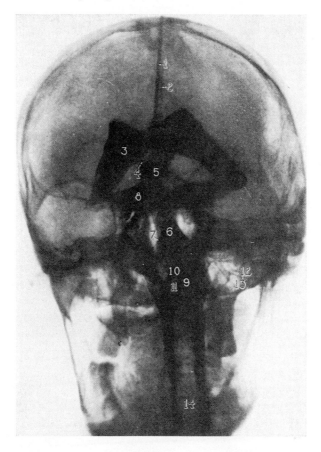

*Figure 78 G.* Antero-posterior view of anatomical specimen shown in preceding figure (from K., 1940). 1: injection needle; 2: fissura longitudinalis cerebri; 3: lateral ventricle; 4: foramen interventriculare; 5: third ventricle; 6: fourth ventricle; 7: recessus superior posterior of fourth ventricle; 8: cisterna ambiens; 9: cisterna magna; 10: region of incisura cerebelli posterior; 11: region of tonsilla cerebelli; 12: left cerebellar hemisphere; 13: sulcus horizontalis cerebelli; 14: intumescentia cervicalis medullae spinalis.

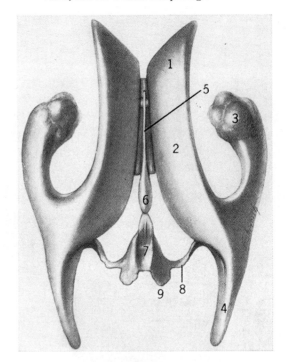

Figure 78H I

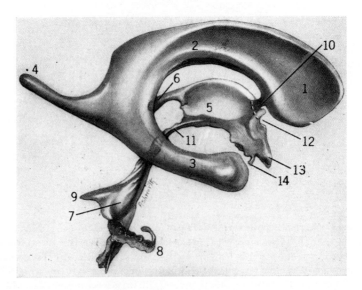

Figure 78H II

Figures 68 to 78 illustrate various aspects of ventricular spaces in the vertebrate brain[63a] and, in addition, a few representative configurational relationships between brain and cranial cavity or, respectively, between brain and head.[63b]

In summary it could be said that the orderly events displayed by the ontogenetic morphogenesis of the vertebrate neural tube at its rostral levels allow the formulation of a reasonably justified and convenient 'fundamental plan' of brain configuration, whereby that organ is roughly subdivided, along a longitudinal rostro-caudal axis, into the *five distinct main portions* enumerated in the accepted nomenclatures (BNA, PNA). Additional aspects of configuration, to be discussed in the following sections (2, 4, 5, 6), combine with, and superimpose upon, this rostro-caudal sequence a dorsoventral sequence of significant longitudinal zonal arrangements. Further complications, which may be disregarded as far as the convenient and practical standard sub-

---

[63a] With regard to figure 72 it should be added that HOLMGREN and VAN DER HORST (1925) emphasize the difference between their figure and that of BING and BURCKHARDT (1905), concluding with the comment (p. 64 loc. cit.): 'Let us hope that in the new editions of textbooks BING's and BURCKHARDT's figure and description will not more be reproduced or referred to'. On the basis of my own first-hand acquaintance with the brain of numerous Anamnia, including Dipnoans, I am inclined to believe that the differences stressed by HOLMGREN and VAN DER HORST are related to a combination of three variables, namely (a) individual and subspecies variations; (b) differences between sagittal and neighboring parasagittal planes as well as between slight degrees in slanting (obliqueness) or straightness of the sections, and (c) slight distortions occurring in the course of processing the material (conservation, fixation, etc.). It seems thus that both I (whose forebrain plane seems to be somewhat oblique and parasagittal) and II provide essentially correct pictures.

[63b] Relevant data on correlations between shape of cranium, respectively cranial cavity, and shape of brain are discussed in papers by DABELOW (1931), HOFER (1952, 1954), and SCHNEIDER (1957). A further reference to the significance of this topic for phylogenetic speculations will be found further below in section 8 on p. 747

---

*Figure 78 H.* Cast of adult human ventricular system (after RETZIUS, from LARSELL, 1939. The labels have been added). I Dorsal view of cast. II Lateral view of cast. 1: anterior horn; 2: body (cella media); 3: inferior horn (impressions of digitationes hippocampi at 3 in fig. I); 4: posterior horn; 1–4: lateral ventricle of telencephalon; 5: third ventricle; 6: suprapineal recess (third ventricle); 7: fourth ventricle; 8: lateral recess of fourth ventricle; 9: bilateral fastigial recesses of fourth ventricle; 10: *foramen interventriculare Monroi:* 11: *aqaeductus Sylvii* (mesencephalic ventricle); 12: shelf of commissura anterior; 13: preoptic recess; 14: infundibulum. Dotted line indicates limit of reduced ventriculus impar telencephali (cf. p. 177 of text).

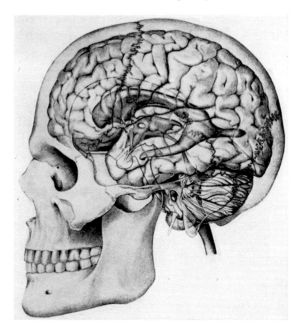

*Figure 78I.* Relationships of skull, brain, and ventricular system in adult man (after
KAUTZKY and ZÜLCH, from HAYMAKER-BING, 1956), for comparison with preceding figures.

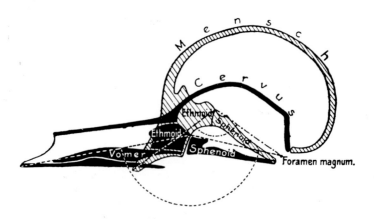

*Figure 78 J.* Superimposed diagrammatic sections through human skull and skull of an
ungulate mammal (Cervus) showing the respective relationships, including the angles
(dotted lines) between cerebral and facial cranium (after BÜTSCHLI, from SELENKA-
GOLDSCHMIDT, 1923).

division of the definitive or adult vertebrate brain is concerned, are introduced by the appearance of a more or less pronounced but essentially temporary transverse segmentation related to the so-called *neuromeres*. Some of these, however, seem to coincide with permanent configurations. The behavior of the neuromeres will be discussed below in section 3.

### C. Disturbances of Morphogenesis: Some Teratologic Aspects

The multiplicity of the factors involved in morphogenesis and the complexities of their interaction patterns which result in the characteristic 'definitive' organic configurations represent an intrinsically 'unreliable', essentially probabilistic system subject to numerous internal and external disturbances.[64] Considering the end-effect displayed by the ontogenetically approximately stable adult organisms exhibiting a highly differentiated and topologically invariant bauplan of body as a whole, and of different organs in particular, one is indeed justified to speak of 'superb architecture and sloppy workmanship' in referring to these 'animated' (i.e. living) products of nature.[65]

It is thus easy to understand that, even within a given species, many *individual variations* of an essentially 'identical' body build should oc-

---

[64] In a formulation based on information theory or *'cybernetics'*, such disturbances represent, of course, *noise*. Again, although the system, composed of a multitude of complex subsystems (cf. e.g. fig. 7 B), many of which are highly susceptible to interference by 'noise', can thus be considered *'unreliable'*, the large number of fairly 'normal' animal and human individuals, substantially in excess of conspicuously 'defective' ones, demonstrates that this 'unreliability' must be qualified as being 'relative' or *statistical*.

Morphogenesis, in the aspect under consideration, like the operations of neural networks, represents a *special case* of the *general mathematical orderliness* displayed by the performance of *very large systems* involving considerable complexity and redundancy of interactions ('connecting lines', 'channels of communication' in 'control flow charts' similar to the diagram of fig. 7 B), including *'correcting'* as well as *'restoring' subsystems*. These latter prevent multiplication of error, which is thus, as it were, kept under control. The generalized mathematical aspects of such intrinsically *unreliable systems* whose behavior is, nevertheless, statistically determinate or probabilistic, and in this respect *fairly 'reliable'*, have been elaborated by J. VON NEUMANN and others (cf. vol. 1, p. 24, of this series). A peculiar Christian theological interpretation, elaborated by ST. AUGUSTINE (345–430 AD), and based on the assumption that the Creator could not make any mistake comparable to the defective work of an unskilled artisan, is quoted further below in footnote 92a on p. 241.

[65] Cf. volume 1, p. 21 of this series.

cur.[66] The human anatomist, who, if competent, has an unusual opportunity to familiarize himself with the manifold configurational and structural variations displayed by one species, namely *Homo* '*sapiens*'(?),[67] might point out, among many others, the following typical instances. In the skeletal system, for example, the sacrum can consist of four, five, or six vertebrae. The last cervical vertebra may have a costal process developed as a cervical rib. Supernumerary carpal bones, or additional sesamoid bones may be present. In the muscular system, even such important muscles as the pectoralis major can occasionally be absent, or odd muscles, such as a musculus sternalis, may be found; a two-headed muscle, e.g. biceps brachii or biceps femoris, can have three heads, and various degrees of connection between adjacent muscles as well as variations in the shape of individual muscles are encoun-

---

[66] In medical and anthropologic terminology, the word '*constitution*' refers to the anatomical (configurational) make-up or to the physiological (functional) habits, or to both, of the individual body. Constitution is presumed to be determined by the genetic, biochemical, and physiologic endowment of the body (heredity), namely the *genotype*, in combination with environmental factors. Interaction of genotype and environment results in the adult *phenotype*.

Everyday casual observation discloses the great variety of constitutional types, and several attempts at classification or standardization have been made. Thus, a *thin* (leptosome, asthenic), a *stout* (pyknic), and an *athletic* (muscular, stenic) habitus can be very roughly described as the three basic *somatotypes*. KRETSCHMER (1922, 1948) made an interesting and at least partially successful attempt to correlate 'mental' aspects, namely 'temperament' and 'character', with body build (cf. the comments in '*Brain and Consciousness*', K., 1957, p. 316 f). One of the earliest endeavors to classify human 'character' is that by THEOPHRASTUS (about 372–287 B.C.) who actually described diverse behavioral patterns and manifestations of mood rather than definite character types. The work of THEOPHRASTUS was later taken up and expanded by LA BRUYÈRE (1645–1696) during the famed or notorious so-called '*Siècle de Louis XIV*'.

A likewise interesting attempt to correlate morphologic typology (of animals as well as plants) with concepts of developmental mechanics and manifestations of growth was made by BRANDT (Grundzüge einer Konstitutionsanatomie, 1931). This author stresses three biological aspects of configuration (biologische Gestaltungsphänomene), which he designates as *Formbildung*, *Wachstum*, and *Differenzierung*.

[67] ENNIUS (239–169 B.C.) offers an amusing comment *(Saturae, fragment 23 ex libris incertis;* also quoted by CICERO, *De Natura Deorum* I, 35, 97):
'*Simia quam similis turpissuma bestia nobis!*
I would, however, suggest, as a variant, the following paraphrase:
*Homo quam similis, turpissima bestia, simiae!*
LUCIUS ANNAEUS SENECA (4 B.C.–65 A.D.) remarks in his *Epistulae morales* (CVII, 7):
'*Et fera nobis aliquo loco occurret et homo perniciosor feris omnibus*'.

tered. Internal organs (viscera) display numerous comparable ranges of variation. As regards the central nervous system, it will here be sufficient to mention the variability manifested by the pattern of gyri and sulci of the cerebral hemispheres, by the configuration and structure of septum pellucidum, by presence or absence of a diencephalic massa intermedia, moreover by cytoarchitectural and other structural or configurational details of multitudinous grisea.

Structures with a branching pattern, such as blood vessels and peripheral nerves, especially cutaneous veins and nerves, show an almost 'infinite' series of different arrangements in their ramifications and anastomoses within a given overall 'design', in the manner of 'theme and variations'. Again, the persisting array of dermal papillae in regular rows, and the corresponding pattern of epidermal ridges, which produce the well-known fingerprints used for identification, display innumerable variations whose individual manifestations are considered unique for all practical purposes. Yet, by an appropriate system of classification (e.g. subsets of loops, whorls, spirals, arches, or composite types) any one among millions of such recorded and filed patterns can be rapidly selected for appropriate comparison (cf. ASHBY's 'mega-pick', discussed in volume 1 of this series, p. 27, footnote 14).

Variations are related to *anomalies* which can be defined as *marked deviations* from an *assumed normal standard*. But, since the range included in such arbitrary standard cannot be rigorously defined, there are many deviations within the broad boundary zone between simple variations and 'true' anomalies. *Situs inversus*, a lateral transposition of the organs, resulting in a mirror image of the normal asymmetry of the viscera, can be considered a fairly rare anomaly, rather than a variation, involving the abdominal viscera, or the thoracic viscera, or both.

Anomalies, again, are related to *malformations*. In these latter, however, defective or faulty development of structures in emphasized as the predominant characterization. Quite evidently, an operationally valid but fictional concept of 'purposeful specification' is thereby introduced.[68]

It is thus generally recognized that 'continuous gradations connect the normal, slightly abnormal and severely abnormal' (AREY, 1954). Anomalies can involve observable spatial configuration, namely form and structure (in the narrower sense, i.e. texture). Other anomalies are functional, involving, e.g. 'faulty' metabolism or other 'physiological'

[68] Cf. K., Mind and Matter (1961), § 78, p.500f.

aspects.[69] Disturbances leading to anomalies may take place at any stage of ontogenetic evolution. As a rule, disturbances occurring during gestation, and therefore becoming evident at birth in man or mammals, are said to result in *congenital* defects. Subsequently originating defects are then considered '*acquired*'. It is claimed that, in man, one live birth in fifty shows an obvious external malformation, the incidence in 'spontaneously' aborted fetuses being much higher, and that some definite anomaly can be found in one out of twelve careful autopsies done on all ages.

With respect to the distribution of anomalies upon the different human organ systems, AREY (1954) quotes the following figures: central nervous system, 60 percent; stomach and intestines, 15 percent; heart and blood vessels 10 percent; bones, muscles and skin, 10 percent; all other systems, 5 percent. Thus, since, at least in man, more than half of all anomalies would accordingly involve the central nervous system, a discussion of teratologic aspects in dealing with the morphogenesis of the neuraxis seems particularly appropriate.

Some pronounced malformations, although compatible with the intrauterine life of a mammalian or human fetus, and therefore not preventing live birth, are incompatible with postnatal life, being loosely classified as *non-viable*. Because of their bizarre or grotesque appearance, the terms *monster* and *monstrosity* have been traditionally used to designate conspicuous anomalies, abnormalities, or malformations, commonly considered to be '*unnatural*' or 'contrary to the order of nature'. Hence, the designation *Teratology*, derived from the Greek (τό) τέρας (sign, portent, prodigy, monster) refers to the branch of biology (anatomy, embryology, pathology, clinical medicine) dealing with anomalies, malformations, and 'monstrosities'.

---

[69] Malformations and anomalies in the morphologic or anatomic sense include evidently only derangements of form and structures as recognizable by gross or microscopic inspection. Functional and metabolic disturbances without such perceptible configurational structural derangements would therefore not pertain to the category under consideration. Anomalies and disturbances of essentially functional type are, among others, *phenylpyruvic oligophrenia, galactosemia, aldosteronism, familial periodic paralysis* (with bouts of sometimes almost complete but transient flaccid paralysis), and at least some forms of *hyperparathyroidism* (tetany), *hyperthyroidism* and *hypothyroidism*. However, since anomalies or 'malformations' of macromolecular structures are involved, some authors, as e.g. TATUM (1961) subsume 'molecular abnormalities detected in any organism from virus to man' under the concept 'congenital malformations'.

The Romans of antiquity, who were, generally speaking, still extremely superstitious despite their cynicism as ruthless empire builders, their ability in assimilating many aspects of Greek culture, and their intellectual capability in establishing a highly sophisticated legal system, included such 'monstrosities' among their '*prodigies*' and '*portents*'. The occurrence of phenomena pertaining to these categories required atonement by intricate official acts of ritual 'purification'. The extant books of Livy's (59 B.C. to 17 A.D.) historical work '*Ab urbe condita*' contain more than 70 passages listing the occurrence of portents, among which numerous births of human or animal 'monsters' are included.[70] During the epochs of the Roman republic, hermaphroditic (gynandromorphic) anomalies, or intersexes, were considered particularly frightening. Human hermaphrodites were put to death, even if fortuitously discovered years after birth: '*Id vero haruspices ex Etruria adciti foedum ac turpe prodigium dicere: exterrem agro Romano, procul terrae contactu, alto mergendum. Vivum in arcam condidere provectumque in mare proiecerunt*' (XXVII, 37, 6); '*Iam animalium obsceni fetus pluribus locis nuntiabantur: in Sabinis incertus infans natus, masculus an femina esset, alter sedecim iam annorum item ambiguo sexu inventus; ... Foeda omnia et deformia errantisque in alienos fetus naturae visa; ante omnia abominati semimares iussique in mare extemplo deportari, sicut proxime C. Claudio M. Livio consulibus deportatus similis prodigii fetus erat,* (XXXI, 12, 6–8)'; '*ex Umbria nuntiatum est semimarem duodecim ferme annos natum inventum. Id prodigium abominantes arceri Romano agro necarique quam primum iusserunt*' (XXXIX, 22, 5).

---

[70] E.g. two-headed pigs or lambs, five-legged colts or lambs, mules or asses with three feet, chicks with three feet, children 'without eyes or noses' or without extremities. A child with 'elephant head' (XXVII, 11, 5: '*constabat*' ... '*cum elephanti capite puerum natum*') may have been a cyclops with proboscis. The otherwise unknown author Julius Obsequens, who lived in the latter Roman Empire about A.D. 350, presumably under the rule of Constantine's (324–337) sons and successors, compiled a '*Prodigiorum liber*', extracted from Livy's *History*. It contains teratologic data of the following sort: '*Teani Sidicini puer cum quattuor manibus et totidem pedibus natus. Urbe lustrata pax domi forisque fuit*'; '*Puer ex ancilla quattuor pedibus manibus oculis auribus et duplici obsceno natus*' ... '*Puer aruspicum iussu crematus cinisque eius in mare defectus*'; '*puer a parte priore alvo aperto ita ut nudum intestinum conspiceretur, idem posteriore natura solidus natus, qui voce missa expiravit*'; '*puer ex ancilla natus sine foramine naturae qua humor emittitur*'; '*mulier duplici natura inventa*'; '*puella biceps, quadrimana, gemina feminea natura mortua nata*', etc. etc.

With regard to the crudity and repulsive depravity which, regardless of highly sophisticated aspects, prevailed in Roman civilization, that subsequently became submerged in the savagery of the great predatory barbarian invasions, one might quote a remark by Saint Bernard de Clairvaux (1090–1153): '*habet mundus iste noctes suas, et non paucas*'.

In the voluminous '*Natural History*' of PLINY (23–79 A.D.), which represents an uncritical miscellany of hair-raising *Mother Goose tales* interspersed with a few actual data, book VII contains some very brief, ambiguous and partly preposterous references to monstrous births of various kinds *( prodigiosa partus )*. As regards intersexes, however, PLINY is somewhat more objective and merely states: '*Gignuntur et utriusque sexus quos hermaphroditos vocamus, olim androgynos vocatos et in prodigiis habitos, nunc vero in deliciis*' (VII, 3, 34). '*Ex feminis mutari in mares non est fabulosum*' (VII, 4, 36).[70a]

In contradistinction to these Roman attitudes, the earlier treatment of this subject matter by ARISTOTLE (384–322 B.C.), although rather brief, seems much more reasonable and sober. ARISTOTLE deals with deviations and malformations in his '*Generation of Animals*'. He states that human offspring which do not take after a human being at all in their appearance can resemble a monstrosity. For that matter, any one who does not take after his parents may really be considered a monstrosity, since, in these cases, nature has in a way strayed from the generic type. The first beginning of a deviation, according to ARISTOTLE, is already the formation of a female instead of a male, although this is a necessity of nature, because a species differentiated into male and female must be kept in being (IV, 3, p. 400 LCL).

He then goes on to say[71]: 'still it is not easy, by stating a single mode of cause, to explain the causes of everything: (1) why male and female are formed, (2) why female offspring often resembles the father, and male offspring the mother, and again (3) the resemblance borne to ancestors, and, moreover, (4) what is the cause why sometimes the offspring of a human being is in no way similar to any of those ancestors and has no human appearance, but only that of an animal, and represents a so-called monstrosity' (IV, 3, p. 416 LCL).

---

[70a] The Greek historian DIODORUS SICULUS (ca. 80–20 B.C.) reports presumably pseudo-hermaphroditic cases of apparent sex changes. Some of these persons are said to have been burnt alive by the authorities (fragment 10–12 of Book 32), but DIODORUS himself, who favors an explanation based on natural events, reproves such atrocious acts resulting from superstitious fear (δεισιδαιμονία).

[71] 'Οὐ ῥᾴδιον δὲ οὐδὲ τρόπον ἕνα τῆς αἰτίας ἀποδιδόντας τὰς αἰτίας εἰπεῖν περὶ πάντων, τοῦ τε γίνεσθαι θῆλυ καὶ ἄρρεν, καὶ διὰ τί τὸ μὲν θῆλυ τῷ πατρὶ πολλάκις ὅμοιον τὸ δ' ἄρρεν τῇ μητρί, καὶ πάλιν τῆς πρὸς τοὺς προγόνους ὁμοιότητος, ἔτι δὲ διὰ τίν' αἰτίαν ὁτὲ μὲν ἄνθρωπος μὲν τούτων δ' οὐθενὶ προσόμοιος, ὁτὲ δὲ προϊὸν οὕτως τέλος οὐδὲ ἄνθρωπος ἀλλὰ ζῷόν τι μόνον φαίνεται τὸ γιγνόμενον, ἃ δὴ καὶ λέγεται τέρατα.'

'Other sorts of monstrosity are characterized by additional parts to their configuration, having additional feet or heads'. 'The explanation of monstrosities is closely similar to that of deformed animals, because a monstrosity represents a kind of deformation' (IV, 3, p. 418 LCL).

Again, monstrosities, which are subsumed under the class of offspring which is unlike its parents, are called *metachoira* (μετάχοιρα), namely strange (τερατῶδες) beings which are affected by lack of a part or presence of an additional part. A monstrosity belongs accordingly to the class of things contrary to nature, although not contrary to nature in her entirety, but only to nature in the generality of cases. Thus, in the phenomena under consideration, the occurrence is contrary to a particular orderliness, but by no means random. It appears therefore less a monstrosity if it is realized that what seems contrary to nature is nevertheless somehow in accordance with nature (as, e.g., when the formal nature fails to control the material nature: IV, 4, pp. 424–426 LCL).

As an aside before reverting to contemporary viewpoints on teratology the amusing literary treatment of the subject by the once much-read amiable German poet JEAN PAUL (JOHANN PAUL FRIEDRICH RICHTER, 1768–1825) might be mentioned. His short novel '*Dr. Katzenbergers Badereise*' depicts the extravagant antics of an eccentric physician, author of the fictional treatise '*De monstris epistula*', who is obsessed with the collecting of teratologic specimens.

In a convenient classification of the phenomena pertaining to teratology, as for instance used by AREY (1954) and here adopted with several modifications and additions, the following categories can be distinguished.

(1) *Developmental failure* involving essentially *qualitative morphogenetic processes (Gestaltungsvorgänge)*. The primordium fails to appear, or does not develop to a significant degree as a whole or in part. *Agenesis* designates complete or substantial failure, resulting in absence of a component (e.g. amelia). *Developmental arrest* includes a gamut of instances where progressive development has remained incomplete.

(2) Partly *abnormal numerical or formal differentiation* of configurated components *within a still essentially normal overall pattern*, such as the *Klippel-Feil* syndrome which involves reduction in number of cervical vertebrae as well as additional anomalies. This category (2) differs from the extremely atypical differentiation subsumed below under (16), and would also include cases of hermaphrodism or pseudohermaphrodism.

(3) *Arrest of normal growth*, involving essentially *quantitative aspects*, i.e. *Wachstumsvorgänge* (e.g. microcephaly).

(4) *Excess of normal growth* (e.g. macrocephaly), essentially the opposite of category 3. The subsequent categories (5) to (18) concern again predominantly *Gestaltungsvorgänge*.

(5) *Splitting, duplication* or *increased number of primordia* (e.g. double ureters or renal pelves, polydactyly, true twinning).

(6) *Failure of primordia to subdivide* (e.g. syndactyly).

(7) *Fusion of primordia* (e.g. horseshoe-kidney).

(8) *Failure of antimeric primordia to fuse in the midline (dysraphic disturbances)*.

(9) *Failure of lobated components to consolidate* (e.g. lobated spleen).

(10) *Failure to atrophy, to obliterate or to disintegrate in the course of normal development*, such that transitory features remain (e.g. double vena cava, cervical fistula, ductus thyreoglossus, anal membrane). It should be recalled that some amount of *cell death* concomitant with the progress of ontogenesis is a significant aspect of *normal morphogenesis*, which latter is characterized by complex interactions of the differentiating tissues.

(11) *Destruction or abnormal disintegration of developing or partly developed structures* caused by nongenetic factors such as circulatory disturbances, infections or other extraneous damage during the course of ontogenesis (e.g. some sorts of porencephaly, of hydranencephaly, and most cases of infantile cerebral palsy).

(12) *Abnormal migration* of cell populations or whole organs (e.g. heterotopias).

(13) *Abnormal placement* (e.g. situs inversus, palatine teeth).

(14) *Abnormal secondary displacements and distortions* (e.g. *Arnold Chiari malformation*, i.e. prolongation of cerebellum caudad into vertebral canal with concomitant displacements of pons and oblongata. Hydrocephalus, in many instances, could likewise be subsumed under this category).

(15) *Abnormalities of shape* involving distorted, but quantitatively (3), (4), essentially 'normal' growth processes (e.g. oxyencephaly, scaphoencephaly, associated with deformation of skull).

(16) *Extremely abnormal or atypical differentiation*. Developmental processes and their results differ here substantially from normal patterns of organization, displaying a random appeearance (e.g. congenital tumors of dysplastic type, teratomas).

(17) '*Atavism*'. Essentially human anomalies or variations which

are interpreted as representing 'ancestral recurrences' such as azygos lobe of lung, normally present in mammalian quadrupeds, or elevator muscle of clavicle, present in climbing primates. The occurrence of a *sulcus lunatus* in the occipital lobe of the cerebral hemisphere might likewise be interpreted in this fashion. However, some malformations due to developmental failures (1), perhaps also involving 'arrest of growth' (2), may lead to configurational patterns characteristic for taxonomically 'lower' vertebrates by interfering with morphogenetic processes common to all vertebrates, and cannot be properly said to represent a true 'atavism' in the accepted sense. Thus, the anomalous development of the human telencephalon occurring in some cases of arhinencephaly and hydranencephaly (cf. further below) may lead to the formation of an everted univentricular telencephalon impar typical for most teleosts (cf. K. and GLOBUS, 1936), but can hardly be interpreted as an *'ancestral occurrence'*. It may be recalled that the telencephalon in presumably all vertebrates originates ontogenetically from a stage of unpaired evagination discussed in the preceding subsection (cf. above pp. 143–158). The designation *'atavism'* becomes here, as well as in other similar instances, quite evidently meaningless, and this term should, therefore, be applied with considerable caution and restraint.

(18) Predominantly *functional, 'idiopathic'* or *biochemical anomalies with morphologic or structural manifestations* (e.g. albinism, amaurotic familial idiocy with several forms,[72] so-called mongolian idiocy, gargoylism or

---

[72] *Tay-Sachs disease* (cerebro-macular degeneration), believed to be a hereditary disorder transmitted as an autosomal recessive trait, is characterized by accumulation of a lipid (ganglioside) in nerve cells of neuraxis and retina. Degeneration of fiber tracts is likewise displayed. The disease seems to be caused by the absence of a specific enzyme, hexosaminidase A, in neuraxis, liver, and other body organs (cf. Science *165:* 698, 1969). As a condition of lipid metabolism, this defect has been grouped with other conditions of that general type (lipidoses), namely *Niemann-Pick disease* (a familial disorder of infantile phospatide metabolism, with anemia, lymphocytosis, enlarged liver and spleen), *Gaucher's disease* (familial splenic anemia), and *Hand-Schüller-Christian disease* (chronic idiopathic xanthomatosis with exophthalmos, diabetes insipidus and defects in the so-called membrane bones). In *Niemann-Pick disease*, cortical nerve cells appear ballooned; histiocytes and reticular elements in all organs are severely involved, the affected cells contain sphingomyelin. In *Gaucher's disease*, the opaque, homogeneous, and often multinucleated cells contain the cerebroside kerasin. In *Hand-Schüller-Christian disease*, whose familial nature is suspected, but as yet not definitely established, lipids in the abnormal cells consist predominantly of cholesterol. Some authors (e.g. MORIARTY and KLINGMAN, 1962) suggest that the *Laurence-Moon-Biedl syndrome* (obesity, hypogenitalism, retinitis pigmentosa, mental deficiency, skull defects, occasionally syndactylism; hereditary nature not fully

*Hurler's disease,*[73] tuberous sclerosis or *Bourneville's disease*).[74] Considering ontogenesis to extend beyond birth, conditions of the following types could be included: hepato-lenticular degeneration, *Huntington's chorea*, some instances of paralysis agitans, myotonia congenita or *Thomsen's disease*, various muscular atrophies and dystrophies, progressive spastic paralyses, amyotrophic lateral sclerosis, *Friedreich's* and similar *hereditary ataxias*, familial myoclonic epilepsy, as well as neurofibromatosis *(Recklinghausen's disease)*. With respect to the correlated neuropathological changes, it may here be sufficient to mention the following. Hyaline, often concentric *Lewy bodies* can be found in the cytoplasm of nerve cells of substantia nigra and other grisea in some cases of paralysis agitans. In familial myoclonic epilepsy, intracytoplasmic '*Lafora bodies*', containing a glycogen-like carbohydrate, are commonly present in *Purkinje cells* and some cells of nucleus dentatus.

(19) Predominantly *functional* or *biochemical anomalies without, or at least with microscopically almost imperceptible respectively negligible structural manifestations* (e.g. alkaptonuria, phenylpyruvic oligophrenia, familial periodic paralysis, myasthenia gravis, congenital insensitivity to pain,[75] and various sorts of color blindness).

As regards the just-mentioned, predominantly functional anomalies subsumed under (18) and (19), constantly increasing lists of such hereditary disturbances become available in connection with a more

---

established), and the *Hallervorden-Spatz syndrome* (to be dealt with further below) are related to, or represent fragmentary forms of, cerebro-macular degeneration. One type of cerebro-macular degeneration (early infantile form), apparently occurring only in Jewish populations, is believed to imply a racial factor.

[73] *Hurler's disease* or syndrome (gargoylism) is a genetically transmitted disorder of mucopolysaccharide metabolism characterized by excessive excretion as well as visceral storage of chondroitin sulfate B and heparitin sulfate, and by excessive storage of gangliosides in the central nervous system. It seems to be caused by a marked deficiency of a specific β-galactosidase isoenzyme (cf. Science *165:* 611, 1969). Development of bone and cartilage is defective, and the head, in particular, assumes a grotesque appearance *(inde nomen)*. The dwarfism is severe, and mental deficiency is usually manifested. The neural lesions include accumulations of ganglioside in swollen, ballooned nerve cells.

[74] Cf. the preceding volume 3, part I, p.781 of this series. Together with neurofibromatosis, cerebelloretinal angiomatosis *(Hippel-Lindau disease)*, and *Sturge-Weber's disease* (cf. footnote 80), tuberous sclerosis has been classified as one of the (hereditary) phakomatoses, characterized by the formation of 'lens-like' tumor masses (φαχός, lens).

[75] Cf. K., Brain and Consciousness (1959, p.183f).

detailed knowledge of metabolism[75a] (cf. STANBURY *et al.*, 1966; HAND-LER, 1970). In general, these anomalies appear to be effects of either the absence of a specific protein (usually an enzyme), or of the presence of a 'defective', nonfunctioning enzyme.

Thus the affected enzymes are tyrosinase in albinism, phenylalanine hydroxylase in phenylketonuria, and ceruloplasmin in hepato-lenticular degeneration *(Wilson's disease)*. Ceruloplasmin, e.g., is a blue, copper-containing alpha globulin of blood plasma, which catalyzes the oxidation of amines, phenols, and ascorbic acid. The amount of ceruloplasmin is substantially reduced in *Wilson's disease*. The hepatic changes are here those of a portal cirrhosis. The brain appears atrophic, particularly in the frontal and parietal telencephalic regions. The corpus striatum and to a lesser extent globus pallidus show 'progressive' neuronal degeneration, which may also occur in diencephalic, mesencephalic, and cerebellar grisea. Degenerative changes of the blood vessels can likewise be detected, and the *Virchow-Robin spaces* often appear distended (cf. HERZ and MEYERS, 1962). In general terms, *Wilson's disease* can be said to be characterized by accumulation of copper in the brain because a normal, copper-carrying protein of the plasma is wanting.

Again, in various hereditary disturbances of glycogen metabolism, designated by terms such as *Andersen's disease, Forbes disease, Gierke's disease, Hers' disease, McArdle-Schmid-Pearson disease*, and *Pompe's disease*, the particular enzymatic defects have been identified with substantial certainty.

Generally speaking, the normal complement of all enzymes is said to be far in excess of metabolic requirements, and the presence of only half that complement in the heterozygous state may remain undetected, since it suffices for 'normal life'.

In accordance with the view that a single 'gene' controls the synthesis of a single enzyme, and that anomalies of enzyme production are the effect of 'faulty', 'aberrant' or 'modified' genes, these latter are

---

[75a] A recent publication (HANDLER, 1970) refers to more than 400 autosomal dominant and at least 175 autosomal recessive heritable human disorders, but that author's remark represents an evident understatement. Although, fortunately, many genetically determined human diseases are relatively rare, more than 1500 such conditions have already been recognized, and new conditions of this type are being reported every year (cf. McKUSICK, 1971). Some seemingly theoretically possible, but still rather fanciful and unconvincing proposals for genetic manipulations in humans and for 'gene therapy' of genetic disease are not infrequently voiced (cf. FRIEDMAN and ROBLIN, 1972).

commonly considered to be 'mutants'. Because of the multitudinous aberrations of human metabolism, it is therefore assumed that mutation occurs continuously in man. Concerning the evaluation of mutations, reference may here be made to the comments included in volume 1, pp. 119, and 131–137 of this series.

Since, except in systems of pure logic, where 'absolutely' valid rigorously consistent formal rules may be postulated, most, if not all definitions are 'false', or at least highly fictional, and arbitrarily inconsistent, it is evident that no system of teratologic classification can be rigorous. Such systems depend, of course, on particular adopted viewpoints such, e.g., as a *morphologic*, a *functional*, or a *causal* one.[75b] Thus, in clinical neurology, a simplified useful system of the following type may be formulated (cf. also MORIARTY and KLINGMAN, 1962).

1. Anomalies due to *duplications*.

2. Defects due to *improper midline closure of neural tube* or associated structures, or both.

3. Defects due to *inhibited development (agenesis* or *hypoplasia)* of the nervous system.

4. *Hereditary defects* that become *manifest later in life (abiotrophies)*.

5. *Faulty development* originating *early in morphogenesis*.

6. *Faulty development* originating *in advanced stages of morphogenesis*.

7. *Hereditary metabolic defects*.

8. *Defects affecting the nervous system secondarily*, including congenital defects of skeletal and other systems.

In a publication dealing with what its author considers to be the 'borderland of embryology and pathology', WILLIS (1958) devotes several chapters to malformations and related anomalies, which he classifies in accordance with the following simplified general grouping.

(1) Malformation of twins, triplets, etc.

(2) Gross malformations of the neural tube and axial skeleton.

(3) Gross malformations of the head end.

(4) Gross malformations of the hind end.

(5) Gross defects of the ventral body walls.

(6) Malformations restricted to particular organs or parts.

(7) Hamartomas and hamartomatous system disorders.

(8) Generalized anomalies of skeletogenesis.

---

[75b] Still another traditional viewpoint, already adopted in some of ARISTOTLE's comments (cf. above p. 217), stresses 'lack' or 'defect' in contradistinction to excess: *monstra in defectu* and *monstra in excessu*.

From the viewpoint of the neuropathologist, OSTERTAG (1956), in his important handbook-chapter on malformations of the nervous system, presents still another suitable overall classification. This author distinguishes:

1. Malformations of the *primitive anlage (including lack or incomplete closure or neural tube and axial skeleton )*.

2. *Spurious aplasias* and developmental disturbances of anatomically demonstrable *exogenous nature*.

3. Malformations caused by *fetal lesions*.

4. *Meningeal disturbances*.

5. *Dysraphic disturbances*[76] and *syringomyelia*.

6. *Systemic malformations*.

In dealing with developmental aberrations or teratologic manifestations, it is obviously necessary, as various authors have pointed out, to distinguish between the *morphogenetic events* leading to faulty end-results, namely the *'formal mechanism'*, and the *causal factors* determining the abnormal morphogenetic events. In the latter case, functional operators are considered the independent variables, of which the formative processes become then the dependent ones.

A 'causal' understanding of malformations is beset with extreme difficulties. In the first place, as pointed out in the subsection on general morphogenesis, no satisfactory 'explanation' of normal ontogenesis, based on rigorous mathematical formulations, can be given at the present time. In the second place, it seems well established that a single, specific malformation may be caused by various entirely different agents and even by different sorts of developmental mechanisms. Thus, as in a many-one transformation, the observed transform (malformation) does not indicate a single operand. Again, one and the same kind of disturbing factor may, depending on different parameters, produce entirely different types of anomalous development. In other words, the 'cause' could here be an operand of a one-many transforma-

---

[76] The term *dysraphism* in the wider sense, as e.g. used by REFSUM (1962) embraces such widely differing conditions as spina bifida, anomalies of the sternum, finger as well as foot deformities, and many others. OSTERTAG (1956), however, in agreement with other authors who include syringomyelia as one manifestation of the dysraphic state, distinguishes *arrhaphic disturbances* (lack of closure), *hyporhaphic disturbances* (failure of secondary closure), and *dysraphic disturbances sensu strictiori* (anomalous closure, *'fehlerhafter Schluss des dorsalen Neuralrohrs')*. The arrhaphic and hyporhaphic disturbances are thus classified under *'Verbildungen der Primitivanlage'*.

tion, where the operand, depending on additional parameters, may go to any one of several possible states.

An important *parameter* is here the *flow of time*. Each developing configuration (whole body, organ, and even organ parts) seems to pass 'through an individual *critical period (or periods)* during which it undergoes accelerated growth and differentiation, and manifests a marked susceptibility to injurious influences brought to bear on it. Other parts may or may not be so scheduled as to be sensitive at that particular moment, to those influences; a part that is sensitive at one particular time is immune to the same influence at other periods of development. The action of such differential susceptibility can be tested in lower forms by altering the external environment. Influences that retard or halt development at a critical period are highly effective in bringing about disturbed, anomalous development' (AREY, 1954).

OSTERTAG (1956) has given considerable attention to the critical periods in the ontogenesis of the human neuraxis. Correlating these periods with the onset of the 'formal mechanisms' leading to malformations, the cited author has provided an extensive, well documented and illustrated tabulation (loc. cit. pp. 370–381) identifying the probable *critical periods* (or periods of teratogenetic determination) for origin of malformations in accordance with his classification, and ranging from the neural plate stage to the tenth lunar month of gestation. In a simplified manner, some pathologists and clinicians distinguish in man or comparable mammals, *qua* critical periods, between *embryonic* and *fetal* pathology. The first stage of ontogenesis, in which the germ layers and primordial organs are formed, followed by the beginning of definite organ differentiation up to the second month in human gestation is said to represent the embryonic period. After the second month in human ontogenesis the developing organism is conventionally and somewhat arbitrarily designated as *fetus*.

From the viewpoint of experimental embryology or 'developmental mechanics', which purports to analyse causal factors, ZWILLING (1955) has elaborated the following teratologic classification:

1. Abnormal initial stimulus.
   a) Initial stimulus absent.
   b) Deficient initial stimulus.
   c) Excessive initial stimulus.
2. Abnormal response of reacting tissues.
   a) Absence of response.
   b) Partial or incomplete response.

    c) Excessive response.

    d) Mechanical interference with response.

3. Abnormality of both initial stimulus and responding tissue.

4. Abnormal differentiation of component tissues.

5. Abnormal growth of structures.

6. Degenerative processes.

    a) Abnormal degeneration.

    b) Excessive 'normal' degeneration.

    c) Failure of 'normal' degeneration to occur.

7. Abnormality of functional activity or regulatory mechanisms.

An amusing causal 'explanation', based on older views of DARESTE (1891), has been propounded by STOCKARD (1920–1921). This author states: 'For the past ten years I have claimed that all types of monsters not of hereditary origin are to be interpreted simply as developmental arrests'. Now, it can hardly be claimed that spontaneous or experimental twinning, particularly studied by STOCKARD, whose actual merits in the investigation of these and related topics are quite undeniable, represent what could be reasonaby called 'developmental arrest'. It should be evident to those readers familiar with logical and semantic problems, that the terms 'developmental arrest' (STOCKARD), 'initial stimulus', 'abnormal response' etc. (ZWILLING), become rather meaningless and highly ambiguous if critically examined. In the aspect under consideration, simplified terms of this sort fail to embody the complex multifactorial aspect of morphogenesis vividly illustrated by the diagram of WEISS (1955) referred to and shown on p. 57-58 (Fig. 7 A).

With respect to a somewhat more relevant *classification of causal events* in the occurrence of malformations, *endogenous* and *exogenous* factors may be roughly distinguished. The former are considered intrinsic to the embryo or fetus, and the latter as acting from the 'outside'. Yet this distinction also easily leads into semantic traps, particularly since, in mammals and man, or even in other viviparous animals, the relationship of maternal organism to the embryonic or 'fetal' one may here remain ambiguous. Strictly speaking, only anomalies of the embryonic genome might be termed endogenous. These anomalies, also called *genotypic* and interpreted as mutations, are hereditary. The sorts of nonhereditary, *phenotypic*[77] anomalies, which may rather closely imi-

---

[77] The term *genotype* subsumes the genetic constitution, presumably 'fixed' in the zygote at the time of fertilization, discounting possible subsequent mutational processes. The term *phenotype* subsumes the general 'outward' appearance of the ontogenetically

tate those presumably caused by mutants are then considered exogenous and were designated by R. GOLDSCHMIDT as *phenocopies*.

*Sensu strictiori endogenous* causative teratologic factors involving the genome are thus represented by mutations, which, again, can be classified as gene mutations and chromosomal mutations.[78] If the 'abnormal' gene under consideration is present at the corresponding locus in both chromosomes of a pair, the *homozygous* state obtains. If a 'normal' gene is matched with an 'abnormal' one, the *heterozygous* state results. In this latter case, either the normal or the abnormal gene may be '*dominant*', the corresponding partner being then '*recessive*'. Some genes, however, are incompletely dominant, and the individual displays an intermediate character. Again, genes may be *sex-linked*, being located in a sex chromosome.[79] Genes not influenced by ordinary environmental factors are called highly '*penetrant*'. A gradual transition to genes with low penetrance is known to exist, such that variations in the environment or in the combinations with other genes may alter the presumed course of development initiated by the gene.

Abnormal conditions recorded as presumably *autosomal dominant* are, e.g., achondroplastic dwarfism, peroneal muscular atrophy, fused fingers, instances of congenital cataract, migraine, types of progressive muscular dystrophy, myotonia congenita *(Thomsen's disease)*, dystrophia myotonica, facio-scapulo-humeral muscular dystrophy, *Hunting-*

---

evolved organism. The phenotype is presumed to result from an interaction of genotype (genes) and environment. Cf. also the comments on the genetic aspect of evolution in vol. 1, p. 115f. of this series.

[78] Cf. volume 1, p. 119 of this series. The term 'spontaneous mutations' refers to 'undetermined origin' of mutations.

[79] In man, 22 pairs of *homologous chromosomes (autosomes)*, and one pair (X, Y) of *sex chromosomes* are present, the diploid complement being thus normally $44 + XY$ in males, and $44 + XX$ in females. Sex-linked conditions are, e.g., hemophilia and color blindness, and *Fabry's disease* (an inherited metabolic deficiency disease involving, *inter alia*, accumulation of a lipid in various tissues and organs). Because of difficulties in the identification of several chromosome pairs displaying very similar configurations, some of the autosomes are classified as pertaining to so-called groups (cf. the groups mentioned below in footnote 80).

Concerning the rapidly increasing new data on chromosome arrangements as well as on other aspects of genetics in general, relevant information may be found in the notable comprehensive 'Textbook of Human Genetics' (1971) by my former associate at Woman's Medical College of Pennsylvania, Prof. MAX LEVITAN, New York, in collaboration with Prof. ASHLEY MONTAGU.

*ton's chorea*, tuberous sclerosis, and perhaps *Sturge-Weber's disease*.[79a] Among conditions classifiable as *recessive*, albinism, color blindness, hemophilia, spinal muscular atrophy of infancy, hereditary ataxias, hepato-lenticular degeneration, and gargoylism *(Hurler's Disease)* could be mentioned. Some of these disorders are *sex-linked recessive* (or x-linked), occurring in males, but carried by apparently unaffected females (red-green color blindness, hemophilia and muscular dystrophy of the *Duchenne type*).

Endogenous causative factors involving the genome but differing in some respects from 'typical gene mutations' may be related to alterations in number and arrangement of chromosomes. Such chromosomal alterations have been observed in mongolism,[80] gonadal dysgenesis *(Turner's syndrome)* and *Klinefelter syndrome* (male hypogonadism with gynecomastia and eunuchoidism).

Essentially *exogenous causative factors* affecting the phenotype can be classified as (1) mechanical; (2) radiational; (3) chemical; (4) deficiencies, and (5) diseases, such as luetic or rubella infections of the maternal organism, transmitted to the embryo or fetus during its development. It is evident that, *qua* causal categories, (2), (3), (4), and (5), display logical or semantic overlap, and might accordingly, at least to a considerable extent, also be subsumed under the generalized concept *'disturbances of metabolism'*.

Again, exogenous congenital anomalies may resemble those that are genetically determined. As regards the well founded view that dif-

---

[79a] Atrophy and calcification of cerebral cortex (usually in one hemisphere) associated with angiomatous vessels and ipsilateral facial naevus. Hemiparesis, hemianopsia, mental deficiency and additional symptoms are commonly present.

[80] In mongolism *(Down's disease)* the chromosome set includes an autosome triplet in the chromosome group 21–22 instead of a pair, and thus consists of 47 instead of 46 chromosomes (cf. JACOBS *et al.*: The somatic chromosomes in mongolism. The Lancet *I:* 710, 1959). In addition, a similar condition seems to involve such single trisomic state in the group 17–18. Moreover, in trisomy 18, conditions such as agenesis of corpus callosum, abnormal inferior olivary nucleus, cerebellar or general cerebral hypoplasia, and other cerebral abnormalities have been recorded (SUMI, 1970). Individuals afflicted with mongolism show, as a rule, symptoms of premature aging. Some investigators suspect that macromolecular cellular events occurring in mongolism might also be related to those which are significant for normal aging. Concerning relevant problems of aging cf. pp. 712–737 of the preceding volume (3, part I, chapter VI, section 9 of this series). A trisomy within the group 13–15 is also known to occur in a type of cyclencephaly involving 'arhinencephaly' and hypoplasia of optic nerves and eyes.

ferent embryonic cell aggregates or tissues have characteristic sensitive developmental periods during which a variety of extrinsic as well as genetic factors may lead to similar congenital defect, the summarizing publication edited by HANDLER (1970) adds the following comment: 'more refined analysis showed some specificity in the response both to different external agents and to different genetic conditions'.

(1) Numerous sorts of anomalies have been recorded in studies of 'developmental mechanics', particularly in lower vertebrates as well as invertebrates, following experimental *mechanical interference* with development by means of excision, destructive electrolysis, centrifugation and related procedures. As regards human anomalies, fibrous amniotic bands were believed to cause intrauterine amputations and other lesions of the fetus, but although such amniotic adhesions may indeed be recorded, contemporary authors doubt their significance for the origin of malformations.[80a]

(2) Depending on its intensity, *ionizing radiation* may have grossly destructive effects on living tissues, leading to severe necrosis, or may cause lasting changes which are not immediately detectable. Subsequent irradiations can then engender *cumulative effects*. It is now well-known that ionizing radiation, such e.g. as x-rays, can induce gene mutations.[80b] Moreover, many experimental studies with irradiation of embryos at definite times have produced malformations in animals, and the susceptibility of human embryos and fetuses to radiation is likewise well established. In chapter V, p. 65 of the preceding volume of this series, the peculiar type of cerebral malformation obtained by RIGGS *et al.* (1956) by prenatal irradiation of rats was briefly pointed out in a discussion of the subependymal cell plate. Additional data on teratologic effects of ionizing radiation are reviewed in the compilation by KALTER (1968).

---

[80a] OSTERTAG (1956), however, on the basis of reasonably well established evidence, assumes the (very rare) occurrence of true amniogenic malformations, but adds, with regard to the more commonly noted amniotic adhesions or bands: '*ihrer Sinnfälligkeit wegen sind diese lange ursächlich überschätzt worden*'.

[80b] On the other hand, some investigators postulate the existence of an 'enzyme mechanism' for the 'repair' of genetic DNA material altered or 'damaged' by radiation or chemicals. Such repair, which might, at least in some instances, counteract radiation and similar effects, is presumed to be accomplished by means of an '*endonuclease*' initiating the rebuilding of the original base sequences in the DNA strand through the action of another enzyme, namely *DNA polymerase*.

(3) Experimental studies have shown that numerous *anorganic and organic substances* can produce anomalous development in lower animals. As regards mammals, injection of *trypan blue*[81] into pregnant rats induced a whole gamut of congenital malformations in the offspring, involving central nervous system, eye, skeletal system, anus, as well as cardiovascular and urogenital systems. In the central nervous system, hydrocephalus, spina bifida, exencephaly, and a condition somewhat similar to *Arnold-Chiari malformation* were recorded. Trypan blue induced developmental defects could also be obtained in mice; rabbits seem likewise to be susceptible.

In offspring of rats, mice, and rabbits, trypan blue does not seem to induce congenital malformations when given to the mother later than about the 9th day of pregnancy, if injections are administered during pregnancy only. The dye thus acts here only during a limited period of ontogenesis, becoming effective between the 8th and 9th day of gestation.

It is reported that trypan blue, injected into pregnant rats and mice, does not appear in the yolk sac fluid of embryos, but does collect in the yolk sac epithelium, and continues to be deposited after the 9th day. The other embryonic tissues remain unstained. In rabbits, trypan blue is found in the yolk sac fluid if injections are made up to the 9th day of pregnancy, but apparently not in the yolk sac epithelium. Injected after the 16th day, however, trypan blue is said to be taken up by the yolk sac epithelium.

This poses the questions why, after the 9th day, trypan blue is not teratogenic in rats and why, on the 9th day, a change of permeability to trypan blue is displayed by the rabbit. There are thus *variations in permeability* related to species difference as well as to the passage of time, and involving problems in some respects comparable to those presented by the behavior of the hemato-encephalic barriers.

The mechanism of the teratogenic action of trypan blue has been much discussed since the original experiments reported in 1948 by GILLMAN *et al.*, but still remains very poorly understood. A detailed discussion of the points at issue is contained in KALTER's review (1968). Three main hypotheses have been formulated: (1) the effects on

---

[81] The use of the acid vital dye *trypan blue*, its toxic effect on the brain parenchyma, and various aspects of its penetration into the neuraxis were discussed in the preceding volume (3, part I, p. 341f) with regard to studies concerning the *hemato-encephalic barriers*.

the embryo are presumed to be the indirect results of disturbances caused by the dye in the maternal organism; (2) dye molecules are presumed to enter the placenta and to interfere with transport of substances necessary for embryonic nutrition, and (3) it is presumed that the dye actually reaches the embryo itself where it interferes directly with developmental processes. The at present available data, which are not sufficiently complete or unequivocal, can, but not must, be interpreted as perhaps favoring the third hypothesis.

Among other chemical compounds whose teratogenic effect upon some mammals has been recorded within the last 20 to 25 years, *alkylating agents*[82] such as *nitrogen mustards* might next be mentioned. Substances of this group were introduced for purposes of gas warfare during World War I. More recently, alkylating agents of this type have been used as antineoplastic drugs. These substances are highly reactive chemicals combining with cellular components and interacting with DNA. It is believed that they thereby prevent the production of new cells. Because of a certain similarity (but by no means identity) with radiation effects, such agents have been loosely called '*radiomimetic*'.[83] Besides nitrogen mustards, related alkylating compounds such as uracil mustards, triethylene melamine, triethylene thiophosphoramide, cyclophosphamide and others are known to be teratogenic for at least some mammals.

The alkaloid *colchicine* (present in *Colchicum autumnale*) can produce congenital malformations in rats, mice, rabbits, and hamsters. It is well-known that this substance, included in the group of '*cytotoxic drugs*', generally arrests mitosis in metaphase; this is concomitant with failure of spindle formation. Bizarre and abnormal nuclear configurations, including polyploidy, may result, and the cells often die.

Another instance exemplifying the toxic action of drugs was exhibited by the disastrous experiences with *thalidomide* (α-phthalimido-glutarimide), a sedative and hypnotic introduced some years ago, and used as medication for pregnant women. It was subsequently found to

---

[82] An *alkyl* is the radical which results when an aliphatic hydrocarbon looses one hydrogen atom. The verb 'alkylate' refers then to the substitution of an aliphatic hydrocarbon radical for a hydrogen atom in a cyclic compound, e.g. introducing a side chain into an aromatic compound.

[83] A very general sort of relationship obtains between the effects of *alkylating agents* (e.g. nitrogen mustards), *colchicine* (briefly referred to further below), *urethan*, *radiophosphorus*, *x-rays* and other radiations, as well as additional agents.

cause severe malformations of the limbs of their offspring. Following the scandal resulting from this tragic catastrophe, thalidomide has been much used, since about 1962, in experimental work, particularly on rabbits, which proved the most susceptible, while mice and rats were affected to a lesser degree. Malformations were also obtained in various primates[84] (several macacus species and unspecified baboons). In some of the experiments with rhesus monkeys, live births were prevented by thalidomide administration.

The malformations obtained in rabbits and other experimental animals included numerous defects of the central nervous system, such as spina bifida, exencephaly and encephalocele, porencephaly, as well as hydrocephalus.[85] In male rabbits, fed with thalidomide, some teratogenic effects on the offspring were also reported.

Various uncertainties remain with respect to the mode of entry of thalidomide into the embryo and to the details of its teratogenic mechanism. The substance is spontaneously hydrolyzed *in vivo* and forms a number of metabolites, believed to be essentially nonteratogenic. Be that as it may, thalidomide is known to enter the embryo, persisting for some time while being converted into its accumulating hydrolysis products, which seem to be retained in the cells. Thalidomide itself is, at any rate, the carrier of the teratogenic material into the embryo (WILLIAMS *et al.*, 1965, as quoted by KALTER, 1968).

Other miscellaneous substances suspected, on the basis of some experimental evidence, to have teratogenic effects in some mammals, are *urethan* (ethyl carbamate) in mice, *hydroxyurea* in rats, *veratrum* (hellebore, containing various alkaloids) in sheep, *Rauwolfia alkaloids* in rats (skeletal defects), *serotonin* (in mice), and *alloxan*[86] (in mice). In addition, diverse purine and pyrimidine antimetabolites, glutamine antime-

---

[84] Spontaneous and experimentally induced congenital malformations in nonhuman primates, including those caused by thalidomide, were reviewed by WILSON and GAVAN (1967). This paper includes a report on original studies with thalidomide in *Macaca mulatta* by the authors of the cited survey.

[85] Although there seems to be little doubt about the teratologic effect of thalidomide in these various experiments, hydrocephalus as well as other defects of the neuraxis may also occur spontaneously in the animals under investigation. Several authors reported some neuraxis defects in their control series (cf. the review by KALTER, 1968).

[86] *Alloxan* (mesoxalyl urea) is a substance which causes diabetes mellitus by rather selectively destroying the insulin-producing β-cells of the pancreas. Alloxan has been widely used to bring about diabetes in experimental animals. HORII (1964) showed that occurrence of alloxan induced teratologic effects could be prevented by insulin injection.

tabolites, steroid substances (e.g. hydrocortisone), and some steroid hormones might be mentioned in this context.

As regard therapeutic agents widely used in medical practice, teratogenic effects of insulin on mice and rabbits have been reported after injections into pregnant animals. Some antihistamines (e.g. meclizine pyrilamine) were likewise shown to manifest such effects, but it is believed that in these teratogenic antihistamines the antihistaminic properties as such are not the significant factors. Even the widely used salicylates were shown to have teratogenic effects on the offspring of salicylate-treated female rats and mice (skeletal malformations and neuraxis defects). There is, however, no relevant evidence that the widespread use of salicylates has caused congenital malformations in man. Antibiotics such as tetracycline, penicillin in combination with streptomycin, and actinomycin D (Dactinomycin) are likewise reported to produce defects in rat offspring.[87]

(4) *Deficiencies (essentially nutritional)*. Numerous experiments, undertaken since about 1933, have shown the teratogenic effect of maternal *vitamin deficiencies* in various mammals (HALE, 1933, WARKANY and NELSON, 1940). Vitamin A deficiency seems to induce effects on eyes and various soft parts of the body. Lack of folic acid seems to affect mostly the skeleton. The teratogenic effect of various folic acid antagonists such, e.g., as aminopterin may be mentioned in this connection. Riboflavin deficiency appears to induce hydrocephalus and occasionally exencephaly in addition to malformations in other systems. Combined vitamin $B_{12}$ (cyanocobolamin)[88] and folic acid deficiency may affect the central nervous system. Pantothenic acid deficiency, investigated by GIROUD et al. (1957) induces considerable developmental defects of the neuraxis. Thus, the telencephalon was shown to remain at the stage of a telencephalon impar with common ventricle. The skeletal system in particularly affected by lack of vitamin D.

Hormonal deficiencies are known to cause dwarfism (hypophysis), cretinism (thyroid), and sexual infantilism (gonads). As regards miner-

---

[87] Adverse allergic and other reactions to tetracycline, penicillin, streptomycin, and actinomycin D, are, of course, well-known to clinicians. Actinomycin D is used as an antineoplastic agent (cf. above the comments on nitrogen mustards and alkylating agents). Again, some plant alkaloids such as vincristine and vinblastine (derived from the periwinkle plant, *Vinca rosea*) are also considered to have antineoplastic properties. These substances, again, were shown to be teratogenic in hamsters, rats, rabbits, and monkeys.

[88] Cyanocobalamin is the antianemia factor of liver extract. It contains cobalt and a cyano (CN) group.

al deficiencies, lack of manganese is believed to induce congenital ataxia. Another type of ataxia ('swayback' of sheep) is apparently caused by
copper deficiency (cf. INNES and SAUNDERS, 1962), which also induces
brain defects in guinea pigs, but does not seem to affect rats. These latter animals, however, appear susceptible to zinc[88a] deficiency.

Severe fasting imposed during gestation in mice (but not in rats)
was also reported to produce anomalies in various systems (cf. KAL-
TER, 1968). Hypoxia of pregnant animals (e.g. exposure to greatly reduced atmospheric pressure) is likewise known to cause various congenital malformations, particularly of the skeleton. There is also some
evidence for the teratogenic effect of hyperoxia.

As in this latter case, taking '*deficient*' to mean '*unsuited*' or '*defective*' in
the sense of '*faulty*', the teratogenic effect produced by excessive doses
of vitamin A may here be arbitrarily subsumed under heading (4) '*Deficiencies*'. Hypervitaminosis A was shown to cause numerous malformations of the brain and some other organs in rats, mice, rabbits, and
hamsters.

*(5) Diseases.* The developmental defects of bones and teeth caused
by *luetic infection* transmitted to the human fetus are well-known. These
defects originate relatively late, namely after the eighteenth week,
when the treponema may first penetrate the disrupted barrier of *Lang-
hans cells* in the placenta. Infection of the mother with *rubella virus* in
early pregnancy may induce anomalies such as microcephaly, deaf-mu-
tism, cataracts, and heart lesions, in addition to thrombocytopenic pur-
pura, hepatosplenomegaly, icterus, anemia, and low birth weight. A
correlation of the type and severity of the lesions with the stage of
pregnancy at the time of infection may obtain. A greater susceptibility
of the early embryo seems established, and when infection occurs during the first 8 weeks of pregnancy, abortion or stillbirth can result. Deformed infants born 6 to 8 months after infection were reported to still
excrete virus in their nasopharyngeal secretions for some length of
time. Even clinically normal infants may excrete virus at birth. Some
experimental studies with rubella and other virus infections of mammals as well as some epizootic observations (e.g. encephalitis infections in cattle and pigs) also disclosed evidence of teratogenic effects
manifested by viruses. *Toxoplasmosis* (caused by a protozoan) is likewise believed to be capable of inducing ontogenetic anomalies.

---

[88a] Inferences from experimental studies suggest that zinc deficiency, at least in rats,
causes impaired DNA synthesis (cf. SWENERTON *et al.* in Science *166:* 1014–1015, 1969).

In evaluating the various established causal factors of teratogenesis, it becomes apparent that among the numerous relevant parameters which must be taken into consideration, the parameter of *time* (with reference to the onset of the diverse formative processes in ontogenesis), and the parameters related to animal *species specificity* (even within taxonomically rather closely related groups such as the mammalian class and mammalian orders) play a very significant role.

As another aside, one might also briefly mention the ancient popular notions concerning various effects of the pregnant mother's mental (emotional) experiences upon the forthcoming offspring ('*maternal impression*', prenatal influences). States of either serenity or fright (including the shock of bodily injury) were credited with favorable or unfavorable influences respectively, and in this latter case even with teratogenic potentialities (e.g. in German: *das Versehen der Schwangeren*). There is little doubt that the generally alleged instances purporting to substantiate effects of this type are based on crude superstition and represent misstatements, distortions of facts, as well as outright forgeries. With regard to this topic AREY (1954) justly comments: 'not only are nervous communications lacking between mother and child, but also the suspected episode usually occurs far too late; that is, long after the fetal body is well laid down at two months'. Nevertheless, in accordance with well substantiated concepts of psychosomatic medicine,[89] organic or functional (physico-chemical) changes correlated with emotional states or disturbances seem indeed to occur. Despite the lack of nervous communication between mother and embryo, respectively fetus, hormonal or unspecified biochemical influences cannot be entirely ruled out. The possibility that some maternal psychosomatic states might, to a still undefined extent, affect the offspring, should therefore not be dogmatically excluded. It appears, however, fully justified to dismiss all uncritical and extravagant claims as figments of the imagination or gross misrepresentations. So far, no case carrying convincing evidence has come to my attention.[89a]

The preceding discussion concerning the *general aspects of teratology*, i.e. of 'faulty' morphogenesis, merely intended to point out the im-

---

[89] Cf. K., Brain and Consciousness, p. 310f (1957).

[89a] A comprehensive study, evaluating 'prenatal determinants' of behavior in animal and man, was recently completed by JOFFE (1969), and contains an extensive bibliography. The author includes a discussion of problems related to maternal stress and maternal emotions, emphasizing the obtaining uncertainties, and pointing out the need for further rigorous methods of investigation.

portancc, thc large number, as well as the complexity of the problems pertaining to that topic. With regard to pertinent *special aspects* particularly related to the neuraxis, only a few relevant types of anomalies or malformations can briefly be considered in this context. The profusion and diversity of the multitudinous dysontogenetic manifestations preclude here a more detailed survey.

The teratologic literature is very extensive. The formal aspects of malformations are perhaps most expertly dealt with in the older publications, of which the handbook edited by SCHWALBE (1906–1927) deserves special mention.[90] It contains a contribution on malformations of the nervous system by ERNST (1909). Contemporary teratologic works dealing with the neuraxis are the publications by OSTERTAG (1956) and KALTER (1968) quoted above. Because of a long tradition, the teratologic nomenclature is extremely intricate, arbitrary, and at times perplexing. Depending on the assumed viewpoint, one of several overlapping sorts of malformation may be emphasized in the designations. Thus, I. G. SAINT HILAIRE (cf. footnote 90) distinguished the facial defects of *cyclocephaly* and *otocephaly*, the former being characterized by approach or by fusion of the orbits and eyebulbs *(cyclopia)* in the midline, and the latter by displacement of the ears in the same direction. Evidently, such displacement of both eyes and ears may be combined in a whole gamut of interrelated patterns, involving, moreover, anomalies of brain (cyclencephaly and arhinencephaly), of external nose (proboscis), and of other parts (jaw, mouth). It is obvious that any attempt at rigorous classification, necessitating a cumbersome terminology, must result in a chaos of highly debatable categories.

Among specific 'anomalies' which, as it were, blend with normal conditions, *twinning, multiple birth,*[91] and related *duplications* shall briefly

---

[90] A still older, very extensive work is that by ISIDORE GEOFFROY SAINT HILAIRE (the son of the more famous ETIENNE GEOFFROY ST. HILAIRE mentioned in vol.1, p.87, and vol.2, pp.283, 319 of this series): *Histoire générale et particulière des anomalies de l'organisation chez l'homme et les animaux, ouvrage comprenant des recherches sur les caractères, la classification, l'influence physiologique et pathologique, les rapports généraux, les lois et les causes des monstruosités, des variétés et vices de conformation, ou traité de tératologie.* 3 vols. and atlas (Baillière, Paris 1832–1837).

[91] It is understood that *monozygotic* true twins, triplets, quadruplets, etc. are here meant, which originate from a single zygote. Fraternal *(dizygotic)* twins, triplets, etc. ('litter-mates') originate from different zygotes. In man (and comparable Placentalia) they are contained within individual, separate chorionic sacs, whose placentae, however, may be closely apposed.

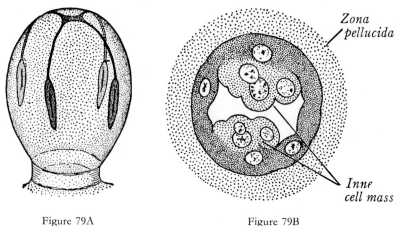

Figure 79A Figure 79B

*Figure 79 A.* Chorionic vesicle of the armadillo, containing four embryos at the stage of the primitive streak (after PATTERSON, from AREY, 1954).

*Figure 79 B.* Hypothetical stage of separate cell masses in a sectioned mammalian blastocyst (after STREETER *et al.*, from AREY, 1954).

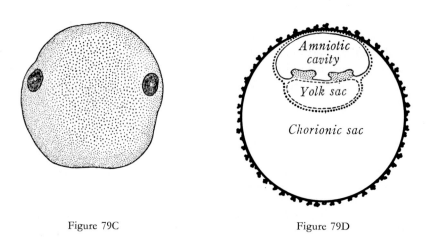

Figure 79C  Figure 79D

*Figure 79 C.* Blastocyst of a sheep, with two embryonic discs, illustrating an early twin embryo (after ASSHETON, from AREY, 1954).

*Figure 79 D.* Hypothetical mammalian twin stage, showing cross-section through two embryonic axes on a single embryonic disc (after STREETER *et al.*, from AREY, 1954).

be considered first. Twinning or multiple embryogenesis may result from separation of early blastomeres, but spontaneous multiple embryogenesis in vertebrates is believed to be caused by (a) the arising of two blastoderms by 'subdivision or segregation of the potential axis', (b) by separate 'organization centers' within a single blastoderm, becoming separate embryonic axes, (c) by secondary fission or budding of a single organization center in a blastoderm.

Normally, the edentate armadillo, *Dasypus*, produces monozygotic quadruplets (Fig. 79 A) assumed to develop in accordance first with mode (b) and then mode (c). In various other instances of monozygotic twinning the development by means of modes (b) or (c) cannot be exactly inferred from mere examination at advanced stages. Figures 79B, C and D illustrate some actual and inferred stages of mammalian twinning. Figure 80, which is self-explanatory, shows an instance of experimental twinning in the urodele Amphibian, *Triton taenia-*

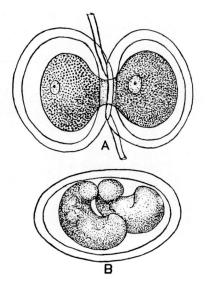

*Figure 80.* Experimental production of a monozygotic twin from a fertilized ovum of *Triton taeniatus*. The first two blastomeres, at the final, but still not quite complete stage of cell division ('dumbbell stage') have been separated by a hair loop (after SPEMANN, from DÜRKEN 1929).

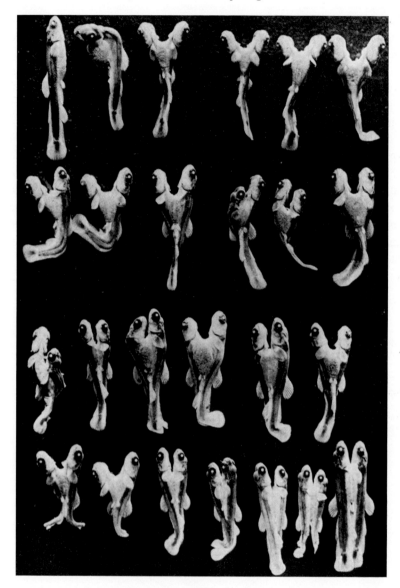

*Figure 81.* 'A series of young trout that started development with a slightly insufficient supply of oxygen. The series begins with an ordinary single individual and passes through increasing degrees of anterior duplicity, shown in the two upper rows. It then continues with specimens showing step after step of completely formed double bodies and tails'. In some other instances, double individuals with unequal components were obtained (from STOCKARD, 1921).

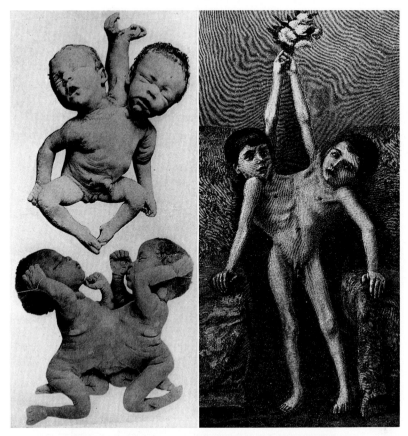

Figure 82                              Figure 83

*Figure 82.* Two degrees of 'duplicity' in human newborns (from STOCKARD, 1921).

*Figure 83.* The double-headed boy *Tocci* (from RANKE, 1923).

*tus*, by purely mechanical interference (cf. the comments on regulative eggs and mosaic eggs in subsection A, p. 8)[92].

Partial duplications, resulting in teratologic monstrosities, and experimentally obtained by STOCKARD (1921) in the trout by means of artificial hypoxia, are illustrated in Figure 81. Figure 82 shows instances

---

[92] In the invertebrate echinoderms any of the first four blastomeres of the sea urchin will form a small but perfect larva if properly isolated. Even blastula halves containing both animal and vegetal cells have been shown to regulate into complete larvae.

of spontaneous teratologic duplicity in man. As regards this latter, a whole gamut of symmetrical conjoined twins (double monsters, '*Siamese twins*', thoracopagus, etc.) and of unequal conjoined twins has been recorded in the literature. The often cited '*Siamese twins*', born in 1811, lived until 1874, thus reaching an age of about 63 years; one twin survived for a short time after the death of his partner. Another, much more pronounced human two-headed malformation, with two conjoined upper trunks including normal arms, but one single lower trunk including abdomen, and a rather normal pair of inferior extremities, except for unilateral polydactyly, remained viable and was exhibited in Europe as the *double-headed boy Tocci* (Fig. 83).

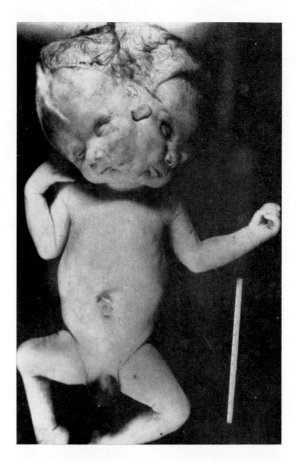

*Figure 84.* Human diprosopus delivered at about full term (courtesy of Dr. F. J. WARNER). Length of ruler: 14 cm.

Malformations in monozygotic twins are said to be more frequent than in ordinary individuals. One of the twins may be an *'acardiacus'* depending on the circulation of its mate. If the disparity in size between twins is considerable, the smaller and usually markedly malformed one is called a *parasite* which may have various bizarre attachments to the host or *autosite*. According to a generally held plausible opinion, duplication of head (complete or incomplete) is due to chorda-mesoderm splitting into two streams during the forward movements of gastrulation through the primitive streak. Doubling of the lower trunk, believed to originate somewhat later, is then assumed to result from a forking divergence of 'retreating' primitive knot and streak.

Anomalous mature human spermatozoa, displaying either two 'heads' or two 'tails', have been recorded and depicted by G. Retzius (1902).

As regards duplications affecting the neuraxis, I had the opportunity to inspect, through the courtesy of Dr. F. J. Warner, Philadelphia, brain and spinal cord of a human *diprosopus ( Fig. 84 ) with a duplex pair of rather normally developed cerebral hemispheres, the actual bifurcation of the neuraxis occurring at the level of mesencephalon ( Fig. 85 ). Spinal cord and brain stem provide here in macroscopic appearance a single, unified stalk for a double forebrain. Few clinical data concerning this case were available, but since the fetus appeared to have been delivered approximately at term, it doubtless was at least alive in utero.* By means of the sophisticated present-day procedures of enforced *medicated survival*, maintaining an 'artificial life', malformations of this type could perhaps be kept viable for an undefined period of time. Adherents of animistic views might ponder over the theological implications concerning the possible presence of two immortal souls in one body, an interesting problem presumably pertaining to the purview of the *Pontifical Academy*.[92a] Serial sections of brain

---

[92a] St. Augustine ( *De Civitate Dei contra Pagamos*, XVI, 8) reports an appropriate case : *'Ante annos aliquot, nostra certe memoria, in Oriente duplex homo natus est superioribus membris, inferioribus simplex. Nam duo erant capita, duo pectora, quattuor manus, venter autem unus et pedes duo, sicut uni homini; et tamdiu vixit ut multos ad eum vivendum fama contraheret'.* St. Augustine comments here briefly *'de monstrosis apud nos hominum partubus'* in connection with the question whether the fabled human races, *'quae gentium narrat historia'*, were descendants of *Adam*, or of the sons of *Noah*, or of other single individuals. St. Augustine's list of such mythical races (one-legged skiopods, cyclopians, neckless men, etc.) follows closely that given by Pliny *( Historia Naturalis*, VII, 2), and also used by Aulus Gellius (123–169 A.D.) in *Noctes Atticae*, IX, 4, but without specifically quoting these authors.

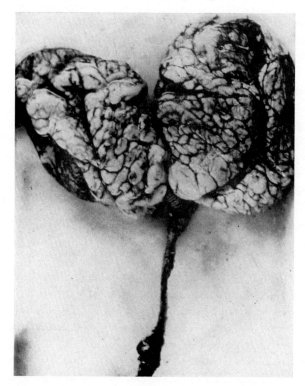

*Figure 85.* Brain and upper spinal cord in human diprosopus of preceding figure (courtesy of Dr. F. J. WARNER).

---

With regard to the individual human monstrous births, the pious Saint remarks: 'far be it from any one to suppose in his folly that the Creator made a mistake, even though one may not know why he acted as he did. He, whose works no one has the right to censure, knows what he has done'. Concerning the above-mentioned fabled monstrous human races, St. AUGUSTINE states: '*non itaque nobis videri debet absurdum ut quem ad modum in singulis quibusque gentibus quaedam monstra sunt hominum ita in universo genere humano quaedam monstra sint gentium*'. Moreover, in the Saint's psychopathic opinion, the Creator might have made human races of this sort expressly to prevent us from thinking that the wisdom by which he moulds the forms of men was at fault in the case of single monsters as if it had been the craft of an unskilled artisan ('*quid si propterea Deus voluit etiam nonnullas gentes ita creare, ne in his monstris, quae apud nos oportet ex hominibus nasci, eius sapientiam qua naturam fingit humanam velut artem cuiuspiam minus perfecti opificis putaremus erasse?*').

It should be recalled that St. AUGUSTINE can be regarded as one of the two main doctrinal pillars of the Roman Catholic Church, the other one being St. THOMAS AQUINAS (1225 ?–1274).

stem, including lower medulla oblongata and adjacent cervical segments of spinal cord displayed at first glance a most interesting pattern
of divergence, respectively convergence manifested by the relevant
tracts and grisea connected with the duplex forebrain. Dr. WARNER has
completed his study of this case and published the results in 1972.
BISHOP (1921) has examined and depicted in considerable detail the
nervous system of a dicephalous 22 mm long pig embryo, whose
head displayed a somewhat higher degree of fusion than Dr. WAR-
NER's human diprosopus. The median eyes were well developed
but closely adjacent in a single palpebral opening, and above a
median conjoined auditory meatus. Eyes and ears on the lateral sides
were normal. A pair of cerebral hemispheres, a diencephalon, and a
mesencephalon were 'normal' for each of the fused heads, and di-
verged rostrally from the conjoined rhombencephalon. Bilateral symmetry and a 'normal' pattern of neural structures were recognizable
throughout the conjoint structures of the neuraxis. The conjoint spinal
cord, although representing a single tube giving off paired spinal
nerves in normal distribution, displayed evidence of incomplete duplication along its ventral portion where a median neural ridge, protruding into the central canal gradually diminished from about the level of
the cervical flexure to the mediastinum. Conjoined and thus unpaired
median spinal nerves, nonbranching and nonganglionated, were seen
to arise at regular intervals from that neural ridge. Despite this barely
suggested duplex aspects of the conjoint spinal cord, the body, including neck, of the embryo appeared grossly as that of a normal single individual attached to one umbilical cord.

Duplications of the spinal cord *(diplomyelia)* has been described in
man and is in many cases associated with *spina bifida* (cf. further below).
The extent of the duplication varies, and diverse degrees of fusion
have been recorded. A bony septum may be present and associated
with other aspects of vertebral column duplication. *Diastematomyelia*
or *pseudoduplication* is a splitting of the spinal cord not always clearly
distinguishable from diplomyelia, especially since borderline cases occur. In diastematomyelia neither half is a complete spinal cord, and the
parts may be unequal. Duplications of the central canal within a single
spinal cord, especially at caudal levels, are likewise occasionally recorded.

Among the *arrhaphic* and *dysraphic* manifestations involving the
neuraxis, *cranioschisis (acrania)* is characterized by an open-roofed skull.
The brain may be reduced to an amorphous vascularized mass, directly

exposed, or covered by a thin membrane continuous with the skin. The eyes are usually rather well developed, the bulb often being still connected to the malformed brain mass. The brain stem, even if exposed, may or may not display some of its typical structural features, as well as the origin of all or some of the cranial nerves. Extreme cases of cranioschisis, although still preserving some amount of amorphous brain mass, are also designated as *anencephaly*, and are usually combined with rachischisis (Fig. 86). *Exencephaly* (Fig. 87) refers to a massive protrusion of malformed brain tissue through a skull defect. The mass may be covered by normal skin, by meningeal tissue, or by a nondescript neuroectodermal layer with or without meningeal components.

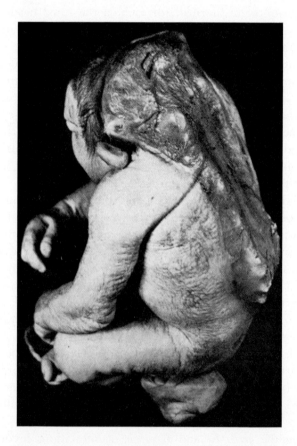

*Figure 86.* Late human fetus with anencephaly and complete spina bifida (rachischisis, cf. fig. 90). The ribbon-like spinal cord is easily identifiable (from RASMUSSEN, 1955).

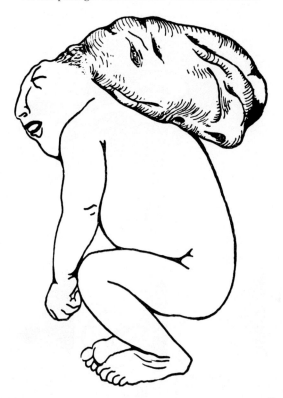

*Figure 87.* Late human fetus with exencephaly (after NAÑAGAS, from RASMUSSEN, 1955).

Lesser degrees of cranioschisis[93] are represented by encephalocystomeningocele, hydrencephalocele, encephalomeningocele, and meningocele (Figs. 88, 89).

Dysraphic disturbances of the *vertebral column* and associated structures, including the *spinal cord*, are posterior and anterior *rachischisis*. The posterior malformation, also called *spina bifida* (posterior), is relatively common and said to occur in about one per thousand births. As

---

[93] *Encephalocystomeningocele* includes dilated meningeal spaces, brain mass, and a ventricular protrusion; *encephalocystocele* includes meninges, brain mass and ventricular space, but without dilated meningeal spaces; *hydrencephalocele* includes meninges, brain mass and dilated ventricular space; *encephalomeningocele* (or simply *encephalocele*) includes brain mass without ventricular space and meninges; *meningocele* includes meninges (usually with dilated spaces) only. The terminology of these conditions is not entirely standardized.

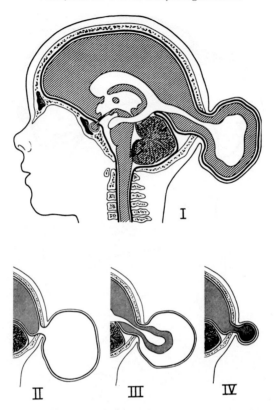

*Figure 88.* Diagrams depicting various degrees of human encephaloceles (from GER-LACH *et al.*, 1967). I Encephalocystocele (hydrencephalocele). II Meningocele. III Encephalocystomeningocele. VI Encephalocele.

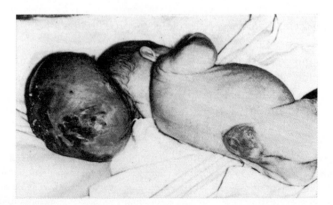

*Figure 89.* Large human occipital encephalocele combined with lumbosacral spina bifida in whose center the area medullo-vasculosa can be seen (from GERLACH, 1967).

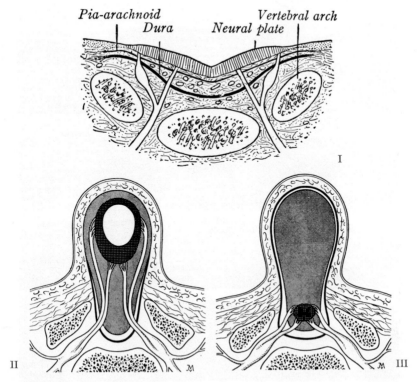

*Pia-arachnoid*          *Vertebral arch*
      *Dura*        *Neural plate*

*Figure 90.* Diagrams illustrating various degrees of human rachischisis *(spina bifida)*. I Spina bifida aperta, i.e. myeloschisis with open neural plate (from AREY, 1954). II Rachischisis with myelocystomeningocele displaying dilated central canal. III Spinal meningocele with essentially normal spinal cord (II and III from GERLACH, 1967).

in the case of cranioschisis, several degrees of severity obtain (Fig. 90). In *myeloschisis* or *spina bifida aperta*, the malformed spinal cord is directly exposed like an early neural plate, but displays, of course, some degree of more advanced histogenetic differentiation.[94] The exposed vascularized neuroectodermal tissue in the midline is designated as *area medullo-vasculosa* (Fig. 91). Laterally, a strip of modified meningeal tissue, the

---

[94] If cystic protrusion caused by accumulated cerebrospinal fluid internally to the medullary plate or closed spinal cord occurs, the malformation is called *myelomeningocele*. The term *myelocystomeningocele* refers to an enlarged cavity within the closed spinal cord as well as an accumulation of meningeal fluid protruding in rachischisis. The term *myelocele* is also used for a protrusion essentially provided by the neural tube alone, which may, however be covered by a thin leptomeningeal layer.

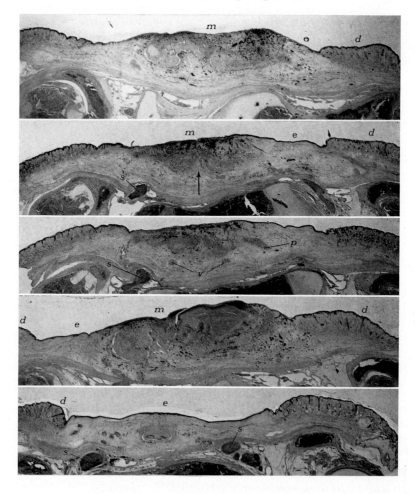

Figure 91 A

*Figure 91.* Cross-sections through a human thoraco-lumbar spina bifida aperta (cf. fig. 90 I) in a 2-month-old infant examined *in vivo* and *post exitum* by the author (from a set of serial sections in my collection; hematoxylin-eosin; A × 4.7, B top × 4.7, B bottom × 9, red. ²/₃). a: adipose tissue overlying closed end of spinal cord with duplex canalis centralis; d: zona dermatica; e: zona epithelio-serosa; m: area medullo-vasculosa; p: dorsal root; s: spinal ganglion; v: ventral root; w: nerve cells of cornu anterius; x: leptomeningeal tissue; y: pachymeningeal tissue; z: periost (perichondrium) of vertebrae; arrow indicates fissura mediana anterior. Second strip from bottom in figure A shows a hemorrhagic cyst in area medullo-vasculosa.

*zona epithelio-serosa*, forms a transition to the normal skin or *zona derma-tica*. Comparing Figures 90, I and 91 with Figures 8, 13, and 48 in volume 2 of this series, it is evident that a topologically identical terminal condition obtains, thus reproducing an invertebrate neural pattern. Yet, it would not be particularly meaningful to speak here of an 'atavism', since a configuration of central nervous system flush with the epidermal ectoderm represents an initial developmental stage in a multitude of taxonomically as well as presumably phylogenetically quite unrelated Metazoa.

Lesser degrees of spina bifida are various forms of meningo-myelocele and meningocele. In the least pronounced instances, rachischisis may not be externally apparent *(spina bifida occulta )*. However, a dimple or reddish discoloration of the skin usually indicates its location, where a fibrous strand connects the meninges with the integument. A

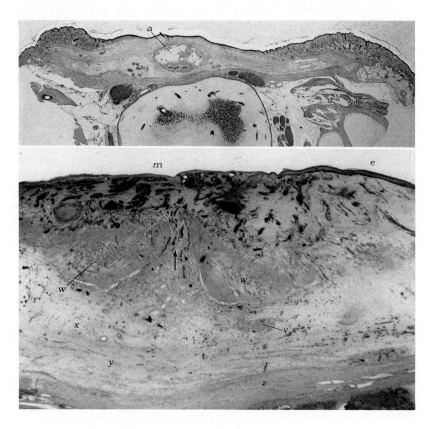

Figure 91 B

tuft of hair may also be present on the overlying skin. Occasionally, a cutaneous fistula or sinus, discharging a clear fluid, leads to the meningeal level. If hair is present within the epithelial duct, the condition known as *pilonidal sinus* obtains, which is interpreted as originating particularly in relation to terminal caudal, 'coccygeal' vestiges of the spinal cord. Some instances might perhaps even be related to the posterior neuropore. Instead of an open duct, a closed *pilonidal cyst* can occur.

In anterior rachischisis or *spina bifida anterior*, a rather uncommon malformation, the bodies of some vertebrae may be defective, and contents of the vertebral canal may herniate into body cavities, particularly into peritoneal cavity. About 30 cases of intrapelvic meningocele have been recorded in man (RASMUSSEN, 1955).

In a recent report, PADGET (1970) attempts to introduce the term *'neuroschisis'* as an allegedly new concept of significance for the understanding of spina bifida, anencephaly and diverse other mammalian defects. According to the cited author, 'any part of the embryonic neural plate or tube may be split open by one or more neural clefts, a neuroschisis as presently first described'. 'Damage to the embryo varies according to when the clefts occur, where they occur, whether they are complete in dividing both inner and outer limiting neural membranes, whether they expand, or close (i.e. 'heal'), and how much their concomitants and sequelae affect neural and other primitive structures'.

With regard to the classifications discussed further above, it is evident that such 'neuroschisis' would be subsumed under AREY's category 11 (cf. p. 2/8), MORIARTY's and KLINGMAN's categories 2 or 5 (cf. p. 222), and OSTERTAG's categories 1 or 5 (cf. p. 223).

While cleft, bleb, or cyst formations which may be designated as 'neuroschisis' could obviously occur in the course of ontogenesis, many of the actual pictures displaying defects of this type can also be interpreted as postmortal changes or artefacts of histologic technical procedure. Even the illustrations of the two supposedly 'unique' cases depicted by PADGET (1970) cannot be considered fully convincing. Nevertheless, in view of the presumed multifactorial origin of malformations, the role of cleft formations, followed by abnormal foldings (which are likewise very frequently artificial postmortal, etc. distortions) could be granted as one of the numerous possibilities. Be that as it may, it remains quite doubtful whether the concepts of *'neuroschisis'* or *'schizencephaly'* (cf. below p. 272 and footnote 103) are significant additions to the many, necessarily vague terminologic attempts at 'explaining' the 'nature' of malformations.

*Syringomyelia* is a condition characterized by cavitation within the spinal cord (*syrinx:* tube, pipe). The cavities may be due to a dilation of the central canal (*hydromyelia*), to a dysraphic insufficient relative reduction of the canal, or to secondary cavitations within intramedullary softening necroses or tumors. These latter, again, can be true neoplasms or glioses of borderline dysplastic type with or without involvement of the central canal (cf. Fig. 109, p. 163 of the preceding vol. 3/I). Most cases of syringomyelia affect the cervical or upper thoracic region of the spinal cord.

Concerning *malformations of the brain*, the term *arhinencephaly*, apparently introduced by Kundrat about 1882, refers to a large group of diversified and, in part, heterogeneous telencephalic anomalies, of which arhinencephaly (absence or aplasia of rhinencephalic structures) may represent only a minor aspect.

Schwalbe and Josephy (1913) have subdivided cases of human arhinencephaly into two groups. In the first, they include brains with a *telencephalon impar*, a malformation now generally referred to as *cyclencephaly*, and closely related to the *cyclopic* anomalies. In the second, they include otherwise normal or nearly normal brains, whose telencephalon is characterized by the (usually bilateral) absence of olfactory bulbs and tracts.[95]

As regards the second group, the developmental defect may be confined to the brain (arhinencephaly *sensu stricto*), or may be combined with multiple anomalies displayed by other systems. Thus, Morton (1947) reported the case of a full-term stillborn male infant whose otherwise grossly quite normal brain lacked olfactory nerves (fila), bulbs, and tracts. The additional anomalies in other systems included colobomata of both irises, polydactyly (six digits on each limb), persistent right and left vena cava superior, abdominal testes, and *Meckel's diverticulum*. The author assumes an interference with normal development between the 5th and 8th weeks of embryonic life. It is evident that the classification of this and similar cases under the simple heading 'arhinencephaly' is highly arbitrary.

---

[95] The term '*rhinencephalon*' is, of course, rather ambiguous and improperly definable. It was omitted from the PNA. In partial accordance with the older nomenclature of the BNA, the human rhinencephalon *sensu strictiori* may, for practical purposes, be restricted to olfactory bulbs, olfactory tracts, olfactory trigone, striae olfactoriae mediales and laterales, as well as substantia perforata anterior. Older authors included various additional structures believed to be 'olfactory centers' (cf. Villiger, 1920). It is intended to include a critical appraisal of the concept 'rhinencephalon' in volume 5, chapter XIII, of this series.

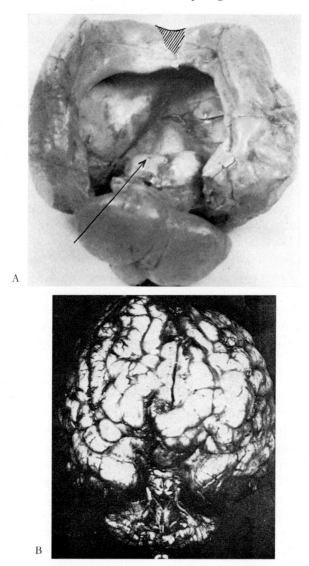

*Figure 92.* Dorsal and basal views of a rhinencephalic brain with cyclencephaly in a human infant surviving a few days *post partum*. A Dorsal view of brain cast. Arrow points to dorsal surface of diencephalon (thalamus); the v-shaped depression between the rostral ends of the ovoid thalami is the free opening of third ventricle into the common telencephalic ventricle; area marked by hatching in frontal region indicates original outline, interrupted by a damage in the cast (from K. *et al*. 1957). B Basal view of original specimen upon removal (from K. and GLOBUS, 1936).

With respect to cases of the first group, displaying a *telencephalon im-par* we (K. and GLOBUS, 1936) have reported a conspicuous instance in a female infant, born at full term and dying on the fourth day. The malformation involved only the brain, which showed an incompletely differentiated telencephalon *(cyclencephaly)* with a large common ventricle and considerable eversion of the malformed pallial wall. Viewed from above, the telencephalon appeared as a crescent or horseshoe-shaped mass, whose dorsal defect was covered by a protruding membranous hydrocephalic sac representing an expanded lamina epithelialis of roof plate (Figs. 92, 93).

The cerebellum was of roughly normal size but exposed owing to the lack of telencephalic occipital lobes. The brain stem (midbrain, pons, oblongata) was, like the cerbellum, grossly normal although slightly undersized. The ventricular system, including the aqueduct, was patent.

As regards the forebrain, the thalami bulged out in the shape of two fairly sizable oval masses separated by a groove corresponding to the third ventricle. Optic tracts and optic chiasm, though somewhat undersized, were of normal shape. The septum pellicidum was absent, and corpus callosum as well as fornix could not be recognized at first sight. On closer examination, however, it was found that the marginal arch of the ventricle at the rostral part of the horseshoe was a rudimentary corpus callosum continuous with the lateral marginal arches corresponding to both crura and fimbria fornicis.

On the basal aspect of the brain the salient feature was the complete bilateral absence of the olfactory bulb and tracts, as well as that of olfactory striae, and of the *diagonal band of Broca*. Because of this defect, which completely agrees with that considered characteristic for arhinencephaly, we classified the case under discussion as pertaining to said category. However, on account of the considerable internal hydrocephalus concomitant with a fairly normal aspect of the skull, this case could also be classified as an instance of hydranencephaly (cf. K. *et al.*, 1957), which will be discussed further below.

The high degree of eversion displayed by the dorsal (pallial) wall of unpaired telencephalon represents a configurational pattern simulating conditions normally obtaining in various lower vertebrates (Anamnia), namely in holocephalians (chimaera), ganoids, teleosts, and dipnoans (cf. Figs. 49 D, 58 III, 59 B, 60, 64 I). Although much less pronounced thant the eversion found in ganoids and teleosts, but rather comparable in degree to that seen in holocephalians and dipnoans, it can, nevertheless, be considered extreme for a human brain, or, generally speaking, a brain of Amniota. Depending on the logical criteria, said condition in man could be called an 'atavism'. Such evaluation, however, would be entirely meaningless as regards 'explanation' since

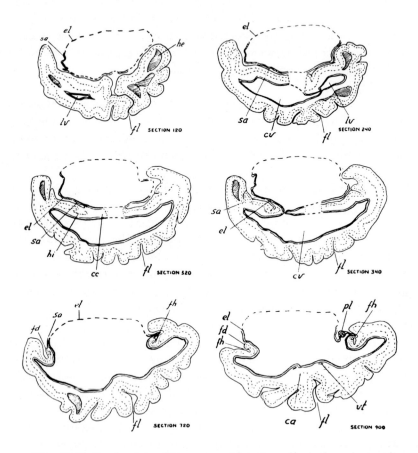

*Figure 93.* Outline drawings of transverse serial sections of brain shown in preceding figure (from K. and Globus, 1936). ac: *aquaeductus Sylvii;* ca: commissura anterior; cc: corpus callosum; cg: lateral geniculate body; ci: capsula interna; cp: colliculi superiores; cs: corpus striatum; cv: common ventricle of the end-brain; el: ependymal lining (the lost ependyma is indicated by the interrupted part of the line); ep: epiphysis; fd: fascia dentata; fh: fissura hippocampi; fl: fissura longitudinalis; gl (?): globus pallidus (?); he: hemorrhages; hi: hippocampal formation; la: lamina affixa; lv: lateral ventricle; mi: massa intermedia; na: nucleus amygdalae; nc: nucleus caudatus; nl: nucleus lateralis thalami; nm: nucleus medialis thalamu; pa: substantia perforata anterior; pl: choroid plexus; pu: putamen; sa: wall of hydrocephalic sac; st: sulcus terminalis; tc: taenia choroidea; th: thalamus; to: optic tract; tt: taenia thalami; vd: ventriculus diencephali; vt: vena terminalis.

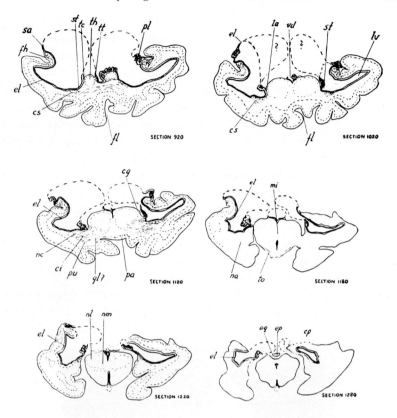

Figure 93 (continued)

the condition is based (a) on the persistence of a configurational stage (telencephalon impar) occurring in the ontogenesis of all vertebrates, and (b) on an apparently secondary eversion perhaps related to the considerable hydrocephalus. Moreover, various degrees of telencephalic eversion occur in various vertebrate forms which could hardly be considered, to any significant degree, as phylogenetically related. Another and different parallelism with conditions found in taxonomically lower forms (marsupials, chiroptera, edentates) concerns the arrested development of the corpus callosum, resulting in the presence of a supracommissural hippocampal formation comparable to that in the cited lower mammals. With regard to the critical periods discussed above, the 'formal genesis' of the malformation may have had its onset about the sixth week of intrauterine development.

*Cyclencephaly*, combined with *arhinencephaly*, is likewise displayed in cases of cyclopia (cyclocephaly). In a critical review of congenital malformations of the brain, METTLER (1947) expresses the opinion that malformations of the cyclencephalic type are correlated with nasal abnormalities such as monorhinic conditions, and that the normal configuration of the telencephalon 'is disclosed to depend upon the successful development of the rhinencephalon'. This view, however, cannot be upheld, although some influence of the differentiating nasal placodes upon the development of olfactory bulbs, as e.g. reported by BURR (1916) in Amphibia, seems indeed to obtain. Yet, the human cases pertaining to the second group of SCHWALBE and JOSEPHY (1913) as mentioned above, and the fact that a predominant telencephalon impar is typical for many Anamnia with bilaterally well developed nasal structures, including olfactory bulbs and tracts, are clearly incompatible with METTLER's overall hypothesis.

In typical extreme human *cyclopia* (Fig. 94), only a single median eye is present, with one optic nerve passing through a single median optic foramen. This nerve may divide into right and left optic tracts at

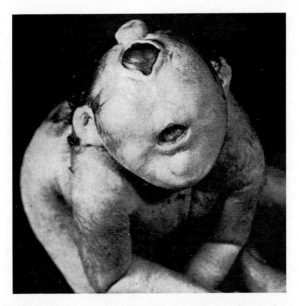

*Figure 94.* Face of human cyclops ('late fetus'). Typical proboscis above single conjoint eye, whose duplex origin by 'fusion' is clearly indicated by dark median line in eyebulb. A small posterior herniation of the defective brain is hidden from view (after RASMUSSEN, 1955).

the base of a cyclencephalic brain. In some cases, an optic nerve seems to be entirely absent, presumably because of secondary disintegration. Hypothalamic commissures are usually rather well developed. An aberrant external nose, in form of a trunk-like *proboscis*, is usually displayed above the single eye, but may be completely missing. Olfactory nerves, bulbs, and tracts are, as a rule, absent. In classifying the malformation, however, arhinencephaly is here traditionally subordinated as concomitant to cyclopia which represent then the main feature. All gradations of eye fusion are recorded, ranging from a truly single eye to imperfect fusion of paired eyes, mere contact, or diminished distance with defects of the separating orbital and other midline structures. Additional variations, as well as complications, including the combination of cyclocephaly and otocephaly, mentioned above (p. 235), have been described.

Relevant data on malformations, respectively developmental anomalies, of the brain in submammalian forms are, on the whole, far more scarce than those concerning mammals and man. However, some significant observations on unilateral anophthalmia in Sauropsidans were recently published by KNOWLTON (1964) and KIRSCHE (1965). KNOWLTON's report deals with a fairly advanced chick embryo displaying, in addition to left anophthalmia and left olfactory defect of arhinencephalic type, various anomalies in the differentiation of diverse grisea along the cerebral neuraxis. KIRSCHE's paper analyzes developmental anomalies in the brain of a lizard *(Lacerta agilis)* with right anophthalmia, detected at about hatching stage, and surviving for about 30 minutes following removal from the egg.

The diagnostic term *hydranencephaly*, used by recent authors, but going back to descriptions by CRUVEILHIER about 1829 and by SPIELMEYER about 1905 (cf. K. *et al.*, 1957, 1959), refers to an essentially congenital condition characterized by a highly deficient development or status of the cerebral hemispheres, combined with hydrocephalus, but concomitant with a fairly normal appearance of head and skull. The intracranial space left by the telencephalic defect is filled out by cerebrospinal fluid, which may be intraventricular or extraventricular, or both. Hydranencephaly, referring to what might be considered a secondary condition, thus subsumes various malformations of different morphologic and etiologic categories, including our case of arhinencephaly (K. and GLOBUS, 1936) briefly discussed above.

The clinical diagnosis of hydranencephaly, although fully justified, and at times the only possible diagnosis *in vivo*, is therefore not particu-

larly satisfactory, since it cannot refer to the different anatomic manifestations nor to the diverse etiologic factors related to hydranencephaly as the common feature.

It may be said that the 'class' hydranencephaly comprises two main
subsets, subsuming two main types, namely (1) the sort caused by intrinsically arrested or defective developmental processes,[96] and (2) the
sort caused by extrinsic destructive encephaloclastic processes, undo-
ing or secondarily modifying the results of normal morphogenesis (including histogenesis). Since the processes subsumed under both
groups can set in at various stages of ontogenesis, and since the processes under groups (1) and (2) are not definable in fully rigorous fashion, intermediate (overlapping) forms of (1) and (2) might perhaps occur.

An unequivocal and typical instance of group (1) is our above-
mentioned case of arhinencephaly; typical instances of group (2) are
two cases reported by LANGE-COSACK in 1944.

Cases of group (2) may be recognized by residues of destructive
and reactive processes. Glial scars or proliferations, diffusely distributed hemosiderin in meninges, choroid plexus, or glial margins justify
the inference of extrinsic destructive processes including hemorrhages.[97]

In cases of group (1), the brain malformation may display features
of arrested and incomplete development, maintaining, however, discounting incompleteness, retardation, and configurational deformation, essentially 'normal' orderly topologic and structural relationships. In other cases of this group, gross distortions and randomness
of structural arrangements obtain, displaying a nearly teratomatous
type of growth.

---

[96] Endogenous genetic involvement, or exogenous nongenetic interferences altering
formative processes at the fundamental morphogenetic field level (cf. vol. 1, p. 231 of this
series). Such interference, although *sensu strictiori* exogenous, is therefore still intrinsic
in contradistinction to extrinsic encephaloclastic processes. However, a logically rigorous
and entirely satisfactory distinction with regard to borderline instances cannot be made
due to the limitations of linguistic models.

[97] LANGE-COSACK (1944) believes that traumatic damage or various circulatory
disturbances play an important role in the genesis of hydranencephaly *(Blasenhirn)*.
This possibility is corroborated by the experiments of BECKER (1949), who produced
vesicular malformations of the prosencephalon, including telencephalon, closely resembling, or identical with, certain forms of hydranencephaly and porencephaly, by blocking
branches of the internal carotid in young dogs. It is evident that the cited authors deal
with group (2) as defined above in the text.

A particularly relevant clinical case of *hydranencephaly* (Fig. 95) which we diagnosed a few weeks after normal birth, subsequently to the onset of decerebrate rigidity symptoms, came to our attention in 1954 (K. *et al.*, 1957, 1959 and additional reports). It concerns a female child, still living[98] and 16 years old at the time of this writing (1970). The brain malformation, visualized by means of pneumoencephalography (Fig. 96), appears roughly comparable to the univentricular telencephalic malformation anatomically studied in our 1936 case of arhinencephaly (K. and GLOBUS, 1936), whose postnatal survival lasted only a few days.

The pneumoencephalograms of our clinical case disclosed a telencephalon reduced to a collapsible flabby sack. Although the presence of a rudimentary pallium with an abortive cortical structure, and some rudiments of basal ganglia, that is, a status more or less similar to the previously studied malformation can be assumed, the child might be considered '*decorticate*' from a functional point of view. The child's behavior appeared 'normal' during the immediate postnatal period, before the 'disturbance' or, or the 'lack' of telencephalic functions could exert an influence on the behavioral pattern, which then began to display a type of decerebrate rigidity.

The child can neither stand nor sit, cannot even raise her head, and is unable to feed herself. Although the pupillary reflex to light and some muscular reactions to sound can be elicited, the eyes neither follow nor avoid light or objects, and there is no convergence. There is absolute lack of even the most elementary behavioral patterns suggesting 'mental' acitivities, and in this respect the child can be likened to an

---

[98] Although, for all practical purposes, the case could be justifiably evaluated as representing a completely '*mindless*' human being not essentially differing from a complex mechanical doll, the mother developed a strong emotional attachment to this child. The mother is a rational and intelligent person who fully understands and realizes the essential features of the situation. Yet, she is convinced that she can, in a fashion, communicate with the child, and that, if only in an exceedingly dim manner, the child recognizes her. On the basis of our own observations, we are inclined to doubt this and to interpret this attitude of the mother as a hyponoic and hypobulic delusion. Be this as it may, the exceedingly meticulous and devoted home care, provided by the mother and the childs's two perfectly normal and rather bright siblings, rendered possible this most unusual survival. The duration of life in a case of severe cerebral agenesis (arhinencephaly with encephalocele) studied by GAMPER (1926) was about 3 months. The child in a case of complete lack of differentiated cerebral hemispheres, reported by EDINGER and FISHER (1913) died before completing 4 years.

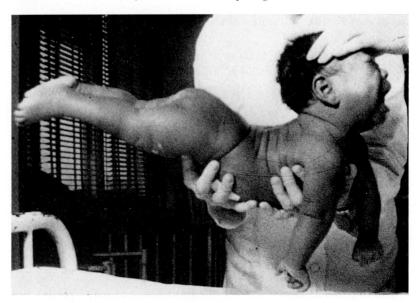

Figure 95 A

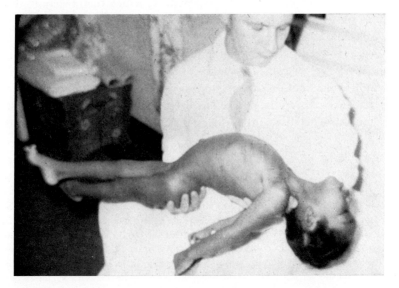

Figure 95 B

*Figure 95.* Manifestation of decerebrate rigidity in hydranencephalic child (from K. *et al.*, 1959). A Several weeks after birth. B After about 3½ years (note 'spontaneous' *Babinski sign* at left margin, now presumably fixed by contracted state of atrophic muscle).

experimentally produced 'thalamus animal' manifesting 'sham emotions' (in this case crying). Alternating periods of 'sleep' and 'wakefulness', that is of rest with closed eyes and of activity, including crying, are manifested. The EEG shows a fairly regular slow activity with high voltage and is not influenced by external optic, acoustic or me-

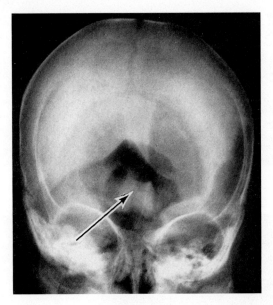

Figure 96 A

*Figure 96.* Pneumoencephalograms of hydranencephalic child shown in preceding figure. A–C several weeks after birth, D–G at about 3¹/₂ years of age. A Anteroposterior view, recumbent position. Arrow indicates ovoid outline of thalamus (cf. fig. 92A); above arrow the air-filled space of velum interpositum (the air is continuous with that in cisterna ambiens and subtentorial space. B Lateral view in erect position of head, showing upper air bubble; occipital 'lobe' is supported by tentorium. C Lateral view in recumbent (supine) position of head, showing upper air bubble in frontal region (A–C from K. *et al.*, 1957; 1: cerebellum; 2: cisterna magna and subtentorial space; 3: cisterna ambiens; 4: cisterna chiasmatis; 5: cisterna interpeduncularis, merging with cisterna chiasmatis, ambiens, and pontis). D Left lateral view with head in semirecumbent (supine) position. E Left lateral view with head in recumbent (prone) position. F Left lateral view with head in semierect position. G Anteroposterior view with head in erect position; purposely underexposed view, to show, by greater contrast, details of flabby, membrane-like pallium, which has partly collapsed on left side, i.e. right side of picture (D–G from K. *et al.*, 1959). 1: cerebellum; 2: subtentorial space; 3: cisterna ambiens; 4: cisterna chiasmatis; 5: cisterna interpeduncularis; 7: optic nerve; 8: cisterna pontis; 9: mastoid cells; 10: cisterna olfactoria; 11: air in region of vallecula cerebri; R: right; L: left; f: rudiment of falx cerebri; l: subluxations of atlas and axis; t: tentorium cerebelli.

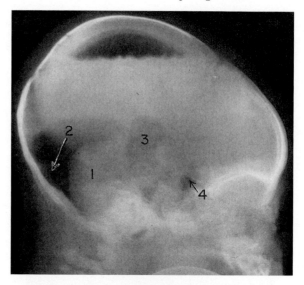

Figure 96 B

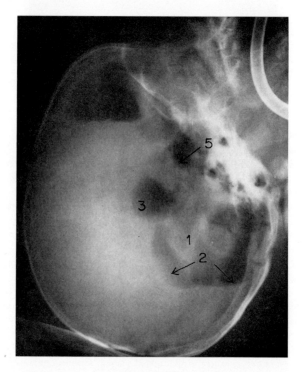

Figure 96 C

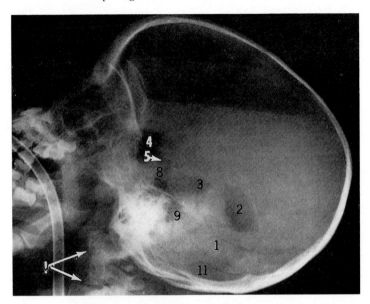

Figure 96 D

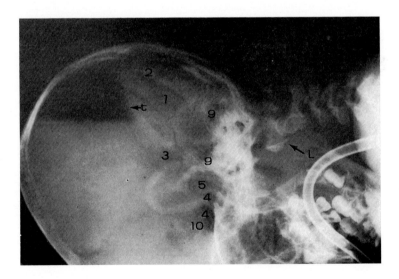

Figure 96 E

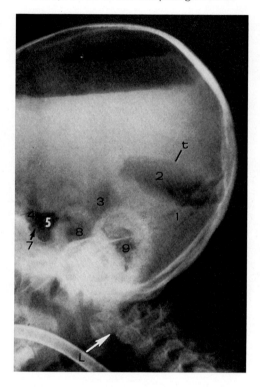

Figure 96 F

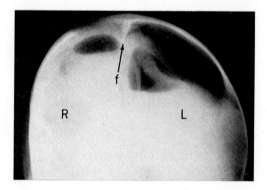

Figure 96 G

chanical stimuli, including 'painful' ones (pinching) which, however, elicit a crying reaction. Nor did rhythmic stimuli of various frequencies show any response-effect in the EEG.

While there cannot be any doubt about the general diagnosis 'hydranencephaly' of 'decorticate' type, it is evident that, without actual gross and microscopic corroboration, our specific diagnosis, namely univentricular telencephalic malformation pertaining to group (1), can only be tentative, although rather probable.[99] An alternate possibility, which, rightly or wrongly, we consider less probable, would be a prenatal encephaloclastic process, esentially restricted to the telencephalon, resulting in softening, resorption, and shrinkage, and corresponding to the cases of group (2) studied by LANGE-COSACK, or somewhat related to a similar case reported by JAKOB (1931). Although the roentgenologic pictures seem to indicate absence of a recognizable corpus callosum rudiment, a rudimentary falx cerebri is present, and an imperfect subdivision into right and left hemispheres cannot be entirely excluded.

*Agenesis*, *aplasia*, and *hypoplasia* of the human *corpus callosum* (German: *Balkenmangel, Balkendefekt*) is a well-known anomaly of the human brain (Fig. 97). According to KIRSCHBAUM (1947), approximately 100 cases have been reported in the available literature during the past hundred years. Absence of the corpus callosum may remain entirely symptomless *qua* clinical manifestations, and then merely represents an incidental finding at necropsy. Since it can thus occur in normal individuals, one might be surprised, as OSTERTAG (1956) remarks, that a relatively large number of individuals without corpus callosum did not show any 'psychic' peculiarity or disorder despite this *prima facie* substantial brain defect. *Mutatis mutandis*, a similar situation obtains with regard to complete cerebellar aplasia, discussed further below. Some sort of 'explanation', in my opinion, is provided by the well supported assumption of considerable redundancy associated with ultrastability (cf. vol. 1, pp. 20–24 of this series).

The anomaly is compatible with long life, up into the 8th decade, but death in infancy and in childhood, perhaps from associated developmental disturbances, has been frequently recorded. In addition to

---

[99] This probability rests on the fact that it is possible to interpret the pneumoencephalographic pictures in rather full conformity with the anatomical features of the case studied in collaboration with Dr. GLOBUS (K. and GLOBUS, 1936).

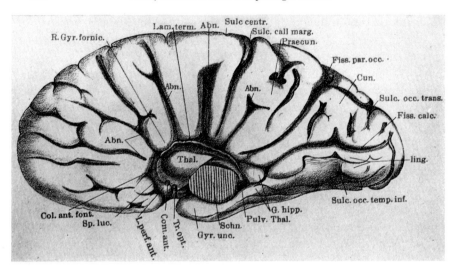

Figure 97 A

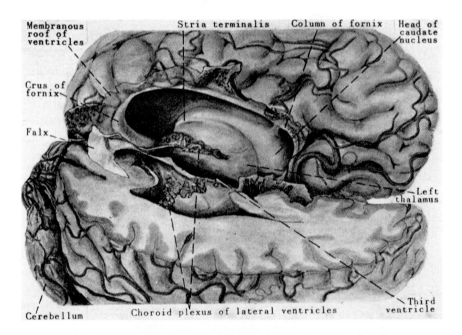

Figure 97 B

the clinically symptomless cases referred to above, callosal agenesis or malformations commonly involve either feeble-mindedness, or lesser 'psychic' disorders of various types, or epilepsy.

Since development of the corpus callosum within the telencephalic lamina terminalis is initiated about the beginning of the third month of intrauterine life, the factors interfering with its formation and growth may be assumed to take effect at approximately this time. Complete lack of corpus callosum entails *absence of a septum pellucidum*. Agenesis of corpus callosum is frequently combined with other anomalies, such as cyclencephaly, cyclopia, porencephaly, and cortical heterotopias. In extreme cases, commissura anterior, hippocampal commissure, and telencephalic choroid plexus are likewise missing.[100] The sulci on the medial aspect of the hemisphere may display a peculiar ('vertical', 'radiating') pattern by slanting off from the fimbria fornicis (cf. Fig. 97). Because a solid transverse cover above the tela chorioidea ventriculi

---

[100] A close inspection of the illustrations in various publications reporting absence of corpus callosum seems, however, to indicate that at least in some such cases a rudiment of corpus callosum, corresponding to its earliest anlage, may be present in the lamina terminalis dorsal to commissura anterior (e.g. presumably in fig. 3 [case 2] of KIRSCHBAUM, 1947). Again, a presumably compensatory hypertrophy of the anterior commissure can be present. In some other cases, abnormal transverse commissures provided by presumably secondary fusions of the two medial hemispheric walls have been described.

---

*Figure 97.* Adult human forebrains with agenesis of the corpus callosum. A Medial aspect of acallosal brain in a case of 'microcephalic' idiocy (after ONUFROWICZ, 1887, and FOREL, 1907). Abn: anomalous radiating sulci, at right angle to ventricular roof ('lamina terminalis', cf. below); col. ant. font. (misprint for forn.): anterior column of fornix; Lam. term.: 'lamina terminalis' (presumably thickened unpaired roof plate, interconnecting, instead of corpus callosum, both hemispheres in the midline, and therefore, quite logically interpreted by FOREL as part of lamina terminalis); R. Gyr. fornic.: components of the dissociated ('zerspaltenen') gyrus fornicatus in the acallosal brain; Schn.: cut surface between diencephalon and mesencephalon; Sp. luc.: 'septum pellucidum' (FOREL, who assumed that the septum was a part of the medial hemispheric wall, quite logically designated the juxtaterminal portion of the paraterminal body, namely the gyrus subcallosus, as 'septum pellucidum'; this latter, however, is an intraterminal portion of $B_4$, not present in this and other similarly malformed brains; the true topologic equivalent of 'septum pellucidum' is here the intraterminal neighborhood between the two anterior columns of the fornix, which represent, in this case, the thickened true telencephalic lamina terminalis; cf. also the discussion in section 6, p. 628); the other abbreviated designations are self-explanatory. B Adult human forebrain with agenesis of corpus callosum. Dorsal part of right hemisphere cut off to expose ventricles and medial surface of left hemispheres (from RASMUSSEN, 1955).

tertii is missing, the roof of the third ventricle may bulge upward and can display a characteristic conspicuous, wide median space between the hemispheres in appropriate ventriculograms, which occasionally, as in some other cerebral malformations, allow for an *intra vitam* diagnosis of anatomical anomalies.[101]

Mesenchymatous disturbances related to callosal aplasia and hypoplasis involve the presence of *meningeal tumor masses*, particularly of *lipomatous* type[102] in the region of the defect or even apparently 'replacing' it. Anomalies of vascular pattern likewise occur on the medial aspect of the hemispheres. Again, a peculiar longitudinal fiber tract *('Balkenlängsbündel')* is commonly found to run along the medial aspect of the hemisphere adjacent to the defect. Although this fiber bundle was already carefully described by W. ONUFROWICZ under FOREL's direction in 1887, more than 10 years before PROBST's subsequent study on *'balkenlose Gehirne'*, some recent authors refer to the *'Balkenlängsbündel'* as *'Probst's bundle'*. FOREL assumed that this tract represents the fasciculus longitudinalis superior, an 'association bundle' which is normally masked by the massive transverse callosal fiber system. I believe that FOREL's interpretation is essentially correct, although, in addition, so-called 'heterotopic' fibers might also be involved. An older summary by FOREL (1907) and a more recent discussion with numerous bibliographic references by OSTERTAG (1956) provide relevant introductions to the complex problems concerning human *'Balkenmangel'*.

---

[101] Some relevant aspects of roentgenologic differential diagnosis are discussed in our report on defects of the human septum pellucidum (HAHN and K., 1930).

[102] Cf. the discussion of neoplastic growth in volume 3, part I, p. 760 and footnote 340 of this series. It might here be added, that, in the cited short summarizing survey based on extensive first-hand acquaintance with the relevant problems, and in accordance with generally accepted views, *medulloblastomas* were classified as *neuroectodermal* neoplasms. In a recent monograph (GULLOTTA, F.: Das sogenannte Medulloblastoma. Springer, Berlin 1967), which has since come to my attention, the author attempts to support and revive an earlier view interpreting medulloblastomas as (mesodermal) sarcomas. It is true that certain types of medulloblastomas, characterized by reticulum fibers, and prone to metastasize, give the appearance of being cerebellar sarcomas, but I would agree with HAYMAKER (1969) in regarding such tumors as variants within the group of medulloblastomas, whose classification as neuroectodermal neoplasms I consider reasonably well established despite the aforementioned objection. On the other hand, there is no doubt that genuine cerebellar sarcomas can occur.

As regards other eutherian mammals, aplasia of corpus callosum was recorded in monkey *(Cebus)* and cat, involving functional disturbances. In a mouse strain, definitely hereditary ('familial') agenesis of corpus callosum without apparent behavioral manifestations has been reported (cf. the review by KIRSCHBAUM, 1947).

Surgical separation of the human hemispheres by practically complete transection of the corpus callosum, as performed for the removal of neoplasms or for the treatment of extreme epileptic conditions, does not seem to affect the 'unity' ('oneness') of conscious experience, nor to be necessarily followed by easily recognizable substantial behavioral changes. This appears to agree with the instances of callosal agenesis that remained apparently 'symptomless' during the affected individual's life. Nevertheless, detailed studies on patients with transected corpus callosum seem to indicate that some differences in 'information processing' may result from an independent activity of the separated hemispheres. There is, however, hardly any justification for the grossly exaggerated statements that, in such cases, each hemisphere manifests 'independent conscious awareness',[102a] and that 'surgery has left these people with two separate minds' (HANDLER, 1970).

As far as the human brain is concerned, the functional significance of the corpus callosum has doubtless a direct bearing on the still poorly understood problems of *handedness, bilateralism,* and *'dominance'* of one hemisphere ('major' vs. 'minor' hemisphere, cf. K., 1957). At present, there are some indications that the two (human) hemispheres may stand 'in a relation of reciprocal specialization', such that, e.g. verbal activities depend more on one hemisphere, and certain nonverbal performances more on the other. 'Higher functions', accordingly, might not be the 'exclusive domain' of a dominant hemisphere (cf. e.g. GAZZANGA, 1965).

---

[102a] Since, in the wider sense, the terms *'consciousness'* and *'awareness'* are synonymous one might wonder what the anonymous neoprimitive authors of that statement intend to express by *'independent conscious awareness'*. There are, of course, as discussed in detail elsewhere (K., 1957, 1961, 1965), numerous different aspects of consciousness. In dealing with questions of this type, it is evidently necessary first to give a sufficiently rigorous operationally valid definition of *'consciousness'*, and second, to define, in likewise manner, its diverse aspects. Under such premises a distinction of *'consciousness'* *('perception')* in general, and *'awareness'* (e.g. meaning-related *'apperception'*) in particular, could perhaps be made. Nevertheless, in general terms, any state of awareness is conscious, and any state of consciousness implies awareness.

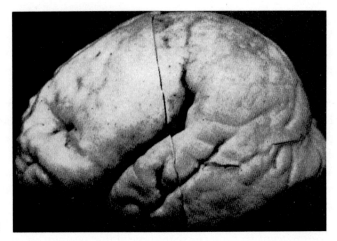

I

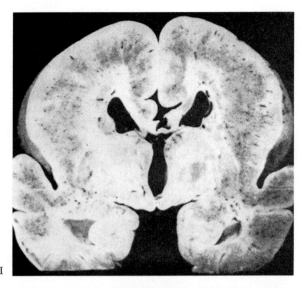

II

*Figure 98.* Cerebral hemisphere of 10-month-old boy, displaying 'agyria' (lissencephaly) and microencephaly (brain weight 700 g). I Lateral aspect. It will be noted that, in addition to a fairly typical fissura cerebri lateralis *(fossa Sylvii)*, the development of some gyri and sulci is faintly suggested. II Approx. frontal section of forebrain along plane shown in I (arrow). The hypoplastic cortex seems to be relatively 'thick', and a claustrum cannot be identified by gross inspection (after DeLange, from Rasmussen, 1955).

It is difficult to ascertain whether hemisphere specialization obtains in other animals, e.g. in mammals such as primates. '*Split-brain*' experiments have been performed on primates. By sectioning the optic chiasma, it becomes possible to restrict optic input from one eye to one (i.e., the homolateral) hemisphere. Interocular transfer of learned pattern discrimination tasks in chimpanzees was thereby found to be dependent on the telencephalic commissures, namely splenium corporis callosi and anterior commissure, the former pathway being the more efficient one (BLACK and MYERS, 1964).

Again, a 'split-brain' monkey with unilateral optic input can be taught, with one eye, that a given signal represents the availability of food, and subsequently, with the other eye, trained to recognize the identical significance of a quite different signal. Thereafter, the animal responds to these signals entirely in accordance with the input from the eye which is left uncovered. Moreover, following unilateral prefrontal lobotomy in a monkey with restricted, homolateral optic input, the animal introduced into a cage containing a snake 'cowered with fear when one eye was uncovered, and completely ignored the snake when the other eye was employed' (HANDLER, 1970).

Among additional relevant anomalies or defects of the human telencephalon, the following can here briefly be considered: *agyria, microgyria, pachygyria, porencephaly, Hallervorden-Spatz syndrome,* and *infantile cerebral palsy.*

*Agyria* or *lissencephaly* is characterized by a localized or generalized failure of the telencephalic pallium to undergo the folding process resulting in a normal pattern of gyri and sulci (Fig. 98). The affected regions of cortex are hypoplastic. In pronounced cases, involving idiocy, the hemispheres of an infant or adult appear like those of an oversized early fetal telencephalon. Although lissencephalic mammalian brains may indeed display an 'ancestral' aspect of the telencephalic surface, one could hardly evaluate human lissencephaly as an 'atavism' without stretching the point at issue.

In *microgyria* the hemispheres show regions with abnormally narrow, rather tortuous gyri and gaping sulci. Supernumerary convolutions can be present. The cortex within the microgyric regions may appear histologically normal.

In *pachygyria*, the gyri are abnormally wide, being separated by inconspicuous sulci, and manifest a rather simple pattern of convolutions. The cortex in the pachygyric regions, which can display a bilateral-symmetric distribution, is commonly hypoplastic.

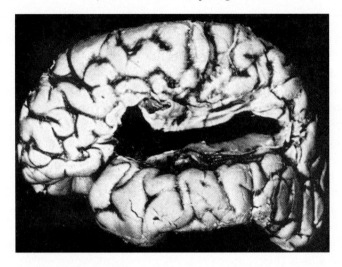

*Figure 99.* Porencephaly, manifested by large defect in lateral wall of left cerebral hemisphere, establishing wide communication between ventricle and leptomeningeal (subarachnoid) space (after BAKER, from RASMUSSEN, 1955).

*Porencephaly* (Fig. 99) is a condition characterized by external cavities in the telencephalic mantle, communicating with ventricles or subarachnoid space, or both. Corresponding skull defects are occasionally present. The sometimes used term 'pseudoporencephaly' refers to
noncommunicating cavities. Cases of porencephaly believed to result
from 'developmental failure', namely from 'agenesis of the cerebral
mantle' in early stages of morphogenesis have been classified as instances of 'schizencephaly' without specific reference to the relevant
particular pathogenesis or 'causal origin'.[103] Such schizencephalic poren-
cephalies may display a bilateral-symmetric pattern. In contradistinction, cases of encephaloclastic porencephaly are presumed to be the result of secondary destructive events, such e.g. as encephalomalacia
caused by circulatory disturbances or inflammatory processes.

The *Hallervorden-Spatz syndrome*[104] involves apparently gradual degenerations of grisea in telencephalo-diencephalic boundary zone and

---

[103] YAKOVLEV and WADSWORTH (1946).

[104] HALLERVORDEN and SPATZ (1922). Homogeneous eosinophilic rounded bodies may
be found within the nerve cells of basal ganglia and some other grisea.

base of mesencephalon (i.e. globus pallidus and substantia nigra). The clinical symptoms, beginning in childhood, include progressive rigidity, athetosis, retardation of speech, and mental deterioration. The condition may be of recessive hereditary nature, but sufficiently conclusive data do not seem to be extant. Some authors regard this syndrome as a neuropathologic rather than a clinical diagnostic term.

*Infantile cerebral palsy* subsumes a variety of crippling conditions characterized by motor disorders, not necessarily nor even commonly involving mental deficiency. Depending on terminologic and statistical criteria, its incidence has been estimated to be in the range of 0.6 to 6 per 1000 births. Clinically, the main forms of this cerebral palsy can very roughly be classified as (1) *choreoathetosis* and related dystonic states; (2) *spastic hemiplegia;* (3) *paraplegia* and *quadriplegia*, such as congenital spastic diplegia *(Little's disease)* and related forms, and (4) *ataxia*.

Quite evidently, these diverse conditions imply a number of different causative factors, including hereditary, 'genetic' ones, as well as various encephaloclastic processes and other exogenous disturbances at different periods of prenatal as well as early postnatal development.

Among these various factors, several sorts of events as well as of neuropathologic states may be enumerated: anoxia, hemorrhages, infections, birth trauma, kernicterus,[105] *status marmoratus*,[106] porencephaly and porencephalic cysts, whose margins frequently display scarred and atrophic convolutions *(ulegyria)*, other related types of so-called 'cystic degenerations', and cerebral hemiatrophy *(lobar sclerosis)*. This latter, again, can result from a variety of causes. The gyri are here small and deformed (secondary microgyria), displaying diffuse gliosis. Mental deficiency is commonly associated with this type of lesion.

*Infantile hemiplegia*, in contradistinction to *congenital hemiplegia*, can occur as a complication of many infectious diseases, such as whooping cough, measles, scarlet fever, diphtheria, chicken pox and others.

---

[105] Cf. volume 3, part I, p.346, of this series. *Icterus gravis neonatorum* is usually the result of a hemolytic disease *(erythroblastosis fetalis)* brought about by antibodies formed in Rh-negative mothers against the Rh-factor of the Rh-positive fetus. *In utero*, the hemolytic bilirubin is presumably excreted into the maternal circulation. If, after birth, the infant organism cannot control the bilirubin excess, kernicterus occurs, causing damage to basal ganglia and other grisea.

[106] Status marmoratus *(état marbré)*, originally described by O. and C.Vogt, is a peculiar mottled appearance of the basal ganglia (putamen and nucleus caudatus). The irregularly scattered whitish spots are due to a perhaps reactive overgrowth of medullated fibers, combined with gliosis and loss of ganglion cells. Arterial occlusion, birth injury, and inflammatory processes have been suggested as causes of the griseal alterations.

Concerning malformations of other subdivisions of the human brain, some *cerebellar anomalies* may here be mentioned. Complete or near complete *agenesis* of the cerebellum (Fig. 100), although rare, has been repeatedly reported (cf. e.g. RUBINSTEIN and FREEMAN, 1940, with further references to literature). In an 11-year-old female epileptic idiot, COMBETTE (1831) described the cerebellum as a mere gelatinous membrane connected with the brain stem by membranous peduncles. In other cases of virtually complete cerebellar agenesis, however, very small flocculi interconnected by a rudimentary nodulus were present, corresponding to a reduced flocculo-nodular lobe (*archicerebellum* of

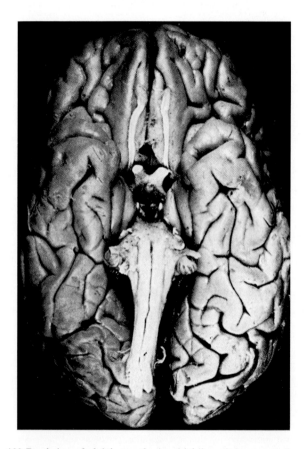

*Figure 100.* Basal view of adult human brain with bilateral almost complete agenesis of cerebellum. A somewhat small flocculus is, however, bilaterally present. Pons and inferior olives 'practically absent' (after FREEMAN, from RASMUSSEN, 1955).

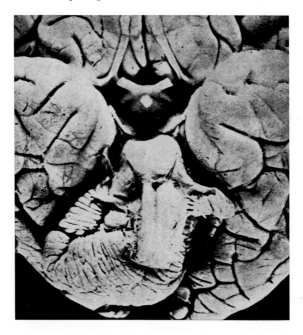

*Figure 101.* Basal view of left unilateral cerebellar agenesis in a child aged three years. Right inferior olive absent, rather small flocculus present (at right in picture) on left side (after STRONG, from RASMUSSEN, 1955).

LARSELL, 1937). Unilateral cerebellar agenesis *(hemiagenesis)* is *relatively* more common (Fig. 101). Agenesis involving most of the vermis, including the fastigial nuclei, is likewise on record.

Complete or partial cerebellar agenesis may be correlated with severe clinical symptoms of various sorts, such as motor incoordination, feeble-mindedness, and epilepsy. On the other hand, even essentially complete cerebellar agenesis can remain entirely symptomless throughout adult life and merely become an incidental finding at necropsy.[107] In some such case, signs of impaired cerebellar function became noticeable during the very last, antemortal period of life. The problems related to the functional compensation for a morphologic and structural defect of this magnitude were discussed by ANTON and ZINGERLE (1914) as well as several other authors (cf. also RUBINSTEIN and FREEMAN, 1940). Personally, I am inclined to believe that vestibu-

---

[107] Cf. above the comments on agenesis and aplasia of corpus callosum.

lar nuclei and reticular formation of rhombencephalon play here an important vicarious role in providing a substitute for cerebellar activities. Cases of congenital unilateral cerebellar defects are said to be commonly symptomless.

*Caudal displacement* of cerebellar mass toward or into the vertebral canal is an anomaly designated as *Arnold-Chiari malformation*, which can also involve displacements or distortions of pons and medulla oblongata. Distorted cerebellar components may become impacted within the foramen occipitale magnum,[108] a substantial part of the medulla oblongata being then located caudal to that foramen. The affected cranial nerve roots are elongated, and interference with the intercommunicating channels of cerebrospinal fluid may cause internal hydrocephalus. Various degrees of this deformity can occur, some of which are combined with cerebellar hypoplasia or heteroplasia, or with spina bifida.

Other sorts of cerebellar anomalies involve agyria, microgyria, macrogyria, various heterotopias, and neoplasias of borderline dysplastic character. The *transitory superficial granular layer* of the cerebellar cortex, contributing to histogenesis, and normally disappearing before the end of the second postnatal year, may remain as a persistent structure, from which neuroectodermal neoplasms can take their origin (K., 1950).[109] This anomaly implies a disturbance of further differentiation, characterized by lack of 'reduction', which becomes anatomically manifest in late (i.e. postnatal) ontogeny, although nothing can be inferred about nature and time of origin of the relevant 'causal' factors.

*Cerebellar neoplasias* of borderline dysplastic character include a number of presumably interrelated conditions described as *gangliocytoma dysplasticum*, cerebellar *hamartoma*, granular cell hypertrophy, and *Lhermitte-Duclos disease* (cf. e.g. AMBLER *et al.*, 1969). In two, apparently familial cases, the cited authors describe hypertrophy of granule cell bodies with hypertrophy and myelination of their axons, distortions of

---

[108] Increased pressure of cerebrospinal fluid caused by the space requirements of intracranial neoplasms may also displace parts of an otherwise normal cerebellum, particularly the cerebellar tonsils, toward or into the foramen occipitale magnum, and result in a *secondary* distortion somewhat similar to the *Arnold-Chiari malformation*. A so-called cerebellar pressure cone of this type is depicted in the author's report on neoplastic transformation of the subependymal cell plate in the floor of the fourth ventricle (subependymal spongioblastoma) in J. Neuropath. exp. Neurol. *6:* 139–151, 1947 (fig. 1, loc. cit.).

[109] Cf. volume 3, part I, p. 781 of this series.

cytoarchitecture, and hypertrophy of affected cerebellar folia. There is some indication that this anomaly could be related to general megalencephaly which displays a diffuse disturbance of cell growth at early stages of ontogenesis. In various conditions of that type, which can also selectively affect diverse particular regions of the brain, there may be excessive cell proliferation of various sorts and degrees, ranging from an approximately normal appearance to diverse pathologic ones, manifesting hypertrophy, or 'embryonic' characters, or predominance of certain cell types (e.g. widespread blastomatous thickenings, diffuse spongioblastomatosis, etc., cf. also K., 1947).

Miscellaneous additional anomalies, which particularly concern the human central nervous system, and should be pointed out in this context, include *congenital hydrocephalus, vascular anomalies,* and *abnormalities in shape as well as size of skull,* concomitant with corresponding deformations of the brain. Following a brief discussion of these topics, and in order to emphasize the extent of the domain pertaining to dysontogenetic manifestations, some comments on *familial demyelinating processes* and on the *degenerative disorders* termed *spastic pseudosclerosis* and *Pick's disease* shall conclude this subsection dealing with anomalies and disturbances of morphogenesis *sensu latiori.*

*Congenital internal hydrocephalus* is an excess of cerebrospinal fluid within the ventricular system, exerting pressure leading to dilation of the ventricles and often considerable enlargement of the head. It is appropriate to distinguish such hydrocephalus from hydranencephaly (discussed further above) in which the fluid is usually not under significantly increased pressure and merely fills out the spaces within a malformed brain ('compensatory' hydrocephalus, hydrocephalus '*ex vacuo*'). Increased pressure may be caused by obstructions of the ventricular system blocking the communications between ventricular and subarachnoid spaces.[110] Congenital hypertrophy of choroid plexuses is

---

[110] Congenital blockage of the communication between fourth ventricle and subarachnoid space through the *foramina of Magendie* and *Luschka,* with concomitant hydrocephalus, is known as the so-called '*Dandy-Walker*' *syndrome,* which may occasionally be accompanied by syringomyelia.

Histologic changes of stretched human brain wall, observed by the author in cases of extreme hydrocephalus, and especially those affecting ependyma and subependymal cell plate, were depicted and briefly discussed in the preceding volume 3, part I, of this series (p. 160, figs. 115A, B, C, D).

Problems concerning the mechanism involved in the resorption of cerebrospinal fluid (e.g. either by small thin-walled vessels of the leptomeninges, or through the so-

also said to be one of a variety of causes (MORIARTY and KLINGMAN, 1962), among which 'impairment of absorption' is also generally suggested. If no obstruction is present, and excessive fluid under pressure is found in ventricular as well as in subarachnoid spaces, *communicating hydrocephalus* obtains. In some cases there are only dorsal obstructions in the subarachnoid spaces above the tentorium, perhaps caused by adhesions following hemorrhages due to birth trauma *(acquired postnatal hydrocephalus)*. A diversity of additional abnormal and exogenous pathological conditions (e.g. *Arnold-Chiari malformation*, inflammatory processes such as meningitis, expanding intracranial lesions, etc.) may be combined with hydrocephalus which then displays only one aspect of the anomalous features, and can be associated with a wide range of clinical manifestations (cf. K. and HAYMAKER, 1950).

As regards the occurrence of hydrocephalus in animals, HENKEL and KIRSCHE (1967) have recently described a case of hydrocephalus internus in the underdeveloped telencephalon of a young *Lacerta agilis*. The cited authors who evaluated it as a case of hydrocephalus *e vacuo*, include in their paper a concise but comprehensive discussion of this and some other cerebral malformations in animals, as well as a relevant bibliography. In accordance with other authors, two main sorts of hydrocephalus are distinguished, namely (1) those related to disturbances of cerebrospinal fluid circulation (hypersecretion, defective resorption, and impeded drainage), and (2) hydrocephalus *e vacuo*, compensating defects of brain substance. SCHERER (1944), in a textbook of comparative mammalian neuropathology, subdivides hydrocephalus *internus* into hydrocephalus *occlusus*, hydrocephalus *ex vacuo*, and *primary* hydrocephalus with excessive fluid production. As regards man, the normal amount of cerebrospinal fluid (about 90–180 cc, cf. below, section 7, p. 699 may, in very extreme cases, be increased to an amount exceeding 10 l.

---

called arachnoid granulations of mammals such as man) shall be briefly referred to in section 7 (Meninges).

As regards the term '*hydrocephalus*', COOPER (1963) quotes the following comment by SMYTH (in 'A treatise on hydrencephalus or dropsy of the brain'. Callow, London 1814). 'I have made a slight change in the Greek name of the disease. That of *Hydrocephalus* only expresses dropsy of the head, whereas *Hydrencephalus* expresses what it really is, dropsy of the brain. To the Greek scholar the difference is obvious; to others it is of no consequence'. Still, since the 'encephalon' is a substantial component of the head, the term 'hydrocephalus', particularly with respect to communicating (external) hydrocephalus, remains reasonably appropriate.

In man, whose neuroanatomy has been investigated with particular intensity, numerous variations and *anomalies of the blood vessels*, supplying and draining the brain, have been recorded. Many of these variations concern different origins of cerebral blood vessels from the basal anastomotic ring (*circulus arteriosus Willisii*, cf. section 7 of this chapter), which may even, in rare cases, remain incomplete (i.e. noncontinuous or 'interrupted'). Additional anomalies comprise angiomas of diverse type, and many sorts of aneurysms, including arteriovenous ones. In the course of ontogenesis, beginning with a rather diffuse network of capillary type characteristic for early embryonic stages, the vascular network of the body in general, and of the neuraxis in particular, undergoes several significant transformations which are susceptible to disturbances at different stages, resulting in variations and anomalies. Many variations are displayed by the cerebral veins, and, among anomalies, arteriovenous aneurysms of *vena magna Galeni*, with aberrant connections from posterior cerebral or choroidal arteries, have been reported (ALPERS and FORSTER, 1945). Variations and anomalies of the dural sinuses are not uncommon. Aneurysms and angiomatous cysts, perhaps originating from embryonic vascular channels, and related to the dural sinuses, are likewise recorded.

As regards abnormalities in the shape of the skull, the obvious ballooning with long persistent and enlarged fontanelles in pronounced cases of congenital hydrocephalus requires no further comments, but some remarks on a number of additional cranial anomalies seem appropriate in this context.

Deformations of the human skull were studied in great detail by 19th century authors such as ECKER, A. RETZIUS, R. VIRCHOW (the famed pathologist, who was also much interested in anthropology), and WELCKER. VIRCHOW, in particular, dealt extensively with a proper classification of skull deformities. A summary of these problems and thereto-related controversies, debated in an extensive literature, is contained in the anthropologic work of J. RANKE (1923), to which reference was made in vol. 1 (p. 87, footnote 21) of this series.

Concerning the classification of *skull deformities*, two main groups might be distinguished, namely fortuitous or secondary ones, caused by external mechanical pressure, and intrinsic ones, related to primary disturbances of morphogenetic processes.

The first group includes prenatal deformities due to diverse events causing pressure against the maternal pelvis or e.g. pressure exerted by multiple pregnancies. Another sort of deformation pertaining to this group is caused by excessive molding during labor and delivery. Depending on the different cephalic presentations (occiput, face, brow, combined with anterior, transverse, and posterior orientation), various patterns of 'congenital' deformation can result, which closely resemble the diverse types of intrinsic ones. As a rule, however, even rather conspicuous deformities caused by the birth mechanisms have the tendency to disappear in the course of postnatal development, but may remain permanent in some cases.

A third subgroup of secondary cranial deformation comprises those caused by postnatal pressure effects, either fortuitous and related to habitual resting positions of the head against hard supporting surfaces, or intentional and related to ethnological practices, as recorded by anthropologists with respect to certain populations (e.g. ancient Peruvians, some Eskimo and Lapplander groups, and aborigines of the New Hebrides, etc.).

The *primary* or *intrinsic cranial deformities* are generally subsumed under the concept of so-called *craniostenoses*. These deformations, which may or may not involve significant reduction of the volume of the cranial cavity, are characterized by premature ossification of the sutures of the cranial vault, of the face, or at the base of the skull. Such conditions are sometimes accompanied by other anomalies, e.g. syndactyly in upper or lower limb, or both. The brain, which can be of rather normal size, but whose shape displays deformations conforming to those of the cranial cavity, may or may not manifest additional anomalies.

Although, *prima facie*, the brain thus appears to have been 'molded' by the shape of the deformed cranial cavity, the morphogenetic interactions between the soft brain mass and its hard osseous capsule are of a more complex nature. The biological plasticity of living bone tissue, due to resorptive and adaptive processes responding to pressure exerted by 'softer' structures upon the physically much 'harder' bone, is well-known. Thus, vessels, nerves, cerebral gyri, and meninges impress grooves, impressiones digitatae, and foveolae granulares upon bony surfaces. An aneurysm of the aorta may cause destruction and resorption of sternum or vertebrae, but not usually of intervertebral disks, ligaments, or skin.

Assuming that premature closure of sutures is a primary factor preventing expansion of the cranial cavity in conformity with a normal growth pattern, it seems likely that the expanding brain affects the shape of the cranium in accordance with various functional *loci minoris resistentiae*, in relation to undefined factors, within the morphogenetic field of the developing skull.[111] As regards microcephaly, implying a brain of pronounced subnormal size, it appears rather probable that factors preventing normal brain development and growth play here the significant primary role.

The following types of so-called *craniostenoses* are commonly distinguished. In *oxycephaly (turricephaly*, German: *Turmschädel)* the skull is high and somewhat tower-shaped. In acrocephaly the configuration is similar, but more pointed, and a crest along the site of sagittal suture is usually present. In *acrobrachyencephaly* the head is, in addition, markedly brachycephalic. A hereditary (heredofamiliar) anomaly, combining features of oxycephaly and acrocephaly with facial deformities,[112] is called *craniofacial dysostosis* or *Crouzon's disease*.

*Trigonocephaly* displays tapering of the skull at the frontal region, *scaphocephaly* is characterized by a subnormal transverse diameter, with somewhat bulging forehead and occipital region. *Cymbocephaly* is occasionally used as a synonym for scaphocephaly, but also designates a slightly bowl-shaped head, with a depression of the surface at the vertex. *Plagiocephaly* refers to an asymmetric skull of 'skewed' appearance, one frontal half and its contralateral occipital half being larger than their antimeric parts.

The neuropathologic concept of *'demyelinating diseases'* subsumes a variety of heterogeneous conditions which display degeneration of medullated fiber systems or fiber tracts as a conspicuous invariant. Axons as well as nerve cells (grey matter) are generally much less affected, the grey matter being only occasionally or incidentally involved. Glial proliferation and hyperplasia (gliosis) is usually the characteristic end-result of chronic demyelinating processes which are therefore also

---

[111] The radio-autographic and biochemical studies initiated by DRATMAN and her collaborators (cf. vol. 1, pp. 144–145 of this series) seem to indicate that thyroid hormones in general, and particularly thyroxin, play a significant role not only for ontogenetic evolution of the neuraxis (DRATMAN, 1967; DRATMAN and K., 1968; DRATMAN *et al.*, 1969), but also for the proper ontogenetic development of skeletal tissues (DRATMAN and K., 1969). The possibility that the factor responsible for the cited abnormal biologic behavior of bone tissue is related to disturbances involving thyroxin metabolism should therefore be taken into consideration.

[112] The facial deformities may include flattening of the bridge of the nose, which displays 'birdlike' prominence, prognathism, and underdevelopment of upper jaw. Exophthalmus, papilledema, optic atrophy and strabismus can likewise be present. 'Intelligence', however, remains usually unimpaired.

designated by the term '*sclerosis*'.[113] Since the causal factors resulting in these diseases are insufficiently known, as well as in most instances rather controversial, 'nosologic entities' have been named in accordance with anatomic distribution or histologic character of the lesions.

With regard to the common feature of *myelin sheath destruction*, clinicians (e.g. SCHUMACHER, 1962) distinguish two main groups of disorders, namely *primary* demyelinating diseases and *secondary* demyelinating conditions. The first group comprises the *cerebral lipidoses* (cf. footnote 72), diffuse sclerosis, multiple (disseminated) sclerosis, acute disseminated encephalomyelitis of postimmunization and postinfectious type, and the numerous diverse forms which can be included under these subgroups. The second group subsumes all or almost all other conditions in which myelin sheath destruction occurs. It therefore includes traumatic *Wallerian degeneration* and degenerations following primary nerve cell loss, heredodegenerative diseases such as *Friedreich's ataxia*, metabolic disorders and deficiencies (e.g. subacute combined sclerosis in pernicious anemia), conditions of anoxia, exogenous and endogenous intoxications, as well as diverse neurotropic virus infections.

There is little doubt that various different pathologic processes may affect the myelin sheaths, which represent the spiral winding of surface membranes, pertaining to supporting cells, around nerve cell axons (cf. volume 3, part. I, p. 423 of this series). Myelin formation and maintenance seem to imply the interactions and activities of both neuronal and supporting elements. It seems likely that a diversity of disturbances involving the metabolism of either neuronal cells or supporting cells, or both, could result in breakdown of myelin.

Among exogenous 'causes' or causal factors, infections, intoxications, allergic processes, deficiencies, metabolic disturbances, and enzymatic processes have been suggested. Experimental demyelination in animals has been produced by a variety of toxic substances including saponin, potassium cyanide, and carbon monoxide. Arsphenamine, sulfonamides, and spinal anaesthetics are also known (on the basis of fairly convincing evidence) to have produced demyelination in human patients, although perhaps rather rarely. Similarly, immunization pro-

---

[113] Cf. section 8, subsection 'Degeneration and Regeneration', particularly with respect to *Wallerian* and related degeneration, p. 667 f., and footnote 302 in volume 3, part I of this series. Gliosis was illustrated (fig. 163) and briefly dealt with in section 3, p. 220, of that volume.

cedures (e.g. against rabies and smallpox) may, in some relatively rare instances, cause an acute disseminated encephalomyelitis.

BRAIN (1962), who likewise classifies as demyelinating diseases only the conditions pertaining to the first group of SCHUMACHER (1962), excluding, however, the cerebral lipidoses, stresses the difficulties of their appropriate classification: 'Apart from the fact that all but the most acute forms of demyelinating diseases are sometimes familial, and that the most acute forms often follow acute infections, especially the exanthemata caused by viruses such as measles, smallpox and vaccination, little is known as to the aetiology of this group of disorders. An aetiological classification is therefore impossible.'

'The attempt to classify them upon a pathological basis encounters the difficulty that although a large number of pathological varieties have been distinguished, they merge into one another to form a continuous series. A purely clinical classification is equally unsatisfactory in that it fails to accommodate transitional forms exhibiting features common to two clinical varieties, which can usually clearly be distinguished. The best available classification is a clinico-pathological one, which is based upon the recognition of transitional forms. Such a classification must be provisional and must be qualified by the recognition of transitional forms' (BRAIN, 1962).

The group of (primary) progressive demyelinating diseases termed *diffuse sclerosis*, and usually occurring early in life includes, in addition to *Schilder's disease* (encephalitis periaxialis diffusa), many types whose classification has remained controversial, and whose terminology is very complicated. A separation into *exogenous* ('inflammatory') and *endogenous* (familial, heredodegenerative) type has been attempted, but cannot be considered entirely satisfactory. Yet, with respect to the present context, the following conditions, which represent the so-called *leukodystrophies*, whose familial nature seem reasonably certain, can be pointed out: *acute infantile type of Krabbe, subacute type of Scholz, adult type of Ferraro, chronic type of Pelizaeus-Merzbacher*. It should be added that the qualification 'hereditary' or 'familial' does not specify whether these conditions arise, as it were 'spontaneously', that is, through the effect of entirely 'endogenous' factors, or whether merely an 'endogenous predisposition' obtains, the actual disease being triggered by undefined but significant 'exogenous' variables.

In classifying hereditary diseases, REFSUM (1962) justly remarks that 'the antithesis of hereditary and nonhereditary characters is a theoretical simplification'. Genetically conditioned susceptibility for numerous exogenous diseases is widely recognized and seems to be well supported by the available evidence.

Among the hereditary diseases of the nervous system, two other conditions could perhaps be included,[114] namely spastic pseudosclerosis and *Pick's disease*.

*Spastic pseudosclerosis*, described by CREUTZFELD and by JAKOB approximately fifty years ago, has its onset between the ages of about thirty and fifty years; its course is rapid, characterized by progressive dementia, dysarthria, spasticity and extrapyramidal symptoms. Death occurs within three months to three years. The neuropathologic lesions comprise atrophy of cerebral gyri, destruction of nerve cells, glial reactions, and demyelination of various fiber systems. Alterations in some important enzymes of brain metabolism within the affected regions have been recently reported. The condition is also known as *Jakob-Creutzfeld* or *Creutzfeld-Jakob disease* (spongiform encephalopathy).

*Pick's disease* or circumscribed cortical atrophy is usually confined to frontal and temporal lobes of the hemispheres. 'Intellectual' functions deteriorate, 'mental dullness' becomes pronounced, epileptic attacks occur in some cases, and general marasmus characterizes the final stages. The disease, which usually sets in between the age of 50 to 60, always ends fatally after a course of about three to twelve years. Despite its 'presenile' onset, senile plaques and *Alzheimer's neurofibrillary change*[115] are generally absent. Arteriosclerosis does not seem to be a factor. Considerable cortical cell loss, particularly in the outer cell layers, and some degree of gliosis represent the main histopathologic changes. The degenerating nerve cells may display cytoplasmic 'inclusion bodies' of presumably nonviral origin.

---

[114] Cf. REFSUM (1962), who refers to the cases indicating familial occurrence of these diseases. However, inoculation of brain biopsy material from a patient diagnosed as a case of *Creutzfeldt-Jakob disease* into a chimpanzee was followed, after 13 months, by the appearance of a subacute, progressive, noninflammatory degenerative brain disease clinically and neuropathologically very similar to the above-mentioned disease in man (GIBBS *et al.*, 1968). The authors of the cited report believe thus that *Creutzfeldt-Jakob disease* was experimentally transmitted to the chimpanzee and is caused by a transmissible agent (virus?). It is, on the other hand, not entirely impossible that the said 'disease' represents a group of closely similar neuropathologic conditions related to a variety of 'causes'.

[115] Cf. volume 3, part I, p. 722 of this series.

## 2. Primary Longitudinal Zones (Floor, Basal, Alar, and Roof Plate)

In the preceding account of morphogenesis, the subsection dealing with the relevant events characterizing the ontogenetic evolution of the vertebrate brain considered mainly the sectionalization manifested by a rostro-caudal sequence of vesicular bulges, displaying transverse, fold-like, respectively ridge-like, external and internal circular or semi-circular boundary regions, oriented at an approximately right angle to the fictional longitudinal axis.

This aspect of the process of differentiation finally leads to a 'definitive' or 'adult' configuration which can be adequately described in terms of the conventional and convenient terminology distinguishing the generally recognized five major brain subdivisions, namely telencephalon, diencephalon, mesencephalon, metencephalon, and myelencephalon.

However, in relation to the three-dimensional aspect of the neural tube, two further criteria of subdivision can be applied, supplementing the concept of rostro-caudal arrangement. These additional morphologic interpretations are based on the recognition of (a) definable *continuous caudo-rostral longitudinal zones* running at an approximately right angle to the dorsoventral plane, and (b) of a *stratification of cellular aggregates* into more or less distinct, roughly concentric (internal, intermediate, external) elementary layers within the transverse planes of the tubular neuraxis or its 'vesicular' derivatives, and approximately orthogonal to imaginary radial axes drawn in such planes.

These two principles of subdivision (a, b) have been established in a systematic manner by His (1888, 1890, 1892, 1893, 1904), who distinguished (a) unpaired *roof plate*, paired *lateral plates* divisible into *alar* and *basal plates*, and unpaired *floor plate*, as well as (b) *ependymal, mantle,* and *marginal layers*.

Both sorts of subdivision were briefly considered, in respect to problems of histogenesis, in section 1 of the preceding chapter V (volume 3, part I). From the viewpoint of morphogenesis, the longitudinal zones of His shall be dealt with in the present section as well as in sections 3, 4 and 5. Data on ontogenetic stratification or *'migrating layers'* were reported in several studies by the author and his collaborators (e.g. K. and von Domarus, 1920; K., 1922, 1931, 1937; Miura, 1933). Some use of this criterion of subdivision was made in tracing the ontogenetic derivation of grisea and in attempting to establish homologies. Because of various uncertainties and ambiguities concerning boundaries and suitable systematization of these variably differentiated 'layers', we subordinated this criterion as being secondary to that of the primary configurational connectedness displayed by emerging grisea ('gestalt-orderliness', K. 1929a; topologic connectedness, K. 1967a). Other authors, notably BERGQUIST (1957), BERGQUIST and KÄLLEN (1953, 1954; KÄLLEN, 1965), HUGOSSON (1957), KAHLE (1951, 1956), RÜDEBERG (1961), and SENN (1968), have devoted particular attention to the stratification of embryonic layers, in or from which grisea and fiber systems respectively delimiting fiber zones originate.[115a] CROSBY and her collaborators (1969; cf. also NAUTA and HAYMAKER, 1969) have furthermore stressed the arrangement into lateral, medial, and periventricular zones, i.e. into external, intermediate, and internal strata, for their subdivision of the hypothalamus at the definitive ('adult') stage. Since, however, the present section is concerned with the primary longitudinal zones of HIS (dorso-ventral sequence), further comments pertaining to stratification (internal-external sequence) shall be included in the subsequent sections concerning the regional pattern of the neural tube, particularly of diencephalon (5) and telencephalon (6).

---

[115a] Stressing the fact that some aspects of stratification and cell populations in the embryonic or developing neural tube differ from those at the adult stage, a 'committee' (Boulder Committee, 1970) has proposed a 'revised terminology'. Four 'fundamental layers' are designated as ventricular, subventricular, intermediate, and marginal zones, whose arrangement at five different stages (A–E) is described. All neurons and 'macroglia' elements are said to be derivable from these developmental zones. In accordance with the terminology adopted in the present treatise, the ventricular zone at stage A seems to be the medullary epithelium. At stage B, ventricular, intermediate, and marginal zone may correspond to ependymal (matrix), mantle, and marginal layers, respectively. Ventricular and subventricular zones at stage D might represent an inner and outer sublayer of periventricular matrix. Intermediate zone and a 'cortical plate' (CP) described at stage D and E refer to a particular instance, namely telencephalic corticogenesis of mammals, and should not, in my opinion, be generalized. I would, moreover, add that the subependymal cell plate, stressed by GLOBUS and myself (cf. vol. 3, part I, chapter V, section I) seems to be a joint derivative of ventricular and subventricular zones in the committee's terminology, while the remainder of ventricular zone becomes the ependymal lining.

In addition, the committee refers to a semantic problem involved by the use of the suffix -blast, said to connote a 'proliferating cell' which therefore cannot be one of 'postmitotic' type. However, if that suffix is taken to indicate an 'immature' cell, which, if required in the aspect under consideration, can be further qualified as labile, stable, or perennial (cf. vol. 3, part I, chapter V, section 9), then an element characterized by the suffix -blast may be proliferating or nonproliferating, respectively intermitotic or postmitotic depending on its specific taxonomic histological classification, and regardless of its immature status.

The arrangement and topologic connectedness of the primary longitudinal zones of His as tubular wall components is especially conspicuous in spinal cord and rhombencephalon, where this configuration is clearly displayed during at least some stages of ontogenesis in all classes of vertebrates. It represents thus a generalized vertebrate morphological feature (Fig. 102). It can easily be seen that, in comparison with the spinal tube, the lateral plates of the rhombencephalon manifest some degree of eversion, correlated with a stretching of the roof plate.

As shown in Figure 102A, this eversion can be of moderate degree, while in some vertebrate forms, at some ontogenetic stages, and in some regions of the rhombencephalon, the eversion can become ex-

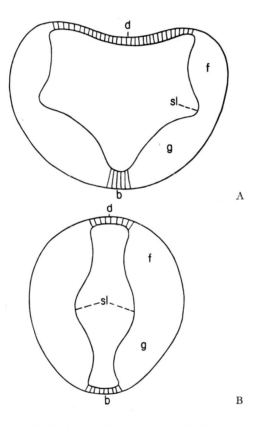

*Figure 102.* The longitudinal *zones of His* in the generalized vertebrate neural tube at their early development stages. A In rhombencephalon. B In spinal cord portion of neuraxis. b: floor plate; d: roof plate; f: alar plate; g: basal plate; f and g: lateral plate; sl: sulcus limitans (from K., 1927).

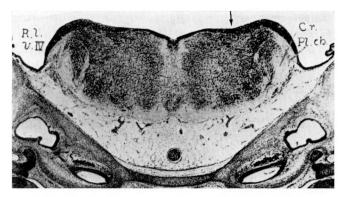

*Figure 103.* Extreme eversion of lateral plates with '*Umstülpung*' of the dorsal portion of the alar plate. Region of recessus lateralis ventriculi quarti (R. l. v. IV) in a human embryo of 28.3 mm length. C.r.: corpus restiforme; Pl. ch.: 'Plica chorioidea' (from HOCH-STETTER, 1929). The added arrow (K.) points to approximate region of poorly defined sulcus limitans.

treme,[116] being combined with inside-out folding (Fig. 103). Such degree of eversion characterizes the so-called rhombic lip as originally described by HIS on the basis of preparations with perhaps additional shrinkage effects exaggerating the manifestations of folding.[117]

---

[116] Cf. the discussion of inversion and eversion on p. 168—176.

[117] Cf. above p. 184 and footnote 58. HOCHSTETTER's contributions to (especially human) brain ontogenesis are of substantial merit for two particular reasons. (1) On the basis of exceptionally well preserved and carefully processed embryonic material, that author published a series of atlases (1919–1929) which provide an extremely valuable documentary material regardless of the at times somewhat pedantic, pedestrian and unnecessarily verbose interpretations in the text. (2) HOCHSTETTER convincingly clarified important points in the ontogenesis of corpus callosum and septum pellucidum in man.

However, for reasons which I was unable to elucidate, HOCHSTETTER seems to display a peculiar animosity toward W. HIS, dismissing, in a most cavalier fashion, many of the outstanding and pioneering contributions of this author by harping on the fact that much of the material used by HIS was evidently poorly preserved, displaying considerable shrinkage and distortion effects. Yet, W. HIS (1831–1904), one of the neuron theory's originators, was an investigator of outstanding acumen, and of far wider outlook than HOCHSTETTER. He may be regarded, in contradistinction to this latter, as a pioneer of the first rank, whose concepts, even if they did not always remain tenable, significantly contributed to further progress in neurobiologic thought based on configurational aspects. RASMUSSEN (1953), in WEBB HAYMAKER's '*Founders of Neurology*' justly comments in this respect: 'Many of HIS's embryologic findings were disputed by HOCHSTETTER, of Vienna, but the fact remains that though his embryologic material was of poor quality he was able to make fundamental discoveries, whereas others with perfect material could not match it with ideas'.

The *roof plate (Deckplatte)* seems to derive from a fusion of the dorsal edges of the neural plate at the groove-stage while changing into that of the neural tube. Although thus arising from the union of a bilateral structure in the region of concrescence, it becomes an essentially unpaired component which may manifest bilateral-symmetric differentiations (e.g. paired choroid plexus arrangements). In some region of the neural tube (spinal cord, 'closed' oblongata, cerebellum, mesencephalon) it appears to provide a glial or ependymal septum, which may become indistinct or entirely unrecognizable at final stages, and particularly in various regions of most vertebrates (e.g. cerebellum and tectum mesencephali). In medulla oblongata, exceptionally in mesencephalon (apparently only in Petromyzon), but commonly in diencephalon and in telencephalon, the roof plate provides the neuroectodermal component of the choroid plexuses. Paraphysis, velum transversum, nonchoroidal membranous rhombencephalic, diencephalic and telencephalic roof walls (e.g. in the telencephalon of Osteichthyes and Choanichthyes) are likewise derivatives of the roof plate.[118] The subcommissural organ (cf. vol. 3, part. I, p. 336), and apparently the epiphysis (including pineal and parapineal organs), also represent differentiations of the epithelial roof plate. Although the participation of this latter in the production of neuronal elements cannot be entirely excluded, very few if any nerve cells (e.g. some dubious pineal elements) seem to originate from that structure. *Mutatis mutandis*, a similar restriction seems to obtain with regard to the floor plate which is considered further below.

The paired *lateral plates (Seitenplatten)* of the neural tube are the main zones providing the bulk of the nervous parenchyma consisting of neuroblastic and spongioblastic derivatives. These plates become subdivided by a longitudinal groove, the *sulcus limitans* of HIS (cf. Fig. 102), into the dorsal *alar plate (Flügelplatte)* and the ventral *basal plate (Grundplatte)*. Although very conspicuous throughout spinal cord and most regions of the deuterencephalon of all examined vertebrate representatives during at least some ontogenetic stages, the identification of this sulcus can become difficult or uncertain at some stages and in some regions of the neuraxis in various vertebrate forms.

As regards the peripheral nerves, the dorsal spinal roots and the branchial nerves are connected with the alar plate, while ventral spinal

---

[118] A detailed study concerning many of these structures *(epitheliale Gebilde am Gehirn der Wirbeltiere)* has been published by Graf HALLER (1922).

roots, third, sixth, and twelfth cranial nerves, respectively spino-occipital roots, connect with, or emerge from the basal plate. Generally
speaking, the efferent neuronal elements concerned with direct output
from the neuraxis seem to originate and to remain located in the basal
plate. The alar plate, on the other hand, contains the grisea mainly concerned with primary input and its further distribution ('primary afferent nuclei'). In some cases, it may also contain the cell bodies of primary afferent neurons (sensory neurons of the 'first order'),[119] such as the
large *dorsal cells of Rohon-Beard* in the larval spinal cord of amphibians
or the vesicular cells of the mesencephalic trigeminal root in tectum
mesencephali of most if not all vertebrates.

    Thus, although both alar and basal plate provide and contain numer-
ous additional correlating grisea, it seems justified, from a 'functional'
viewpoint, to characterize the alar plate as the *'primary afferent zone'* and
the basal plate as the *'primary efferent zone'*. Some authors (e.g. AREY,
1954, and many others) accordingly designate the alar plate as 'sensory', and the basal plate as 'motor'. From a more rigorous viewpoint,
however, these terms seem less suitable, since the word 'sensory' retains mentalistic implications, and the term 'motor' does not, strictly
speaking, subsume efferent fibers carrying impulses for secretion.

    The unpaired *floor plate (Bodenplatte)* represents the ventral midline
neighborhood of the neural tube, and originally consists of medullary
epithelium (cf. vol. 3/I, chapter V, section 1). At subsequent early ontogenetic stages, it is still not possible to determine to which extent this
ventral part of the developing neuraxis is 'non-neuronal' (cf. also
KINGSBURY, 1920–1921). During further growth and differentiation
the floor-plate then becomes clearly demarcated, consisting essentially
of spongioblastic elements forming a rather narrow shelf of ependymal
elements ('primitive' or 'ependymal' spongioblasts). These cells, respectively their processes, may reach from the internal to the external
surface of the tube. Throughout its longitudinal extent in spinal cord
and rhombencephalon, the floor plate remains at first relatively or
completely free of neuronal elements, although it may subsequently
become transformed, at least in some regions, by the inclusion of either migrating or perhaps autochthonous neuroblasts, respectively

---

[119] The classification of neuronal units, respectively nerve fibers, into afferent and
efferent sequences of the first, second, and third respectively higher order was discussed
in chapter V (volume 3/I, pp. 272–274). A brief comment on classification is also included
in section 7, pp. 544–545 of that chapter.

nerve cells. Thus, at least in some Amphibian larvae, as shown by COGHILL (1929) in Amblystoma, commissural neurons seem to derive from the floor plate and were designated as *'floor plate cells'*. These internuncial elements are here of particular functional significance for certain early larval swimming reflexes which are discussed in chapter VIII of the next volume. In addition, the floor plate appears to develop, in Amblystoma and perhaps also some unspecified other forms, from a ventral midline ridge of the neural plate, resting directly upon the notochord, and described by BAKER (1927) as the *'neural keel'*.

With regard to these four fundamental zones, the vertebrate neural plate can be conceived as a bilaterally symmetric configuration, whose unpaired midline region becomes the floor plate, and whose paired lateral halves become the lateral plates, which subsequently differentiate into basal and alar plates. At the neural tube stage, the roof plate results from the dorsal midline fusion of lateral plates. Although in this respect of paired origin, it may be considered a secondarily essentially unpaired median component, which may, however (as e.g. in telencephalic choroid plexuses, and to a lesser extent in other plexus formations), still display paired differentiations.

The neuraxis can thus be interpreted as a tubular organ with four (two paired and two unpaired) wall components, and closed by a dorsal seam[120] with rostral as well as caudal neighborhoods of terminal fusion. This interpretation involves the question of rostral and caudal termination of the four wall components and, in addition, that of defining a suitable conceptual longitudinal axial line.

As regards the spinal tube, the available evidence seems to indicate that all four zones extend to the caudal end, whose closure, involving posterior neuroporus and canalis neurentericus, was dealt with above in section 1 (p. 105). In the spinal tube, the two unpaired zones become substantially modified, being reduced to dorsal and ventral septal structures which, in many instances, may be quite inconspicuous.

The caudal end of the neural plate is evidently not a 'stabilized' structure, but can be assumed as represented, in the 'dynamic' living state, by a neighborhood of proliferating cells originally rostral to the dorsal lip of the blastopore or to the primitive pit.

---

[120] However, such closure by a dorsal seam does evidently not take place in those instances where the neural tube or portions of it develop by differentiation and organization of a more or less solid cell mass (cf. vol. 3, part I, p. 12 and particularly figures 16, 18 loc. cit.).

The question of the rostral termination of the longitudinal zones is of much greater import, since it involves an interpretation of the anterior portion of the brain. Because of various difficulties, it has led to several different conceptions.

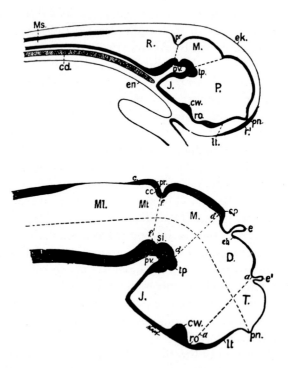

*Figure 104.* Two diagrams showing the embryonic subdivisions of the vertebrate brain according to Kupffer. The lower figure indicates the longitudinal axis of the neural tube as traced by this author. aa: telencephalo-diencephalic boundary according to Kupffer (the basal part of this boundary corresponds to sulcus intraencephalicus anterior; in accordance with my own concept, this line should curve toward the bulge at the right of 'ro', which bulge represents the torus transversus; the dorsal part of the boundary, however, agrees with my own views); c: cerebellum; cc: 'commissura cerebellaris'; cd: notochord; ch: commissura habenulae; cp: commissura posterior; cw: chiasmatic ridge; dd: diencephalo-mesencephalic boundary; e: epiphysis; e': paraphysis; ek: integumentary ectoderm; ff: mesencephalo-rhomboencephalic boundary; lt: lamina terminalis; pn: processus (respectively recessus) neuroporicus; pr: plica rhombo-mesencephalica; pv: 'plica encephali ventralis'; r': 'unpaired' olfactory placode; ro: recessus praeopticus; si: sulcus intraencephalicus posterior; D: diencephalon; J: infundibular region; M: mesencephalon; Ml: myelencephalon; Ms: medulla spinalis; Mt: metencephalon; P: prosencephalon; R: rhombencephalon; T: telencephalon.

With respect to the lines of closure of the neural tube and the designation of '*seams*', HIS (1892) recognized three '*Säume*' corresponding to what he called '*ursprüngliche Nahtlinien*', namely a self-evident dorsal one, a rostral one, and a ventral or 'neurochordal' one.[121] GORONOWITSCH (1893) introduced the term '*sutura terminalis anterior*' for the rostral seam.

KINGSBURY (1920–1921) justly points out that the neural plate was thus interpreted as a completely paired structure , consisting of two separate halves united through concrescence and joined together by the floor plate. KINGSBURY (loc. cit.), although differing from the views of HIS concerning the rostral extent of the longitudinal zones, accepts that author's concept of sutura neurochordalis. In contradistinction to HIS and others who adopted this term, I do not believe that the distinction of a neurochordal suture is justified, since I interpret the neural plate as an unpaired configuration, primarily continuous in the midline, which arises rostral to the dorsal lip of the blastopore of invaginated gastrulae, and rostral to the primitive streak of blastoderms (even if this streak is regarded as having arisen by 'concrescence').

As far as the neural plate is concerned, and discounting the instances where the tubular neuraxis originates from the organization of a solid cell cord (cf. above, p. 291, footnote 120), one might even merely recognize a single dorsal suture with a rostral and a caudal end-region. Since, however, the rostral region of closure, involving formation of the *lamina terminalis*, is of particular significance, it seems justified to stress the morphogenetic importance of that region by a special designation (e.g. '*rostral suture*').[122]

If the rostral end of an imaginary axis of the neural tube is assumed to be located at the spot of ultimate anterior closure, and if this spot in-

---

[121] HIS (1892) stated: '*Alle drei Säume des Medullarrohres entsprechen ursprünglichen Nahtlinien. Am längsten ist die dorsale Naht (d. N.) bekannt. Die ventrale oder neurochordale Naht ist zur Zeit noch von manchen Seiten bestritten*'. '*Ungenügend gewürdigt ist auch die vordere oder frontale Endnaht.*' '*Sie entsteht durch Verbindung der vorderen Bänder der Medullarplatte und nimmt bei allen Wirbeltieren eine durchaus selbständige und charakteristische Stellung ein.*' '*Die Wand des Medullarrohres ist in den an die Nahtlinien anstossenden Strecken im Allgemeinen dünner, als in den beiden Seitenwänden, die verdünnte Strecke der ventralen Röhrenwand ist die sogenannte Bodenplatte, die der dorsalen die Deckplatte. Die verdünnten Nahtstrecken der vorderen Naht liegen im Boden des dritten Ventrikels und in der Lamina terminalis.*'

[122] Thus, although HIS (1892) speaks of '*drei Säume*', he could also have distinguished four by including a 'caudal' suture closing posterior neuropore and neurenteric canal. The concept of a 'caudal suture' would even seem more justified than that of an assumed so-called 'neurochordal seam'.

deed corresponds to the 'recessus neuroporicus' displayed by various vertebrate brains, and located within the telencephalic lamina terminalis, then a tracing of that axis as postulated by KUPFFER (1906) and shown in Figure 104 could be upheld on quite logical grounds.

It is evident that the question concerning the rostral end of the neural plate previously to the formation of lamina terminalis by anterior fusion, differs from the question of the rostral end of the longitudinal axis of the neural tube at the final stages of anterior closure immediately preceding, and partly concomitant with, differentiation of telencephalon impar.

The studies by HIS (1893), JOHNSTON (1909) and KINGSBURY (1920–1921, 1922) led to the conclusion, which agrees with my own observations on a diversified vertebrate material, that the brain portion of the neural plate ends anteriorly with a terminal ridge *(Basilarleiste* on *Endleiste* of HIS). This structure may be regarded as a precursor of the chiasmatic ridge. The rostralward-proceeding dorsal closure of the neural tube, resulting in the formation of the lamina terminalis as a midline structure, ends thus basally at the rostral border of that ridge. A transverse depression between lamina terminalis and chiasmatic ridge becomes the preoptic recess. At the preceding earlier stage, a transverse furrow, called 'primitive optic furrow' or 'primitive infundibulum' (KINGSBURY, 1922), running across the median plane, may delimit the caudal boundary of chiasmatic ridge. KINGSBURY, essentially following JOHNSTON (1909), interprets this furrow as 'continuous with the 'retinal foveae' and later the optic vesicles. However, subsequently to the actual evagination of optic vesicles following differentiation of a recognizable prosencephalon, the lumen of the optic stalk, which obliterates to become the sulcus intraencephalicus anterior, connects with the transverse groove between rostral border of chiasmatic ridge and lamina terminalis, namely with the floor of preoptic recess.

These events, in general, occur before the prosencephalon differentiates into telencephalon and diencephalon through evagination of a telencephalon impar, combined with median forward eversion of lamina terminalis, whereby the torus transversus is formed and the lamina terminalis becomes subdivided into a telencephalic and a diencephalic portion. This latter represents the anterior wall of preoptic recess.

Reverting to the problem concerning the rostral extent of the primary longitudinal zones, HIS (1893) made an attempt to project subsequent brain regions upon the open vertebrate medullary plate as shown in Figure 105. The interpretation of this author concerning the

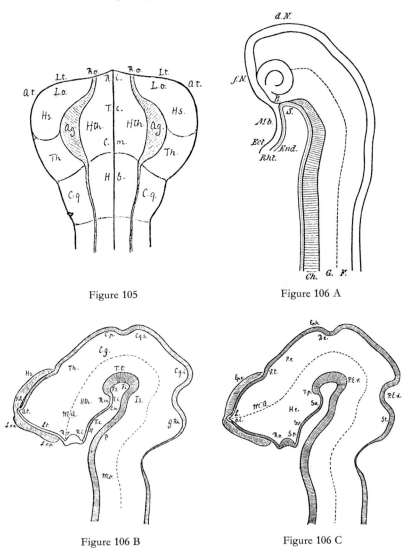

Figure 105

Figure 106 A

Figure 106 B

Figure 106 C

*Figure 105.* Attempt according to His (1893) of projecting the various brain regions upon the open and flattened medullary plate of a Selachian (from His, 1893). This figure should be compared with figures 107 and 111. Ag: locus of eye vesicle; At: locus of angulus terminalis; Cm: locus of mammilary body: Cq: locus of tectum mesencephali; Hb: locus of tegmental ridge; Hs: locus of telencephalic pallium; Hth: locus of hypothalamus; Lo: locus of 'lobus olfactorius' (olfactory bulb); Lt: locus of lamina terminalis; Ri: locus of recessus infundibuli; Tc: locus of tuber cinereum; Th: locus of 'thalamencephalon'. The locus of optic chiasma is assumed to lie between Ro and Ri. The sequence of the loci Ri, Tc, Hth, Cm, Hb is presumed to represent the basal plate.

rostral end of floor, basal and alar plates are depicted in Figures 106, 107 B and 108 B. JOHNSTON's concept is shown in Figures 107 C and 108 C. In both interpretations, the preoptic recess approximately indicates the locus of the rostral end of the neural plates, although, in HIS' view, the 'recessus basilaris' ('primary recessus infundibuli') represents a transitory end preceding formation of chiasmatic ridge.[123] In contradistinction to this view, SCHULTE and TILNEY (1915) assumed that the brain plate did not extend farther than the mammillary region (Figs. 107 A and 108 A). Accordingly, not only the lamina terminalis, but the entire floor of the prosencephalon would be formed by closure of the 'rostral suture'. Neither the subsequent investigations of KINGS-BURY (1920–1921, 1922) nor my own observations on early vertebrate embryonic stages are in agreement with this interpretation. On the

---

[123] In figure 108 B KINGSBURY has not correctly indicated HIS' concept concerning the rostral end of floor plate. It is evident (cf. fig. 106 B) that HIS traced this plate as far as the '*Basilarleiste*' or '*Endleiste*' which, by concrescence, forms the chiasmatic ridge. The primary infundibulum of HIS (Recessus infundibuli s. basilaris) is not the infundibulum indicated in figure 108 B, but the transverse furrow indicating the caudal border of chiasmatic ridge.

---

*Figure 106.* Diagrams of HIS, indicating that author's concept of the rostral end of the embryonic neural tube. A Relationships of bent brain tube to notochord, *Seessel's pouch*, and pharyngeal membrane (from HIS, 1892). B: basilar ridge; dN: dorsal seam; Ect: ectoderm; End: entoderm; fN: frontal (rostral) seam; F: alar plate; G: basal plate; Mb: stomodeum; Rht: stomodeal membrane; S: *Seessel's pouch*. The bend between dN and fN indicates approximate locus of angulus terminalis.

B Longitudinal brain axis (corresponding to sulcus limitans, and its assumed rostral end) at a later stage of human ontogenesis. At: angulus terminalis; Cg: locus of geniculate bodies; Cm: corpus mammillare; Cqi, Cqs: corpora quadrigemina of tectum mesencephali: Cp: locus of pineal body; Ei: eminentia interpeduncularis; Fch: fissura chorioidea; Fi; fossa interpeduncularis; Grh: genu rhombencephali; Hs: hemisphere of telencephalon; Hth: hypothalamus; If: locus of (secondary) infundibulum; Is: isthmus rhombencephali; Loa: 'lobus olfactorius impar'; Lop: 'lobus olfactorius posterior'; Lt: lamina terminalis; Mo: medulla oblongata; Ri: (primary) recessus infundibuli s. recessus basilaris; Rm: recessus mammillaris; Ro: recessus (prae)opticus; P: locus of pons; Tc: tuber cinereum; Th: thalamus; Tt: 'tuber tegmentale' (tegmental ridge); dotted line: sulcus limitans.

C KUPFFER's concepts, as interpreted by HIS, and drawn in the brain profile of the preceding figure B. De: synencephalic region of diencephalon; Epe: 'epencephalon' (pallium telencephali); Eph: epiphysis; He: hypencephalon (hypothalamic region); Loi: 'lobus and recessus olfactorius impar'; Pe: parencephalic region of diencephalon; PEd: plica encephali dorsalis; PEv: plica encephali ventralis; Ro: recessus opticus; Sd: saccus dorsalis hypencephali; Sv: saccus ventralis hypencephali; Sp: sinus postopticus; Tp: tuberculum posterius; Vt: velum transversum (B and C from HIS, 1893).

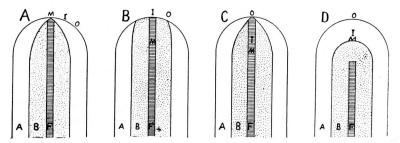

*Figure 107.* Schema to illustrate the interpretations of SCHULTE and TILNEY (A), HIS (B), JOHNSTON (C), and KINGSBURY (D) of the fundamental plan of the brain plate, in terms of primary longitudinal zones (from KINGSBURY, 1922). A: alar plate; B: basal plate; F: floor plate; I: primitive (primary) infundibulum; M: mammillary recess; N: noto-chord; O: preoptic recess.

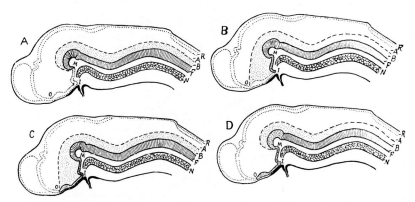

*Figure 108.* Schema to illustrate the interpretations shown in preceding figure, but with respect to the closed brain tube (from KINGSBURY, 1922) R: roof plate; other designa-tions as in figure 107.

other hand, SCHULTE's and TILNEY's findings concerning the rostral end of sulcus limitans are of substantial importance and could be cor-roborated by KINGSBURY as well as by myself. SCHULTE and TILNEY were apparently the first authors pointing out that sulcus limitans, as-sumed by HIS to be continuous with the *sulcus hypothalamicus Monroi*, and terminating in the preoptic recess,[124] actually seems to end with a

---

[124] This interpretation by HIS is easily understandable because of secondary fusions between sulci at certain stages, and in certain forms. HIS may have indeed observed a continuity between sulcus limitans, sulcus diencephalicus ventralis, and sulcus intraen-cephalicus anterior, whereby a linear sequence corresponding to that author's concept is indeed provided (cf. further below in section 5).

ventral convexity at the mesencephalo-prosencephalic boundary in the region of mammillary recess.

STREETER (1912, 1933), however, essentially retained the view of HIS concerning the rostral end of basal plate. Yet, STREETER's actual findings, reported in 1933, rather clearly show the rostral end of sulcus limitans in accordance with the views expressed by SCHULTE and TIL-NEY, KINGSBURY, and ourselves (Fig. 136).

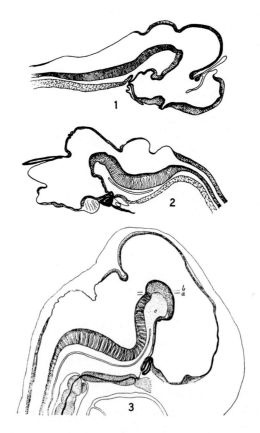

*Figure 109.* Reconstructed approximately midsagittal sections through embryonic brains of a Selachian, a Sauropsidan, and a mammalian, showing rostral end of floor plate (from KINGSBURY, 1920–21). 1: *Squalus acanthias* of 40-mm total length; 2: chick of $7^1/_2$ days incubation; 3: calf embryo of 23-mm length. The floor plate is indicated by the parallel, somewhat irregular lines representing its ependymal or spongioblastic 'fibers'. Labelings in 3 not relevant to present discussion.

The investigations of KINGSBURY (1920–1921, 1922) on embryos of sharks, birds, and mammals (Figs. 109, 110, 111) showed that the floor plate cannot be traced farther rostrad than the rhombencephalo-mesencephalic boundary, and seems to end at the fovea isthmi (cf. p. 182). KINGSBURY is inclined to accept, with some modification, HIS' concept of a sutura neurochordalis, and considers floor plate and noto-chordal plate as primarily coextensive, thus recognizing epichordal and prechordal portions of the brain.

However, while such distinct brain portions doubtless can be de-fined in the vertebrate neuraxis, the notochord does nevertheless ex-tend further rostrad than the level of fovea isthmi and may reach the mammillo-infundibular region (cf. Figs. 48 C, 51 A, 53 C, 55 C, 69 A).

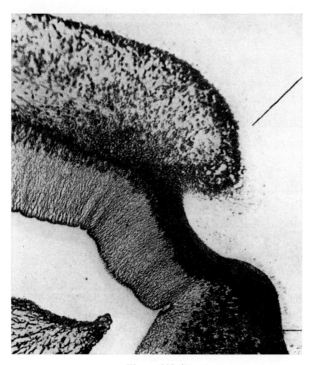

Figure 110 A

*Figure 110.* Photomicrographs of median plane of brain tube floor at plica encephali ventralis and fovea isthmi, to be compared with reconstructions in preceding figure (from KINGSBURY, 1921–22). A *Squalus acanthias* of 40 mm ($\times$67.5); only the region between the two lines is median. B Calf embryo of 23-mm length ($\times$37.5). Both figures reduced $^7/_{10}$.

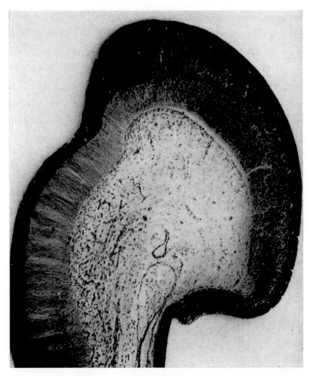

Figure 110 B

Yet it may, on the other hand, lose its close contiguity to floor plate in the region of fovea isthmi.[125]

Moreover, discounting the entirely epichordal location of the neuraxis in Amphioxus (Figs. 43–45) as irrelevant for detailed comparison

---

[125] KINGSBURY (1934), referring to investigations by GAGE (1917), SUNDBERG (1924), and ALOISI (1932), stresses that a noticable amount of glycogen is displayed in the raphé-neighborhood of the floor plate *('septum medullae')*. This glycogen-containing raphé does not seem to extend rostrally beyond the fovea isthmi (cf. also VAAGE, 1969, who uses the term 'nonproliferating cell column', 'NPC', in referring to the floor plate). The notochord is likewise rich in glycogen (concerning the presence of this substance in the CNS, cf. also vol. 3/I, chapter V, p. 137 of this series). While SUNDBERG and KINGSBURY seem to consider notochord and glycogen-containing notochord as coextensive, VAAGE (1969) justly remarks that the notochord in the chick (and, according to my observations, presumably in most or many vertebrates) terminates rostrally in the mammillary region. VAAGE thus states that from the relevant ontogenetic stages onward, the notochord and the glycogen-containing 'ventral raphé' are not coextensive: 'Consequently, no correlation seems to exist between the notochord and the glycogen-containing raphé as supposed by SUNDBERG (1924) and KINGSBURY (1934)' (VAAGE, loc. cit.).

with conditions in true vertebrates, the rostral extent of notochord in these latter becomes not only secondary *qua* adult condition, but also greatly varies from stage to stage and in different forms (cf. also HIS, 1892 and KINGSBURY, 1921–1922). The assumed particular relationship between the evolution of notochord and that of neuraxis, although *prima facie* not improbable, remains still poorly elucidated and cannot yet be formulated in a satisfactory manner. This question is, moreover, related to various, at least at present, quite undecidable embryologic problems concerning the so-called 'prechordal plate', the so-called 'preoral entoderm' and diverse obscure aspects of the ontogenesis of the vertebrate head as well as of inferred phylogenesis.

Again, as regards KINGSBURY's view concerning the rostral end of floor plate at the fovea isthmi, I would prefer, on the basis of my own observations, a formulation stating that said region corresponds to the gradual disappearance of floor plate within an undefined rostrally adjacent neighborhood. Looking at Figure 110B, from KINGSBURY's

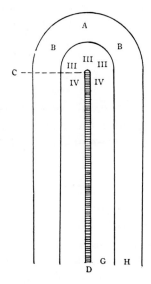

*Figure 111.* Diagram to illustrate KINGSBURY's interpretation of the rostral portion of the neural plate (from KINGSBURY, 1920–21). A: region of retinal areas; B: region of 'olfactory lobes' and 'cerebrum' (i.e. telencephalon); C: rostral end of floor plate or 'sutura neurochordalis' (fovea isthmi in the neural tube); D: floor plate ('sutura neurochordalis'); G: basal plate; H: alar plate; III: locus of nucleus n. oculomotorii; IV: locus of nucleus n. trochlearis. This diagram, which is doubtless better founded than the early attempt by HIS, shown above in figure 105, should also be compared with figures 107 and 112.

original material, it will be seen that the 'fibrous' structure of floor plate seems to 'taper out' at levels rostral to fovea isthmi. Although, at some stages and in some forms, a rather sudden termination at fovea isthmi may be displayed, the rostralward extending gradual 'fading out' of floor plate is not infrequently manifested.

Another important result of the cited investigations by KINGSBURY was the conclusion, based on additional and rather convincing evidence, that the basal plate, as already suggested by SCHULTE's and TILNEY's study (1915) does not extend beyond the mesencephalo-prosencephalic boundary, and that the sulcus limitans, if properly recognizable, ends in the region of the mammillary recess (Figs. 107D, 108D, 111). Except for the aforementioned boundary region, the entire prosencephalon and its derivatives, namely telencephalon and diencephalon, must therefore be considered as formed by the rostral extension of paired alar plate and essentially unpaired roof plate. KINGSBURY, who comments on comparable features of roof and floor plates in his paper of 1920–1921, does, however, not specifically refer to the rostral end of roof plate in his communication of 1922 which discusses the fundamental plan of the vertebrate brain.

According to KINGSBURY (1922), the outstanding feature, as far as the brain tube is concerned, becomes manifested by the great expansion of its dorsal portion, 'effecting a rotation, as it were, roughly around the regions of preaxial mesoderm as a center or axis. This brings thus, dorsal regions ventral, ventral aspects more dorsal, more caudal material cephalad – thus reversing the primitive sequence, with the more rostral expansion primitively dorsal. Certain of these rotations are more extreme in the embryo than in later stages'.

'Little need be added to what was previously said upon the effect of the interpretation upon the relations of alar plate, basal plate, and sulcus limitans in the prechordal portion of the neural tube. The boundary between them as primary sensory and motor zones must necessarily be here indefinite and perhaps indeterminate. The boundary clearly lies, I think, in the region of the mammillary recess. The question, however, possesses no embryological bearing and I gladly leave it to the consideration of neurologic workers' (KINGSBURY, 1922).

The results of our own studies (K. 1924, 1926, 1927, 1929a, b, 1931, 1934, 1936, 1937, 1939, 1956; GERLACH 1933, 1947; MILLER, 1940; MIURA, 1933) are in essential agreement with most of the interpretations by KINGSBURY. As indicated in Figure 112, we assume that, rostral to the mesencephalo-prosencephalic, respectively mesencephalo-diencephalic boundary zone, the entire 'neuronal' wall of the brain tube should be regarded as a derivative of alar plate. To this, of course, must be added, as mentioned above, the persisting derivatives of the roof plate. As regards the concept of sutures propounded by His and other authors, it seems therefore convenient to distinguish a *dorsal suture*, involving the *roof plate*, from an anterior or *rostral suture* involving the fusion of *alar plate* (Figs. 105, 106, 107 D). This latter suture provides the *lamina terminalis*. Although at early stages, and at rostral levels, a distinction of roof plate from alar plate cannot always be made with sufficient clarity, the *angulus terminalis* of His (cf. above p. 150), if recognizable, seems to indicate a convenient boundary between the dorsal suture (of roof plate) and the rostral suture (of alar plate) forming the lamina terminalis.

Concomitantly with the differentiation of prosencephalon into diencephalon and telencephalon, various complicating features arise in connection with the transitory so-called neuromeric stage, involving semivesicular components and their transformations. These events

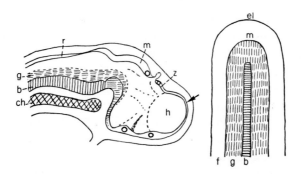

*Figure 112.* Rostral end of longitudinal zones of His in essential agreement with KINGSBURY's views, and in conformity with my own interpretation (slightly modified from KUHLENBECK, 1927). b: floor plate; ch: notochord; el: terminal or basilar ridge; f: alar plate; g: basal plate; h: telencephalon; m: (at left) mesencephalon; (at right): end of basal plate in diencephalo-mesencephalic boundary region; r: rhombencephalon; z: diencephalon. The curved line from the self-evident velum transversum to torus transversus with anterior commissure (likewise not labeled) is the telo-diencephalic boundary. The three broken lines directed toward this boundary indicate, in dorso-basal sequence, the loci of sulcus diencephalicus dorsalis, medius, and ventralis. The lines related to chiasmatic ridge (not labeled) are sulcus intraencephalicus in recessus praeopticus (rostral) and locus of sulcus lateralis infundibuli (sulcus lateralis hypothalami posterioris). The broken line within mesencephalon, caudal to sulcus diencephalicus dorsalis, indicates locus of sulcus lateralis mesencephali. Arrow indicates telencephalic roof plate.

will be considered below in section 3. As regards the diencephalic and telencephalic alar plate, two specific fundamental longitudinal zonal systems, characterizing diencephalon respectively telencephalon, and differing from the longitudinal alar plate zones in deuterencephalon (mesencephalon and rhombencephalon), subsequently become manifest. These zonal systems, of which the diencephalic one arises in superposition upon a fading transverse neuromeric segmentation into parencephalon and synencephalon, represent a definitive arrangement in numerous Anamnia, but are further modified by considerable additional differentiations in Amniota. The morphologic significance of this arrangement, which was extensively studied in our own investigations, will be discussed in sections 5 and 6. Supplementary evidence concerning the assumed rostral end of the longitudinal zone to which the term 'basal plate' can be appropriately applied shall likewise be included in section 5.

KIRSCHE (1967), who started out from a different methodologic approach, has elaborated a concept of *postembryonic matrix zones* of the vertebrate brain based on processes of histologic differentiation including 'matrix consumption'. The cited author pointed out that development and localization of these matrix zones have significant relationships to the morphological zonal pattern configuration outlined in our own studies. Additional investigations undertaken in KIRSCHE's Institute by KRANZ and W. RICHTER (1969, 1970) as well as by SCHULZ (1969) deal with various aspects of this approach whose further developments must be awaited.

In this connection it should also be again recalled (cf. above p. 286), that SENN (1968, 1969, 1970), on the basis of studies especially concerning the reptilian brain, has particularly emphasized the relevance of an orderliness of stratification within the wall of the neuraxis. This author regards the principle of stratification in a central-peripheral sequence within the tube as a third general type of neuraxial configuration in addition to the dorso-ventral sequence of longitudinal zones, and to the rostro-caudal sequence of main subdivisions.


### 3. Neuromery; Definitive Main Subdivisions


MARCELLO MALPIGHI (1628–1694) was perhaps the first author known to have recorded three cerebral vesicles (presumably prosencephalon, mesencephalon, and rhombencephalon) in the developing chick (1673). TIEDEMANN (1781–1861), who refers to MALPIGHI,[125a] likewise described cerebral vesicles in the human fetus (1816). K. E. VON BAER (1828) is generally credited with the first systematic description of a segmentation manifested by a regularly occurring sequence of vesicular or bulging configurations in the chick's developing neural tube. Other early investigators, such as REMAK (1855) and DURSY (1869), likewise recorded the 'segments' displayed during ontogenetic development of the neuraxis. MIHALKOVICS (1877) used the designation 'folds' ('*Falten*'), and ORR (1887) seems to have introduced the term '*neuromeres*'. This author particularly emphasized the lateral ('dorso-ventral') constrictions, correlated with corresponding ridges on the internal surface of the tube.

A thorough analysis of the problems related to the morphogenetic significance of neuromery was undertaken by v. KUPFFER (1906) in HERTWIG's '*Handbuch der Entwicklungsgeschichte*'. KUPFFER based this account on his own extensive and pioneering studies of the ontogenetic development of the vertebrate neural tube, and on a critical review of the then already extensive literature characterized by numerous disagreements among the investigators of neuromeric segmentation.

KUPFFER distinguished a *primary neuromery*, displayed by apparent segmentation of neural plate or neural groove before closure, from *secondary neuromery* which becomes fully manifested after complete closure of the brain tube. As is well-known, the closure of the cerebral tube proceeds, as a rule, anteriorly in a caudorostral direction, and posteriorly from the original closure in an undefined region of rhombencephalon, 'downward' in a rostrocaudal direction. This rostrocaudal closure continues along the spinal tube, whose differentiation is accompanied by that of somites and notochord, respectively axial skeleton, in accordance with a rostrocaudal axial morphogenetic gradient (*Differenzierungsgefälle*, cf. K., 1930) recorded by a number of authors. Because the available, rather contradictory data concerning primary neuromery still remain dubious, being insufficiently documented by unambiguous findings, and, moreover, not particularly relevant to the problems here

---

[125a] In vol. 2 of a five-volume work '*Marcello Malpighi* and the Evolution of Embryology', ADELMANN (1966) has included the original Latin text, with English translation, of MALPIGHI's treatises '*De formatione pulli in ovo*', and '*De ovo incubato observationes*'. Previously, ADELMANN (1942) had published, with translations and comments, a facsimile reproduction of 'The embryologic treatises of *Hieronymus Fabricius ab Aquapendente*' (1533–1619) of Padua, who was the teacher of WILLIAM HARVEY (1578–1657).

under consideration, only the significant questions pertaining to secondary neuromery shall be discussed in this section.

With respect to this neuromery, KUPFFER regarded the following statements as generally valid:

1. At the time when the brain tube becomes rostrally detached from the epidermal ectoderm and a rhombo-mesencephalic boundary is recognizable, at least six neuromeres of the rhombencephalon can be distinguished caudally to that boundary. In some forms, 7 rhombomeres seem to be present. In the urodele amphibian *Amblystoma*, however, the second and the third of these neuromeres are not separated from each other.

2. The first rhombencephalic neuromere provides the cerebellum. The second is primarily connected with the maxillo-mandibular ganglion of the trigeminal nerve, the fourth with the acustico-facial nerve root complex, the fifth with the glossopharyngeal nerve. The third does not display a nerve-connection.

3. About the same time period, five neuromeres, separated by external grooves, can be recognized in the region of prosencephalon and mesencephalon. If the paired (i.e. bilaterally symmetric) groove, at which dorsally the posterior commissure becomes first recognizable, is taken as a boundary mark, then three of these neuromeres pertain to prosencephalon, and two are mesencephalic.

In concluding these statements, KUPFFER (loc. cit.) cautiously remarks: '*Alles weitere erscheint noch unsicher, so namentlich die Herleitung aller dieser Neuromeren einzeln für sich von ebensoviel primären Neuromeren der Neuralplatte resp. des massiven Neuralstranges. Es empfiehlt sich daher, diese Neuromerie als sekundäre von der primären zu unterscheiden. Die sekundäre Neuromerie des Hirnes verstreicht danach, ist überhaupt am Boden des Vorder- und Mittelhirngebietes nicht so ausgesprochen wie am Dache und an den Seitenwänden*'.

The results of subsequent investigations by KAMON (1906), MEEK (1907), GRÄPER (1913), SCHUMACHER (1928), HALLER (1929), and various others can be interpreted as essentially substantiating KUPFFER's conclusions, with, however, a few relatively minor differences. Thus, KAMON (1906), who studied the ontogenesis of the chick's brain, saw two distinctive mesencephalic segments only at a very early stage, the mesencephalon then becoming a single vesicle. As regards the rhombencephalon, he found the glossopharyngeal nerve related to the sixth rhombomere. MEEK (1907), in his paper on neuromery of the gull *(Larus fuscus)*, referred rhombomeres 5 and 6 to n. abducens, rhombomere

7 to n. glossopharyngeus, and rhombomere 8 to n. vagus. GRÄPER (1913), who investigated the rhombomeres of chick, sheep and pig embryos, referred rh 5 to n. abducens, rh 6 to n. glossopharyngeus, and rh 7 to n. vagus. SCHUMACHER (1928), in the European lapwing *(Vanellus cristatus)*, describes a single mesencephalic vesicle and a sequence of five rhombomeres which represent KUPFFER's rhombomeres 2 to 6. He admits, however, two additional, less clearly demarcated bulges *('Erhebungen')* corresponding to KUPFFER's rh 1 and rh 7. SCHUMACHER's prosencephalic neuromeres are comparable to those of KUPFFER.

Subsequently to an investigation of the diencephalic longitudinal zonal pattern in Anamnia (K., 1929b), which becomes predominant at stages characterized by fading or disappearance of the two transitory neuromeres parencephalon and synencephalon, it was necessary to deal with the problem of transitory neuromery in the course of our studies on reptilian, avian, and mammalian embryos (K., 1931, 1935, 1936, 1956; MIURA, 1933).

It became apparent that, while the diencephalic longitudinal zones of Anamnia represent, generally speaking, a rather simple 'permanent' configurational arrangement which may conveniently be studied at stages displaying the definitive brain subdivision, these zones are clearly recognizable in Amniota only at transient key stages of ontogenetic evolution. At these stages, the longitudinal arrangement becomes superimposed upon, and supplants, the fading transverse neuromeric parencephalic and synencephalic segmentation. In the course of these events, certain stages reveal the zonal features of the diencephalon as distinctly as it appears in many adult Anamnia. In Amniota, however, this 'fundamental' or 'primitive' pattern rather rapidly breaks up into many separated diencephalic grisea. Without reference to the transient key stages mentioned above, it becomes difficult if not impossible to recognize and to trace the original longitudinal zones in the Amniote diencephalon.[126]

Following these studies, the morphologic significance of secondary neuromery was then evaluated (K., 1935) from a more generalized viewpoint, summarizing our own observations on the segmentation of

---

[126] Although the differentiation of the telencephalic longitudinal zonal pattern of Amniota (cf. section 6) does not involve the disappearance of transitory neuromeres, the tracing of these zonal derivatives in the Amniote telencephalon involves, *mutatis mutandis*, very similar problems requiring the study of appropriate transient 'key stages'.

the vertebrate brain tube, and reviewing the most pertinent data available in the literature (cf. the tabulations Figs. 113, 114).

In accordance with these data it could be stated that essentially homologous segments *('neuromeres')* of the developing neural tube seem to occur in all vertebrates. Likewise, a fairly well definable topographic relationship of cerebral nerves to some of these 'neuromeres' could be assumed as commonly obtaining. Despite complications by the successive arising, in rostral and caudal direction, of new segments, some of which are transitory, and thus subsequently again disappear, reasonably well definable boundaries or rather boundary regions *(topologic neighborhoods)* between the definitive brain subdivisions can be traced to several of these 'neuromeric' configurations (K., 1935, 1954, 1956; MIURA, 1933). In addition to the morphologic homology of early vertebrate brain segments, a *promorphologic homology*, which applies to comparison within one and the same organism, namely in this case *homodynamy (metameric homology)* can be admitted as obtaining with regard to, and within, at least some sets of these 'neuromeric' segments[127] (K., 1935b, 1967a).

Subsequent further detailed investigations on vertebrate neuromery were undertaken by BERGQUIST and KÄLLEN (e.g. BERGQUIST, 1952, 1956, 1964a, b, 1966; BERGQUIST and KÄLLEN 1953a, b, 1954; KÄLLEN, 1953; KÄLLEN and LINDSKOG, 1953). In an extensive and valuable study on the morphology of the diencephalon in Anamnia, BERGQUIST (1932) had made use of numerous fairly early stages of forebrain development, many of which are characterized by complex transitory patterns related to the disappearance of certain neuromeric segmentations and their superposition as well as replacement by the definitive longitudinal pattern. In our own studies on the diencephalic pattern in Anamnia (K., 1929b; GERLACH, 1933, 1947; MILLER, 1940), we had particularly considered the definitive fundamental longitudinal zones. Thus, rather late embryonic as well as (preferably young) adult stages were especially dealt with. Divergences between the conclusions, interpretations, or concepts propounded by BERGQUIST and those expressed in our communications are presumably mainly related

---

[127] The questions pertaining to this neuromeric homodynamy are briefly discussed further below on p. 327f.

Übersichtstabelle der sekundären Neuromerie

| | v. KUPFFER, 1906, Wirbeltiere, allgemein | KAMON, 1906, Hühnchen | MEEK, 1907, Larus fuscus | GRÄPER, 1913, Hühnchen und Säugetiere | SCHUMACHER, 1928, Vanellus cristatus | HALLER, 1929, Wirbeltiere |
|---|---|---|---|---|---|---|
| Vorderhirn | 1 Telencephalon | 1 Telencephalon | 1 Olfactorius | | 1 Telencephalon | B I |
| | 2 Parencephalon | 2 Parencephalon | 2 Opticus | | 2 Parencephalon | B II |
| | 3 Synencephalon | 3 Synencephalon | 3 — | | | |
| | | | | | 3 Synencephalon | B III |
| Mittelhirn | 1 | 1 | 1 III | | 1 | B IV |
| | 2 | 2 | 2 IV | | | B V |
| Rautenhirn | 1 Kleinhirn | 1 Kleinhirn | 1 Kleinhirn | 1 | 1 | |
| | 2 V | 2 V | 2 V | 2 V | 2 V | |
| | 3 — | 3 — | 3 — | 3 — | 3 — | |
| | 4 VII/VIII | 4 VII/VIII | 4 VI/VIII | 4 VII/VIII | 4 VII/VIII | |
| | 5 IX | 5 — | 5 } VI | 5 VI | 5 — | |
| | 6 | 6 IX | 6 } | 6 IX | 6 IX/X | |
| | 7 | | 7 IX | 7 X | 7 | |
| | | | 8 X | | | |

Figure 113. Tabulation of secondary neuromery in accordance with the observations of various authors up to 1929 (from K., 1935b).

Übersichtstabelle zur Beurteilung der Neuromeren

| Phylogenetische Bewertung | Encephalomer | Nervenbeziehungen | Bleibende Hirngliederung |
|---|---|---|---|
| Archencephalon | Telencephalon | I | Hemisphären |
| | Parencephalon | II (?) | Zwischenhirn |
| | Synencephalon | | |
| | M 1 | III (?) | Mittelhirn |
| | M 2 | IV (?) | |
| | R 1 | | Kleinhirn |
| Deuterencephalon | R 2 | V | |
| | R 3 | V (?) | |
| | R 4 | VII/VIII | Medulla oblongata |
| | R 5 | VI (?) | (im weiteren Sinne) |
| | R 6 | IX | |
| | R 7 | X | |

*Figure 114.* Evaluation of neuromery as propounded in 1935 on the basis of the author's studies (from K., 1935b).

to the difference in our respective theoretical and methodological approaches.[128]

As regards the question of neuromeric segmentation of the brain tube, BERGQUIST (1964b, 1966) distinguishes five different phases, namely (1) *proneuromery;* (2) *interneuromery I;* (3) *neuromery;* (4) *interneuromery II,* and (5) *postneuromery.* The 'neuromeric' phases are said to be characterized by a prominent transverse segmentation, which appears less distinct 'or is even quite extinguished' during the two interneuromeric phases. During the postneuromeric phase, a longitudinal ar-

---

[128] The term 'methodological' subsumes here, *inter alia,* the emphasis on certain relevant stages in ontogenetic evolution. With respect to the diencephalon of Amniota, we have likewise especially considered the longitudinal zonal system. This required particular attention to those stages in which these zones became superimposed upon a fading neuromeric segmentation and still display a rather simple configuration easily comparable with that in Anamnia (reptiles, K., 1931; birds, K., 1936; mammals, MIURA, 1933).

rangement of cell populations in the wall of the neural tube also appears, which, according to Bergquist (1964b) represents 'the so-called *His-Herrick longitudinal cell columns*'. In this manner, a '*squared pattern*' is said to arise. These squares were designated as '*Grundgebiete*' by Bergquist (1932), and later as '*migration areas*' (Bergquist and Källen, 1953a, 1954). From said areas, the various grisea are assumed to derive by migrations during the successive so-called migration phases I and II, followed by gradual differentiation. Figures 115–117 illustrate the concepts formulated by Bergquist and Källen.

A still more recent, and very detailed study on the segmentation of the primitive neural tube in chick embryos, including histochemical and autoradiographic investigations, was published by Vaage (1969). The intricate system of subdivisions described by this author is shown in Figure 118 A. The terminology and notations used by Vaage are

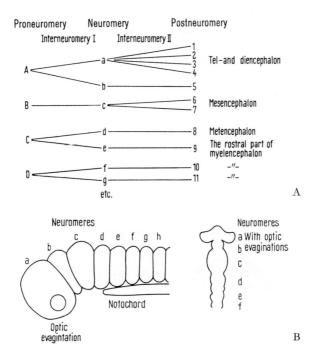

*Figure 115 A*. Diagram of proneuromery, neuromery, and postneuromery, with interneuromeric phases I and II according to Bergquist's concepts (from Bergquist, 1964b).

*Figure 115 B*. Neuromeres a–f according to Bergquist (from Bergquist, 1964b; at left, read optic evagination instead of evagintation).

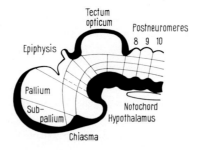

*Figure 116.* Migration area pattern according to BERGQUIST (from BERGQUIST, 1964b)

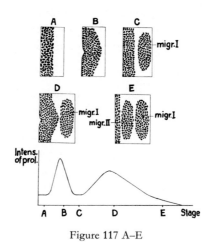

Figure 117 A–E

Figure 117 F

*Figure 117.* BERGQUIST and KÄLLEN's concept of migration layers (from KÄLLEN, 1955). A–E Diagrams depicting the formation of successive migration layers, correlated with changes in mitotic activity illustrated by the curve. F Diagram showing the assumed relationship between a migration layer and two adjacent migration areas. The diagram is said to represent a cross-section through the brain wall, with the ventricular surface at bottom.

adapted to, but somewhat modified from, those of previous authors. *Encephalomeres (prosomeres, mesomeres, rhombomeres)*, and *myelomeres* (of the spinal tube) are distinguished. In contradistinction to BERGQUIST and KÄLLEN, VAAGE did not observe any 'reduction' of the neuromeres during the 'interneuromeric phases' of the cited authors. He states that the neuromeres subdivide and are transformed during the early developmental stages. Thus, 'new encephalomeres are formed by successive subdivisions of the preceding neuromeres. In this way eight prosomeres, two mesomeres and eight rhombomeres are created'. Hence, with reference to the concepts of BERGQUIST and KÄLLEN, there are two different recent points of view concerning both the segmentation pattern of the brain tube and the number of neuromeres.

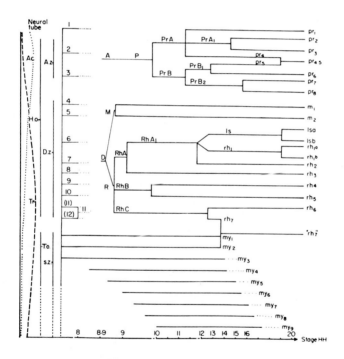

*Figure 118 A.* Diagram illustrating VAAGE's concept of 'neuromery' (from VAAGE, 1969). The subdivisions of archencephalon (A), deuterencephalon (D) and spinal cord (sz, my) are shown at different stages of development in the notation HH8 etc. of HAMBURGER and HAMILTON (1951). In addition, the diagram displays the assumed relationships between the neural tube and the 'primary neuromery' of HILL (1900), indicated by the vertical sequence 1–12, the archencephalic (Az), the deuterencephalic (Dz), and the spinochordal (sz) zone of LEHMANN, the head (Ho) and trunk (To) organizer of SPEMANN, the activating (Ac) and the transforming (Tr) principles of NIEWKOOP, as indicated on the left.

According to VAAGE, an increasing number of neuromeres is formed by successive subdivisions of the preceding ones. 'According to the view advocated by BERGQUIST and KÄLLEN, on the other hand, three different series of neuromeres occur' (VAAGE, loc. cit.). It should, however, be noted that concerning the diencephalic neuromeres, whose transitory nature I have particularly emphasized, VAAGE remarks: 'The diencephalic neuromeres disappear shortly after their formation. According to my observations some of the neuromeric sulci in chick

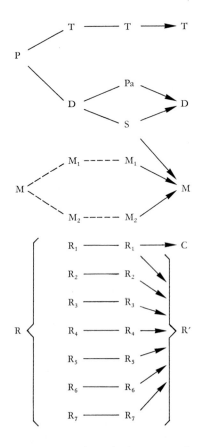

*Figure 118B.* Diagram illustrating the author's concept of cerebral tube neuromery. C: cerebellum; D: diencephalon; M: mesencephalon; $M_1$, $M_2$: transitory mesencephalic segments; P: prosencephalon. Pa: parencephalon; R: rhombencephalon; R': adult rhombencephalon exclusive of cerebellum; $R_{1-7}$: transitory rhombomeres; S: synencephalon.

can be traced stage by stage without the preceding transformation and disappearance into permanent sulci'.[129]

With respect to events in the spinal tube, which are of secondary interest in the aspect here under consideration, VAAGE expresses his views as follows, in general accordance with previous authors such as McCLURE, ZIMMERMANN, NEAL, LILLIE, and KINGSBURY, whose pertinent publications are listed in his bibliography: The segmentation starts rostrally in the prospective spinal cord and spreads caudalward in one wave of bulges. The myelomeres are thus formed in a rostrocaudal sequence in accordance with the generally accepted embryological differentiation pattern. The disappearance of the myelomeres likewise proceeds rostrocaudalwards subsequently to the formation of the spinal nerves.

The *overall* interpretations of VAAGE (1969) are formulated in the following conclusions:

'A. The patterns of segmentation are different in the cerebral tube and the spinal tube.

a) At early ontogenetic stages the encephalomeres and the myelomeres have a histological structure conforming to that of the neuromeres of ORR.

b) The cerebral tube differentiates initially in a caudorostral direction, the spinal tube in a rostrocaudal direction.

c) At later stages segmentation proceeds differently in the prosencephalon, the rhombencephalon and the spinal tube.

---

[129] VAAGE cites here as such 'permanent' grooves sulcus intraencephalicus anterior and posterior ('fovea isthmi'), s. 'thalamicus' ventralis, s. 'thalamicus' medius, and recessus metathalamicus of HALLER (1929). According to my own interpretation, the sulcus intraencephalicus anterior, related to the obliteration of the originally hollow optic stalk, should not be evaluated as a 'neuromeric' sulcus or structure. Likewise, I would not regard sulcus intraencephalicus posterior (fovea isthmi) as a 'neuromeric' groove remnant. As regards sulcus diencephalicus medius and ventralis, as well as recessus metathalamicus, namely synencephalic recess respectively sulcus synencephalicus, we have indeed traced the complex relationships, through a number of relatively rapidly changing transformations, between these sulci and the lumina of diencephalic neuromeres (parencephalon and synencephalon, cf. K., 1931, 1935, 1936, 1956; MIURA, 1933). This would essentially agree with some of VAAGE's interpretations concerning his findings. Thus, VAAGE (1969) comments: 'BERGQUIST and KÄLLÉN (1953 a, b, 1954, 1955) describe the occurrence of three or even four successive waves of ventricular sulci varying both in location and morphology. Hence the sulci cannot be used as landmarks during the early ontogenesis (BERGQUIST, 1964). In my material some of the sulci may be traced continuously from their early beginnings and to their final positions in the brain wall.'

d) Homologous neuromeres presumably occur in all vertebrates.

B. A topographical relationship exists between the cranial nerves and the neuromeres. A similar relationship presumably exists in other vertebrates.

C. The intracerebral boundaries develop from interneuromeric or neuromeric structures present at early developmental stages.

a) The first interprosomeric boundary to develop extends caudal to the infundibular region, the second extends rostral to the optic nerve.

b) The pros-mesencephalic boundary is located caudal to the posterior commissure.

c) The mes-rhombencephalic boundary runs between the oculomotor and the trochlear nucleus.

d) The rhombo-spinal boundary lies at the level of the rostralmost somite.

D. The notochord extends rostralwards to the boundary between di- and mesencephalon. The floor plate, the raphe-containing glycogen and the non-proliferating cell column terminate just caudal to the meso-rhombencephalic boundary. The isthmic fovea occurs rostrally in the rhombencephalon. The $m_2$-sulcus turns into the boundary sulcus between the mes- and rhombencephalon.'

As regards the detailed new observations disclosed by the investigations of BERGQUIST, KÄLLEN, and VAAGE, supplemented by the interpretations which these authors have propounded, I consider none of these data sufficiently apposite, respectively convincing, for a significant modification of my previous views on vertebrate brain ontogenesis and 'neuromery' as included in the subsection 1 B on morphogenesis of the vertebrate neuraxis (p. 141) and in the first part of the present section 3. However, some comments on, and references to the data provided by the above-mentioned authors necessarily require incorporation in my further discussion of this subject matter.

It seems quite evident that the definite configuration of the vertebrate brain tube is the end-result of complex morphogenetic events characterized by the occurrence and transformation, in a temporal and spatial sequence, of vesiculo-tubular segment-like subdivisions which may be called *neuromeres*. A reasonably adequate understanding of these events must therefore be based on an appropriate conceptualization of said neuromeres, that is, on a pertinent formulation of the 'neuromery problem'. This complex problem, in turn, presents a number of different aspects, which can be subsumed under three main headings, namely (1) *semantic problems;* (2) *causal problems,* and (3) *morphologic problems.*

The semantic problems include questions of (a) interpretation; (b) of definition, and (c) of appropriate, reasonably unambiguous terminology.

Concerning, e.g. only moderately distinctive bulgings of the developing neural tube, perhaps combined with additional, still less distinct relief features, should, or should not a *two-vesicle stage* of the brain tube, displaying archencephalon and deuterencephalon, be recognized? As elaborated in the subsection on morphogenesis of the neuraxis, I have

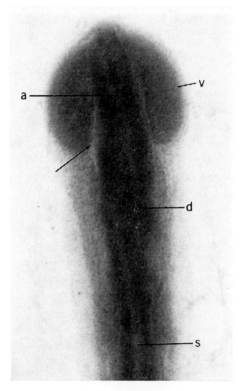

Figure 119 A

*Figure 119.* Embryonic brain tubes displaying a configuration which can be evaluated as displaying the 'two-vesicle stage'. A Embryo of Teleost, *Coregonus clupeiformis*, of about 7.5-mm length (carmine stain, ×55, red. ²/₃). a: archencephalon; d: deuterencephalon; s: spinal cord; v: eye vesicle arising from archencephalon; arrow indicates approximate boundary between a and d. B Chick embryo of 24 h, before complete rostral closure of brain tube. The eye vesicles appear still included in the outline of the primordial archencephalon. Other designations as in A (carmine stain, ×64, red. ²/₃). C Ventral view of neural tube in chick embryo at stage HH-9 (from VAAGE, 1969). Although this author dis-adopted the interpretation recognizing a two-vesicle stage (cf. Figs. 47 I, 119), while BERGQUIST (1966) states: '*das Zwei-Blasen-Stadium besteht nicht*'. VAAGE (1969), however, considers archencephalon and deuteroencephalon as the two primary encephalomeres, which he even recognized *in vivo*, and explicitly confirms the interpretations of the older authors, particularly of KUPFFER (1906), with regard to these 'vesicles' or subdivisions, quoting the following statement by KUPFFER: '*Der vordere kugelige Abschnitt der Vögel differenziert sich in die Augenblasen und das bleibende Vorderhirn, der hintere begreift das Mittel- und Rautenhirn zusammen*'. With respect to the authors (BERGQUIST, KÄLLEN and others) rejecting the morphological division of the 'cere-

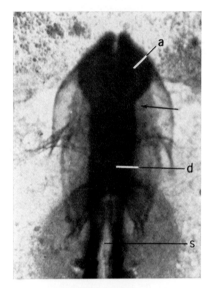

Figure 119 B

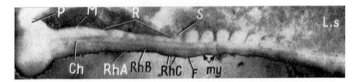

Figure 119 C

tinguishes mesencephalon (M) from rhombencephalon (R), I would still regard this configuration as a 'two-vesicle stage' displaying archencephalon (P) and deuterencephalon (M and R). ch: notochord; F: 'interneuromeric furrow'; L.s: 'longitudinal sulcus'; my: myelomere; Rh A–C: rhombomeres; S: somite.

bral rudiment' into an archencephalic and deuterencephalic portion, VAAGE (loc. cit.) comments: 'As far as I am aware, these authors have not examined the two vesicle stage'. Personally, however, I believe that they may indeed have examined it, but that they used different semantic (in the wider sense) criteria. Evidently any such criteria are both arbitrary as well as permissible, and depend upon one's concepts of 'economy of thought' (in MACH's terminology).

Again, with respect to the *three-vesicle stage*, I would distinguish prosencephalon, mesencephalon, and rhombencephalon at the stage of Figures 120A and D, which I consider essentially tripartite (cf. also Fig. 47 II), regardless of lesser additional relief features indicating the beginning of further segmentation processes. BERGQUIST's (1964) interpretation of comparable stages is shown in Figure 115B, while Figures 120B and C depict similar stages (displaying a somewhat more advanced segmentation than in my own Figure 120A), as evaluated by KUPFFER (1906). STREETER's (1933) interpretation is shown in Figure 121.

Likewise, concerning the *five-vesicle stage* (Fig. 47 III), I would distinguish telencephalon, diencephalon, mesencephalon, metencephalon (cerebellum), and myelencephalon at stages such as depicted in Figures 129–133, 138–140, despite further secondary segmentations into parencephalon, synencephalon, and a series of rhombomeres. Because of the intrinsic limitations encumbering human language, combined with the somewhat blurred aspect of complex morphogenetic events, the rigorous definition of the subdivisions at the two-, three-, and five-vesicle stages is essentially restricted to what KANT[130] called explanation of an empirical, concrete term. Such 'definition' applies to referential language providing a 'symbol' or 'referent' for a certain pattern of sensory percepts, and simply meaning: 'this is what we call that'.

---

[130] *Kritik der reinen Vernunft. II. Transcendentale Methodenlehre* (1781, 1787), cf. KUHLENBECK, 1968 (Some Comments on Words, Language, Thought, and Definition). Empirical, concrete terms, representing 'mental' abstractions of certain patterns of sensory percepts, involve POINCARE's *qualités pures*, and cannot be adequately defined in syntax language by logical rules alone, but require explanation in referential language by sensory perceptual demonstration ('*this is that*'). Almost never can it be clearly determined whether, in using the symbol designating a certain object, one refers at one time to more and another time to fewer of the qualities of features (which can be rather numerous) of that object. As regards the difficulties of interpretation and classification, it will here suffice to mention the disagreements concerning cytoarchitectural parcellation or taxonomic classification

On the other hand, if we ask, e.g., *'what is a neuromere?'*, or 'What is a rational terminology applicable to neuromeric configuration?', fairly rigorous formal definitions based on arbitrarily chosen postulates can be given in terms approximating those of syntax language.

In this respect, one may obviously define neuromeres as representing distinctive, reasonably relevant[131] transverse segmentations of the neural tube displayed in the course of ontogenesis. Yet, it is evident that, *qua* generally recognized definitive and distinctive conventional main brain subdivisions, not all 'neuromeres' are of equivalent anatomical value.

The primordial telencephalic and mesencephalic vesicles which become recognizable at rather early ontogenetic stages, correspond each to one of the major 'adult' brain parts. Likewise, the first rhombomere essentially corresponds to the definitive cerebellum. Again, at the early two-vesicle stage, the archencephalic (prosencephalic) and the deuterencephalic vesicle correspond each to major distinctive subdivisions which can still be made in the adult, namely forebrain and combined meso-rhombencephalon.

On the other hand, parencephalon and synencephalon, recognizable mesencephalic neuromeres $M_1$ and $M_2$, as well as the rhombomeres caudal to the first one, including the 'myelomeres', can be regarded as quite transitory, being replaced by zonal systems superimposed upon the transverse 'segmentation', and becoming, as it were, 'fused' into a more or less unified diencephalon, mesencephalon, rhombencephalon, and spinal cord, respectively.

---

(cf. K., 1954, 1967a). Paraphrasing remarks by CRAIK and by J.B.S.HALDANE (cf. K., *Brain and Conciousness*, 1957, pp. 48–49) it could be said that the precise definitions of blurred and therefore confusing morphogenetic events do not, in fact, cover very much of the range even of sensory experience. Rigorous and strictly logical terminologies will only work for material that has certain highly abstract properties (e.g. operationally well-definable *numerical* relationships), which are rather less frequently and much less completely exemplified in the 'real ('actual') world than scientists or logicians would like us to believe.

[131] One can here, of course, again ask: *what is relevant*, respectively reasonable, or, distinctive? The only further answer is here reference to *'intuition'* as discussed in §48, p. 195f. and § 64, p.293 of the author's monograph *'Mind and Matter'* (1961). In the aspect under consideration, this would merely imply that the observer using the cited questionable terms may assume that at least some undefined other observers, with whom he communicates by means of his report, will experience comparable percepts and relate these in a similar fashion to the verbal symbols under consideration.

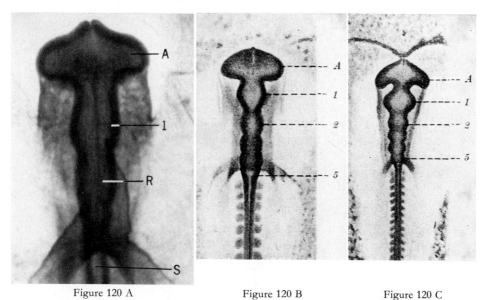

Figure 120 A                    Figure 120 B                    Figure 120 C

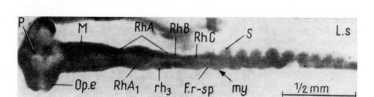

Figure 120 D

*Figure 120.* Embryonic brain tubes which can be evaluateds a displaying the 'three-vesicle-stage'. A Chick embryo of 33 h (carmine stain; ×66; rcd. ²/₃). The anterior neuropore is still open; the eye vesicles, evaginated further laterad than in Fig. 120B, are still included in the overall outline of prosencephalon with paired optic vesicle evagination; R: rhombencephalon; S: spinal cord (not yet closed). B Sparrow with 10 somites. C Sparrow with 15 somites (B and C after KUPFFER, 1906). A: archencephalon (prosencephalon); 1: mesencephalon; 2, 5: beginning segmentation of rhombencephalon. D Ventral view of neural tube in chick embryo of stage HH-9-10 (from VAAGE, 1969). M: mesencephalon; my: myelomere; Op.e: optic evagination; P: prosencephalon; Rh and rh: rhombomeres. By extending the domain of RhA relatively far caudalward, VAAGE implies that the rhombomeres rh arise through 'subdivision' of those designated as Rh. F.r-sp: 'rhombo-spinal boundary and furrow'; L.s: 'Longitudinal sulcus'.

Accepting the aforementioned 'vesicle-concept', and, moreover, designating all 'major' transverse segmentations as *'neuromeres'*, one might thus distinguish *permanent* from *transitory ones*, and consider 'transitoriness' as a special feature. It would then be permissible to use the term *neuromeres sensu strictiori* for the latter, and *neuromeres sensu latiori* for those representing permanent subdivisions.

Accordingly, a classification of cerebral neuromeres *sensu latiori* could again comprise those of the *first order* (archencephalon, deuterencephalon), those of the *second order* (prosencephalon, mesencephalon, rhombencephalon), and those of the *third order* (telencephalon, diencephalon, cerebellum). The neuromeres *sensu strictiori*, characterized by

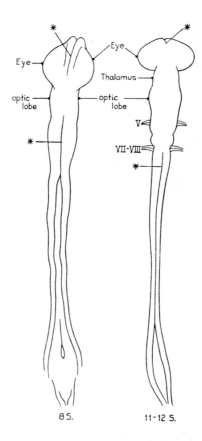

*Figure 121.* Outline drawings of embryonic chick brains as interpreted by STREETER (from STREETER, 1933). Left: 8-somites stage, right 11- to 12-somites stage. Asterisk: closure incomplete.

their *transitory* nature, would then comprise *parencephalon, synencephal-on, mesencephalic neuromeres* $M_1$ *and* $M_2$, and *all rhombomeres caudal to* $rh_1$.[132]

This classification of neuromeres in accordance with my own observations and concepts embodies most of KUPFFER's relevant interpretations, from which it only slightly differs with regard to the basal telo-diencephalic boundary, provided, in KUPFFER's interpretation, by the sulcus intraencephalicus anterior, and in my interpretation by torus transversus and commissura anterior.

In contradistinction to the interpretations of BERGQUIST and KÄLLEN, and of VAAGE, I do not evaluate various transitory embryonic configurations, which could be considered secondary, and may even, in some instances represent accidental distortions, respectively artefacts, as 'neuromeres'.

With regard to the concepts of BERGQUIST and KÄLLEN[133] (Figs. 115, 116), it can be seen that their 'proneuromeres' A and B correspond to prosencephalon and mesencephalon in the terminology which I have adopted, while C and D together represent the rhombencephalon at the three-vesicle stage. Again, 'neuromere' *a* seems to be prosencephalon at a stage when the synencephalon *(b)* is beginning to differentiate. According to my own experience, however, the synencephalon may, as a rule, become clearly recognizable at a period when the telencephalon likewise becomes rostrally distinctive from the original prosencephalon, whose remainder thereby represents the diencephalon composed of parencephalic and synencephalic portions. BERGQUIST's neuromere C corresponds to the mesencephalon, and those designated as *d, e,* etc., represent the rhombomeres.

With respect to the 'postneuromeres', 1 presumably corresponds to the telencephalon, while 2, 3, 4, *in toto* may be regarded as the parencephalon, but represent additional subdivisions which do not agree with my interpretations. The 'postneuromeres' 6, 7 ($M_1$ and $M_2$), and 8, 9, etc. (rhombomeres 1, 2, etc.), on the other hand, evidently correspond to the segmentation described by KUPFFER and others, including myself.

---

[132] Thus, SCHUMACHER (1928), not without some justification, designates KUPFFER's rhombomere 2 as his rhombomere 1, since he evidently noticed that KUPFFER's first neuromere, namely the cerebellar one, somewhat differs in its behavior from those of the caudal sequence beginning with KUPFFER's $rh_2$.

[133] In earlier papers, BERGQUIST and KÄLLEN used a slightly different neuromeric notation for 'proneuromeres' (1–6) and 'neuromeres' (I–XI). Figure 8 in BERGQUIST's 1956 paper brings a comparison of their original and their subsequent notations.

Concerning the terminology and the findings of VAAGE (cf. Figs. 115, 118 A, 120 D, 137), I am in essential agreement with his subdivision A, D (archencephalon, deuterencephalon, two-vesicle stage) and P, M, R (prosencephalon, mesencephalon, rhombencephalon, three-vesicle stage). As regards the subsequent neuromeric parcellation of VAAGE, however, I am unable to interpret that author's PrA and PrB with sufficient certainty, but pr1 and pr2 seem to represent the telencephalon, pr3 appears related to the eye vesicle evagination which I include in the basal part of parencephalic segment, pr 4, 5, 6 may likewise pertain to parencephalon, and pr 7–8 presumably correspond to the synencephalon.[134] The mesencephalic and rhombencephalic segments described by VAAGE are at least roughly comparable to, if not, in most instances, identical with, those that I previously enumerated in accordance with KUPFFER and others.

In contradistinction to the diagrams of BERGQUIST (Fig. 115), and VAAGE (Fig. 118 A), which both end at a stage of considerable and mutually disagreeing parcellation, my own diagram (Fig. 118 B) attempts to correlate the 'neuromeric vesicles', as I can best interpret these in part somewhat blurred and 'squirming' configurations,[135] with the distinctive 'permanent' (adult) subdivisions of the vertebrate neuraxis.

Again, as already mentioned above, BERGQUIST and KÄLLEN describe three different 'waves of segmentation', resulting in 'proneuromeres', 'neuromeres', and 'postneuromeres'. Thus, there are two 'interneuromeric stages' ('interneuromery I and II'). VAAGE, however, states that no reduction of the 'neuromeres' takes place during these 'interneuromeric' phases, but that an increasing number of 'neuromeres' is provided by successive subdivisions of the preceding ones. Differentiation of the brain tube is said to proceed initially in a caudorostral direction, while the spinal tube differentiates in a rostro-caudal direction.

---

[134] In comparison with STREETER's (1933) interpretation (figs. 121, 136), it can be seen that my own interpretation of details respectively 'parcellation' of relief features, which is, on the whole, quite compatible with STREETER's somewhat simplified concepts, stands, as it were, 'midway' between the interpretation of this latter author, and the intricate concepts based on the features as shown in the interpretations by BERGQUIST, KÄLLEN, and VAAGE.

[135] Although my interpretation differs in various respects from that by either of these cited authors, it can easily be seen that some sort of generalized, if hazy, overall agreement seems to obtain.

According to my own observations, the telencephalic vesicle originates as a rostral unpaired evagination of the original prosencephalic vesicle, from which the synencephalic vesicle arises as a more or less concomitant but perhaps predominantly subsequent expansion, the region between telencephalon and synencephalon becoming the parencephalon.

In the course of these events, that is, while the telencephalon first evaginates from the prosencephalon, whose remainder, namely the diencephalon, becomes further subdivided into parencephalon and synencephalon, the demarcation between these three segments may seem at times rather inconspicuous. Such intermediate stages of forebrain morphogenesis may perhaps correspond to BERGQUIST's concept of 'interneuromery'.

The subdivision of mesencephalon into $M_1$ and $M_2$ remains, at best, rather indistinct and somewhat uncertain as to definite timing. The segmentation of rhombencephalon sets in rather early, and appears to proceed by increasing degrees of tube wall constrictions, which do not seem to involve further subdivisions of the originally formed rhombomeres, in a rostro-caudal direction. In other words, I believe that the rhombomeres arise as a progressive direct segmentation of the rhombencephalon, but not by successive subdivision of preceding neuromeres.

Following this elaboration on the different facets of the semantic aspects pertaining to the 'neuromery problem' (definition, appropriate terminology, and interpretation), a brief consideration of the *causal* aspects may be appropriate.

Several factors, enumerated and discussed in the publications of BERGQUIST (1952, 1964, 1966, as well as others) and VAAGE (1969), seem here to be involved. Since 'neuromeres' can easily be detected in the living condition (developing ova of diverse Anamnia, blastoderms of chick, etc.), their explanation as shrinkage effects caused by postmortal changes and methods of histological processing has been discounted. I believe, nevertheless, that some features of 'neuromeric' configuration are frequently modified, exaggerated or distorted by changes of this type, and that many of the 'odd', 'aberrant', confusing and 'complex' aspects of 'neuromery', described by various authors and in part related to the numerous differences in opinion and interpretation concerning these configurations, are indeed postmortal or fixation (etc.) 'distortion effects'.

Some authors have suggested traction produced by the flexures of the brain tube, by the outgrowing nerves, or by the somites in the trunk region. Again, the existence of proliferation maxima which produce localized expansions of the neural tube has been assumed. BERGQUIST and KÄLLEN have particularly maintained the significance of proliferation centers. (*'Grundgebiete'*, *'migration areas'*). Generally speaking, I believe that this approach, emphasizing the effects of proliferative activity, is fully justified, although I remain sceptical with respect to the presently possible formulations in terms of rigorously describable interactions, and to some of the proffered detailed descriptive interpretations.

According to VAAGE (1969), the concept of archencephalon and deuterencephalon has assumed a special significance through the experimental investigations undertaken by various authors quoted in his monograph. These experiments are said to show that the differentiation of the neural anlage is governed by two different 'induction systems' in the roof of the archenteron. The one system ('activating system') is presumed to induce the formation of prosencephalic configurations. The other system ('transforming system') allegedly causes the formation of mesencephalic, rhombencephalic, and spinal cord structures. It is claimed that both in amphibians and birds, the prechordal mesoderm induces prosencephalic structures exclusively. In evaluating such conclusions derived from experimental embryology, VAAGE (loc. cit.) suggests that the archencephalon and deuter(o)encephalon of birds are 'early morphological expressions of the activating principles and the transforming principles respectively'. The cited author then expresses the belief that the induction experiments to which he refers confirm observations leading to an interpretation of mesencephalon and rhombencephalon as epichordal brain regions, and of prosencephalon as a prechordal domain.

From the morphologic viewpoint stressed in our own approach, it could be stated that the distinction of archencephalon and deuterencephalon represents a concept which can be based on actual configurational facts, and is justifiable by form-analytic, topologic considerations whose validity does not depend on, and in fact is not, in the aspect under consideration, logically related to, the sorts of physicochemical interactions causing the motion of masses resulting in the given patterns.[136]

---

[136] Thus e.g., the motions of masses resulting in the formation of a neural tube by he folding of a plate into a groove, followed by closure into a tube by a dorsal suture, differ considerably from those occurring during organization of a solid cell cord or of a nondescript cell mass (e.g. within the tail-bud) into a tube. Yet, at a given tube stage, the

Configurational orderliness must evidently be the result of 'vital' (but in a 'nonvitalistic' sense, cf. p.67f.) interactions within the substratum. Theoretically, like all other physical phenomena, such interactions can be considered definable in mechanistic, rigorous mathematical terms. As regards the multitudinous difficulties thereby involved, an attempt has been made to point out some of the relevant problems in the introductory subsection, on morphogenesis, of the present chapter VI. Without intending to question the justification of experimental embryology or the value of the data obtained by this approach, I believe that, so far, the formulations based on said methods represent merely a vague, hazy, and highly debatable first approximation, which is anything but 'rigorous' if viewed according to the standards of the exact experimental natural sciences. Morphology, on the other hand, admittedly operates with vaguely circumscribed types of form, which, nevertheless, can be analyzed within a combinatorially rather exactly determined topologic framework. It would, therefore, be an unjustified overestimation, if devotees of experimental embryology, by implication or innuendo, should evaluate their procedures as more 'exact' than, as 'superior' to, or as conflicting with, the classical morphologic approach.

The third main aspect of the neuromery-problem, subsuming the *morphologic questions*, comprises the following principal topics: (a) to which extent does this sequence of segments involve truly 'metameric' i.e. promorphologically homologous ('homodynamic') configurations; (b) how are these segments related to the 'definitive' subdivisions of the neuraxis, and (c) what is the relation of these segments to the sequence of cranial nerves.

As regards topic (a), distinctive segments arranged along the main longitudinal body can be evaluated as *'homodynamic'* or as true metameres if they display an identical bauplan by exhibiting the same basic subsets *('Grundbestandteile')*.

In this respect, telencephalon and parencephalon, both consisting of roof plate and alar plate, might be considered comparable metameres. Likewise, the mesencephalic segments $m_1$ and $m_2$ (if present) provided by (ordinarily fused) roof plate, alar plate, and basal plate, can thus be evaluated as metameres. The synencephalon seems to in-

resulting configuration pattern of the neuraxis (floor plate, lateral plates, sulcus limitans, roof plate) can be evaluated as morphologically identical, regardless of the different formative processes.

clude the end of the rostrally disappearing basal plate, and poses a dubious question, which may be arbitrarily solved by considering it a segment *sui generis*, transitional between parencephalon and mesencephalon. The rhombomeres display an identical bauplan, corresponding to that of spinal tube, consisting of roof, alar, basal, and floor plates, thus representing, in this aspect, true 'metameres'.

It is therefore permissible to classify the above-mentioned 'recognizable' and 'definable' vesicle-like segments of the embryonic vertebrate brain tube as pertaining to three main groups, with an 'intermediate' or 'transitional' segment.

1. *Archencephalic segments:* telencephalon and parencephalon (alar and roof plates).

2. *Rostral deuterencephalic segments:* synencephalon and mesencephalon, respectively $m_1$ and $m_2$, if displayed (basal, alar, and roof plates). The synencephalon may be here considered a special, transitional segment, representing an open neighborhood characterized by overlap.

3. *Caudal deuterencephalic segments:* rhombomeres (floor, basal, alar, and roof plates). The isthmus region, although, in my opinion, not forming a comparable 'segment' in the here accepted sense, may likewise be evaluateed as a neighborhood intermediate or transitional between 3 and 2 (cf. further below).

If the significant '*Grundbestandteile*' characterizing the morphologic pattern of deuterencephalon are presumed to be alar and basal plates (roof plate being regarded as a differentiation of alar plate, and floor plate as midline differentiation of basal plate), then rostral and caudal deuterencephalic may be considered 'homodynamic', and likewise metamerically homologous with the spinal tube.

It is noteworthy that, before the significance of the longitudinal zones became generally stressed following the studies of His and of various Anglo-American authors (GASKELL, C. J. HERRICK, JOHNSTON and others), GORONOWITSCH (1888) made the following remark in his investigation on the brain of the ganoid *Acipenser:* '*Das Mittelhirn ist ein vollständiges Segment des primitiven Gehirnrohres und ist als ein Abschnitt des Rückenmarkrohres homodynam aufzufassen. Das sekundäre Vorderhin ist eine lokale sackförmige Ausstülpung des Gewölbeteiles des Gehirnrohres und hat deshalb einen vollständig anderen morphologischen Wert.*'

On the other hand, GORONOWITSCH (loc. cit.) stated in a further comment: '*Das primitive Vorderhirn ist homodynam einem Abschnitte des Rückenmarks, wie das* GÖTTE *auffasste*'. This latter interpretation, however, does not appear acceptable. The 'primitive forebrain', namely that transitory configuration following rostral closure of the neural tube, and from which eye vesicles, telencephalon ('secondary forebrain') and parencephalon differentiate, appears to be a derivative of alar plate (and roof plate) only. It should, therefore, not be considered metamerically homologous with the spinal tube.

The embryonic segments of the vertebrate brain tube, which, according to the terminology here adopted and defined on p. 322, were designated as neuromeres *sensu latiori* of the third order, and as neuromeres *sensu strictiori*, are illustrated, in representatives of diverse classes, by Figures 122–140. Concerning individual segments, a few comments should be added, which will include references to the above-mentioned topics (b) and (c).

1. *Telencephalon.* HENRICH (1896), a student of KUPFFER, discusses, with reference to the older literature, the question whether the first anlage of telencephalic hemispheres represents a paired or unpaired

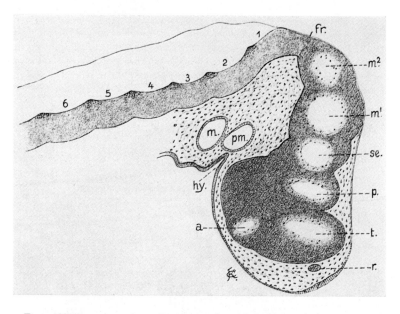

*Figure 122.* Neuromeres in an Acanthias embryo (Selachian) of 10- to 11-mm length (after KUPFFER, 1906). a: optic vesicle stalk; fr: 'fissura' rhombo-mesencephalica; hy: hypophysial invagination *('Anlage')*; m: mandibular cavity ('preotic somite'); m¹, m²: mesencephalic neuromeres $m_1$ and $m_2$; p: parencephalon; pm: premandibular cavity ('preotic somite'); r: peripheral section *('Anschnitt')* through bottom of unpaired olfactory pit; se: synencephalon; 1–6: rhombomeres (unusually well developed in basal portion of neural tube).

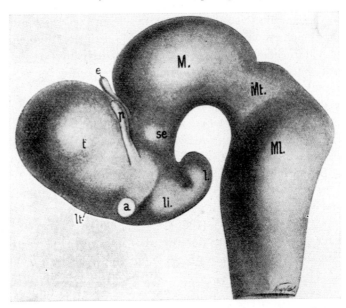

*Figure 123.* External view of the brain tube in an Acanthias embryo of 25 mm (after KUPFFER, 1906). a: optic vesicle stalk; e: epiphysis; l: 'lobus posterior' hypothalami; li: 'lobus inferior' hypothalami; lt: lamina terminalis at approximate region of telo-mesence-phalic boundary; M: mesencephalon; Ml: myelencephalon; Mt: metencephalon; p: par-encephalon; se: synencephalon; t: telencephalon.

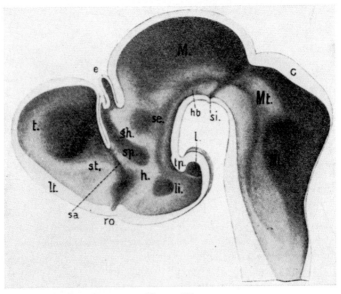

Figure 124

configuration. He reaches the conclusion, that the (ontogenetically) primordial telencephalon, namely the epencephalon *sensu latiori* in KUPFFERS terminology, is neither paired nor unpaired, but tripartite: '*es findet von Anfang an mit Beginn der Abgrenzung des Epencephalon vom Parencephalon eine Dreiteilung statt, in einen unpaaren, mittleren und rudimentär bleibenden Teil, das unpaare Epencephalon s. str., und in zwei paarige zu mächtiger Entfaltung gelangende birnförmige Ausstülpung der Seitenwände, die Hemisphärenblasen*'.

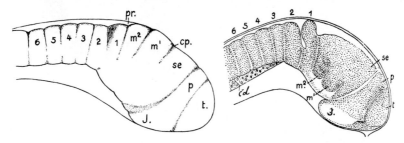

*Figure 125.* Segmentation of the brain tube in a 20-day-old embryo of the Teleost *Salmo purpuratus* (after HILL from KUPFFER, 1906; designations according to KUPFFER). cd: notochord; cp: locus of commissura posterior; J: 'infundibulum' (hypothalamic region related to parencephalic segment); $m_1$, $m_2$: mesencephalic segments; p: parencephalon; pr: 'plica rhombo-mesencephalica'; se: synencephalon; t: telencephalon; 1–6: rhombomeres. Although STREETER (1933) regards HILL's 'neuromeric' outlines with scepticism, the segmentation as indicated in this figure corresponds rather well with my own observations and interpretations. In general, however, I found this segmentation basally much less conspicuous than laterally respectively dorsally.

*Figure 124 A.* Internal view of Acanthias embryo brain tube depicted in preceding figure 123 (after KUPFFER, 1906). c: cerebellum; e: epiphysis; gh: anlage of ganglion habenulae (epithalamus); h: 'regio hypencephalica' (approximately dorsal hypothalamus); hb-tegmental ridge (basal plate); l: 'lobus posterior' hypothalami; li: 'lobus inferior' hypo: thalami; lt: lamina terminalis; M: mesencephalon; Ml: myelencephalon; Mt: metencephalon; ro: preoptic recess; se: synencephalic recess; si: sulcus intraencephalicus posterior; sa: sulcus intraencephalicus anterior (due to a probable shrinkage effect, KUPFFER has traced this sulcus further dorsalward than warranted by my own observations; in KUPFFER's interpretation, this sulcus would reach the attachment of velum transversum, and be continuous with recessus posterior of hemisphere); sp: recessus parencephalicus; st: 'striatum' (basal part of telencephalon); t: telencephalon; the sulcus limitans, not labeled, is clearly recognizable to the right of 'se'; although perhaps somewhat exaggerated by shrinkage and blurred at its rosto-basal termination, this aspect of sulcus limitans corresponds essentially to my own observations; in crossing the lamina terminalis, the lead 'sa' there approximately passes through the telo-diencephalic boundary of my interpretation; the external indentation between preoptic recess and chiasmatic ridge, also visible in figure 123 presumably represents a shrinkage effect.

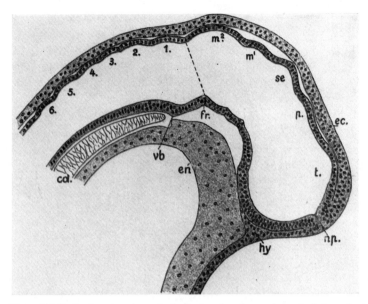

*Figure 126.* Sagittal section through the brain tube of an embryonic *Salamandra* (urodele amphibian) combined from two sections in the approximate median plane (from KUPFFER, 1906). cd: notochord; ec: ectoderm; en: entoderm; fr: 'fissura rhombo-mesencephalica; hy: adenohypophysial anlage; $m_1$, $m_2$: mesencephalic segments; np: locus of anterior neuropore; p: parencephalon; se: synencephalon; t: telencephalon; vb: 'ventral rhombencephalic flexure'; 1–6: rhombomeres.

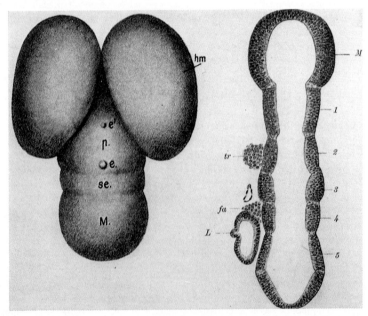

Figure 127 A, B

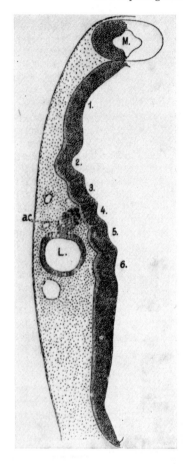

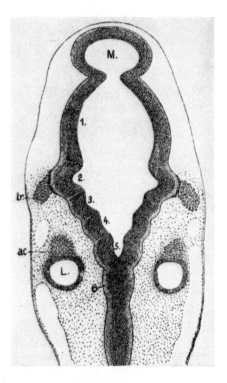

Figure 128 A                              Figure 128 B

*Figure 128*. 'Horizontal' sections through the rhombencephalon of a druck embryo (Anas, avian Sauropsidan) at 4th day of incubation (from KUPFFER, 1906). Section B is slightly *('um drei Schnitte')* more ventral than section A. ac: ganglion of eighth nerve; L: otic vesicle; M: mesencephalon; tr: trigeminal root and ganglion; 1–6: rhombomeres.

*Figure 127*. Neuromeres as displayed in the embryonic brain tube of reptilian Sauropsidans (from KUPFFER, 1906). A Embryonic brain of *Anguis fragilis* (cf. also fig. 54). B 'Horizontal section through mesencephalon and rhombencephalon of *Coronella austriaca*. e: epiphysis; e': paraphysis; fa: acustico-facialis ganglion; hm: telencephalon; L: otic vesicle (anlage of labyrinth); M: mesencephalon; p: parencephalon; se: synencephalon; tr: trigeminal ganglion; 1–5: rhombomeres. Compare A with fig. 49 B.

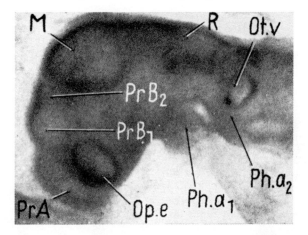

*Figure 129.* Lateral view of brain tube in living duck embryo of stage HH-13 (from
VAAGE, 1969). M: mesencephalon; Op.e: optic evagination; Ot.v: otic vesicle; Ph. a$_1$, Ph.
a$_2$: pharyngeal arches 1 and 2; Pr A: 'prosomere A' (telencephalon, K.); Pr B$_1$: 'prosomere
B$_1$' (parencephalon, K.); Pr B$_2$: 'prosomere B$_2$' (synencephalon, K.); R: rhombencephalon.

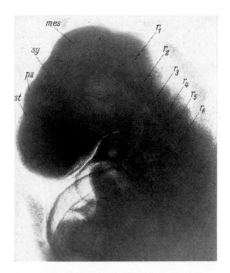

*Figure 130.* Head of duck embryo at 72 h of incubation (total mount, carmine, from
K., 1936). mes: mesencephalon; pa: parencephalon; r$_{1-6}$: rhombomeres; st: sulcus telo-
diencephalicus; sy: synencephalon.

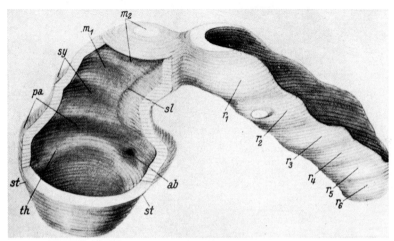

*Figure 131.* Plate model of brain in a duck embryo of 72 h incubation time (from K., 1936). ab: optic vesicle stalk; $m_1$, $m_2$: mesencephalic segments; pa: parencephalon; $r_{1-6}$: rhombomeres; sl: sulcus limitans; st: indication of sulcus telo-diencephalicus; sy: synencephalon; th: torus hemisphaericus.

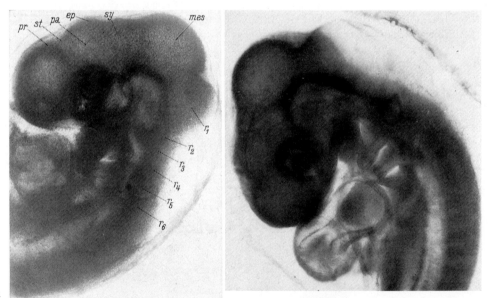

A                                                                                                          B

*Figure 132.* Heads of avian embryos displaying neuromery (whole mounts, carmine stain). A Duck embryo at 85 h of incubation (from K., 1936). B Chick embryo at 72 h of incubation displaying configuration very similar to A. ep: epiphysis; mes: mesencephalon; pa: parencephalon; pr: paraphysis; $r_{1-6}$: rhombomeres; st: sulcus telo-diencephalicus; sy: synencephalon. The olfactory placode (not labelled) is clearly seen in A, lens and otic vesicle (at level of $r_6$ in A) are distinctly displayed in B, which is kept unlabeled.

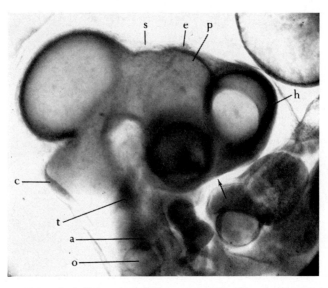

*Figure 133.* Head of chick embryo at four days (96 h) of incubation (whole mount, carmine stain). a: ganglion of acustico-facialis complex; c: cerebellum; e: epiphysis; h: tel-encephalon; o: otic vesicle; p: parencephalon; s: synencephalon; t: trigeminal ganglion. Arrow indicates locus of basal telo-diencephalic boundary. The extensive development of mesencephalic tectum opticum (not labeled) is conspicuous. Within the telencephalon, the common ventricle (ventriculul impar s. communis telencephali) is clearly recognizable.

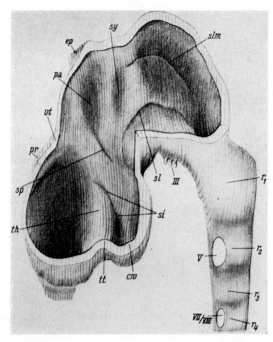

Figure 134

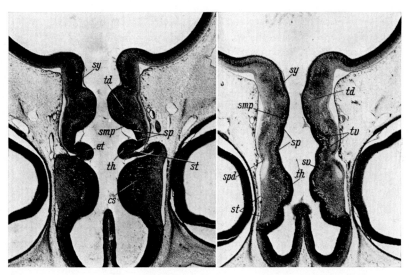

Figure 135 A                              Figure 135 B

*Figure 135.* Three sections from a 'horizontal' (rostro-caudal 'frontal') series through the prosencephalon of a chick embryo at 5 days and 20 h of incubation (from K., 1936, cf. also fig. 14, 1936). A is most dorsal, C most basal, C passing through torus transversus. The apparent secondary median fusions of hemispheres in B and C near bottom result from oblique sections *('Flachschnitte')* through rostral parts of lamina terminalis (cf. fig. 222). The sections show a transformation stage of parencephalon and synencephalon as well as differences in cellular density and indications of so-called 'migration layers'. Even a cursory inspection clearly indicates that considerable regional differences preclude a schematic classification of 'migration layers' as attempted by several authors. cs: basal ganglia of telencephalon; et: eminentia thalami (ventralis); smp: sulcus diencephalicus medius, pars posterior; sp: sulcus parencephalicus; spd: sulcus peduncularis (dorsalis); st: sulcus telencephalo-diencephalicus; sv: sulcus diencephalicus ventralis; sy: recessus synencephalicus; td: thalamus dorsalis; th: torus hemisphaericus; tv: pars ventralis thalami (thalamus ventralis); z: connection of sulcus diencephalicus ventralis and sulcus parencephalicus.

---

*Figure 134.* Plate model of brain in a duck embryo at 120 h of incubation, showing details of prosencephalo-mesencephalic ventricular relief (from K., 1936). cw: chiasmatic ridge; ep: epiphysis; pa: parencephalic recess; pr: paraphysis; $r_{1-4}$: rhombomeres; si: sulcus intraencephalicus anterior; sl: sulcus limitans; slm: sulcus lateralis mesencephali; sp: sulcus parencephalicus; sy: synencephalic recess; th: torus hemisphaericus; tt: torus transversus; vt: velum transversum; III: oculomotor rootlets; V: trigeminal root; VII/VIII: root of facial and acoustic nerve.

The primordial telencephalon, as referred to by HENRICH, corresponds approximately to the stage shown in Figures 131 and 141. However, seen in cross-section (Fig. 142), such stage can hardly be called tripartite. It is, moreover, a question of arbitrary semantics, whether one should consider it paired or unpaired, and I prefer to call it *unpaired bilateral symmetric*. The anlage of the dorsal ridge, which is the epencephalon *sensu strictiori* of HENRICH, pertains to the roof plate, from which subsequently paraphysis and telencephalic choroid plexus arise.

At a later stage (Fig. 143) the tripartite stage stressed by HENRICH then becomes evident. Yet, with regard to the large common telencephalic ventricle, I have also evaluated this configuration as a *telencephalon impar* (cf. section 1C of this chapter). Although telencephalon and parencephalon may be considered serially homologous at initial stages, a divergent differentiation subsequently becomes manifest. Telence-

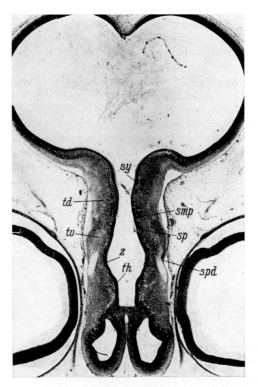

Figure 135 C

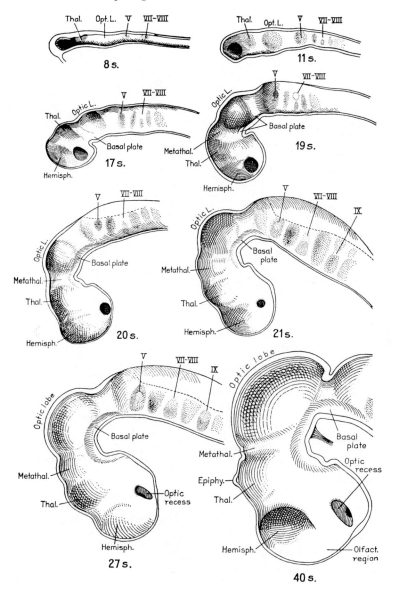

Figure 136 A

*Figure 136.* STREETER's interpretation of 'neuromeric' segmentation in the chick brain at different stages of development. Based on dissections of the brain tube, enlarged approximately 25 diameters in A, and as indicated in B, s referring to number of somites (from STREETER, 1933).

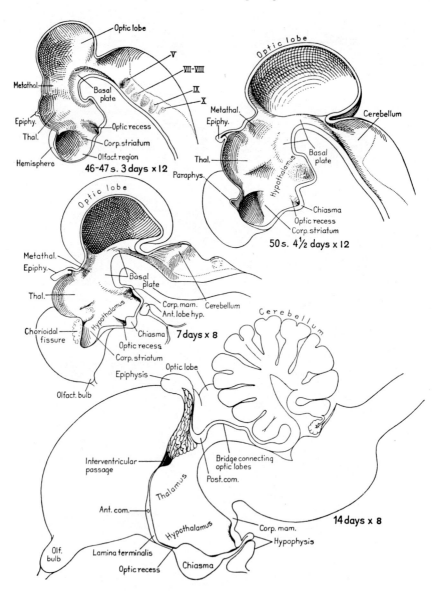

Figure 136 B

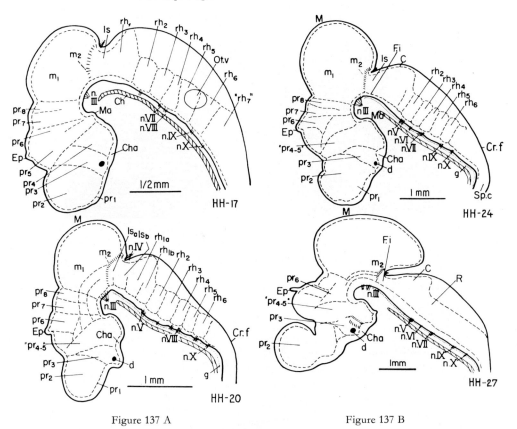

Figure 137 A                                    Figure 137 B

*Figure 137.* VAAGE's interpretation of 'neuromeric'segmentation in the chick brain at stages HH-17 to HH-27 (from VAAGE, 1969). C: cerebellum; Ch: notochord; Cha: chiasmatic ridge; Cr. f: cervical flexure; d: 'ductulus opticus' (entrance into eye vesicle stalk); Ep: epiphysis; F.i: fovea isthmi; g: 'glycogen containing raphe'; I.s., I.s$_a$, Is$_b$: 'rhombencephalic segments' (isthmus region); M: mesencephalon; m$_1$, m$_2$: ' mesomeres' 1 and 2; Ma: 'mammillary recess'; nIII etc.: cranial nerve roots; Ot.v: otic vesicles; pr$_1$ etc.: 'prosomeres'; R: rhombencephalon; rh$_1$ etc.: rhombomeres; Sp.c: spinal cord.

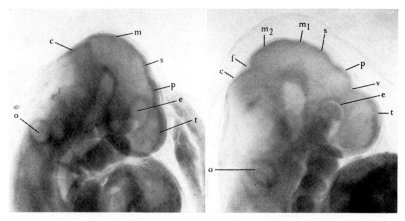

Figure 138                                Figure 139

*Figure 138.* Head of 4-mm pig embryo (SUS, artiodactyle ungulate mammal), showing mammalian prosencephalic neuromery (total mount, carmine stain). c: cerebellum; e: optic vesicle of right side (left one as blurred shadow); m: mesencephalon; o: right otic vesicle (left one in blurred overlapping outline); p: parencephalon; s: synencephalon; t: telencephalon.

*Figure 139.* Head of 6-mm pig embryo (total mount, carmine stain) to be compared with preceding figure. f: rhombo-mesencephalic groove; $m_1$, $m_2$: faintly indicated mesencephalic segmentation; v: velum transversum; other abbr. as in figure 138.

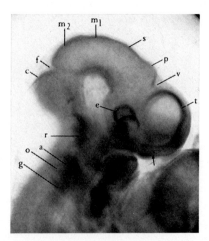

*Figure 140.* Head of 10-mm pig embryo (total mount, carmine stain; compare with sections shown in figs. 56 C–D). a: facial-acoustic ganglion; f: glosso-pharyngeal ganglion; r: trigeminal ganglion; arrow indicates approximate locus of basal telo-diencephalic boundary; the common telencephalic space (ventriculus impar telencephali) is conspicuously visible.

phalon and parencephalon originate zonal systems of dissimilar morphologic value, the parencephalic zonal system extending caudad into
the synencephalon. It is evident that telencephalic and diencephalic
(parencephalo-synencephalic) zonal systems are not 'homodynamic'
(promorphologically homologous). The telencephalo-diencephalic
boundary has been dealt with in section 1, and shall again be considered in sections 5 and 6. As regards the relationship of 'neuromeres'
to cranial nerves, it seems rather evident that the telencephalic vesicle
is associated with the nervus olfactorius. Many details of this relation-

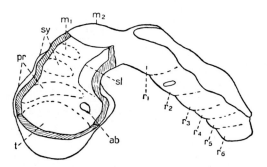

*Figure 141.* Reconstruction of the brain tube in a duck embryo of 72 h, showing some
details of the prosencephalic and mesencephalic ventricular relief (from K., 1935b, cf. also
fig. 131). ab: opening of eye vesicle stalk; $m_1$, $m_2$: midbrain segments; pr: parencephalon;
sl: sulcus limitans; sy: synencephalon; $r_{1-6}$: rhombomeres; t: telencephalon.

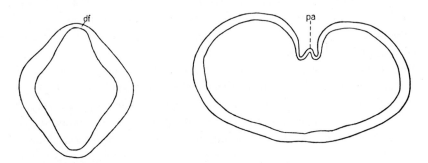

*Figure 142.* Cross section through the early embryonic telencephalon shown in preceding figure (from K., 1935b). df: 'dorsal ridge' (so-called epencephalon of HENRICH,
1896).

*Figure 143.* Cross-section through telencephalon impar in a chick embryo of 87 h
(from K., 1935b). pr: paraphysial rudiment.

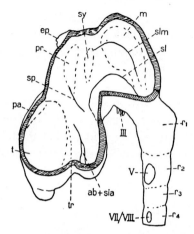

*Figure 144.* Reconstruction of brain tube in a duck embryo of 120 h, showing some details of prosencephalic and mesencephalic ventricular relief (from K., 1935a). ab + sia: eye vesicle stalk groove being transformed into sulcus intraencephalicus anterior; ep: epiphysis; m: mesencephalon; pa: locus of paraphysial rudiment; pr: parencephalic recess; $r_{1-4}$: rhombomeres; sl: sulcus limitans; slm: sulcus lateralis mesencephali; sp: sulcus parencephalicus transversus; sy: synencephalon; t: telencephalon; tr: torus transversus.

ship, however, particularly those referring to the interactions of onto-genetic events during ontogenesis, still appear insufficiently clarified. The configurations described by BERGQUIST (1952) as '*neuromerode*' or '*neurod o*', and by VAAGE (1969) as $pr_1$ at his stages HH-14 and following ones, may perhaps represent early aspects of the olfactory bulb while becoming differentiated from the remainder of telencephalon. The nervus terminalis shall be discussed in the parts of this treatise dealing with peripheral nerves and with the olfactory system. Concerning this thin nerve, which seems to have relationships with the lamina terminalis at the telencephalo-diencephalic boundary, I was unable to obtain pertinent findings at relevant stages of early ontogenesis. It should also be stressed that neither n. olfactorius, n. terminalis, n. parietalis nor optic 'nerve' are configurations serially homologous with the deuterencephalic cranial nerves.

    2. *Parencephalon.* This conspicuous segment originates dorso-cau-dally the epiphysial structures as derivatives of the roof plate. Within the lateral walls (alar plate) arises the diencephalic longitudinal zonal system, which, unrestricted by segmental boundaries, extends into the synencephalon. Figure 144 shows an early manifestation of the events

related to the development of the zonal system, namely the sulcus parencephalicus transversus, characteristic for these morphogenetic processes in the sauropsidan brain.[137] Rostro-basally, the parencephalon
retains its connection with the optic vesicle stalk, whose region of origin is involved in the formation of sulcus intraencephalicus anterior.
In this respect, the optic 'nerve' which actually represents a brain tract,
since the retina must be considered a prosencephalic derivative, may
be considered related to the parencephalon (cf. the tabulation of
Fig. 114).

3. *Synencephalon*. This segment consists of roof, alar, and basal
plates. Its ventricular wall commonly displays the rostral convexity of
the here terminating sulcus limitans. The synencephalon may thus be
considered the most anterior of the segments 'homodynamic' with the
spinal tube and could therefore be included in the deuterencephalon.

However, being a region of transition between archencephalon and
deuterencephalon, it should preferably be evaluated as a segment of
distinctive morphologic value. During ontogenesis, the rostral portion
of synencephalon, whose alar plate cell masses are included in the diencephalic longitudinal system, becomes 'absorbed' into the diencephalon, which develops by this 'fusion' of parencephalon and synencephalon. The caudal portion of synencephalon 'amalgamates', or 'joins up',
with the mesencephalon. Thus, although for practical purposes of terminology, parencephalon and synencephalon may be said to represent
an early embryonic 'diencephalon', this convenient and quite acceptable simplification of complex ontogenetic events must accordingly be
qualified.

Strictly speaking, the diencephalon results from the combination of
two 'nonhomodynamic' neuromeres, the archencephalic parencephalon and the greater (rostral) portion of the 'hybrid' deuterencephalic synencephalon, which represent a transitory segmentation, replaced by
the diencephalic longitudinal zonal system. The 'vesicular' lumen
within the alar plate region of synencephalon becomes the recessus synencephalicus which may display connection with sulcus diencephalicus dorsalis rostrally, and caudally with sulcus lateralis mesencephali. In

---

[137] Further details on the complex transformations occurring during the development
of the diencephalic zonal system of Sauropsida can be found in my publications of 1931,
1935, and 1936, and were recently confirmed by SENN (1968). Some pertinent aspects are
also included in section 5 of the present chapter.

different vertebrate forms, these relationships of recessus synencephalicus appear to be somewhat variable.

STREETER (1933), who expressed considerable scepticism about the significance and the diverse interpretations of neuromery, described the synencephalon as 'methathalamus'. This latter designation, however, does not seem desirable, since it might be mistaken for the homonymous BNA term, which has a quite different significance.[138]

4. *Midbrain segments.* While some authors, particularly also KUPFFER (1906), and, among the recent ones, BERGQUIST (1964) and VAAGE (1969), recognized two mesencephalic neuromeres, SCHUMACHER (1928), STREETER (1933), and other investigators recorded only one single midbrain segment. On the basis of my own observations I believe that, although, depending on ontogenetic stage and particular vertebrate form, only a single mesencephalic vesicle may be clearly displayed, the tendency toward a subdivision into segments $M_1$ and $M_2$ can be noticed. Some of my own findings are here rather similar to those depicted by HALLER (1929).

MEEK (1907) considered these two segments as representing an oculomotor and a trochlearis segment. Yet, the vesicular or segmentai aspect is far less prominent in the basal plate region, where these nuclei arise, than in the domain of alar plate.[139] Nevertheless, it is not impossible that MEEK's hypothesis is justified, and that a relationship between the rudimentary, perhaps inconstant mesencephalic segmentation, and the said nuclei does obtain. VAAGE (1969), however, with reference to various supporting findings recorded in the literature, expresses the opinion that the nucleus trochlearis is situated in the rhombencephalon. Since the peculiar trochlear nerve presents several poorly clarified morphologic problems, whose discussion will be included in section 3 of the following chapter VII, and a secondary relationship, through rostral migration, of nucleus IV to 'segment' $M_2$ cannot be excluded, the apparently differing views of MEEK and of VAAGE *(et al.)* are perhaps not necessarily mutually exclusive.

---

[138] The metathalamus of the BNA subsumes 'corpus geniculatum' mediale et laterale. Critical comments concerning this subdivision were given in a paper summarizing our observations on the development of corpus geniculatum mediale in the rabbit (K., 1935a).

[139] In contradistinction to VAAGE (1969), who interpretes the fovea isthmi (sulcus intraencephalicus posterior) as a remnant of $M_2$, I am therefore inclined to assume that this fovea is of different origin and represents an 'independent' configuration sui generis.

5. *Rhombomeres.* As regards these segments, my own observations agree well with those reported by KAMON (1906) and SCHUMACHER (1928). The first rhombomere (recorded but not designated as such by SCHUMACHER) seems related to the cerebellar anlage, while the second is connected with the trigeminal root. The third rhombomere could not be linked with a nerve root but might still pertain to the domain of the developing trigeminal grisea. The fourth rhombomere appears connected with the roots of the acoustico-facial complex. I was unable to notice, with sufficient certainty, a nerve root connection of the fifth rhombomere, but it seems possible, as GRÄPER (1913) believed, that this segment corresponds to the nervus abducens. In contradistinction to the nerves V, VII/VIII, IX and X, the VIth nerve represents a ventral root, originating in a basal and medial location whose relationship to the rhombomeric segmentation is not easily recognizable. The segment boundaries, which are most clearly displayed in the dorsolateral regions of the rhombencephalon, become rather indistinct basomedial-ward.

The sixth rhombomere appears related to the glossopharyngeal root, and the seventh to that of the vagus. It remains, however, questionable whether these overall relationships of rhombomeres to cranial nerve roots indicate a primary, fundamental orderliness of segmentation with phylogenetic significance or should be considered incidental and secondary manifestations resulting from an interplay of diverse ontogenetic processes. The discrepancy between KUPFFER's interpretation, referring the glossopharyngal root to the fifth rhombomere, and that of other observers, who traced this nerve root to the sixth, might well be due to actual 'fluctuations' or 'irregularities' occurring in the course of the just-mentioned interactions.

Reverting to the definitive main subdivisions of the 'adult' vertebrate brain (Figs. 67, 145) as recognized in accordance with the generally accepted standard nomenclature, it could be stated that this terminology appears both convenient and justifiable. All five major subdivisions, namely telencephalon, diencephalon, mesencephalon, metencephalon (cerebellar segment), and myelencephalon can be evaluated as 'units' characterized by particular configurational features.

Two slightly different logical procedures can be followed in describing the ontogenetic evolution of which this definitive brain pattern is the end-result. Such descriptions, representing operationally useful semantic thought models, are required for an understanding of, or an 'insight' in, form, structure, function, and pathology of the neuraxis.

As was attempted to show within the section on morphogenesis (section 1, subsection B), it still seems permissible and appropriate to base one of these models on the traditional concepts of two-, three-, and five-vesicle stages, 'downgrading', as it were, the significance of the still poorly elucidated and controversial transitory neuromery. This latter, whatever it may imply, nevertheless does obtain, and the traditional model under consideration thus represents no more than a rather useful first approximation.

For a somewhat more advanced approach, it seems therefore necessary to consider, as far as possible, the relevant behavior of *sensu strictiori* neuromeric configuration. However, because of the numerous problems and uncertainties, including those of proper interpretation,

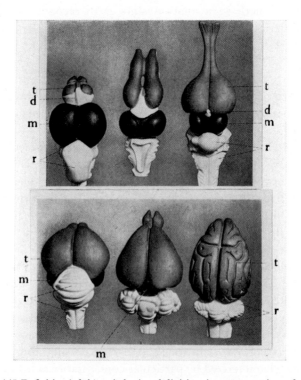

*Figure 145.* Definitive (adult) main brain subdivisions in representatives of various vertebrate classes (Courtesy of Prof. H. SPATZ, after models by SOMMER; cf. also fig. 67). Top row from right to left: Teleost (trout), amphibian (frog), reptile (crocodile); bottom row: bird (pigeon), lissencephalic mammal (rabbit, lagomorph), gyrencephalic mammal (dog, carnivore). d: diencephalic roof; m: mesencephalon; r: rhombencephalon; t: telencephalon.

this second procedure, at least at the present time, likewise remains an approximation which is perhaps, in some respects, even less satisfactory than the first one.

With regard to the boundaries between the definitive subdivisions, it should again be stressed that such limits are here not provided by

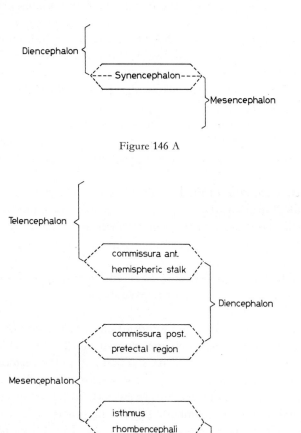

Figure 146 A

Figure 146 B

*Figure 146.* Two diagrams illustrating transitional status of synencephalon (A) and zones of transition representing telo-diencephalic, diencephalo-mesencephalic, and mesen-cephalo-rhombencephalic boundaries (boundary regions) (B).

geometric lines, but by topologic neighborhoods representing zones of transition which may be relatively narrow or relatively wide. The telencephalo-diencephalic boundary, indicated by commissura anterior and vestiges or derivatives of sulcus, respectively torus, telo-diencephalicus, is fairly sharp. Yet, a mingling of diencephalic (particularly hypothalamic) cell masses in the bed griseum of commissura anterior obtains.[140]

The diencephalo-mesencephalic boundary zone, indicated by commissura posterior, is wider and less well defined, involving the pretectal region. It seems best to consider the posterior commissure as pertaining to both subdivisions, while, despite overlap, the derivation of pretectal grisea from either diencephalic, or mesencephalic fundamental longitudinal zones can be traced with a fairly reasonable degree of accuracy. The mesencephalo-rhombencephalic boundary zone, in turn, is represented by the relatively wide transitional region designated as isthmus.

The two diagrams of Figure 146 attempt to illustrate, in a necessarily simplified manner, the concept of boundary zones and the peculiar transitional significance of the synencephalic neuromere.

### 4. The Deuterencephalon and the Transformations of its Primary Zonal System

On the basis of the data and interpretations discussed in the preceding sections, the deuterencephalon, consisting of mesencephalon and rhombencephalon, can be conceived as that part of the vertebrate brain tube whose wall includes basal, alar and roof plates, and which is connected with the roots of the cranial nerves III to XII (respectively spino-occipital nerves of Anamnia). The rostral portion of the deuterencephalon, namely the mesencephalon, seems to lack a floor plate component, which is present in rhombencephalon and spinal cord. The ventral portion of the lateral plate, to which the concept 'basal plate' can be applied, appears to end within the deuterencephalo-archencephalic (mesencephalo-prosencephalic or mesencephalo-diencephalic) boundary region.

---

[140] The peculiar boundary features resulting from the developmental events involving the hemispheric stalk are considered in sections 4 and 5.

It may be justified to consider the deuterencephalon as kathomologous with the rostral part of the spinal cord in Amphioxus, namely
with the region caudally adjacent to the apical 'brain' vesicle of that
Acranian. Again, from a morphologic viewpoint, the deuterencephalon can be regarded as promorphologically homologous (homodynamic), that is to say in this case serially homologous, with the spinal cord.

The hereby implied, and actually quite clearly recognizable similarities with the spinal tube, displayed by the deuterencephalic tube, nevertheless decrease, that is, become progressively less and less distinct
along a caudo-rostral direction. Again, although the mesencephalon
can justly be considered to represent the rostral portion of the deuterencephalon, this latter, because of the gradually increasing caudo-rostral 'gradient' of dissimilarity which becomes steep in the isthmus region, thereby comprises two subdivisions with somewhat different
morphologic patterns, namely rhombencephalon and mesencephalon.
Finally, with regard to the special differentiation of alar plate providing the cerebellum, a further subdivision of rhombencephalon into
metencephalon and myelencephalon can be made in accordance with
traditional terminology (BNA, PNA).

The transformations, displayed in the course of ontogenetic evolution by two of the primary longitudinal zones of the deuterencephalon,
namely roof plate and floor plate, have been pointed out in section 2.
The present section 4 shall deal mainly with the transformations affecting alar and basal plates, which represent, in the aspect here under consideration, the more relevant constituents. These adjoining two fundamental pattern elements, whose junctional open boundary zone corresponds, at the ventricular surface, to the sulcus limitans, can be rather
easily recognized and distinguished in many embryonic as well as adult
stages of the vertebrate brain, although, depending upon ontogenetic
stages or vertebrate forms, the sulcus limitans may display considerable variations of distinctiveness.

After the studies of His had established the morphologic significance of the primary longitudinal zones dealt with in section 2 of the
present chapter, the investigations of GASKELL (1886, 1889), OSBORN
(1888), STRONG (1895), C. J. HERRICK (1899), JOHNSTON (1902), and
others, led to a further subdivision of these primary zones, based on
the concept of *functional nerve components*, and characterized by a distinction of 'somatic' and 'visceral' attributes involving additional subordinated categories. The term 'somatic' is here supposed to subsume functions 'concerned with the adjustment of the body to its environment',

while '*visceral*' is said to refer to those functions 'concerned with the internal adjustments of the body' (HERRICK, 1931). It is evident that both 'somatic' and 'visceral' cannot be defined in a satisfactorily rigorous manner and represent rather hazy, ambiguous and 'slippery' abstractions which may easily lead into semantic traps.[141] Nevertheless, there is little doubt that, as a first approximation, these notions, and the concept of functional longitudinal zones, respectively fiber tracts, elaborated upon this theoretical basis, have been most valuable and helpful in further studies on the organization of the neuraxis.

GASKELL, HERRICK, and other authors concerned with the '*doctrine of nerve components*' pointed out that the peripheral nerves emerging from the central neuraxis comprise a variety of fibers with different functional significance. Thus, e.g., the vagus, glossopharyngeal, and facial nerves carry *afferent* fibers from taste buds,[142] from entodermal mucous membranes (at least nerves X and IX), from the ectodermal outer skin, and, in fishes as well as in some larval or even adult (mainly aquatic) amphibians, afferent fibers from the lateral line system.[143] Upon entering the neuraxis, these various sorts of fibers are segregated

---

[141] HERRICK (1922) attempted to answer the question '*what are viscera?*' and to provide relevant definitions which I do not find very convincing. However, a gleam of diffidence seems to have crossed that author's mind, since he concludes his paper with the following statement: 'in a functional analysis of the animal body three classes of organs must be recognized: (1) visceral, (2) somatic, and (3) ambiguous or transmutant cases, in origin belonging to one of the two primary types but secondarily transformed wholly or in part into the other. The third class cannot be eliminated or ignored, for organisms are not static, but are ever in flux and old materials may be transformed and put to new uses quite at variance with any formal rules which we may lay down in our logical system'. DART (1922) justly points out *the misuse of the term 'visceral'* in a detailed analysis of the difficulties and inconsistencies involved by the use of this term with respect to skeletal, muscular, and nervous systems. In contradiction to HERRICK and others, he considers olfaction to be 'a thorough-going somatic, ectodermal and exteroceptive sense'. TRETJAKOFF, about 1913, had previously also evaluated olfactory sensory elements as '*somatisch-rezeptorisch*'. Needless to say, I have always, since my early neurobiologic studies, considered olfaction to be an exteroceptive 'sense'.

[142] In some fishes, taste buds are present in the outer skin, being essentially innervated by the VIIth nerve. In the catfish *Ameiurus (Siluroid Teleost)*, e.g., these cutaneous taste buds are distributed over practically the entire body surface as well as over the 'barblets'.

[143] Further details concerning the lateralis system, the cutaneous gustatory system, and the morphologic significance of cranial as well as spinal nerves are discussed in sections 1, 2, and 3 of the next chapter (VII).

from each other and redistributed, being gathered, as it were according to their provenance or 'functional' significance, into several well defined longitudinal bundles with descending and ascending fibers (Figs. 147–150). These bundles, which are located in topologically distinctive neigborhoods of the alar plate, contain therefore fibers of a particular category pertaining not to one, but to a group of cranial nerves, and represent afferent fibers of the first order. In addition, said fiber tracts are accompanied or surrounded by grisea displaying a more or less 'columnar' arrangement, and containing correlated neurons of the second order, from which further connections of the second order arise.[144]

---

[144] The relative *specificity of receptors ('peripheral analyzers')* for certain sorts of stimuli, and the distribution of the resulting signals, in accordance with such specificity (which can be conceived as normally corresponding to an *'invariant'*), upon particular 'collecting' grisea, pertaining to a 'functionally' specific system, represent elementary factors of the complex and hierarchically organized 'logical mechanism' dealing with *'abstraction of invariants'*. At higher hierarchical levels, such, e.g. as related to cortical activities in mammals and man, this mechanism involves pattern registration and recognition, as well as symbolization and higher 'memory' functions.

In a paper on *'Physiologie der Hirnrinde'* (Jahrb. d. Max-Planck-Gesellschaft 1968, 62–89), CREUTZFELDT stresses the self-evident fact that only those optic environmental aspects which are registered by the retina can be processed by the brain. It is, of course, well-known that the retina, being a peripheral portion of the brain, already performs analyzing functions of high order (cf. e.g. the mechanisms studied by J. Z. YOUNG in the optic ganglion of cephalopods, which corresponds to grisea of the vertebrate retina, as discussed in vol. 2, pp. 231–237 of this series). Because an apparently diffuse and mixed distribution, leading from 'simple' to 'complex' and 'hypercomplex' receptive fields, obtains in the mammalian visual cortex, CREUTZFELDT takes issue with the conclusion *'dass eine hierarchische Ordnung im Nervensystem besteht mit zunehmender Abstraktion bestimmter Reizparameter in getrennten Kanälen'*. He then adds: *'eine Parallele zum Problem der Mustererkennung von Computern bietet sich an, und ein Vorgang ähnlich der Invariantenerkennung wird vermutet. Diese Parallele ist wahrscheinlich falsch'*. As stated above, I believe that this comparison (*'Parallele'*) is not only justified, but rather self-evident.

It can be maintained that the mathematical and logical orderliness obtaining in computer mechanisms is likewise displayed by neuronal mechanisms. This, quite obviously, includes registration of, and operations with, *invariants*. I consider oversimplified models, such as the application, by McCULLOCH and PITTS, of *Boolean algebra* in SHANNON's notation to be highly valuable, despite the evident shortcomings of their elementary formulation. Further progress must be sought by formulations expressing the basic features of these models in probabilistic and statistical terms. The semantic approaches of ASHBY, V. NEUMANN, and RASHEVSKY seem to me rather useful.

CREUTZFELDT, moreover, denies the justification of attempts to correlate the available data on neural coding with the phenomena of consciousness, implying that such, at the time necessarily generalized and oversimplified attempts, represent *'voreilige Anthropomorphismen'* and teleologic interpretations. He oddly concludes by quoting the once

As regards the *efferent* components of cranial nerves, the fibers innervating the skeletal and eye musculature leave the neuraxis in root bundles comparable to ventral roots of the spinal cord (except for the peculiar root of the IVth or trochlear nerve), and are considered to be 'somatic'. Their grisea of origin ('nuclei') represent an interrupted 'somatic' 'nuclear' (or cell) column, continuous with the ventral horn of spinal cord.

The efferent fibers innervating the branchial musculature and its derivatives, moreover the 'autonomic' preganglionic fibers of the 'vegetative nervous system', leave the neuraxis as components of branchial nerves (VII, IX, X, XI) which are comparable to dorsal roots of the spinal cord. The preganglionic fibers of the oculomotor nerve (III) provide here another exception. All these fibers are considered to be 'visceral'. Again, their grisea of origin are arranged in one (Anamnia) or two (Amniota) interrupted cell columns, approximately continuous with the lateral horn or the dorsal part of ventral horn of spinal cord. Both the 'somatic' and the 'visceral' efferent cell columns display, like the afferent tracts and their grisea, a topologically distinctive and definable longitudinal zonal arrangement within the basal plate (cf. Figs. 150–154).

The *'doctrine of nerve components'*, widely accepted by American and other neurobiologic authors, and accordingly stressed by American texts on embryology (e.g. AREY, 1954) and neuroanatomy (e.g. RANSON, 1943), distinguishes, in its most extreme formulation, as given by HERRICK (1931), eight different categories, subsumed under the concepts *'somatic'* and *'visceral'*, combined with the qualifications *afferent* and *efferent*. These categories, which can be assigned to fibers, tracts, neurons, grisea, cell columns, and wall portions of the neural tube, are enumerated as follows.

---

well-known saying of pre-World War I French revanchists and warmongers: '*N'en parlons jamais, mais il faut savoir que nous y pensons toujours*'.

Unless, adopting the proverbial policy of the apocryphal ostrich, one chooses to ignore the actual facts of life, it becomes unavoidable to formulate a logical or semantic correlation of physical neural events ( *N-events* ) with mental events, i.e. with consciousness ( *P-events* ). The imputation that 'anthropomorphic' or 'teleologic' concepts are thereby required, is completely and obviously unjustified if, in accordance with rigorous concepts of neurological epistemology, physical (behavioristic), and mentalistic (psychic) terms are properly used and restricted to their respective domains. Besides, how can any significant progress be made unless the problems are widely discussed. I would thus counter CREUTZFELDT's strange quotation by citing the Spanish adage '*Hablando se entiende la gente*'.

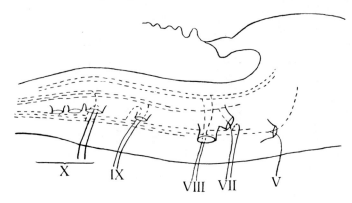

Figure 147 A

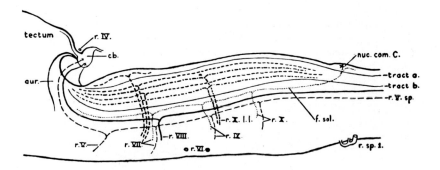

Figure 147 B

*Figure 147.* Diagrams of longitudinal root fiber bundles in the amphibian rhombence-phalon. A Arrangement in a 'generalized' amphibian larva (partly based on concepts by HERRICK, and modified after K., 1927). The Roman numerals represent cranial nerves. Top bundle vestibulo-lateral system ('special somatic afferent'); middle bundle 'general and special visceral afferent' (tractus solitarius); lower bundle cutaneous exteroceptiv-'general somatic afferent' (radix descendens et ascendens trigemini). B Diagram of the central courses of the sensory components of cranial nerves V to X of larval Amblystoma seen as projected upon the lateral surface of the medulla oblongata (from HERRICK, 1948). The general cutaneous component is drawn in dashed lines, the vestibular in thick unbroken lines, the visceral gustatory in dotted lines, the three lateral line VII roots in thick dash lines, the two lateral line X roots in dot-and-dash lines, and correlation tracts a and b in thin continuous lines. aur: cerebellar auricle; cb: cerebellum; nuc. com. C.: 'nucleus commissuralis' of CAJAL; r.: radix.

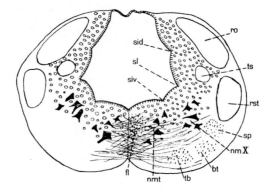

*Figure 148.* Semidiagrammatic cross-section through medulla oblongata of the urodele amphibian, Triton, at the level of nucleus motorius vagi, showing the location of the
three main longitudinal root fiber tracts (ro, ts, rst) indicated in figure 147 A (from K.,
1927). Although HERRICK, in figure 147B, separates the vestibular from the lateral fibers,
these two components may, for practical purposes, here be considered as forming a single
distinctive common tract system. bt: tractus bulbo-tectalis; fl: fasciculus longitudinalis
medialis; nm: efferent nucleus of vagus; nmt: nucleus reticularis ('motorius') tegmenti;
ro: radix nervi octavi; rst: radix descendens trigemini; sid: sulcus intermedius dorsalis;
siv: sulcus intermedius ventralis; sl: sulcus limitans; sp: tractus spino- and bulbo-tectalis
( ?); tb: tractus tecto-bulbaris ( ?); ts: tractus solitarius.

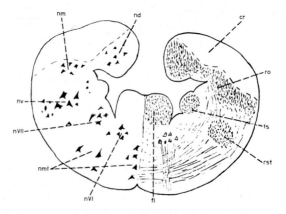

Figure 149 A

*Figure 149.* A Semidiagrammatic cross-section through the rhombencephalon of the
Selachian Acanthias at the level of efferent abducens and facial nuclei (in part after KAP-
PERS, from KUHLENBECK, 1927). B Diagram, comparable to preceding figure, and indicating topologic connectedness of zonal arrangements (from K., 1927). Both figures should
also be compared with figure 148. cr: crista cerebellaris; fl: fasciculus longitudinalis medialis; nVI: nucleus of abducens; nVII: efferent nucleus of facial nerve; nd: nucleus dorsalis

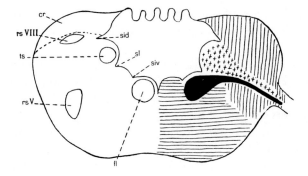

Figure 149 B

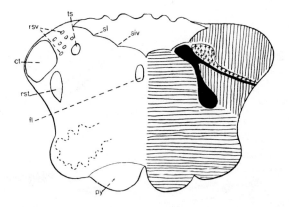

*Figure 150.* Diagram, comparable to that of figure 149 B, indicating topologic connectedness of zonal arrangement in a cross-section through the human medulla oblongata
(from K., 1927). ct: corpus restiforme (inf. cerebellar peduncle); py: pyramid (tractus cor-
tico-spinalis); rsv: radix descendens vestibularis; other designations as in figure 149. It can
be seen that, in comparison to the arrangement in Selachians (and other Anamnia), the intermedioventral zone has two subdivisions, namely the medial, paraventricular pars interna (preganglionic n. dorsalis vagi et glossopharyngei), and the ventrolateral, pars externa
(n. ambiguus, 'branchiomotor').

nervi octavi; nm: nucleus medialis nervi octavi; nmt: nucleus reticularis ('motorius') tegmenti; nv: nucleus ventralis nervi octavi; ro, rs VIII radix nervi octavi; rs V, rst: radix
descendens nervi trigemini; sid: sulcus intermedius dorsalis; siv: sulcus intermedius ventralis; sl: sulcus limitans; ts: tractus solitarius; horizontal hatching: ventral zone and undifferentiated basal plate; black: intermedioventral zone and its root fiber channel; crosses:
intermediodorsal zone (tractus solitarius and correlated grisea) with its root fiber channel;
vertical hatching: dorsal zone; octavus stands here for 'octavolateralis'.

1. *General somatic afferent.* Related to general exteroceptive innervation of the integument and deep proprioceptive innervation of muscles and tendons.

2. *Special somatic afferent.* Related to the innervation of highly differentiated sense organs, such as cochlea, vestibular apparatus, and lateral line organs. The visual organ and its neuronal connections are likewise included in this category.

3. *General somatic efferent.* Related to the innervation of the 'general skeletal musculature' of the body.

4. *Special somatic efferent.* Related to the innervation of 'highly specialized somatic muscles', namely, the external eye muscles and 'part of the tongue muscles'.

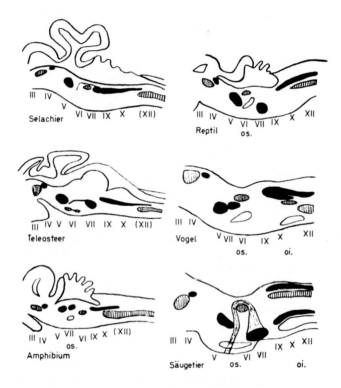

*Figure 151.* Diagrams showing interrupted 'columnar' arrangement of efferent cranial nerve grisea ('nuclei') in diverse vertebrates (modified after Kappers, from Kuhlenbeck, 1927). Vertical hatching: ventral zone ('somatic' efferent); black: intermedioventral zone ('visceral' efferent); oi: inferior olivary complex (presumably also present, although not indicated, in all forms here included); os: superior olivary complex (presumably present in anuran Amphibia, but perhaps not in other Anamnia).

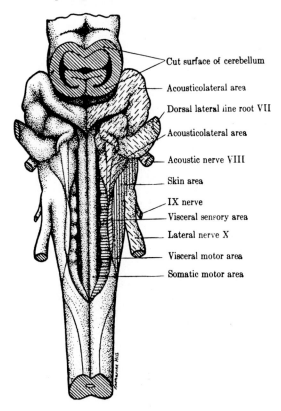

Cut surface of cerebellum

Acousticolateral area

Dorsal lateral line root VII

Acousticolateral area

Acoustic nerve VIII

Skin area

IX nerve

Visceral sensory area

Lateral nerve X

Visceral motor area

Somatic motor area

Figure 152 A

*Figure 152.* Interrupted 'columnar' arrangement of cranial nerve grisea projected upon floor of fourth ventricle in the Selachian Acanthias, in an urodele amphibian larva (Amblystoma), and in adult man. A *Squalus acanthias* (from HERRICK, 1931). B Amblystoma (modified after HERRICK, from K., 1927). n: nucleus; nm: nucleus 'motorius'. C Adult man. cu: nucleus cuneatus; gr: nucleus gracilis; Vm: cell column of mesencephalic trigeminal root; Vp: griseum of descending ('spinal') trigeminal root; Vs: trigeminal main 'sensory' nucleus; Vt: 'motor' trigeminal nucleus; VII, IX, X s: griseum of tractus solitarius; VIII v: vestibular grisea; IX, X, XI a: nucleus ambiguus; IX, X d: nucleus dorsalis (alae cinereae); vertical hatching: octavus system (dorsal zone); fine dots: trigeminus and posterior funiculi system (dorsal zone); oblique hatching: system of tractus solitarius; coarse dots: ventral zone; crossed hatching: intermedioventral zone, pars interna (preganglionic); horizontal hatching: intermedioventral zone, pars externa ('branchiomotor').

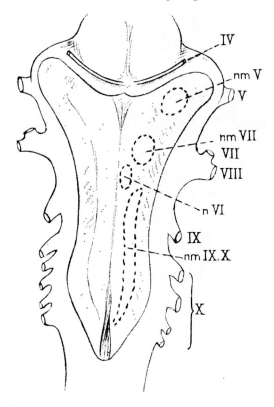

Figure 152 B

5. *General visceral afferent.* Related to the innervation of 'visceral mucous surfaces without highly differentiated sense organs'.

6. *Special visceral afferent.* Related to the innervation of 'specialized sense organs serving the senses of taste and smell'.[145]

7. *General visceral efferent.* Related to the innervation of 'unstriped musles, heart muscle, glands'. This category includes the preganglionic fibers arising within the central neuraxis.

---

[145] It should here be recalled that speaking of the 'senses' of taste, smell, and sight, as well of 'sensory organs' or 'sensory fibers' in general, evidently involves 'mentalistic terms' derived from conscious experience (consciousness). In a truly behavioristic terminology based on a rigorously logical language it would therefore be necessary to speak of chemoreceptor input I (olfaction) and II (taste), optic input, and generally of input or afferent fibers etc. Since such, in the aspect under consideration, excessive rigor is not only connected with various semantic difficulties but would clearly display undue and

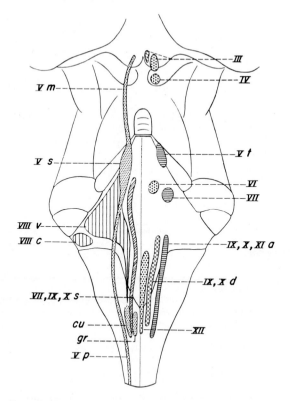

Figure 152 C

8. *Special visceral efferent.* Related to the innervation of 'highly specialized striated muscles of a different origin (both embryologically and phylogenetically) from the striated trunk muscles'.

A somewhat shorter classification, omitting a distinction of 'general somatic' and 'special somatic' efferent categories, was adopted in the following manner by AREY (1954).

---

unnecessary pedantry, the traditional terminology may be used with the required caution, and ignoring the hidden mentalistic implications. A consistent behavioristic approach should regard all organic functions as intrinsically identical with the behavior of non-biological physico-chemical systems or of artificial mechanisms. In the highly useful and unavoidable behavioristic semantic model *(fiction* in VAIHINGER's sense) the actually obtaining differences are here based on complexity and substratum (with or without organic 'metabolic' activities) but not on 'essence' or 'quiddity'.

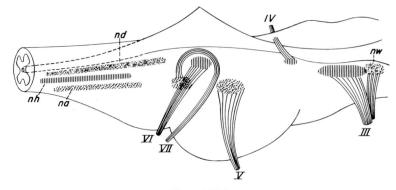

Figure 153 A

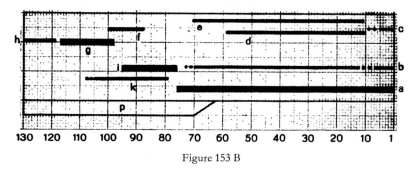

Figure 153 B

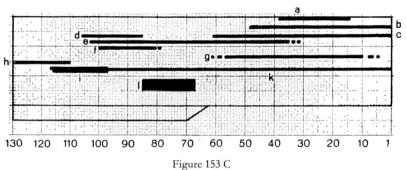

Figure 153 C

*Figure 153.* Interrupted 'columnar' arrangement of rhombencephalic cranial nerve grisea of man in lateral projection. A Diagram referring to adult man, for comparison with figure 151 B. na: nucleus ambiguus column, to which facial and trigeminal motor grisea likewise pertain; nd: nucleus alae cinereae; nh: nucleus hypoglossi; nw: nucleus of *Edinger-Westphal;* nd and nw pertain to pars interna of intermedioventral zone; grisea of ventral zone indicated by vertical hatching. B Diagram showing actual length of efferent griseal columns and olivary nuclei in the human newborn, as determined by a wax plate reconstruction. Each of the 130 plates, representing four 30 μ sections, corresponds thus to an actual thickness of 120 μ (from Hɪĸɪᴊɪ, 1933). a: n. olivaris inferior; b: n. ambiguus

A. Afferent (or sensory).

1. *Somatic afferent.*

   (a) *General* (fibers ending chiefly in the integument).

   (b) *Special* (fibers from the sensory epithelia of the eye and ear).

2. *Visceral afferent.*

   (a) *General* (sensory fibers from the viscera).[146]

   (b) *Special* (fibers of smell and taste).

B. Efferent (or motor).

   1. *Somatic efferent* (fibers ending on skeletal muscle).

   2. *Visceral efferent.*

(a) *General* (fibers ending about autonomic ganglion cells which, in turn, control smooth muscle, cardiac muscle and glandular tissue).

(b) *Special* (cranial nerve fibers terminating on the striated musculature derived from branchial arches).

In the *spinal cord*, however, the 'functional components' which here display a lesser degree of diversity, are simply classified as somatic afferent, visceral afferent, visceral efferent and somatic efferent (cf. Fig. 155). These categories essentially conform to the corresponding 'general' ones of deuterencephalon and its cranial nerves.

Although the configurational arrangement, upon which the 'doctrine of functional nerve components' is superimposed, can be regarded as a reasonably well established and demonstrable fact, the complex terminology involving a subdivision into the above-mentioned eight or seven 'functional' categories remains encumbered by a number of semantic ambiguities and by the implication of some insufficiently clarified actual relationships.

---

[146] AREY apparently includes here the afferent innervation of the vasculature. But should blood vessels be considered viscera? It will also be seen that AREY 'sidesteps' the semantic problem related to proprioceptive innervation. HERRICK, however (cf. above), at least attempts to deal with some aspects of this question.

---

column; c:n. accessorii dorsalis; d: n. hypoglossi; e:n. dorsalis vagi et glossopharyngei (alae cinereae); f:n. abducentis; g:n. motorius trigemini; h:n. loci caerulei; i:n. motorius facialis; k: n. olivaris superior complex; p: pons.

   C Diagrams showing actual length of afferent griseal columns and of area postrema in the human newborn (procedure identical with that of fig. 152B, from HIKIJI, 1933). a: area postrema; b: n. gracilis; c: n. cuneatus; d: n. vestibularis sup.; e: nn. vestibularis medialis et vestibularis spinalis; f: n. vestib. lateralis; g: nn. tractus solitarii; h: n. radicis mesencephalicae trigemini; i: main sensory n. of trigeminus ('convolutio trigemini'); k:n. radicis descendentis (spinalis) trigemini; l: nn. cochleares.

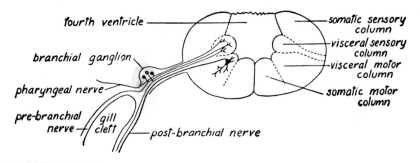

Figure 154 A

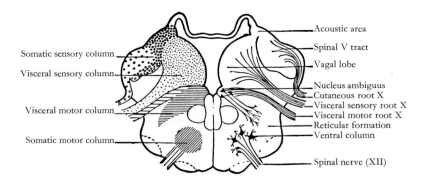

Figure 154 B

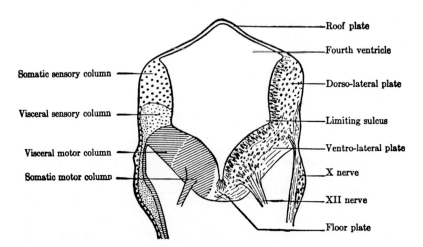

Figure 154 C

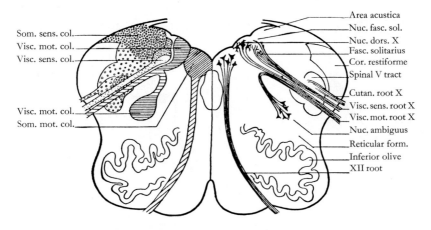

Figure 154 D

*Figure 154.* Functional columnar, respectively zonal arrangement in the medulla oblongata of fishes and man as depicted by HERRICK (from HERRICK, 1931). A Cross-section
through oblongata of the Selachian Acanthias. B Cross-section through oblongata of the
Teleost *Prionotus carolinus.* C Cross-section through oblongata of a 10.2-mm human embryo (HERRICK designates here alar plate as 'dorso-lateral plate', and basal plate as 'ventrolateral plate'). D Cross-section through adult human oblongata.

It is evident that, for an adequate formulation of functional categories, SHERRINGTON's (1906) concepts of *exteroceptive, interoceptive,* and
particularly *proprioceptive*[147] stimuli or neural signals are, in various respects, preferable to the concepts '*somatic*' and '*visceral*'. Although 'somatic afferent' and 'visceral afferent' can be considered partly equivalent to exteroceptive and interoceptive, the qualifier proprioceptive
must then be distributed upon both categories of the *Gaskell-Herrick
terminology.* This cannot be done in a sufficiently convenient and satisfactory manner.

As regards the allegedly 'visceral sense' of *olfaction,* I believe, with
DART and others, that it is clearly related to an 'adjustment of the body
to its environment', and should therefore be classified as 'exteroceptive', respectively 'somatic'.[147a]

---

[147] Cf. vol. 1 (pp. 13–14) of this series.

[147a] One might, moreover, insist that all input, exteroceptive, interoceptive, and
proprioceptive, causes 'internal adjustments of the body'. However, the qualifiers 'primarily' and 'secondarily' might here perhaps provide a semantic 'loophole'.

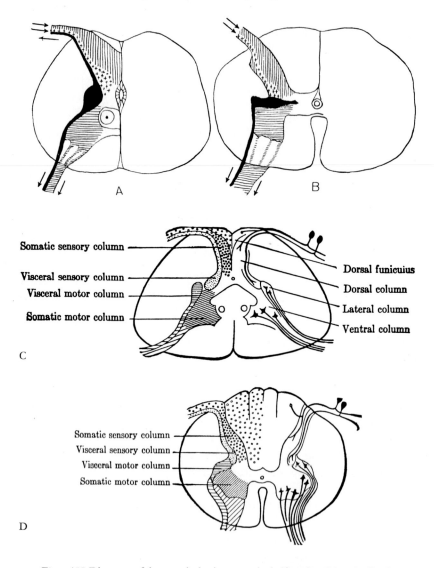

*Figure 155.* Diagrams of the morphologic, respectively 'functional' longitudinal zones in the spinal cord of Anamnia (Teleosts) and Amniota (Mammalia, man). A Generalized spinal cord of Teleost. B Generalized spinal cord of man. Vertical hatching: dorsal zone; crosses: intermediodorsal zone; black: intermedioventral zone; horizontal hatching: ventral zone; the encircled dot in A represents Mauthner's fiber. C Spinal cord of the Teleost, Menidia. D Human spinal cord (A and B from K., 1927; C and D from HERRICK, 1931; although diagrams of this sort represent a very crude first approximation, disregarding various complex 'secondary' rearrangements, they remain useful as indicating some relevant aspects of the obtaining configurational orderliness).

Concerning the *'gustatory sense'*, which is mediated through 'taste buds' located in oral cavity and pharynx, the junction line between entoderm and ectoderm runs anteriorly to the row of vallate papillae (in man and at least various other mammals). It can thus be admitted that most of the taste buds are entodermal, although some lingual taste buds seem to be *ectodermal*. Nevertheless, one might concede, in this respect, that gustatory functions are interoceptive as well as visceral. Yet, the cutaneous taste buds of fishes, mentioned above in footnote 142, are clearly ectodermal, receiving, moreover, input from the external environment, and must, therefore, be evaluated as exteroceptive 'sense organs'. Their classification as 'visceral' is obviously *Procrustean*. The exteroceptive 'gustatory' or chemoreceptor fibers, however, join a common 'general' *and* 'special' 'visceral' fiber tract and griseal column of the deuterencephalon.

Again, 'fibers from the sensory epithelia of the *ear*', including, in higher forms, *cochlear* and *vestibular* ones, as well as those of the *lateral line* system of Anamnia, are classified as 'special somatic afferent'. Now, the cochlear fibers are evidently essentially exteroceptive, while the vestibular ones can reasonably be evaluated as proprioceptive. With regard to the lateral line system, its classification as exteroceptive rather than proprioceptive seems quite justified. In the deuterencephalon, however, vestibular and lateralis components are closely joined *qua* configurational arrangements, the vestibularis system being fairly well segregated from the cochlear tracts or grisea.

With respect to other proprioceptive systems, HERRICK classifies 'deep proprioceptive sensibility' of muscles and tendons as 'general somatic afferent'. In the deuterencephalon, however, the only system reasonably well identified as proprioceptive *qua* cranial musculature is the *mesencephalic root of the trigeminus* with its vesicular cells (neurons of the first order), which is related to the muscles of mastication whose motor innervation has been termed 'special visceral efferent'. Thus, in a consistent terminology, the corresponding proprioceptive fibers of these muscles should be called 'special visceral afferent'.

Concerning the mesencephalic root of the trigeminus, a number of additional poorly clarified problems obtain. It is believed that this system also deals with input (pressure?) from teeth and gums, besides containing proprioceptive fibers from other cranial nerves such as oculomotor, trochlear, abducens, facial, and vagus (cf. e.g. PEELE, 1961). Accordingly, some of these (facial, vagus) fibers should be 'special visceral afferent' while others (in eye muscle nerves) should be 'special somatic afferent'. Pressoreceptor [147b] vagus and glossopharyngeal fibers from carotid sinus and aortic arch, which could be evaluated as 'general visceral afferent', together with chemoreceptor fibers, and therefore presumably 'special visceral afferent' components of these nerves, however, seem to join the combined 'general and special visceral' system (tractus solitarius and its grisea). All this, at best, involves a rather awkward and confused terminology, which might, of course, delight a lawyer's heart.

Further unclarified problems, awaiting their proper solution in accordance with the doctrine of 'functional nerve components', are related to the presence and connections of nerve cells scattered along the extracerebral (and perhaps intracerebral) course of the eye muscle nerves. Their proprioceptive function has been suspected. Whether such cells are displaced elements of the mesencephalic root of the trigeminus, and whether this latter is exclusively proprioceptive, still remain moot questions. Likewise, on the basis of not entirely conclusive observations, the presence of proprioceptive cells within the efferent nuclei of eye-muscle nerves has been suggested. This would represent an unusual (but of course not 'impossible') exception to the general rule implying a location of 'all' (i.e. 'most') primary afferent cellular elements (perikarya) in alar plate.

Reverting to the well corroborated 'factual' morphologic basis for the doctrine of *'functional nerve components'*, the following orderliness is manifested by the configuration of the central neuraxis.[148]

1. *Primary efferent grisea* ('centers') are generally located in the basal plate, while the *primary afferent grisea* are generally located in the alar plate.

2. The primary centers or grisea of origin for those *efferent fibers* which leave the neuraxis with *dorsal roots* are located in the dorsal portion of the basal plate, frequently closely ventral to sulcus limitans, i.e. dorsally to the more ventral grisea originating the efferent fibers of ventral roots for body and eye musculature. In the spinal cord, grisea of origin for preganglionic fibers are, as a rule, located in the dorsal portion of basal plate. This may be related to the fact that, in Anamnia, some of these fibers join the dorsal root. In Amniota, however, these

---

[147b] One could evidently also argue that such 'deep' pressoreceptor fibers should be classified as 'proprioceptive'.

[148] The peripheral relationships of functional nerve components are considered in the sections of chapter VII dealing with the relevant peripheral nerves.

preganglionic fibers generally leave with the ventral roots. This, of course, might be a secondary, 'phylogenetic adjustment'. In the mesencephalon, the preganglionic grisea of the oculomotor nerve are likewise essentially located dorsally to those supplying the striated musculature, although both sorts of fibers leave the neuraxis by a ventral root.

3. The primary afferent deuterencephalic grisea for *cutaneous innervation* and their fiber tract are located more dorsally than primary grisea and fiber tract for *interoceptive, gustatory, vascular chemoreceptor* and *vascular pressoreceptor* innervation. The grisea and the fiber tract of these systems lie rather closely dorsal to sulcus limitans[149] (tractus solitarius and its grisea).

4. In the rhombencephalic portion of deuterencephalon, the primary afferent grisea (and their root fiber bundles) pertaining to the *statoacustico-lateralis systems* have, as a rule, the most dorsal position of the here enumerated configurations.

In addition to the distinction of basal and alar plate, a further morphologic subdivision into the following four main longitudinal zones can thus be made with respect to the rhombencephalon (K., 1926, 1927).

I. *Ventral zone (Ventralgebiet)*. Efferent nuclei for innervation of striated muscles by the nerves III, IV, VI, and XII (respectively spino-occipital) nerves.

II. *Intermedioventral zone (Intermedioventralgebiet)*. Efferent nuclei for branchial musculature and their derivatives and for preganglionic fibers (nerves V., IX., X., XI.). In Anamnia, both the efferent elements for branchial musculature and the preganglionic elements appear more or less intermingled within this zone. In Amniota, however, the 'branchiomotor' elements assume a distinct, more lateral or external position, while the preganglionic ones are separately gathered in a more medial or internal position. Thus, in these latter forms, the intermedioventral zone consists of two subdivisions, namely

(A) *Pars interna* ('n. dorsalis' vagi et glossopharyngei, 'ala cinerea').

(B) *Pars externa* (n. motorius V, VII, n. ambiguus IX, X, n. XI).

III. *Intermediodorsal zone (Intermediodorsalgebiet)*. Tractus solitarius with its grisea (n. tr. solitarii and n. parasolitarius).

---

[149] It should be recalled that, because of the eversion displayed by the rhombencephalon, and depending on the degree of such eversion, morphologically ventral and dorsal relationships may become topographically medial and lateral ones (cf. figs. 167 B, 169, 172, 176, 180).

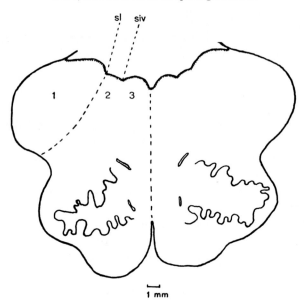

*Figure 156.* Cross-section through adult human medulla oblongata indicating approximate extent of alar plate (1) and basal plate (between the two dotted lines). 2: open neighborhood of intermedioventral zone; 3: open neighborhood of ventral zone; sl: approx. location of sulcus limitans; siv: approx. location of sulcus intermedioventralis (from K., 1929a).

---

*Figure 157.* BECCARI's interpretation of the rhombencephalic longitudinal arrangement in Anamnia (A) and Amniota (B) at different levels (from BECCARI, 1943). Explanations for figure 157 A. 1: pars superior, lateralis et medialis (K.) of dorsal zone; 2: pars inferior et lateralis (K.) of dorsal zone; 3: intermediodorsal zone; 4: intermedioventral zone; 5: ventral zone; V–X: levels of corresponding cranial nerve roots (spino-occipital nerve roots of Anamnia, comparable to Amniote XII, and mesencephalic root of V are omitted); a: 'general somatic afferent' fibers; b: branchiomotor fibers (a and b of V); c: 'somatic' efferent fibers of VI; d: lateralis fibers; e: 'somatic' afferent cutaneous fibers; f: gustatory fibers; g: branchiomotor fibers (d–g of VII); h: vestibular fibers; i: lagena fibers of VIII, precursors of Amniote cochlearis system; k: 'somatic' afferent cutaneous fibers; l: gustatory fibers; m: branchiomotor fibers (k–m of IX); n: lateralis fibers; o: 'somatic afferent cutaneous fibers; p: 'visceral' afferent fibers; q: branchiomotor and preganglionic fibers (n–q of X). Explanations for figure 157 B. 1–5: same as in A; 6, 7: dorsal, respectively ventral horn of upper cervical spinal cord (compare with fig. 155); VII–X: same as in A; XIS: spinal accessory; XII: hypoglossus; a: 'somatic' afferent cutaneous fibers; b: gustatory fibers; c: preganglionic salivary fibers of VII; d: branchiomotor fibers (a–d of VII); e: 'somatic' afferent cutaneous fibers; f: gustatory fibers; g: 'visceral' afferent (pharyngeal); h: preganglionic (salivatory) fibers; i: branchiomotor fibers (e–i of IX); j: 'somatic' afferent cutaneous fibers; k: 'visceral' afferent fibers (from gastro-intestinal tract); l: preganglionic

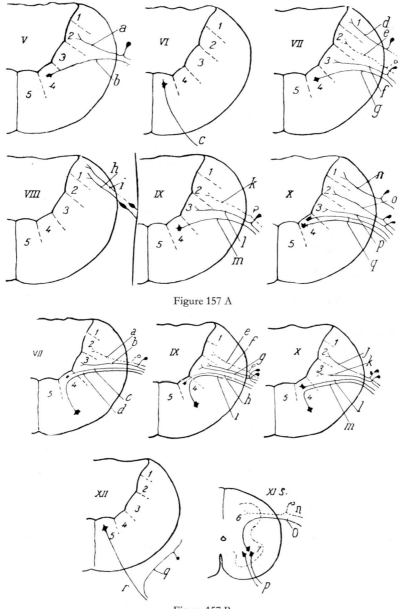

Figure 157 A

Figure 157 B

efferent fibers; m: branchiomotor fibers, e.g. for pharynx and larynx (j–m of X); n: rudimentary dorsal ganglion of spinal accessory; o: 'branchiomotor' fibers for mm. trapezius
and sternocleidomastoideus of spinal accessory; p: first spinal ventral roots; q: 'somatic'
afferent fibers of *Froriep's ganglion* of XII; r: 'somatic' efferent lingual fibers of XII.

IV. *Dorsal zone ( Dorsalgebiet )*

(1) Afferent nuclei and root fiber tracts of stato-acustico-lateralis systems, (2) radix s. tractus spinalis trigemini, main afferent trigeminal nucleus. The two components listed under (1) and (2) have somewhat different locations, and the dorsal zone is thus further subdivided into

(A) *Pars superior, lateralis et medialis,* related to (a) vestibulo-lateralis system. This region becomes somewhat segregated from an additional subregion pertaining to

(b) cochlearis system in forms where present.

(B) *Pars inferior et lateralis,* related to the exteroceptive innervation whose primary fibers are gathered in tractus s. radix spinalis V. (with its accompanying grisea), or reach the main sensory nucleus of n. V.

This general configurational arrangement (K., 1926, 1927) which was also recognized by BECCARI (1943, p. 86f.), is very clearly displayed in the rhombencephalon of adult Anamnia (cf. Figs. 148, 149, 157 A) and can still be identified in Amniota, including man (cf. Figs. 150, 156, 157 B). The boundary neighborhoods between the four main longitudinal zones are frequently indicated, on the ventricular surface, by a system of sulci comprising *sulcus limitans* and two additional, roughly parallel grooves in alar and basal plates. In the former,

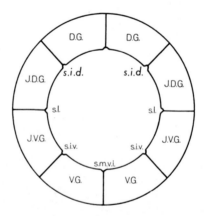

*Figure 158.* Diagram, based on the configuration of the oblongata in the Dipnoan Protopterus, but valid as a general schema indicating the morphologic pattern of the vertebrate rhombencephalon (after GERLACH, 1933). D.G.: dorsal zone; J.D.G.: intermediodorsal zone; J.V.G.: intermedioventral zone; V.G.: ventral zone; s.i.d.: sulcus intermediodorsalis; s.i.v.: sulcus intermedioventralis; s.l.: sulcus limitans; s.m.v.i.: sulcus medianus ventralis internus rhombencephali. The floor plate, shown in figure 164, has been omitted as not relevant for the purpose of the diagram.

*sulcus intermediodorsalis* (or intermedius dorsalis) approximately corresponds to the boundary between dorsal and intermediodorsal zone. This latter, in turn, is roughly demarcated from the intermedioventral zone by the sulcus limitans. The *sulcus intermedioventralis* (or *intermedius ventralis*) approximates a ventricular boundary 'line' between intermedioventral zone and ventral zone (cf. Figs. 148, 149, 156). Concerning these two last mentioned zones *(Gebiete)*, it should be stressed that they represent open topologic neighborhoods within the basal plate, and thus, taken together, not necessarily preempting the entire basal plate. Accordingly, certain grisea, such as components of nucleus reticularis tegmenti, might be considered derivatives of basal plate in general, not classifiable as subdivisions of one of the two specialized zones. A similar qualification seems perhaps justified with respect to some alar plate structures. Nevertheless, discounting the cerebellum, which could be evaluated as a special rostral differentiation[150] of dorsal zone and alar plate in general, the morphologic pattern or bauplan of the vertebrate rhombencephalon may be indicated by the diagram of Figure 158, elaborated by GERLACH (1933) on the basis of our investigations.

As regards the *cerebellum*, it will here be sufficient to state that this alar plate configuration of the rostral rhombencephalon essentially consists of a transverse, median portion, the *primordial corpus cerebelli* (cerebellar plate), and the paired lateral *'auricles'*. These latter, in many Anamnia, are 'ear-like' folds surrounding the lateral recess of fourth ventricle. Such auricle includes an inferior or posterolateral fold, caudally continuous with acustico-lateral area, and a superior or anteromedial fold, continuous with its antimere through, or along, the caudal edge of corpus cerebelli.

---

[150] Some comments on cerebellar bauplan were included in section 1B. The highly specialized features of cerebellar ontogenesis and configuration shall be dealt with in chapter X of volume 4, and their details can be ignored as being irrelevant for the generalized formulation here under consideration. In this respect, however, the peculiar ontogenetic development of the cerebellar cortex might be mentioned. It involves, besides the subependymal matrix, a transitory superficial granular layer (cf., e.g., K., 1950), which has been shown to be present in numerous different vertebrates. As regards other aspects of cerebellar ontogenesis and phylogenesis, the following publications should also here be pointed out: RÜDEBERG's attempt (1961) to derive the cerebellar nuclei of vertebrates from so-called 'migration layers', and LARSELL's comprehensive posthumous monograph (1967) on the cerebellum from Myxinoids to birds. JANSEN's and BRODAL's *Handbuch* volume (1958) on the human cerebellum likewise contains numerous pertinent data on mammalian and submammalian cerebellar comparative anatomy.

Generally speaking, it could be said that the auricles of Anamnia are closely related to the vestibular, respectively vestibulo-lateralis system, while the primordial corpus has relationships to spino-cerebellar and therewith associated input. If the corpus cerebelli is well developed, its caudalmost portion, continuous with the lateral, auricular component, represents a morphologically and perhaps functionally distinctive unit, the *flocculo-nodular lobe*, linked to the vestibularis system, the *flocculus* of Amniota being a derivative of the Anamnian auricle. The *nodulus* can be evaluated as the midline neighborhood interconnecting the two flocculi. The cerebellum is then accordingly subdivided into a (secondary) *corpus* and a *flocculo-nodular lobe*, the ('*vermian*') *nodulus* being, according to LARSELL, a 'new feature' displayed by Amniota (particularly birds and mammals).

In mammalians, a portion of the corpus cerebelli becomes closely related to certain regions of neocortex, by way of connections mediated through pontine grisea. Roughly speaking, this portion, which expands into paired lateral configurations, corresponds to the *cerebellar hemispheres* of macroscopic anatomy. The median neighborhoods of the mammalian cerebellum are conventionally described as the *vermis*.

EDINGER (1910) introduced the phylogenetic terms *palaeocerebellum* and *neocerebellum*. The former subsumes vermis and flocculo-nodular lobe, the latter designates the hemispheres, and was meant to subsume the portions of cerebellum predominantly linked with neocortex.

LARSELL (1951) subdivided the mammalian cerebellum into *archicerebellum* (flocculo-nodular lobe with vestibular connections), *palaeocerebellum* (portions of cerebellum with spino-cerebellar and other connections), and *neocerebellum* (portions with predominantly neocortical connections). This author justly emphasized that there are no sharp conceptual boundaries between the relevant stages of presumptive phylogenetic development (as inferred from observable taxonomic series). This is quite obvious, since all the topologic neighborhoods corresponding to said subdivisions are already present in the most 'primitive' or rather 'simplest' cerebellar configuration consisting of auricle and primordial median corpus, as displayed by many Amphibia.

There is, nevertheless, some justification for LARSELL's convenient subdivision into archicerebellum, palaeocerebellum and neocerebellum. Concerning this latter, however, it should moreover be pointed out that certain regions of 'vermis' are presumably 'neocerebellar', and that certain regions of 'hemispheres' are considered 'palaeocerebellar'. An objection against the prefixes '*palaeo-*' and '*archi-*' as judiciously employed for the cerebellum by LARSELL could yet be raised on the grounds that these terms tend to perpetuate a misleading *cliché* commonly applied to a classification of cerebral cortex, basal ganglia, and thalamus. Comments on the evaluation of these telencephalic and diencephalic configurations will be found in sections 5 and 6 of the present chapter.

Since a typical neocortex appears to be an exclusive characteristic of Mammalia, the presence of a pons and of a neocerebellum in the here adopted sense seems likewise restricted to this vertebrate class. This interpretation does not contradict the claim, by JANSEN and BRODAL (1958), and others, that prosencephalic cerebellar connections, mediated by 'rudimentary' pons-like grisea, may obtain in birds, and perhaps even other submammalian vertebrates. Depending on arbitrary semantic rules and on the degree of logical rigor in their interpretation, one could here either speak of pontine and neocerebellar kathomologa, or of 'analoga', if the functional significance of prosencephalic fiber connections is stressed.

As regards cerebellar fiber connections, it might also be added here that vestibular input does not seem entirely restricted to the archicerebellum or flocculo-nodular lobe, but reaches additional regions, namely rostral vermian portions such as the lingula and adjacent neighborhoods of mammals. Relevant data concerning these topics as well as cerebellar cortex and the internal cerebellar grisea (nuclei) are to be dealt with in chapter X of volume 4.

In the *mesencephalon*,[151] the essential features of the deuterencephalic bauplan are still recognizable, although somewhat modified, particularly by the incursion of optic fibers into the dorsal portion of alar plate *(tectum mesencephali, tectum opticum)*. These exteroceptive fibers,

---

[151] OSTERTAG (1956, p.292) quotes the following statement by an opinionated embryologist denying the justification of evaluating the mesencephalon both as an ontogenetic brain vesicle and as a significant subdivision of the neuraxis: '*Im Schrifttum findet sich häufig die Beschreibung eines Stadiums «der drei primären Hirnbläschen»: Prosencephalon, Mesencephalon und Rhombencephalon. Diese Gliederung geht auf Befunde an Vogelkeimlingen (Hühnchen) zurück. Bei Vögeln ist oft (nicht bei der Ente) tatsächlich das spätere Mittelhirngebiet als Folge einer ontogenetischen Heterochronie früh mächtig entwickelt. Ein selbständiges Mittelhirnbläschen fehlt aber den meisten übrigen Wirbeltieren, insbesondere den Säugern und dem Menschen. In der Tat beruht die Entfaltung des Mittelhirngebietes bei Vögeln zunächst auch nur auf einer dorsal lokalisierten Proliferation des Materials übergeordneter optischer Gebiete (Tectum). Die seitlichen und basalen Teile dieses Gebietes sind echtes Rhombencephalon (Tegmentum). Eine innere Strukturgrenze zwischen Tegmentum mesencephali und Rhombencephalon existiert auch im aus-*

arising from the retinal ganglion cells, can be classified as essentially of the third order (1. neuron: rod, respectively cone, 2. neuron: bipolar cell, 3. neuron: retinal ganglion cell). Their incursion into tectum mesencephali is especially massive or preponderant in those vertebrates whose vision is well developed *and* depends mainly on the mesencephalic mechanisms.

The tectum mesencephali, however, represents a correlating griseum furthermore receiving considerable nonoptic input from rhombencephalon and spinal cord. In addition, the mesencephalic root of the trigeminus with its vesicular cells of the first order is located in a deep layer near the ventricular surface, often extending as far mediad as the midline neighborhoods.

In submammalian forms, particularly in Anamnia, fibers of the eighth nerve system ('lemniscus lateralis', including cochlear respectively lateralis input, where these systems are present) especially reach a more ventral portion of alar plate, which may bulge into the mesencephalic ventricle, bilaterally forming the *torus semicircularis* with its griseum (nn. mesencephali lateralis). The torus semicircularis becomes delimited from tectum opticum by a frequently very conspicuous groove, the *sulcus lateralis mesencephali*.

In mammals, this griseum expands in a caudo-dorsal direction, becoming the paired *colliculus inferior* tecti mesencephali. A tendency toward such displacement can be seen in some reptiles. In this respect, the torus semicircularis of submammalian forms, with its grisea (n. lateralis mesencephali and subdivisions), may be evaluated as a ventral (essentially nonoptic portion) of 'tectum mesencephali, displaced by *'internation'* in REMANE's sense (cf. above, p. 72).

With the exception of some Anamnia (particularly teleosts characterized by highly developed interoceptive and exteroceptive gustatory systems), and displaying a 'nucleus gustatorius secundarius', a well defined intermediodorsal zone cannot be recognized in the mesencephal-

---

*gebildeten Zustand nicht. Die besondere und verfrühte Anlage des Tectums steht mit der bevorzugten Ausbildung der Augen bei Vögeln in ursächlichem Zusammenhang'.* Nevertheless, OSTERTAG (loc. cit.), enumerating several evident reasons, rightly retains the '*Mittelhirn*'-concept. On the basis of the data discussed in the present chapter, it seems hardly necessary to refute the quoted statement which cannot be taken seriously. Merely a cursory glance at many of the relevant illustrations of the present chapter should be sufficient to show that, even at early ontogenetic stages, the mesencephalon represents a (deuterencephalic) subdivision *sui generis* throughout the entire craniote vertebrate series.

ic alar plate. Moreover, a corresponding sulcus intermediodorsalis does not become discernible on the ventricular surface.

In conformity with the considerable expansion and the peculiar differentiation of the mesencephalic alar plate, the sulcus limitans assumes a comparatively ventral position (Figs.159–162). Although in many instances poorly defined, it can be identified with reasonable certainty at some ontogenetic stages (occasionally including the 'adult' configuration) of many or perhaps most vertebrates. The mesencephalic basal plate, in addition to rostral differentiations or components of the griseum known as n. motorius s. reticularis tegmenti, contains the efferent nuclei of two segmental deuterencephalic nerves, namely of III and IV. The perikarya of the oculomotor, innervating extrinsic eye musculature, form a griseum representing the ventral zone *(Ventralgebiet)* of basal plate, and the locus containing the preganglionic elements corresponds to the intermedioventral zone *(Intermedioventralgebiet)*. A sulcus intermedioventralis is poorly and rarely indicated, but can nevertheless be noticed in some forms and at some stages. In its 'definitive' arrangement, the efferent griseum of nervus trochlearis is located in the ventral zone of mesencephalic basal plate, caudal to the oculomotor complex.[152]

Although the mesencephalon can indeed be considered to represent a 'true' (rostral) portion of the deuterencephalon (*'echtes Deuterencephalon'* but not *'echtes Rhombencephalon'*, cf. footnote 151), it does represent a significantly modified subdivision, characterized by a different configuration of its alar plate and a presumable lack of floor plate. Our interpretation of the mesencephalic morphologic pattern is shown in the diagrams of Figure 163. With regard to the concept of 'functional nerve components' and to the interpretation of its basic facts in predominantly morphologic terms (dorsal zone, intermediodorsal zone, etc.), I believe that these formulations, which refer to characteristics of the series of cranial nerves III to XII, can only be applied, with reasonable justification, to spinal cord and deuterencephalon. It seems evident that the archencephalic nerves (I, nervus terminalis, nervus parietalis, and so-called nerve II) are of an entirely different morphologic

---

[152] The morphologic interpretation of nervus trochlearis and its nucleus has posed a still poorly understood problem of interpretation, mentioned above on p. 346 of section 3, and also briefly dealt with on pp. 99–100, and 195–196 of my *Vorlesungen* (K., 1927). This question will again be pointed out in chapter VII, sections 3 and 4.

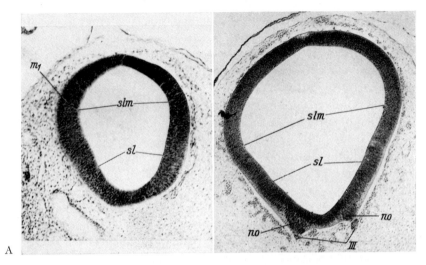

A                                                                     B

*Figure 159.* Early developmental stages of the Sauropsidan mesencephalon (from K., 1936). A Cross-section, duck embryo of 72 h. B Cross-section, duck embryo of 120 h. Figure 159 corresponds closely to Hugosson's figure 21 (1957) depicting this configuration in stage HH-18 (65–69 h) of a chick embryo. $m_1$: first mesencephalic neuromere; no: anlage of oculomotor nucleus; sl: sulcus limitans; slm: sulcus lateralis mesencephali; III: root fibers of oculomotor.

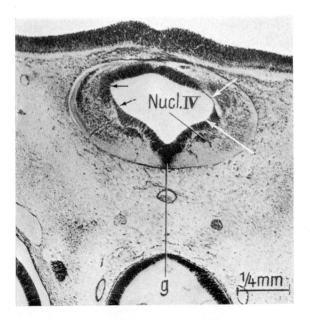

*Figure 160.* Cross-section through caudal mesencephalon in a chick embryo of stage HH-27. Left side is more rostral than right side, which is close to the isthmus rhombencephali (after Vaage, 1969; the arrows have been added to indicate my interpretation). g: rostral end of floor plate ('glycogen-containing raphe'); upper arrows: sulcus lateralis mesencephali; lower arrows: sulcus limitans.

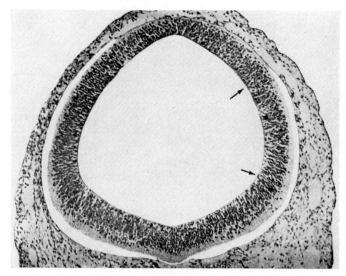

Figure 161 A

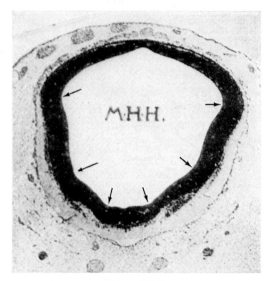

Figure 161 B

*Figure 161.* Cross-sections through the human mesencephalon at early embryonic stages. The arrows have been added to the original figures. A Embryo of about 4 weeks, less than 10 mm (from His, 1904). The sulcus lateralis mesencephali (upper arrow, right) is hardly recognizable, being merely suggested; lower arrow: sulcus limitans. B Embryo of 15.3 mm (from Hochstetter, 1929); upper arrows: sulcus lateralis mesencephali; middle arrows: sulcus limitans; lower arrows: sulcus intermedioventralis.

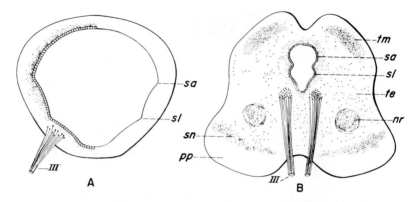

*Figure 162.* Semidiagrammatic cross-section through the human mesencephalon in an embryo (A) of about 15 mm (6–7 weeks) and in the adult (B). It should be stressed that the ventricular outline of the adult's cerebral aqueduct is subject to considerable secondary variations and deformations which commonly prevent an accurate identification of the relevant sulci. nr: nucleus ruber tegmenti; pp: pes sive basis pedunculi; sa: sulcus lateralis mesencephali; sl: sulcus limitans; sn: substantia nigra; te: tegmentum; tm: tectum mesencephali; III: n. oculomotorius.

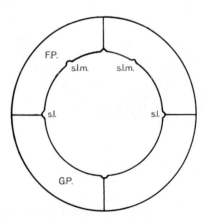

*Figure 163.* Diagram, based on the mesencephalic configuration in the Dipnoan Protopterus, but valid as a general schema indicating the morphologic pattern of the vertebrate mesencephalon (after GERLACH, 1933). G.P: basal plate; F.P.: alar plate; S.L.: sulcus limitans; S.L.M.: sulcus lateralis mesencephali.

character, quite apart from the essentially different diencephalic and telencephalic longitudinal zonal arrangements.

Concerning the significance of the various sulci approximately suggesting ventricular boundaries of the longitudinal zonal systems, it can
be claimed that a sufficiently consistent correlation of these sulci with
the corresponding zones obtains, which justifies their evaluation as
configurations providing '*contributory evidence*' in favor of the zonal bauplan concept. Such interpretation does not seem invalidated by the
possibility that the formation of these sulci may be 'caused' by a multiplicity of factors, some of which may be 'accessory' and unrelated to
the zonal configurational orderliness. The variability in the manifestation of these sulci, including the presence of accessory and atypical
ones (cf., e.g., K. and NIIMI, 1969), was repeatedly pointed out. Their
possible distortion by postmortal and fixation effects must also be kept
in mind. Nevertheless, their overall correlation with the zonal pattern
can be regarded as well established.

The deuterencephalic bauplan, as defined above in morphologic
terms, gradually begins to become recognizable at various successive
ontogenetic stages of Anamnia, and is particularly conspicuous in
young 'adult' specimens of numerous relatively 'undifferentiated' or
'generalized' forms. Amniota may display, at rather early embryonic
periods, a configuration very similar to the definitive aspect of this
bauplan as manifested in Anamnia (cf. Figs. 164–166). *Prima facie*, it
seemed possible to suppose that the grisea pertaining to or located
within the various longitudinal zones (Figs. 167–169) might originate
*in situ*.

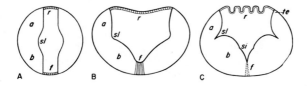

*Figure 164.* Diagrams of the fundamental *zones of His* characterizing the spinal cord
and deuterencephalic morphologic pattern, as displayed in spinal cord (A) and oblongata
(B) of early human embryos (4–5 weeks), compared with the arrangement in the adult Cyclostome Petromyzon (C). a: alar plate; b: basal plate; f: floor plate; r: roof plate; si: sulcus intermedioventralis; sl: sulcus limitans; te: region of 'taenia' medullae oblongatae (roof-
plate attachment).

However, a number of observations disclosed various substantial migrations and displacements of cell masses in the course of ontogenetic evolution. One of the most significant first reports concerning these ontogenetic events is that by Essick (1907, 1909, 1912) who disclosed evidence, confirmed by numerous other authors, that several grisea, which, in the definitive arrangement, are located within the domain of the basal plate, consist of cellular elements derived from the alar plate. Subsequently to their dorsal origin and first differentiation, these cells appear to migrate ventralward into the basal plate (Fig. 170) where they become organized as griseal aggregations.

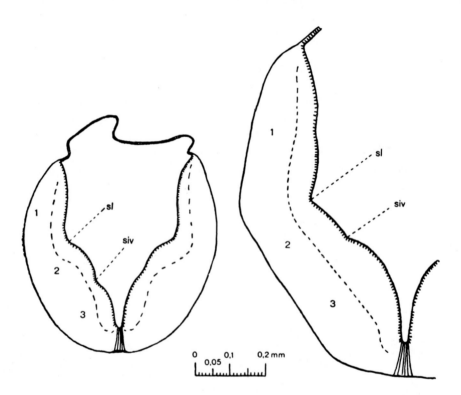

*Figure 165.* Actual outline, showing cross-section through oblongata of Petromyzon larva (Ammocoetes, left), compared with a section through oblongata of a 10.2-mm human embryo (right, after His) with sulcus intermedioventralis and perhaps intermediodorsalis which is faintly indicated, but not labeled, in alar plate (from K., 1929a; the scale refers only to Ammocoetes at left). 1: alar plate; 2: intermedioventral zone; 3: ventral zone; siv: sulcus intermedioventralis; sl: sulcus limitans.

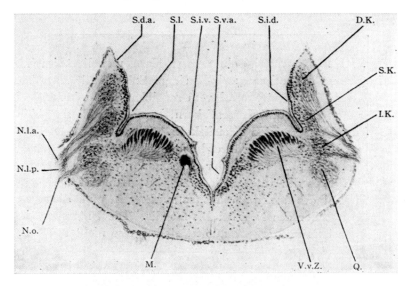

*Figure 166.* Cross-section through rostral oblongata of adult Cyclostome, *Entosphenus japonicus*, showing arrangement of sulci and cell groups (from SAITO, 1930). D.K.: dorsal nucleus of laterialis; I.K.: nucleus nervi vestibularis; M: *Müller cell* no. 7; N.l.a.: nervus laterialis anterior; N.l.p.: nervus laterialis posterior; N.o.: nervus octavus; Q.: radix descendens trigemini; S.d.a.: sulcus dorsalis accessorius; S.i.d.: sulcus intermediodorsalis, S.i.v.: sulcus intermedioventralis; S.K.: intercalated nucleus of laterialis (the intermediodorsal zone can be seen as a narrow cell strip medially to this nucleus); S.l.: sulcus limitans; S.v.a.: sulcus ventralis accessorius; V.v.Z.: rostral intermedioventral cell column.

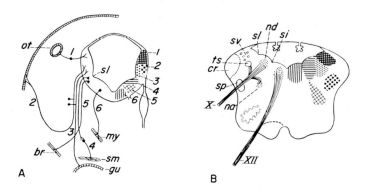

*Figure 167.* Diagrammatic cross-sections showing the relationship of longitudinal griseal and fiber zones at an early embryonal stage (approx. between 10 and 12 mm) and (B) in the adult configuration of the human brain (combined on the basis of findings or interpretations of HIS, HERRICK, RANSON, and others, and of the author's original observations). 1: dorsal zone (pars superior et lateralis); 2: dorsal zone (pars inferior et lateralis); 3: intermediodorsal zone; 4: intermedioventral zone (pars interna); 5: intermedioventral

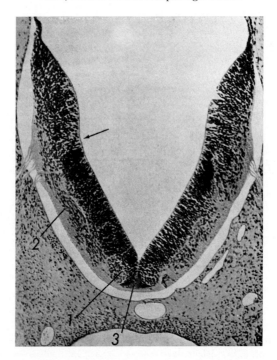

*Figure 168.* Cross-section through the oblongata at the glossopharyngeal root level in a human embryo of about 4 weeks, and somewhat less than 10 mm (from HIS, 1904). According to this author's interpretation: '*Eintritt sensibler Glossopharyngeusfasern in den Tractus solitarius, links Austritt motorischer Fasern der Accessoriusreihe aus dem Mark*'. '*Der Hypoglossuskern ist jederseits als gesonderte Masse der Mantelschicht erkennbar, aber es liegen keine austretende Wurzelfasern im Schnitt.*' Arrow: sulcus limitans; 1: ventral zone (n. XII); 2: intermedioventral zone; 3: floor plate. The designations have been added (K.) to HIS' unlabeled original figure.

zone (pars externa); 6: ventral zone; br: branchiomeric muscles; cr: restiform body; gu: entodermal gut; my: 'myomeric' muscles; na: nucleus ambiguus; nd: n. dorsalis vagi (preganglionic cell column of X, IX, and VII); ot: otic vesicle; sl: sulcus limitans; si: sulcus intermedioventralis; sm: smooth muscles ('viscera', cardiovascular system, including striated cardiac muscle); sp: radix descendens trigemini et eius nucleus; sv: radix descendens vestibularis et eius nucleus; ts: tractus solitarius with its grisea.

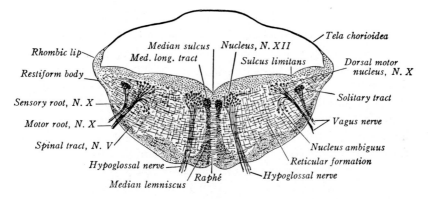

*Figure 169.* Cross-section through developing human oblongata at about two months of intrauterine life, displaying a configuration intermediate between that of A and B in preceding figure 167 (after AREY, 1954).

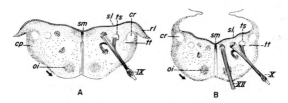

*Figure 170.* Diagrammatic cross-sections through the embryonic human oblongata of approximately 9 weeks (about 30 mm) at the levels of 'rhombic lip' (A) and of so-called 'ponticulus' (B). The migration of cells from the embryonic corpus pontobulbare in the alar plate is indicated by arrows. cp: corpus pontobulbare; cr: corpus restiforme; oi: inferior olivary complex; rl: 'rhombic lip'; sl: sulcus limitans; sm: sulcus medianus ventriculi quanti; ts: tractus solitarius; tt: radix descendens ('spinalis') trigemini; IX–XII: cranial nerves. The 'ponticulus' is dorsal to cr.

Such grisea are particularly the nuclei pontis of mammals,[153] located in a basal portion of metencephalon, the pars basilaris pontis, which represents a bulging mass, also containing longitudinal pyramidal tract bundles and other fiber connections, and is, as it were, ventrally super-

---

[153] JANSEN and BRODAL (1958) refer to 'pontile rudiments' in birds, and express the opinion that the nuclei pontis are not exclusively mammalian structures but might also be present in lower vertebrates. This is of course possible, but in view of the rather definite correlation of typical neocortical grisea, *specific* for mammals, and the characteristic pontine nuclei, it remains doubtful to which extent '*pontile primordia*' in submammalian forms represent comparable grisea. The relevant questions remain insufficiently elucidated, and, for practical purposes, a typical pons may still be evaluated as a mammalian morphologic attribute.

imposed upon the basal plate components. The nuclei arcuati of the mammalian myelencephalon, which can, at least in part, be interpreted as caudal pontine nuclei, also pertain to this group of grisea. Within the basal plate proper of metencephalon, the superior olivary complex of vertebrates with a cochlear system seems likewise, either wholly or to a large part, to be provided by such migrating elements of the alar plate. In addition, the inferior olivary complex of vertebrates, located in the basal plate of the myelencephalon, might perhaps also be formed, at least in part, by cell migrations from the alar plate.

As regards the apparently rather massive migration process of the pontine grisea which appear to originate in the region of the so-called 'rhombic lip' of the alar plate, 'aberrant' cellular aggregates, whose displacement by ventral migration seems to have remained partial or incomplete, represent the *corpus ponto-bulbare* of ESSICK (1907)[154] in the adult rhombencephalon.

Of greater import, in the aspect here under consideration, namely with respect to ventral zone and intermedioventral zone of the 'definitive' bauplan, are further early migrations reported by TELLO (1922), BECCARI (1923), WINDLE (1933), and others. According to these observations it appears possible that, during an early embryonic stage, corresponding to 10 mm or less length in man (about 4 or 5 weeks of age), at least some of the 'visceral' efferent neuroblasts of the branchial nerves (V, VII, IX, X, XI) originate in the basal plate near the midline, medially to the 'somatic' efferent neuroblasts, or from a common nuclear column shared by these latter cells. Within a very short period the branchial nerve neuroblasts seem to migrate into their subsequent lateral position which (in man) they may reach at approximately 6 weeks of age. Figures 171 and 172 show AREY's (1954) interpretation of this migration process.

It is, however, difficult to interpret properly the relevant data, which still require further clarification. Whether the migrating neuroblasts are mostly branchiomotor, as seems to be the case, or whether they also include preganglionic elements, is not yet sufficiently detectable. It is not possible to state with certainty whether this assumed early ontogenetic migration is also characteristic of all 'primitive' lower forms (Anamnia) or whether it represents a cenogenetic feature in

---

[154] Further data concerning pontine grisea, superior olivary complex (including nuclei of trapezoid body, and inferior olivary complex will be included in volume 4, chapter IX of this series.

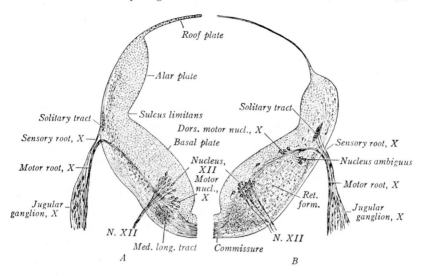

*Figure 171.* Semidiagrammatic cross-sections indicating AREY's interpretation of griseal migrations in the embryonic human oblongata, at the hypoglossal level, between the (approx.) 10- and 12-mm stages (after AREY, 1954).

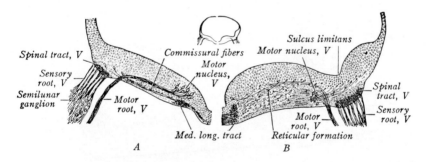

*Figure 172.* Semidiagrammatic cross-sections (with an orientation sketch, above), indicating griseal migrations in the embryonic human rhombencephalon at root entrance level of trigeminus, between the (approx.) 6-mm and 11-mm stages (after AREY, 1954).

'higher' forms. The possibility of neurobiotactic influences[155] exerted by the fasciculus longitudinalis cannot be excluded as a plausible factor, since KAPPERS and his associates have shown considerable shifting or displacements of motor cranial nuclei in the vertebrate series, sub-

---

[155] Neurobiotaxis in the qualified sense as elaborated in section 7 of chapter V (volume 3, part I) of this series.

stantiating the concept of neurobiotaxis. Again, some observations (KIMMEL, 1940) could be interpreted to indicate a 'homogeneous' efferent cell column subsequently separating into a medial ('somatic') efferent ventral zone and a lateral ('visceral') efferent intermediolateral zone.

It might, however, be pointed out that in Amphioxus, according to BONE (1960), the viscero-motor elements of the spinal cord, whose neurites emerge through the dorsal roots, are located in the most basal portion of the neuraxis ventrally to the neuronal elements believed to have 'somatic-motor' function. Amphioxus, however, displays in other respects highly peculiar arrangements, which preclude significant or convincing comparisons with Craniota for purposes of phylogenetic speculations. Thus, the so-called ventral or 'somatic-motor' root fibers of Amphioxus appear to be not nerve fibers but extensions of muscle fibers which provide neuromuscular junctions, at the surface of the spinal cord, with intraspinal longitudinal motor fiber systems (FLOOD, 1966, Cf. chapter VII; sections 1 and 2 of the present volume).

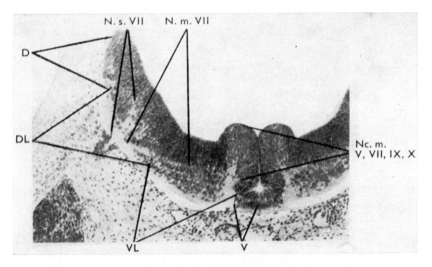

*Figure 173.* Transverse section through the embryonic medulla oblongata of the Selachian Torpedo at the 17-mm stage and on a level with the facial nerve, as interpreted by HUGOSSON. The motor V, VII, IX, X nucleus (Nc.m. V, etc.), and the motor and sensory facial fibers are said to be seen (from HUGOSSON, 1957). D: columna dorsalis; DL: col. dorsolateralis; N.m. VII: efferent facialis fibers; N.s. VII: afferent facialis fibers; V: columna ventralis; VL: col. ventrolateralis. It is evident that HUGOSSON includes here the domain of floor plate into what he designates as 'ventral column'.

During ontogensis of the craniote Vertebrate *spinal cord*, and as far as I could ascertain on the basis of my own observations, the 'somatic-motor' elements seemingly differentiate, *ab initio*, in the ventral zone of the basal plate, and the 'visceral efferent' elements in that plate intermedio-ventral zone.

In order to obtain further data concerning the early embryonic cell column, HUGOSSON (1957) undertook a detailed investigation on the formation of cranial nerve nuclei within the entire medulla oblongata (including pons) of several different vertebrates (Anamnia and Amniota). HUGOSSON attempts to correlate his findings with the concepts of 'Grundgebiete' and 'migration areas' elaborated by BERGQUIST and KÄLLEN, discussed further above in section 3.

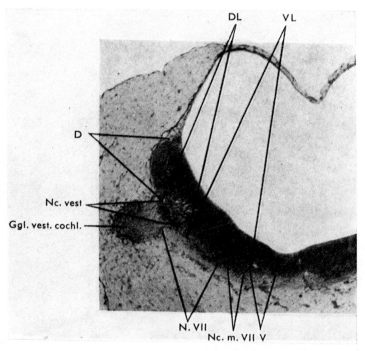

*Figure 174.* Transverse section through the embryonic medulla oblongata of the Sauropsidan chick at stage HH-21-22 (3.5 days) and on a level with the vestibulo-cochlear ganglion, as interpreted by HUGOSSON. The anlagen of the vestibular nuclei and the motor facial nucleus are said to be seen (from HUGOSSON, 1957). D: columna dorsalis; DL: columna dorsolateralis; Ggl: vest. cochl.: ganglion vestibulo-cochleare; N.VII: nervus facialis; Nc.m. VII: nucleus motorius facialis; Nc. vest: 'nucleus vestibularis'; V: columna ventralis; VL: col. ventrolateralis. In contradistinction to the preceding figure, HUGOSSON's columna ventralis does not here seem to include the domain of floor plate.

HUGOSSON (1957) reports that the development of migration layers results in the formation of four embryonic longitudinal cell columns present during developmental stages of the rhombencephalic neural tube, designated by the cited author as *ventral, ventrolateral, dorsolateral,* and *dorsal* (Figs. 173–175). These columns, forming 'longitudinal bulges on the mantle layer of the tube', are said to originate at a time when the postneuromeres of BERGQUIST and KÄLLEN begin to disappear. The ventral and dorsal columns are described as the first to develop, an ill-defined 'intermediate column' subsequently becoming differentiated into ventrolateral and dorsolateral column.

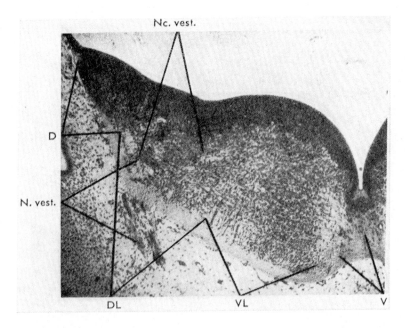

*Figure 175.* Transverse section through the embryonic medulla oblongata of mammalian man at the 13.5-mm stage and on a level with the vestibular root, as interpreted by HUGOSSON. The section is said to show the 'vestibular nucleus' (from HUGOSSON, 1957). D: columna dorsalis; DL: col. dorsolateralis; N. vest.: nervus vestibularis; Nc. vest.: 'nucleus vestibularis' (I would interpret this neighborhood as the primordium of n. vest. lat., n. vest. med., and perhaps n. of the descending vestibular root. K.); V: columna ventralis; VL: columna ventrolateralis. Here, again (cf. fig. 173), HUGOSSON's columna ventralis includes the domain of floor plate, in which scattered cellular elements (autochthonous or migrated?) can be seen.

According to HUGOSSON (loc. cit.) the 'somato-motor' grisea develop from the ventral column, the 'viscero-motor' from the ventrolateral
column, the 'viscero-sensory' from the dorsolateral column, and the
'somato-sensory' from the dorsal column. Although, *prima facie*, this
embryonic configuration and its differentiation seem to imply a certain
degree of correspondence between HUGOSSON's embryonic cell columns and the four major subdivisions of the *Gaskell-Herrick functional
zonal system*, HUGOSSON emphasizess that such interpretation cannot be
upheld.

Thus, the cited author claims that both 'somato-motor' and 'viscero-
motor' elements of the oculomotor complex develop from the ventral column, and that the 'somato-sensory' vestibular nuclei arise from
the dorsolateral column. The embryonic columns are not interpreted
to have any definite functional value, or to represent the anlagen of the
adult functional columns, 'but must rather be looked upon as building
units that provide material for the nuclear structures. Not until later on
in development, when the nuclei with their cell zones of uniform
function have formed into rows, do the four functional columns of
the adult stage appear. Terms such as 'somato-motor', 'viscero-motor', etc. should, therefore, not be applied to the early embryonic
stages, neutral designations such as 'ventral column', 'ventro-lateral
column' etc. being preferable' (HUGOSSON, 1957). In accordance with
concepts of BERGQUIST and KÄLLEN, the quoted author stresses that
his embryonic cell columns are related to longitudinal zones of high
mitotic frequency. He believes that these longitudinal proliferation
zones are formed 'through an influence exerted by the notochord, this
effect being probably 'in the nature of a general mitotic stimulation'.
'When the functionally similar nuclei, with their associated cell zones,
are ranged in four rows, there forms a new generation of columns
with strictly functional properties' (HUGOSSON, loc. cit.). Moreover,
this author interprets his findings as contradicting the presumed order
of development upon which BOK based his theory of neurobiotactic
stimulogenous fibrillation (cf. vol. 3, part I, p. 552 of this series).

The difficulties in evaluating 'static' histologic pictures of ontogenetic stages in terms of cell migrations are evidently considerable,[156]

---

[156] Such difficulties of interpretation are well illustrated by the quite unconvincing,
although not altogether unreasonable attempt of DART and SHELLSHEAR (1921, 1921–1922)
to show that the vertebrate spinal cord neuroblasts providing motoneurons for striated
musculature arise from mesodermal 'indifferent cells' of the primitive somites and become

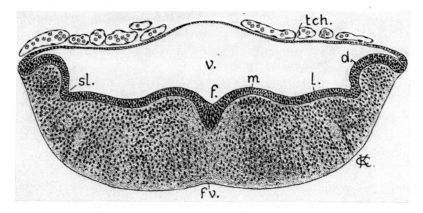

*Figure 176.* Cross-section through the oblongata in an 'older' embryo of the Selachian Scyllium at the 'trigeminal level' (from KUPFFER, 1906). d.: 'dorsal longitudinal ridge' (presumably dorsal zone); f.: 'sulcus longitudinalis medialis'; fv.: 'fissura mediana ventralis'; l.: 'lateral longitudinal ridge' (presumably intermediodorsal zone); m: 'medial longitudinal ridge' (presumably ventral zone); sl: sulcus limitans (probably; the alar plate elements doubtless extend ventrolateralwards beyond this sulcus); tch.: lamina epithelialis of plexus chorioideus; v.: fourth ventricle. The groove between the l and m neighborhoods is presumably sulcus intermedioventralis.

and lead to discrepancies of interpretations by different authors. In his monograph, HUGOSSON (1957) refers, *inter alia*, to some objections by VRAA-JENSEN (1956), who regarded most of the vestibular grisea as derivatives of the 'dorsal column'. I am here inclined to agree with this latter author.

Be that as it may, and although various other details of HUGOSSON's interpretation seem to me somewhat unconvincing, I would agree with that investigator's general conclusions, namely that the definitive morphologic pattern (bauplan), which is particularly well displayed in the deuterencephalon of some Anamnia, becomes clearly manifested relatively late in ontogeny, and that, at early stages, complex morpho-

---

secondarily incorporated into the neural tube. Actually, a somewhat comparable, although in other respects different neuromuscular relationship seems indeed to obtain in Amphioxus, where the ventral roots have been recently described as provided by processes of muscular elements connecting with the spinal cord. A similar mode of innervation, whereby *'the muscle goes to the nerve'*, is well-known to be displayed by invertebrate Nematoda. Further references to this kind of neuromuscular relationship will be made in chapter VII, sections 1 and 2, and in chapter VIII (vol. 4), section 3.

genetic and histogenetic events take place, whose description and interpretation could require still unclarified concepts differing from
those relevant for a formanalytic approach in the evaluation of their 'final', more or less 'static', transforms.

Nevertheless, one might still contend that, with respect to both
morphologic and 'functional' significance, the early distinction of *alar*
and *basal plate*, as based on the views of His, retains its validity. These
plates can be considered to represent a *primary zonal system* which undergoes subsequent transformations.

On the basis of the hitherto recorded observations it thus still remains uncertain to which extent the 'definitive' zonal arrangement of
the deuterencephalic vertebrate bauplan does or does not arise by differentiation *in situ* or through secondary cell '*migrations*'. Both sorts of
events are not entirely mutually exclusive, and a preliminary survey of

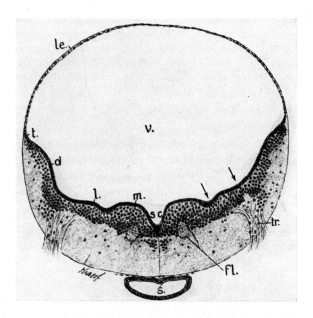

*Figure 177.* Cross-section through the oblongata in a 4-week-old embryo of the Ganoid Acipenser at the trigeminal root level (from Kupffer, 1906). f l.: fasciculus longitudinalis medialis; le.: lamina epithelialis (roof plate); t.: taenia of fourth ventricle (attachment
of roof plate); tr.: trigeminal root; s.: saccus vasculosus (diencephalon); sc: 'sulcus centralis longitudinalis' (median sulcus of fourth ventricle). Other abbr. as in preceding figure. The two arrows, added to Kupffer's original fig., indicate sulcus limitans region (lateral) and sulcus intermedioventralis (medial).

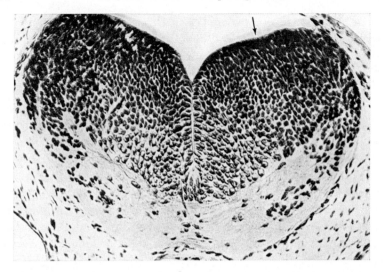

*Figure 178.* Cross-section through the oblongata in an embryo of the Teleostean trout of 6.5 mm (from His, 1904). The differentiation is less advanced than in the stages represented by the two preceding figures. An arrow pointing to region of sulcus limitans has been added (K.). The cellular pattern seems to indicate a basalward migration from the alar plate, although some of the scattered ventral elements, despite their transverse orientation, possibly related to fiber growth, may have originated from the basal plate.

the available data seems to indicate that both these morphogenetic 'tendencies' seem to obtain.

In this respect, the ontogenetic development of the rhombence-phalon appears to differ from the doubtless somewhat less complex morphogenesis of the spinal tube. As regards *histogenesis* (volume 3, part I, section 1) in contradistinction to *morphogenesis*, a substantial amount of the significant data was obtained from investigations, by various authors, of the spinal cord's differentiation. The relevant observations on these differentiation processes in deuterencephal-

---

*Figure 179.* Cross-section through the oblongata in an embryo of the mammalian rabbit of about 25.5 mm (from Ziehen, 1906). Labels and interpretations in accordance with the present treatise have been added to the unlabeled original figure. Arrow: region of sulcus limitans, 1: alar plate; 2: anlage of n. dorsalis vagi; 3: anlage of hypoglossal nucleus; the unlabeled basolateral cell clusters represent the anlage of the inferior olivary complex.

*Figure 180.* Cross-section through the oblongata in a fetus of the Prototherian (Monotreme) mammalian Echidna of about 22 mm (from Ziehen, 1906). Added designations as in preceding figure. The anlage of inferior olivary complex has here a more medial location than displayed in the section from the rabbit embryo.

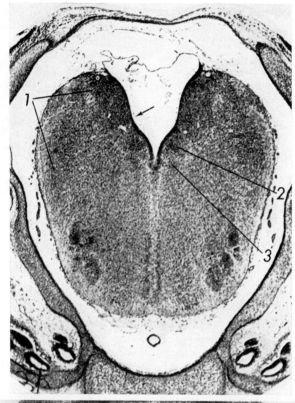

179

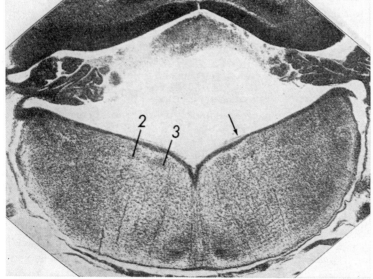

180

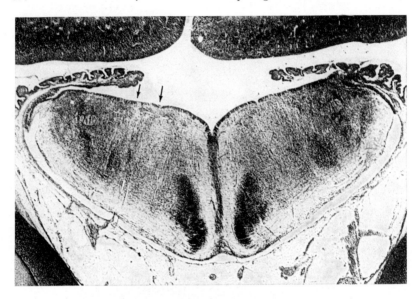

*Figure 181.* Cross-section through the oblongata in the mammalian human embryo of about 4 weeks (from His, 1891). Added arrows: sulcus limitans region (lateral), sulcus intermedioventralis region (medial). The cell group medial to sulcus limitans is presumably anlage of the glossopharyngeal portion of (preganglionic) n. dorsalis vagi (n. alae cinereae); the cell group medial to sulcus intermedioventralis may be the anlage of n. praepositus hypoglossi; 'rhombic lip' eversion recognizable at lateral edge of alar plate.

on, diencephalon and telencephalon appear to indicate that, with respect to the brain, these particular events proceed in a manner not essentially different from the sequences of histogenetic evolution displayed by the spinal cord. It seems, however, that some of the initial cell migration processes inferred on the basis of radioautographic studies exhibit certain regional differences concerning the site of DNA synthesis in matrix elements, and perhaps characteristic for cerebellum,[157] tectum mesencephali, diencephalon, and telencephalon (YoHIDA, 1967).

---

[157] Referring to *cerebellar histogenesis*, which is characterized by the peculiar transitory superficial granular layer, YOSHIDA (1967), whose radioautographic study was undertaken in chick embryos, gives the following account: 'In the mantle layer of the velum medullare posterius of the cerebellum, germinal pool is formed by the undifferentiated cells migrating from the ependymal layer of this area, and from this germinal pool embryonic granular layer elongates along the external limiting membrane. From these findings, it is concluded that the origin of embryonic granular layer is in the subependymal germinal pool of the velum medullare posterius.'

As regards the morphogenesis of the deuterencephalon, involving the demarcation of cell zones and grisea, particularly in the rhomben-cephalon, figures 176–181 show some developmental stages, recorded by KUPFFER (1906), HIS (1891, 1904), and ZIEHEN (1906) in several different vertebrates.

## 5. The Diencephalon and its Secondary Zonal System

On the basis of pioneering and fundamental studies primarily concerning the ontogenesis of the human brain, HIS (1893) recognized three main, more or less longitudinal subdivisions of the diencephalon, namely, in dorso-basal sequence, epithalamus, thalamus, and hypothalamus. In an investigation of the amphibian brain, C. J. HERRICK (1910) described four distinct subdivisions, *viz.*, epithalamus, thalamus dorsalis, thalamus ventralis, and hypothalamus, whose boundaries are approximately indicated by fairly constant sulci within the wall of the third ventricle. By this further subdivision involving the thalamus, HERRICK thus introduced the significant concept of a ventral thalamus[158] distinguishable from the dorsal one.

In accordance with JOHNSTON's investigations and interpretations,[159] (1902, 1909) it seemed at first possible that these four subdivisions might somehow correspond to the pattern correlated with the four 'functional nerve components', comprising somatic afferent, visceral afferent, visceral efferent, and somatic efferent longitudinal zones as displayed in spinal cord and deuterencephalon (cf. the preceding section 4). Again, the findings of HERRICK (1910), HERRICK and OB-ENCHAIN (1913), and DROOGLEVER FORTUYN (1912) suggested that a diencephalic zonal arrangement, similar to the configuration displayed

---

[158] Since the thalamus ventralis of Mammalia, and particularly of primates, including man, is far less developed than the thalamus dorsalis, becoming relatively inconspicuous at the adult stage, it is understandable that HIS did not consider the ventral thalamus to represent a 'separate' morphologic entity distinguishable from dorsal thalamus. In contradistinction to the condition in Amniota, however, the thalamus ventralis of Anamnia is a griseal configuration of substantial relative size.

[159] JOHNSTON (1902), however, although extending the somatic motor division into the forebrain, reaching almost the extreme rostral end of the telencephalon (exclusive of olfactory bulb), expressed scepticism about the recognition of the other zones at levels rostral to the rhombencephalon.

by amphibians, could be detected in reptiles, cyclostomes, and mammals. However, these first attempts at establishing the relevant homologies remained inconclusive, inexact, and in part contradictory, since the available data still precluded a more accurate formulation.

In a series of studies extending from 1924 to 1969, and in collaboration with SUMI (1926), SAITO (1930a, b), MIURA (1933), GERLACH (1933, 1947), MILLER (1940, K. and M., 1942, 1949), HAYMAKER (K. and H., 1949), KUMAMOTO-SHINTANI (1959), NIIMI (K. and N., 1969), and CHRIST (1969), a systematic analysis of the morphological pattern in the vertebrate diencephalon (K., 1924, 1929b, 1931, 1936, 1937, 1956) was undertaken. We reached the conclusion that the subdivision of the diencephalic grisea into four longitudinal zones as displayed by amphibians is not limited to this class but can also be detected in all classes of fishes, in reptiles, birds, and mammals. By establishing the relevant homologies on a consistent formanalytic basis, it could be shown that this configurational arrangement is an intrinsic morphologic feature of the vertebrate diencephalon, and may be evaluated as 'completely' differing from the above-mentioned four 'functional' zones of deuterencephalon and spinal cord. Investigations by JEENER (1930), PAPEZ (1935), GILBERT (1935), WARNER (1942, 1969), and others provided additional evidence and independent confirmations concerning the vertebrate diencephalic morphological pattern.

Amplifying the results of KINGSBURY (1920, 1922), which had been preceded by SCHULTE's and TILNEY's report (1915) on the development of the cat's neuraxis, our studies also led to the conclusion that the deuterencephalic and spinal basal plate of HIS with its accompanying sulcus limitans ends rostrally, in all examined vertebrates, at the mesencephalo-diencephalic boundary region within the topologic neighborhood of mammillary recess.[160]

While the longitudinal zonal pattern of the diencephalon is rather easily recognizable in most adult lower vertebrates (Anamnia), it is difficult to identify it in adult higher vertebrates (Amniota: reptiles, birds, mammals), because this simple configuration breaks up during ontogenetic evolution and differentiates into more or less complex pat-

---

[160] This region of the posterior hypothalamus is sometimes also designated as 'infundibular' or 'primary infundibular' recess, which, however, does not correspond to the 'primitive infundibulum' of KINGSBURY (1922) mentioned on p. 294 in section 2. The relevant terminology is not sufficiently standardized. Although the posterior inferior hypothalamic region is frequently qualified as 'infundibular', the term 'infundibulum' should perhaps be restricted to a portion of the neurohypophysis (cf., e.g., K., 1970).

terns of 'separate' diencephalic grisea. Nevertheless, in these higher vertebrates, some transitory stages of embryonic development always display the diencephalic zonal arrangement as distinctly as in Anamnia. However, particularly in birds and mammals, the most striking manifestation of this pattern-identity is restricted to some key stages of ontogeny.

As pointed out in section 3 (p. 307 *et passim*) the ontogenetic development of the secondary zonal system characteristic for the diencephalon becomes clearly manifest at undefined stages corresponding to the fading or disappearance of the parencephalic and synencephalic neuromeric segmentation. Three longitudinal sulci, variable as to first appearance and persistance in the various vertebrate classes, orders, and even species, roughly separate the longitudinal zones on the ventricular wall, and represent, within a certain degree of approximation, fairly reliable limiting sulci, despite variability as regards accessory sulci or extension into the domain of other sulci.[161]

Keeping in mind these restrictions and qualifications, the following generalization seems justified on the basis of sufficiently constant correlations. Sulcus diencephalicus dorsalis (subhabenularis), which may caudally merge with sulcus synencephalicus or sulcus lateralis mesencephali,[162] or both, indicates the ventral boundary of epithalamus (habenular griseum). Conversely, it corresponds to the dorsal boundary zone of thalamus dorsalis, whose ventral 'limiting sulcus' is sulcus

---

[161] These other sulci are: remnants of sulcus synencephalicus, sulcus lateralis mesencephali (these two may become continuous with each other), sulcus limitans, sulcus (lateralis) hypothalami (posterioris), also less rigorously designated as sulcus lateralis infundibuli, and sulcus intraencephalicus anterior (which represents a remainder of the optic stalk evagination).

[162] It should here be emphasized that the *epithalamus*, as interpreted by our form analysis of the diencephalic zonal systeme, is restricted to the *habenular grisea* with their parencephalic habenular commissure, and does not extend caudalward beneath (i.e. basally to) the synencephalic commissura posterior, even if the sulcus diencephalicus dorsalis (or s. subhabenularis) merges with recessus or sulcus synencephalicus. This latter groove is generally located within the (dorsal thalamic) pretectal region. The most dorsal diencephalic neighborhood at these levels comprises the so-called *posthabenular region* (*posthabenuläres Zwischenhirngebiet*, pertaining to *thalamus dorsalis*. Frequently, a conspicuous dorsalward extension of sulcus diencephalicus dorsalis separates epithalamus from posthabenular dorsal thalamic neighborhoods. Sulcus synencephalicus may caudally become continuous with sulcus lateralis mesencephali, which indicates a 'boundary' between tectum opticum and torus semicircularis of 'lower vertebrates'. This sulcus was discussed in the preceding section 4.

diencephalicus medius, basally to which the thalamus ventralis extends. This latter is 'delimited' from the hypothalamus by sulcus diencephalicus ventralis.

Sulcus diencephalicus ventralis, which corresponds to the sulcus hypothalamicus (BNA, PNA) *sive Monroi* (BNA) of the adult human brain, may join either the rostral convexity of sulcus limitans or display connections with dorsal end of sulcus hypothalami (posterioris), or both. More rarely, sulcus diencephalicus medius can likewise display a similar connection with rostral convexity of sulcus limitans.

Since the diencephalic longitudinal zones (respectively their 'boundary grooves') develop rostrally to the end of sulcus limitans, being, as it were, roughly perpendicular or 'radial' to the convexity of that sulcus (cf. Fig. 112), they can be interpreted as derivatives of the alar plate. As regards these observable morphologic manifestations, it can be reasonably assumed that certain configurational invariants, compatible with caudo-rostral as well as baso-dorsal gradients and transitions, must somehow be physico-chemically 'encoded' in the cellular substratum providing the configurations.[163]

However, because some authors (e.g. RICHTER, 1965, following SPATZ and KAHLE, as shown in Figure 182, which should be compared with Figs. 107 and 108)[163a] have essentially retained the original concept of HIS concerning rostral end of basal plate, respectively sulcus

---

[163] Since, despite exaggerated and premature claims by *'macromolecular biologists'*, nothing certain or definite is known with respect to the morphologically significant 'coding system', the question may be left open whether the significant 'code-groups' (implying negentropy) are 'located' exclusively in the 'genome' of the individual cells, or are manifested by 'factors' pertaining to the specific interactions of multicellular arrays.

Seen from another viewpoint, the manifestation of configurational orderliness evolving in the course of ontogenesis (and phylogenesis) on the basis of its encoded variety, is properly expressed by GOETHE's poetic formulation:

'*Geprägte Form, die lebend sich entwickelt.*'

(*Urworte, Orphisch*)

GOETHE likewise intuitively recognized topologic invariance (cf. vol.1, chapter 3, pp. 194–195 of this series, which he experienced as *'ein geheimes Gesetz'*:

'*Alle Gestalten sind ähnlich, und keine gleichet der andern,*

*Und so deutet das Chor auf ein geheimes Gesetz;*

*Auf ein heiliges Rätsel.*'

(*Die Metamorphose der Pflanzen*)

[163a] RICHTER (1965) is even inclined to include a rostral extension of floor plate into the hypothalamus.

limitans (discussed above in section 2), it seems here perhaps appropriate to summarize the arguments in favor of the interpretation which we have adopted on the basis of our comparative and embryologic studies.

It should first be stressed that the term 'basal plate', like the other terms referring to wall portions of the tubular neuraxis, is an abstraction derived from the classification of certain observable features, and thus represents, in the same manner as other 'taxonomic' terms, no more than a 'fiction' introduced by the observer and describer. It does not imply that there exists, figuratively speaking, the *Platonic idea* of a basal plate (etc.) such that, in a more naive anthropomorphic formulation, a certain wall portion might state 'I am the basal plate'.

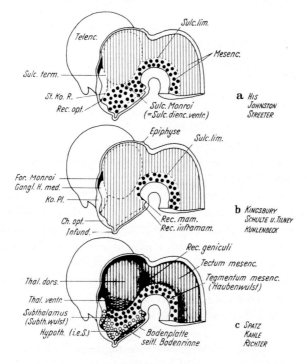

*Figure 182.* Diagram, based on the brain of a human embryo of 24 mm, and comparing different concepts concerning rostral end of basal plate, respectively sulcus limitans (from RICHTER, 1965). Vertical hatching: alar plate; large dots: basal plate; circles: 'hypothalamus *sensu strictiori* with relations to floor plate'; small dots: thalamus ventralis as 'transitional zone'; Gangl. H. med.: medial 'ganglionic hill'; Rec. geniculi presumably corresponds to recessus synencephalicus in the interpretation of the present treatise (K.); St. Ko. R.: *Stielkonusrinne* i.e. sulcus intraencephalicus anterior of the present treatise (K.); other designations self-evident.

Nevertheless, there are, on the other hand, substantial logical and semantic, as well as factual (actual) reasons for a classification of basal plate in accordance with our viewpoint. First, one might, semantically and logically, define the basal plate as that ventral wall portion of the neural tube which provides efferent grisea whose neurites emerge in metamerically homologous, serially arranged dorsal (branchial or spinal) and ventral nerve roots. This wall portion is, on the ventricular surface, and in a fairly constant manner, delimited from the alar plate by the sulcus limitans. In addition, the basal plate contains a number of different grisea, all of which need not necessarily be evaluated as 'motor' or 'efferent', but which are ontogenetically and topographically in close spatial connection with the 'true' efferent grisea.

Second, *qua* actual findings, substantially numerous careful observations disclose that a rostralward convex groove, basally ending in the neighborhood of the mammillary recess, anteriorly and ventrally to tuberculum posterius, is very frequently and clearly displayed at some ontogenetic stages or even in the definitive adult configuration of many vertebrates pertaining to all classes from fishes to mammals. In most cases, this groove can be identified, with reasonable certainty, as the anterior end of sulcus limitans.

Moreover, the cell masses enclosed by this rostral end of sulcus limitans appear, in frequent instances, as a cytoarchitecturally definable rather 'compact' griseal configuration, which includes the rostral end of the oculomotor nuclear column (with its *Edinger-Westphal nucleus* where present), the *interstitial nucleus of Cajal*, and additional cell groups. Because of its striking aspect (cf., e.g., K., 1937, 1954, K. and MILLER, 1942) we have designated this griseum as the *tegmental cell cord* (or cell plate). It corresponds to the prerubral tegmentum of PAPEZ (1940). Conspicuous aspects of that configuration, and its close relationship to the assumed rostral end of sulcus limitans are displayed in suitable sagittal sections such as shown in Figures 76 III, 213 and 237A–C.

Finally, it can be maintained that the tegmental cell cord represents the rostral end of that griseal complex defined above as basal plate, whose significant criterion is the relationship to the efferent components of metameric cranial and spinal nerves.

Concerning the alar plate, the following definitions seem permissible. As a first approximation, it can be said to represent that wall portion of the neural tube located between basal plate (sulcus limitans) and roof plate, or, in regions lacking a clearly recognizable roof plate, the wall portion dorsal to basal plate. From a functional viewpoint, it represents that part of the neural tube wall which contains correlating grisea ('afferent centers' of the first order) directly related to peripheral 'sensory' input through spinal, and branchial nerves (including VIII) as well as optic and olfactory nerves.[164]

This of course, does not exclude the presence of other grisea, some of which might also be defined as partly or essentially 'motor'. Thus the nucleus vestibularis lateralis *(Deiters)*, with its 'efferent' or 'motor' vestibulo-spinal tract, is rather clearly a derivative of the alar plate and can be evaluated as still located in this latter. Moreover, the 'motor' neocortex of mammals (with cortico-bulbar and cortico-spinal tracts, i.e. 'upper motor neurons'), originates in the dorsal (incontestably alar plate) wall of the telencephalon, being part of the dorsal brain 'mantle' (pallium).

Again, some afferent fibers of the first order doubtless make direct (e.g. 'reflex') connections with motoneurons in the basal plate of the spinal cord, and similar connections, either with motoneurons, preganglionics, or other basal plate derivatives (e.g. reticular formation) can be reasonably presumed to obtain in the brain stem.[165] In addition, some still poorly elucidated efferent connections within 'sensory' nerves (VIII, II) reaching labyrinth structures or retina, and the possibility that preganglionic fibers are included in the prosencephalic nervus terminalis, might here be mentioned. There is also some incompletely clarified evidence that optic[166] input fibers from the retina directly reach rostral basal plate grisea in the mesencephalon.

It could, of course, also be maintained, on semantic and logical grounds, that the subdivision, into basal and alar plate, being related to spinal and cranial serially homologous (metameric) nerves, applies only to spinal cord and deuterencephalon. Accordingly, neither basal nor alar plate would be distinguishable in prosencephalon and its derivatives.

This formulation, however, seems somewhat extreme. Because the roof plate can clearly be seen to extend rostralward as far as telence-

---

[164] *Nervus terminalis*, and *n. parietalis* (K., 1927, p. 75, footnote 3) must also be included, but present not entirely understood special cases.

[165] Some *Golgi impregnation* pictures can be interpreted as substantiating this view.

[166] The optic *'nerve'*, originating in a peripheral *brain portion* (retina) is, of course, in this respect not quite comparable to other, genuine *peripheral*, nerves.

phalon, and in view of the close relationship (connectedness) of alar
and roof plate, in combination with other arguments discussed in the
preceding definitions, it seems preferable to consider the diencephalic
and telencephalic neural brain wall as rostral portions of the alar plate.
One might, of course, evaluate this prosencephalic tube portion as a
'modified' alar plate. Yet, the deuterencephalic alar plate becomes rostralward already significantly 'modified' in the mesencephalon.[167]

In addition to the antimeric lateral wall regions providing paired,
bilaterally symmetric grisea with their nerve cells and fiber connections, the vertebrate diencephalon includes certain (essentially 'unpaired') *dorsal* and *basal midline configurations*. The former comprise, depending on the diverse taxonomic forms, (1) a thin epithelial roof
*(lamina epithelialis)* which may protrude as a *(secondary)* *'parencephalon'*,
also designated as *Dorsalsack* or *Zirbelpolster;* (2) a *choroid plexus,* and
(3) the *epiphysial structures;* moreover, in the synencephalic, diencephalo-
mesencephalic boundary region, (4) the *subcommissural organ.* All
these configurations can be regarded as differentiations of the roof
plate.

The *epiphysial formations* (Fig. 183 A), between habenular and posterior commissure, derive from the caudal part of the parencephalic
neuromere. In many lower vertebrates, the epiphysis is a rather complex saccular structure of various relative and absolute size, or a comparatively small vesicle. This latter may be attached to a fairly long
stalk resting on the 'dorsal sac'. Again in some instances (e.g. in amphibian Gymnophiona), it is represented by a short epithelial vesicle

---

[167] Taking a rather skeptical viewpoint, one could also cite SHAKESPEARE's well-known
lines from *Romeo and Juliet* (act II, scene 2):

'*What's in a name! that which we call a rose*
*By any other name would smell as sweet;*
*So Romeo would, were he not Romeo call'd,*
*Retain that dear perfection which he owes*
*Without that title…*'

Paraphrasing this quotation, one might then say that the configurational arrangement
of the prosencephalic wall and of its derivatives displays a recognizable pattern which
involves both conspicuous invariants and considerable differences. These actual data are
not affected by whatever names the observer chooses to give them.

Yet, although, being *'fictions', 'all classifications are false',* some are less false than
others, and the here adopted conceptualization is very definitely maintained as the most
thought-economical or suitable one with respect to a desirable 'semantically operational'
'insight' concerning the morphologic pattern of the vertebrate brain.

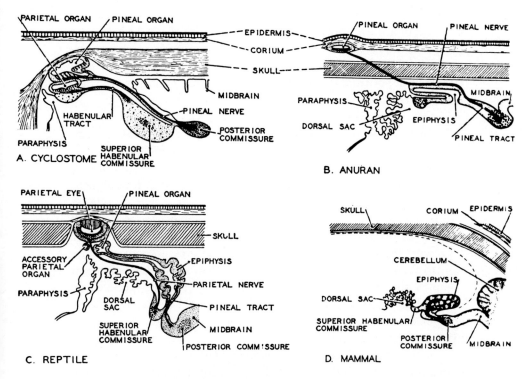

*Figure 183 A.* Epiphysial configurations in diverse vertebrates as seen in diagrammatic median longitudinal sections (redrawn from Oppel, after Studnička, from Neal and Rand, 1936). A: Cyclostome; B: anuran Amphibian; C: Reptile; D: Mammal. The 'parietal organ' or eye in this illustration corresponds to the parapineal organ respectively eye of the nomenclature adopted in the text.

within the roof of the synencephalic recess.[168] In mammals, the epiphysis is commonly a relatively small, cone-like solid cellular appendage *(conarium)* with very short stalk. It should be added that an epiphysis is apparently entirely missing or exceedingly small in various adult vertebrates, such as the selachian *Torpedo,* the reptilians Alligator and Crocodilus, and in some mammalians including Dasypus (Edentate), moreover whales, and the elephant.

In Cyclostomes such as Petromyzon, the pineal complex displays two median *'parietal'* eyes, the better developed upper *'pineal organ'*

---

[168] Because of the transformations involving fading of neuromeric segmentation and development of a permanent sulcus or recessus synencephalicus, this relationship is not incompatible with the caudal parencephalic origin of epiphysial anlage.

being located dorsally to, i.e. above the less developed '*parapineal org-an*'. The pineal eye appears to be connected with the right habenular ganglion, the parapineal eye with the left one. Whether this arrange-ment should be interpreted as indicating a reduction from a bilateral (side by side) or from a serial (rostro-caudal) manifestation of the par-ietal eyes remains a moot question.

Some anuran amphibians possess an epiphysis and a pineal organ (eye), while in some reptilians, an epiphysis and a parapineal organ (eye) are found. Small bundles of nerve fibers connecting these parietal eyes with the habenular region are designated as *nervus parietalis*. A short transitory stage, recognizable during ontogenetic development of the epiphysis in man and some other mammals, seems to indicate that two separate but closely adjacent primordia may have secondarily fused. It can thus be assumed that the 'solid' mammalian pineal body represents the kathomologon of both parietal eyes. In contradistinc-tion to these photosensitive vesicular eyes formed by the epiphysial complex, the histologic structure of the 'solid' vertebrate (particularly mammalian) epiphysis or *pineal body ('pineal gland')* suggests a still in-completely understood endocrine function which is presently studied by diverse investigators.

Further data and details concerning the parietal eyes and the pineal body, not relevant to the present discussion of the vertebrate dience-phalic bauplan, shall be included in chapter XII of volume 5.

The *basal midline configurations* of the diencephalon represent hy-pothalamic derivatives and can accordingly be considered differentia-tions of the alar plate. These formations are (1) the *supraoptic crest* of Amniota; (2) the *neurohypophysial complex* found in all classes of verte-brates, and (3) the *saccus vasculosus* of numerous fishes. Data on the su-praoptic crest are included in volume 3, Part I, chapter V, section 5 of this series.[168a] Although hypophysis and saccus vasculosus shall be

---

[168a] In an interesting paper on the ependymal lining of the ventricles in the brain of the Holocephalian Selachian *Chimaera monstrosa*, BRAAK (1963) has described an '*organon vasculosum praeopticum*' which evidently pertains to the group of '*circumventricular organs*' and might possibly represent a kathomologon of the Amniote *supraoptic crest*, being, however, topographically in close relation to the rostral edge of chiasma opticum. Unfortunately, BRAAK's paper, which has no apparent direct bearing upon the problems of morphologic zonal pattern relationship, escaped our attention and was not cited in our study of *Chimaera* and *Callorhynchus* (K. and NIIMI, 1969), nor in chapter 5, section 5, vol. 3/I of the present series. Some scattered incidental observations suggest that homologa or kathomologa of BRAAK's '*organon*' might be more or less distinctively displayed by other Anamnia.

dealt with in chapter XII of volume 5, a few comments on these structures seem appropriate in connection with the present general discussion of the vertebrate diencephalic bauplan.

Generally speaking, the vertebrate *hypophysis* consists of two main subdivisions with essentially different origin, namely of *adenohypophysis* (lobus buccalis, lobus glandularis), and of *neurohypophysis* (lobus nervosus, pars nervosa). The anlage of adenohypophysis originates from the ectoderm of the stomodeum by the infolding of an epithelial pocket known as *Rathke's pouch*. The neurohypophysis, on the other hand, develops, as a derivative of the neuraxis, within the posterior inferior part of the hypothalamus, that is, from a neighborhood of hypothalamic floor caudal to the ridge of optic chiasma.[169]

Both adenohypophysis and neurohypophysis display further subdivisions (Figs. 183 B–C), of which only those of pars nervosa need to be mentioned in this context. As a rule, the vertebrate lobus nervosus consists of *median eminence, infundibular stem*, and *infundibular lobe*. The median eminence (eminentia mediana, less rigorously also termed eminentia medialis) represents that part of the neurohypophysis which is directly continuous with the non-hypophysial parenchyma of the hypothalamus caudally or, respectively, caudo-basally to the chiasmatic ridge. In many Amniota, the eminentia mediana forms a bulge on the external brain surface, and is separated from the hypothalamus *sensu stricto* by a shallow groove (sulcus tubero-infundibularis or sulcus hypophysio-hypothalamicus).

The infundibular stem is an often somewhat indistinctly delimited portion, which provides a stalk-like connection between eminentia mediana and infundibular process. This latter, also known as the neural lobe in the narrower sense, is the most distal part of the neurohypophysis.

A vascular plexus (mantle plexus) between neurohypophysis and adenohypophysis, which is present in cyclostomes, selachians and osteichthyes, may, in various gnathostome fishes, display an arrangement characterized by a *hypophysio-portal circulation*. More or less distinctly differentiated hypophysio-portal vessels occur in amphibians and dipnoans. This portal system is still more complex in Amniota. Generally speaking, the primary net of the hypophysio-portal system seems to be

---

[169] The comparative anatomy and the presumable phylogenetic evolution of the hypophysis are discussed by WINGSTRAND (1966).

located in the eminentia mediana, and the flow in the portal vessels is assumed to be directed toward the adenohypophysis. A portal circulation of this type appears to be a rather constant feature of vertebrates, establishing a link between neuraxis, of which the neurohypophysis is a part, and the adenohypophysis (cf. also GREEN, 1966).

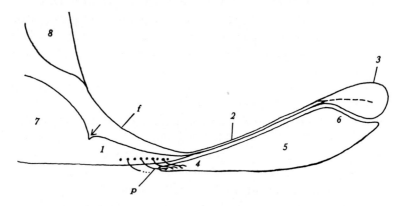

*Figure 183 B.* Semidiagrammatic midsagittal section through the pituitary complex of the amphibian (Gymnophione), Schistomepum thomense, displaying the various subdivisions of the hypophysis (from K., 1970). 1–3: neurohypophysis; 4–6: adenohypophysis; 1: eminentia mediana; 2: infundibular stem; 3: infundibular process; 4: pars tuberalis; 5: pars distalis; 6: pars intermedia; 7: ridge of supra- and postoptic commissures (chiasmatic ridge); 8: caudal part of hypothalamic wall basal to mammillary region; f: membranous roof of infundibular recess; p: hypophysio-portal system; arrow: internal hypophysio-hypothalamic groove.

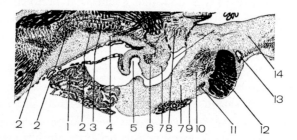

*Figure 183 C.* Paramedian sagittal section through the diencephalo-mesencephalic region of the Plagiostome, *Scyllium canicula*, displaying saccus vasculosus (from KAPPERS, 1921). 1: saccus vasculosus; 2: various tegmental decussations; 3: root fibers of n. oculomotorius; 4: nucl. interpeduncularis; 5: recessus lobi inferioris hypothalami; 6: nucl. sacci vascularis; 7: hypothalamus; 8: thalamus ventralis; 9: adenohypophysis; 10: basal forebrain bundle; 11: part of postoptic decussations; 12: optic chiasma; 13: preoptic recess; 14: wall of preoptic recess.

The *saccus vasculosus* appears to be a 'sensory' ependymal organ occurring in numerous gnathostome fishes[170] (Elasmobranchs, Ganoids, Teleosts). Its presence in Choanichthyes (Latimeria, Dipnoans) has been suspected, but remains somewhat doubtful.[171] The saccus vasculosus, located in the postero-inferior region of the hypothalamus between hypophysial complex and so-called mammillary recess, is a highly vascularized, and relatively thin-walled, 'membranous' expansion of the third ventricle, which may display a large number of folds (Fig. 183 C). In many adult forms, the saccus lumen can be seen to remain in communication with the third ventricle, but the available data are not sufficient to exclude, in some instances, the occurrence of a saccus which, by secondary closure, might become completely separated from the ventricular lumen.

The saccus vasculosus is lined by a highly modified ependyma consisting of supporting (and perhaps 'secretory') elements, and of *neurosensory epithelial cells*. These latter are provided, on the ventricular side, with hair-like processes, terminating in small knobs, and protruding into the sac's lumen. At the opposite end of the cell, a neurite originates. Bundles of these nerve fibers form nonmedullated tracts connecting with diencephalic grisea. DAMMERMAN (1910) introduced the term '*Krönchenzellen*' (*crown cells, cellules en couronne*) for the neurosensory cells of saccus vasculosus, whose structure seems to indicate that they must be subsumed, together with olfactory cells, retinal rods, and retinal cones, under category V (*neuroepithelial nerve cells*) of the tentative classification given in the preceding chapter (volume 3, part I, p. 90) of this series. In addition to the fiber system *originating* from the crown cells of the saccus vasculosus, this latter also *receives* nonmedullated fiber bundles originating in diencephalic grisea.

The functional significance of saccus vasculosus has not be elucidated with any certainty. Its structure suggests registration of condi-

---

[170] DORN (1955) brings an extensive list of fishes in which a saccus vasculosus has been recorded.

[171] MILLOT and ANTHONY (1965) describe a rudimentary saccus vasculosus in Latimeria: *on observe accolées à, ou plus ou moins incluses dans l'extrémité postérieure de l'hypophyse, des vésicules irrégulièrement divisées et plissées, tapissées par un épithélium épendymaire, qui semblent ne pas faire directement partie de la pars nervosa et pourraient être des vestiges du sac vasculaire. Nous ne saurions l'affirmer avec certitude, n'ayant pu mettre en évidence dans ces vésicules les cellules à couronne caractéristiques'.* The cited authors likewise point out that, in other instances where the saccus vasculosus is rudimentary, it becomes reduced to what appears as '*une annexe de l'hypophyse plus ou moins reconnaissable*'.

tions related to the ventricular fluid, DAMMERMAN (1910) interpreted the saccus as a sense organ for pressure (water depth registration). This might include conditions such as blood pressure, or gas content of water, blood, or liquor. In addition, a 'secretory activity', and an undefined role in the 'dynamics of cerebrospinal fluid circulation' have been suspected.

Although the topologic neighborhood, from which the saccus vasculosus of fishes derives, is evidently present within the postero-inferior wall of the hypothalamus in all vertebrates, a particular wall configuration which could justifiably be considered a saccus homologon does not seem to be displayed by amphibians or amniota. With regard

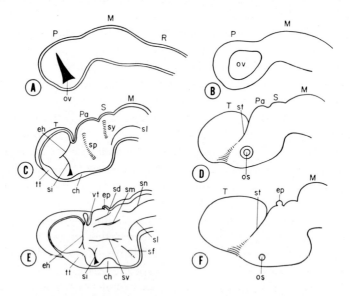

*Figure 184.* Diagram of early telo-diencephalic differentiation as seen in developing larvae of tailed Amphibia (after K. and HAYMAKER, 1949). A Stage of three primary brain vesicles, ventricular view. B External view of A.C Stage of transitory diencephalic neuromeres, ventricular view. D external view of C.E stage of diencephalic longitudinal zones, ventricular view. F external view of E. ch: chiasmatic ridge; eh: posterior evagination of telencephalon; ep: epiphysis; M: mesencephalon; os: optic stalk; ov: optic vesicle; P: prosencephalon; Pa: parencephalon; R: rhombencephalon; S: synencephalon; sd: sulcus diencephalicus dorsalis; sf: sulcus lateralis thalami posterioris ('sulcus lateralis infundibuli'); si: sulcus intraencephalicus anterior; sl: sulcus limitans; sm: sulcus diencephalicus medius; sn: sulcus lateralis mesencephali; sp: groove of parencephalic neuromere ('sulcus parencephalicus'); st: sulcus telo-diencephalicus; sy: groove of synencephalic neuromere ('recessus sive sulcus synencephalicus'); T: telencephalon; tt: torus transversus; vt: velum transversum.

to this topic, MILLOT and ANTHONY (1965) comment as follows: '*Le sac vasculaire est strictement propre aux Poissons: les homologies que l'on a essayé d'établir à son sujet chez les autres Vertébrés, Homme y compris, sont plus que contestables. L'organe est d'ailleurs très inégalement représenté dans la série des Poissons*'.

Reverting to the diencephalic zonal system as displayed by *Anamnia*, and becoming well defined following fading or disappearance of parencephalic and synencephalic neuromeric segmentation, Figure 184, based on our investigations, illustrates their early differentiation. This diagram, referring to tailed Amphibia, likewise applies to Anamnia in general.

*Cyclostomes* can be evaluated as very 'primitive' vertebrates. Although their brain is, in many respects, correspondingly 'simple' (JOHNSTON, 1902; SAITO, 1930a; HEIER, 1948), a number of complicating 'pattern deformations', particularly in Myxinoids (JANSEN, 1930) introduce features whose interpretation presents considerable difficul-

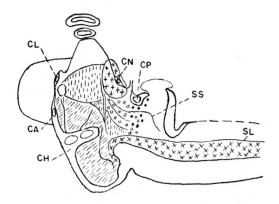

*Figure 185.* Semidiagrammatic reconstruction of longitudinal zones in the diencephalon of adult Petromyzon or Entosphenus (from K., 1956). Crosses: epithalamus; circles: thalamus dorsalis; black dots: pretectal component neighborhoods of dorsal thalamus; vertical hatching: thalamus ventralis; oblique hatching: hypothalamus; x: basal plate of deuterencephalon (including tegmental cell cord; CA: commissura anterior; CH: chiasmatic ridge; CL: commissura pallii; CN: commissura habenulae; CP: commissura posterior; SL: sulcus limitans; SS: sulcus lateralis mesencephali continuous with sulcus synencephalicus; the vertical sulcus indicated by double line, and extending from epithalamus into thalamus ventralis, is the 'pseudosulcus' of *Saito*; the other sulci ('boundary sulci', hypothalamic sulci, and accessory ones) are easily identifiable by their respective position and are unlabeled in order not to overload the diagram.

ties. Several different opinions concerning morphologic significance of the diencephalic ventricular wall portions have been expressed (HEIER, 1948; ADAM, 1956).

Figure 185 illustrates the arrangement of the diencephalic zones, according to our interpretation, in adult Petromyzon or in the very similar *Entosphenus japonicus*. Figure 63 I, included further above in section 1 of this chapter, shows, at the larval or Ammocoetes stage of Petromyzon, a lesser degree of 'pattern deformation', and a diencephalic ventricular wall configuration corresponding more closely to that of other Anamnia.

On the other hand, comparing Figure 185 with Figures 190 and 203, it can be seen that, in the midsagittal plane, the orientation of diencephalic floor to sagittal brain stem axis displays here a degree of dorsalward rotation not manifested in the other Anamnia.[172] The presence of a peculiar dorso-basal sulcus, designated as *pseudosulcus* by SAITO (1930a) and shown in our figure, may be correlated with, and is perhaps a result of, this obtaining morphologic 'rotation'. The general 'pattern deformation' could also be interpreted as a 'compression' of prosencephalic configurations in the rostrocaudal axis.[173] These displacements result in a peculiar expansion of ventral thalamus rostral to the habenular grisea and dorsal to the hypothalamic recess (Fig. 185). The topologic neighborhood represented by that expanded ventricular wall region corresponds to the eminentia thalami ventralis of Amphibia,[174] and is closely related to the hemispheric stalk, which will be discussed further below in the present section. Two arrays of cells, pertaining to thalamus ventralis and hypothalamus (preoptic recess), extend laterad and connect with the periventricular cell aggregates of the caudal paired hemispheric evagination, near the polus posterior. The connecting cell bands represent *massa cellularis reuniens superior* (thalamus ventralis) and *massa cellularis reuniens inferior* (hypothalamus) of RÖTHIG (1923) and K. (1924). Figures 186–189 illustrate additional aspects of the diencephalic zonal system. Further data concerning our concepts

---

[172] Cf. the discussion remarks to ADAM's paper. It should also be mentioned that, in this respect, non-negligible differences between species, and apparently also individual variations within one and the same species seem to obtain.

[173] 'Distortions' correlated with 'compressions' of that type are, e.g., very evident in the brain configuration of adult *Chimaera* (cf. K., and NIIMI, 1969).

[174] Other authors (cf. HEIER, 1948) designate this configuration as 'primordium hippocampi' and interpret it as pertaining to the telencephalon.

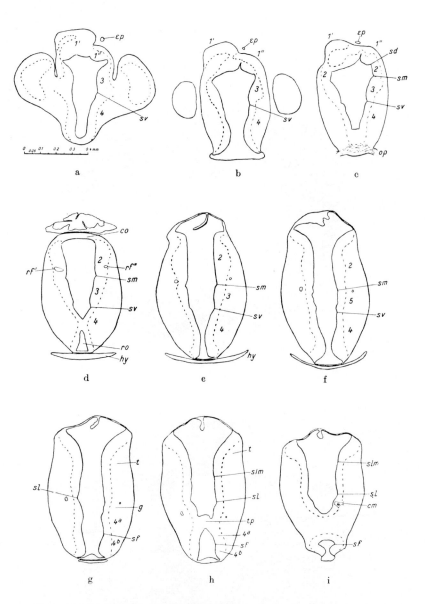

*Figure 186.* Cross-sections through diencephalon of a 6.3-mm Ammocoetes (from K., 1929b). 1: epithalamus; 1': right ganglion habenulae; 1": left ggl. hab.; 2: thalamus dorsalis; 3: thalamus ventralis; 4: hypothalamus; 4a: pars superior hypothalami; 4b: pars inferior hypothalami; 5: boundary region of thalamus ventralis and mesencephalic basal plate; cm: *cell of Müller;* co: commissura posterior; ep: epiphysial stalk; g: basal plate of mesencephalon; hy: adenohypophysis; op: optic nerve and chiasma; rf: fasciculus retroflexus (rf': right; rf": left); ro: recessus postopticus; sf: sulcus lateralis hypothalami; sl: sulcus limitans; slm: sulcus lateralis mesencephali; sm: sulcus diencephalicus medius; sv: sulcus diencephalicus ventralis, t: tectum mesencephali; tp: tuberculum posterius.

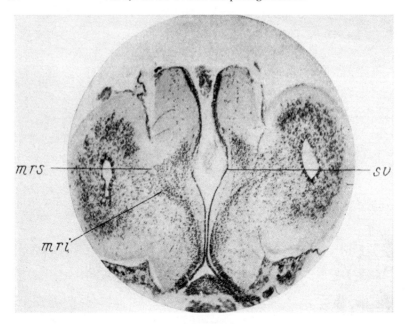

Figure 187

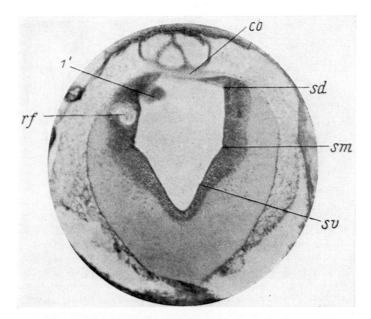

Figure 188

will be found in two of the cited publications (K., 1929b, 1956). Details pertaining to configuration and cytoarchitecture of the grisea, to
fiber connections, as well as to other diencephalic structures (pineal
and parapineal organs, hypophysis) are not relevant to the aspect here
under consideration and shall be dealt with in chapter XII of volume 5.

The arrangement of the diencephalic zonal system in *Selachians* is illustrated by Figures 190–192. This is rather similar to the configuration obtaining in *Ganoids* such as Amia (Figs. 193, 194). However, the
diencephalon of some *Holocephalian Plagiostomes* (Chimaera, Callorhynchus) is characterized by a peculiar pattern deformation, whereby the
preoptic recess becomes greatly elongated and transformed into a
quasi-membranous stalk containing the fiber systems connecting telencephalon and diencephalon (cf. K. and NIIMI, 1969). Yet despite some
additional distortions related to what could be called a compression of
diencephalic and mesencephalic neighborhoods into each other, the
typical zonal features of the diencephalon still remain clearly recognizable.

Notwithstanding complications resulting from the considerable
degree of differentiation and diversity of forms displayed by *Teleosts*,
the fundamental features of the vertebrate diencephalic pattern are
preserved in these fishes, as indicated by Figures 195–198.

In *Latimeria chalumnae*, identified as a *Coelacanth*, pertaining to the
class Choanichthyes,[175] to which the order Dipnoi is likewise assigned,

---

[175] Cf. vol. 1, p. 59 of this series. While Crossopterygians were tentatively considered
to represent an 'order' of Choanichthyes, YOUNG (1955) uses the term Crossopterygii as
a class synonymous with Choanichthyes, and adopts a subdivision of this class into three
orders: (1) extinct Osteolepidoti, (2) Devonian-recent Coelacanthini, (3) Devonian-recent
Dipnoi.

---

*Figure 187.* Cross-section showing cellular arrangements in the diencephalon of Petromyzon fluviatilis at the level of preoptic recess and hemispheric stalk (from K., 1929 b).
mri: massa cellularis reuniens inferior; mrs: m.c.r. superior; sv: sulcus diencephalicus ventralis; the wall region above sv is interpreted as eminentia thalami ventralis; the fig. also
shows the posterior recesses of the telencephalic ventricles, with dorsal (pallial) and basal
cell masses.

*Figure 188.* Cross-section showing cellular arrangements in the diencephalon of a 36-
mm Ammocoetes at level of posterior commissure and (ventrally) postoptic commissures
(from K., 1929b). 1': posterior tip of right gangl. habenulae; other designations as in figure 186.

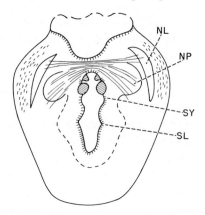

*Figure 189*. Semidiagrammatic cross-section through pretectal region (diencephalo-mesencephalic boundary region) of Petromyzon (modified after K., 1956). NL: nucleus lentiformis mesencephali; NP: nucleus commissurae posterioris; SL: sulcus limitans; SY: sulcus synencephalicus; the subcommissural organ, indicated by the region of prominent, high columnar ependyma, displays, in Petromyzon, two distinct, lateral and medial ridges (*'Leisten'* of ADAM, 1956); the space above the commissure represents part of the mesence-phalic ventricle, with its epithelial roof plate.

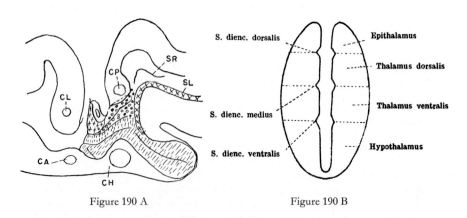

Figure 190 A          Figure 190 B

*Figure 190 A*. Semidiagrammatic reconstruction of longitudinal zones in the dience-phalon of adult Acanthias (from K., 1956). SR: sulcus lateralis mesencephali; other abbr. as in figure 185.

*Figure 190 B*. Diagram of Selachian diencephalic bauplan as seen in cross-section (slightly modified after GERLACH, 1947).

the typical vertebrate diencephalic zonal system has been identified by
MILLOT *et al.* (1964), and is shown in Figure 199.

In *Dipnoans*, particularly in young adult forms, e.g. *Protopterus*
(Figs. 200–202) the diencephalic zonal pattern is rather clearly manifested.[176] The periventricular arrangement of the cell masses, and the
'limiting' grooves are here, on the whole, more similar to conditions
prevailing in Amphibia than to those illustrated in Latimeria. However, the diencephalon of Ceratodus, as depicted by HOLMGREN and VAN
DER HORST (1925) doubtless displays some features vaguely resembling the ventricular outlines in Latimeria.

The diencephalic zonal system of *Amphibia* is illustrated by Figures 203–208. It will be recalled that HERRICK's findings (1910) concerning the brain morphology in this vertebrate class introduced the
concept of a thalamus ventralis, amplifying the basic subdivisions propounded by HIS. Generally speaking, the cellular arrangements obtaining in Urodeles and Gymnophiones show a lesser degree of cytoarchitectural differentiation than displayed by many Anurans. Figure 206 B,
illustrating fairly typical Anuran conditions, indicates indeed a stratification into external (ld, lv) and internal thalamic cell masses. These latter, again, manifest a narrow, dense ependymal and subependymal
ventricular lining, and an indistinctly stratified outer sublayer. The hypothalamus, in addition to ependymo-subependymal lining, merely
displays an outer sublayer of the 'internal cell mass', whose density increases basalward. Although the thalamic cell grouping might here be
interpreted as loosely corresponding to the concept of so-called migra-

---

[176] In the Protopterus material used for our Breslau studies and reconstructions (K.,
1929b; GERLACH, 1933), the velum transversum was apparently either greatly reduced or
secondarily distorted, respectively damaged. Thus, our model (fig. 2, GERLACH, 1933) did
not show the boundary between parencephalon and paraphysis. Moreover, this material
had not been processed to show cytologic details clearly distinguishing these two structures. DORN (1957), who subsequently studied the structure of the paraphysis in that
Dipnoan, therefore correctly noticed that we erroneously included the paraphysis (which
we could not identify) into our parencephalon. On the basis of additional material and
observations, I have indicated a particularly well developed velum transversum, separating
telencephalic paraphysis from diencephalic (secondary) parencephalon in the diagrammatic
figure 200. DORN's figure 1 (1957) shows a configuration characterized by a very long
paraphysis, a shallow fold of velum transversum, and a short saccus dorsalis (secondary
parencephalon). The stalked epiphysis seems to rest in the dorsal concavity of velum transversum. These, and other findings (cf. above, fig. 72 with the comments on Ceratodus in
footnote 63a seem to indicate that non-negligible individual variations with some degree
of pattern distortion might occur in the diverse species of Dipnoans.

tion layers, the hypothalamic grouping in this instance, and the periventricular arrangements shown in Figures 205 and 206 A, do not corroborate the assumption that the concepts of migration layers, or of
'matrix phases' can be adopted as overall valid criteria for the establishment of homologies. Rather, both concepts can be evaluated as referring to special cases, and as the formulation of an orderliness which
may or may not apply, with considerable variations, to particular regions of the neuraxis in different vertebrate forms.

Proceeding, within the taxonomic series, from Anamnia to *Amniota*, it can be seen that the grisea in the adult *Reptilian diencephalon*
(Figs. 209 A–D) display a configuration whose overall arrangement is
prima facie comparable with that in amphibians (Figs. 205, 206).

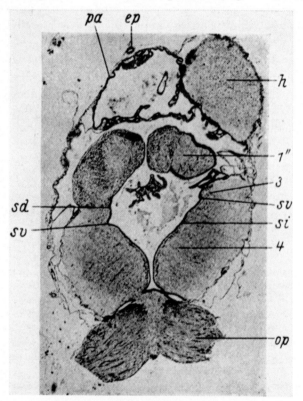

Figure 191 A

*Figure 191.* Cross-sections through the diencephalon of a young (adult) Acanthias
(from K., 1929a). Section A is slightly more rostral than B. h: caudal tip of telencephalon;
other abbr. as in figure 186.

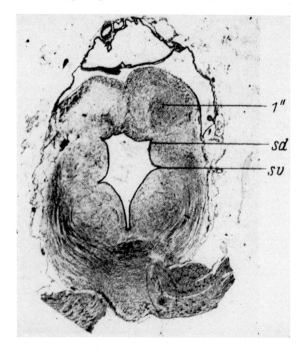

Figure 191 B

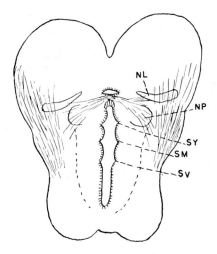

*Figure 192.* Semidiagrammatic cross-section through caudal diencephalon of Acanthias (from K., 1956). SM: sulcus diencephalicus medius; SV: sulcus diencephalicus ventralis; other abbr. as in figure 189.

Yet, because of the more complex griseal differentiation obtaining in reptilians, a mapping of various adult diencephalic neighborhoods upon the simpler amphibian longitudinal zonal system presents several difficulties. The adult configuration displayed by the reptilian cell aggregates does not provide sufficiently valid evidence for identification of some neighborhoods as pertaining, *qua* derivatives, to a specific longitudinal zone. Also, the diencephalic sulci which, to some extent, rep-

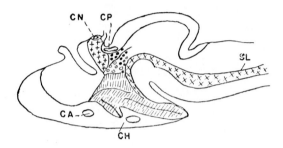

*Figure 193.* Semidiagrammatic reconstruction of longitudinal zones in the diencephalon of a young (adult) Amia (from K., 1956). Abbr. as in figures 190 and 185.

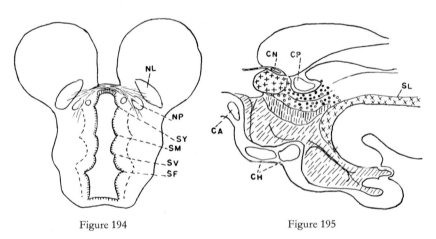

Figure 194                              Figure 195

*Figure 194.* Semidiagrammatic cross-section through caudal diencephalon of Amia (from K., 1956). SF: sulcus lateralis infundibuli (s. lat. hypothalami posterioris); other abbr. cf. figures 191 and 189.

*Figure 195.* Semidiagrammatic reconstruction of longitudinal zones in the diencephalon of a young adult specimen of the Teleost Corydora (from K., 1956); abbr. as in figures 190 and 185.

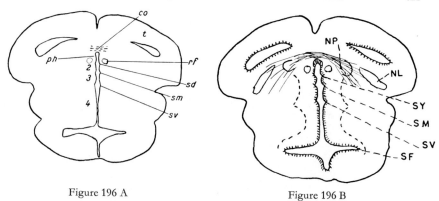

Figure 196 A                    Figure 196 B

*Figure 196 A*. Outline of cross-section through caudal diencephalon in a 66-mm embryo of the Teleost Anguilla (eel) at level of commissura posterior (from K., 1929a). ph: posthabenular diencephalic neighborhood, pertaining to pretectal region of dorsal thalamus; sd: sulcus synencephalicus, rostrally continuous with sulcus diencephalicus dorsalis; other abbr. as in figures 186 and 191.

*Figure 196 B*. Diagrammatic cross-section based on actual outline drawing of preceding figure, but indicating diencephalic (NP) and mesencephalic (NL) griseal components of pretectal region (from K., 1956). Abbr. as in figures 189, 191, and 194.

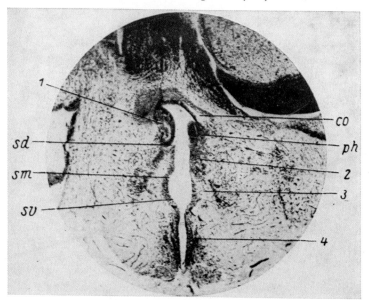

*Figure 197*. Cellular arrangements and ventricular configuration in the diencephalon of the adult Teleost, *Carassius auratus*, as seen in a cross-section passing dorsally through commissura posterior and basally through postoptic commissures (from K., 1929b). ph: transition of ganglion habenulae (epithalamus) to posthabenular diencephalic neighborhood (thalamus dorsalis); other abbr. as in figures 186, 188, 196A.

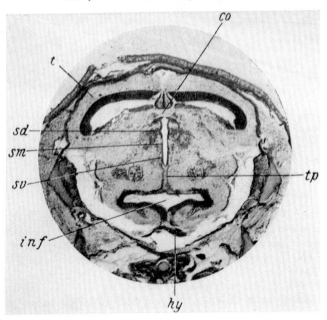

*Figure 198.* Cellular arrangements and ventricular configurations in the caudal diencephalon of an advanced Anguilla (eel) embryo (66 mm), at the level of tuberculum posterius, i.e. slightly caudal to the levels of figure 196 (from K., 1929b). inf: third ventricle space in lobi inferiores hypothalami, loosely called 'infundibulum'; sd is here sulcus synencephalicus, rostrally continuous with s. dienc. dorsalis; the lead tp indicating neighborhood of tuberculum posterius ends on tegmental cell cord; other abbr. as in figures 186, 188, 196A and 197.

*Figure 199.* Cross-sections through the diencephalon of the Crossopterygian Coelacanth, Latimeria (from MILLOT *et al.*, 1964). Cell arrangements shown on right side, main fiber tracts on left. c. hab.: habenular commissure; ep.: epiphysis; f. opt.: tractus opticus; f. retr.: fasciculus retroflexus; f.t.b.: basal forebrain bundle and its components (1: 'striotubercular'; 2–5: 'strio-hypothalamic); f. th. hyp.c.: fasc. thalamo-hypothalamicus caudalis; g. hab.: ganglion habenulae (epithalamus); hypoth.: hypothalamus; s.d.d.: sulcus diencephalicus dorsalis; s.d.m.: sulcus diencephalicus medius; s.d.v.: sulcus diencephalicus ventralis; s. th. d.: 'sillon du thalamus dorsal' (sulcus synencephalicus *mihi*); str. m.: stria medullaris; t. post.: tuberculum posterius; th.d.: thalamus dorsalis and subdivisions (d.: *'partie dorsale'*, v.: *'partie ventrale'*); th. v.: thalamus ventralis. Lower s.d.m. in a should read s.d.v.

*Figure 200.* Semidiagrammatic reconstruction of longitudinal zones in the diencephalon of a young adult specimen of the Dipnoan, Protopterus annectens, (from K., 1956). Abbr. as in figures 195, 190, 185; the added arrow indicates the velum transversum separating the paraphysis from the parencephalic roof plate ending at commissura habenulae.

resent 'limiting sulci' indicating zonal boundaries, are very poorly defined, and quite indistinct in many adult reptilian forms. A further
complication arises from the occurrence of an additional ventricular
sulcus, designated as 'pseudosulcus' (K., 1931), which can be very deep
in some groups (e.g. Chelonia), and which obliquely crosses the system

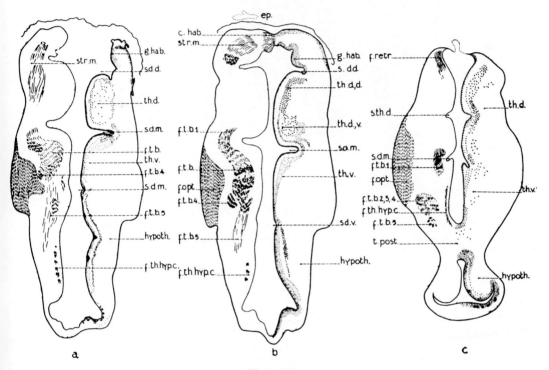

Figure 199

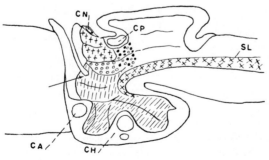

Figure 200

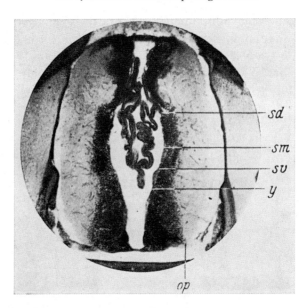

*Figure 201.* Cross-section through the diencephalon of Protopterus annectens at a caudal level of ganglion habenulae and preoptic recess (from K., 1929b). op: optic nerve; y: caudal accessory groove of sulcus intraencephalicus anterior system; other abbr. as in figures 191 and 186.

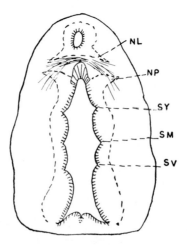

*Figure 202.* Semidiagrammatic cross-section through the caudal diencephalon of Protopterus annectens (from K., 1956). The extensive choroid plexus included within the ventricle (cf. fig. 201) has been omitted. Abbr. as in figures 187, 192, 194, 196B).

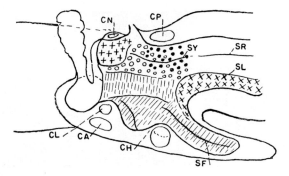

*Figure 203.* Semidiagrammatic reconstruction of longitudinal zones in the diencephalon of a 'generalized' urodele amphibian, based on reconstructions from Salamandra, Triton, and Amblystoma brains (from K., 1957). Abbr. as in figures 185, 190, 193, 195, 200. An essentially identical configuration also obtains in the anuran brain, and, with only slight modifications, as shown by figure 204, in the Gymnophione brain.

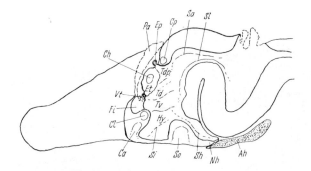

*Figure 204.* Diagram of midsagittal section through the brain of the amphibian Gymnophione Schistomepum, showing features of third ventricle. The diencephalic sulci have not been labeled in order to avoid overloading with leads, but can be easily identified (from K. *et al.,* 1966). Ah: adenohypophysis (cf. fig. 183 B); Ca: commissura anterior; Cl: commissura pallii ('commissura hippocampi'); Ep: epiphysis; Et: epithalamus (nucl. s. gangl. habenulae); Fi: foramen interventriculare; Hy: hypothalamus; Nh: neurohypophysis (cf. fig. 183 A); Pa: paraphysis; Sa: sulcus lateralis mesencephali; Sh: sulcus lateralis hypothalami (inferioris, also 's. lat. infundibuli'); Si: sulcus intraencephalicus anterior; sl: sulcus limitans; So: supraoptic commissure(s); Td: thalamus dorsalis; Tdp: synencephalic and pretectal subdivisions of thalamus dorsalis; Tv: thalamus ventralis; Vt: region of vestigial velum transversum; Ch: commissura habenulae.

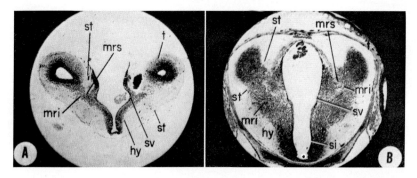

*Figure 205.* Cellular arrangement in the amphibian forebrain at level of hemispheric stalk and preoptic recess (from K., 1929b). A In the urodele Siredon pisciformis. B In a metamorphosing larva (24 mm) of the anuran *Rana fusca.* hy: hypothalamus (preoptic recess); mri: massa cellularis reuniens inferior; mrs: massa cellularis reuniens superior; st: sulcus telo-diencephalicus; si: sulcus intraencephalicus anterior system (posterior branch); sv: sulcus diencephalicus ventralis; above this, in figure A, a well developed eminentia thalami (cf. Petromyzon, fig.187); t: telencephalon near posterior pole (cf. Petromyzon, fig. 187), in figure B, the posterior pole (not labeled) ist shown caudally to the end of its ventricle.

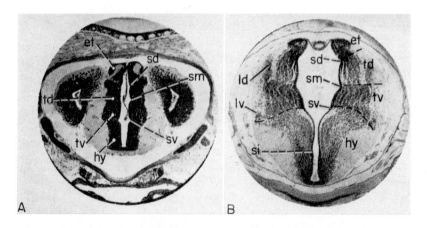

*Figure 206.* Cellular arrangements in the amphibian forebrain at levels displaying all four diencephalic longitudinal zones (from K., 1929b). A Young adult of the urodele Spelerpes fuscus. B Young adult of the anuran Pelobates fuscus. et: epithalamus (n.s. gangl. habenulae); ld: dorsal lateral geniculate nucleus; lv: ventral lateral geniculate nucleus; sd: sulcus diencephalicus dorsalis; sm: sulcus diencephalicus medius; td: thalamus dorsalis; tv: thalamus ventralis; other abbr. as in figure 205.

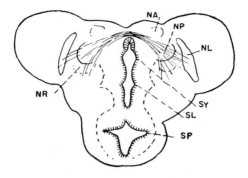

*Figure 207.* Semidiagrammatic cross-section through the diencephalo-mesencephalic boundary zone of the anuran amphibian Rana. NA: area praetectalis; NR: primordium of n. praetectalis; other abbr. as in figures 189, 192, 194, 196B, 202. The high epithelium indicated dorsally beneath the fibers of the posterior commissure in this and the other here-mentioned figs. represents the subcommissural organ.

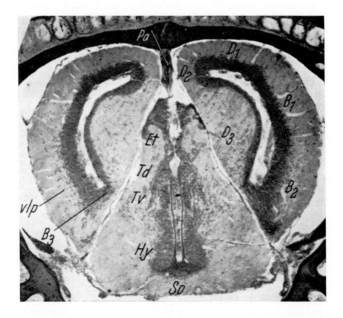

*Figure 208.* Cross-section through forebrain of the Gymnophione amphibian Schistomepum at rostral level of supraoptic commissures and caudal to hemispheric stalk (from K. *et al.*, 1966). $B_{1-3}$, $D_{1-3}$: telencephalic zonal system; Et: epithalamus (gangl. habenulae); Hy: hypothalamus; Pa: paraphysis (caudal end); Td: thalamus dorsalis; Tv: thalamus ventralis; vlp: area ventrolateralis posterior (basal 'cortex' or 'precortex').

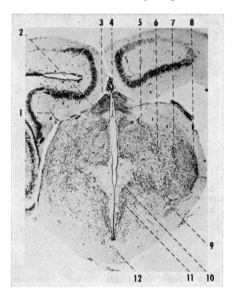

*Figure 209 A.* Cross-section through the diencephalon of adult Lacerta vivipara at rostral level of postoptic commissures (from FREDERIKSE, 1931). 1: medial wall of telencephalon; 2, 3: lateral and medial habenular nucleus; 4: n. dorsomedialis; 5: n. rotundus; 6: n. dorsolateralis; 7, 8: inner and outer cell layer of corpus geniculatum laterale ventrale; 9: n. ventromedialis and n. ventrolateralis; 10, 11: subdivisions of n. paraventricularis hypothalami; 12: postoptic (supraoptic) commissures with scattered interstitial cells; 4, 5, 6 are derivatives of thalamus dorsalis, and 7, 8, 9 of thalamus ventralis (interpretation of grisea according to K., 1931).

of zonal diencephalic sulci, becoming in part superimposed upon or combined with, two of these sulci, namely sulcus diencephalicus medius and ventralis. The term 'pseudosulcus' was introduced because this groove does not represent a 'limiting sulcus' *('Grenz furche')* in the here adopted sense. It can be interpreted as a transform of the parencephalic neuromeric cavity related to a folding process connected with the development of thalamus dorsalis.[177]

---

[177] *'Eine tiefe Furche…, die vom vorderen Abschnitt des Sulcus diencephalicus medius schräg basal- und kaudalwärts über den Thalamus ventralis hinweg zieht. Sie stellt keine echte Grenz furche dar, sondern wird als Faltungserscheinung betrachtet, die mit der Entwicklung des Thalamus dorsalis einhergeht, und daher als Pseudosulcus bezeichnet'* (K., 1931). The reptilian pseudosulcus is presumably not morphologically identical (homologous) with SAITO's pseudosulcus in the cyclostome diencephalon, despite some general analogies in the features of both sulci.

However, at certain stages of ontogenetic evolution, concomitant with a fading out of the parencephalic and synencephalic segmentation, the fundamental diencenphalic zones and, to some extent, the 'limiting' diencephalic sulci are clearly recognizable in a manner directly comparable, by one-to-one mapping, with the corresponding configurations of Amphibia.

Figures 210 A–H, and 211 A–C show the configuration of the longitudinal zonal system in cross-section at key stages of Lacertilian embryos. Figure 212 illustrates the corresponding ventricular relief,

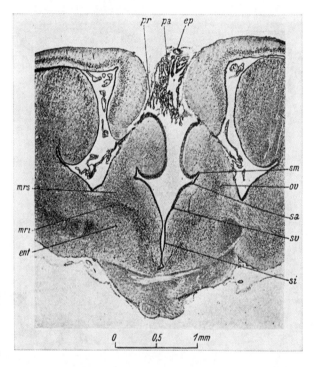

*Figure 209 B*. Cross-section through the forebrain of an adult Chrysemys (Chelonian) at level of hemispheric stalk and rostral end of optic chiasma (from K., 1931). ent: ant. entopeduncular nucleus (hypothalamus); ep: epiphysis; mri: massa cellularis reuniens inferior (hypothalamus); mrs: massa cellularis reuniens superior (thalamus ventralis); ov: nucleus ovalis (thal. ventr.); pa: paraphysis; pr: (secondary) parencephalon; sa: sulcus accessorius int. of pseudosulcus (sulc. parencephalicus system); si: sulcus intraencephalicus anterior; sm: sulcus diencephalicus medius; sv: sulcus diencephalicus ventralis; the unlabeled groove at the bottom of lateral ventricles is the sulcus terminalis; this figure, showing above sm a large eminentia thalami (dorsalis), should be compared with figure 187 (Petromyzon).

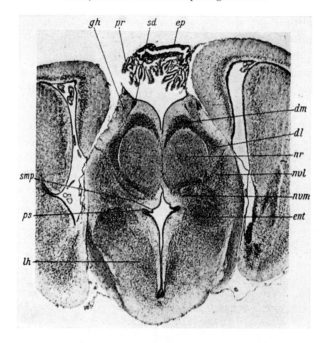

Figure 209 C

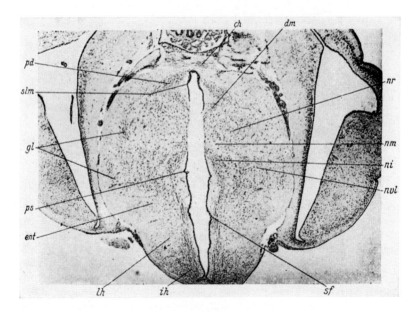

Figure 209 D

which, if compared with that of Amphibia, is complicated by presence and course of the pseudosulcus. Figure 213 displays the diencephalic zones in a parasagittal section and, moreover, shows the rostralward convex end of the deuterencephalic basal plate. It can be seen that, as mentioned above (p. 400), the diencephalic longitudinal zones appear arranged in a roughly 'perpendicular' orientation to the convex end of basal plate (respectively sulcus limitans).

Figures 214 A–D and 215 A, B illustrate the diencephalic configuration in embryos of other reptilian orders, namely of Chelonia and Crocodilia. Comparable and topologically identical zonal arrangements in the embryonic Ophidian brain *(Natrix sipedon)* were depicted and described by WARNER (1942). The parencephalic sulcus (pseudosulcus), although present and identified, was less pronounced in *Natrix sipedon* than in similar stages of Lacertilia, Chelonia, and Crocodilia as illustrated in this section.

In his study concerning diencephalic and mesencephalic ontogenetic development of *Lacerta sicula*, SENN (1968) described the differentiation of embryonic *elementary layers*, which essentially correspond to stratification patterns stressed as 'migration areas' or migration layers by BERGQUIST and KÄLLEN (1953a, 1954) and by HUGOSSON (1957).

As interpreted by SENN (1968), this laminar pattern, formed by cellular and fibrous arrangements arising in the process of early organization, provides primordia in or from which nuclei and further fiber layers differentiate. According to our own observations, a tendency toward stratification may indeed be displayed by the cell masses at significant key stages (cf. Figs. 211 A, B, C). An external cell layer, from

---

*Figure 209 C.* Cross-section through the forebrain of Chrysemis at the level of caudal end of optic chiasma and rostral beginning of postoptic commissures. At left, caudal end of hemispheric stalk, at right, the section plane passes caudally to that stalk (from K., 1931). dl, dm: n. dorsolateralis and n. dorsomedialis; nm: n. medialis s. subrotundus (thalami dorsalis); nr: n. rotundus (thalami dorsalis); nv: n. ventralis (griseum thalami ventralis); ps: pseudosulcus; other abbr. as in preceding figure.

*Figure 209 D.* Cross-section, caudal to hemispheric stalk and to postoptic commissures, through the forebrain of an adult *Testudo graeca* (from K., 1931). ch: commissura habenulae; gl: nucleus s. corpus geniculatum laterale (upper lead: pars dorsalis, thalamus dorsalis; lower lead: pars ventralis, thalamus ventralis); ih: nucleus ventralis hypothalami; lh: nucleus lateralis hypothalami; ni: nucleus ventralis internus s. medialis (thal. ventr.); nvl: n. ventralis, pars lateralis (thal. ventralis); pd: n. posterodorsalis; sf: sulcus lateralis infundibuli (s. lat. hypothalami posterioris); slm: sulcus synencephalicus, caudally continuous with sulcus lateralis mesencephali; other abbr. as in the two preceding figures.

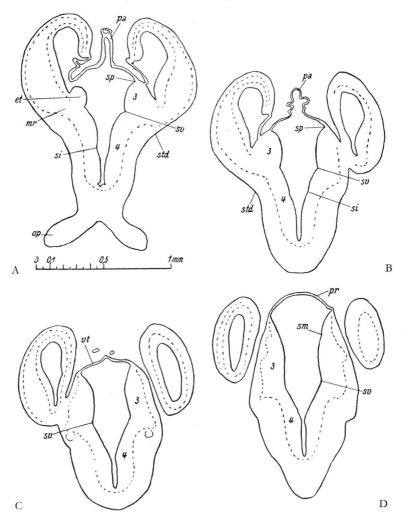

*Figure 210.* Outlines of cross-sections through the forebrain in a 22 mm long embryo of *Lacerta agilis* (from K., 1931). 1: epithalamus; 2: thalamus dorsalis; 3: thalamus ventralis; 4: hypothalamus; cp: commissura posterior; ep: epiphysis; et: eminentia thalami (ventralis); gp: 'geniculatum praetectale' (n. praetectalis; thal. dors.); hy: adenohypophysis; mr: massae cellulares reunientes in hemispheric stalk; ncp: nucleus commissurae posterioris, op: optic nerve and chiasma; pa: paraphysis; pr: (secondary) parencephalon; ps: pseudosulcus; sd: sulcus diencephalicus dorsalis; sf: 'sulcus lateralis infundibuli'; si: sulcus intraencephalicus anterior, dorsal branches; sl: sulcus limitans; slm: sulcus synencephalicus (in G), sulcus lateralis mesencephali (in H); sm: sulcus diencephalicus medius; sma: sulcus dienc. med., pars anterior; smp: sulcus dienc. med., pars posterior; sp: sulcus paraphysio-thalamicus; std: sulcus telencephalo-diencephalicus; sulcus diencephalicus ventralis; vt: velum transversum.

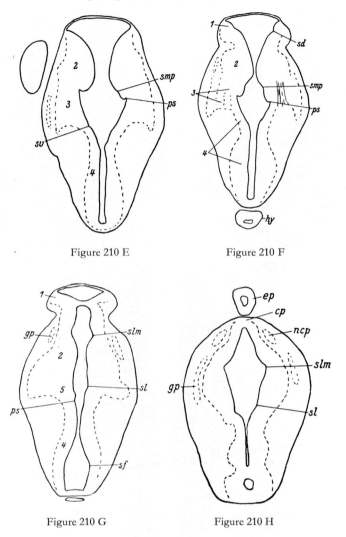

Figure 210 E          Figure 210 F

Figure 210 G          Figure 210 H

which, in particular, the components of ventral and dorsal geniculate body differentiate, can be roughly distinguished from a wider internal cell layer. This latter, again, may show additional, often ill-definable sublayers.

The innermost sublayer includes ependyma and a narrow subependymal cell plate, both being derivatives of the primary (histogenetic) matrix as discussed in section 1 of chapter V (volume 3, part I of this series). The wider outer sublayer of the internal layer represents a mor-

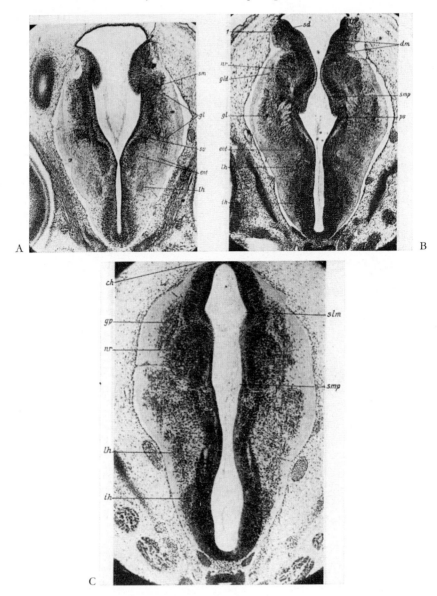

*Figure 211.* Cross-sections through the diencephalon of a 22-mm embryo of *Lacerta agilis*, showing the cellular arrangements (from K., 1931). dm: n. dorsomedialis; ent: nucleus entopeduncularis (anterior) hypothalami; gl: corpus geniculatum laterale ventrale; gld: corp. gen. lat. dorsale; other abbr. as in figures 209 and 210; section A shows posterior pole of telencephalon at left.

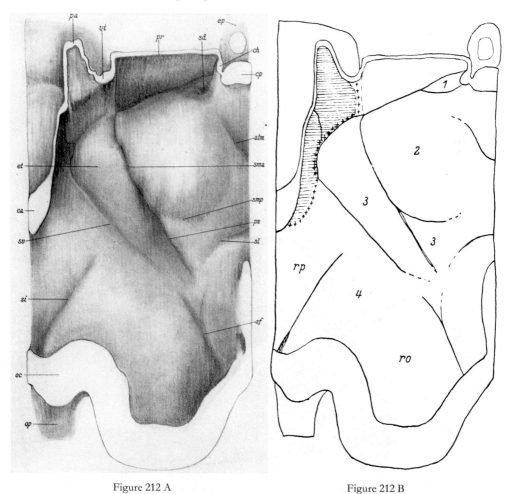

Figure 212 A                         Figure 212 B

*Figure 212.* Ventricular aspect of the developing diencephalon in a 22-mm embryo of *Lacerta agilis* (from K., 1931). A Wax plate reconstruction. B Outline sketch. ca: commissura anterior; ch: commissura habenulae; et: eminentia thalami (ventralis); oc: optic chiasma; op: optic nerve; ro: recessus postopticus (posterior hypothalamus); rp: recessus praeopticus (anterior hypothalamus); horizontal hatching in B represents rudimentary ventriculus impar telencephali, whose boundary against diencephalic third ventricle is indicated by crosses; other abbr. as in preceding reptilian figs.

phogenetic matrix for different grisea (nuclei), and, depending upon additional criteria as well as upon different taxonomic forms, allows for a still more detailed 'parcellation' qua stratification.

SENN (1968) has proposed an overall laminar systematization applicable to the Lacertilian diencephalon and mesencephalon. He distinguishes three main strata subdivided into zones, as tabulated on p. 11 of his publication. The periventricular strata of SENN correspond to the internal cell layer of our terminology, and comprise 5 zones (1–5). SENN's central strata are subdivided into two zones (6 and 7), of which 7 roughly corresponds to our external cell layer, and 6 may represent the transition, characterized by a lesser cellular density between inter-

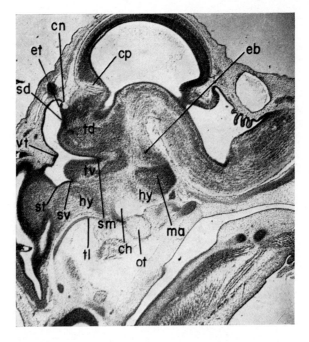

*Figure 213.* Paramedian sagittal section through brain of a 27-mm embryo of *Lacerta agilis*, showing longitudinal diencephalic zones and rostral end of basal plate (from K., 1931, and CHRIST, 1969). ch: optic chiasma; cn and cp: commissura habenulae respectively posterior; eb: rostral end of basal plate; hy: hypothalamus; ma: anlage of mammillary grisea in hypothalamus; ot: optic nerve; sd: junction of sulcus diencephalicus dorsalis and sulcus synencephalicus, continuous with sulcus lateralis mesencephali; sm and sv: sulcus diencephalicus medius respectively ventralis; st: sulcus terminalis reaching commissural plate of lamina terminalis (telo-diencephalic boundary); td: thalamus dorsalis; tl: sulcus telo-diencephalicus; tv: thalamus ventralis.

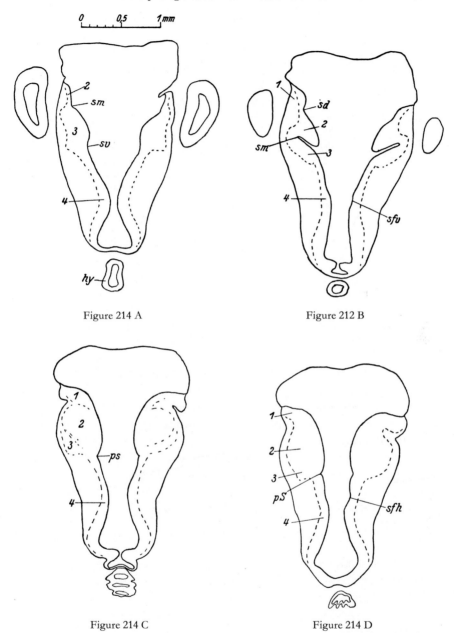

Figure 214 A

Figure 212 B

Figure 214 C

Figure 214 D

*Figure 214.* Outlines of cross-sections through the forebrain in an embryo of *Chelone midas* (length of head only: 7 mm), caudal to postoptic commissures (from K., 1931). sfv: anterior dorsal branch of sulcus lateralis infundibuli; other abbr. as in figure 210.

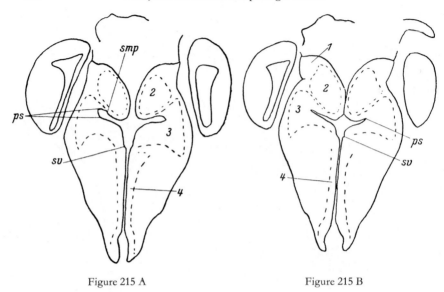

Figure 215 A                    Figure 215 B

*Figure 215.* Outlines of cross-sections through diencephalon of an embryo of *Crocodilus biporcatus* (length of head only: 11.8 mm). The plane of section is slightly oblique, being dorsally somewhat more rostral, and basally somewhat more caudal (from K., 1931).

nal and external cell layer. SENN's superficial strata, including 7 zones[178] (8–14) essentially correspond to the zonal or marginal layer, consisting predominantly of fibers, but which may include superficial cellular aggregates presumably derived from the outer part of our external cell layer.[179]

As regards the dorsal midline structures of the diencephalon, it should also be mentioned that, in some reptiles, an accessory *telencephalic commissura pallii posterior* can take its course through the roof plate within the caudal, diencephalic leaf of velum transversum (cf. Fig. 63 II).

---

[178] The very detailed lamination of the superficial strata (marginal layer) proposed by SENN refers particularly to a stratification displayed in the tectum opticum (cf. figs. 20, 35, and 36 of SENN, loc. cit.).

[179] In the stratification introduced by HIS (1888) and essentially based on embryonic stages of the spinal cord, the matrix or ependymal layer corresponds roughly to the innermost sublayer of our internal layer, the mantle layer (HIS) to the outer sublayer of internal cell layer *and* to external cell layer, which may be considered a further differentiation of the internal one. The marginal layer of HIS would then correspond to SENN's superficial strata.

The configuration of grisea in the diencephalon of *birds* at the definitive (adult) stage manifests a higher degree of differentiation and complexity (Fig. 216) than in the reptilian forms which are likewise Sauropsida. Nevertheless, a dorso-basal sequence of distinctive neighborhoods, suggesting a zonal arrangement, is clearly recognizable. Yet, except for the most dorsal (epithalamic, habenular) and the most basal (obviously hypothalamic) grisea, the derivation of most cell groups from either thalamus dorsalis, thalamus ventralis or hypothalamus is by no means evident and cannot be inferred with sufficient certainty by inspection of the adult pattern.

At key stages of ontogenesis, however, both the longitudinal diencephalic zonal systems and the configuration of sulci are displayed in a manner permitting easy one-to-one mapping upon corresponding conformations in embryonic reptilians and in most adult Anamnia. The avian diencephalic longitudinal zonal system, which becomes clearly manifested toward the end of what might be termed the 'formative

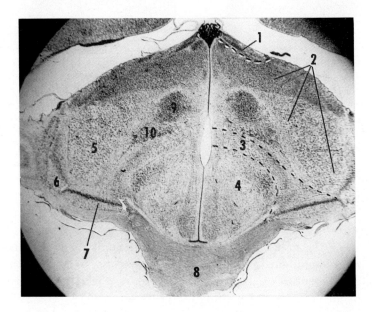

*Figure 216.* Cross-section through diencephalon of newly hatched chick at level of optic chiasma, showing the definitive ('adult') avian griseal configuration (*Nissl stain*, ×23, red. ²/₃; modified after K., 1937). 1: epithalamus; 2: thalamus dorsalis; 3: thalamus ventralis; 4: hypothalamus; 5: n. rotundus; 6: corp. geniculatum laterale, pars dorsalis; 7: corp. genic. lat., pars ventralis; 8: optic chiasma; 9: n. ovoidalis; 10: n. subrotundus.

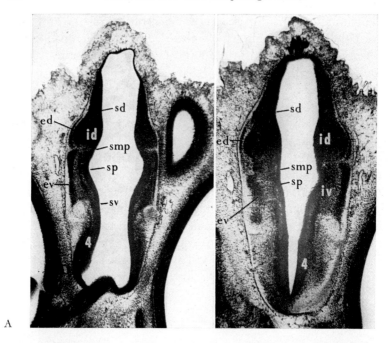

A                                                                          B

*Figure 217.* Cross-sections through the diencephalon in a chick embryo of 5 days and
16 h of incubation (hematoxylin-eosin stain, ×40, red. ²/₃; modified after K., 1937). ed,
ev: external cell layer of dorsal respectively ventral thalamus; id, iv: internal cell layer of
dorsal respectively ventral thalamus; sd: sulcus diencephalicus dorsalis; smp: sulcus diencephalicus medius posterior; sp: sulcus parencephalicus ('pseudosulcus'); sv: sulcus diencephalicus medius. A relatively cell-free, narrow 'zona limitans' in recognizable, particularly on the right side. Section A displays, at right, the caudal tip of telencephalon; section B
about 0.28 mm caudal to A.

phase'[180] reaches its most distinct pattern configuration at the beginning of the organogenetic phase and gradually loses this distinctiveness because of considerable shifting processes involving the griseal
primordia.

During the initial stages in the differentiation of nuclear masses
within the longitudinal zones, a remarkable similarity to the conditions
obtaining in reptiles can be noticed. This similarity includes the formation of an external and an internal cell layer (cf. Figs. 217–220) as
discussed above in connection with the reptilian diencephalic morphogenesis. Thus, the reptilian homologues of most avian dience-

---

[180] Cf. vol. 1, p. 230 of this series.

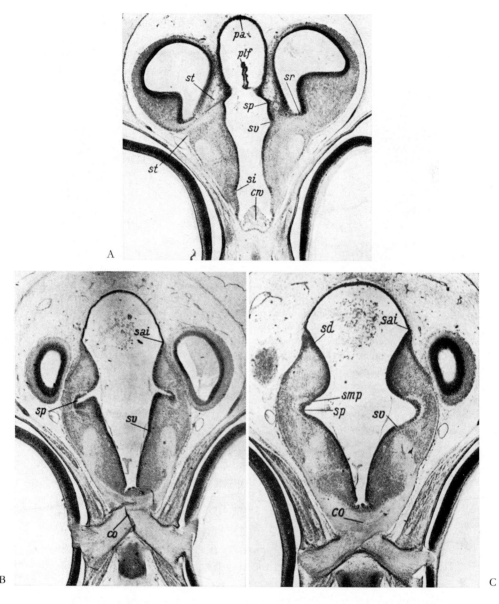

*Figure 218.* Cross-sections through the forebrain in a chick embryo of 7 days of incubation (from K., 1936). A At level of hemispheric stalk. B Just caudal to hemispheric stalk. C At level of telencephalic polus posterior. co: optic chiasma; cw: rostral end of chiasmatic ridge; pa: (secondary) parencephalon (roof plate); plf: fold of choroid plexus; sai: 'sulcus (internus) taeniae thalami'; sd: sulcus diencephalicus dorsalis; si: sulcus intraencephalicus anterior; smp: sulcus diencephalicus medius (posterior); sp: sulcus parencephalicus ('pseudosulcus'); sr: sulcus terminalis; st: sulcus telo-diencephalicus; sv: sulcus diencephalicus ventralis.

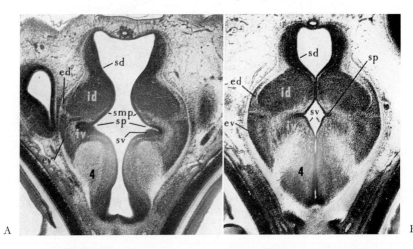

A                                                                          B

*Figure 219.* Cross-sections through the diencephalon in a chick embryo of 7 days of incubation (hematoxylin-eosin, ×33, red. ²/₃; modified after K., 1937). A Through preoptic
region. Within zone 4: sulcus intraencephalicus anterior. At left and right, caudal end
of telencephalon. B Through level of optic chiasma, about 0.24 mm caudal of A. Abbreviations as in figure 218. The epiphysial anlage, not labeled, is easily identified on the parencephalic roof in figures 217B, 219, and 220.

phalic nuclei can be mapped with a reasonable degree of certainty or
at least probability.[181]

The particularly extensive development of the thalamus dorsalis of
birds, as compared with Reptilia and Anamnia, is conspicuous, and
may be correlated with a transitory sulcus accessorius thalami dorsalis
(Fig. 220B), which seems to delimit the neighborhood of n. ovoidalis from that of n. subrotundus (cf. Fig. 216). However, a similar
groove, although much more faintly, can be detected in some reptilian
ontogenetic stages. Figures 217 to 220 illustrate, in cross-sections, the
diencephalic zonal system of chick embryos at 5 days and 16 h, 7 days,
and 8 days of incubation. At some stages (cf. Figs. 217A and 219B),
a relatively cell-free, narrow '*zona limitans*' between dorsal and ventral
thalamus can be noticed. Some such limiting zones apparently correspond to neighborhoods in which fiber bundles develop and become

---

[181] Since the present chapter is essentially concerned with questions pertaining to the
overall morphologic pattern, a more detailed discussion of the homologies in the vertebrate
diencephalon, and particularly of the numerous diencephalic grisea in the brain of Amniota,
must be deferred to chapter XII in volume 5 of this series.

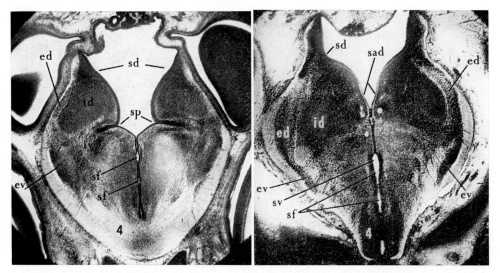

Figure 220 A                                        Figure 220 B

*Figure 220.* Cross-sections (A, B) through the diencephalon in a chick embryo of 8 days of incubation (hematoxylin-eosin, $\times 31$, red. $^2/_3$; modified from K., 1937). A At level of supraoptic and postoptic commissures and of caudal part of optic chiasma. B At level of posterior hypothalamus, about 0.56 mm caudal to A. 1: epithalamus; 2: thalamus dorsalis; 3: thalamus ventralis; 4: hypothalamus; gd: dorsal lateral geniculate body; gv: ventral lateral geniculate body. sad: sulcus accessorius thalami dorsalis; sf: branches of sulcus lateralis thalami posterioris (s. lat. infund.) system; remnants of sulc. dienc. medius and s. dienc. ventralis (not labeled) can be seen at the ventricular boundary of the respective longitudinal zones.

aggregated. Fiber systems of this type may form *medullated 'laminae'* separating different grisea.

A mapping of the diencephalic zonal system, in a chick embryo of 5 days, upon the ventricular wall is shown in Figure 221 A. The sulcus parencephalicus (pseudosulcus) and its transformations[182] correlated with the development of the longitudinal zones is illustrated by Figures 221 B–224.

---

[182] With regard to these rather substantial configurational changes, one may indeed quote the lines
'...*Gestaltung, Umgestaltung*
*Des ewigen Sinnes ewige Unterhaltung.*'
(GOETHE, *Faust II*)

The griseal configuration in the adult *mammalian diencephalon* is illustrated by Figure 225, which shows the arrangement in a Metatherian and in a 'lower' Eutherian mammal. Again, as in birds, a dorso-basal zonal 'stratification' is clearly evident upon inspection, but the boundaries 'separating' dorsal thalamic, ventral thalamic, and hypothalamic derivatives, although vaguely suggested, cannot be convincingly established on the basis of the definitive (adult) pattern, except for the rather obvious basal limit of epithalamus (habenular grisea). In

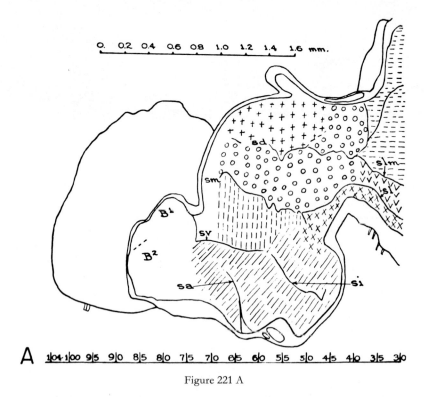

Figure 221 A

*Figure 221 A–C.* Mappings, based on graphic reconstruction, of the diencephalic longitudinal zones upon the ventricular wall in chick embryos at 5, 7, and 8 days of incubation (from K., 1939). Crosses: epithalamus; circles: thalamus dorsalis; vertical hatching: thalamus ventralis; oblique hatching: hypothalamus; horizontal hatching: tectum mesencephali; v-notation: torus semicircularis mesencephali; x-notation: basal plate; neighborhoods within broken lines in B and C: pretectal grisea; $B_1$, $B_2$ in A: telencephalic zones; sa: sulcus intraencephalicus anterior; sd: s. dienc. dors.; si: s. lat. hypothal. posterioris (s. lat. infund.); sl: s. limitans; slm: s. lat. mesencephali; sm: s. diencephalus. medius; sma: s. dienc. med., pars ant.; smp: s. dienc. med., pars. post.; sp: s. parencephalicus ('pseudosulcus'); sv: s. dienceph. ventralis.

comparison with the adult avian configuration (Fig. 216) it will be noticed, furthermore, that the vague bilateral (antimeric) 'stratification'
of diencephalic grisea displays a curved 'laminar' arrangement with
dorsal concavity in the depicted mammals, but a 'lamination' with dorsal convexity in birds. This latter appears evidently correlated with the
ventrolateral shift of the avian tectum mesencephali, in contradistinction to the dorsomedial location of this configuration in mammals.

At key stages of mammalian ontogenetic development, the arrangement of the fundamental diencephalic longitudinal zones, including their limiting sulci, and accessory ones such as sulcus intraencephalicus anterior and sulcus lateralis hypothalami posterioris ('sulcus
lateralis infundibuli'), is very conspicuous (cf. Figs. 226 to 230) and

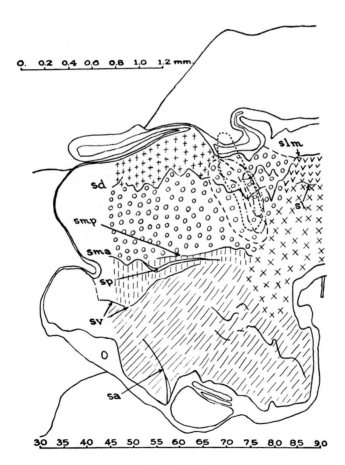

Figure 221 B

permits a reasonably unequivocal one-to-one mapping upon the corresponding neighborhoods of adult simple Anamniote brains, or of embryonic Sauropsidan brains. A relatively cell-poor *zona limitans intrathalamica* between dorsal and ventral thalamus (cf. Figs. 227, 228) as well as, in some instances, a similar but less pronounced zona *limitans hypothalamica* (Fig. 227) can be recognized at some of these stages. Zona limitans intrathalamica becomes the *fasciculus thalamicus* (FOREL's field $H_1$), and zona limitans hypothalamica develops into *fasciculus lenticularis* (FOREL's field $H_2$) of the adult mammalian diencephalon.

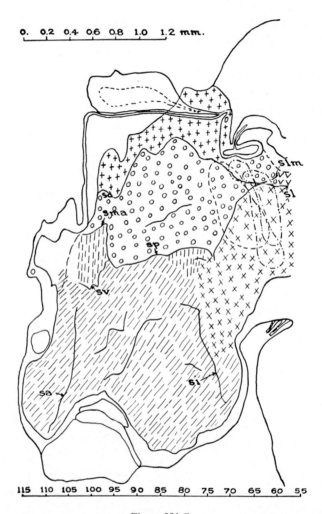

Figure 221 C

In our joint studies attempting to establish the morphologic significance of the vertebrate diencephalic pattern, MIURA (1933) demonstrated a particularly evident display of the zonal system in rabbit embryos at the stage of 15-mm length, which thus, as regards the rabbit, falls within what may here be called the 'key stage period'. Comparable

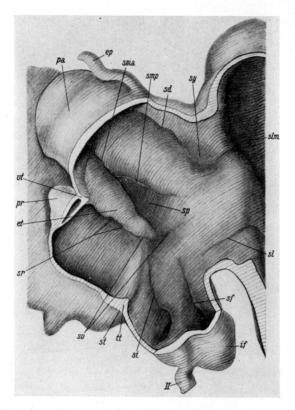

*Figure 222.* Diencephalic ventricular relief in a chick embryo at 5 days and 20 h of incubation (wax plate reconstruction, from K., 1936). Legends for figures 222–224. II: optic nerve; V: trigeminal root (on right external surface of rhombencephalon); cs: telencephalic basal ganglia; ep: anlage of epiphysial structures; et: eminentia thalami (ventralis); if: 'infundibular' (posterior hypothalamic) neighborhood; ltt: neighborhood of lamina terminalis telencephali; pa: (secondary) parencephalon (roof plate); plf: choroid plexus folds; pr: paraphysial anlage; rpc: recessus postcommissuralis (dorsal to commiss. post.); sai: sulcus taeniae thalami (internus); sar: sulcus paraphysio-thalamicus; sd: s. dienc. dorsalis; sf: s. lat. hypothalami posterioris (s. lat. 'infundibuli'); si: s. intraencephalicus anterior; sl: s. limitans; slm: s. lat. mesencephali; sma: s. dienc. medius ant.; smp: s. dienc. med. posterior; sp: s. parencephalicus ('pseudosulcus'); sr: s. terminalis; st: s. telo-diencephalicus; sv: s. dienc. ventralis; sy: s. synencephalicus; tt: torus transversus; vt: velum transversum; x: sulcus accessorius thalami dorsalis.

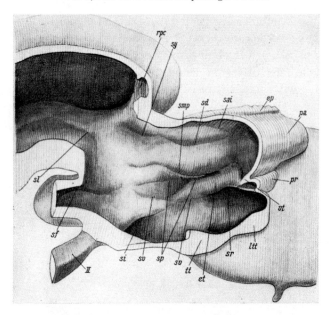

*Figure 223.* Diencephalic ventricular relief in a chick embryo at 6 days of incubation (wax plate reconstruction, from K., 1936). Abbr. as in figure 222.

stages of pig embryos [182a] are shown in Figures 234 and 235. In the embryonic human brain, the distinctly recognizable diencephalic zonal configuration may clearly show sulcus diencephalicus medius and ventralis at the eight-week stage. At the eleven-week stage, the sulcus diencephalicus medius seems to vanish, while the sulcus diencephalicus ventralis has been retained, and sulcus diencephalicus dorsalis may appear (cf. Fig. 236). A noteworthy feature in the differentiation from the eight-week to the eleven-week stage is the drawing together of dorsal thalamus and hypothalamus in the region of the sulcus diencephalicus ventralis,[183] and the lateral displacement of the thalamus ventralis which provides nucleus reticularis thalami, corpus geniculatum

---

[182a] As regards *Metatheria* (marsupials), WARNER (1969) has recently studied the ontogenetic development of the diencephalic zonal system in Australian marsupials, particularly in *Trichosurus vulpecula*, but including some stages of *Perameles nasuta* and *Phascolarctos phascolomys*.

[183] This groove becomes the *sulcus hypothalamicus* (PNA, *sulcus hypothalamicus Monroi*, BNA) of the adult human brain.

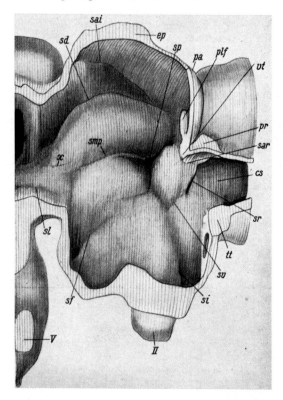

*Figure 224.* Diencephalic ventricular relief in a chick embryo at 7 days of incubation (wax plate reconstruction, from K., 1936). Abbr. as in figure 222.

laterale ventrale, and zona incerta.[184] At the eleven-week stage, the medial part of the thalamus ventralis has already become dorso-basally flattened. This change in the configuration of the thalamus ventralis appears to be a distinctive feature in the ontogeny of all Amniote vertebrates (Reptiles, Birds, Mammals). The further development of the human (and generally the mammalian) diencephalon is characterized by an increasing predominance of the gradually expanding dorsal thalamus and a continued lateral displacement and narrowing of the ventral thalamus.

The dorsal part of the hypothalamic cell zone in mammals and man assumes a distinctive character as a result of a marked accentuation of

---

[184] The zona incerta is included in the so-called 'subthalamus' as understood, on the basis of morphologically somewhat dubious criteria, by various authors.

its lateral, or *entopeduncular* subdivision that develops a substantial relationship with the extrapyramidal motor system (K. and HAYMAKER, 1949; K., 1954). Although this differentiation is much less pronounced

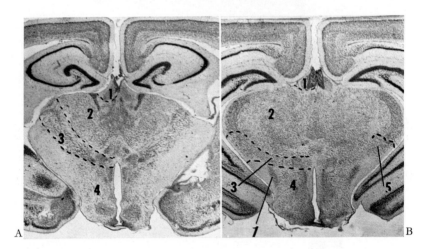

*Figure 225 A.* Cross-section through the diencephalon in the adult Marsupial mammal *Dromiciops australis* at a level caudal to optic chiasma. 1: epithalamus; 2: thalamus dorsalis; 3: thalamus ventralis; 4: hypothalamus. In this, and the following two figures, it will be noted that the ventricular space between the antimeric thalami dorsales and ventrales is obliterated by a secondary fusion, involving disappearance of the ependymal lining. Such fusion is typical for a large number of, if not for most, mammals, and to a lesser extent (massa intermedia) also occurs in perhaps 30 percent of human brains (*Nissl stain*, ×17, red. ²/₃).

*Figure 225 B.* Cross-section through diencephalon of *Dromiciops*, further caudal than in preceding figure. 5: corpus geniculatum laterale, pars ventralis (pars dorsalis above griseum 5); 7: *corpus subthalamicum* (n. subthalamicus) *Luysii* (dorsal hypothalamus). 1–4 as in preceding figure.

---

*Figure 225 C.* Cross-section through the diencephalon in the adult mouse caudal to optic chiasma (hematoxylin-eosin, ×22, red. ²/₃). 1–4 as in preceding figures.

*Figure 226.* Cross-section through the prosencephalon of a 15-mm rabbit embryo at the level of hemispheric stalk (from MIURA, 1933). ae: area cellularis externa; ai: area cellularis interna; ep: epiphysial anlage; mri: massa cellularis reuniens, pars inferior; mrs: m. cell. reun., p. superior; sd: sulcus diencephalicus dorsalis; sm: sulcus dienc. medius; sma: stria medullaris thalami; st: sulcus terminalis; stt: tractus strio-thalamicus (lateral forebrain bundle); sv: sulcus diencephalicus ventralis. The medial wall of lateral ventricle between st and choroid plexus folds (not labeled) becomes the lamina affixa. Figures 226–230 pertain to the same 15-mm series.

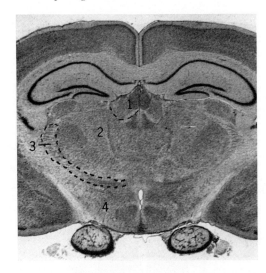

225 C

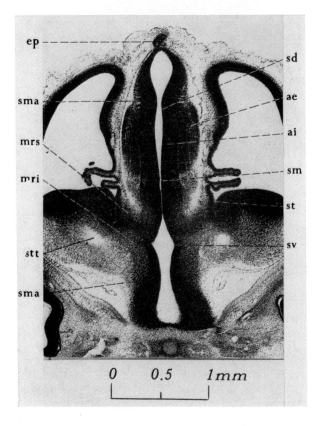

226

in submammalian forms, the application of rigorous comparative criteria nevertheless permits the conclusion that the dorsal part of the mammalian hypothalamic cell zone (a medial neighborhood of which pertains to the anlage of the periventricular hypothalamic nuclei) is homologous with the dorsal part of the less differentiated hypothalamic zone of Amphibia and other Anamnia.

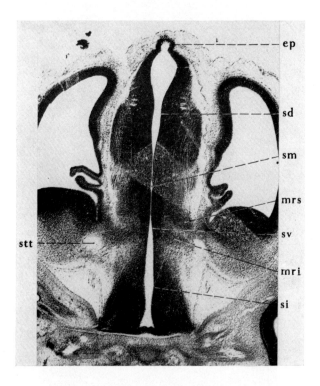

*Figure 227.* Cross-section through embryonic rabbit prosencephalon slightly caudal to preceding figure (from MIURA, 1933). si: sulcus intraencephalicus anterior; above the lead to sm, the zona limitans intrathalamica can be recognized; the lead to sv passes through the unlabeled zona limitans hypothalamica; other abbr. as in preceding figure.

---

*Figure 228.* Cross-section through embryonic rabbit prosencephalon just caudal to hemispheric stalk (from MIURA, 1933). rsp: recessus synencephalicus; tro: optic tract; zit: zona limitans intrathalamica; other abbr. as in preceding figures.

*Figure 229.* Cross-section through embryonic rabbit prosencephalon at level of posterior commissure (from MIURA, 1933). cop: commissura posterior; thd: thalamus dorsalis; thv: thalamus ventralis; other abbr. as in preceding figures.

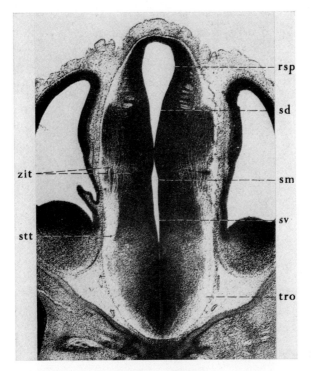

rsp

sd

zit

sm

sv

stt

tro

228

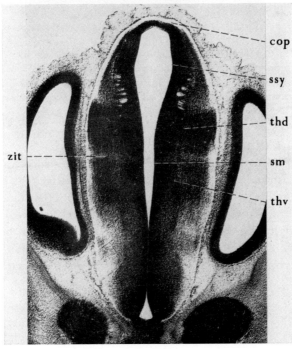

cop

ssy

thd

zit

sm

thv

229

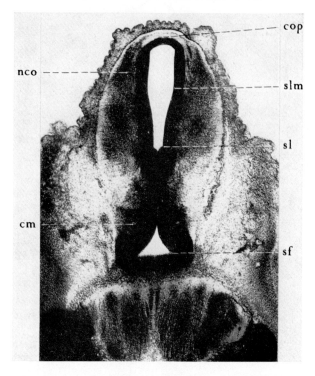

*Figure 230.* Cross-section through embryonic rabbit diencephalon at level of tuberculum posterius (from MIURA, 1933). cm: anlage of mammillary body; nco: nucleus commissurae posterioris; sf: sulcus lateralis hypothalami posterioris ('s. lat. infundibuli'); sl: sulcus limitans; slm: approximate junction of sulcus lateralis mesencephali and synencephalicus (cf. fig. 240); other abbr. as in preceding figures.

Granted that the dorsolateral entopeduncular structures of mammals and man are much more elaborate and impressive than the homologous neighborhoods of the submammalian hypothalamus, there is little justification for the concept of an additional '*subthalamic*' fundamental longitudinal zone.[185] Such a basic concept would be justified only if it were applicable to the entire vertebrate series. This, however, is doubtless not the case. The entire hypothalamic diencephalic zone

---

[185] As already indicated above in the preceding footnote 184, the nonstandardized term 'subthalamus' subsumes, in the interpretation of some authors, a variety of structures which are definitely not 'hypothalamic' in the morphologic sense (cf. K., 1948).

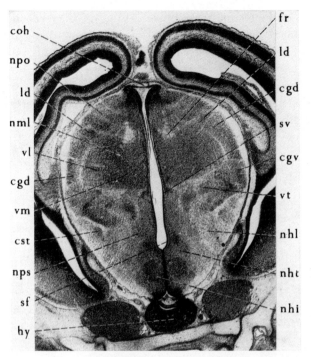

coh
npo
ld
nml
vl
cgd
vm
cst
nps
sf
hy

fr
ld
cgd
sv
cgv
vt
nhl
nht
nhi

*Figure 231*. Cross-section through the prosencephalon of a 30-mm rabbit embryo caudal to hemispheric stalk, and showing further differentiation of diencephalic zonal system
(from MIURA, 1933). cgd: corpus geniculatum, pars dorsalis; cgv: corpus geniculatum,
pars ventralis; coh: commissura habenulae; cst: nucleus subthalamicus *(corpus subthalamicum Luysii);* fr: fasciculus retroflexus; hy: adenohypophysis; ld: nucleus lateralis dorsalis;
nhi: 'nucleus hypothalamicus inferior'; nhl: nucleus lateralis hypothalami; nht: 'nucleus
intermedius hypothalami'; nml: 'nucleus mediolateralis'; npo: nucleus posterior thalami;
nps: 'nucleus posterior superior hypothalami'; rt: nucleus reticularis thalami; sf: 'sulcus
lateralis infundibuli'; vl: nucleus ventralis lateralis; vm: nucleus ventralis medialis; epithalamus consists of lateral and medial habenular nucleus, not labeled, above npo; thalamus dorsalis comprises cgd, ld, nml, npo, vl, vm; thalamus ventralis consists of cgv and rt;
hypothalamus contains cst, nps, nhl, nht.

including its dorsolateral entopeduncular portion,[186] can be evaluated
as a single vertebrate 'bauplan-grundbestandteil'.[187]

---

[186] The mammalian rostral entopeduncular grisea include the *globus pallidus*, which thus
represents a diencephalic, hypothalamic derivative, 'protruding', through the hemispheric
stalk, into the telencephalon. The caudal entopeduncular grisea give origin to the nucleus
subthalamicus *(corpus subthalamicum Luysii).*

[187] A summary of our views concerning the vertebrate diencephalic bauplan, with
particular reference to the hypothalamus, was prepared for HAYMAKER's, ANDERSON's and

While, at the 30 mm-stage of the rabbit, the third ventricle is present as a single cleft-like space lined by ependyma, this space has become obliterated at the level of thalamus dorsalis by secondary fusion
of its two ventricular surfaces, whereby the *adhaesio interthalamica*
(PNA) or so-called *massa intermedia* (BNA) is formed. At first, two parallel dense cellular rows still indicate the original ependymal lining
with its subependymal cells (Fig. 232). Subsequently the *nuclei reu-*

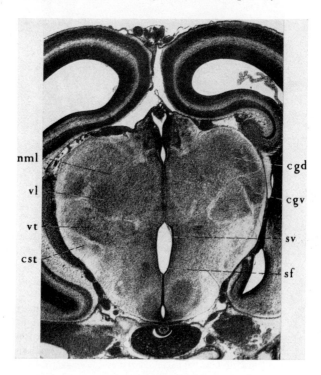

*Figure 232.* Cross-section through the prosencephalon of a 45-mm rabbit embryo caudal to hemispheric stalk, and displaying diencephalic griseal differentiation processes subsequent to those at the 30-mm stage (from MIURA, 1933). Beginning of intrathalamic fusion leading to formation of massa intermedia. In addition, relatively cell-free limiting
zones are noticeable between ventral and dorsal thalamus, between ventral thalamus and
hypothalamus, between reticular nucleus and ventral geniculate body (right) and between
n. ventralis lateralis and dorsal geniculate body (left). Slightly oblique section, left side
being somewhat more caudal than right one. Abbr. as in preceding figures.

NAUTA's treatise on the hypothalamus (1969), and included in CHRIST's chapter 'Derivation
and Boundaries of the Hypothalamus, with Atlas of Hypothalamic Grisea (loc. cit. 1969,
pp. 13–60).

*nientes*, which, at least in part, may appear as unpaired midline nuclei, become differentiated within the region of fusion.

In the human diencephalon, such midline fusion occurs only with a frequency of about 30 percent, the resulting massa intermedia or *commissura mollis* being much less extensive than in most lower mammals. Small aberrant types of massa intermedia, e.g. within the hypothalamus, can occasionally be seen.

As regards a *stratification* into *embryonic elementary layers*, the findings in our own material disclosed a relatively late manifestation of this tendency. Thus, while such lamination was only very faintly suggested in some regions at the 15-mm-stage of rabbits (cf. Figs. 226–230), it became more conspicuous at the 30- and 45-mm-stages. There is little doubt that nucleus habenularis lateralis of epithalamus, the dorsal and ventral thalamic components of corpus geniculatum laterale, and the nucleus posterior appear to be derivatives of our *'external cell layer'* *(area cellularis externa* of MIURA, 1933). Likewise, the lateral hypothalamic grisea, including also entopeduncular grisea, and subthalamic nucleus, seem to differentiate from that external cell layer. The origin of

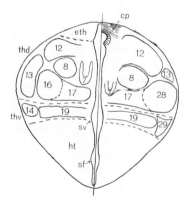

*Figure 233*. Diagram indicating griseal primordia in the embryonic rabbit diencephalon at a period roughly intermediate between MIURA's 30- and 45-mm stages. Based on our Breslau material jointly studied with Dr. MIURA (from K., 1935). Left half, still showing epithalamus, somewhat rostral to right half, which is caudal to end of epithalamic zone. cp: commissura posterior and subcommissural organ; eth: epithalamus (gangl. habenulae); ht: hypothalamus; sf: 'sulcus lateralis infundibuli'; sv: sulcus diencephalicus ventralis; thd: thalamus dorsalis; thv: thalamus ventralis; 7: nucleus parafascicularis; 8: 'n.mediolateralis'; 12: n. posterior thalami; 13: dorsal lateral geniculate body; 14: ventral lateral geniculate body; 16: n. ventrolateralis; 17: n. ventromedialis; 19: n. reticularis thalami; 28: dorsal medial geniculate body; 29: ventral medial geniculate body.

most other hypothalamic grisea, as well as the nucleus reticularis of ventral thalamus, however, can be traced to formative events within the morphogenetic matrix of the *internal cell layer (Area cellularis interna* of MIURA, 1933).

In the thalamus dorsalis, on the other hand, except for the above-mentioned differentiation of corpus geniculatum laterale dorsale and nucleus posterior, the formative events appear to be somewhat more complex. Subsequently to definite appearance of the anlagen of dorsal lateral geniculate and posterior grisea, the remnants of external cellular area seem to merge with internal cellular area, blending into a diffuse

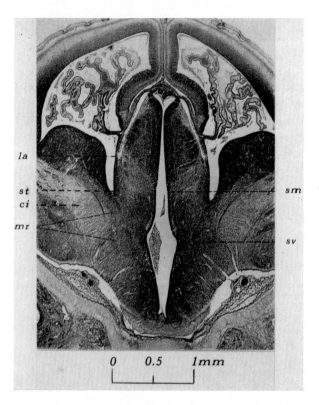

*Figure 234 A.* Cross-section through prosencephalon of a 50-mm pig embryo *(Sus domesticus)* at the level of hemispheric stalk and optic chiasma (slightly modified after K., 1930). ci: capsula interna (component of differentiated lateral forebrain bundle); la: lamina affixa; mr: massa cellularis reuniens (upper lead: of thalamus ventralis, becoming bed nucleus of stria terminalis; lower lead: of hypothalamus, matrix of globus pallidus and ant. interpeduncular nucleus); sm: approximate neighborhood of sulcus diencephalicus medius; st: sulcus terminalis; sv: sulcus diencephalicus ventralis.

morphogenetic matrix. Within this substratum, the nuclear masses (grisea) emerge by *differentiation, condensation* (aggregation, clustering), combined with *rarefaction* of cell groups *in situ*, as well as by growth of neuropil, and by *proliferation*, which may perhaps essentially involve the spongioblastic elements. The sequences of this whole process display a superficial resemblance to crystallization or to flocculation within a colloidal system, and has been described in our reports (MIURA, 1933; K., 1954, also K., 1937, in birds). As regards some of the main definitive grisea of mammalian thalamus dorsalis, their topologic relationships in *statu nascendi* is indicated by the diagram of Figure 233, based on the rabbit series jointly studied by MIURA (1933) and myself (K., 1935).

In the sagittal planes (Figs. 237 A–C, 238) the longitudinal arrangement of the fundamental diencephalic zones and the rostrally

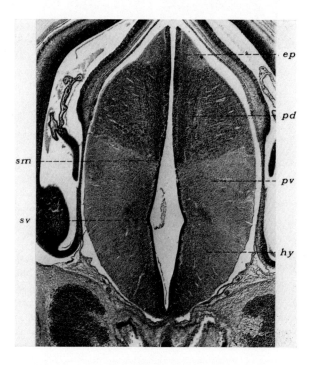

*Figure 234 B.* Cross-section through prosencephalon of a 50-mm pig embryo caudal to hemispheric stalk (from K., 1930). ep: epithalamus (ganglion habenulae); hy: hypothalamus; pd: thalamus dorsalis; pv: thalamus ventralis; sm: sulcus diencephalicus medius; sv: sulcus diencephalicus ventralis.

convex anterior end of the basal plate become conspicuously evident. The ventricular aspect of this configuration, including its relationship to the more or less pronounced ventricular grooves, is illustrated by Figures 239 and 240.

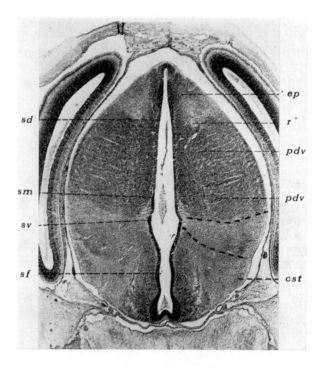

*Figure 234C.* Cross-section through prosencephalon of a 50-mm pig embryo caudal to preceding figure and close to posterior poles of hemispheres (slightly modified from K., 1930). cst: probably anlage of nucleus subthalamicus; ep: epithalamus (the caudal portion of commissura habenulae, not labeled, is recognizable); pds: dorsal part of thalamus dorsalis; pdv: ventral part of thalamus dorsalis (before the availability of additional data from mammalian key stages, which Miura, 1933, provided under the author's direction, this neighborhood was, in 1930, erroneously interpreted as thalamus ventralis); rf: anlage of fasciculus retroflexus; sd: sulcus diencephalicus dorsalis near fusion with sulcus synencephalicus; sf: 'sulcus lateralis infundibuli'; sv: dorsal end of widened sulcus diencephalicus ventralis; at this caudal level, the paraventricular neighborhood of thalamus ventralis (not labeled, but approximately outlined between the two broken lines) has become considerably narrowed.

Summarizing the course of events in griseal differentiation, it could
be stated that the original *primary* or true *histogenetic matrix* (cf. vol. 3/I,
p. 31, of this series) provides, as it were, by cell migration, the external
and internal cell layers, while remaining a gradually diminishing inner-

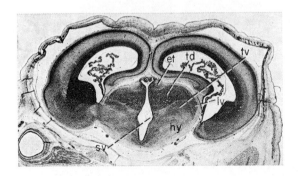

*Figure 235.* Slightly oblique cross-section through the prosencephalon of a 70-mm pig
embryo (after K., 1948). At left, caudal end of hemispheric stalk with recognizable stria medullaris (not labeled). Sulcus terminalis and lamina affixa, dorsally attached to fimbria of hippocampal anlage, are easily identifiable (cf. fig. 234 A). At right, the section plane passes
caudally to hemispheric stalk. et: epithalamus; hy: hypothalamus; lv: ventral lateral geniculate body; sv: sulcus diencephalicus ventralis; td: thalamus dorsalis; tv: thalamus
ventralis; at bottom, postoptic commissures perhaps overlapping with caudal end of optic
chiasma, may be seen.

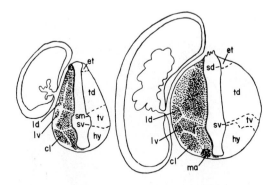

*Figure 236.* Diagram of the diencephalic zones as seen in cross-sections through the
human prosencephalon in embryos of about 25- (left) and 47-mm length (right). Based on
observations by HOCHSTETTER (1919) and K. (from K., 1948). cl: anlage of nucleus subthalamicus *(corpus subthal. Luysii)*; et: epithalamus; hy: hypothalamus; ld: dorsal lateral
geniculate griseum; lv: ventral lateral geniculate griseum; ma: anlage of mammillary grisea; sd: sulcus diencephalicus dorsalis; sv: sulcus diencephalicus ventralis; td: thalamus
dorsalis; tv: thalamus ventralis.

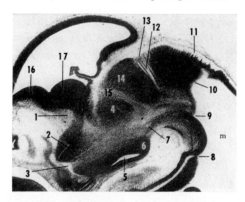

*Figure 237 A.* Slightly oblique sagittal section through the brain of a 15-mm rabbit embryo, showing longitudinal arrangement of diencephalic zonal system (slightly modified after MIURA, 1933). 1: commissural plate; 2: preoptic recess; 3: anlage of optic chiasma; 4: thalamus ventralis; 5: posterior ('infundibular') recess of hypothalamus; 6: anlage of mammillary grisea (2, 5, and 6 represent the hypothalamus); 7: rostral end of deuterencephalic basal plate; 8: fovea isthmi; 9: sulcus limitans; 10: sulcus lateralis mesencephali; 11: commissura posterior; 12: fasciculus retroflexus; 13: epithalamus; 14: thalamus dorsalis; 15: zona limitans intrathalamica; 16: lateral ganglionic hill ($B_1$ of telencephalon) 17: medial ganglionic hill ($B_2$); m: mesencephalic ventricle.

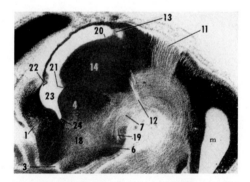

*Figure 237 B.* Sagittal section passing through velum transversum (slightly modified from MIURA, 1933). 18: hypothalamus; 19: tractus mammillo-tegmentalis; 20: sulcus diencephalicus dorsalis; 21: sulcus diencephalicus medius; 22: velum transversum; 23: *cavum Monroi* (ventriculus impar telencephali); 24: sulcus diencephalicus ventralis; other designations as in preceding figures.

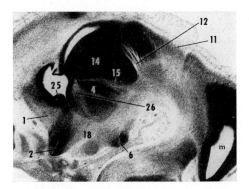

*Figure 237C.* Sagittal section somewhat lateral to B (unpublished figure from Breslau material studied by MIURA and the author). 25: torus transversus; 26: fascic. mammillo-thalamicus; other designations as in preceding figures.

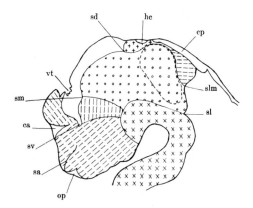

*Figure 238.* Mapping of diencephalic zones and of their relation to tegmental cell cord and tectum mesencephali in a rabbit embryo of 15 mm, based on joint studies by MIURA, MILLER, and the author (from K. and MILLER, 1942). Crosses: epithalamus; circles: thalamus dorsalis; dotted outline within thalamus dorsalis: neighborhood of dorsal thalamic pretectal primordia; vertical hatching: thalamus ventralis; rostrodorsal-caudoventral oblique hatching: telencephalic portion of lamina terminalis (commissural plate); caudo-dorsal-ventrorostral oblique hatching: hypothalamus; horizontal hatching: alar plate of mesencephalon; x: tegmental cell cord; ca: commissura anterior; cp: commissura posterior; hc: habenular commissure; op: optic chiasma; sa: sulcus intraencephalicus anterior; sd: sulcus diencephalicus dorsalis; sl: sulcus limitans; slm: sulcus lateralis mesencephali; sm: sulcus diencephalicus medius; sv: sulcus diencephalicus ventralis; vt: neighborhood of velum transversum.

most sublayer of the internal one, whose ultimate remnants *may* persist as subependymal cell plate. In this respect, only the reduction of the primary matrix represents the process of *'matrix exhaustion'* or *'Matrix-aufbrauch'* pointed out as particularly conspicuous in mammalian telencephalic corticogenesis (K. and v. DOMARUS, 1920; K., 1924). Griseal differentiation, on the other hand, can be said to involve cellular displacements and rearrangements within the *secondary* or *morphogenetic matrix*, in contradistinction to 'matrix exhaustion'.

The various degrees of condensation and rarefaction displayed by the secondary matrix correspond essentially to the concept of *'matrix phases'* propounded by KAHLE (1951, 1956, 1958, 1969), and are directly correlated with the configuration of the regionally differing griseal crytoarchitectural pattern. The cited author, however, subsumes under

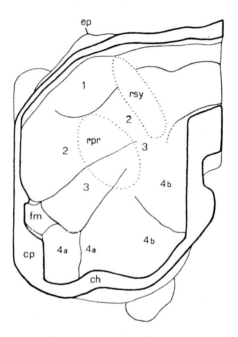

*Figure 239 A.* Outline drawing of diencephalic ventricle and its relief features in a 15-mm rabbit embryo (from MIURA, 1933). 1: epithalamus; 2: thalamus dorsalis; 3: thalamus ventralis; 4: hypothalamus (4a: preoptic portion; 4b: postoptic or 'infundibulo-mammillary portion); ch: chiasmatic ridge; cp: commissural plate of lamina terminalis; ep: epiphysial anlage; fm: ventriculus impar s. communis telencephali, including foramen interventriculare Monroi (the midline protrusion above the f of fm is the remnant of velum transversum); rpr: recessus parencephalicus; rsy: recessus synencephalicus. Compare with figure 239 B.

his concept both 'exhaustion' of the primary, and subsequent rarefaction within the secondary matrix (cf. Fig. 335). Nevertheless, both
processes, which substantially differ from each other, are not necessarily correlated (cf. e.g. Figs. 226–232). Moreover, the differentiations
involving the secondary matrix are structural and do not affect the topologic significance of the relevant *grundbestandteil neighborhood*. Thus,
the matrix phases as conceived by KAHLE cannot be used as criteria in
establishing morphologic homologies within the vertebrate series.

In the course of mammalian embryonic and fetal development, the
thalamus dorsalis becomes the most bulky component of the diencephalon both through increase in wall thickness and dorsoventral
width. Concurrently with these growth processes the original tubular
configuration of the diencephalon undergoes marked changes, resulting in the considerable enlargement of the original region of continuity between diencephalon and telencephalon provided by the comparatively exiguous *primary hemispheric stalk* of lower vertebrates.

This broad zone of continuity between telencephalon and diencephalon constitutes the human and mammalian *(secondary)* *hemispheric
stalk*, characterized by a groove, the *sulcus terminalis*, within the floor of

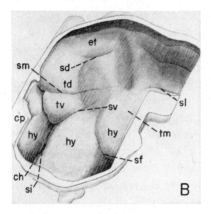

*Figure 239 B.* Wax plate reconstruction of diencephalic ventricular relief in a 15-mm
rabbit embryo (slightly modified after MIURA, 1933, and to be compared with preceding
fig. A). ch: neighborhood of chiasma opticum; cp: commissural plate; hy: hypothalamus;
sd: sulcus diencephalicus dorsalis; sf: 'sulcus lateralis infundibuli'; si: sulcus intraencephalicus anterior; sl: sulcus limitans (dorsally to which sulcus lateralis mesencephali, not
labeled, can be recognized); sm: sulcus diencephalicus medius; sv: sulcus diencephalicus
ventralis; tm: neighborhood corresponding to rostral end of tegmental 'cell plate' (or 'cell
cord'); tv: thalamus ventralis.

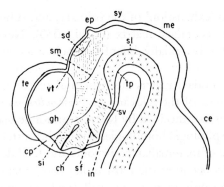

*Figure 240.* Diagrammatic outline of the brain in a human embryo of between 10- and 11-mm length (6th week), as seen from ventricular surface in midsagittal section (from K., 1948). Circles: epithalamus; vertical hatching: thalamus dorsalis; horizontal hatching: thalamus ventralis; oblique hatching: hypothalamus; x: deuterencephalic basal plate (tegmental cell cord); ce: cerebellar plate; ch: chiasmatic ridge; cp: commissural plate with torus transversus (telo-diencephalic boundary); ep: epiphysial anlage; gh: ganglionic hills (primordium of corpus striatum); in: infundibular neighborhood; me: mesencephalon; sd, sf, si, sl, sm, sv as in preceding figures; sy: synencephalon; te: telencephalon (cerebral hemisphere); tp: neighborhood of tuberculum posterius; vt: velum transversum.

the lateral ventricle.[188] Through this zone, containing the topographically complicated region of striatothalamic junction, pass numerous fiber systems, comprising the components of the internal capsule and the various thalamic stalks.

The analysis and understanding of the developmental events in the diencephalo-telencephalic border zone have met with considerable difficulties. A number of earlier authors, including HIS (1889) assumed the occurrence of a *secondary fusion* or adhesion between the lateral surface of the diencephalon and that part of the hemispheric wall which corresponds to the so-called corpus striatum. Until fairly recently, this view could still be encountered in neurologic literature and in some texts on embryology. However, studies on the development of the striatothalamic junction disclosed that such secondary fusion does not occur, and that the broad zone of transition between the striatum and the diencephalon derives from the expansion, by essentially intussusceptional growth, of the primary hemispheric stalk (SCHWALBE, 1880; GOLDSTEIN, 1903; HOCHSTETTER, 1919; K., 1927; 1954).

---

[188] The relationship of sulcus terminalis to lamina affixa and sulcus telencephalodiencephalicus is discussed further below on p. 468 and illustrated by figure 241.

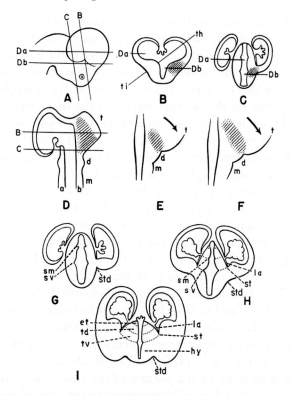

*Figure 241.* Diagrams illustrating the development of hemispheric stalk, telencephalo-diencephalic boundary neighborhoods, and fundamental diencephalic longitudinal zones in the embryonic human brain. A–F as seen in a fetus of 11 mm crown–rump length (about 5–6 weeks), G at about 7 weeks, H at about 8 weeks, and H at about 12–13 weeks (from K., 1951; A–F after K., 1927; G–I modified after HOCHSTETTER on the basis of original observations, and after K., 1927). A Key figure showing planes of the transverse sections (B, C) and of the dextrosinistral longitudinal sections (Da, Db). D, E, and F show the fan-like widening of the hemispheric stalk, whereby a portion of the lateral thalamic surface becomes a posterior surface. The telencephalo-diencephalic boundary neighborhood in the hemispheric stalk is indicated by oblique hatching. a: plane corresponding to line Da; b: plane corresponding to line Db; d: diencephalon; et: epithalamus; hy: hypothalamus; la: lamina affixa; m: mesencephalon; sm: sulcus diencephalicus medius; st: sulcus terminalis in floor of lateral ventricle; std: sulcus telo-diencephalicus; sv: sulcus diencephalicus ven-tralis; td: thalamus dorsalis; ti: portion of sulcus telo-diencephalicus corresponding, on the external surface, to torus hemisphaericus on internal (ventricular) surface; th: torus hemisphaericus (whose median neighborhood is the torus transversus of lamina termina-lis); tv: thalamus ventralis. It will be noted that, in C and G, the left half of the picture shows a plane caudal to hemispheric stalk, while the plane of the right half still passes through the caudal portion of this stalk. In figure I a rostral portion of lateral thalamic sur-face, coextensive with hemispheric stalk (cf. G, H) tends, to become a dorsal surface, such as finally obtains in the definitive (adult) stage.

In the human brain, at about the sixth week of intrauterine life (11-mm-length), the hemispheric vesicle is demarcated from the rostral end of the diencephalic ventricle by a low ridge in the floor of the primitive interventricular foramen. This is the *torus hemisphaericus* (Fig. 241 B), which represents the rostral part of the hemispheric stalk and corresponds on the external surface to the *telencephalo-diencephalic groove*. Right and left grooves converge rostrad to join in the midline on the external surface of the lamina terminalis. The thickened median portion of the lamina terminalis corresponding to this groove and to the midline junction of right and left torus hemisphaericus is the *torus transversus* of v. KUPPFER (1906) or the *commissural plate*, in which anterior commissure and corpus callosum develop. As the telencephalic vesicles expand in a posterior direction by caudal paired evagination, the primary hemispheric stalk thickens and broadens but continues to retain its original connection with the hypothalamic and in part ventral thalamic lateral wall of the diencephalon. On its external and ventral aspect the hemispheric stalk contains the *telencephalo-diencephalic sulcus*, while on its dorsal and ventricular aspect it includes a groove formed by the floor of the caudal part of the lateral ventricle (Fig. 241 C, H, I). This groove is the *sulcus terminalis*, bounded laterally by the ventrolateral portion of the telencephalic wall in which the ganglionic ridges develop, and medially by the portion of the thin *lamina epithelialis* of the roof plate of the telencephalon ventrad to the choroid fissure. This portion of the lamina epithelialis is adjacent to the lateral diencephalic wall from which it is separated by a limb of sulcus telo-diencephalicus (Fig. 241 G). Subsequently this lamina epithelialis fuses with the lateral diencephalic wall to become the *lamina affixa* (Fig. 241 I). The *extraventricular* sulcus telo-diencephalicus is thereby obliterated, being, as it were, replaced in this region by the closely adjacent *intraventricular* sulcus terminalis. This latter, along the corresponding portion of hemispheric stalk, now indicates the telo-diencephalic boundary. The part of the ganglionic ridge which develops into the body and tail of the caudate nucleus thus adjoins the lateral wall of the thalamus covered by the lamina affixa but remains demarcated from the thalamus by the sulcus terminalis.

The further development of the brain is marked by a growth of the hemispheres which greatly exceeds that of the diencephalon. The *lateral surface* of the diencephalon dorsal to the hemispheric stalk, that is, the part which originally projected upward between the hemispheres, flattens and becomes a *dorsal* instead of a lateral surface (Fig. 241 I).

The hemispheric stalk at the same time broadens by means of a fanlike lateral expansion (Fig. 241 E, F). *Pari passu* with this growth process the original lateral surface of the diencephalon caudal to the hemispheric stalk rotates gradually to face caudolaterad and finally becomes a *caudal* instead of a *lateral* surface (Fig. 241 F, C, D). The resulting disappearance of the lateral surfaces of the embryonal diencephalon, a process which puzzled many investigators and prompted the hypothesis of a secondary telodiencephalic fusion of these surfaces, is therefore only apparent. The lateral surfaces remain as surfaces, but their orientation and relative extension become altered by the process of development resulting in the thickening of the hemispheric stalk and in the overgrowth and overexpansion of the cerebral hemispheres.

With the formation of the temporal lobes and their poles, the sulcus terminalis, accompanied by its neighboring structures, tail of caudate nucleus, the lamina affixa, vena terminalis, and stria terminalis, curves around the hemispheric stalk and extends as far as the tip of the inferior horn of the lateral ventricle.

The formation of body and tail of the caudate nucleus takes place in the wall of the lateral ventricle along the whole extent of the striatothalamic junction outlined by the sulcus terminalis. These histogenetic processes leave behind embryonal remnants in the wide subependymal cell plate adjacent to the sulcus terminalis. We have stressed the significance of these remnants for the formation of striatothalamic neuroectodermal brain tumors (GLOBUS and K., 1942, cf. also vol. 3/I, pp. 59–63 and 781–783, of this series).

As regards the rostral delimitation of the fundamental diencephalic longitudinal zones from the longitudinal zonal system of the telencephalon, which shall be discussed in the next section, the following relationships obtain. Within the hemispheric stalk and the lamina terminalis, the cell masses of the anterior *hypothalamus*, namely preoptic grisea and anterior entopeduncular components, are in direct contiguity with basal telencephalic components, although their boundaries are indicated by zones of transition (*massa cellularis reuniens, pars inferior* of adult Anamnia and of embryonic Amniote brains, bed nucleus of commissura anterior and of stria teminalis)[189] or by conspicuous cytoarchitectural limits (between globus pallidus and striatum).

---

[189] Concerning the reptilian brain, this was particularly stressed by SENN (1968). The behavior of the *massae cellulares reunientes* in the course of embryonic development of the human brain was described by SCHNEIDER (1950). In his investigation on the ontogenesis

The *thalamus ventralis* likewise displays some continuity provided by *massa cellularis reuniens, pars superior* of Anamnia, but this continuity becomes, on the whole, practically negligible in adult Amniota. The grisea of thalamus dorsalis and epithalamus remain more or less completely disconnected from those of the telencephalon's zonal system. Considering the ontogenetic events involved in the formation of the hemispheric stalk, one might therefore, with some reservations concerning massa cellularis reuniens, pars superior, essentially agree with SENN's statement (1968) that the embryonic development clearly demonstrates a close connection between hypothalamus and hemisphere: 'there is no cytoarchitectonic connection between the hemisphere and the thalamus or epithalamus'. This, however, does not justify the contention of GRÜNTHAL (1952) who interprets the hypothalamus as the primordium of the prosencephalon.[190] Rather, the prosencephalon should be conceived as the overall archencephalic anlage, from which telencephalon *and* diencephalon derive. In the course of the formative events resulting in a delimitation of the diencephalon, the hypothalamus becomes *then* secondarily established as a particular diencephalic derivative, concomitantly with the differentiation of the three other longitudinal zones.

With regard to the *overall differentiation of the diencephalon* in the Vertebrate taxonomic or presumed phylogenetic series, it could be stated that, in *Anamnia*, epithalamus and hypothalamus are relatively well

---

of the basal ganglia this author confirmed that *massa cellularis reuniens, pars inferior* contains the anlage of globus pallidus (cf. also K., 1924). Although there can hardly be any doubt about the hypothalamic origin of that griseum, I would somewhat more cautiously state that *pars inf. of massa cell. reun.* includes at least *part* of the pallidum-anlage.

[190] GRÜNTHAL (1952) states: '*das Vorderhirn entsteht nach dem Prinzip der Ausstülpung, ohne eindeutige Längs- oder Segmentärgliederung. Als Urabschnitt des Vorderhirns hat der Hypothalamus zu gelten, der sich sehr frühzeitig, zusammen mit dem Epithalamus am oralen Ende des Neuralrohres sackförmig vorwölbt. Dieser primäre Vorderhirnabschnitt hat seine produktivste Epoche auf phylogenetisch und ontogenetisch früher Stufe. Aus dem hypothalamischen Zwischenhirn bilden sich sehr bald als vorgeschobene, exponierte Organe durch Ausstülpung: oral der Paläocortex (Riechrinde) mit dem medialen Ganglienhügel (Pallidum), lateral das Sehorgan und kaudal das Corpus mammillare, welches durch Verdickung der Seitenwände des Recessus mammillaris entsteht*'.

Quite apart from the rather well documented evidence that the medial ganglionic hill ($B_1$) does not correspond to the 'pallidum', GRÜNTHAL's account ignores all details of the complex events resulting in the formation of telencephalon and diencephalon with their respective longitudinal zones (cf. also the comments in K., 1956).

developed, being, by and large, on a par with the thalamus. Within this latter, morevoer, the thalamus ventralis, again, is roughly on a par with thalamus dorsalis. In *Amniota*, on the other hand, the thalamus dorsalis becomes proportionally larger and more extensively differentiated. The epithalamus displays here, both *qua* volume and differentiation, a relatively minor status, while the hypothalamus, although less ostensibly than the dorsal thalamus, manifests 'progressive' features. In higher mammals such e.g. as man and primates, the thalamus dorsalis represents the most 'prominent' griseal configuration of the diencephalon.

Since the present section 5 is essentially concerned with the developmental events involving the secondary longitudinal zonal system of the vertebrate diencephalon in respect to the overall morphologic pattern, a detailed discussion of the multitudinous diencephalic grisea has been omitted. In dealing with these cell groups a comprehensive and critical consideration of their complex fiber tract connections becomes necessary. These topics, including relevant data pertaining to the final stages of ontogenetic differentiation into distinctive nuclei, pertain to the subject matter of volume 5 in the 'special part' (vols. 4 and 5) of this series. It should also be added that the results of recent studies on *'chemodifferentiation'* in the rat's diencephalon (EITSCHBERGER, 1970) suggest that events which can be described in terms of 'chemodifferentiation' are related to the demonstrable production of particular enzymes. The ontogenetic development of the enzyme pattern thus represents a significant but still poorly understood factor during 'maturation'. 'Chemodifferentiation', quite evidently, involves not only developmental events in the diencephalon, but concerns the ontogenesis of the entire organism.

## 6. The Telencephalon and its Secondary Zonal System

The investigations of ELLIOT SMITH (1901, 1907, 1910), CAMPBELL (1905), BRODMANN (1909), C. and O. VOGT (1919), v. ECONOMO and KOSKINAS (1925) and various other authors led to the mapping of numerous distinctive cortical as well as subcortical grisea in the human and mammalian telencephalon, whose cortex can be said to include the following major subdivisions (Fig. 242).

(1) A medial *hippocampal formation* of relatively simple and dense stratification; (2) an adjacent wider, somewhat more stratified *parahip-*

*pocampal cortex;* (3) a dorsolateral well stratified isocortex or '*neo-cortex*'[190a] which can be described as hexalaminar with various modifications; (4) a basilateral *cortex of piriform lobe* with lesser stratification, and (5) a rather loosely arranged *basal cortex* (tuberculum olfactorium, cortical amygdaloid nucleus).

The internal grisea[190b] comprise (a) the complex of *corpus striatum* (nucleus caudatus and putamen, where clearly separated); (b) the *claustrum;* (c) the *amygdaloid complex*, and (d) the *preterminal* and *paraterminal complex sensu latiori.*

---

[190a] The term '*neopallium*' was coined and introduced in 1901 by ELLIOT SMITH to designate 'one of the three histologic formations which constitute the true pallium (of REICHERT)'. ELLIOT SMITH then added: 'as it is the latest of these to reach the height of its development, we may call it the "new pallium", or, if the hybrid term be permissible, "neopallium" in contradistinction to the "old pallium" of the Sauropsida and the earlier Vertebrata, which is chiefly formed of the other two pallial areas'. These two latter were defined as the hippocampal formation and 'part only of the pyriform lobe'. The term '*neocortex*' as subsequently generally used in neurologic literature, refers thus to the cortex of the 'neopallium'. EDINGER then applied the term '*archicortex*' to the hippocampal formation, and KAPPERS designated the piriform lobe of mammals, together with adjacent regions of basal cortex, as '*palaeocortex*'. ELLIOT SMITH (1910), to whom the introduction of the term '*archipallium*' had been ascribed, repudiates, in a '*disclaimer*', the responsibility for that designation as employed by contemporary authors, and suggests that the term 'archipallium' be dropped.

The term '*pallium*' (mantle, cloak) was already used by BROCA (1824–1880) in contradistinction to 'rhinencephalon' (cf. VILLIGER, 1920) and also by REICHERT (1811–1884) about 1859. In the latter author's sense, 'pallium' referred to the thinner dorsal part of the embryonic human brain as contrasted to the thicker basal 'Stammlappen', and, in the adult human brain, to the 'mantle' of cortex surrounding the 'striatal thickening'. The designation '*pallium*' was retained by both BNA and PNA without an intelligible specification, and divergent views concerning its exact limits or its morphological significance have been expressed in the literature (cf., e.g., RETZIUS, 1896; VILLIGER, 1920; KIESEWALTER, 1925). Few authors, in fact, give a clear definition of their concepts concerning 'pallium' ('*Mantelteil der Hemisphäre*') and '*basis*'. REICHERT (1859) even included the floor of the *fossa Sylvii*, i.e. the insula with its 'neopallial' isocortex, into the basal '*Stammlappen*'. According to ELLIOT SMITH (1910), 'the greater part of the pyriform lobe is not pallium in REICHERT's sense'. This, however, is due to the fact that he includes a part of the basal cortex (area ventralis anterior) not comprised within the swelling of tuberculum olfactorium, as well as the entire area ventrolateralis posterior (cortical nucleus amygdalae) into his 'pyriform lobe'. My own definition of '*pallium*' and '*basis*' (K., 1929a) will be made clear, as I hope, in the present section 6. As regards my definition of '*piriform lobe*', which I consider to be part of the pallium cf. p. 580.

[190b] It will be recalled that, in accordance with our interpretation, the globus pallidus is a hypothalamic, diencephalic griseum (cf. above, footnote 186).

Evaluated from a more comprehensive viewpoint of comparative neurology, this subdivision with its further detailed parcellations posed additional basic problems related to a determination of morphologic homologies for the telencephalic grisea throughout the entire vertebrate series. In a phylogenetic interpretation, the manifestation of homologies permits a tentative tracing of the evolution of the telencephalon from fishes to mammals.

Among fundamental studies dealing with the here relevant questions, those by L. EDINGER (1905, 1908, 1911), ELLIOT SMITH (1910, 1919,) JOHNSTON (1906, 1909), HERRICK (1910), KAPPERS (1920, 1936), HOLMGREN (1922, 1925), and BECCARI (1943) are of particular significance.[191] In attempting to investigate these problems on the basis of concepts established by the *Gegenbaur school* of comparative anatomy in JACOBSHAGEN's interpretation, the author (K., 1921), at the suggestion of his teacher MAURER, started out with a mapping of the pertinent grisea in the relatively uncomplicated or 'unspecialized' urodele amphibian brain. Paraphrasing a remark of LUDWIG EDINGER, it could be said that the brain of some amphibians, especially of Urodeles, if perhaps not in all respects the simplest vertebrate brain,[192] does indeed display the most clear-cut manifestation of the fundamental configurational pattern characteristic for the neuraxis of all vertebrates. Thus, in particular, it provides a most suitable frame of reference[193] for an understanding of both comparative anatomy and presumptive phylogenetic evolution of the vertebrate forebrain (i.e. diencephalon and telencephalon).

---

[191] Further references to the numerous additional contributions by the quoted authors and other investigators can be found in the cited publications.

[192] In this respect, the brain of larval *Petromyzon* (Ammocoetes), despite certain 'pattern distortions', might be considered a suitable competitor or 'candidate' for such evaluation. Thus, HEIER (1948) based his monograph 'Fundamental principles in the structure of the brain' on the pattern obtaining in *Petromyzon fluviatilis*. HERRICK (1948), on the other hand, dealt particularly with the urodele Amblystoma as a 'paradigm' of the vertebrate brain.

[193] Namely, a frame of reference for mappings of adult Amphibian telencephalic neighborhoods upon those of both other Anamnia, and embryonic Amniota. Also, conversely, for mappings from other Anamnia and from embryonic Amniota upon Amphibia. The mappings from Amniota upon Amphibia, and *vice versa*, require, however, in order to be sufficiently unambiguous, ontogenetic key stages of Amniota within certain ranges of development.

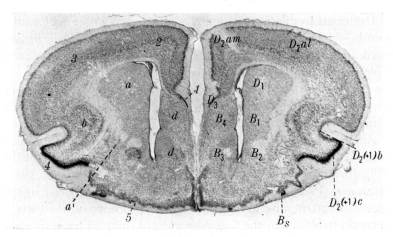

Figure 242 A

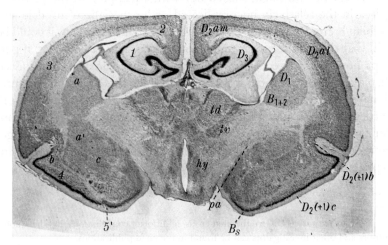

Figure 242 B

*Figure 242.* The main telencephalic grisea in the relatively simple brain of a Metatherian mammal *(Dromiciops australis,* marsupial). A Cross-section through rostral paired evaginated hemisphere. B Cross-section, at level of hemispheric stalk, through caudal paired
evaginated hemisphere. *(Nissl-stain,* ×17, red. ³/₅). 1: hippocampal formation; 2: parahippocampal cortex; 3: hexalaminar cortex ('neocortex'); 4: cortex lobi piriformis; 5: basal cortex (area ventralis anterior); 5′: basal cortex (area ventrolateralis posterior, whose
topographically secondarily medial position is morphologically and topologically 'basilateral'). a: corpus striatum (nucleus caudatus); a′: corpus striatum (putamen); b: claustrum; c: amygdaloid complex; d: pre- and paraterminal complex; hy: hypothalamus; pa:
globus pallidus; td: thalamus dorsalis; tv: thalamus ventralis. D and B symbols on right
side represent the topologic notation explained in text.

Impressed with the results of BRODMANN's work concerning a pattern of distinguishable mammalian cortical areas, the author (K., 1921) initially studied the urodele telencephalon in order to ascertain whether a differentiation of the pallium into distinctive, cytoarchitecturally characterized areas, perhaps homologous with particular griseal regions of mammals, could be found in simpler, Anamniote submammalian brains as displayed by Amphibia.[194] In addition, this investigation involved a mapping of all telencephalic areas of the urodele brain, especially since mammalian cortical differentiation did not appear restricted to the so-called 'pallium'. Moreover, because this latter term, as used in the literature, seemed rather ambiguous and arbitrary, it was also attempted to provide suitable morphologic criteria for a well-definable standardized distinction of '*pallium*' and '*basis*'.

Following an extension of these studies to Anuran and Gymnophione amphibian brains (1921, 1922), the more generalized problems concerning vertebrate corticogenesis and the homologies of cell aggregates in the vertebrate telencephalon were investigated (K., 1922, 1924). Taking into account the relevant stages of telencephalic ontogenesis, it finally became possible to identify the significant features of a fundamental longitudinal telencephalic pattern *(bauplan)*, displayed in the entire vertebrate taxonomic series, and characterized by a topologic configuration of arc-wise connected neighborhoods manifesting a 'hierarchy' of form values such as *Formbestandteile* and *Grundbestandteile* (K., 1929a, 1967). These morphologic, respectively topologic, concepts became, of course, also applicable to the pattern features manifested by deuterencephalon and diencephalon, as dealt with above in sections 4 and 5 of the current chapter.

The presently extant *Cyclostome* Anamnia (Agnatha) comprise two groups classifiable as orders, namely *Petromyzonts* and *Myxinoids*. Because of numerous pecularities or 'distortions', the forebrain of the latter order presents rather difficult problems of interpretation (cf. above, section 1B, p. 191).

The relatively simple telencephalon of *Petromyzonts*, which displays both rostral and caudal paired evagination with a greatly reduced, practically negligible telencephalon medium s. impar (Figs. 63 I, 243)

---

[194] This was formulated as '*die Frage, ob eine Differenzierung des Palliums in einzelne, auf Grund der Schichtungsverhältnisse abgrenzbare, den Rindenfeldern homologe Bezirke nicht schon dem Auftreten eines eigentlichen Cortex cerebi vorangeht, oder ob diese Differenzierung erst einer zunächst einheitlichen Rindenbildung entspringt*' (K., 1921, p.464).

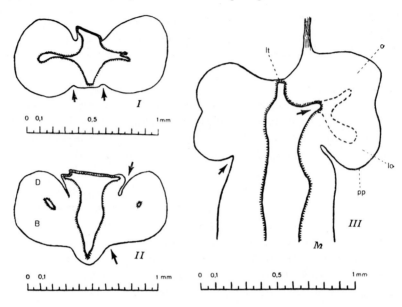

*Figure 243 A.* Morphologic aspects of the telencephalon in Ammocoetes specimens of about 12 cm length, representing an advanced 'larval' stage of Petromyzon (from K., 1929a). I Cross-section at level of interventricular foramen. II Cross-section through caudal evagination of telencephalon. III Approximately horizontal section through forebrain. D, B: telencephalic 'Grundbestandteile'; lo: lobus hemisphaericus; lt: lamina terminalis; o: bulbus olfactorius; pp: polus posterior; added arrows indicate sulcus telo-diencephalicus; in I, the basal median neighborhood between the antimeric sulci is a rostral portion of recessus praeopticus; in III, right arrow points to region of torus.

consists of a rostral *bulbus olfactorius* and a caudally adjacent *lobus hemisphaericus* ('lobus olfactorius'). The cell populations in the hemispheric lobe show a rather diffuse, essentially periventricular arrangement with apparently random clusters and irregular, disjoint stratifications. Depending on rather subjective criteria, it may or may not be evaluated as forming a '*praecortex*'[195] (Fig. 243 B). In young adults, the peri-

---

[195] HERRICK (1910) formulated the following *definition of cortex cerebri: 'correlation tissue developed as superficial gray matter within the dorsal (pallial) walls of the cerebral hemispheres'.* The author (e.g. K., 1922), elaborating on HERRICK's definition, pointed out that, in the aspect under  consideration, the designation 'superficial' should imply the location of a griseal sheet peripheral (i.e. abventricular) to its fiber pathways. Because such cortex develops not only in the *pallial,* but also in the *basal* telencephalic wall, the *cortex cerebri* represents thus a correlation tissue developed as superficial gray matter *within the walls of the telencephalon* (omitting the restriction 'pallial').

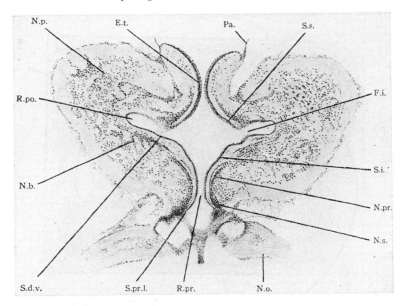

*Figure 243 B.* Cross-section through the forebrain of adult *Entosphenus japonicus* at the level of interventricular foramina (from SAITO, 1930). E.t.: eminentia thalami; F.i.: foramen interventriculare; N.o.: nervus opticus; N.b.: 'nucleus basalis' (B); N.p.: 'nucleus pallialis' (D); N. pr. nucleus praeopticus (hypothalamus); N.s.: 'nucleus supraopticus' (hypothalamus); Pa.: parencephalic roof (epithelial roof plate); R.pr.: recessus praeopticus; R.po.: recess of lateral ventricle; S.d.v.: accessory sulcus related to sulcus diencephalicus ventralis near caudal rim of interventricular foramen; S.pr.l.: 'sulcus praeopticus lateralis'; S.s.: 'sulcus suprainterventricularis' (related to sulcus diencephalicus ventralis system); s.i.: 'sulcus infrainterventricularis'.

Although spread out as a cellular sheet along the external telencephalic wall surface, and separated from the ependyma, respectively from periventricular matrix or periventricular grisea by a (generally but not necessarily medullated or 'white') fiber layer, the cortical griseum is not necessarily located externally to all of its fiber connections, since extrinsic superficial fiber systems may persist in the marginal or 'zonal' layer.

With respect to these definitions and concepts, telencephalic periventricular griseal arrangements in vertebrate forms without 'true' cortex, but displaying various degrees of dispersal or 'migration' toward the surface, and manifesting some tendency toward stratification (e.g. a *dispersal layer* or '*Schwärmschicht*') could be designated as *precortical*, representing a '*praecortex*'. In addition to the telencephalic cortex cerebri, superficial gray matter of cortical type is, of course, also developed as cortex cerebelli and cortex-like gray matter of the tectum mesencephali (particularly in submammalian forms). These cortices shall be dealt with in volume 4 of the present series.

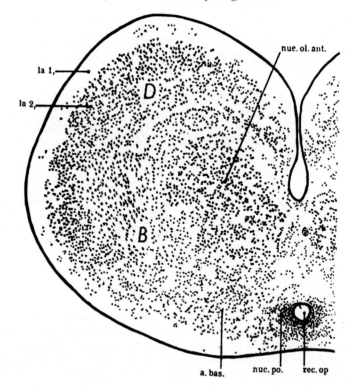

Figure 244 I

*Figure 244.* Cross-sections through the forebrain of Myxine glutinosa. I At rostral lev el of recessus preopticus. II At level of ganglion habenulae and commissura postoptica (after JANSEN, 1930,, with added B and D symbols referring to the topologic bauplan concept, and differing interpretations, in accordance with the present treatise, added in parenthesis to the explanations of abbreviations). B: telencephalic basis; D: telencephalic pallium; a. bas.: 'area basalis telencephali' (in I presumably massa cellularis reuniens inferior, in II presumably lateral part of hypothalamus, K.); com. po.: commissura postoptica; hab: ganglion habenulae; la 1–5: layers of telencephalic cell groups; nuc. ol. ant.: 'nucleus olfactorius anterior' (this interpretation as a telencephalic cell group seems somewhat dubious; it is perhaps a rostral extension of the so-called 'primordium hippocampi', i.e. of thalamus ventralis); nuc. po.: nucleus praeopticus; pr. hip.: 'primordium hippocampi' (rostral extension of thalamus ventralis as interpreted in present treatise; the loose cell groups between habenular ganglion and thalamus ventralis correspond perhaps to the rostral end of thalamus dorsalis); rec. op.: recessus praeopticus; s. bas. ant.: 'sulcus basalis anterior' (basal portion of sulcus telo-diencephalicus).

ventricular grouping is usually more compact (Fig. 187). The wall of
the telencephalic tube dorsal to the ventricle can be designated as the
'*pallium*', the corresponding basal wall being the '*basis*'. The grisea
within these walls accordingly represent two definable neighborhoods
with open boundaries, here representing the telencephalic *grundbestandteile* which shall be designated as D (dorsal, pallial), and B (basal).
Taking the olfactory bulb with its formatio bulbaris, characterized by
olfactory glomeruli, as a single unit, the *bauplan* of the Petromyzont
telencephalon can be expressed by the simplified formula O, D, B.

     In order to interpret the arrangements in adult *Myxinoids*, whose
telencephalic ventricle is obliterated (Figs. 244 I, II), it would be nec-

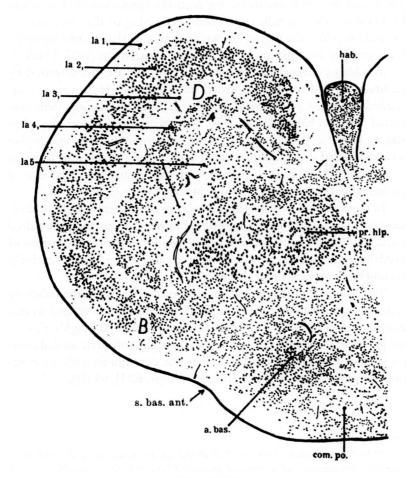

Figure 244 II

essary to examine key stages of the relevant ontogenetic transformations. Although CONEL (1931) and HOLMGREN (1946) have succeeded
in procuring highly interesting embryonic material of *Bdellostoma* and
of *Myxine glutinosa*, respectively, the data recorded by the cited authors
are not sufficiently complete for a proper understanding of the significant morphogenetic processes, and do not permit an unequivocal mapping (without inter- and extrapolation) of the Myxinoid grisea upon the
D and B neighborhood of Petromyzonts. A very tentative mapping,
reinterpreting the telencephalic grisea of *Myxine* as described by JAN-
SEN (1930), is indicated in the explanation to Figures 244 I and II. Concerning the so-called 'primordium hippocampi' of earlier writers on
Myxinoids and Petromyzonts, my own interpretation of that neighborhood as a diencephalic component, pertaining to thalamus ventralis, and including the region known as eminentia thalami of Amphibia,
is in complete agreement with the views of HOLMGREN (1922, 1946).

*Selachians (Elasmobranchs)*, like Cyclostomes, are characterized by
an inverted, and paired evaginated telencephalon. However, in contradistinction to Cyclostomes (or Agnatha), a fairly large unpaired evaginated telencephalon impar (Fig. 245), which persists at the adult
stage, obtains in Selachians. Elasmobranchs, moreover, do not display
a caudal paired evagination of the telencephalon, which thus lacks, in
these forms, a true polus posterior of the hemisphere related to a
paired caudal extension of lateral ventricle.

In the rostral paired evaginated portion of the hemispheres, the olfactory bulb extends commonly laterad rather than rostrad, the paired
portion of lobus hemisphaericus being then located medially instead of
caudally to the bulb, and protruding, as a rule (with a 'polus anterior'),
rostrad beyond the stalk of the bulb[196] (Fig. 246).

The pallial or D-neighborhoods are located dorsal to the olfactory
stalk's evagination, and the basal or B-neighborhoods ventral to this
evagination (Fig. 245). Caudalward, the antimeric B-neighborhoods
are joined in the lamina terminalis; likewise, the antimeric medial components of D-neighborhoods may become contiguous within the supraneuroporic portion of that lamina (cf. Figs. 62 II, 64 III).

---

[196] In Holocephalians (*Chimaera*, Callorhynchus), however, whose forebrain configuration also differs in other respect from that of 'true' Selachians, the olfactory bulb
represents the rostral extremity of the paired hemisphere.

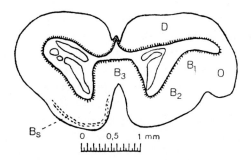

*Figure 245*. Slightly oblique cross-section through the telencephalon in a 60 mm long embryo of the Selachian *Mustelus laevis* (from K., 1929 a). It can be seen that the entrance into olfactory bulb ventricle is here located so far caudolaterally as to lie in the transverse plane of telencephalon impar (cf. figs. 246 and 247).

The telencephalic cell populations show a fairly dense but rather narrow periventricular layer, followed by a layer of low density, peripheral to which a partly interrupted, and in some regions or instances uninterrupted dense cell 'plate' of cortical type is located. This lamina is separated from the external surface by a 'zonal' layer of low cell density (Fig. 247). With reference to the concepts of periventricular griseum, praecortex, and cortex cerebri (cf. footnote 195) one could thus say that the Selachian telencephalon might manifest grisea of rudimentary cortical, or at least corticoid, type.

The cortical arrangements obtain both in pallium (D) and basis (B), being actually more pronounced in the basis.[197] Again, at certain rather advanced ontogenetic stages, this cortical differentiation is generally more conspicuous than in the older adults, where the cellular arrangements tend to assume a somewhat 'diffuse' aspect.

---

[197] The basal cortical plate represents the '*area basalis superficialis*' corresponding to the '*area ventralis anterior*' (cortex olfactoria, tuberculum olfactorium) and, posteriorly, to the '*area ventrolateralis posterior*' (true periamygdalar cortex, cortical amygdaloid nucleus) of mammals. In many instances a much thinner or almost lacking periventricular layer is correlated with the extent of the greatest compactness in the Selachian area basalis superficialis. It should be added that the area ventrolateralis posterior is both topologically and topographically ventrolateral in 'lower' vertebrates, but assumes, topographically, a rather ventromedial position in mammals, becoming displaced by the considerable expansion of the pallium. Nevertheless, because of its topologic connectedness, the term area ventrolateralis posterior was retained. It might, of course, also be simply called area ventralis (or basalis) posterior.

In contradistinction to the telencephalon of Petromyzon, in which a further regional subdivision of pallial and basal cell masses is not sufficiently evident, it seems possible to recognize more or less distinctive griseal regions displaying a longitudinal zonal pattern, both in the dorsal and the basal wall of the lobus hemisphaericus of Selachians. This pattern of pallial and basal zones is suggested by slight differences of cell population density and 'cytoarchitecture' as well as by a fairly consistent or constant manifestation of ventricular sulci.

The *pallial* (dorsal, respectively dorsomedial) grisea can be roughly subdivided into three neighborhoods, namely into a *lateral* ($D_1$), an *intermediate* or *dorsal* ($D_2$), and a *medial* ($D_3$) region. This was already pointed out by HOLMGREN (1922), who designated the lateral area as 'pyriform cortex', the intermediate one as 'general pallium', and the medi-

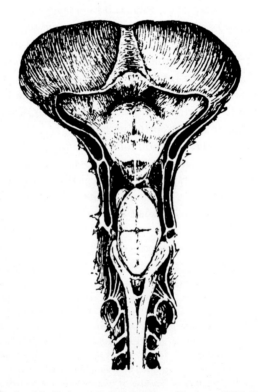

*Figure 246.* Brain of Scyllium canicula *in situ* (from KAPPERS, 1921). The lateral position of olfactory bulb in relation to polus anterior of hemispheric lobe is conspicuous.

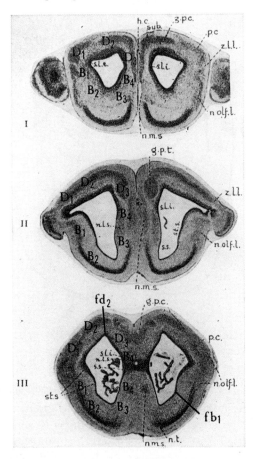

*Figure 247.* Cross-sections through the paired evaginated portion of the telencephalon in an advanced (80 mm long) embryo of Acanthias (after HOLMGREN, 1922, from KAPPERS, 1947). I Rostral to olfactory stalk. II At level of olfactory stalk. III Caudal to olfactory stalk, at level of lamina terminalis in which recessus neuroporicus (not labeled) is clearly recognizable. g.p.c.: 'general pallium cortex'; g.p.t.: 'general pallial thickening'; h.c.: 'hippocampal cortex'; n.l.s., n.m.s.: 'nucleus lateralis septi', resp. 'n. medialis septi'; n. olf. l.:'nucleus olfactorius lateralis'; n.t.: 'nucleus taeniae'; p.c.: 'pyriform pallium'; s.l.e., s.l.i.: 'sulcus limitans externus' (fd$_1$) resp. 's.l. internus' (fd$_4$); s.s.: 'sulcus septalis' (fb$_3$); st.s.: 'striatal swelling' (B$_1$); z.l.l.: 'zona limitans lateralis'. Designations in accordance with my own notation D$_{1-3}$, B$_{1-4}$, and fb$_1$ have been added. B$_4$′ indicates supraneuroporic, caudodorsal extension of B$_4$ (cf. fig. 248B).

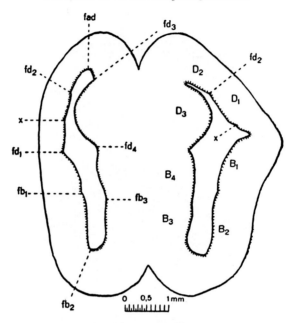

*Figure 248 A.* Cross-section through telencephalon of a 25 cm long young Acanthias (from K., 1929a). x: accessory ventricular sulcus within wall neighborhood D$_1$.

al portion as 'hippocampal cortex'.[198] Without reference to the abbreviated topologic notation, the pallial corticoid grisea could be designated as area (respectively cortex) lateralis, dorsalis, and medialis pallii.

Within the basal grisea a dorsolateral neighborhood B$_1$ which is ventromedially adjacent to the olfactory bulbar stalk, may slightly protrude into the ventricle. It corresponds essentially to the region described as nucleus olfactorius lateralis by HOLMGREN (1922). The contiguous ventrolateral wall sector, commonly characterized by a particularly dense portion of the cortical cell plate, represents neighborhood B$_2$ in the proposed topologic terminology. B$_2$, in turn, is contiguous with neighborhood B$_3$, which includes the ventral sector of the medial wall of the hemisphere, and may likewise display a dense structure of

---

[198] E.g. in HOLMGREN's figures 10, 11, 12, 18, 19 (loc.cit.). In his figures 15 and 17B, I would interpret HOLMGREN's *h.c.* as the supraneuroporic component of B$_4$. In HOLMGREN's figures 15, 16, 17B, 21, I would interpret his *g.p.t.* ('general pallial thickening') as D$_3$ on the basis of my observations on original material and of additional findings by GERLACH (1947).

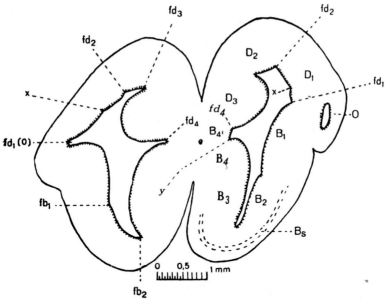

*Figure 248 B.* Cross-section through telencephalon of a 26 cm long Acanthias (modified after K., 1929a). At left, entrance into olfactory stalk, $fd_1(O)$ being the extension of $fd_1$ into olfactory stalk ventricle. At right, a caudal portion of olfactory stalk ventricle is shown. Neuroporic recess (not labeled) can be seen in lamina terminalis. The rostral portion of this latter, protruding into the paired evaginated telencephalon, is also shown in figures 247 III and 248 A. $B_{4'}$: supraneuroporic portion of $B_4$ (cf. fig. 247 III); y: accessory sulcus; x: cf. preceding figure.

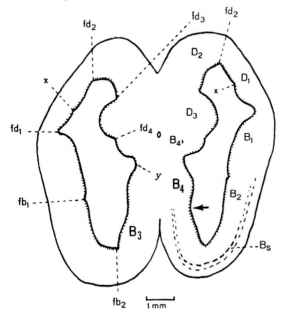

*Figure 248 C.* Cross-section through the telencephalon of adult Acanthias at level of neuroporic recess (slightly modified after K., 1929a). x, y: cf. preceding figure; neuroporic recess not labeled, but clearly recognizable.

the basal cortical plate. The density of this latter decreases in the dorso-medial and dorsolateral directions. Caudalward, in the region of lamina terminalis, the contiguous antimeric $B_3$ components provide the median floor of telencephalon impar.

Again, dorsally to $B_3$, the medial wall of the paired evaginated hemisphere contains a distinctive neighborhood $B_4$ which becomes adjacent to the medial pallial neighborhood $D_3$. The dorsal portion of $B_4$ represents thus a zone of connectedness between pallial ($D_3$) and basal ($B_4$) components of the rostral paired evaginated and inverted hemisphere. Although small caudal portions of $B_4$ may extend into both dorsal and basal wall of telencephalon impar, the formation of a neighborhood $B_4$ appears intrinsic to rostral paired inverted evagination, and does not seem to occur in forms with completely everted lobus hemisphaericus (Ganoids, Teleosts, and 'Crossopterygian' Latimeria).

Again, without direct reference to the topologic notation, the basal grisea might be designated by the self-explanatory terms *nucleus basilateralis superior* ($B_1$), *basilateralis inferior* ($B_2$), *basimedialis inferior* ($B_3$), *basimedialis superior* ($B_4$), and *cortex basalis* with further subdivisions as deemed appropriate[198a] (e.g. nucleus basimedialis superior internus et externus, etc.).

Concerning the *ventricular sulci*, which, to a significant extent, appear correlated with boundary regions between the telencephalic griseal wall neighborhoods, the following terminology was suggested (K., 1929a). A groove approximately indicating the basal limit of $D_1$ against $B_1$ is designated as $fd_1$. A more dorsal groove, $fd_2$, is approximately located between $D_1$ and $D_2$. Sulcus $fd_3$ corresponds to an open boundary between $D_2$ and $D_3$, while a medial pallio-basal boundary between $D_3$ and $B_4$ is suggested by sulcus $fd_4$.

---

[198a] E.g. distinguishing the periventricular cell groups as 'nuclei' from the peripheral 'cortical' cell plate. Instead of '*nucleus*' and of '*basilateralis*' etc., one could also use the terms '*griseum*' and '*basolaterale*' respectively '*basilaterale*', thus griseum periventriculare aut corticale basolaterale superius et inferius, basomediale etc. inferius et superius.

Again, because, not only in Selachians, but apparently in all gnathostome vertebrates, the caudal portions of the grisea comprising nucleus basimedialis inferior and superior display conspicuous preterminal and terminal relationships to the lamina terminalis, these grisea may be subsumed under Elliot Smith's concept of *paraterminal body sensu latiori*. Said grisea are likewise subsumed under the rather vague and ambiguous concept 'septum'. The term *paraterminal body sensu strictiori* might preferably be restricted to griseum basimediale superius ($B_4$) and its derivatives (cf. also K., 1969).

As regards the basal telencephalic wall, sulcus $fb_1$ may be seen between $B_1$ and $B_2$, and sulcus $fb_2$ between $B_2$ and $B_3$. Within the basomedial wall, ventricular groove $fb_3$ can occur between $B_3$ and $B_4$ (cf. Figs. 248 A–C).

Of these ventricular sulci, $fd_1$ and $fd_4$, related to lateral, respectively medial pallio-basal boundary zones, seem perhaps to be the most constant and commonly most pronounced ones. The others may, somewhat more often, be poorly displayed or entirely missing. Further complications arise in connection with the inversion of the rostral paired evaginated hemisphere, such that a dorsal and a ventral folding of the telencephalic wall provides, within the lateral ventricle, a dorsal and ventral angle *(concavity, Ventrikelumschlagsfurche)*, which can but need not correspond, and may thus obtain in addition, to some of the above-mentioned 'limiting sulci' (Fig. 249 A).

Not only in Selachians, but apparently in all Gnathostome vertebrate forms with rostrally paired evaginated and inverted telencephalon, this folding of the ventricular wall seems to manifest some degree of (even individual) variation, whereby the angulus dorsalis may coincide with $fd_2$ or $fd_3$, or remain an independent groove between both (fad, furrow of angulus dorsalis). *Mutatis mutandis*, angulus ventralis ventriculi may coincide with $fb_2$ or become an independent groove fav, medially or laterally parallel to $fb_2$. In addition, further accessory ventricular sulci can be manifested, some of which might, of course, be fixation and shrinkage artefacts.[199]

Taking into consideration the configuration of both grisea and ventricular groove pattern, GERLACH (1947) has mapped the general aspect of the telencephalic zonal system in Selachians as shown by Figure 249 B.

---

[199] The author has, for many years, studied this behavior of grooves in the ventricular wall of numerous representative forms pertaining to all major vertebrate groups, keeping in mind the evident occurrence of artefacts. Nevertheless, rather constant findings in apparently well preserved material, including both removed brains and specimens serially cut *in situ craniale*, seem to justify the inference that, despite a considerable range of variations, an orderly and morphologically significant system of ventricular grooves obtains. Since the spatial distribution of genetically determined specific griseal configurations appears correlated with substantial displacements of material, including the moulding of a tubular structure, it seems justified to assume that the manifestations of grooves bears a definite relation to pattern orderliness resulting from morphogenetic events. Despite much greater 'randomization' of secondary (accessory) sulci and gyri on the human and primate telencephalic surface, the 'constant' orderliness of the overall surface relief pattern, and its overall correlation with distinct cortical grisea can hardly be denied.

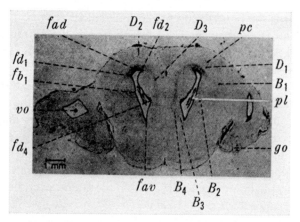

*Figure 249 A.* Cross-section through the telencephalon of adult *Galeus canis* (after GERLACH, 1947). go: olfactori glomeruli; pc: 'precortical' plate; pl: choroid plexus of lateral ventricle; vo: ventricle of olfactory bulb.

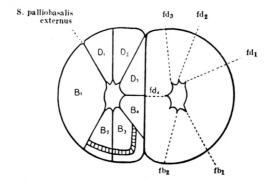

*Figure 249 B.* Diagram of Selachian telencephalic zonal pattern (bauplan) as mapped by GERLACH for Galeus (from GERLACH, 1947).

Using the subscript *s (superficial)* to denote a pronounced cortical differentiation, and indicating the relationship of such cortex to its respective periventricular longitudinal zone by a line (signifying separation from the matrix region) in the manner $\frac{B_1}{B_s}$, the bauplan of the Selachian telencephalon accordingly can be expressed by the general formula[200]

$$O, D_1, D_2, D_3, \frac{B_1, B_2, B_3,}{B_s,}, B_4.$$

---

[200] By arbitrarily omitting the notation $\frac{D_1}{D_s}$ etc., it is intended to stress the more pronounced 'cortical' aspect along and within the basal zones.

The telencephalon in *Holocephalians* such as Chimaera and Callor-hynchus, recently re-examined with regard to the topologic bauplan concept (K. and NIIMI, 1969) differs in several respects from the telencephalon of Squalidae and Rajidae. In particular, the caudal part of the Chimaeroid hemisphere is differentiated as an unpaired evaginated and slightly *everted* telencephalon impar of a type comparable to that of the Dipnoan *Protopterus* (to be discussed further below). The anterior, paired evaginated portion of the lobus hemisphaericus, located caudally to the rostral olfactory bulb, likewise differs in its internal (ventricular) and external shape from that of the other orders of Elasmobranchs (Fig. 250). Nevertheless, despite these considerable pattern deformations and distortions, the general bauplan formula for the Selachian telencephalon also applies to that of Chimaeroids.

In comparison with Cyclostomes, it is evidently possible to map the Selachian neighborhoods O, $(D_1 \leftrightarrow D_3) = D$, and $(B_1 \leftrightarrow B_4) = B$ upon the corresponding Cyclostome neighborhoods, O, D, B. Since this involves, for the olfactory bulb, a one-to-one transformation or mapping, special orthohomology evidently holds for O. In the case of D and B, however, because the transformations are many-one, or, conversely, one-many mappings, special kathomology obtains.[201]

Generally speaking, the telencephalon of *Osteichthyes* is characterized by a substantial eversion of its lobus hemisphaericus, which commonly represents an unpaired rostral evaginated telencephalon impar. In various forms, some degree of likewise everted, but, of course paired, caudal evagination may occur (cf. Fig. 61 A). As a rule, only the olfactory bulbs and, where present, their stalks are rostral paired evaginated configurations.

In evaluating the basic telencephalic components or *Grundbestandteile* of these brains, it is first necessary to realize that, because of the 'rotation' intrinsic to eversion, the D-region becomes displaced dorsolaterad, respectively laterad, such that neighborhood $D_3$, which is the most medial pallial area of the inverted telencephalon, assumes the most lateral position in the everted pallium (cf. Fig. 60). Evidently, a sulcus $fd_4$ cannot be present in an everted telencephalon, since it

---

[201] Previously to the detailed elaboration of his topologic homology concept (K., 1967, cf. also section 1 A, p. 64f. of this chapter), the author (K., 1929a) had considered the transformations $D \leftrightarrow (D_1, D_2, D_3)$, and $B \leftrightarrow (B_1, B_2, B_3, B_4)$, because of an assumed 'internal' differentiation of D, respectively B, as representing instances of 'complete' or 'orthohomology' in GEGENBAUR's and JACOBSHAGEN's sense.

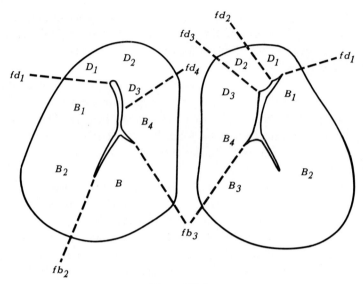

Figure 250 A

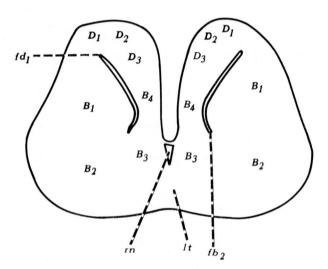

Figure 250 B

*Figure 250.* Cross-sections through the telencephalon of adult Chimaera Colliei (simplified after K. and NIIMI, 1969). A Rostral to lamina terminalis. B Through rostral portion of lamina terminalis. C Near caudal end of inverted paired hemispheres. D At rostral level of everted telencephalon impar. fo: 'fornix'; S: falx membranacea (septum membranaceum) of roof plate.

represents a groove delimiting pallium from basis within the medial wall of the rostral paired evaginated and inverted lobus hemisphaericus. The sulcus between everted pallium (neighborhood $D_3$ of alar plate), and attachment of lamina epithelialis (roof plate) is not morphologically or topologically comparable to $fd_4$, which is a groove *within* the (inverted) alar plate.

Again, eversion of the Osteichthyan brain may be accompanied by considerable thickening of the telencephalic wall, particularly in an in-

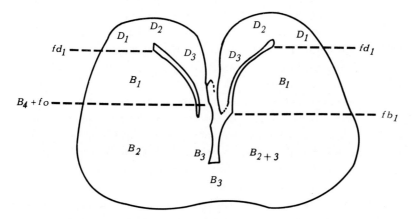

Figure 250 C

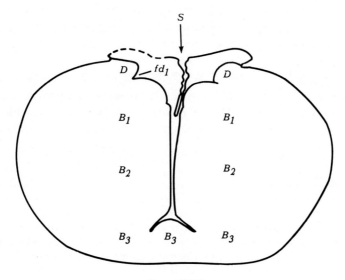

Figure 250 D

termediate region comprising the adjacent neighborhoods of pallium and basis as, e.g., especially in Teleosts, or the telencephalic wall may remain relatively thin as, e.g., in Polypterus. Moreover, the extent of lateral eversion (i.e. the degree of eversion) likewise displays, in accordance with different taxonomic forms, a wide range of variations.

In attempting to map the telencephalic wall neighborhoods, respectively their grisea, of the everted telencephalon of Osteichthyes upon the corresponding neighborhoods in Selachians and amphibians, a number of difficulties arise because of the peculiar griseal differentiation in the everted brain.[202] One of the main problems involves the recognition of an open boundary homologous to that between the pallial (D) and the basal (B) components of the inverted fish and tetrapod brains. Moreover, limiting grooves comparable to those of Selachians and amphibian brain are often lacking or poorly displayed in various forms of the everted Osteichthyan brain. There are, however, in some of these latter forms, at least in some regions of their telencephalon, cell-poor 'limiting' zones which may provide additional clues for a conceptual separation of certain grisea. Yet, these limiting zones are not entirely constant as regards their number, respectively their presence between particular griseal neighborhoods.

Concerning the 'limiting' grooves, a particular sulcus, separating a set of dorsolateral respectively lateral grisea from a set of medial, respectively basal ones, can, in our opinion, be considered to represent the sulcus $fd_1$. In many forms, the groove $fb_2$ of our interpretation, 'separating' the more dorsally located basal neighborhood $B_1$ from the more basal ones $B_2$ and $B_3$, is particularly conspicuous. In addition, some forms, especially Carassius auratus and Corydora, display both a roughly tripartite griseal subdivision and a suggestion of grooves within the pallium, apparently corresponding to the neighborhoods $D_1$, $D_2$, $D_3$ with their related grooves $fd_2$ and $fd_3$. Figures 251–259, illustrating the telencephalic configuration in a variety of

---

[202] Thus, NIEUWENHUYS (1966) comments: 'the forebrain' (apparently meaning the telencephalon) 'of teleosts cannot be directly compared with that in terrestrial vertebrates'. The author (K.), on the other hand, believes that a reasonably valid comparison by means of topologic mapping can be accomplished. It should, however, be added that NIEUWEN-HUYS' methodologic concepts as well as his interpretation of the everted Osteichthyan telencephalon (discounting the details related to a definition of the pallio-basal or D–B boundary) are essentially identical with those of my more than 40-year-old studies (K., 1924, 1929).

Ganoids and Teleosts, show the different basal and pallial neighborhoods with or without correlated 'limiting' sulci.

In some *Teleostean* forms, particularly in those whose telencephalon
comprises a caudal paired evagination or at least a fairly expanded caudal portion of the telencephalon, an additional, but transverse lateral
sulcus, connecting at roughly a right-angle with groove $fd_1$, may be
present, resulting in a T- or Y-shaped pattern. This is the '*sulcus ypsiliformis*' of GOLDSTEIN and VAN DER HORST, discussed and depicted by
KAPPERS (1921). If present (Figure 263 C), this groove can be interpreted as limiting the rostral portion of the pallium from a caudal one.

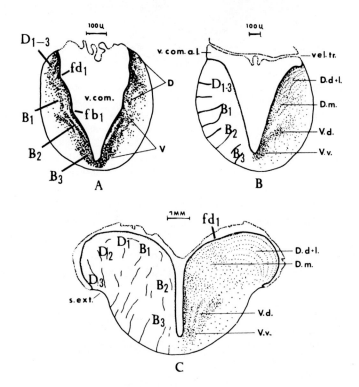

*Figure 251.* Cross-sections through the middle of the telencephalon in the Ganoid Acipenser ruthenus (after NIEUWENHUYS, 1963, modified by addition of my interpretation).
A Embryo of 14 days. B Embryo of 64 days. C. Adult. D: 'area dorsalis telencephali';
D.d.+l.: 'nucleus occupying the dorsal and lateral part of D; D.m.: 'medial part of D'; s.
ext.: 'sulcus externus'; v. com.: ventriculus communis; v. com. a.l.: 'angulus lateralis of
ventriculus communis'; V: area ventralis telencephali; V.d.: dorsal part of V.; V.v.: ventral part of V. The notations $D_{1-3}$, $B_{1-3}$, $fd_1$ and $fb_1$, added to NIEUWENHUYS' illustration,
indicate my interpretation.

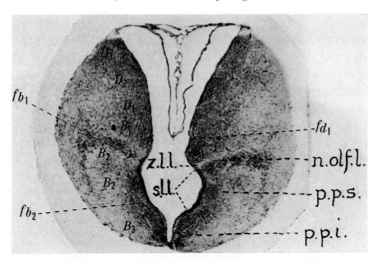

*Figure 252.* Cross-section through the telencephalon of the Ganoid Lepidosteus (gar-pike) at 'some distance' rostral to anterior commissure (after HOLMGREN, 1922, with added designations indicating my interpretation). n.olf.l.: 'nucleus olfactorius lateralis'; p.p.i.: 'corpus precommissurale pars inferior'; p.p.s.: corpus precommissurale pars superior; s.l.l.: 'sulcus limitans lateralis'; z.l.l.: 'zona limitans lateralis'. Although HOLMGREN point-ed out only one zona limitans ('lateralis') between $B_1$ and $B_2$, it can easily be seen that a sim-ilar limiting zone obtains between the upper ($B_{2'}$) and the lower part of $B_2$.

It does, however, not seem to be correlated with a significantly recog-nizable cytoarchitectural gradient within the longitudinal pallial zones.[202a]

Among *Ganoids*, the African Palaeoniscoid Polypterus is supposed to show various 'primitive' features, which, however, may well be

---

[202a] Since KAPPERS *et al.* (1936) had depicted an 'anterior limb of sulcus ypsiliformis' as a deep fissure corresponding to our $fd_1$, MILLER (1940) retained the term 'sulcus ypsili-formis' as a synonym for the anterior part of our $fd_1$. A true sulcus ypsiliformis was not seen in our specimens of Corydora.

---

*Figure 253.* Cross-section through the telencephalon of the Ganoid Amia calva (hema-toxylin-eosin, approx. ×38, red. $^2/_3$). A At rostral level of anterior commissure. B At rostral level of recessus praeopticus. C At slightly more caudal level, where pallium (D) and basis (B) become conspicuously demarcated by a depression (x) corresponding to the locus of $fd_1$ and perhaps related to the stem of a rudimentary 'sulcus ypsiliformis'; this de-marcation of 'pallium' and 'basis' is somewhat similar to that obtaining for Chimaera in fig-ure 250D. ca: commissura anterior; $fd_1$: locus of barely suggested groove $fd_1$; rp: recessus praeopticus; x: depression at locus of $fd_1$; y: attachment of roof plate to 'pallium'.

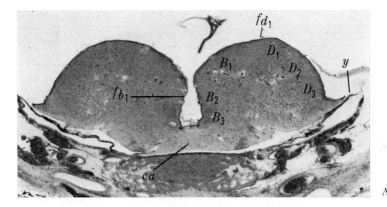

A

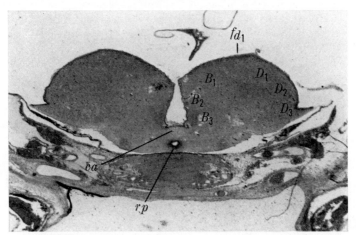

B

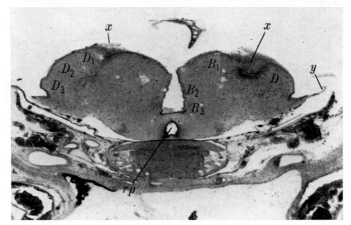

C

Figure 253

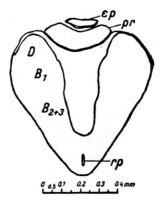

*Figure 254.* Cross-section through the forebrain in a 26 mm long embryo of the Iso-spondylous Teleost Salmo at a level of lamina terminalis including the rostral end of the hypothalamic (diencephalic) preoptic recess (from K., 1929b). ep: epiphysial stalk; pr: (probably) parencephalic roof; rp: recessus praeopticus.

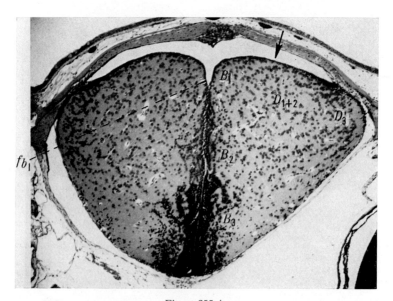

Figure 255 A

*Figure 255.* Cross-sections through telencephalon and forebrain of the adult Esocid Teleost, Trutta (trout). A Rostral to anterior commissure. B Through level of anterior commissure. C At level of preoptic recess (hematoxylin-eosin, about ×110, red. $^2/_3$). Arrows in A and B indicate approximate lateral boundary of $B_1$, which becomes, caudal to lamina terminalis, assimilated into a more or less joint oval nuclear mass formed by $B_1$ and $D_{1+3}$. ca: anterior commissure; rp: preoptic nucleus around obliterated recessus preopticus.

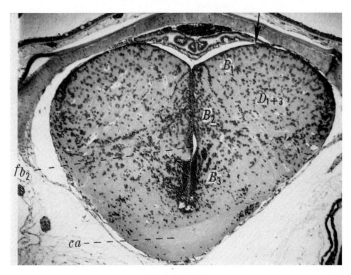

Figure 255 B

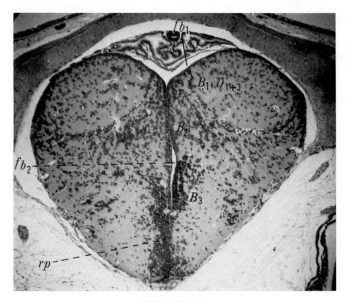

Figure 255 C

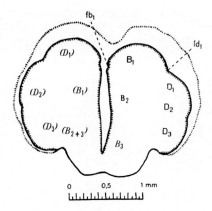

Figure 256 A

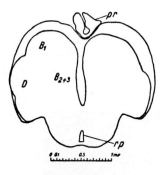

Figure 256 B

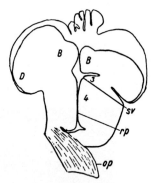

Figure 256 C

*Figure 256.* Cross-sections through the forebrain in the adult Cyprinoid Teleost, Carassius auratus (A modified after K., 1929a, B and C from K. 1929b). A Section closely anterior to rostral end of anterior commissure. On the left side, the notation in parentheses
conforms to HOLMGREN's interpretation of the Osteichthyan lobus hemisphaericus (cf. fig.
278, I). B Section at level of rostral end of preoptic recess. C Section at level just rostral to
paired (everted) caudal evagination. op: optic nerve respectively chiasma; rp: preoptic recess; sv: sulcus diencephalicus ventralis; 3: rostral end of thalamus ventralis; 4: hypothalamus; the groove between 3 and B can be interpreted as sulcus terminalis telencephali. On
left side of A and in accordance with terminology and interpretation of HOLMGREN, ($B_1$):
'nucleus olfactorius lateralis'; ($B_{2+3}$): 'corpus precommissurale', 'pars superior' (2) and
'pars inferior' (3); ($D_1$): 'pyriform pallium'; ($D_2$): 'general pallium'; ($D_3$): 'hippocampal
pallium'.

quite 'aberrant'. Relevant studies of its forebrain are those by HOLM-GREN (1922), and NIEUWENHUYS (1963, 1966 and other publications). In the definitive state, Polypterus (Figs. 260, 261) displays a rather uniform wall portion dorsally to a groove delimiting that region from basal grisea which I would interpret as representing $B_2$ and $B_3$. I be-

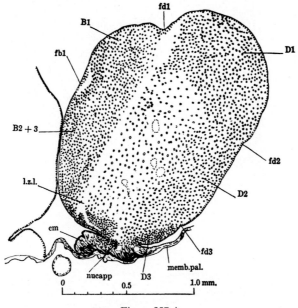

Figure 257 A

*Figure 257.* Cross-sections through the telencephalon and forebrain of the adult Siluroid Teleost, Corydora (from MILLER, 1940, with slight modifications of terminology in B, C, and D). A Rostral cross-section near attachment of olfactory stalk. B Caudal cross-section rostral to anterior commissure. C Cross-section at a rostral level of anterior commissure. D Cross-section through forebrain at level of preoptic recess and optic chiasma. bt: approximate region of junction between external pallio-basal sulcus and basal telo-diencephalic sulcus; cm: medial crus of olfactory stalk; co: optic chiasma; l.z.l.: lateral zona limitans (a somewhat comparable medial zona limitans between $B_1$ and $B_2$ can be noticed in cross-sections B, C, and D); memb. pal.: 'membranous pallium' (roof plate); nca: bed nucleus of commissura anterior (essentially a $B_3$ derivative); nen: (anterior) nucleus entopeduncularis here differentiated as massa cellularis reuniens, pars inferior s. hypothalamica); nm: 'mediocaudal nucleus' (of $D_3$); nucapp: 'nucleus olfactorius anterior, pars precommissuralis'; nup: preoptic nucleus; nut: 'nucleus taeniae' (perhaps pertaining to $D_3$ and not comparable with 'nucleus taeniae' of inverted Anamniote telencephalon). sa: so-called 'somatic area'; s.acc.: 'sulcus accessorius ($fb_2$)'; st: 'sulcus terminalis telencephali'; $B_1$ da: anteromedial part of $B_1$; $B_1$dp: dorsolateral subdivision, posterolateral part; $D_1$c: central subdivision; $D_1$e: subependymal subdivision (*mutatis mutandis* also for $D_2$ and $D_3$).

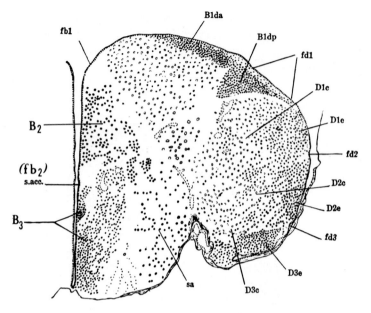

Figure 257 B

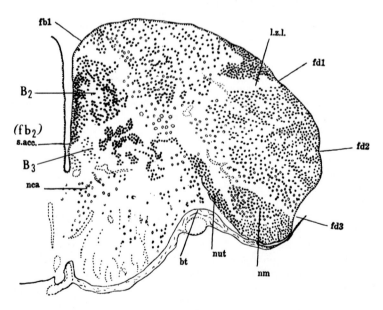

Figure 257 C

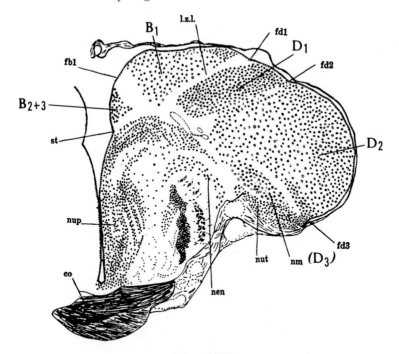

Figure 257 D

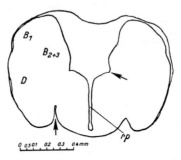

*Figure 258*. Cross-section through the forebrain in a 66 mm long embryo of the eel (Anguilla), an Apodan Teleost perhaps related to the Clupeids (from K., 1929b). rp: recessus praeopticus; upper arrow indicates 'sulcus terminalis', lower arrow points to external sulcus telo-diencephalicus.

lieve, therefore, that this groove should be evaluated as $fb_1$, and that the $B_1$ neighborhood has been secondarily 'absorbed', respectively 'included' into the D-complex. This seems to be suggested by findings in embryonic and some adult stages depicted by Nieuwenhuys *et al.* (1969) and Holmgren (1922) as shown, with my interpretation, in Figures 260 and 261. One could here assume, during morphogenesis, a 'blurred' differentiation of the area $B_1$, which constitutes an intermediate or 'transitional' open neighborhood between the D- and B-compo-

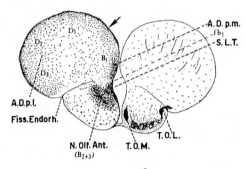

Figure 259 A

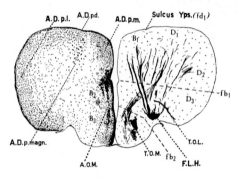

Figure 259 B

*Figure 259.* Cross-section through the telencephalon of the Mesichthyan Teleost, Gasterosteus aculeatus (stickleback), related to the Esocids according to a current classification, but formerly included under the Acantopterygii (after Nieuwenhuys, 1959, modified by addition of my interpretation). A Through rostral part of lobus hemisphaericus. B Through more caudal portion, but still rostral to anterior commissure. A.D.p.d.: 'pars dorsalis of area dorsalis'; A.D.p.l.: 'pars lateralis of area dorsalis'; A.D.p.magn.: 'pars magnocellularis of area dorsalis'; A.D.p.m.: 'pars medialis of area dorsalis'; A.O.M.: 'area olfactoria medialis'; Fiss. Endorh.: 'fissura endorhinalis' (external pallio-basal sulcus mihi, cf. fig. 257B, C); F.L.H.: lateral forebrain bundle; N. Olf. Ant.: 'nucleus olfactorius anterior; S.L.T.: 'sulcus limitans telencephali'; Sulcus Yps.: 'sulcus ypsiloniformis'; T.O.L.: 'tractus olfactorius lateralis; T.O.M.: 'tractus olfactorius medialis'. The other designations, which are self-explanatory, are my additions. Arrow in A indicates approximate boundary of $D_1$ and $B_1$, roughly corresponding to locus of groove $fd_1$, if present.

nents.[203] In consideration of the complex morphologic aspects exhibited by the Osteichthyan *skull* components, whose elements greatly vary as regards presence or reduction, one could also suppose that such 'fluctuations' occur with respect to *telencephalic grisea*. Accordingly, the configuration of the lobus hemisphaericus in Polypterus might be expressed by the two alternate bauplan formulae $D_{1+3} + B_1$, $B_2$, $B_3$, or assuming that $B_1$ has become essentially eliminated, $D_{1+3}$, $B_2$, $B_3$.

Concerning the D-components, adherents of the *Herrick, Crosby, Humphrey and Schnitzlein school of comparative neurology* have propounded an interpretation which, at least from the topologic viewpoint, appears to me inappropriate. Apparently because, in the inverted brain, the so-called 'primordium hippocampi' ($D_3$) is the most dorsomedial pallial griseum, a dorsomedial griseum of the everted Osteichthyan telencephalon is considered to represent said 'primordium hippocampi' (cf. FORTUYN, 1961) as shown in Figure 262. According to my own interpretation, this griseum (in the sunfish, *Eupomotis*) would not even pertain to the pallium, but presumably represents $B_1$.[204]

---

[203] Reference to this problem was made in volume 1, p.184 of the present series. It was there stated that NIEUWENHUYS' interpretation of my $B_1$ as a part of D might possibly be more justified. However, upon careful review of the data and rechecking of Osteichthyan material, I am inclined to uphold my original interpretation and 'to stick to my guns'. One could here add that (1) the region $B_4$ likewise represents, in the inverted brain, an intermediate neighborhood between basis and pallium, displaying, despite topologic invariance, a diversity of topographic configurational (shape) relationships with the adjacent neighborhoods $D_3$ and $B_3$. Again (2), the intermediate pallial neighborhood $D_1$, closely contiguous with $B_1$, becomes 'absorbed', respectively 'included', by 'introversion' or 'internation', into the basal neighborhood complex $B_1$ and $B_2$ of Amniota, as discussed further below. This morphogenetic behavior seems to indicate a somewhat *'labile'* character of the griseal neighborhoods $B_1$ and $D_1$ at the 'transition' from 'pallium' to 'basis', $B_1$ becoming 'pushed' toward the 'pallium' in Osteichthyes, and $D_1$ toward the 'basis' in Amniota.

[204] In another Teleost *(Perca flavescens)*, CROSBY et al. (1966, figs. 1B and C, loc. cit.) even identify a 'gyrus dentatus' (or 'primordial hippocampus, dorsomedial part') and a 'cornu Ammonis' (or 'primordial hippocampus, dorsal part'), comparable with the corresponding grisea of the mammalian hippocampal formation. In a topologic interpretation, however, I would evaluate the Teleostean 'gyrus dentatus' of the cited authors as $B_2$, and their 'cornu Ammonis' as $B_1$.

Although the *Chicago-Michigan-Alabama school of comparative neurology* has propounded, in its extensive studies, a very large number of minutely detailed griseal homologies applying, in terms of the mammalian brain, to the entire vertebrate series, I have been unable to find, in any of the publications of that school, a definite statement and rigorous

As regards the Osteichthyan basal neighborhoods $B_1$ and $B_2$, considerable regional differences in one and the same form, as well with respect to different forms, seem to obtain. This is manifested by a fairly uniform griseal arrangement $B_1 + B_2$, by a separation of both, or by a differentiation of $B_2$ and $B_3$ into further distinct cell groups.[205]

Without reference to the topologic notation, the formulation of a generally applicable and standardized terminology for the cell groups of the Osteichthyan brain, conforming to the designations suitable for Selachian, Dipnoan and amphibian grisea, involves two problems. First, the evident displacements related to *eversion*, and second, the very high and diversified *degree of griseal differentiation* particularly obtaining in the large group of Teleosts. Because the present section concerns merely the topologic features of the secondary telencephalic zonal system within the framework of the overall morphologic brain pattern, no detailed attempt at dealing with this problem can be made in this context. Further data on nuclear configuration and fiber connections of the Osteichthyan telencephalon shall be discussed in chapter XIII of volume 5 of this series. Nevertheless, for a preliminary formulation, the following adaptable terms, allowing further qualifications or subdivisions, can be suggested. *Griseum basale inferius* for derivatives of $B_3$, *griseum basale intermedium* for those of $B_2$, and *griseum basale superius* for those of $B_1$. The pallial components might then comprise *griseum mediale* ($D_1$), *dorsale* ($D_2$), et *laterale* ($D_3$) pallii. Obviously, the term 'nucleus', with further qualifications, might be substituted. Again it is evident that, in the everted pallium, $D_3$ assumes a lateral position, and $D_1$ a medial one, in contradistinction to the inverted pallium, in which $D_3$ becomes a griseum mediale, and $D_1$ a griseum laterale.

---

elaboration of the criteria for the concept of homology as used for the identification of grisea in the investigations by the members of that group. It is evident that no useful discussion of the relevant problems is possible unless the concepts under consideration are clarified by intelligible and unequivocal definitions.

[205] In our tentative first studies of the Osteichthyan telencephalic zonal system (K., 1929a; MILLER, 1940) we therefore chose to designate the set of neighborhoods basal to our sulcus $fb_1$ with the noncommittal term $B_{2+3}$. Although the reasons given (on pp. 26–27, K., 1929a) for the identification of $fd_1$ were actually valid for the limited material then at my disposal, and do not hold for forms with poorly displayed or missing ventricular sulci and nondescript griseal groupings, further studies concerning the mapping of Osteichthyan telencephalic neighborhoods have led me to uphold, on topologic grounds, my original identification of the pallio-basal boundary in these forms.

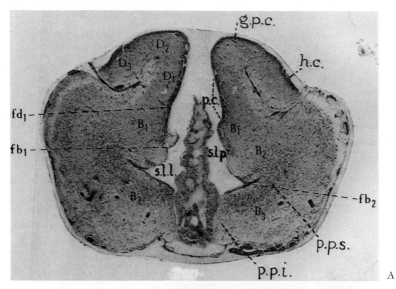

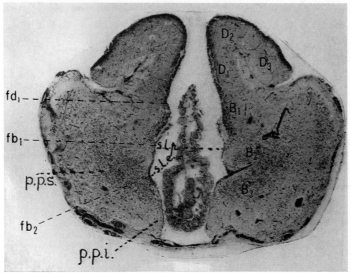

*Figure 260.* Cross-sections through the telencephalon in the Palaeoniscoid Ganoid, Polypterus (after HOLMGREN, 1922, modified by addition of my interpretation). A Section through about middle part of lobus hemisphaericus. B Section somewhat caudal to level of preceding figure. g.p.c.: 'general pallium cortex'; h.c.: 'hippocampal cortex'; p.c.: 'pyriform cortex'; p.p.i.: 'corpus precommissurale pars inferior'; p.p.s.: 'corpus precommissurale pars superior'; s.l.e., s.l.l.: 'sulcus limitans lateralis'; s.l.p.: 'sulcus limitans pallii'.

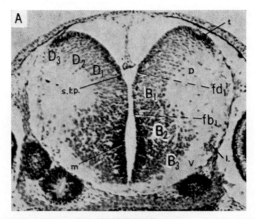

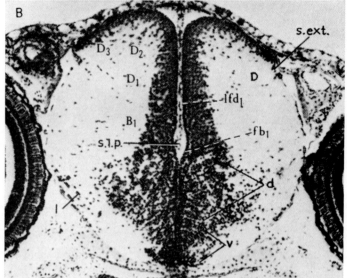

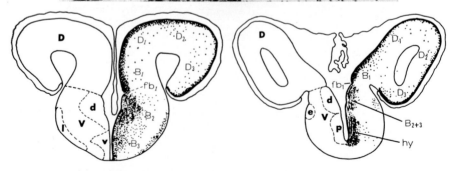

Figure 261 C                              Figure 261 D

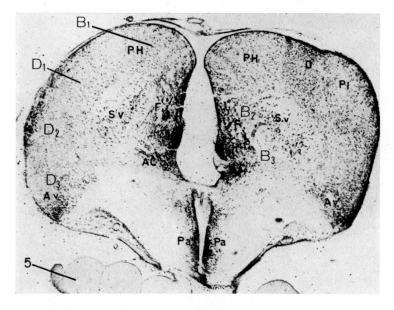

*Figure 262.* Cross-section through the forebrain of the sunfish, Eupomotis gibbosus, at a rostral level of diencephalic preoptic recess and caudal end of anterior commissure (after FORTUYN, 1961, modified by addition of my interpretation). Ac: 'bed nucleus of anterior commissure'; Av: 'amygdaloid complex, ventrolateral nucleus'; $B_{1, 2, 3}$: basal neighborhoods of telencephalic zonal system; D: 'dorsal area'; $D_{1, 2, 3}$: pallial neighborhoods of telencephalic zonal system; F: 'pars fimbrialis of primordium hippocampi'; Pa: 'parvocellular nucleus of preoptic area (preoptic recess)'; PH: 'primordium hippocampi, main part'; PI: 'piriform arca'; SV: 'caudoventral component of stria medullaris'.

*Figure 261.* Cross-sections through telencephalon of Polypterus (A, B after NIEUWENHUYS, BAUCHOT and ARNOLD, 1969; C, D after NIEUWENHUYS, 1966, modified by addition of my interpretation). A Telencephalon in a larva of 8.8-mm length. B Telencephalon in a larva of 17-mm length. C Telencephalon of adult Polypterus. D Forebrain of adult Polypterus at rostral level of diencephalic preoptic recess. D: area dorsalis telencephali or pallium according to NIEUWENHUYS; $D_1, D_2, D_3$ pallial longitudinal neighborhoods according to my interpretation; $B_1, B_2, B_3$: basal longitudinal zones according to my interpretation; V: area ventralis telencephali or subpallium according to NIEUWENHUYS; d: 'dorsal subpallial nucleus'; e: 'nucleus entopeduncularis'; $fb_1$, $fd_1$: sulci of telencephalic zonal system according to my interpretation; l: 'lateral subpallial nucleus'; $lfd_1$: locus of faded groove $fd_1$; m: 'matrix'; p: nucleus praeopticus; s. ext.: 'sulcus externus'; s.l.p.: 'sulcus limitans pallii lateralis' (it will be seen that, in my interpretation, the groove indicated by NIEUWENHUYS *et al.* as s.l.p. in A is not the same as that so indicated in B); v: 'ventral subpallial nucleus'.

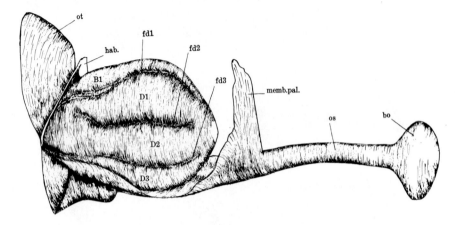

Figure 263 A

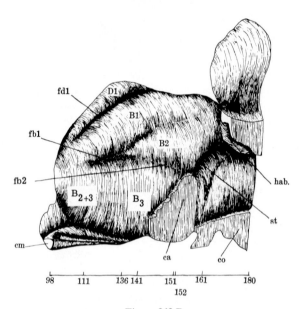

98    111    136 141    151    161    180
                        152

Figure 263 B

*Figure 263.* Wax-plate reconstructions of the Teleostean telencephalon. A Lateral view of the telencephalon in Corydora paliatus (after MILLER, 1940). B Medial view of telencephalon in Corydora (slightly modified after MILLER, 1940). C Lateral view of telencephalon in Monopterus albus (after VAN DER HORST, from KAPPERS, 1921). bo: bulbus olfactorius; ca: commissura anterior; co: optic chiasma; hab: ganglion habenulae; memb. pall.: epithelial roof plate of telencephalon; os: olfactory stalk; ot: optic tectum (tectum mesencephali); st: 'sulcus terminalis telencephali'; the designations pertaining to telencephalic zonal system in A and B are self-explanatory; septum (in C) presumably $B_1$.

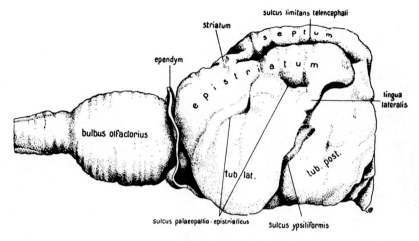

Figure 263 C


Reverting to the topologic aspect, the *bauplan* of the *Osteichthyan telencephalon* can be expressed by the general formula

$$O, D_1, D_2, D_3, B_1, B_2, B_3.$$

Emphasizing the reversal due to eversion, one could, of course, also write

$$O, D_3, D_2, D_1, B_1, B_2, B_3.$$

Moreover, taking into account forms with rather compact arrangements of basal cell groups, $B_{2+3}$ might be substituted for $B_2$, $B_3$ (cf. K., 1929a, MILLER, 1940). Figures 263 A, B show a wax plate reconstruction displaying the longitudinal arrangement of the fundamental telencephalic zones in the brain of the Teleost Corydora. A similar reconstruction of the telencephalon in the Teleost *Monopterus albus*, after VAN DER HORST, is depicted by Figure 263 C.

The generally extinct *Crossopterygians*, of which *Latimeria* is considered to be a surviving, recent form, are evaluated as 'phylogenetically related' both to Dipnoans and to the first tetrapod Anamnia. Yet, from the data published by MILLOT and ANTHONY (1965, 1966) it can easily be seen that the telencephalon of Latimeria, although displaying considerable pattern deformation, and thus manifesting an 'aberrant' aspect, is more closely related to the type obtaining in Ganoids or Teleosts.

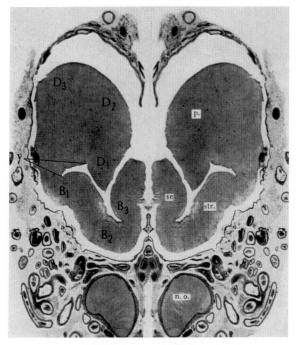                    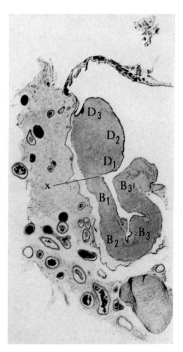

Figure 264 A                                        Figure 264 B

*Figure 264.* Two cross-sections through the telencephalon of Latimeria (from MILLOT and ANTHONY; A, 1966; B, 1965; modified by my added interpretation). p.: pallium; se: 'septum'; str.: 'striatum'; n.o. optic nerve; x: probably groove $fd_1$; y: grooves probably representing a modified $fd_1$ system; $B_{3'}$: it is doubtful whether this dorsal part of $B_3$ should or should not be evaluated as a true $B_4$ neighborhood.

Discounting the olfactory bulb, which is described as being located in the nasal capsule and connected with the lobus hemisphaericus by a long olfactory stalk, almost the entire telencephalon of Latimeria is un-paired evaginated and everted, thus representing a telencephalon medium sive impar, shown in Figure 264. At the rostral end of lobus hemisphaericus, however, a relatively small rostral paired evaginated, but likewise everted portion seems to protrude beyond the anterior end of lamina terminalis.[205a]

---

[205a] Since the author had no opportunity to study original material of the brain in Latimeria, his evaluation of the telencephalon in this peculiar form is based on the figures shown in MILLOT's and ANTHONY's publications (1965, 1966).

The thick pallium, corresponding to our D-zones ($D_{1-3}$), manifests, however, a considerable *internation* or *introversion* in the region which may mainly include the $D_1$ neighborhood of our terminology. Again, the so-called 'septum', which can be interpreted as essentially our $B_3$ zone, forms a prominent bulging ridge on both sides of the midline. It includes a basal and a dorsal part. Depending on arbitrary criteria, the dorsal portion of $B_3$ could or could not be evaluated as $B_4$.[206]

In cross-sections (Fig. 264), the lumen of the ventriculus impar or communis telencephali displays a number of diverticula. The dorsolateral diverticulum is bounded by the attachment of lamina epithelialis, the subpallial diverticulum, whose bottom grooves presumably correspond to manifestations of our sulcus $fd_1$, is approximately located between pallium and basis, and the basal diverticulum, perhaps our sulcus $fb_2$, is found within the B-neighborhoods. At different transverse levels, additional parallel sulci can ben seen. Besides these paired diverticula, an unpaired median one is seen in the midline between the bulging $B_3$ protrusions. Despite pattern distortion, the comprehensive bauplan formula O, $D_3$, $D_2$, $D_1$, $B_1$, $B_2$, $B_3$ ($B_4$ ?) can be said to hold for Latimeria.

Generally speaking, the *Dipnoan* telencephalon consists of a rostral bulbus olfactorius and a caudal lobus hemisphaericus. This latter, in turn, is rostrally paired evaginated and *inverted*, and caudally[206a] unpaired evaginated with some degree of *eversion*. A caudal paired evagination, with polus posterior, does not obtain. The Dipnoan telencephalon shows thereby similarities with that of Elasmobranchs, and, as regards *eversion*, with the telencephalon impar of Holocephalians. Again, concerning the cellular arrangements, some similarities with those displayed by both the amphibian and the Selachian telencephalon could be said to obtain.

In *Protopterus* and *Lepidosiren* (Figs. 265, 266, 267) the pallial and the basal wall of lobus hemisphaericus are characterized by the pres-

---

[206] If medial connectedness with $D_3$ in the rostral paired evaginated and inverted hemisphere is an essential characteristic of zone $B_4$, then such neighborhood does not obtain in Latimeria. If $B_4$ is defined as merely a distinctive and substantial dorsal subdivision of $B_3$, then a $B_4$ neighborhood could be described in Latimeria.

[206a] The term 'caudally' denotes here merely the relationship between two portions (anterior and posterior) of a *rostrally* evaginated telencephalon, and does not imply the meaning of 'true' caudal evagination referred to in the next sentence. Any *unpaired* telencephalic evagination whose *floor* is provided by the expanded lamina terminalis must necessarily be directed rostralward.

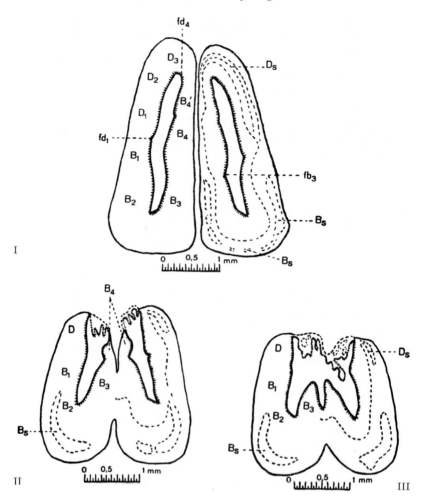

*Figure 265.* Three outline cross-sections through the telencephalon of Protopterus annectens, indicating the neighborhoods of the telencephalic zonal system (from K., 1929a).

ence of a cortical or corticoid cell plate, separated from the periventricular griseum by a relatively cell-free layer. The corticoid arrangement
is more pronounced in the pallium than in the basis, and again, more
distinct in very young adults than in older ones.[207]

---

[207] Both dorsal and basal corticoid plates display a tendency toward disruption into
'islets' of cell aggregates, not unlike those obtaining (as *islands of Calleja*) in the basal
cortex (cortex olfactoria, tuberculum olfactorium) of mammals. The dorsal corticoid
layer tends to join the periventricular layer at both lateral and medial pallio-basal boundaries. A similar tendency may be shown by the basal corticoid layer in the $B_1$ region.

The neighborhoods $D_1$, $D_2$ and $D_3$ do not show distinct cytoarchitectural boundaries, neither do the basal neighborhoods $B_1$, $B_2$ and $B_3$. Nevertheless, thickness and density gradients of cell aggregates are suggested. However, a detailed parcellation into numerous 'nuclei' or 'area', such as described by RUDEBECK (1945) and other authors, can

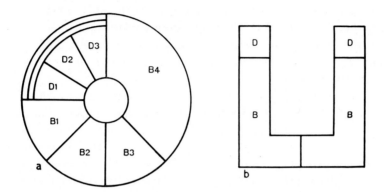

*Figure 266.* Diagrams outlining the telencephalic zonal system of Protopterus in cross-sections (from GERLACH, 1933). a At level of rostral paired evaginated and inverted hemisphere. b At level of unpaired evaginated and slightly everted hemisphere, approximately corresponding to figure 265 III.

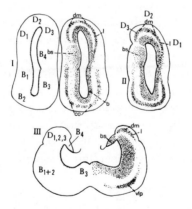

*Figure 267.* Cross-sections through the telencephalon of Lepidosiren (from K., 1927, slightly modified by addition of the topologic notation). I Level at rostral third of lobus hemisphaericus. Il Slightly rostral to lamina terminalis. III At level of lamina terminalis and telencephalon impar with slight eversion. b: 'nucleus basalis'; bs: nucleus basimedialis superior; co: 'cortex olfactoria'; dm: 'cortex dorsomedialis' ($D_{3+2}$); l: 'cortex lateralis' ($D_1$); vlp: area ventrolateralis posterior.

hardly be regarded as very convincing. Neighborhood $B_4$ is represented by a rather conspicuous and large dorsomedial cell group with a
ventral and a dorsal subdivision. Both are partly coextensive with
slight laterad archings into the ventricular lumen. The connectedness
relationships of this subdivided dorsomedial longitudinal zone to pallium and basis at the levels of telencephalon impar (cf. Figs. 265, 266)
unequivocally establish its identity as neighborhood $B_4$. ELLIOT SMITH
(1908) and HOLMGREN (1922) justly interpreted this wall region as 'corpus paraterminale', respectively septum or fimbrial portion of septum,
namely as configurations pertaining to the $B_4$ zone of the topologic telencephalic pattern.[208]

The ventricular grooves are, in general, indistinctly evolved, except for $fd_1$, $fb_3$ and to some extent $fd_4$. Relatively cell-poor zonae limitantes (lateralis and medialis) as noticed by HOLMGREN (1922) may occur in correlation with $fd_1$ and $fd_4$, respectively. The just-quoted author's sulcus septalis represents our $fb_3$. The angular grooves of the
paired lateral ventricle (Ventrikelumschlagsfurchen fad and fav) may or
may not closely coincide with $fd_4$ dorsally and $fb_2$ ventrally.

On the whole, the African *Protopterus* and the South American
*Lepidosiren* show a rather similar overall shaping of the telencephalon,
while the Australian *Ceratodus* (Fig. 268) displays, as regards its shape,
a somewhat different transformation of the topologically invariant
Dipnoan telencephalic bauplan. The wall of the main, paired evaginated portion of lobus hemisphaericus appears here more 'flabby' than in
the two other forms. It includes, as already mentioned in section 1 B on
general morphogenesis, a component of roof plate (lamina epithelialis)
in its medial wall (Fig. 58 C). This structure was described as 'lingula
interolfactoria (BING and BURCKHARDT, 1905; HOLMGREN and VAN
DER HORST, 1925) or septum ependymale (NIEUWENHUYS and HICKEY,
1965). In the dorsal $B_1$ region, an external sulcus or fold corresponds to
a bulging, without thickening, of the 'flabby' wall into the ventricle.
The cellular arrangements, including a tendency toward corticoid

---

[208] Because of its dorsal position and a superficial similarity with the so-called
primordium hippocampi ($D_3$-neighborhood) of amphibians, the author, during his early
comparative studies about fifty years ago, interpreted at first the $B_4$ zone of Dipnoans as
being homologous to the amphibian 'area medialis pallii' (K., 1922), but, upon more
detailed analysis, adopted ELLIOT SMITH's and HOLMGREN's interpretation which may be
considered conclusive. H.N.SCHNITZLEIN and E.C.CROSBY (The telencephalon of the
lungfish, *Protopterus*. J. Hirnforsch. *9:* 105–149, 1967), however, still interpret our $B_4$
neighborhood as 'cornu ammonis' and 'gyrus dentatus', i.e. 'area medialis pallii'.

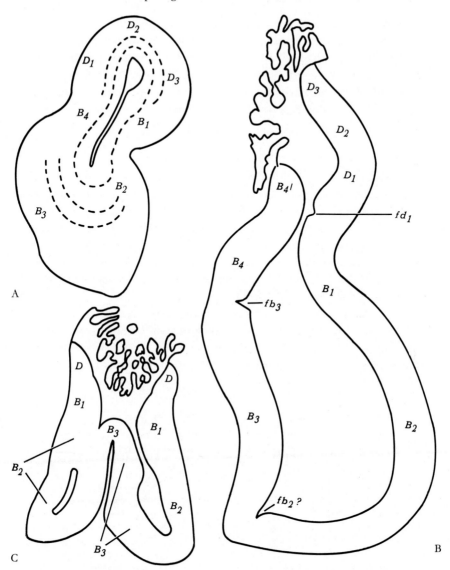

*Figure 268.* Three cross-sections through the lobus hemisphaericus of Ceratodus (A B, modified after NIEUWENHUYS and HICKEY, 1964; C modified after HOLMGREN and VAN DER HORST, 1925, all reinterpreted in accordance with topologic neighborhood notation). A Through rostral part of lobus hemisphaericus. B At somewhat more caudal level of rostrally paired evaginated hemisphere. C At level of lamina terminalis and telencephalon impar with slight eversion, which is somewhat more pronounced than that shown in Lepidosiren (fig. 267, III).

'laminae' or 'clusters', are similar to those in Protopterus and Lepidosiren, but apparently somewhat more 'blurred' or 'nondescript'.[209]

Emphasizing the endbrain configuration of Protopterus and Lepidosiren, and taking into consideration the definite tendency toward corticoid plates displayed by Ceratodus, the bauplan of the Dipnoan telencephalon can be expressed by the formula

$$O \; \frac{D_{1+2+3}}{D_s}, \; \frac{B_1, B_2, B_3}{B_s}, \; B_4.$$

A diagram of the Dipnoan telencephalic pattern, as mapped with respect to Protopterus by GERLACH (1933), is shown in Figure 266. Again, in a purely descriptive terminology of the telencephalic grisea, the designations medial, dorsal (intermediate) and lateral parts of dorsomedial corticoid 'cell plate' or of pallial periventricular cell layer might be used. For the basal components, *'nucleus' basilateralis superior* ($B_1$), *basilateralis inferior* ($B_2$), *basimedialis inferior* ($B_3$), *basimedialis superior* ($B_4$) and basal cortex ($B_s$) seem appropriate.

The *Amphibian telencephalon* is, in some respects, similar to that of Selachians, being rostrally paired evaginated and inverted, and displaying a likewise inverted telencephalon medium s. impar, which is quite conspicuous although of relatively moderate dimension. The amphibi-

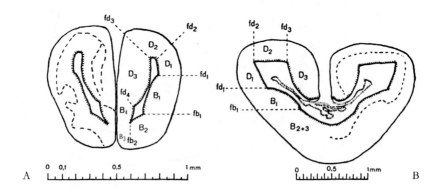

*Figure 269.* Cross-sections through the telencephalon of urodele amphibians (from K., 1929a). A Rostral paired evaginated hemisphere of Spelerpes fuscus. B At level of telencephalon impar of Triton.

---

[209] Unfortunately, I have been unable to procure, or to examine, original material of Ceratodus (*Neoceratodus forsteri*).

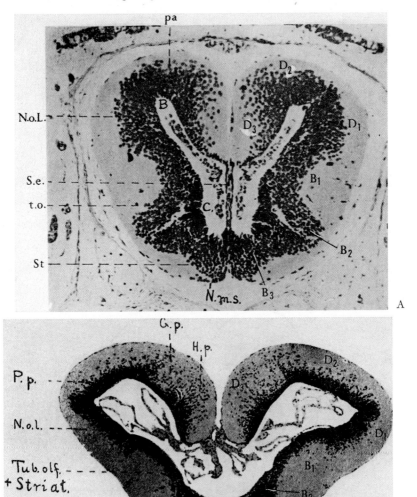

*Figure 270.* Cross-sections through the telencephalon of urodele amphibians (from SÖDERBERG, 1922, modified by addition of topologic neighborhood notation). A Telencephalon of advanced, 2-cm Triton larva rostral to interventricular foramen. B Telencephalon impar of adult Salamandra. B: 'dorsal groove', interpreted by SÖDERBERG as delimiting 'pallium' from 'subpallium' and as fd₂ by myself; C: 'ventral groove', interpreted by SÖDERBERG as delimiting 'striatum' from 'septum', and as fb₂ by myself; G.p.: 'general pallium'; H.p.: 'hippocampal pallium'; N.m.s.: 'nucleus medialis septi'; N.o.l.: 'nucleus olfactorius lateralis; Pa: 'pallium'; P.p.: 'pyriformal pallium'; S.e.: 'septum ependymale'; St., Sept.: 'septal parts'; Striat.: 'Striatum'; t.o., Tub. olf.: 'tuberculum olfactorium'.

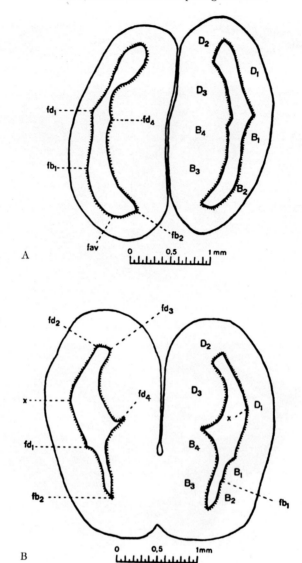

*Figure 271.* Cross-sections through paired evaginated telencephalon of the frog, Rana fusca (from K., 1929a). Section B, caudal to A, includes a rostral neighborhood of lamina terminalis; ✕: accessory sulci.

an hemispheres differ from the Selachian and Dipnoan ones by the presence of a caudal paired evaginated and inverted component with polus posterior, and containing a paired recessus posterior ('posterior horn') of the lateral ventricle. Also, in contradistinction to true Selachians, whose olfactory bulb has a lateral position, the amphibian bulbus olfactorius represents the rostral end of the telencephalon.

As regards the distribution of cellular elements within the telencephalic wall, there is little tendency toward a corticoid arrangement, although in certain larval stage, particularly of Anura, an abortive and transitory lamination of periventricular cell masses is faintly suggested. In adult Gymnophiona, however, a ventral basilateral $(B_2)$ rudimentary corticoid formation, and a still less pronounced, but 'near-corticoid' arrangement in the posterior dorsomedial area $(D_3)$ can be seen.

The pallial or D-region displays three cytoarchitecturally differing neighborhoods, which are particularly distinct in Urodeles (Figs. 269, 270). Three layers can be recognized, (1) a dense periventricular layer; (2) a much less dense dispersal layer *(Schwärmschicht*, K., 1921), and (3) a cell-poor molecular or zonal layer containing the bulk of the fiber systems. The three neighborhoods, area lateralis pallii $(D_1)$, area dorsalis pallii $(D_2)$ and area medialis pallii $(D_3)$ are characterized by the relative thickness of these layers, of which (1) decreases, and (2) increases in passing from $D_1$ to $D_3$. A zona limitans occurs between $D_3$ and $B_4$ in all three orders; another such zone, between $D_1$ and $B_1$, obtains in most Anurans (Figs. 271, 272).

The groove $fd_1$ is generally quite distinct. In Anurans, however, $fd_2$ and $fd_3$ are frequently less well recognizable, and their manifestation may be blurred by the pronounced dorsal ventricular angle (fad). Again, groove $fd_4$ is quite distinct in Urodeles and Anurans; it can be recognized in some Gymnophione larvae (Fig. 273) but tends to disappear in adults because of the considerable internation of $D_3$ and $B_4$, whereby a single bulge is formed, as pointed out further below.

The basal or B-region can be subdivided into $B_1$, $B_2$, $B_3$ and $B_4$. Grooves $fb_1$, $fb_2$, and $fb_3$ may, in some instances, be rather poorly suggested; $fb_2$ can coincide with the angulus ventralis ventriculi (fav). Especially in Urodeles, $B_2$ and $B_3$ tend to form a single griseum at levels of telencephalon impar. The caudal paired evagination of the hemisphere contains, in Urodeles and Anurans, essentially the pallial neighborhoods $D_1$–$D_3$ with a minor and variable caudal extension of the conjoint basal Grundbestandteile $B_{1+3}$. In Gymnophiona, however, the neighborhoods $B_1$, $B_2$, $B_3$ remain rather distinctly differentiated at these caudal levels. The telencephalon of Gymnophiona, moreover, displays a peculiar medial concavity of the hemispheric ventricle, correlated with a high degree of inversion. The medial concavity corresponds to a considerable bulge jointly formed by the adjacent longitudinal zones

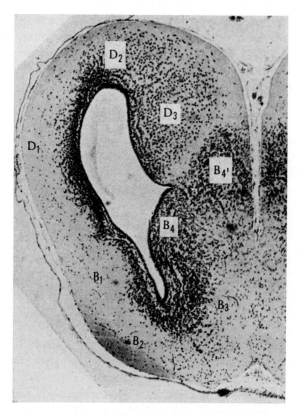

Figure 272 A

*Figure 272.* Cross-sections through the telencephalon in anuran Amphibia (A and B reinterpreted after KIESEWALTER, 1928, C modified after SÖDERBERG, 1922, from K., 1927). A Rostral paired evaginated hemisphere of adult frog. The ventricular grooves (not labeled) can easily be identified by comparing with figure 271 A and B. B Caudal paired eva-ginated hemisphere of frog. The left half of the section passes just caudal to interventricular foramen. C Telencephalon impar in a frog larva of 3.5 cm length. b: basal grisea ($B_{1, 2, 3}$); bs: supraforaminal portion of nucleus basimedialis superior; d: area dorsalis pallii ($D_2$); l: area lateralis pallii ($D_1$); m: area medialis pallii ($D_3$); nc: 'nucleus commissurae hippo-campi' (a diencephalic griseum representing a rostral component of eminentia thalami ven-tralis); pr: nucleus praeopticus (hypothalamus); vlp: area ventrolateralis posterior (exter-nal portion of $B_{2+3}$); the midline portion of $B_3$ in the lamina terminalis forms an interstitial griseum of commissura anterior; $B_4'$, $B_4''$: pars fimbrialis septi.

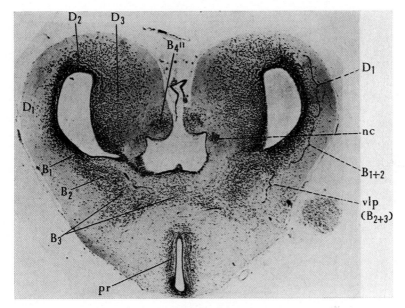

Figure 272 B

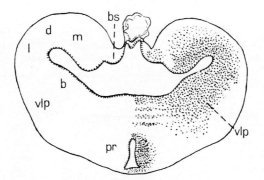

Figure 272 C

$D_3$ and $B_4$, whose 'limiting sulcus' $fd_4$ disappears, but whose zona limitans remains conspicuous (Figs. 273–275). A caudal part of $B_4$ commonly extends into the dorsomedial wall of the telencephalon impar and even as far as rostral levels of the caudally paired evaginated hemispheres. This cell group, particularly noticeable in Anurans, is the so-called '*pars fimbrialis septi*' (cf. Fig. 272).

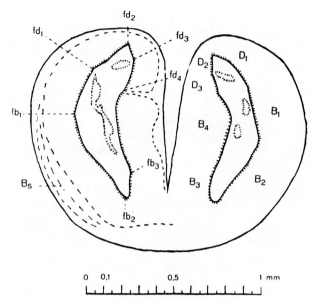

*Figure 273.* Cross-section through the telencephalon in an advanced larva of the Gymnophione Ichthyophis glutinosus (from K., 1929a).

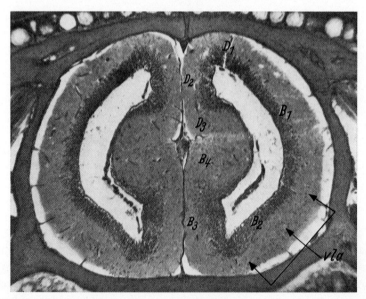

*Figure 274.* Cross-section through the telencephalon of the Gymnophione Siphonops annulatus slightly rostral to lamina terminalis and interventricular foramen (after K. *et al.*, 1966). vla: area ventrolateralis anterior ($B_s$, co).

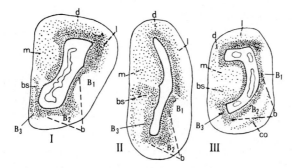

*Figure 275.* Cross-sections through the rostral paired evaginated telencephalon in the three orders of extant Amphibia (from K., 1927, with some added designations). I Salamandra (urodele). II Frog (anuran). III Siphonops (Gymnophione). b: region of basal grisea ($B_1$, $B_2$, $B_3$); bs: nucleus basimedialis superior ($B_4$); co: cortex olfactoria; d: area dorsalis pallii ($D_2$); l: area lateralis pallii ($D_1$); m: area medialis pallii ($D_3$, so-called 'primordium hippocampi').

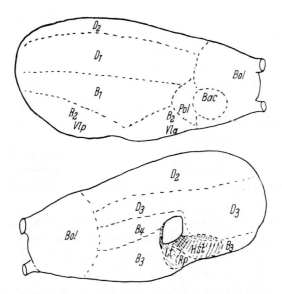

*Figure 276.* Mapping of the telencephalic longitudinal zones upon lateral and medial aspect of the hemisphere of the Gymnophione Amphibian, Siphonops (from K. *et al.*, 1966). Bac: bulbus olfactorius accessorius; Bol: bulbus olfactorius; Hst: hemispheric stalk; Lt: lamina terminalis; Pol: nucleus postolfactorius lateralis; Rp: preoptic recess; vla: area ventralis (or ventrolateralis) anterior; vlp: area ventrolateralis posterior.

On the basis of the cellular arrangements obtaining in the amphibian hemispheres, the telencephalic bauplan of the three extant orders can be expressed by the following formulae.

Urodeles $\quad$ $O, D_1, D_2, D_3, B_1, B_2, B_3, B_4$ respectively $(B_{3+4})$

Anurans $\quad$ $O, D_1, D_2, D_3, B_1, B_2, B_3, B_4$

Gymnophiona $O, D_1, D_2, D_3, \dfrac{B_2}{B_s}, B_3, B_4.$

A mapping of these longitudinal zones for Gymnophiona is shown in Figure 276. For a descriptive terminology of grisea, consistent with the topologic bauplan notations, the following designations, which, if required, allow for further subdivisions by appropriate qualifying additions, can be suggested. In the pallium: area lateralis pallii $(D_1)$, area dorsalis pallii $(D_2)$, area medialis pallii $(D_3)$. In the basis: nucleus basilateralis superior $(B_1)$, nucleus basilateralis inferior $(B_2)$, nucleus basimedialis inferior $(B_3)$, nucleus basimedialis superior $(B_4)$. Transitional neighborhoods between olfactory bulb and lobus hemisphericus may be designated as 'nucleus olfactorius anterior' in general agreement, but discounting some details, with views of HERRICK. Such transitional lateral and medial neighborhoods include, inter alia, the grisea which we described as 'nucleus postolfactorius medialis', respectively 'lateralis' (cf., e.g., K. et al., 1966).

Also, a manifestation of basal cortical or corticoid arrangements in the $B_2$ zone (and adjacent parts of $B_1$ or $B_3$) of the rostral paired evaginated hemisphere may be designated as area ventralis anterior. A more caudal cellular arrangement of this type, at levels of telencephalon impar and of caudal paired evagination, can be termed area ventrolateralis posterior. These formations are rather conspicuous in Gymnophiona, but may likewise be recognized, although somewhat faintly and variably outlined, as rather transitory features in advanced Anuran larvae.

Concluding this topologic evaluation of the griseal neighborhoods displayed in the telencephalon of Anamnia ('lower vertebrates') a comparison with earlier basic concepts proposed by HERRICK (1910) and by HOLMGREN (1922) seems perhaps expedient.

Figure 277 A discloses that, at the time of his important contribution to forebrain morphology in Amphibia, HERRICK (1910) distinguished four main regions of the amphibian telencephalon, namely two dorsal and two ventral ones. The former included the dorsomedial 'primordium hippocampi' and the 'pars dorsolateralis', while the basal regions comprised 'pars ventromedialis' and 'pars ventrolateralis'. As can be seen from Figure 277 II, HERRICK attempted, moreover, to correlate these four telencephalic regions with the four diencephalic zones, one of which (thalamus ventralis s. 'pars ventralis thalami') he had then identified and described as a previously not recognized significant morphologic component. This attempt at telencephalo-diencephalic correlation, however, could not be upheld on the basis of either configurational aspects or of specific fiber connections.

Among other authors who had reported on the distribution of cell masses in the amphibian telencephalon, without, however, introducing new criteria for a relevant morphologic subdivision, the following should be particularly mentioned: GAUPP (1899), RÖTHIG (1912), BINDEWALD (1914), and KAPPERS and HAMMER (1918). References to their publications may be found in volume 2 of KAPPERS' '*Vergleichende Anatomie des Nervensystems*' (1921) and in the papers by KUHLENBECK (1921) and SÖDERBERG (1922).

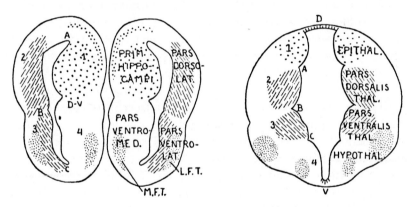

Figure 277 I                    Figure 277 II

*Figure 277.* HERRICK's concept of telencephalic and diencephalic longitudinal zones of amphibians and of their assumed correspondence (from HERRICK, 1910). I Diagrammatic cross-section through the frog's telencephalon 'in front of the lamina terminalis'. II Diagrammatic, slightly oblique cross-section through the diencephalon in Urodela and Anura. The numbers 1, 2, 3, 4, and the letters A, B, C mark corresponding structures in figures I and II, 'the two figures being designed to illustrate the way in which the cerebral hemispheres have been formed by the lateral evagination of the walls of the neural tube'. In figure I: A: dorsal angle of hemisphere; B: zona limitans lateralis and fissura endorhinalis; C: ventral angle of hemisphere; D–V: zona limitans medialis and fissura limitans hippocampi; L.F.T.: lateral forebrain tract; M.F.T.: medial forebrain tract. In figure II: A: sulcus diencephalicus dorsalis; B: sulcus diencephalicus medius; C: sulcus diencephalicus ventralis; D: roof plate; V: floor plate.

Subsequent investigations concerning the configurational connectedness of cytoarchitecturally distinctive grisea (K., 1921, 1922) disclosed three rather distinctive neighborhoods in the amphibian pallium, namely area medialis, dorsalis, and lateralis. SÖDERBERG (1922), in accordance with views of HOLMGREN (1922) likewise subdivided the amphibian pallium into three parts, namely lateral 'pyriform pallium', dorsal 'general pallium', and medial 'hippocampal pallium'. Although, *prima facie*, this would correspond to the subdivision into area lateralis ($D_1$), area dorsalis ($D_2$) and area medialis ($D_3$), SÖDERBERG's 'pallium' included only area dorsalis (her 'pyriform pallium') and area medialis, the dorsal part of which was her 'general pallium', the basal part being her 'hippocampal pallium' (cf., e.g. Fig. 29, p. 113 of her contribution), while area lateralis pallii (K., 1921) was interpreted as 'subpallial' and termed 'nucleus olfactorius lateralis'.

In his later contributions, as summarized in a monograph on the brain of the tiger salamander, HERRICK (1948) adopted a subdivision of the pallium into 'hippocampal pallium', 'dorsal pallium', and 'pyriform pallium', corresponding to our area medialis, dorsalis, and lateralis, respectively. Although using the term nucleus olfactorius dorsolateralis as synonymous with primordium piriforme, he thus, like myself (K., 1921), evaluated this area as pallial.

An appropriate further subdivision of HERRICK's (1910) ventrolateral and ventromedial wall portions of the telencephalon proved to be a much more difficult task because of the rather nondescript cellular arrangements for which, moreover, a highly confusing nomenclature had been introduced in the literature. Major factors for this confusion were the various attempts, based on vague 'intuition', to recognize, in the amphibian and generally the Anamniote hemispheres, configurations such as 'corpus striatum', 'septum', 'nucleus caudatus', 'amygdala', 'epistriatum', and lobus piriformis, characteristic for mammalian or ('epistriatum') reptilian brains. After a number of similar attempts and frustrations[210] in following these precedents, the approach based on a delimitation of topologic neighborhoods[211] and the study of key stages

---

[210] These attempts, however, may be regarded as representing a succession of increasingly useful partially valid 'first approximations'. One might also liken these attempts with the struggle to solve a difficult *'jigsaw puzzle'*.

[211] Like *Monsieur Jourdain*, who did not realize that he was talking prose (MOLIÈRE's *'Le Bourgeois Gentilhomme', acte II, scène 6)*, I did not realize that I was following elementary topologic procedures. At that time, more than 40 years ago, I was mathematically rather

in Amniote ontogenesis (K., 1925, 1929a) finally resulted in what I believe to be a reasonably valid interpretation since it can now be supported by fairly rigorous topologic criteria involving the connectedness of relevant neural tube wall neighborhoods. HERRICK's ventrolateral region is thereby conceived as consiting of $B_1$ (nucleus basilateralis superior) and $B_2$ (nucleus basilateralis inferior). The ventromedial region, inappropriately designated as 'septum', is conceived to include $B_3$ (nucleus basimedialis inferior) and $B_4$ (nucleus basimedialis superior).[212]

HOLMGREN (1922) published a significant contribution to forebrain morphology in lower vertebrates (Anamnia) referring to Cyclostomata, Selachians (including Holocephalians), Dipnoans, and to the peculiar everted telencephalon of Polypterus, and of other Osteichthyes, omitting, however, Amphibians, which had been dealt with by his pupil SÖDERBERG (1922).

The pattern of the paired or unpaired evaginated and inverted lobus hemisphaericus, as conceived by HOLMGREN, is shown in Figure 278. As regards the inverted Selachian brain, I fully agree with HOLMGREN's *subdivision*, although disagreeing with his *interpretation* of the lateral pallial area as 'pyriform pallium'.[213] Moreover, HOLMGREN's subdivision of the basis seems to lump together $B_2$ and $B_3$ as 'tubercu-

---

naive, and, moreover, particularly impressed by some aspects of '*gestalt*' *theory*. Paraphrasing MOLIÈRE, I could now say: '*Par ma foi, il y a plus de quarante ans que je commençai a faire de la topologie, sans que j'en susse rien*'.

[212] $B_{3+4}$ represent thus the so-called 'septum' *sensu latiori*, $B_3$ corresponding to 'nucleus accumbens septi' and the $B_4$ derivatives to the 'septum' *sensu strictiori*. $B_4$, however, does not seem to become differentiated in the unpaired evaginated, everted lobus hemisphaericus of Osteichthyes, since it develops as a dorsomedial neighborhood of connectedness between $D_3$ and $B_3$, requiring an inverted evagination with an at least significant degree of rostral paired hemispheric configuration. ELLIOT SMITH's term (1903) '*paraterminal body*' is, both *sensu latiori* ($B_3+B_4$) and *strictiori* ($B_4$), far better than the indiscriminately used designation 'septum'. Preferably, the application of the term 'septum' should be restricted to a particular mammalian configuration (cf. K., 1969).

[213] As will be elaborated further below, the neighborhood $D_1$ of Anamnia should *not* be considered homologous to the mammalian lobus piriformis, respectively to its cortex. An evaluation of key stages in Amniote telencephalic ontogenesis leads to the conclusion that $D_1$, by '*internation*' or '*introversion*', becomes included into the basal nuclei complex of Sauropsida and Mammalia, and that the mammalian piriform lobe cortex represents essentially a lateral derivative of neighborhood $D_2$ with only a minor contribution of migrated dorsolateral respectively lateral $D_1$ elements.

lum olfactorium'. HOLMGREN's nucleus olfactorius lateralis is here evidently our $B_1$ or nucleus basilateralis superior, and his 'nucleus taeniae' appears to be a caudal component of $B_2$, corresponding to our 'area ventrolateralis posterior'.

As regards HOLMGREN's concept of the everted telencephalon, the *diagram* of Figure 278 seems likewise, *prima facie*, and discounting the implied interpretation of 'pallium pyriforme', to be in substantial agreement with my own views. In contradistinction to HOLMGREN, however, I would evaluate his corpus precommissurale, pars inferior as $B_3$ in the everted telencephalon, while HOLMGREN indicates it as corresponding to the 'nucleus lateralis septi' ($B_4$) characteristic for the inverted one.

Again, in his *actual descriptions* of the Osteichthyan lobus hemisphaericus, HOLMGREN (1920) generally interprets the neighborhood which, in my opinion, represents $B_1$, that is, what should be the nucleus olfactorius lateralis of his *diagram* (Fig. 278), as pallium, pars medialis (i.e. as his 'pyriform pallium' or my $D_1$). What HOLMGREN actually designates as 'nucleus olfactorius lateralis' in these forms (i.e. as $B_1$ according to the diagrammatic pattern of Figure 278) is, in my interpretation, a dorsal subdivision of $B_2$ while HOLMGREN's corpus pre-

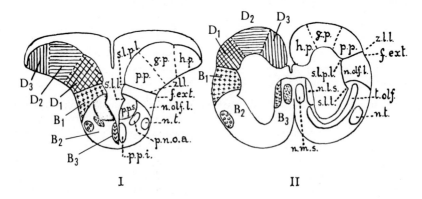

*Figure 278.* HOLMGREN's concepts of the everted (I) and of the inverted (II) telencephalon in lower vertebrates (from HOLMGREN, 1922, modified by addition of my topologic notation). f.ext.: 'fissura' *seu* 'fovea externa'; g.p.: 'general pallium'; h.p.: 'hippocampal pallium'; n.l.s.: 'nucleus lateralis septi'; n.m.s.: 'nucleus medialis septi'; n.olf.l.: 'nucleus olfactorius lateralis'; n.t.: 'nucleus taeniae'; p.n.o.a.: 'nucleus olfactorius anterior, pars precommissuralis; p.p.: 'pyriform pallium'; p.p.i.: 'corpus precommissurale pars inferior'; p.p.s.: 'corpus precommissurale pars superior'; s.l.l.: 'sulcus limitans lateralis'; s.l.p.l.: 'sulcus limitans pallii lateralis'; t.olf.: 'tuberculum olfactorium'; z.l.l.: zona limitans lateralis. In II, n.m.s. and n.l.s. might pertain to $B_4$.

commissurale, pars superior seems to represent the main, ventral sub-division of that neighborhood.[213a]

Alternate designations for Osteichthyan telencephalic grisea pro-posed by HOLMGREN (1920, cf. also KAPPERS et al., 1936, p. 1260, Fig. 544) are: primordium pallii, pars dorsolateralis for cell masses per-taining to neighborhoods $D_1$ and $D_2$ of our terminology, primordium pallii, pars lateralis for cell populations of $D_3$, and primordium pallii, pars medialis for the neighborhood which we interpret as $B_1$.

It should be pointed out that both pallial and basal neighborhoods of Osteichthyes display numerous variations resulting from greater or lesser differentiation of cell groups pertaining to a given topologic neighborhood. Thus, in some forms, the entire pallium is cytoarchitec-turally rather uniform, while in other forms two, three, or more differ-ent cellular areas become differentiated. This also applies to the basal (B) cell masses. In addition, the neighborhood $B_1$ may show a tendency to be drawn toward the pallium. Considering the thereby resulting dif-ficulties of interpretation, there obtains, nevertheless, a rather close general similarity between the two different telencephalic pattern mod-els for gnathostome Anamnia proposed by HOLMGREN (1922) and my-self (1924, 1929a).

Before proceeding with an appraisal of the fundamental telence-phalic zonal system in *Amniota*, it should be stressed that an appropri-ate understanding of the configurational aspects displayed by the telen-cephalon in *Anamnia* must be regarded an essential prerequisite for a morphologically and phylogenetically relevant evaluation of the telen-cephalic components or grisea of Amniota.[214] Summarizing the topo-

---

[213a] I might add that, in my first attempt at a homologization of telencephalic cell masses throughout the vertebrate series, and shortly before elaborating a more rigorous bauplan concept, I had tentatively adopted HOLMGREN's interpretation of the pallio-basal boundary in the teleostean telencephalon (K., 1924). Figure 11 *(Cyprinus auratus)* in the cited publication was in full accord with HOLMGREN's overall viewpoints concerning the cell masses in the everted Osteichthyan lobus hemisphaericus. For further reference, and in comparison with my present concept (K., 1927, 1929a), I have added HOLMGREN's interpretation, indicated by terms of my notation enclosed in parentheses, on the left side of the current figure 256 A. A look at this figure will easily explain why, despite a different *actual* evaluation of neighborhoods in the Osteichthyan telencephalon, HOLMGREN's *diagram*, shown further below in figure 278 I, is compatible not only with that author's view but also with the diverging one propounded by myself.

[214] Thus, numerous extensive as well as detailed studies and publications on develop-ment or structure of the human and mammalian prosencephalon by neurologists not sufficiently acquainted with the brain morphology of Anamnia suffer from considerable weaknesses seriously impairing the value of their generalized conclusions.

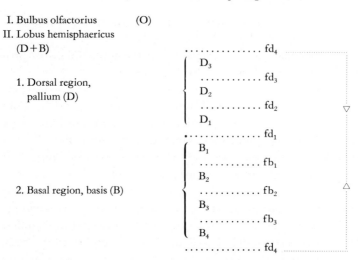

I. Bulbus olfactorius       (O)

II. Lobus hemisphaericus
    (D + B)

                  .............. $fd_4$

1. Dorsal region,
    pallium (D)

$D_3$

........... $fd_3$

$D_2$

........... $fd_2$

$D_1$

.............. $fd_1$

$B_1$

........... $fb_1$

$B_2$

2. Basal region, basis (B)

........... $fb_2$

$B_3$

........... $fb_3$

$B_4$

.............. $fd_4$

*Figure 279 A.* Tabulation showing the general bauplan formula of the paired evaginated and inverted Gnathostome Anamniote Telencephalon. In the unpaired evaginated portion, $fd_4$ dissappears, being basally replaced by a midline neighborhood with or without a median sagittal groove.

I. Bulbus olfactorius (O)

II. Lobus hemisphaericus
    (D + B)

                         attachment of roof plate
               .............. (sulcus taeniae)

1. Dorsal region,
    pallium (D)

$D_3$

........... $fd_3$

$D_2$

........... $fd_2$

$D_1$

.............. $fd_1$

$B_1$

........... $fb_1$

2. Basal region,
    basis (B)

$B_2$

........... $fb_2$

$B_3$

.............. basal midline

*Figure 279 B.* Tabulation showing the general bauplan formula of the Gnathostome Anamniote telencephalon with unpaired evaginated and everted lobus hemisphaericus. The bauplan formulae of this and of the preceding tabulation denote an 'ameboid' or 'molluscoid', i.e. 'squirming' topologic space conceived as a frame of reference *('Bezugssystem'*, *'Bezugsmollusk')*, and which allows for 'agglutinations' or 'fusions' such as $D_{1+2+3}$, $D_{1+}B_1$, $B_{2+}B_3$, etc.

|     | Selachians | Osteichthyes | Dipnoans | Amphibians |
|-----|------------|--------------|----------|------------|
| O | bulbus olfactorius | bulbus olfactorius | bulbus olfactorius | bulbus olfactorius |
| $D_1$ | griseum laterale pallii (cortex lateralis) | griseum mediale pallii (nucleus medialis) | griseum laterale pallii (cortex lateralis) | area lateralis pallii |
| $D_2$ | griseum dorsale pallii (cortex dorsalis) | griseum dorsale pallii (nucleus dorsalis) | griseum dorsale pallii (cortex dorsalis) | area dorsalis pallii |
| $D_3$ | griseum mediale pallii (cortex medialis) | griseum laterale pallii (nucleus lateralis) | griseum mediale pallii (cortex medialis) | area medialis pallii ('primordium hippocampi') |
| $B_1$ | nucleus basilateralis superior | nucleus basalis superior | nucleus basilateralis superior | nucleus basilateralis superior |
| $B_2$ | nucleus basilateralis inferior | nucleus basalis intermedius | nucleus basilateralis inferior | nucleus basilateralis inferior |
| $B_3$ | nucleus basimedialis inferior | nucleus basalis inferior | nucleus basimedialis inferior | nucleus basimedialis inferior |
| $B_4$ | nucleus basimedialis superior | absent | nucleus basimedialis superior | nucleus basimedialis superior |
| $B_5$ | cortex olfactoria | absent | cortex olfactoria | cortex olfactoria (area ventralis ant., area ventrolateralis posterior) |

*Figure 279C.* Synopsis of telencephalic grisea in Gnathostome Anamnia, correlated with hauplan-neighborhoods.

logic connectedness of neighborhoods and cell aggregates[215] of the endbrain in gnathostome fishes and in amphibians the following tabulations (Fig. 279 A–C) can be given.

---

[215] The significant components under consideration are populations of neuronal *perikarya* representing sets *(Grundbestandteile)* and subsets *(Formbestandteile*, with further secondary subsets) of topologic neighborhoods. The perikarya pertaining to such populations or subpopulations are interconnected by their own cell processes (fibers) and by those of neurons in the other ones (discounting intrinsic connections within a given neighborhood). Since it seems well substantiated that, in various forms, topologically 'identical' neighborhoods may display a variety of different fiber connections, these latter cannot be said to represent a rigorous criterion for the establishment of morphologic homology. Moreover, if fiber connections are taken to provide evidence for the establish-

The telencephalon in all three classes of *Amniota* (reptiles, birds, mammals) is rostrally and caudally paired evaginated as well as inverted. The unpaired evaginated telencephalon medium (s. impar), conspicuous during early ontogenetic development, becomes, for practical purposes, negligible at the definitive (adult) stage (cf. Fig. 63, II, III). In birds and mammals, the caudally paired portion of the hemisphere has the tendency to 'curve' or to 'rotate' in a ventral or even rostroventral direction around the hemispheric stalk (cf. Figure 305). This tendency, which is also slightly suggested in some Reptilia and even amphibian Gymnophiona, has been termed '*Endhirnbeuge*' (K., 1972, p. 82, cf. also K., 1924, Fig. 29, loc. cit.) or, by Spatz (1966), '*Hemisphärenrotation*'. It leads, in higher mammals, to the formation of a typical *temporal lobe*, whose rostral tip manifests a temporal 'pole' in addition to the posterior or occipital pole of the hemisphere. A lateral telencephalic surface region, becoming, as it were, 'internated', surrounded, and gradually 'covered' by pallial components *(opercula)* participating in the 'telencephalic bend' represents the *insula Reili* of 'higher' mammals. The depression within the concavity of the 'semicircular' bend corresponds here to the *fossa Sylvii*, containing the insula as its 'floor'.

---

ment of 'homologies', it becomes first necessary to define the *criteria* for the recognition of '*identical' fiber tracts*. Do fiber tracts in different brains manifest 'sameness' because taking their course through 'identical' regions or because they interconnect 'identical' grisea ? In the first case, the significance of their origin respectively termination remains open, and a tract or channel is merely identified by the position of its main 'trunk'. In the second case, we have an evident *petitio principii:* tracts are the 'same' because they interconnect the 'same' regions, and said regions are the 'same' because they are interconnected by the 'same' tracts.

Again, one could give a functional definition of a tract as being concerned with the transmission of signals of a particular sort (e.g. output or input, 'motor' or 'sensory' or related to specific sense organs, etc.). This, however, does not provide any useful information for the configurational evaluation of a neighborhood related to such fiber tract. Nevertheless, since, on the other hand, a certain degree of general orderliness is indeed manifested by fiber connections, this orderliness can be evaluated as a secondary, but *nonrigorous auxiliary criterion*. In cases where, because of indistinct neighborhood boundaries, a topologic mapping becomes ambiguous, particularly constant and thereby recognizable fiber systems, of which some features are clearly identifiable, might thus provide some nonrigorous '*circumstantial evidence*' for an appropriate mapping of grisea with insufficiently definable topologic value (cf. also the comments on this topic in vol. 1, pp. 294–295 of the present series). It should, however, be kept in mind that, in general, the commonly described fiber connections are far less completely elucidated than apparently assumed by most authors.

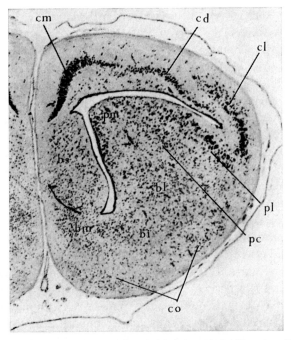

*Figure 280.* Cross-section through the telencephalon of adult Lacerta agilis somewhat rostral to lamina terminalis (from KIESEWALTER, 1925, with added labels according to my interpretation). bi: nucleus basilateralis inferior ($B_2$); bl: nucleus basilateralis superior ($B_1$); bm: nucleus basimedialis inferior ($B_3$); bs: nucleus basilateralis superior ($B_4$); pc: nucleus epibasalis centralis; pl: nucleus epibasalis lateralis; pm: nucleus epibasalis medialis (pc, pl and pm represent EDINGER's epistriatum). cd: cortex dorsalis; cl: cortex lateralis; cm: cortex medialis; co: cortex olfactoria (cd, cl, cm are pallial cortices, co represents the basal cortex).

The telencephalic griseal configuration displayed by a relatively simple adult reptilian brain such as that of *Lacerta*[216] (Fig. 280) might be an appropriate departure in dealing with the problems at issue. Looking at cross-sections through the hemisphere, in this or in other reptilian forms, the perhaps most conspicuous feature is a considerable thickening of the lateral wall, protruding into the ventricle. EDINGER (1908) regarded this thickening as pertaining to the '*hyposphaerium*', i.e.

---

[216] *Lacertilia* and *Ophidia*, being considered as representing two suborders, are commonly subsumed under the order *Squamatae*. The telencephalon of Lacerta (Lacertilian infraorder *Scincomorpha*) could be evaluated as the reptilian endbrain which displays the greatest resemblance to that of lower mammals. However, Iguana, which, together with

to the *basis*, and designated its dorsal ventricular ridge, including its grisea, as '*epistriatum*'. The wall portion forming the *roof* of the hemispheric ventricle contains three well defined cortical plates, *cortex lateralis*, *cortex dorsalis*, and *cortex medialis*, interpreted by EDINGER as homologous with neocortex (neopallium), cornu Ammonis, and fascia dentata (gyrus dentatus) of mammals in the just mentioned sequence. Along the surface of the ventral hemispheric wall, a less pronounced basal cortical plate (cortex olfactoria) can be seen, whose dorsolateral part would, accordingly, represent the mammalian cortex lobi piriformis, and whose ventral part would then correspond to the 'tuberculum olfactorium' of mammals. ELLIOT SMITH (1910), JOHNSTON (1915), CROSBY (1917), and KAPPERS (1921), however, interpreted the lateral cortex (EDINGER's neopallium) as cortex lobi piriformis, the neopallial primordium being, according to CROSBY, an intermediate area located between cortex lateralis and cortex dorsalis (in the alligator). ELLIOT SMITH (1919), moreover, considered the 'hypopallium', i.e. EDINGER's 'epistriatum', to be a pallial derivative.

Because a comparison of adult Amphibia with adult Reptilia seemed to suggest a mapping fully consistent with EDINGER's views, I initially adopted the interpretation of this pioneering author, although I believed, in partial agreement with ELLIOT SMITH, that the reptilian 'epistriatum' was formed by components pertaining to both lateral pallial and adjoining basal regions of Amphibia (K. and KIESEWALTER, 1922).

---

*Anolis, Phrynosoma* and some other forms, is classified as pertaining to the infraorder *Iguania* included in the suborder *Lacertilia*, manifests diverse odd pattern peculiarities of the telencephalon. The reports by KAPPERS and THEUNISSEN (1908) and others indicate, in Iguana, a conspicuous thickening of the $D_1$ zone with rather diffuse cellular arrangements somewhat reminiscent of the crocodilian and avian type. Moreover, in the superpositio lateralis of Iguana, a large extent of the dorsal cortical lamina is covered by the lateral one. The Rhynchocephalian *Sphenodon*, considered a 'relic' or 'living fossil', manifests a variety of lesser telencephalic pattern differences if compared with *Lacerta*. In contradistinction to the telencephalon of Scincomorph *Lacerta*, that of *Chelonia*, who are supposed to have retained a 'primitive organization', is somewhat less similar to the endbrain of lower Mammalia. The *Crocodilia*, subsumed under the 'superorder' *Archosauria*, from which birds may have evolved, display aspects of telencephalic configuration which indeed show a definite resemblance to the griseal configuration obtaining in the avian hemisphere. A tendency toward such configuration could nevertheless have independently originated in diverse reptilian groups. Yet, all extant Reptilia manifest an overall telencephalic pattern which is easily comparable with respect to the different orders and which can be regarded as morphologically 'identical'.

Further observations made while attempting to obtain additional evidence for the establishment of the relevant homologies suggested that configurations of the adult or at least quite late larval Anamniote neuraxis could rather easily be compared with configurations displayed during certain key stages of Amniote ontogenesis (cf. Fig. 281). Study of pertinent Sauropsidan and mammalian ontogenetic stages disclosed that, in order to obtain comparisons which maintain properly corresponding sequences of neighborhoods with identical topologic connectedness, both the so-called lateral ganglionic hill in the human or mammalian telencephalon, and the 'epistriatum' anlage of reptiles should be mapped upon the area lateralis pallii (neighborhood $D_1$) of the amphibian hemisphere. Again, the so-called medial ganglionic hill must then be mapped upon the amphibian basal components ($B_1$ and $B_2$), namely nucleus basimedialis superior et inferior (K., 1925). ELLIOT SMITH's interpretation of the reptilian 'hypopallium' or epistriatum as a pallial derivative became thus substantiated on the basis of conclusive (i.e. topologically valid) embryologic findings. Because it became furthermore evident, by inspection and one to one mappings, that the relationships of 'epistriatum' to cortex lateralis in reptiles are topologically identical with those of claustrum to cortex lobi piriformis in key stages of mammalian ontogenesis, it seemed likewise evident that EDINGER's interpretation of reptilian cortex lateralis as pertaining to the neopallium could not be upheld.

The evaluation of key stages in the ontogeny of the Amniote telencephalon led thus to the conclusion that the neighborhood designated as area lateralis pallii ('nucleus epibasalis', K., 1924; $D_1$, K. 1929a) of amphibians provides, by internation, the epistriatum or hypopallial ridge of reptiles, respectively a part of the basal griseal complex of birds and mammals. Area dorsalis pallii ($D_2$), which remains pallial, represents therefore the neighborhood giving origin to the lateral and the dorsal cortical lamina of reptiles. Thus, the reptilian lateral cortex had to be interpreted as homologous to the mammalian piriform lobe cortex.[217] An adjacent region of dorsal cortex, as it were transitional

---

[217] To some extent, this agrees with SÖDERBERG's interpretation (1922) of area dorsalis pallii ($D_2$) in amphibians as 'pyriform pallium' (cf. fig. 270B). In contradistinction to SÖDERBERG, however, I consider neighborhood $D_2$ to represent a 'general pallium', which provides the lateral and the dorsal cortical plate of reptiles, respectively piriform lobe cortex, neocortex, and parahippocampal cortex of mammals. Both general pallium and hippocampal pallium of SÖDERBERG, as shown in figure 270B, designate, in my evaluation, neighborhood $D_3$ (so-called 'primordium hippocampi').

between this latter and the lateral cortex, can then be regarded as the neighborhood from which the mammalian neocortex originates, i.e. as 'neocortical primordium' in essential agreement with the view of CROSBY (1917). The adjacent extent of dorsal cortex, medial to this 'primordium neopallii', being intercalated between 'neocortical' and 'hippocampal' neighborhoods, accordingly represents a griseum homologous to mammalian parahippocampal cortex. In contradistinction to CROSBY (1917), however, ELLIOT SMITH (1910) stated that, in reptiles, there is 'no structure to which the designation "neopallium" can be

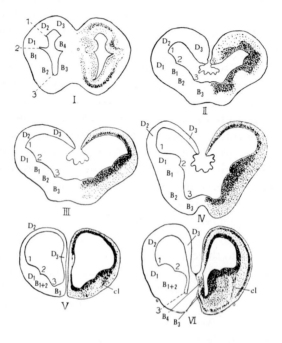

*Figure 281.* Early concept of fundamental telencephalic longitudinal zones (modified, by substituting topologic notations for pallia land basal neighborhoods, from K., 1925–1926). I Selachian (young adult specimen of Acanthias). II Urodele amphibian (young adult specimen of Triton); III Embryonic stage of the reptilian Lacerta showing beginning internation $D_1$; IV Avian embryo (chick of 7 days incubation). V Human embryo of about 38-mm length (approx. between 9–10 weeks). VI Same embryo as in V, somewhat more caudal section, at rostral end of lamina terminalis. On right side of V and VI, the anlage of the claustrum, deriving from $D_1$, and corresponding to reptilian nucleus epibasalis lateralis, is indicated. 1: 'sulcus epibasalis superior' ($fd_2$) 2: 'sulcus epibasalis inferior' ($fd_1$); 3: 'angulus ventralis ventriculi lateralis' (fav, corresponding, in these instances, essentially with $fb_2$). In the original figures, $D_1$ was labeled a, $B_1$ was $b_1$, and $B_2$ as well as $B_3$ were comprised under $b_2$; $B_4$ was recognized as a separate region designated as C.

properly applied'. The cited author, nevertheless, qualified this statement [217a] by the following remark: 'The sensory fibres newly admitted to the hemisphere make their way to the dorsal aspect of the corpus
striatum, and, as the immediate effect, the cortex in the neighborhood
takes on a sudden and rapid growth. This leads to great confusion in
the structure of the hemisphere at the pallio-striate junction, the exact
details of which have not yet been properly elucidated'. ELLIOT SMITH,
moreover, admits 'that the disturbance produced is a sign of the first
stage of the labour attending the birth of the new pallium', and, in his
Figure 17 (loc. cit.) designates the region of CROSBY's 'primordium
neopallii' as 'the earliest rudiment of the neopallium'. I believe not
only that the evaluation of neighborhood $D_2 b$ as neopallial primordium is justified, but also that the topologic analysis of brain wall
neighborhoods in terms of the zonal pattern (K., 1929a) has significantly clarified the difficulties which ELLIOT SMITH recognized and
pertinently characterized as a 'great confusion in the structure of the
hemisphere at the pallio-striate junction'. One could also speak of a
considerable morphogenetic disturbance, at the open boundary of
neighborhoods $D_2$ and $D_1$, correlated with the 'internation' of $D_1$.

The interpretation of the reptilian medial cortical lamina, which derives from the neighborhood $D_3$ (area medialis pallii of amphibians), as
corresponding to the mammalian hippocampal formation (cornu Am-

---

[217a] Included in his second *Arris and Gale lecture* (1910). In his third lecture (1910),
ELLIOT SMITH adds then this further comment: 'I disagree with most other writers upon
this subject in maintaining that no true neopallium is found in the reptilian brain. In
making this statement I am fully aware of the fact that (for the reasons explained below)
I am recanting my own previously expressed opinion; but renewed examination of the
anatomy of the brain in the reptile has convinced me that there is no structure yet differentiated, either from the pallium or from the corpus striatum, which can be homologised
with any degree of exactitude with the neopallium. (It is necessary to make the explanatory
reservation, to which I referred in my last lecture, that the subiculum hippocampi is a
transition region between the hippocampal and neopallial formations, and is not true
neopallium, even though it is convenient to class it with the latter for merely descriptive
purposes. The part of the reptilian cortex called neopallium in my former publications is
subiculum; so long as we fail to discriminate between the latter and the true neopallium
we cannot logically deny the existence of this structure in reptiles.)'. The term *'subiculum'*
as here used by ELLIOT SMITH in agreement with other older authors (cf. also VILLIGER,
1920, p. 50 and fig. 125, p. 131) refers to the *parahippocampal cortex*. Subsequent authors on
cytoarchitectonics (e.g. M. ROSE and others), however, apply the term subiculum to that
wider and more loose portion of the hippocampal cell plate (HE of v. ECONOMO) forming
a transition between hippocampal and parahippocampal cortex, but, in my opinion, still
pertaining to the *cornu Ammonis*.

monis and gyrus dentatus s. fascia dentata) appears reasonably well substantiated. In many reptilian forms the medial cortical lamina displays a distinction between a dorsal region with larger cells (pars macro- or magnocellularis), and a ventral region with smaller elements (pars micro- or parvocellularis), the former comparable to mammalian cornu ammonis, and the latter to gyrus dentatus. It should nevertheless be stressed that the thus obtaining morphologic homologies do not, *per se*, imply an identical or analogous functional significance which depends on details of fiber connections and synaptic mechanisms, including biochemical parameters.

To some extent, these morphogenetic conclusions (K., 1924, 1925, 1929a) are in agreement with the results of independent and simultaneous studies on brain morphology in higher vertebrates published by HOLMGREN (1925), who recognized the pallial origin of the dorsal ventricular ridge (reptilian hypopallium) and distinguished, in the Amniote pallium *sensu strictiori*, three regions, namely lateral 'pyriform pallium', dorsal 'general pallium', and medial 'hippocampal pallium'.[218]

HOLMGREN's reptilian 'general pallium' consists, in my interpretation, of the above-mentioned relatively minor lateral 'neopallial primordium', and of a more extensive parahippocampal cortex. It should, moreover, be recalled that, in Anamnia, HOLMGREN's 'general pallium', corresponding to the neighborhood $D_2$ of my notation, includes, according to my interpretation, the primordia of piriform pallium, neopallium, and parahippocampal pallium, while HOLMGREN's 'pyriform pallium' of Anamnia is our neighborhood $D_1$, which, on rather conclusive ontogenetic evidence, can be considered to provide the 'epistriatum' of reptiles, respectively a portion of the basal grisea in birds, and of the corpus striatum (caudate nucleus) in mammals.

It is evident that the morphogenetic events and configurational aspects allowing for a comparison of adult (or near-adult) telencephalic stages in Anamnia with certain transitory ontogenetic key stages in Amniota can be interpreted according to *Meckel's 'law of reduction'*, or, by cautious phylogenetic extrapolations, according to *Haeckel's 'biogenetic principle'* (cf. vol. 1, p. 240f. *et passim* of this series).

---

[218] HOLMGREN based his concepts upon embryologic studies in the reptilian *Chelonia*, and in the mammalian rodent, *Mus*. He elaborated, in addition to his pallial subdivisions as quoted above in the text, a rather complex terminology involving what I would consider an excessive parcellation of nondescript and in part transitory cell groups.

Again, this interpretation in terms of a morphologic homology concept based upon the topologic connectedness of neural tube wall neighborhoods does not disagree with an appraisal concerning their function as may be reasonably inferred on the basis of the still incompletely understood details of Sauropsidan telencephalic fiber connections. Moreover, even within orders of one and the same class (reptiles, birds) differences as regards particular fiber connections of topologically identical grisea doubtless obtain. Such differences *qua* fiber tracts and synaptic connections become rather substantial if reptiles are compared with birds, or, more generally, Sauropsidan representatives with mammals.

Thus, with respect to the here proposed formulation of an overall telencephalic pattern displayed by all Amniota, the distribution of fiber tracts cannot, in general, be evaluated as either corroborating or contradicting the said formulation (cf. also footnote 215 on p. 531). Studies on fiber connections of reptilian telencephalon undertaken by KRUGER and BERKOWITZ (1960) as well as by others, have used electrophysiological procedures and methods of degeneration. The data yielded by these techniques indicated that the dorsal cortical lamina receives optic and other 'somatic' sensory impulses. This input appears to be diffusely distributed, and related to a type of projection differing from that obtaining in mammals. The grisea of the dorsal thalamus seem to have connections essentially restricted to the so-called 'striatal' basal telencephalic grisea. As regards the lateral ('piriform') and the medial ('hippocampal') cortical laminae, these experimental investigations disclosed that both cortices receive olfactory input, thus confirming well substantiated previous conclusions resulting from the anatomical study of fiber tracts. Olfactory input into lateral cortex and into medial cortex was seen to differ somewhat with respect to the recorded type of potentials. It seems evident that the results based on experimental investigations of this sort have no bearing on problems of *morphologic homology* and cannot be of any significant help for attempts to trace a detailed sequence of phylogenetic stages. Much the same can be said concerning similar studies on the Avian telencephalon, such e.g. as our own experiments related to distribution of optic input (K. and SZEKELY, 1963). The assumption that the prosencephalon of recent Sauropsidans displays a type of phylogenetic evolution considerably differing from that manifested in mammals appears rather obvious.

As regards some specific ontogenetic stages of *reptilian telencephalic development*, Figure 282 A shows an early spatial sequence of wall neigh-

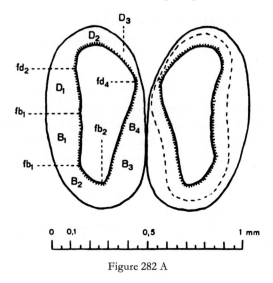

Figure 282 A

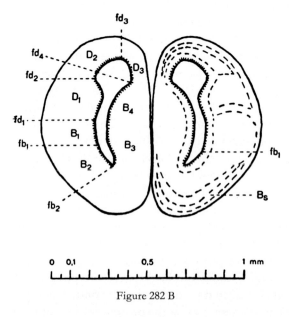

Figure 282 B

*Figure 282.* Cross-section through paired evaginated reptilian hemisphere indicating the telencephalic wall neighborhoods. A Lacerta agilis embryo of 4-mm head length, with one branchial cleft (from K., 1929a). B Lacerta agilis embryo of 5-mm head length (from K., 1929a).

borhoods, comparable to that, e.g., in adult anuran Amphibia
(Fig. 271). The cell populations still display here an essentially periventricular arrangement. Several ventricular sulci are moderately well indicated. At the slightly more advanced stage of Figure 282 B, the differentiation of a cortical plate within the neighborhoods $D_2$ and $D_3$ becomes recognizable. Neighborhood $D_1$ exhibits a moderate protrusion
into the ventricular lumen, thus displaying an early manifestation of
internation or introversion. The cell masses in the dorsal portion of $D_1$
blend, along a zone of close contiguity, with the ventrolateral cell populations of $D_2$.

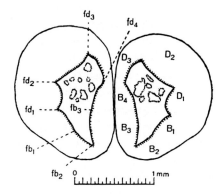

*Figure 283.* Cross-section through embryonic reptilian hemispheres with onset of conspicuous internation of neighborhood $D_1$ (embryo of Crocodilus biporcatus, from K.,
1929 a).

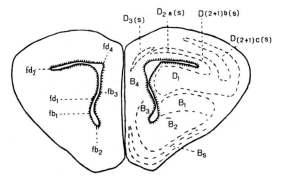

*Figure 284.* Cross-section through hemispheres of a young adult Lacertilian reptile, indicating definitive differentiation of telencephalic neighborhoods (Lacerta agilis; from K.,
1929 a).

Along the basal wall surface pertaining to neighborhoods $B_1$ to $B_3$, a cortical plate $B_s$, separated from the wider periventricular cell masses, has become differentiated. With the exception of $fb_3$, all ventricular 'limiting sulci' can be identified. Figure 283 shows a very similar onto-genetic stage in the telencephalic morphogenesis of a *Crocodilus* embryo, displaying the significant wall neighborhoods and correspond-ing ventricular sulci, including $fb_3$, which is faintly suggested.

Because of the *introversion* or *internation* manifested by the zone $D_1$ of reptiles (and of Amniota in general), this longitudinal zone becomes a component of the *basal ganglia*, being thereby 'shifted' into the ventral hemispheric wall. Thus, in contradistinction to Anamnia with invert-ed telencephalon, such as Selachians and amphibians, in which the dor-sal hemispheric wall or 'pallium' is provided by $D_1$, $D_2$, and $D_3$, the '*pallium*' (EDINGER's *episphaerium*) of Amniota includes only the neigh-borhoods $D_2$ and $D_3$. This difference can be considered to represent a major morphologic distinction between the Amniote and the (invert-ed) Anamniote telencephalon. One might also distinguish a *pri-mary* ($D_{1-3}$), '*pallium*' *of Anamnia*, from a *secondary* one ($D_{2-3}$) *of Amni-ota*. Conversely, the *basis (hyposphaerium* of EDINGER) in *Anamnia* is provided by neighborhoods $B_{1-4}$, while that of *Amniota* includes the neighborhoods $D_1$ and $B_{1-4}$.

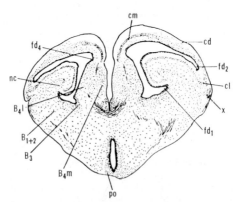

*Figure 285.* Cross-section through the forebrain in an advanced embryo of the Rhyn-chocephalian reptile Sphenodon (from HINES, 1923, modified by substituting, where re-quired, my own terminology). cd: cortex dorsalis; cl: cortex lateralis at junction with nu-cleus epibasalis lateralis; cm: cortex medialis; nc: nucleus epibasalis centralis (derivative of $D_1$); po: nucleus praeopticus (hypothalamus); x: so-called 'nucleus olfactorius lateralis' of HINES, i.e. dorsal edge of basal cortex (cortex olfactoria); $B_4l$ internal portion of $B_4$ or nu-cleus basimedialis superior (so-called 'nucleus lateralis septi'); $B_4m$ external portion of neighborhood $B_4$ or n. basimedialis superior (so-called 'nucleus medialis septi').

At further stages of development, as, e.g. indicated by Figures 284 and 285, the three pallial cortical plates, *cortex medialis, cortex dorsalis,* and *cortex lateralis* approach their definitive arrangement (cf. Fig. 280), and only a reduced layer of more or less scattered periventricular cells remains. The lateral edge of medial cortical plate, and the medial edge of the lateral one are 'split' from, and superposed upon, the corresponding, more internally located edges of the dorsal cell plate by a 'superpositio medialis' respectively 'lateralis'. The lateral edge of the dorsal cell plate may retain a transitory connection with $D_1$ cell groups, while a similar, more lateral connection of $D_1$ with the ventral edge of cortex lateralis usually remains permanent. These connections appear to be related to the close link between $D_2$ and $D_1$ as neighborhoods jointly pertaining to the original 'pallium' of Anamnia. In addition, these connections indicate that the lateral portion of the dorsal cortical plate, and apparently the entire extent of the lateral cortical plate, although essentially deriving from $D_2$, include a component of dorsolateral $D_1$ cellular elements.

However, the connections between cortex dorsalis and cortex lateralis, as well as those of both with the neighborhood $D_1$, manifest considerable variations between taxonomic reptilian orders (cf. Figs. 285, 288, 289, 290) as well as lesser, but still rather conspicuous individual variations within a given species. These differences can be interpreted as morphogenetic fluctuations related to the complex, multifactorial developmental mechanisms involved in the displacement and definitive arrangement of cell populations. A blurred aspect of in part nondescript and transitional cellular groupings and 'migrations' thereby obtains, exhibiting a degree of vagueness or haziness precluding excessive schematic parcellation and the drawing of sharp, linear boundaries.[219]

---

[219] These '*fluctuations*' (cf. also the present chapter's subsection 1C, *Disturbances of Morphogenesis,* p. 211), may be regarded as manifestations of the biologic 'maxim' '*superb architecture and sloppy workmanship*'. Evidently, in attempting to obtain a reasonably valid conceptual and semantic formulation of the configurational problems here involved, the intrinsic *vagueness* of the observable data should be taken into consideration. A purported '*precision*' based on excessive parcellation and *linear boundaries* becomes therefore highly '*inaccurate*'. In order to be as rigorous and 'accurate' as the given circumstances permit, morphology must be based on *topology*, which, guided by conspicuous singularities or even in free construction, places into the manifold *a vaguely localized but combinatorially exactly determined framework*. Topology has the peculiarity that questions pertaining to its domain may, under certain circumstances be decidable, although the continua here concerned cannot be given exactly, but only vaguely, as is always the case in actuality (H. WEYL). Cf. vol. 1, p. 292 of this series, and p. 66 of the present volume.

Within the '*hypopallial ridge*' formed by $D_1$, the cell masses begin to display a differentiation into a lateral portion, *nucleus epibasalis lateralis*, and into a medial periventricular portion, *nucleus epibasalis medialis*. Both grisea are in continuity with each other, and a third, somewhat less dense cell group, *nucleus (epibasalis) centralis* becomes basally segregated from the dorsal hypopallial cell masses.[220] The continuity of nucleus epibasalis lateralis with the cortical derivatives of $D_2$ were pointed out in the foregoing discussion.

Within the ventrolateral, basal wall of the hemispheres, cell condensations topologically identical with nucleus basilateralis superior ($B_1$) and nucleus basilateralis inferior ($B_2$) of Amphibia become recognizable. The ventromedial hemispheric wall likewise displays distinctive primordia of nucleus basimedialis inferior ($B_3$) and superior ($B_4$). These designations for grisea are thus applicable for Reptilia as well as for Amphibia. The differentiation of a corticoid cell plate along the external surface of the basal neighborhoods (particularly $B_2$ and $B_3$) has made further progress.

Of the ventricular grooves, $fd_2$, providing the dorsal ventricular boundary of the hypopallium ($D_1$) has become a limiting sulcus between 'secondary' 'pallium' and 'basis'. The grooves $fd_1$, $fb_1$, $fb_2$, $fb_3$ and $fd_4$ remain, in general, still identifiable.

Although the primordial telencephalic grisea manifest an unmistakable longitudinal zonal arrangement, low rostro-caudal gradients within one and the same zone superimpose, upon this pattern, secondary further subdivisions. These latter, however, such, e.g. as the rostral and caudal portions of $D_1$, briefly discussed further below, can be evaluated as not significant in the aspect here under consideration, and shall again be dealt with in chapter XIII of volume 5.

Figures 286 and 287 show the wall neighborhoods in ontogenetic stages of the alligator's telencephalon as described by KÄLLEN (1951). It can easily be seen that this author, in avoiding my notation, to which he does not refer,[221] nevertheless proposes a subdivision almost identical, as far as the unquestionable and morphologically relevant aspect is

---

[220] Before obtaining, on the basis of appropriate ontogenetic stages (K., 1929a) convincing evidence for the origin of nucleus centralis from neighborhood $D_1$, I was in doubt whether this cell group should be considered a derivative of the hypopallial ($D_1$) or of the basal ($B_1$) hemispheric wall (K., 1924, 1927).

[221] Although not mentioned in his text, the relevant contribution (KUHLENBECK, 1929a) is included in the bibliography of KÄLLEN's paper (1951).

concerned, with my analysis of 1929. KÄLLEN e.g. substitutes $d^{II}$ for $D_2$, $d^{III}$ for $D_1$, $b^{II}$ for $B_2$, $b_m$ for $B_3$, and $C_m$ for $B_4$. KÄLLEN's groove *s.d.* is evidently $fd_2$, and his *s.a.* is our $fd_1$.

Insofar as, in the aspect here under consideration, rather dubious and irrelevant further parcellations are concerned, KÄLLEN's additional notations designate subdivisions of the just-mentioned Grundbestandteile. Such parcellation, if actually required, is, moreover, also im-

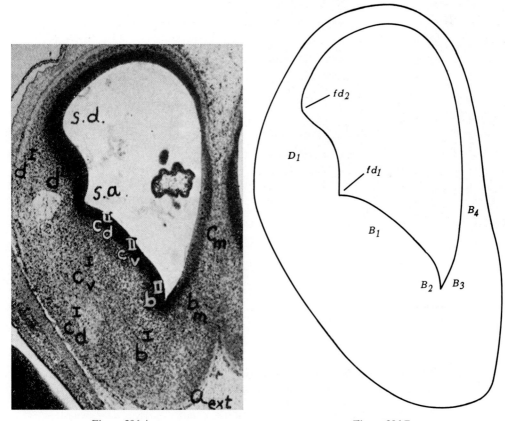

Figure 286 A                    Figure 286 B

*Figure 286.* Cross-section through the hemispheres in embryos of Alligator mississipiensis, showing KÄLLEN's 1951 adaptation of my 1929 notation. A Telencephalon of a 13.5-mm embryo, rostral to 'septum ependymale (from KÄLLEN, 1951). B Outline sketch of cross-section in preceding figure, indicating topologic neighborhoods in accordance with my 1929 notation. C Telencephalon of a 15-mm embryo, rostral to 'septum ependymale' (from KÄLLEN, 1951). D Outline sketch of cross-section in preceding figure, indicating topologic neighborhoods in accordance with my 1929 notation.

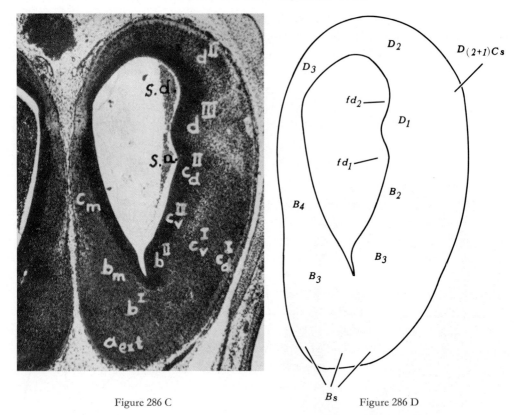

Figure 286 C                                    $B_s$        Figure 286 D

plicitly and explicitly expressible in terms of the original $D_{1-3}$ and $B_{1-4}$
notation (K., 1929a).[222]

Concerning the definitive (adult) configuration of grisea in the reptilian telencephalon (Figs. 280, 288–292), the medial cortical lamina

---

[222] Two incidental short references to my pattern concept (K., 1929a) in another
publication by KÄLLEN (Lunds Universitets Arsskrift, N.F. Avd.2, Bd.47, Nr.5, 1951)
do not indicate that said concept is significantly based on the investigation of key stages
in Amniote *ontogeny*, nor that a substantial part of KÄLLEN's own results is merely provided
by translating my original notation into a set of different, and perhaps less suitable,
symbols. The first reference by KÄLLEN (p.5, loc. cit.) merely states that I attempted to
draw up a bauplan in JACOBSHAGEN's sense for the telencephalon (K., 1929a) and for the
diencephalon (K., 1929b), being followed by GERLACH (1947). The second reference
(p.21, loc. cit.) reads: 'KUHLENBECK described in a series of papers the vertebrate forebrain,
especially the amphibian and the reptilian forebrains, and summarized his idea of the
homologies between the nuclei in a number of «Bauformeln» (1929a). He then used the
ventricular furrows as marks, after which the homologies were settled. His studies were
chiefly carried out on adult material'.

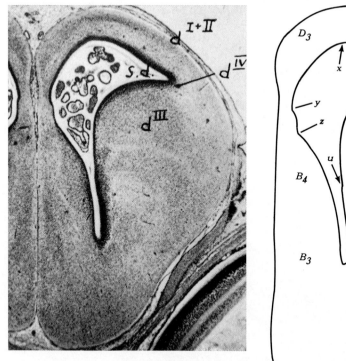

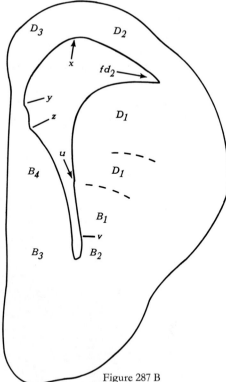

Figure 287 A                                Figure 287 B

*Figure 287.* Cross-section through the telencephalon in a 24-mm embryo of Alligator mississipiensis, showing KÄLLEN's 1951 modified adaptation of my 1929 concepts. A Cross-section 'a short distance rostral to the *foramen Monroi*' (from KÄLLEN, 1951). B Out-line sketch of KÄLLEN's cross-section, indicating topologic neighborhoods in accordance with my 1929 notation. The neighborhood $D_1$ shows a differentiation into dorsal and ventral portion, this latter being the nucleus centralis (cf. Figure 290 showing the adult configuration). u: probable locus of $fd_1$; v: probable locus of $fb_1$, $B_1$ and $B_2$ being pre-sumably somewhat reduced by the expansion of $D_1$ somewhat similar to that in birds; x: perhaps $fd_3$; y, z: perhaps duplication of $fd_4$ related to expansion of $D_3$ and its interaction with $B_4$. Note that $d^{III}$ in Figs. 286 C and 287 A is designated as $d^I$, $d^{II}$ in Fig. 286 A.

can be interpreted as a derivative of $D_3$, dorsal and lateral cortical lami-nae being essentially differentiations of neighborhood $D_2$, with some contributions, by $D_1$, to the most lateral part of dorsal lamina and pres-umably to the entire extent of lateral lamina.

The pattern formula for the main cortical subdivisions of the rep-tilian telencephalon may accordingly be given by the expression

$$\overline{\phantom{--}}, \overline{\phantom{--}}, \overline{\phantom{--}}, \overline{\phantom{--}}.$$
$$D_{3s} \quad D_2a_s \quad D_2(_{+1})b_s \quad D_{2+1}c_s$$

With lesser emphasis on the contribution by zone $D_1$, and in a somewhat more simplified notation, this could also be expressed in the form

$$D_{3(s)}, D_{2a(s)}, D_{2b(s)} D_{2(+1)c(s)};$$

where, in still more abbreviated fashion, and with reference to the mammalian pattern discussed further below, $D_3$ denotes *hippocampal formation*, $D_{2a}$ *parahippocampal cortex*, $D_{2b}$ '*primordium neocorticis*', and $D_{2(+1)c}$ *cortex lobi piriformis*.

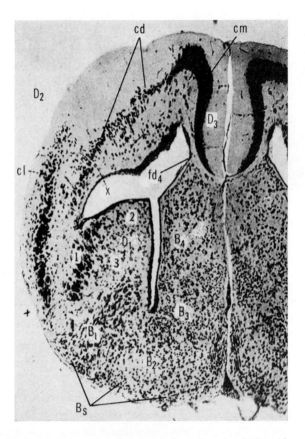

*Figure 288.* Cross-section through the hemisphere of an Amphisbaenid reptile, probably Lepidosternon microcephalon (from JAKOB and ONELLI, 1911, who mistakenly identified this reptile as a Gymnophione; modified by labeling in accordance with my own interpretation). cd: cortex dorsalis; cl: cortex lateralis; cm: cortex medialis; x: region of so-called superpositio lateralis; 1: nucleus epibasalis lateralis; 2: nucleus epibasalis medialis; 3: nucleus epibasalis centralis.

The grisea within the dorsal or 'hypopallial' ridge represent derivatives of the 'internated' or 'introverted' $D_1$ neighborhood. In order to
avoid the ambiguities inherent in the term '*epistriatum*', the designation
*nucleus epibasalis* was suggested for these cell masses (K., 1924, 1929a).
The following definitive subdivisions of that griseum can be recognized: (1) *nucleus epibasalis lateralis*, in many instances a rather dense cell
plate, laterad contiguous or continuous with the basal end of lateral
cortical lamina; (2) *nucleus epibasalis medialis*, which is a medial and in
part ventromedial paraventricular cell plate, more or less continuous
with the lateral subd vision, and (3) *nucleus epibasalis centralis*, a somewhat less dense and more diffuse cell group within the basal concavity
displayed by the cell populations (1) and (2). These grisea, together,
constitute the complex of *nucleus epibasalis anterior*.

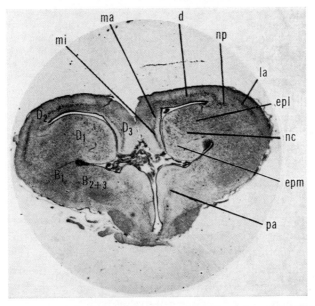

*Figure 289.* Cross-section through the forebrain in the adult tortoise, *Testudo graeca*,
displaying the topologically invariant reptilian telencephalic pattern in a manner differing
from that of Lacertilia (from K., 1925; the topologic notation on left side has been subsequently added). epl: nucleus epibasalis lateralis; la: cortex lateralis; ld: cortex dorsalis; ma:
cortex medialis, pars macrocellularis; mi: cortex medialis, pars microcellularis; nc: nucleus epibasalis centralis; np: so-called 'primordium neopallii'; pa: nucleus entopeduncularis
anterior hypothalami (corresponding to mammalian globus pallidus); the optic tract on
both sides of the hypothalamus and the lateral forebrain bundle, darkly stained on left side
by the carmine stain, are readily recognizable.

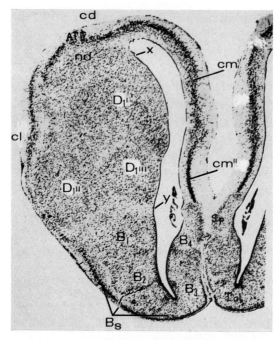

*Figure 290.* Cross-section through the telencephalon of the adult alligator (after ROSE, 1935, labeled in accordance with my interpretation). cd: cortex dorsalis; cl: cortex lateralis; cm$_{1'}$: cortex medialis, pars dorsalis; cm$_{1''}$: cortex medialis, pars ventralis; nd: nucleus diffusus subcorticalis; x: groove fd$_2$; y: groove fd$_1$; D$_{1'}$: nucleus epibasalis dorsalis; D$_{1''}$: nucleus epibasalis centralis lateralis (s. accessorius); D$_{1'''}$: nucleus epibasalis centralis. Sp: 'septum pellucidum' (ROSE); Tol: 'tuberculum olfactorium' (ROSE).

A caudal portion of D$_1$, which may be separated from the rostral one by a transverse groove, represents (4) the *nucleus epibasalis posterior*. In certain reptilian forms, particularly in Lacertilians, this griseum (Figs. 291, 292) displays an oval, semi-elliptic to elliptic arrangement of the cell plate provided by the posterior portions of nucleus epibasalis lateralis and medialis. These cell masses were also designated as '*nucleus sphaericus*' by EDINGER, or as '*archistriatum*' by KAPPERS (1921). The posterior epibasal complex, which commonly includes some scattered caudal elements of nucleus centralis, can also be conceived as a compound griseum, subsuming the caudal portions of (1), (2), and (3). However, despite the obvious subdivisions into an anterior and a posterior griseum epibasale, the entire epibasal (or hypopallial) complex may be evaluated, on the basis of its ontogenetic development, as representing a single morphologic Grundbestandteil corresponding to the longitudinal zone $D_1$.[223]

Within the ventrally adjacent hemispheric wall, two or more or less distinct cell condensations with ill-defined boundaries are commonly

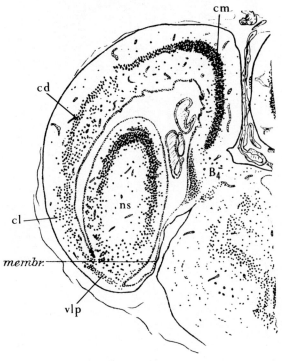

Figure 291 A

*Figure 291.* Cross-sections through the forebrain of Lacerta showing the posterior epibasal and basal griseal complex forming the so-called 'amygdala' of reptiles (A from KIE-SEWALTER, 1922; B from KIESEWALTER, 1928; labeling and interpretation of both sections in accordance with present viewpoints). The sections pass approximately through level of commissura pallii posterior, B being slightly more caudal than A, and showing partial obliteration of lateral ventricle. cd: cortex dorsalis; cl: cortex lateralis; cm: cortex medialis; membr.: membranous wall of telencephalon (lamina epithelialis of roof plate); nc: nucleus centralis (posterior); nl: posterior part of n. epibasalis lateralis forming lateral sector of n. sphaericus; nm: posterior part of n. epibasalis medialis forming medial sector of n. sphaericus; ns: nucleus sphaericus; vlp: area ventrolateralis posterior of basal cortex; x: caudal part of $B_{1+2}$ included in the posterior basal and epibasal complex.

---

[223] This was pointed out (K., 1924, Folia anat. japon. *2:* 243) with the statement: '*Nach meiner Auffassung ist vielmehr ein besonderes, bei den Amphibien pallial ausgebildetes Feld (Area lateralis pallii), in das Innere der Hemisphäre versenkt und bildet den Nucleus epibasalis,*

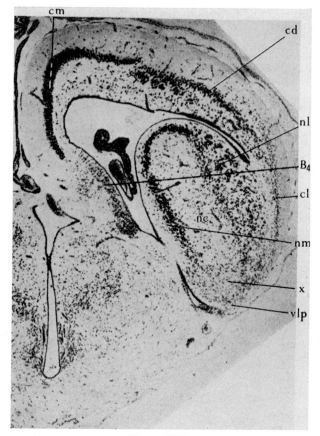

Figure 291 B

recognizable, the dorsal group being *nucleus basilateralis superior*, derived from $B_1$, and the ventral one, derived from $B_2$, *nucleus basilateralis inferior*. *Medially* to the ventricular lumen, the basal hemispheric wall displays *nucleus basimedialis superior* in neighborhood $B_4$, and *nucleus basi-*

---

*der vom oralen bis zum kaudalen Ende ein einheitliches Feld darstellt und nur infolge der caudal-ventralen Krümmung eine gerade Pars anterior und eine caudal-ventralwärts gekrümmte Pars posterior zeigt'*. As item 5 of his summary, KÄLLEN (1951b, p. 345) proffers the conclusion: 'The anlage of the hypopallium is primarily homogeneous: the division into a hypopallium anterior and a hypopallium posterior is a secondary process of minor morphological significance'. Discounting the moot question of 'minor' or perhaps somewhat more relevant morphological significance, and preferring anterius respectively posterius in connection with hypopallium, I would consider KÄLLEN's quoted 1951 affirmation concerning the unity of the hypopallial anlage as identical with my own statement of 1924 (cf. also the preceding footnote 222).

*medialis inferior* in neighborhood $B_3$. These four basal nuclei are directly comparable, by one-to-one mapping, with the corresponding grisea in amphibians.

Along the external surface of the neighborhoods $B_1$ to $B_3$, a corticoid plate $B_s$ can be easily recognized despite its nondescript cellular arrangement and vague outline. It may extend dorsally, with or without interruption, to the external (medial) surface of $B_4$. Caudalward, the corticoid plate reaches the planes of the posterior epibasal complex (cf. Fig. 291), and thus includes an '*area ventralis anterior*' and an '*area ventrolateralis posterior*' (cf. footnote 197 on p. 481). Comparing this basal 'cortex' with the three pallial cortices, it is significant that, in the differentiation of these latter, the periventricular matrix becomes essentially exhausted by abventricular migration, while the basal corticoid plate is formed by a 'migration layer' of relatively lesser proportion, such that the main griseal masses retain their original contiguity with the ventricular lining. These two different modes of ontogenetic corti-

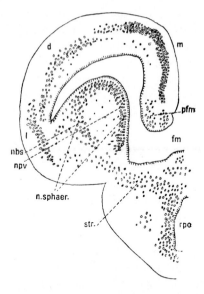

*Figure 292.* Cross-section through the forebrain of Lacerta at level of foramen interventriculare, showing 'nucleus sphaericus' (from K. and KIESEWALTER, 1922). d: cortex dorsalis; fm: foramen interventriculare; l: cortex lateralis; m: cortex medialis; nbs: nucleus epibasalis lateralis; npv: nucleus epibasalis paraventricularis s. medialis; n. sphaer.: nucleus sphaericus; pfm: 'pars fimbrialis septi' (supraforaminal portion of n. basimedialis superior; rpo: recessus praeopticus; str: caudal basal grisea ($B_{1+2+3}$) connected with preoptic hypothalamic grisea through massa cellularis reuniens inferior.

cal development represent relevant characteristics for a morphogenetic distinction between cortex pallii and cortex basalis not only in reptiles but also in mammals.

At posterior levels, particularly within the caudal paired evaginated portion of the reptilian hemisphere, the *posterior epibasal complex* and the *adjacent caudolateral basal cell groups* form an aggregate which is comparable to the *amygdaloid complex* of mammals, and has been described under the collective term '*amygdala*' (e.g. CURWEN, 1939). This is doubtless justifiable on valid morphologic grounds.

However, this comparison tends to disregard significant differences between Sauropsidan and mammalian basal ganglia resulting from a fusion of the $D_1$ and $B_{1-2}$ cell masses in mammals. Nevertheless, discounting a more detailed parcellation, and neglecting some aspects of diverging development, the reptilian *nucleus epibasalis posterior sensu latiori* (i.e. including its nucleus centralis) roughly corresponds, by kathomology, to the main mammalian amygdaloid group (*nucleus amygdalae beta*, K., 1924). Caudal portions of $B_1$ and $B_2$, in like manner, [224] may correspond to components of the mammalian *nucleus amygdalae gamma*, while the corticoid area ventrolateralis posterior represents the mammalian *cortical amygdaloid nucleus* (*nucleus amygdalae alpha and delta*, cf. also K. 1924, 1927; CURWEN, 1939).

Again, the telencephalon of Crocodilia, such as, e.g. the alligator (Fig. 290) displays a general and rather uniform thickening of its entire lateral ventricular wall, in which the basal ($B_{1+2}$) subdivisions, the epibasal grisea ($D_1$) and the lateral part of the dorsal cortical lamina, together with a peculiar subcortical griseum (nucleus diffusus subcorticalis) of $D_2$, assume a configuration closely related to that obtaining in the avian telencephalon. This similarity agrees with other morphologic features suggesting that Crocodilia represent a Sauropsidan group related to the Archosauria and thereby (perhaps rather distantly) to birds.

Summarizing the data on the differentiation of the telencephalic longitudinal zonal system in Reptilia, and neglecting a further parcellation of $B_4$ (including a dorsal extension of $B_s$) the following pattern formula can be given:

---

[224] CURWEN (1939) subdivided the amygdala of the South American lizard *Tupinambis* into seven nuclear masses, whose relationship to the three main groups of my classification (K., 1924, 1927, 1929a) were pointed out. The nucleus centralis of CURWEN's terminology corresponds to caudal portions of $B_{1+2}$ (nucleus basilateralis superior et inferior).

$$O. \; \overline{\phantom{D_{3(s)}}}_{D_{3(s)}}, \; \overline{\phantom{D_{2a(s)}}}_{D_{2a(s)}}, \; \overline{\phantom{D_{2(+1)b(s)}}}_{D_{2(+1)b(s)}}, \; \overline{\phantom{D_{2+1c(s)}}}_{D_{2+1c(s)}}, \; D_1, \; \frac{B_1, B_2, B_3,}{B_s} B_4.$$

The *avian telencephalon* displays, at early stages of ontogenesis (Fig. 293, 294) a zonal configuration easily comparable with that obtaining in Amphibia. Subsequently (Figs. 295, 296, 299), the internation of the neighborhood $D_1$ becomes increasingly manifest, and a *zona limitans* separating $D_1$ from $B_1$ is recognizable. Although, on the whole, a distinction can be made between a dense periventricular layer (formed by the inner primary matrix with its added secondary matrix), [225] and a less dense outer 'migration layer', various irregularly arranged condensations in the outer layer occur, and preclude the drawing of welldefined boundaries. Some of these condensations are nondescript and transitory, suggesting 'fluctuating' morphogenetic events, while others seem to represent precursors of definitive grisea (cf., e.g., Figs. 296 and 300).

Figure 296 shows, in the telencephalon of an 8-day-old chick embryo, a conspicuous thickening formed by the lateral part of $D_2$, and superimposed upon the 'hypopallial ridge' $D_1$. This thickening is related to the differentiation of two particular grisea within the avian pallium, namely *nucleus diffusus dorsalis (superficialis)* and *nucleus diffusus dorsolateralis (internus)*, which correspond to the lateral end of dorsal cortical lamina, and to the subjacent nucleus diffusus subcorticalis in the alligator (Fig. 290). Although predominantly derived from the lateral part of $D_2$, the nucleus diffusus dorsolateralis seems to include some elements originating in the adjacent boundary zone of $D_1$.

At these early stages, all of the 'limiting sulci' $fd_1$ to $fd_4$ and $fb_1$ to $fb_3$ can be identified, although variations in the distinctiveness of one and the same groove at diverse stages, and similar disparities between different grooves at a given stage obtain. Sulci $fd_1$ and $fd_2$, approximately indicating the limits of the epibasal zone $D_1$, are relatively pronounced, while $fd_3$, $fd_4$ and $fb_3$ commonly become merely suggested. An additional sulcus $fd_x$ within the neighborhood $D_2$ corresponds roughly to the dorsomedial limit of the pallial nuclei diffusi. At later stages of development, the rostral end of this groove becomes the anterior portion of the dorsal ventricular angle (fad). In the adult avian telencephalon, the lateral boundary of the nuclei diffusi is indicated on the dorsal surface of the hemisphere's rostral half by a conspicuous

---

[225] Cf. vol. 3, part I, p. 31 of this series.

longitudinal groove known as the *vallecula*, related to a slight sagittal prominence or *pallial torus* formed by said grisea and by the adjacent parahippocampal cortex. At progressively caudal levels, the dorsolateral recess of the hemispheric ventricle extends more and more laterad, covering the dorsal epibasal nucleus, and the vallecula thereby disappears (Fig. 301).

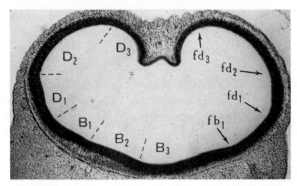

Figure 293 A

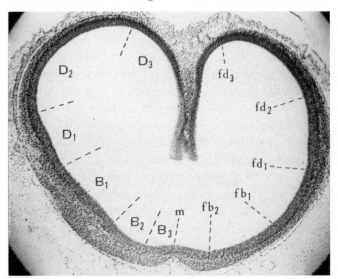

Figure 293 B

*Figure 293.* Cross-sections through the hemispheres of chick embryos, showing the fundamental telencephalic zonal neighborhoods (from K., 1938, slightly modified). A Chick embryo of 87 h of incubation, level of pars impar telencephali. B Chick embryo of 120 h, telencephalon impar (m: midline ridge of $B_3$). C Chick embryo of 5 days and 16 h, level of rostral paired evaginated telencephalon.

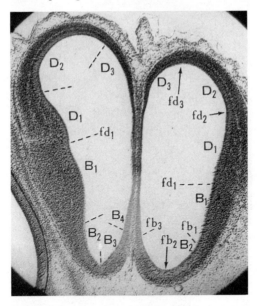

Figure 293 C

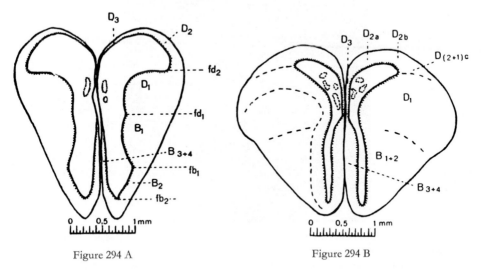

Figure 294 A                               Figure 294 B

*Figure 294.* Cross-sections through the hemispheres of chick embryos at 6 and 9 days
of incubation (from K., 1929a). A Embryo of 6 days. B Embryo of 9 days.

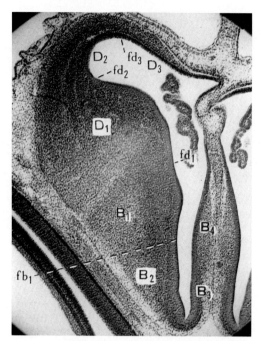

*Figure 295.* Cross-section through the paired hemispheres of a chick embryo at 7 days of incubation, showing cellular density patterns within the telencephalic zonal neighborhoods (from K., 1938).

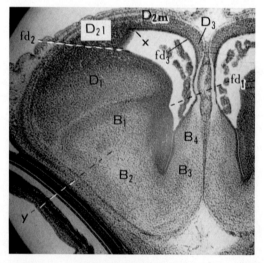

*Figure 296.* Cross-section through the hemisphere of a chick embryo at 8 days of incubation, slightly rostral to lamina terminalis, showing cellular density patterns (from K., 1938, slightly modified). $D_{2l}$: lateral portion of $D_2$; $D_{2m}$: medial portion of $D_2$; x: accessory groove between subdivisions of $D_2$; y: probably locus of rostral extension of diencephalic (hypothalamic) nucleus entopeduncularis anterior.

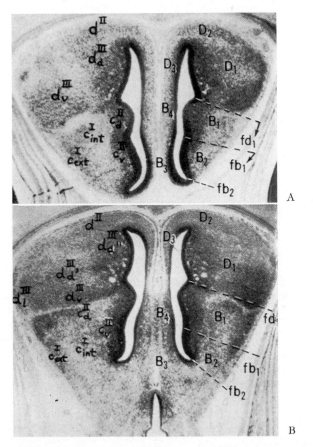

*Figure 297.* Cross-sections through the hemispheres of an embryo of Columba at the 23-mm stage. (After KÄLLEN, 1953, modified by adding my notation on right side.) A At level rostral to lamina terminalis. B At level of lamina terminalis, showing diencephalic preoptic recess of hypothalamus near bottom of section.

Figure 297 shows transverse sections through the telencephalon of a 23-mm embryo of *Columba* as interpreted by KÄLLEN (1953), and roughly comparable to a stage intermediate between that of Figures 295 and 296. It can again be seen that the cited author's designations essentially represent a transcription of the notation which I previously introduced[226] (K., 1929 a, 1938), $d^{II}$ being $D_2$, $d^{III}$ being $D_1$, $c_d^{II}$ being

_____

[226] In his paper on the ontogenesis of the avian forebrain, KÄLLEN (1953) does not refer (either in text or bibliography) to my preceding and rather similar publications on that topic (K., 1929a, 1938). Cf. also above, footnotes 222–223.

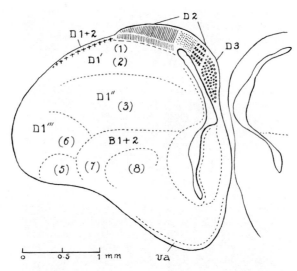

*Figure 298.* Cross-section through the telencephalon of a chick embryo at 10 days of incubation (from K., 1938). Circles represent hippocampal formation; x and dots denote subdivisions of parahippocampal formation; hatching stands for nuclei diffusi of $D_2$ (n. diff. dorsalis et n.d. dorsolat.); crosses indicate lateral corticoid area; $D_1$, etc.: subdivisions of $D_1$ as indicated by numerals 1–3, 5, 6; 1: nucleus epibasalis dorsalis, pars superior; 2: n. epib. dors., pars inferior; 3: n. epibasalis centralis; 5: n. epibasalis caudalis; 6: n. epibasalis centralis accessorius; 7: nucleus basalis; 8: nucleus entopeduncularis anterior (presumably a diencephalic, hypothalamic derivative); va: area ventralis anterior (basal cortex, $B_s$).

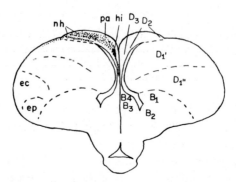

*Figure 299.* Cross-section through the forebrain of the pigeon, Columba. The diencephalic recessus praeopticus and the rostral part of the hemispheric stalk are seen basally (after K., 1929a and 1957). ec: n. epibasalis centralis accessorius ('ectostriatum'); ep: n. epibasalis ('epistriatum' of EDINGER, 'archistriatum' of KAPPERS); hi: 'hippocampal cortex'; nh: nn. diffusi (morphologic homologon of 'neocortical' neighborhood in Mammalian hemisphere); pa: parahippocampal cortex.

$B_1$. The symbols $c_v^{II}$ stand for $B_2$ in Figure 297a and for a ventral portion of $B_1$, overlapping with $B_2$, in Figure 297B. The additional notations refer to some of the various nondescript condensations of cell masses mentioned further above.

At 8 to 9 days of incubation, the chick's telencephalon, which, during earlier stages, manifests a rather generalized Sauropsidan pattern, begins to display, in cross-sections, an unmistakably avian outline (Figs. 294, 296). At 10 days of incubation (Fig. 298) the adult griseal differentiation becomes clearly indicated. This definitive pattern of the avian telencephalon is characterized by a relatively minor development of cortex cerebri in contrast to the expansive enlargement of the lateral internal grisea or basal ganglia, which derive particularly from $D_1$, but also from $B_1$ and $B_2$ (cf. Fig. 299).

As regards the grisea homologous to the reptilian *cortical formations*, the avian derivatives of $D_3$ and of a medial portion of $D_2$ show a rather typical cortical structure, which is, moreover, somewhat more 'advanced' than in reptiles with respect to stratification and further subdi-

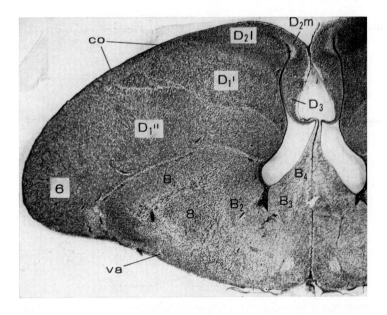

*Figure 300.* Cross-section through the telencephalon of a newly hatched chick, showing cellular pattern, for comparison with figures 298 and 299 (modified after K., 1938). $D_{2m}$: medial, parahippocampal subdivision of $D_2$; $D_{2l}$: lateral subdivision of $D_2$; co: lateral corticoid formation ($D_{1+2}$, respectively $D_{2+1}$); va, 6 and 8 as in figure 298.

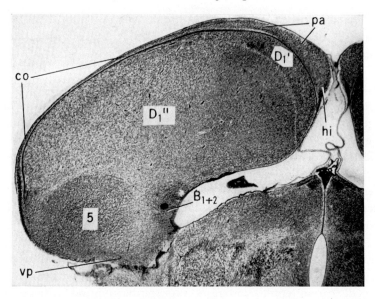

*Figure 301.* Cross-section through the prosencephalon of the newly hatched chick, showing cellular configurations of telencephalic neighborhoods in the posterior (caudal paired evaginated) part of the hemisphere (modified from K., 1938). co: lateral corticoid formation; hi: 'hippocampal' cortex; pa: parahippocampal cortex; vp: area ventrolateralis posterior (caudal part of basal cortex); 5: nucleus epibasalis caudalis ('epistriatum', 'archistriatum').

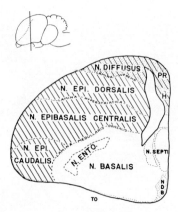

*Figure 302.* Diagrammatic representation of telencephalic centers of a newly hatched chick in transverse plane whose inclination to longitudinal brain axis is shown by inset sketch (from JONES and LEVI-MONTALCINI, 1958). H: 'hippocampal' cortex; NDB: 'nucleus' of *Broca's diagonal band;* N. ENTO.: n. entopeduncularis; PR: parahippocampal cortex; TO: area ventralis anterior.

visions. Investigations by CRAIGIE (1932, 1941) have provided significant data concerning morphology and cytoarchitecture of the avian
cerebral cortex.

With respect to the ontogenetic development of these grisea,
neighborhood $D_3$ gives origin to a cortical lamina homologous to cortex medialis of reptiles and to the mammalian hippocampal formation.
The narrowing ventral tip or edge of this lamina corresponds to gyrus
dentatus, and the wider dorsal portion to cornu Ammonis, the entire
formation being somewhat similar to the precommissural hippocampus of mammals, which is a cell band decreasing ventrad in thickness
as well as in size of cells. Nevertheless, neither the evident morphologic homology of avian '*hippocampus*' or region H (K., 1938) and mammalian hippocampus, nor the just-mentioned superficial structural similarities imply an identical functional significance.

A distinctive and rather clearly laminated cortical griseum originating from the medial part of $D_2$, and adjacent to the region H can be
identified by its topologic location, as a *parahippocampal cortex* homologous with the major, medial extent of the dorsal cortical lamina in rep-

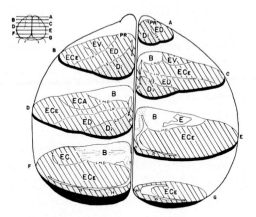

*Figure 303.* Transverse sections through the hemispheres of a newly hatched chick.
The levels of the sections are indicated in the inset (after JONES and LEVI-MONTALCINI,
1958). B: nucleus basalis; Di: nuclei diffusi; E: nucleus entopeduncularis; EC: nucleus
epibasalis caudalis; ECA: nucleus epibasalis centralis accessorius; ECE: nucleus epibasalis
centralis; ED: nucleus epibasalis dorsalis; EV: nucleus epibasalis ventrolateralis; h: hippocampal area; LH: 'lamina hyperstriatica'; LMD: 'lamina medullaris dorsalis'; LS: 'lamina medullaris suprema'; P: parahippocampal area; PR: 'nucleus praepyriformis'.

tiles.[227] Its morphologic significance can be expressed by the formulation $\overline{D_{2a(s)}}$.

Within a more lateral subneighborhood of $D_2$, namely $D_{2b}$, from which the lateral extremity of the reptilian dorsal cortical lamina ('primordium neopallii') derives, a diffuse noncortical griseum takes its origin, and displays two subdivisions, namely an external (superficial) *nucleus diffusus dorsalis* and an approximately coextensive internal *nucleus diffusus dorsolateralis*.

The internal subdivision originates at the boundary of $D_2$ and $D_1$, apparently including some elements of $D_1$ and constitutes thereby a subset $D_{(2+1)b}$. The conjoint nucleus diffusus (dorsalis et dorsolateralis) can thus be evaluated as morphologically homologous with reptilian 'primordium neopallii' and mammalian neocortex. Again, no functional analogy to this latter is hereby implied, such analogy being, moreover, ruled out by the conspicuous difference in structure (cytoarchitectonics) and in relative size. Of the two nuclei diffusi, the external one (dorsalis) corresponds to the *hyperstriatum accessorium* of KAPPERS, HUBER, and CROSBY, and the internal one (dorsolateralis) essentially represents the *nucleus intercalatus hyperstriati* of these authors, but may perhaps also include some of the cell masses which they interpret as *hyperstriatum dorsale*.

Nucleus diffusus dorsolateralis and to some extent also n.d. dorsalis are continuous with a *lateral corticoid region*, derived from the lateral portion of $D_2$ and from elements of $D_1$. In the rostral part of the hemisphere this corticoid region $D_{(2+1)c}$ or $D_{2(+1)c}$ represents a narrow superficial cell band separating components of the epibasal $(D_1)$ complex from the brain surface (Figs. 298, 300). In the *caudal part of the hemisphere* (Fig. 307), $D_{(2+1)c}$ differentiates as a rather distinct, but narrow *corticoid* of 'cortical' cell plate, becoming partly coextensive with the nucleus epibasalis caudalis. On the basis of its topologic relationships the entire lateral corticoid respectively 'cortical' avian formation $D_{(2+1)c}$ can be evaluated as being homologous with the lateral cortical plate of reptiles.

---

[227] The author (K., 1938) tentatively distinguished in the chick four subdivisions ($H_1$ to $H_4$) of the hippocampal cortex, and likewise four subdivisions (a–d) of the parahippocampal cortex. This parcellation, however, can be disregarded as irrelevant in the general aspect here under consideration. Topics of this sort, related to the specific structure of the adult avian telencephalon, and to various additional studies such as those by CRAIGIE (1932, 1941, etc.), DURWARD (1934), STINGELIN (1958 and others), shall be dealt with in chapter XIII of volume 5 of the present series.

The basal cortex which may be designated by the notation $B_s$, $B_{(1+2)_s}$ respectively $B_{(1-3)_s}$, or $B_{(1-4)_s}$, shows a very indistinct differentiation, being a nondescript band of cells along the external surface of the B-neighborhoods. This rudimentary condition presumably corresponds to the decrease in morphologic and functional significance of the olfactory components, as evidenced by the relatively minor development of the olfactory bulb. Nevertheless, as in reptiles and some Anamnia, an area ventralis (anterior) and an area ventrolateralis posterior can be recognized. A small medial subdivision of this latter area has been termed '*nucleus taeniae*'.

With regard to the *noncortical*, respectively *nonpallial, secondarily basal*, i.e. '*hypopallial*' region and to the *primary basal cell masses* (exclusive of basal cortex), $D_1$, besides taking a relatively minor part in the differentiation of nucleus diffusus dorsolateralis $D_2(+_1)_b$, and lateral corticoid region $D_{(2+1)_c}$, provides the (secondary) matrix for the following grisea:

1. *Nucleus epibasalis dorsalis, pars superior* (so-called hyperstriatum dorsale *autorum*).

2. *Nucleus epibasalis dorsalis, pars inferior* (so-called hyperstriatum ventrale).

3. *Nucleus epibasalis centralis* (so-called neostriatum).

4. *Nucleus epibasalis ventrolateralis*, a rostral differentiation.

5. *Nucleus epibasalis caudalis* (so-called archistriatum), a caudal differentiation.

6. *Nucleus epibasalis centralis accessorius* (so-called ectostriatum), a lateral differentiation.

Of these grisea, 1, 2, and 4 may be considered kathomologous to the conjoint griseum comprising nucleus epibasalis lateralis and medialis of reptiles, and 5 to the reptilian nucleus epibasalis posterior complex, which includes a component of nucleus centralis. Nuclei 3 and 6 correspond to the major, anterior part of reptilian nucleus centralis, while the avian nucleus epibasalis caudalis (5) seems to be formed by components derived from posterior portions of central as well as dorsal epibasal nuclei. In the *alligator*, however, a more or less distinctive lateral subdivision of nucleus centralis, topologically homologous to the avian nucleus centralis accessorius (6), can be noticed.

7. The cell masses of the avian neighborhoods $B_1$ and $B_2$ merge in the formation of a common *nucleus basalis* (so-called palaeostriatum augmentatum of KAPPERS or mesostriatum of EDINGER) and also provide, directly below the external brain surface, the (secondary)

matrix for the *basal cortex* $B_s$ as mentioned above. Transitory (cf. Fig. 300) and occasionally permanent differences in cellular density between dorsal and ventral portions of nucleus basalis may occur.

8. In addition, a dorsal and rostrolateral derivative of the hypothalamus, namely the *nucleus entopeduncularis anterior* (so-called palaeostriatum primitivum) protrudes through the hemispheric stalk into the $B_{1+2}$ region along and within the caudal telencephalic part of the lateral forebrain bundle. Some elements derived from $B_1$ and $B_2$ may become 'assimilated' in the formation of this griseum, which, however, represents essentially a diencephalic, hypothalamic cell mass, homologous to nucleus entopeduncularis anterior of reptiles and to the globus pallidus complex of mammals.

9. The avian neighborhood $B_3$ differentiates into *nucleus basimedialis inferior* (so-called nucleus accumbens septi) pertaining to the ventral paraterminal complex.

10. The avian neighborhood $B_4$ becomes *nucleus basimedialis superior* representing the dorsal paraterminal complex (so-called septum *sensu strictiori*). Both $B_3$ and $B_4$ commonly take part in the formation of area ventralis anterior ($B_s$), whose medial and dorsomedial extension includes the nucleus of the diagonal band.

The lateral hypopallial and basal grisea of the Sauropsidan hemisphere display thus a definite dorso-basal band-like stratification, manifested in *reptiles* by nucleus epibasalis ($D_1$), centralis (also $D_1$), basimedialis superior ($B_1$), and basimedialis inferior ($B_2$). The *avian telencephalon*, however, becomes characterized by an additional stratification of the $D_1$ components, which is, moreover correlated with a similar stratification of $D_{2b}$ (nucleus diffusus dorsalis and n. diffusus dorsolateralis). On the other hand, the avian neighborhoods $B_1$ and $B_2$ tend to merge into a common nucleus basalis.[228]

The differentiation and stratification within the avian $D_1$ neighborhood is correlated with the appearance of more or less distinct *medullary laminae*, which separate the griseal subdivisions. The main medullary laminae (Fig. 302) are (a) *lamina medullaris suprema*, between n. epi-

---

[228] In mammals, as discussed further below, the bulk of the secondary matrix in $D_1$, $B_1$ and $B_2$ merges to form a more or less homogenous *corpus striatum (nucleus caudatus* and *putamen)*. Only *nucleus epibasalis lateralis* (which provides the *claustrum*), and *the posterior epibasal complex sensu latiori*, taking part in the differentiation of the mammalian *amygdaloid complex*, remain excluded from the corpus striatum. A relatively minor external component of $B_1$ and $B_2$ (as well as $B_3$ and in part $B_4$) provides the basal cortex *(area ventralis anterior, nucleus of the diagonal band*, and *amygdalar area ventrolateralis posterior)*.

basalis dorsalis and nn. diffusi; (b) *lamina hyperstriatica*, between n. epibasalis centralis and n. epibasalis dorsalis, and (c) *lamina medullaris dorsalis*, between n. basalis and n. centralis; the early anlage of this latter
medullary lamina is the zona limitans between $D_1$ and $B_1$ mentioned
above on p. 555. Additional *accessory medullary laminae* can be found,
which display variations with respect to orders and species, as well as
some individual variations.[229]

Again, variations of this sort are generally correlated with lesser or
greater distinctness of griseal subdivisions, such as the separation of
nucleus epibasalis dorsalis into superior (1) and inferior (2) parts, or
the demarcation within the nuclei diffusi ($D_{2b}$) whose cell masses merge
mediad with the parahippocampal cortex. This latter, in turn, replaces
the nuclei diffusi in the caudal third of the hemisphere (cf. Figs. 301
and 305). Variations concerning subdivisions of nucleus centralis likewise obtain (cf. Figs. 300 and 306).

JONES and LEVI-MONTALCINI (1958) have investigated the embryonic origin of fiber tracts in the chick's telencephalon, using a silver
impregnation technique in addition to cell stains, and obtained significant new data concerning these fiber systems. The cited authors, moreover, confirmed my conclusions concerning the subdivisions of the
avian cerebral hemisphere and adopted my nomenclature for its grisea.
Figures 302 and 303 show the telencephalic cell masses as described by
the cited authors, who also provided, for a suitable comparison of the
terminology, the tabulation reproduced in Figure 304.

JONES and LEVI-MONTALCINI (1958) evaluate the medullary laminae as important boundaries between nuclei, providing convenient
natural subdivisions of the basal ganglia into well defined areas. While
they stand out conspicuously in preparations processed by cell stains,
the cited authors found these laminae to be barely visible in silver-im-
pregnated material, adding the following comments: 'Examination of
the embryonic telencephalon makes it apparent that the laminae are
basically junctions of cell populations of different densities. This is particularly evident in the case of the lamina medullaris dorsalis which ap-

---

[229] Thus, a *'lamina medullaris suprema'* may separate nucleus diffusus dorsolateralis from
nucleus diffusus dorsalis, and a *'lamina hyperstriatica accessoria'* may be found between pars
superior and inferior of dorsal epibasal griseum. Because of such variations and of the
resulting difficulties in identifying homologous cell masses by extrapolation and inter-
polation in default of available observations on a series of sufficiently continuous ontogenetic stages, identical terms used in various publications of the avian telencephalon do
not necessarily always indicate identical cell masses.

pears at 5 days of incubation, long before fibers enter in this area. At this stage a prominent sulcus marks the boundary between the basal and the epibasal primordia of the lateral telencephalic wall.' This sulcus represents, of course, the topologically significant groove $fd_1$ (K., 1929 a) which separates, in Anamnia, the primary 'pallium' from the primary 'basis'. Thus, in Amniota, an 'originally' pallial component ($D_1$), becoming '*hypopallial*', joins, by internation, the 'secondary' basis of the Amniote hemisphere. Again, according to JONES and LEVI-MONTALCINI (loc.cit.), 'Below this sulcus lie the loosely arranged cells of the basal centers; immediately opposite and rostral to it are the epibasal centers in a dense population of cells closely pressed against one another. Shortly after the two laminae, hyperstriatica and suprema di-

Nomenclature of the major nuclei of the corpus striatum in birds

| EDINGER, WALLENBERG and HOLMES, 1903 | ROSE, 1914 | HUBER and CROSBY, 1929 | KUHLENBECK, 1938, and present investigation |
|---|---|---|---|
| Cortex frontalis | B | hyperstriatum accessorium | n. diffusus dorsalis |
| Frontalmark | A | n. intercalatus hyperstriati | n. diffusus dorsolateralis |
| Hyperstriatum | C | hyperstriatum dorsale | n. epibasalis dorsalis, pars superior |
| | D, $D_1$ | hyperstriatum ventrale (dorso-ventrale) (ventro-ventrale) | n. epibasalis dorsalis pars inferior |
| | $G_1$, G, $G_2$, L $G_3$ | neostriatum { ventrale, intermed., dorsale } | n. epibasalis centralis { pars medialis, pars posterior } |
| Parolfactory lobe | R | n. basalis | n. epibasalis ventrolateralis |
| Ectostriatum | S | ectostriatum | n. epibasalis central. accessorius |
| Epistriatum | K | archistriatum | n. epibasalis caudalis |
| Mesostriatum | H | paleostriatum augmentatum | n. basalis |
| N. entopeduncularis | J | paleostriatum primitivum | n. entopeduncularis |

*Figure 304.* Tabulation comparing nomenclature and notations used by different authors for major telencephalic grisea of the avian brain (from JONES and LEVI-MONTALCINI, 1958).

vide the epibasal complex into three large areas; their formation again precedes the differentiation of fiber tracts and offers them a convenient pathway for outgrowing. We came therefore to the conclusion that if they play a role in directing the outgrowth of fiber tracts, they do not represent a significant component of the nervous organization of the telencephalon'.

This conclusion by JONES and LEVI-MONTALCINI seems essentially compatible with my own estimate concerning the significance of fiber tracts for the morphologic evaluation of grisea (cf. above, footnote 215).

Nevertheless, since the laminae (or zonae limitantes) indeed offer a convenient pathway for the outgrowing of fiber tracts (cf. above), said laminae, generally speaking, and in accordance with the views of KAPPERS *et al.* (1936), could be conceived as representing 'collecting fields' for fiber systems of the lateral forebrain bundle and its associated tracts.[230]

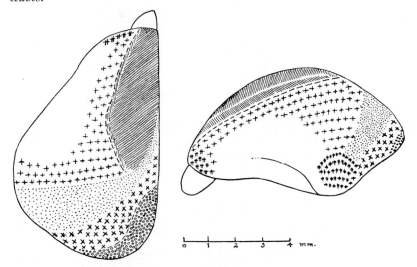

*Figure 305.* Maps, based on graphic reconstructions from serial sections and showing extent of pallial cortex respectively cortical homologa in chick embryo of 19 days of incubation. At this stage, the definitive overall configuration is already manifested (from K., 1938). Crosses with two vertical bars: lateral corticoid field homologous to anteroventral mammalian piriform lobe cortex; crosses with two horizontal bars: lateral corticoid field homologous to a posterior portion of mammalian piriform lobe cortex; other symbols same as for figure 298.

---

[230] The complex, still poorly understood and insufficiently analyzed fiber systems of the avian hemisphere shall be considered in chapter XIII of volume 5.

For comparison with the chick's telencephalon as shown in Figures 300 and 301, Figure 306 illustrates the definitive arrangement of the grisea in the hemisphere of a Passeriform bird *(Parus)*. Supplementing the aspect displayed by cross-sections, Figure 305 shows a mapping of the 'cortical' and 'corticoid' regions upon the chick's dorsal and lateral telencephalic surfaces at an advanced ontogenetic stage. Figure 307 displays the longitudinal arrangement of telencephalic zones in the hemispheres of adult goose (Anseriform, *Anser*) and crow (Passeriform Corvid, *Corvus* sp). Figure 299 illustrates, in a semidiagrammatic cross-section, the zonal arrangement in the telencephalon of an adult pigeon (Columbiformes).

Summarizing the discussion of the telencephalic longitudinal zonal system in birds, and neglecting the parcellations of $D_1$ as well as those

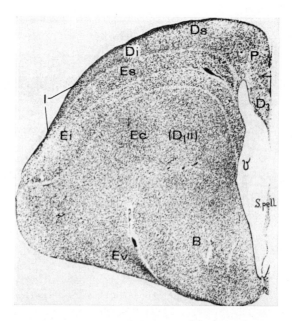

*Figure 306*. Cross-section through the hemisphere of the titmouse, Parus sp. (after Rose,, 1935, modified by substituting my own notation). B: nucleus basalis ($B_{1+2}$); Di: nucleus diffusus dorsolateraliis; Ds: nucleus diffusus dorsalis; Ec: nucleus epibasalis centralis; Ei: nucleus epibasalis dorsalis, pars inferior; Es: nucleus epibasalis dorsalis, pars superior; Ev: nucleus epibasalis ventrolateralis (all n. epibasalis subdivisions are considered derivatives of $D_1$); l: lateral corticoid area ($D_{2+1}$ derivative); P: parahippocampal cortex; S. pell.: 'Septum pellucidum' (Rose); v: lateral ventricle.

of $B_3$ and $B_4$, the following generalized pattern formula may be given for the avian hemisphere:

$$O, \frac{}{D_{3(s)}}, \frac{}{D_{2a(s)}}, D_{2b}, D_{(2+1)c}, D_1, \frac{B_{1+2},}{B_s} B_{3+4}.$$

With a slightly different emphasis, one could also use the expressions

$$D_{2(+1)b'}, D_{2(+1)b''}, \frac{}{D_{(2+1)c(s)}}$$

for the nuclei diffusi and the lateral 'corticoid' and 'cortical' areas. As regards the further differentiations of $D_1$, the terms $D_{1d}$, $D_{1b}$ and $D_{1k}$ would stand for nucleus *epibasalis dorsalis, centralis* (b='basal') and *caudalis*, respectively, allowing, moreover, for further subdivisions, such as $D_{1bl}$ for the rather conspicuous nucleus epibasalis centralis accessorius ('ectostriatum'). Although the elaboration of topologic bauplan-formulae appears definitely helpful in attempts to trace morphologic homologies, an excessive elaboration of this method would result in a notation with an inordinate complexity, comparable to the extreme formalism of pasigraphy or 'logistics',[231] and thus defeat its useful purpose.[232] Concerning the bauplan-formulae given in text and tabulations, the flexibility of the adopted notation may be pointed out. Depending on the aspect under consideration, more detailed or more concise as well as modified expressions become available. Again, it should be added that, in expressions such as, e.g. $B_{1+2}$ or $B_{(1-3)}$ the sign + denotes 'and', while the sign – denotes 'to', i.e. 'from $B_1$ to $B_3$'.

At early stages of ontogenesis, the *mammalian telencephalon* displays a configuration of its wall neighborhoods easily comparable with that of both the rostral paired evaginated and the unpaired evaginated lobus hemisphaericus of amphibians (Fig. 308). In accordance with a considerable expansion of the (secondary) pallial wall ($D_2$ and $D_3$), the neighborhood $D_1$ assumes a relatively ventral position (Fig. 309). With increasing thickening of the (secondary) lateral basis ($D_1$ and $B_{1+2}$), however, the neighborhoods comprising that region may again as-

---

[231] Cf. vol. 1, pp. 287–289 of this series.

[232] The morphologic pattern formulae for the vertebrate telencephalon are comparable to the likewise useful *'flower diagrams'* and *'flower formulae'* of botanists. A short comment on this botanical notation, which also implicity involves a topologic approach, will be found on p. 603.

sume a relatively greater dorsal extension. At such stages, as e.g. illustrated by Figure 310, the configuration of mammalian telencephalic wall neighborhoods is quite similar to that of advanced reptilian embryos (cf. Figs. 281, 282) and allows a homeomorphic mapping of the main longitudinal zones. The grooves $fd_2$ and $fd_1$ are usually well rec-

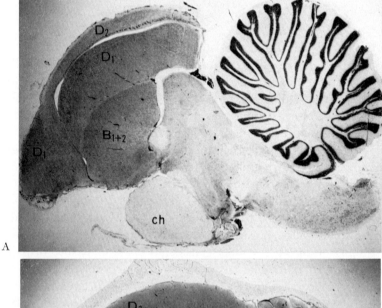

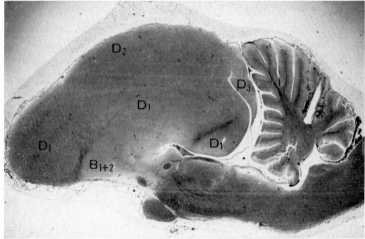

*Figure 307.* Sagittal sections through adult avian hemispheres, showing overall configuration of main longitudinal zonal telencephalic neighborhoods. A Hemisphere of crow (*Corvus* sp.). B Hemisphere of goose (*Anser* sp.).

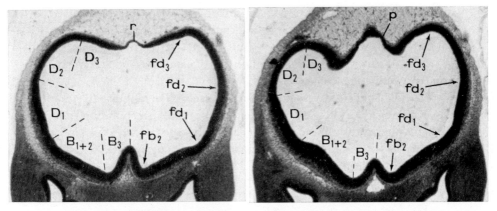

*Figure 308.* Cross-sections through the early unpaired evaginated hemispheres in a 13-mm pig embryo, showing the fundamental telencephalic zonal neighborhoods. Level A is slightly rostral to level B. The configuration is closely similar to that displayed at comparable Sauropsidan stages (cf. fig. 293A) and to that obtaining in Anamnia (cf. fig. 269B, 270B). p: parencephalon (diencephalon); r: roof-plate of telencephalon.

ognizable. Sulcus $fb_2$ commonly coincides with the basal ventricular angle groove fav. The ventricular continuity of $B_1$ and $B_2$, which becomes suggested during ontogenetic sequences in Sauropsidans, seems to be more or less primarily given in mammals, eliminating, as a rule, $fb_1$ (Figs. 311–313).

The fusion of the (primary and secondary) matrix cell masses of neighborhoods $D_1$ and $B_{1+2}$ may be considered a characteristic feature of mammalian telencephalic ontogenesis, leading to the formation of a more or less unified *corpus striatum*.[233]

During the initial stages of this morphogenetic process, $D_1$ represents the so-called '*lateral ganglionic hill*', separated from the '*medial ganglionic hill*' ($B_{1+2}$) by groove $fd_1$ (Figs. 309–313). Concomitantly with

---

[233] From a phylogenetic viewpoint, one could say that a *primary stratification* of the $D_1$, $B_1$, $B_2$ neighborhoods, *obtaining in reptiles* (n. epibasalis, n. centralis, n. basimedialis superior, n. basimedialis inferior) *increases* by further splitting of $D_1$ *in birds*, concomitantly with a considerable expansion of the $D_1$ to $B_2$ wall region, and *decreases (by fusion) in mammals*. Per contra, the reptilian pallial cortical differentiation remains relatively unchanged in birds, while greatly expanding and assuming major proportions in mammals. It is doubtful whether an occasional groove between bed of stria terminalis and body or tail of human nucleus caudatus, described by KAPPERS as 'fissura palaeo-neostriatica' and also noticed by myself (K., 1925b) corresponds to either $fd_1$ or to an unusual manifestation of $fb_1$. Rather, such groove may represent a variation within the adult configuration.

further development, fd₁ gradually disappears, being first levelled at rostral levels while still identifiable at caudal ones (Figs. 320A, B). Sulcus $fd_2$ persists at the definitive (adult) stage and then represents the ventricular groove between lateral border of caudate nucleus and (secondary) pallial ventricular wall.

The medial neighborhoods $B_3$ and $B_4$ are well distinguishable, but do not reach a thickness comparable to that of the corresponding lateral hemispheric wall. In the various forms and stages which I examined, the groove $fb_3$ was rarely present, and if so, only faintly indicated. The

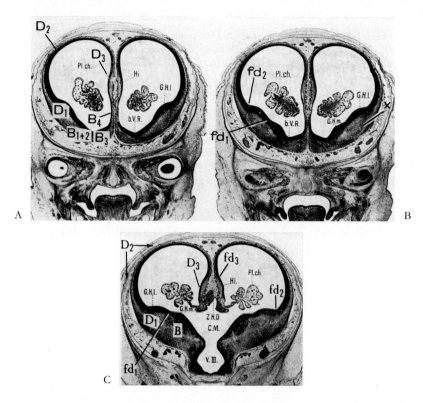

*Figure 309.* Cross-sections, in rostro-caudal sequence, through hemispheres and forebrain of a 25.1-mm human embryo (from HOCHSTETTER, 1919, with some added designations). b.V.R.: 'basale Vorderhirnrinne' (angulus ventralis ventriculi, fav); C.M.: 'cavum Monroi'; g.H.l.: 'lateral ganglionic hill' ($D_1$); G.H.m.: 'medial ganglionic hill' ($B_{1+2}$); Hi: Hippocampus ($D_3$); Pl.ch.: choroid plexus; V. III: preoptic recess (hypothalamus); x: cell condensation including primordium of claustrum; Z.H.D.: 'Zwischenhirndach' (parencephalic roof).

secondary matrix of neighborhoods $B_1$, $B_2$, and $B_3$, while retaining its periventricular location, differentiates, along its external surface, a narrow '*cortical*' or '*corticoid*' *layer* $B_s$, which provides the tuberculum olfactorium (cortex olfactoria with so-called '*islets of Calleja*', or 'area ventralis anterior'), the nucleus of Broca's 'diagonal band' (related to $B_3$), and the 'true' periamygdalar cortex ('nucleus amygdalae corticalis' of mammals, or area ventrolateralis posterior of Amphibia and Reptilia, which becomes displaced medialwards in mammals).

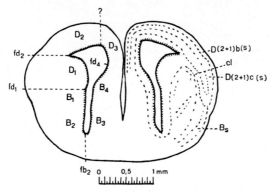

*Figure 310.* Cross-section through paired evaginated hemispheres of a 40-mm mole (Talpa sp.) embryo near rostral end of lamina terminalis (from K., 1929a). cl: locus of claustrum anlage.

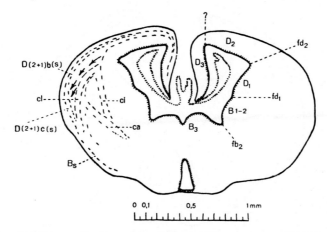

*Figure 311.* Cross-section through hemispheres of 30 mm opossum (*Didelphys* sp.) embryo at level of lamina terminalis. The already reduced telencephalon impar is now the region of interventricular foramen. The preoptic recess (not labeled) appears at bottom (from K., 1929a). ca: anlage of commissura anterior; ci: anlage of capsula interna; cl: anlage of claustrum; ?: perhaps fad (angulus ventriculi dorsalis) or $fd_3$.

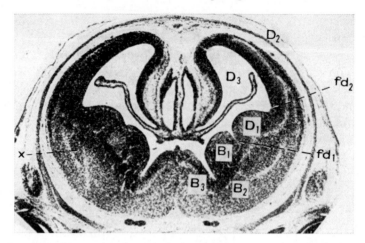

*Figure 312.* Cross-section through the hemispheres of an embryo of the hedgehog (*Erinaceus* sp.) at level of lamina terminalis. The ontogenetic stage corresponds approximately to that of the opossum in the preceding figure (from GROENBERG, 1902, interpreted in terms of my notation). x: cell condensation including anlage of claustrum (on both sides the cellular arrangement is somewhat disrupted by damage to the preparation).

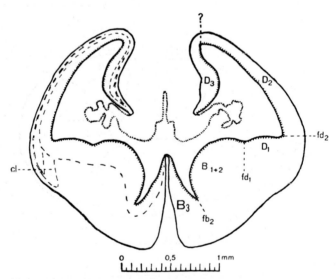

*Figure 313.* Cross-section through the hemispheres of a 40-mm pig *(Sus* sp.) embryo at fairly rostral level of lamina terminalis (from K., 1929a). cl: anlage of claustrum; ?: fad or fd₃ or combination of both.

The developing mammalian telencephalon is, moreover, characterized by a peculiar, dorsally convex rostro-caudal curvature of the lobus
hemisphaericus around the hemispheric stalk, whereby the olfactory
bulb assumes a pronouncedly basal position (Fig. 314). This so-called
hemispheric bend or 'rotation' *('Endhirnbeuge'*, *'Hemisphärenrotation')*,
already mentioned above on p. 532 as partially or faintly manifested in
submammalian forms, results in a corresponding curvature of the telencephalic longitudinal zonal system, which runs relatively straight in
Amphibia (cf. Fig. 276).

Because of the ventral concavity correlated with the dorsal convexity of the hemispheric bend, the $D_1$ and $B_{1+2}$ zones appear, in 'horizontal' sections (Fig. 315 A, B), to be rostrally and caudally 'surrounded'
by the $D_2$ and $D_3$ neighborhoods. Moreover, the rostral extremity of
the precommissural hippocampal cortex, which originates from the
matrix $D_3$, secondarily becomes displaced, as the rostral bend of the
curvature increases, over the anterior extremity of the $B_3$ and $B_4$ neighborhoods (Fig. 320 A). Without a study of intermediary stages permitting to follow this displacement and the corresponding matrix migrations, one might erroneously assume an origin of basal precommissural hippocampus components from $B_4$ and even $B_3$.

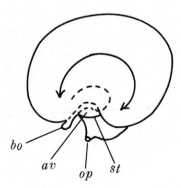

*Figure 314.* Sketch illustrating the curvature of the telencephalon in higher mammals
and man. Based on a figure by HOCHSTETTER (1919) depicting the brain of a 38-mm human
embryo. av: area ventralis anterior (external aspect of basis telencephali; bo: olfactory
bulb; op: optic nerve; st: basal portion of sulcus telencephalo-diencephalicus; the strip between the lower and middle broken lines corresponds to anterior part of piriform lobe,
coextensive with stria olfactoria lateralis; middle broken line corresponds to limen insulae;
locus of isocortical insula between upper and middle broken line; bent line with arrowheads indicates curved 'axis' of 'hemispheric rotation' caused by the embryonic growth
processes.

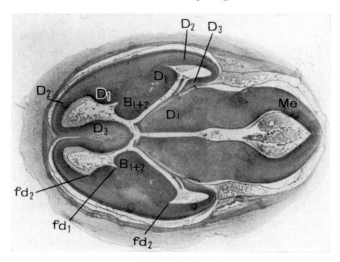

Figure 315 A

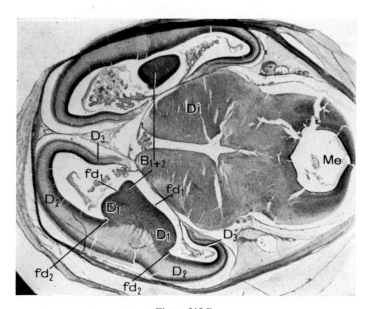

Figure 315 B

*Figure 315.* 'Horizontal' sections through the brain of cat embryos, showing the effect of the hemispheric bend upon the configuration of telencephalic neighborhoods. A Embryo of 8-mm head length. B Embryo of 12-mm head length. Di: diencephalon; Me: mesencephalon.

The *cortex pallii* of mammals can be said to originate essentially from the neighborhoods $D_2$ and $D_3$, the former being the most extensive one. In contradistinction to the *basal cortex* $B_s$, the formation of true cortex pallii is correlated with *matrix exhaustion*, which shall be discussed further below on p. 610 in dealing with some aspects of corticogenesis. In respect to its matrix, the major part of cortex pallii may therefore be termed *hologenic*, while the basal cortex can be characterized by the term *merogenic*. ROSE (1926) had proposed the designations cortex totoparietinus and cortex holoparietinus, respectively (cf. also further below).

The *hippocampal cortex*, comprising a *precommissural*, a *supracommissural*, and a major *postcommissural* portion, originates from neighborhood $D_3$. A fully differentiated hippocampal formation, occurring only in mammals, consists of *cornu Ammonis* ('hippocampus *sensu stricto*' of ELLIOT SMITH, 1901) and of a distinctive *gyrus dentatus* ('fascia dentata'). An undifferentiated hippocampal formation is a cortical band which merely displays a more or less distinct gradient between an outer (topologically dorsal or parasubicular) neighborhood corresponding to *cornu Ammonis*, and an inner (topologically ventral) neighborhood corresponding to gyrus dentatus. The precommissural hippocampus of mammals is essentially undifferentiated, the supracommissural hippocampus is differentiated in a few forms (e.g. Prototheria and Metatheria) but undifferentiated respectively rudimentary in most Eutherian mammals, becoming a narrow fringe of parahippocampal cortex as well as part of the indusium. The postcommissural portion is fully differentiated in apparently all mammals.

The extensive neighborhood $D_2$ includes a medial, *parahippocampal region* $D_{2am}$, from which the *parahippocampal cortex* differentiates, and a main, *lateral region* $D_{2al}$, from which most of the hexalaminar 'neocortex' *sensu strictiori* ('isocortex' of O. and C. VOGT) originates. In the ventrolateral portion of $D_2$, however, namely in a region of continuity between $D_2$ and $D_1$, two subzones $D_{2+(1)b}$ and $D_{(2+1)c}$ can be noticed, in which cortical differentiation involves not only elements of the matrix $D_2$ but also some (basalward increasing) participation of elements derived from $D_1$.

These two subzones include the *regio insularis* $D_{2+(1)b}$, whose cortex can still be classified as (transitional) *hexalaminar neocortex*, and the rostral extent of the *piriform lobe* (the so-called prepiriform area), namely the neighborhood $D_{(2+1)c}$, whose cortex may be classified as an 'allocortex' in the terminology of the VOGTS. Said piriform lobe cortex represents most of the so-called 'palaecortex' of KAPPERS, being, moreover, homologous to the *lateral cortical lamina* of reptiles (cf. also Figs. 242 and 280). With respect to their matrix, which is both pallial and hypopallial (i.e. secondarily basal), the cortices of *regio insularis* and *regio piriformis anterior* are formed partly by matrix exhaustion of $D_2$ and partly by a relatively minor matrix participation of $D_1$ through migration of peripheral elements. ROSE (1928) therefore designated these cortical formations as *cortex pallio-striatalis* sive *bigenitus*.[234] With regard to *matrix exhaustion* (K., 1922, 1929a), the insular isocortex and the anterior cortex of lobus piriformis thus represent a formation transitional between *hologenic* and *merogenic* cortex and may accordingly be designated as *hekaterogenic*.

In the caudal paired evaginated portion of the hemisphere, and in connection with the posterior hemispheric bend involving an anteriorly concave, basalward and rostralward directed curvature of the hippocampal formation, a caudobasal region of parahippocampal cortex becomes contiguous with the posterior end of the anterior piriform areas. The caudal part of the mammalian piriform lobe thereby includes a transition to, and a region of, *parahippocampal cortex*. This is illustrated by Figure 316, depicting a 'simple' Eutherian mammalian hemisphere in comparison with a reptilian one.

It will be seen that on its external surface, the *lobus piriformis* (or pear-shaped lobe) in a relatively simple hemisphere of a 'lower' mammalian brain is delimited from the neopallium by the *sulcus rhinalis lateralis*, which can be roughly subdivided into an anterior and posterior portion. Ventrally, the piriform lobe is rostrally delimited from the basal cortex of area ventralis anterior by the *sulcus endorhinalis*, related to the stria olfactoria lateralis. Caudally, the limit between piriform lobe and basal cortex of nucleus amygdalae ('area ventrolateralis posterior') is given by a variably distinctive '*sulcus periamygdalaris*', generally more or less continuous with the caudal end of sulcus endorhinalis.

---

[234] In one of his publications, ROSE had attacked my studies on a personal level which, as I then thought, required a reply in kind, leading to an exchange of immature unpleasantries in vols. 66, 67, 68 of the *Anatomischer Anzeiger* (1929). The controversy was ended with the following closing remark by the editor, Prof. v. EGGELING (then my esteemed 'Chef'): '*nachdem die Erwiderung des Herrn* HARTWIG KUHLENBECK *auf den Angriff des Herrn* MAXIMILIAN ROSE *eine zweimalige Darlegung der gegenseitigen Ansichten und Auffassungen nach sich gezogen hat, halte ich den Stand der Angelegenheit für genügend geklärt, um allen Lesern des Anat. Anz. eine eigene Stellung zu ermöglichen. Ich schliesse damit die Auseinandersetzung*' (Anat. Anz. *68:* 200, 1929).

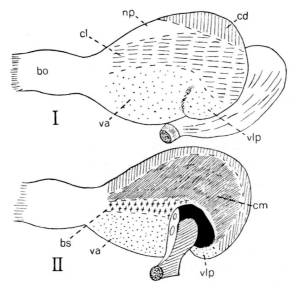

Figure 316 A

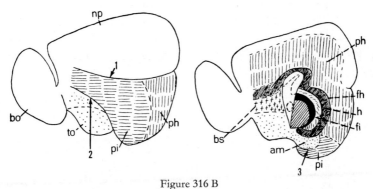

Figure 316 B

*Figure 316.* Comparison of reptilian and simple mammalian Eutherian telencephalic regions in surface projection (from K., 1927). A Hemisphere of the tortoise *(Testudo* sp.) in lateral (I) and medial (II) view. bo: bulbus olfactorius; bs: nucleus basilateralis superior (B$_4$); cd: cortex dorsalis; cl: cortex lateralis; cm: cortex medialis; np: pars lateralis corticis dorsalis (so-called 'primordium neopallii'); va: area ventralis anterior; vlp: area ventrolateralis posterior. B Hemisphere of the hedgehog (Erinaceus sp.) in lateral (left) and in medial (right) view. am: cortex of nucleus amygdalae (area ventrolateralis posterior); bo: bulbus olfactorius; bs: nucleus basimedialis superior; fh: fissura hippocampi; fi: fimbria of fornix; h: hippocampal formation; np: 'neopallium'; ph: parahippocampal cortex; pi: cortex of piriform lobe; to: tuberculum olfactorium (the extent of basal cortex B$_s$ is indicated by stipples, cf. also fig. 316A); 1: sulcus rhinalis lateralis; 2: sulcus endorhinalis; 3: locus of sulcus periamygdalaris, if present.

In accordance with this interpretation, the entire lobus piriformis, whose cortex derives from $D_2$ with a contribution from $D_1$, may be evaluated as pertaining to the 'pallium'.

As regards the basal cortex ($B_s$), it should be kept in mind that the so-called *tuberculum olfactorium*, particularly conspicuous in macrosmatic mammals, is merely a bulging region *within* the area ventralis anterior, from which it is not separated by any sharp delimitation or transition. Such typical 'tuberculum' is, for instance, hardly if ever clearly recognizable in man. The *anterior* basal cortex extends laterally as far as the *sulcus endorhinalis*, medially along the surface of the paraterminal body, and caudally as so-called *substantia perforata anterior* toward the telo-diencephalic boundary. The *posterior* basal cortex (area ventrolateralis posterior of lower vertebrates) is, of course, the above-mentioned cortical portion of the amygdaloid complex. In man, where these 'rhinencephalic' structures are closely crowded together at the base of the hemisphere (cf. Figs. 317 and 348), the anterior part of the sulcus rhinalis lateralis at the *'limen insulae'* is inconspicuous or completely lacking, and the sulcus endorhinalis becomes blurred by the indistinct *stria olfactoria lateralis*, being represented by a faint double groove along both edges of that stria. A good illustration, based on my sketch of 1927, is given by NAUTA and HAYMAKER on p. 150 (as Figure 4–10) in the treatise of HAYMAKER *et al.* on the hypothalamus (1969).

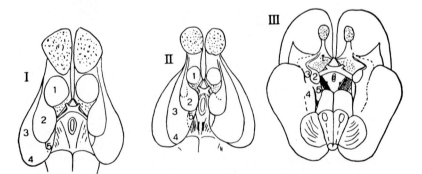

*Figure 317.* Basal view of mammalian telencephalic configurations (from K., 1927). I Oppossum *(Didelphys marsupialis)*. II Rabbit. III Human embryo of about 5 months. The differences in actual brain size are here disregarded. 1: area ventralis anterior; 2: area ventrolateralis posterior; 3: lobus piriformis, pars anterior; 4: lobus piriformis, pars posterior; 5: basal tip of postcommissural hippocampus.

A lateral cell proliferation and migration from the dorsal secondary matrix of $D_1$ results in the formation of a train of elements, commonly appearing as a vaguely outlined condensed cell plate basally continuous with the ventral edge of the differentiating piriform lobe cortex (Figs. 318–320). Dorsally, this cellular band, which may manifest indistinct further splitting or irregular alignments of cell clusters, becomes continuous with the external 'migration layer' of the secondary matrix in the lateral region of $D_2$.

Except for this indistinct dorsal boundary zone, the bulk of said cellular plate, extending laterad from $D_1$, constitutes the *anlage of the claustrum*. Its configurational relationship, recorded by ELLIOT SMITH (1919), LANDAU (1923), myself (K., 1924, 1925a), and FAUL (1926), indicates that the mammalian claustrum is topologically identical, i.e.

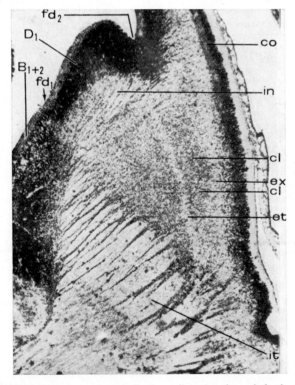

*Figure 318.* Oblique, approximately 'horizontal' section through forebrain of a 4-cm human fetus, showing anlage of claustrum (from LANDAU, 1923, with added $D_1$ and $B_{1+2}$ notation). cl: anlage of claustrum; co: anlage of insular cortex; et: capsula externa; ex: capsula extrema; in: capsula interna, anterior limb; it: capsula interna, posterior limb.

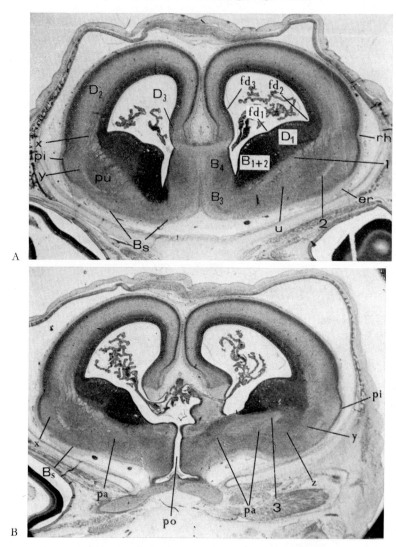

*Figure 319.* Cross-sections through the forebrain in a 7-cm pig embryo, showing primordia of telencephalic grisea. A At rostral end of lamina terminalis. B At level of preoptic recess. er: locus of sulcus endorhinalis; pa: anlage of globus pallidus; pi: cortex of piriform lobe; po: preoptic recess; pu: anlage of putamen; rh: sulcus rhinalis lateralis; u: less dense area in anlage of putamen, which should not be confused with anlage of globus pallidus (pa) whose diencephalic (hypothalamic) origin was recently followed step by step in man by RICHTER (1965); x: migrating elements pertaining to matrix of insular isocortex and of piriform lobe cortex; y: cell masses related to claustrum anlage; z: locus of amygdaloid complex anlage; 1: anlage of capsula interna; 2: anlage of capsula externa; 3: anlage of stria terminalis system, lateral to capsula interna fiber bundles.

homeomorph with the *nucleus epibasalis lateralis* of reptiles. This homology was pointed out, on the basis of mutually somewhat differing morphologic considerations, by ELLIOT SMITH (1919) and by myself (K., 1924, 1925a).

Although disagreeing with some (perhaps essentially semantic) details of ROSE's interpretation (cf. K., 1929a, pp. 44–45), this author's evaluation of insular and 'prepiriform' cortex as *cortex 'bigenitus'* or as

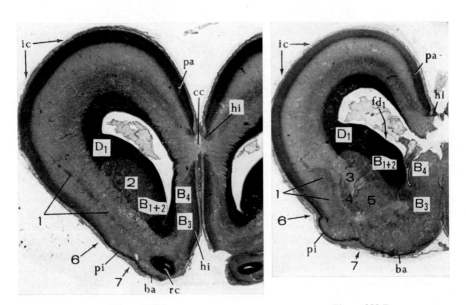

Figure 320 A                    Figure 320 B

*Figure 320 A, B.* Cross-sections through telencephalon respectively forebrain in an 8-cm pig embryo, showing primordia of telencephalic grisea. A At rostral end of corpus callosum, showing caudal end of obliterating rhinocoele. The secondary overlap of precommissural hippocampus ($D_3$) with $B_3$ and $B_4$ neighborhood is due to the shifting correlated with progressive hemispheric bend, and becomes, moreover, exaggerated by the obliqueness of the section's plane. 1: cell band derived from $D_1$ but containing at dorsal extremity $D_2$ components, and pertaining to matrix of insular isocortex, cortex of lobus piriformis, and claustrum; 2: area of lesser density representing secondary matrix of nucleus caudatus; 3: anlage of capsula interna; 4: dense portion of putamen anlage; 5: less dense portion of putamen anlage; 6: locus of sulcus rhinalis lateralis; 7: locus of sulcus endorhinalis; ba: basal cortex (area ventralis anterior including tuberculum olfactorium); cc: anlage of corpus callosum; hi: precommissural hippocampus ($D_3$), secondarily displaced over $B_{3+4}$ neighborhoods; ic: isocortex anlage; pa: parahippocampal cortex anlage; pi: piriform lobe cortex anlage; rc: rhinocoele (ventricle of olfactory bulb and tract).

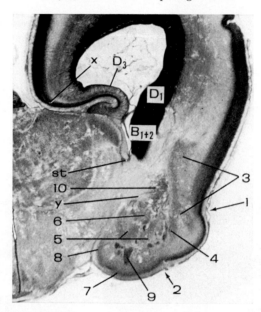

*Figure 320 C.* Cross-sections through forebrain in an 8-cm pig embryo at level of caudal paired hemispheric evagination, showing primordia of amygdaloid complex. 1: sulcus
rhinalis (lateralis), corresponding to boundary (gradient) between isocortical insular cortex
and allocortex of piriform lobe; 2: locus of (posterior) sulcus endorhinalis, corresponding
to boundary (gradient) between cortex of piriform lobe and cortical amygdaloid 'nucleus'
(basal cortex, area ventrolateralis posterior); 3: anlage of claustrum; 4: intermediate cell
group representing transition between claustrum and main amygdaloid complex; 5: subdivisions of main amygdaloid complex (n. amygdalae beta); 6: nucleus amygdalae gamma
(n. amygd. centralis); 7: nucleus amygdalae alpha (cortical amygdaloid nucleus, basal cortex); 8: nucleus amygdalae delta (medial accessory nucleus); 9: clusters of undifferentiated
cells pertaining to matrix of main amygdaloid complex; 10: caudal end of putamen anlage,
medial to which scattered elements of globus pallidus, not visible in the photograph, are
located (y); x: approximate boundary (gradient) between hippocampal formation ($D_3$) and
parahippocampal cortex (derived from $D_2$); y: locus of caudal portions of globus pallidus.


'hekaterogenic' (mihi) seems well substantiated and agrees with my own
findings. KAHLE (1969), on the other hand, assumes that the human insular cortex derives exclusively from a pallial region corresponding[235]
to the basolateral edge of $D_2$, and that the 'prepiriform cortex' originates from a 'striatal' matrix equivalent to $D_1 + (B_{1+2})$. According to

---

[235] KAHLE (1969) does not use my notation.

this assumption[236] the insular cortex would represent a subset $D_{2b\,(s)}$ instead of $D_{2\,(+1)\,b\,(s)}$, the 'prepiriform cortex' being then a subset $D_1 + B_{(1+2)\,a\,(s)}$ instead of $D_{2\,(+1)\,c(s)}$, namely a hypopallial, basal cortex closely related to the tuberculum olfactorium $B_{1+2\,(s)}$ and associated corticoid structures (nucleus of diagonal band, cortical amygdaloid nucleus).

According to my observations in man and other mammals, however, the cortical plate identifiable as lateral edge of isocortex and as 'prepiriform cortex' is connected, *ab initio*, with both the basal part of $D_2$ and the dorsal part of $D_1$ by continuous trains of immature cellular elements.

An internal condensation of those elements which originate from $D_1$ forms the *claustral anlage*, while cell groups along the external border of this primordium provide an indistinct layer related to the cortical cell plate, and gradually disappearing, by incorporation with this latter, during a succession of stages in which the capsula extrema becomes identifiable as a cell-poor layer of 'zona limitans' type. Concomitantly, the anlage of capsula externa and also that of capsula interna develop. Particularly in man, but also in other mammals, the 'insular' border of isocortex tends to become displaced basalward, covering, as it were, the surface of the basal cell masses, while still receiving, as suggested by clearly observable oblique 'streams' of cells, elements from the secondary matrices of $D_2$ and $D_1$.

It is, of course, doubtless justified to state that the dorsal part of insular cortex receives less $D_1$ elements than the basal one, and, again, that the differentiating insular cortex, *in toto*, receives a lesser contribution from $D_1$ than the 'prepiriform cortex'.

Stressing the preponderant role of zone $D_2$ in the differentiation of these cortices, it might furthermore appear permissible to conceive both insular cortex and cortex lobi piriformis as essentially (but not wholly) derivatives of $D_2$. Thus, in the purposely oversimplified diagram of Figure 10B on p. 187, volume 1 of this series, the entire 'neocortex' including thereby the insular cortex, was subsumed under the

---

[236] KAHLE (1969), in elaborating his concept, also quotes conclusions reached by FEREMUTSCH (1962). The reader interested in further details concerning the cortical subdivisions in man and mammals as proposed by these authors may be referred to KAHLE's monograph and its bibliography. Some of these questions shall be dealt with in chapter XV of volume 5, and are not relevant to the comparative approach of the present volume, concerning the overall pattern. The concept of five different fundamental mammalian cortical regions based upon this approach is concisely summarized in the monograph 'Brain and Consciousness' (K., 1957, p. 200) and in our paper on the rabbit's cortical pattern (K. *et al.*, 1960).

symbol $D_2$, and even the still much less negligible $D_1$ contribution to cortex lobi piriformis was minimized by a question mark meant to indicate the uncertainty in defining its amount.

The topologic connectedness of the *claustrum anlage* in mammals is identical with that of the *lateral epibasal matrix* ($D_1$) in Sauropsida, particularly reptiles. It is evident that, in these latter forms, the *nucleus epibasalis lateralis*, although an independent derivative of $D_1$, has, by contiguity and continuity, genetic relationships with both the lateral cell plate and the lateral edge of the dorsal cell plate (CROSBY's '*primordium neopallii*'). Said portion of the dorsal cell plate can be evaluated, on reasonably valid grounds, as topologically corresponding to the mammalian insular isocortex. In lower mammals, the *claustrum* is mainly coextensive with the 'prepiriform cortex', while in higher mammals, particularly in man, where this cortex is reduced and 'pushed' basomediad at the adult stage, the claustrum then becomes mainly coextensive with the insular cortex.

FAUL (1926), who also reviewed *in extenso* previous findings by DE-VRIES (1910) and LANDAU (1923), reached the following conclusions concerning the claustrum: (1) that it does not belong to the striatum at any stage of its development; (2) that the source of its cells is the same as that of the cortex; and (3) that at no stage of its development the human claustrum does have relations with the cortex, but that it is an independent derivative.

My own conclusions, based on previously published (K., 1924, 1925a, 1929a) and on unpublished subsequent studies, could be formulated in the following manner: (1) the *claustrum* can be conceived as representing an independent mammalian '*basal ganglion*' homologous to *nucleus epibasalis lateralis* of Reptiles; (2) the source of its cells is the lateral portion of $D_1$ and (3) although originating from a primordium which contributes to the formation of insular and 'prepiriform' cortices, it can be regarded as 'an *independent derivative*'.

The complexity of the relevant morphogenetic events, and the intrinsic semantic inadequacy of verbal models is evidenced by the fact that, depending upon interpretation of semantic ambiguities, FAUL's formulation may be considered both compatible and incompatible with my own.[237]

------

[237] Evidently, if, by '*striatum*', adult nucleus caudatus and putamen, respectively their early recognizable distinctive anlage are meant, then the *claustrum* does not 'belong to the striatum at any stage of its development'. If, however, the early embryonic $D_1 + B_{1+2}$

The *claustrum* is laterally separated from the cortex by fiber systems constituting the *capsula extrema*, and medially from the striatum by a similar *capsula externa*. In man, the cellular 'stream' originating from adjacent portions of $D_2$ and $D_1$, whose medial $D_1$-component includes the claustral primordium, is clearly recognizable at ontogenetic stages of about 6 weeks (roughly 12-mm length). The bauplan formula of the claustrum can be expressed by the notation $D_{1dl}$, where dl stands for 'dorsolateral'.

As regards the other main internal grisea or *'basal ganglia'* of the mammalian telencephalic hemispheres, the unified secondary matrix $D_1+B_{1+2}$, exclusive of the previously discussed lateral cell migration from $D_1$ related to cortical and claustral morphogenesis, provides the anlage of *corpus striatum*. This latter consists of a medial portion, the *nucleus caudatus*, protruding into the telencephalic ventricle whose lateral wall it provides, and of a lateral portion, the *putamen*, separated from the nucleus caudatus by the massive fiber tracts of the *capsula interna*. Both parts are rostrally conjoint, and also retain additional connections through continuous strands which appear as 'stripes' in sections. The term corpus striatum, coined in human anatomy, refers to this particular feature. In some forms, e.g. in the mouse and rat, no definite separation into caudate nucleus and putamen obtains, since the fiber systems corresponding to the capsula interna traverse the corpus striatum in scattered small bundles.

Generally speaking, the entire *corpus striatum complex* may be evaluated as a morphologic unit displaying, moreover, a rather uniform cytoarchitectural structure. Its bauplan formula can be expressed as $(D_1+B_{1+2})$ m, l, where m (medial) denotes nucleus caudatus, and l (lateral) refers to putamen. Nevertheless, the region adjacent to the inferi-

---

zones are considered to subsume the *'striatal anlage'*, then, of course, the claustrum must be considered as ontogenetically related, respectively 'belonging', at this stage, to the 'striatum'. Second, since $D_1$ presumably contributes to the formation of insular and 'prepiriform' cortex, the 'source' of claustral cells is indeed the same neighborhood from which the cortex obtains (some of its) cellular elements. On the other hand, these cortices are, to a major degree, (also) derivatives of $D_2$. Is, therefore, FAUL's conclusion (2) 'true' or 'false'? The reader may provide his own answer. Thirdly, since FAUL assumes a common source of cortical and claustral cells in his conclusion (2), is he justified or not justified to state 'that at no stage of its development the claustrum does have relations with the cortex'? Again, the reader is here invited to decide this question in accordance with his own views.

or ventricular angle, contiguous with the medial griseum derived from
$B_3$, and bordering on the basal cortex $B_s$, may show some structural
peculiarities and was designated as '*fundus striati*' by BROCKHAUS (1942).
Moreover, within the zone of transition between basal neighborhoods
($B_{1+2}$) of corpus striatum and bulbus respectively tractus olfactorius,
vaguely outlined, rather nondescript cell populations can be subsumed
under the term '*nucleus olfactorius anterior*', briefly referred to above on
p. 524. An additional relatively small and variable griseum, which apparently represents a basilateral differentiation of the striatum, is the
nucleus ansae peduncularis *(nucleus basalis of Meynert*, n. substantiae in-
nominatae, n. subputaminalis) in man and some other mammals.
These various details concerning the corpus striatum and its 'para-
striatal' grisea are not relevant to the present generalized discussion
and shall be dealt with in the special part of this series (vol. 5,
chapter XIII).

The *globus pallidus* (BNA, PNA), although generally still considered, together with the putamen, as pertaining to the telencephalic
so-called nucleus lentiformis of standard anatomical nomenclature, is
doubtless a diencephalic dorsolateral hypothalamic derivative, as already recognized by STRASSER (1920) and particularly stressed by
SPATZ (1921, 1925). This diencephalic origin was subsequently observed by several other authors (e.g. K., 1924, 1954; SCHNEIDER, 1950,
RICHTER, 1965) and can now be regarded as established beyond reasonable doubt (cf. also above, section 5, p. 455, footnote 186, of this
chapter). Because of 'fluctuations' or 'blurring', and the lack of 'precision' obtaining in morphogenetic processes (cf. also subsection 1C,
p. 211) this does not exclude the possibility that the external, dorsolateral portion of globus pallidus 'receives' or 'assimilates' an essentially
negligible amount of elements from the basal secondary $B_2$ respectively
$B_{1+2}$ matrix.

The *nucleus amygdalae* (BNA) or *corpus amygdaloideum* (PNA) of human anatomical nomenclature is a compound griseum corresponding
to the posterior epibasal complex of reptiles (cf. above p. 554) and
consisting of several cell groups allowing for additional parcellations.
In an overall morphologically relevant subdivision (K., 1924, 1927,
1929a) the following main griseal regions ('nuclei') can be distinguished (Fig. 320C, 322):

1. *Nucleus amygdalae gamma* (n. centralis of KAHLE, 1969),
continuous with the basomedial portion of putamen (striatum) and
presumably mainly derived from caudal $B_{1+2}$ components of the

$D_1 + B_{1+2}$ complex. It was also designated as *'dorsaler Nebenkern'* (K., 1924).[238]

2. *Nucleus amygdalae beta* or *main amygdaloid complex (Mandelkernhauptkomplex*, K., 1924, 1927; nucleus amygdalae basalis and n. lateralis of KAHLE, 1969). This griseum, which can be further subdivided, corresponds to the posterior epibasal nucleus of reptiles, and represents a differentiation of the caudal $D_1$ neighborhood pertaining to the joint $D_1 + B_{1+2}$ matrix. Nondescript cellular groups, comparable to scattered griseal 'islets', connect lateral components of the main amygdaloid complex derived from the $D_1$ neighborhood with caudoventral cell aggregates of the *claustrum*. In many instances it becomes highly arbitrary whether such cell groups should be evaluated as pertaining to the claustrum or to the amygdaloid complex.

3. *Nucleus amygdalae alpha* or *cortical amygdaloid nucleus* (nucleus amygdalae corticalis of KAHLE, 1969). It corresponds to the area ventrolateralis posterior of the inverted Anamnian hemisphere and of the reptilian telencephalon. In mammals, however, because the expansion of isocortex results in a ventrad shifting of 'prepiriform cortex', this area has assumed a more basal and even to some extent ventromedial posterior position. The most medial portion of the cortical amygdaloid nucleus, which may be contiguous to the baso-rostral extremity (i.e. the 'end') of the curved postcommissural hippocampal formation, is commonly differentiated as a somewhat distinctive griseum, the medial accessory nucleus (*nucleus amygdalae delta*, *'medialer accessorischer Kern'*, K., 1924, 1927; nucleus amygdalae medialis of KAHLE, 1969, and other authors). The bauplan formula for nucleus amygdalae alpha with its accessory griseum can be expressed by the notation $B_{(1+2)k(s)}$ or $B_{(s)k}$, where k denotes 'caudal'. A generalized formula for the entire amygdaloid complex is provided by the expression $(D_1 + B_{1+2})_k$.

---

[238] This component was originally conceived as corresponding to a caudal part of reptilian nucleus centralis. However, in 1924, before obtaining additional ontogenetic evidence concerning morphogenesis of the Amniote basal ganglia, I had left the question open whether the reptilian nucleus centralis pertained to the epibasal or to the basal complex, but favored a basal origin. Subsequently (K., 1929a) the nucleus centralis could be identified as a derivative of $D_1$, while the *'dorsaler Nebenkern'* apparently derives from a more basal (i.e. $B_{1+2}$) matrix. It should be added that the term 'nucleus amygdalae *centralis*' used by KAHLE and other authors does not involve any reference or relationship to the Sauropsidan hypopallial nucleus (epibasalis) *centralis*. The terminology referring to subdivisions of the amygdaloid grisea is complicated as well as confusing. Further details concerning these grisea shall be discussed in chapter XIII of volume 5.

The internal grisea within the basal portion of the medial hemispheric wall can also be classified as *basal ganglia sensu latiori*. Because of their relationship to the lamina terminalis, ELLIOT SMITH (1910 and other publications) designated these cellular aggregates as the *paraterminal body*.[239] This very appropriate term, or the alternate designation '*preterminal body*', can be taken as subsuming both the neighborhoods $B_3$ and $B_4$ of my notation.

The internal griseum derived from $B_3$ is homologous to the *nucleus basimedialis inferior* of submammalian forms and should be designated by the same term. In mammals, however, this nucleus, continuous with the basal portion of corpus striatum, and forming part of the '*fundus striati*', has also been called *nucleus accumbens* (e.g. OBENCHAIN, 1925) or nucleus *parolfactorius accumbens* (K., 1924, 1927).

The cell masses derived from the neighborhood $B_4$ provide the *nucleus basimedialis superior*, homologous to that nucleus of submammalian vertebrates with inverted telencephalon. This griseal complex, which may also be conceived to represent either the paraterminal body *sensu strictiori* or the dorsal part of that body *sensu latiori*, is commonly designated as '*septum*', but this latter term frequently also includes undefined portions of $B_3$ cell aggregates. The mammalian nucleus basimedialis superior displays a number of vaguely circumscribed cell groups. Detailed parcellations of that griseum have been attempted by ANDY and STEPHAN (1966, 1968), and others.

The *basal cortex* $B_s$ extends medialward and dorsalward over $B_3$ as far as, and apparently including the neighborhood $B_4$. It becomes related to the fibers of *Broca's diagonal band*, whose 'interstitial nucleus' it provides, and is dorsally continuous, if not identical, with what is generally described as '*nucleus medialis septi*'.

---

[239] ELLIOT SMITH (1910) comments: 'the paraterminal body presents a relationship to the hippocampus analogous to that which the striate body has to the rest of the cortex'. 'Just as the projection fibres of the neopallium (as well as many other tracts ascending to it) traverse the corpus striatum in the mammalian brain, so in all vertebrate classes the corpus paraterminale forms a matrix for fibres passing to and from the hippocampus; and part of the body in question often becomes prolonged as a fringe along the ventral edge of the reptilian and amphibian hippocampus, in a position analogous to that occupied by the fimbria in the mammalian brain'. This 'fringe', of course, is the '*pars fimbrialis septi*' in amphibians (cf. fig. 272), mentioned above on p. 521. It was also called '*pars supraforaminalis septi*'. A griseum of this type occurs also in Selachians (cf. p. 486). Although ELLIOT SMITH speaks of 'all vertebrate classes', the relationships in the everted telencephalon, where $D_3$ assumes a lateral position, are evidently quite different.

In man, some primates, and in some other mammals with large telencephalon, a non-negligible portion of the paraterminal body *sensu strictiori* ($B_4$) becomes intraterminal, i.e. included into the stretched lamina terminalis, and forms the structure known as '*septum pellucidum*' (BNA, PNA) of man.[240] Further comments on the development and morphologic significance of the human septum pellucidum and its cavity will be found further below (p. 627) in a short discussion concerning the mammalian lamina terminalis and its commissures. It will here be sufficient to state that the mammalian *griseum basimediale superius* includes a *preterminal*, a *terminal*, and a *supraterminal* portion, the latter being the '*indusium verum*' of ELLIOT SMITH.

Summarizing the relevant data on the differentiation of the telencephalic longitudinal zonal system in mammals, the following bauplan formula can be given:

$$O, \overline{D_{3(s)}}, \overline{D_2am_{(s)}}, \overline{D_2al_{(s)}}, \overline{D_{2(+1)}b_{(s)}}, \overline{D_{(2+1)}c_{(s)}}, D_1dl, D_2+B_{(1+2)}d,l, \overline{B_{(1+4)(s)}}, B_3, B_4.$$

As regards the significant ventricular grooves, only $fd_2$ and $fb_2$ seem to be constant, the former representing the lateral boundary of nucleus caudatus, and the latter becoming the inferior ventricular angle *fav*.

If, in more generalized terms, the Amniote telencephalon is compared with that of Anamnia, the bauplan of the former can be easily derived from the morphologic pattern displayed by the latter, although quite definite changes become obvious. With respect to the 'limiting grooves', sulcus $fd_2$ assumes major significance by indicating the ventricular boundary of (secondary) pallium and (secondary) basis in the lateral wall of the Amniote hemisphere. Groove $fd_1$ is here reduced to a 'secondary' sulcus which, particularly in mammals, may completely disappear.

---

[240] According to ELLIOT SMITH (1910), HUXLEY, in his 1864 lectures, introduced the general term '*septal area*' for the noncortical basomedial ganglionic masses. ELLIOT SMITH then states: 'It is a signal instance of HUXLEY's perspicuity that he came to realise the important morphological fact that the septum lucidum is merely the upper greatly stretched portion of the small and, in the human brain, quite in significant band of ganglionic matter in front of the anterior commissure, to which ZUCKERKANDL many years afterwards gave the name "gyrus subcallosus". ... HUXLEY grouped the septum lucidum and gyrus subcallosus together and called the whole "septal area". The septal area is the surface of the ganglionic mass to which I have given the name "paraterminal body" – "paraterminal" because in the lowlier vertebrates, in which its upper part is not yet modified to form the septum, its outstanding feature is its relationship to the lamina terminalis'.

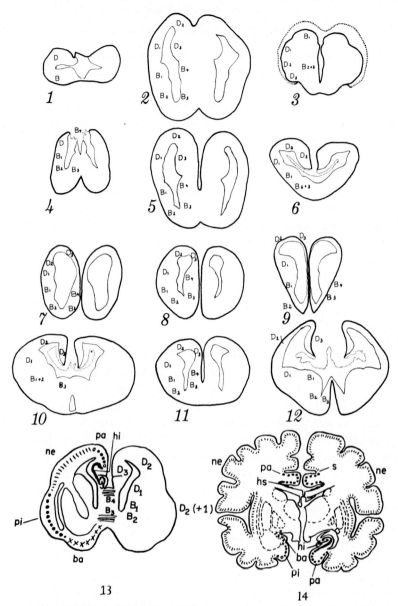

*Figure 321.* Synoptic representation of the topologic invariant neighborhoods in the telencephalon of different vertebrates. 1 Petromyzon. 2 Selachian. 3 Teleost. 4 Dipnoan. 5 Anuran amphibian. 6 Urodele amphibian. 7 and 8 Embryonic reptiles. 9 Embryonic bird. 10, 11, 12 Embryonic mammals. 13 Adult Marsupial mammal, comparing topologic notation with morphologic designations. 14 Adult man, left at level of amygdaloid nucleus, right at level of postcommissural hippocampus. The morphologic designations correspond to the topologic notation as shown in 13. ba: basal cortex; hi: hippocampal cortex; hs: supracommissural rudiment in man; ne: neocortex; pa: parahippocampal cortex; pi: cortex of piriform lobe; s: sulcus cinguli.

This is correlated with an internation or introversion of wall neighborhood $D_1$, whose matrix thereby becomes fused with that of the basal neighborhoods $B_1$ and $B_2$. In Anamnia, on the other hand, the cell masses of $D_1$ provide a significant component of the cytoarchitectural $D_1$ or pallial region. Moreover, the neighborhood $D_2$, which, in Anamnia, displays a rather undifferentiated cytoarchitecture, and can be regarded as an essentially 'single' unit, becomes particularly differentiated in Amniota, reaching its highest degree of differentiation in mammals by providing the diversified as well as extensive regions of *isocortex* and *parahippocampal cortex*. A comparison of Figures 321, 322 and 323 will indicate the difference between the concept elaborated by KAPPERS *et al.* (1936), who did not take into consideration the changes correlated with an internation or introversion of $D_1$, and my own interpretation based on topologic connectedness. Figures 324 and 325 show an adaptation of KAPPERS' and CROSBY's concepts elaborated by ROMER (1950) and included in an elementary textbook on the vertebrate body. These diagrams may be compared with Figures 316 and 321.

More recently, in a 'symposium' on 'comparative and evolutionary aspects of the vertebrate central nervous system', sponsored by the *New York Academy of Sciences*, NORTHCUTT (1969) has presented a concept of the evolutionary fate of what he designates as the '*prototypic columns*' in the amphibian telencephalon. It can easily be seen, by looking at NORTHCUTT's illustration, reproduced here as Figures 326 A and B, that his notation is, like that of KÄLLEN (1951, 1953), a rather close adaptation of the notation which I had introduced in 1929.[240a] Thus, P I represents $D_3$, P IIa and P IIb are the medial and lateral subdivisions of $D_2$, P IIIa and P IIIb being subdivisions of $D_1$. NORTHCUTT's BI+BII correspond to $B_1$, $B_2$, $B_3$, while his BIII represents $B_4$. As

---

[240a] Although NORTHCUTT (1969) briefly cites my papers of 1929 and 1938, no reference is made to the notions introduced by these contributions, which concerned an invariant as well as specific pattern of fundamental vertebrate telencephalic *longitudinal zones* and the principle of a generally valid morphological respectively topologic *notation* applicable to this zonal system. NORTHCUTT merely states that he has divided the reptilian telencephalon into six *longitudinal phylogenetic columns* (prototypic columns) 'based on the ontogenetic evidence of KÄLLEN' (1951) and on experimental and architectonic evidence of adult nuclear groups. One might perhaps wonder, whether both cited authors, to use an expression coined by SCHOPENHAUER, have not 'secreted' (*sekretiert*) the findings and conclusions, dating back to 1929, of a predecessor.

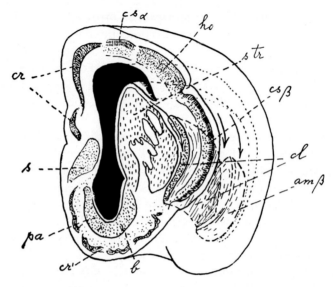

Figure 322

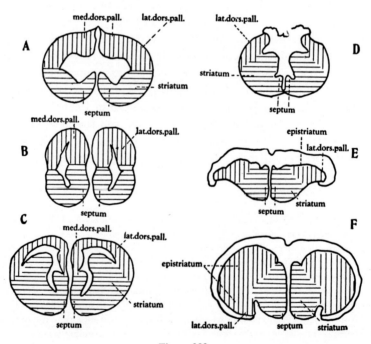

Figure 323

regards differences of interpretation, I believe, in contradistinction to
Northcutt, that the piriform lobe cortex of mammals derives from
both $D_2$ (P IIb) and $D_1$ (P IIIb). Likewise, the lateral fringe of isocortex
(insular cortex) doubtless has ontogenetic relationships to $D_2$ as well as

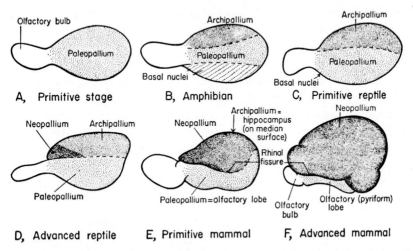

*Figure 324.* Diagrams purporting to show progressive phylogenetic differentiation of
the vertebrate cerebral hemisphere in lateral view according to Romer's interpretation
(from Romer, 1950).

---

*Figure 322.* Diagram indicating my early concept of relationships between claustrum,
main portion of amygdaloid complex, and corpus striatum in a generalized 'lower' mammalian brain (from K., 1924). Except for a further subdivision of 'b' and a substantial participation of this neighborhood ($B_{1+2}$ of subsequent notation), this diagram is fully compatible with my present views. amβ: main amygdaloid complex ('n. amygdalae beta'); b: 'nucleus basalis'; csβ: cortex lobi piriformis ('cortex striatus beta'); cl: claustrum; cr: hippocampal formation ('cortex rudimentarius'); cr': cortex basalis ('cortex rudimentarius');
csα: parahippocampal cortex ('cortex striatus alpha'); ho: isocortex ('cortex homogeneticus', 'neocortex'); pa: 'nucleus parolfactorius accumbens'; s: nucleus basimedialis superior
('nucleus septi'); str: corpus striatum; the terms 'cortex rudimentarius', 'cortex striatus',
'cortex homogeneticus', which I have now discarded, were based on, and adapted from
Brodmann's concepts.

*Figure 323.* Schematic representation of the different developmental types of telencephalon in various vertebrates according to the interpretation of Kappers, Huber, and
Crosby, for comparison with the figure 321. While A, B, and E would be essentially
compatible with my interpretation, C fails, in my opinion, to express the internation of
neighborhood $D_1$, while D and F differ from my interpretation as regards the extent of the
pallium (from Kappers *et al.*, 1936). A Petromyzon. B Amphibia. C Reptilia. D Holoce-
phali. E Holostei. F Teleostei.

$D_1$ (P IIIa and P IIIb). Also, it will be noted that Northcutt does not take into consideration the peculiar significance of the claustrum, respectively lateral epibasal nucleus. Despite some obvious similarities which can be recognized by comparing the relevant illustrations, my own concept therefore differs in several significant respects from the interpretation by the cited author. Looking at figure 326B, it can, in particular, be noted that Northcutt seems to derive the isocortex of Marsupials from $D_1$ (his P III).

It might, moreover, be added that the *homology concept* stressed in the quoted symposium is defined in terms of phylogeny and said to mean 'common ancestry'. It is hardly necessary to repeat here the considerations in favor of a purely formanalytic homology definition independent of phylogenetic speculations and based on elementary notions of topology, as elaborated in volume 1 of this series.

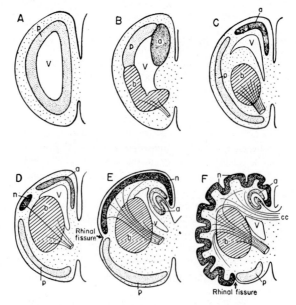

*Figure 325.* Diagrams purporting to show phylogenetic stages in the evolution of vertebrate cerebral cortex and 'corpus striatum' according to Romer's interpretation (from Romer, 1950). A 'Primitive stage'. B 'Stage seen in modern amphibians'. C 'More progressive stage, in which basal nuclei have moved to interior, and pallial areas are moving toward surface'. D 'Advanced reptilian stage'. E 'Primitive mammalian stage'. F 'Progressive mammal'. a: 'archipallium'; b: 'basal nuclei'; cc: 'corpus callosum'; n: 'neopallium'; p: 'paleopallium'; V: ventricle.

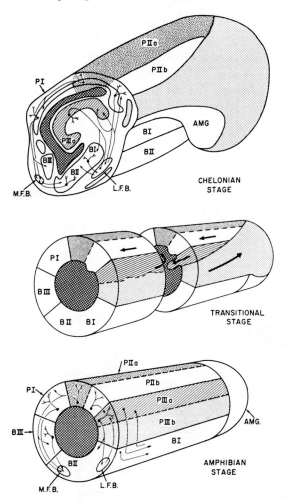

*Figure 326 A.* Schema purporting to be a representation of the left telencephalic hemi-spheres of supposed ancestral amphibian stage leading to the 'Chelonian stage' of recent turtles (from NORTHCUTT, 1969). P I: dorsal column or field; P IIa: dorsomedial compo-nent of dorsal pallial column; P IIb: dorsolateral component of dorsal pallial column; P IIIa: dorsal ventricular ridge; P IIIb: ventral component of lateral pallial column; B II: ventral basal column; B III: medial basal column; L.F.B.: lateral forebrain bundle; M.F.B.: median forebrain bundle. NORTHCUTT's so-called 'Chelonian stage' should be compared with my 'generalized lower mammalian stage' of 1924 shown in figure 322.

Although, in the cited symposium, the participants dealing with the problem of homology chose to sidestep the here relevant issues, a gleam of diffidence occasionally seems to cross their mind, since allusions to the problem of circular reasoning (i.e. *petitio principii*) can be found in their discussions, but the logical cogency of this predicament is conveniently ignored. Instead, vague phylogenetic qualifications, such as field, discrete, patristic, and cladistic homologies are made use of. Readers interested in these elaborations on phylogenetic homology are referred to the volume published by the *New York Academy of Sciences* (1969).

A general bauplan of the Amniote telencephalon, taking into consideration the arcwise connectedness of wall neighborhoods and their relationships to some of the ventricular grooves, might be convenient-

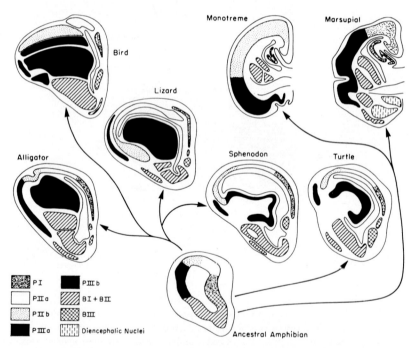

*Figure 326 B.* Schema purporting to be a representation of the 'prototypic columns' of the ancestral amphibians through modern tetrapods. It is suggested that sphenodons, crocodilians, lizards and birds represent '*cladistic field homologies*', and that turtles and mammals represent independent '*cladistic lines*' which are more closely related to each other than to other living tetrapods (from NORTHCUTT, 1969).

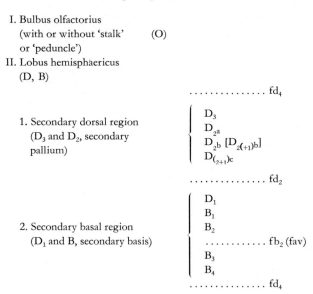

I. Bulbus olfactorius
  (with or without 'stalk'      (O)
  or 'peduncle')
II. Lobus hemisphaericus
  (D, B)

                                    ............. fd$_4$

1. Secondary dorsal region          D$_3$
   (D$_3$ and D$_2$, secondary       D$_{2^a}$
   pallium)                          D$_{2^b}$ [D$_{2(+1)^b}$]
                                     D$_{(2+1)c}$

                                    ............. fd$_2$

                                     D$_1$
                                     B$_1$
2. Secondary basal region            B$_2$
   (D$_1$ and B, secondary basis)    ........... fb$_2$ (fav)
                                     B$_3$
                                     B$_4$

                                    ............. fd$_4$

*Figure 327 A*. Generalized pattern formula of Amniote telencephalon.

ly expressed by the tabulation of Figure 327 A. An additional tabulation (Figures 327 B) provides a comparison of the main griseal components within the Amniote hemisphere.

The introduction of a *formanalytic symbolism* (K., 1929a), which subsequently proved to be an elementary *topologic notation* applicable to
vaguely circumscribed wall neighborhoods within a tubular configuration, places these nevertheless reasonably well distinguishable neighborhoods into a combinatorially exactly determined framework (K.
1967). It introduces thereby a set of logically rigorous designations for
morphologically identical grisea within the neuraxis of the entire vertebrate series from fish to man.

This notation, which, at first glance, might seem unduly abstract as
well as unwieldy and intricate in its application to 'higher vertebrates',
has not become generally accepted.[241] Yet, I maintain that, in view of
the obtaining intrinsic difficulties and complexities of the subject mat-

---

[241] It was, however, used by my collaborators (cf., e.g., GERLACH, 1933, 1947; and
MILLER, 1940). The partial use of this notation with modified symbols by KÄLLEN (1951,
1953) and by NORTHCUTT (1969) was pointed out further above.

|  | Reptiles | Birds | Mammals |
|---|---|---|---|
| O | bulbus olfactorius | bulbus olfactorius | bulbus olfactorius |
| $D_3$ | cortex medialis | cortex medialis | hippocampal formation |
| $D_{2am}$ | cortex dorsalis, pars medialis | cortex dorsomedialis | parahippocampal cortex |
| $D_{2al}$ | cortex dorsalis, pars intermediolateralis | nucleus diffusus dorsalis | isocortex |
| $D_{2(+1)}b$ | cortex dorsalis, pars lateralis |  | isocortex, regio insularis |
| $D_{(2+1)}c$ | cortex lateralis | nucleus diffusus dorsolateralis with lateral corticoid area | 'prepiriform' cortex |
| $D_1$ | nucleus epibasalis lat. et med. | nucleus epibasalis with subdivisions | claustrum $(D_1l)$ |
|  | nucleus (epibasalis) centralis | nucleus centralis with subdivisions | corpus striatum |
|  | nucleus epibasalis posterior | nucleus epibasalis caudalis | part of amygdaloid complex |
| $B_1$ | nucleus basilateralis superior |  |  |
| $B_2$ | nucleus basilateralis inferior | nucleus basalis | corpus striatum |
| $B_3$ | nucleus basimedialis inferior | nucleus basimedialis inferior | nucleus basimedialis inferior |
| $B_4$ | nucleus basimedialis superior | nucleus basimedialis superior | nucleus basimedialis superior |
| $B_s$ | basal cortex | basal cortex | basal cortex |

*Figure 327 B.* Synoptic tabulation of main grisea in the Amniote telencephalon.

ter, it affords a much better approach than the conventional procedures, which, in addition to considerable disagreements between the
interpretations of the diverse authors, have led to a highly confusing,
and, I believe, in part even misleading, terminology.

With respect to the problem under consideration, it seems therefore appropriate to point out that botanists, in the study of plant morphology,[242] elaborated, long ago, *diagrams* and *formulae* for the map-

---

[242] The noted anatomist, histologist and cytologist MARTIN HEIDENHAIN of Tübingen
(1864–1949) likewise stressed, in his monograph '*Die Spaltungsgesetze der Blätter*' (1932),
the significance, for general theoretical morphology, of problems specifically pertaining
to botany. As regards Vertebrate comparative anatomy, it is hardly necessary to point

ping of configurations displaying various degrees and aspects of differentiation. It is evident that this procedure by botanists, introduced and employed without explicit reference to topology, is, nevertheless, likewise *topologic*. Although plant structures, as a rule, are somewhat more clearly delimited from each other than grisea within the vertebrate neuraxis, we have, both in the tubular neuraxis and in the flowers of Angiospermae or in the stem of Pterydophyta, a segregation of definable neighborhoods resulting in specific patterns of arcwise connectedness. Figure 328, from FITTING (1931) shows, in diagrammatic fashion, the *progressive configurational differentiation* of conducting systems formed by complex permanent tissues (xylem and phloem) in the stem of Pterydophyta.

As regards *flowers*, characterizing Angiosperms, and constituting complex configurations provided by an array of 'organs' involved in reproduction, the following components, which may be evaluated as

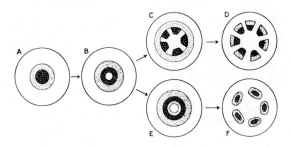

*Figure 328.* Diagram showing progressive differentiation of conducting systems in the stem of Pterydophyta (from FITTING, 1931). The arrows are meant to indicate an inferred phylogenetic sequence. White: cortex and pith (marrow); white with black dots: phloem; black with white dots: xylem; A: Ferns, generalized; B: Schizacea; C: Osmunda; D: Ophioglossum; E: Marsilia; F: Aspidium Filix mas.

---

out the common use of *dental formulae*, such as $i_3^3$, $c_1^1$, $p_4^4$, $m_4^4$ for a postulated Mammalian dental *bauplan*, or the abbreviated notations $\frac{2\,1\,2}{2\,1\,2}$ and $\frac{2\,1\,2\,3}{2\,1\,2\,3}$ for human deciduous respectively permanent dentitions. Various authors, particularly L. BOLK in his *dimer* theory, and RÖSE in his *concrescence* theory, have elaborated hypotheses concerning dental evolution on the basis of these pattern models. Again, with respect to the Vertebrate muscular system, my esteemed old friend SEIHO NISHI, in his fundamental morphologic contributions, has made use of relevant *bauplan formulae* of the type:

    $D_3\,D_2\,D_1\,L$ *( Dorsolateralis trunci)*

    $V_1$ *(Obliquus sup., obl. inf., rectus trunci)*

    $V_2$ *( Longus trunci, hypothetical)*.

*Grundbestandteile*, are commonly recognized. (1) The *calyx*, an outer circle of frequently green leaves, called sepals. (2) The *corolla* (corona, crown), an inner circle of frequently white or colored leaves called petals, which may be separate (choripetal flowers) or fused (sympetal flowers). The circles (1) and (2) are collectively called *perianth*. If both circles display the same aspect, they are subsumed under the term *perigon*. (3) The *androeceum*, consisting of male reproductive organs (stamens) inside the petals. A stamen includes a stalk (filament) and a pollen bearing anther. (4) The *gynaeceum* or pistil, a female reproductive organ, located in the center of the flower. It is either single or compound, consisting of fused pistils. The pistil comprises a basal ovary, a

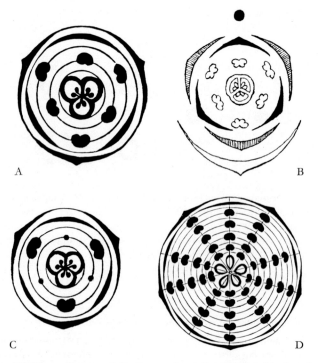

*Figure 329.* Flower diagrams, indicating configurational relationships based on pattern orderliness, and elaborated by botanists for a suitable morphologic classification of Angiospermae. Further explanations in text. A Generalized flower diagram (from Hansen, 1914). B Diagram of pentacyclic Monocotyledone flower, Lilium (from Karsten, 1931). C Diagram of Monocotyledone iris (from Hansen, 1914). D Diagram of Dicotyledone Ranunculacea Aquilegia (from Hansen, 1914).

stalk (style), and an enlargement (stigma) at the top. With respect to
the attachment of the other organs at the central neighborhood (recep-
tacle), the position of the ovary may be superior, inferior, or intermediate. A perfect flower contains androeceum and gynaeceum, an imperfect flower either one but not both.[243]

The *flower diagrams* indicate the configuration of organs by *concentric
circles*, the corolla leaves being usually marked with a median external
crest; androeceum and gynaeceum are represented as cross-section
outlines. Missing or reduced components become indicated by dots
(or crosses), and a dot outside the diagram stands for the inflorescence-
(or mother-) axis. Figure 329 illustrates several such diagrams.

In a still more *abstract mapping*, the diagrams have been substituted
by *flower formulae* using an appropriate *notation*. K stands here for calyx,
C for corolla, and P for perigon; A designates the androeceum, and G
the gynaeceum. Figures affixed to the letters indicate the number of
form elements within each Grundbestandteil. Additional numbers of
circles displayed by a Grundbestandteil are indicated with the sign $+$.
The sign 0 refers to missing components, and a large, more or less indefinite increase in the number of elements is expressed by the sign $\infty$.
Inclusion in parenthesis (...) indicates fusion of the components, and
duplication is marked by a superscript number. The notations G below
or above a line ($\frac{}{G}$, $\frac{G}{}$) refer to an inferior respectively a superior gynaeceum. A hyphen – is used in case of variable numbers of compo-
nents.

Thus, the generalized flower diagram of figure 329 A can be expressed by the formula K3, C3, A3, 3, G(3). With a minor modification,
namely P3$+$3, A3$+$3, $\frac{G}{}$ (3), it would apply to most Liliaceae, as
shown in the slightly different diagram of Figure 329 B. This latter includes two structures not pertaining to the flower itself, namely the
cross-section of the inflorescence axis above, and of the so-called 'supporting leaf' below. An Iridaceous flower, depicted by figure 329 C
corresponds to the formula P3$+$3, A3$+$0, $\frac{}{G}$(3). Figure 329 D shows a
Ranunculaceous flower with the formula K5, C5, A$\infty$, $\frac{G}{}\infty$.

It is evident that such *bauplan formulae*, particularly well applicable
to circular organic configurations consisting of arcwise connected

---

[243] Perfect flowers are hermaphroditic, and imperfect ones, also called 'dicline flowers',
are of single sex. Monoecious plants carry dicline flowers of both sexes on one and the
same plant. Dioecious plants carry staminate flowers and pistillate flowers on separate,
male, respectively female plants. Perfect flowers occur only in various Angiosperms.

neighborhoods and displayed by the flowers of Angiosperm Spermatophyta as well as by the vertebrate neuraxis, are helpful in expressing the topologic invariants pertaining to a diversity of forms with dissimilar shapes and differing degrees of differentiation.

In this respect, the vertebrate telencephalon, despite the complexity of its considerable differentiation culminating in mammals and man, retains a topologically relatively simple circular bauplan. In the unpaired evaginated hemisphere (telencephalon impar s. medium), a complete (closed) bilateral symmetric wall circle is formed by the components B, D, and R (epithelial roof plate). In each rostral paired evaginated hemisphere, a complete (closed) asymmetric wall circle is formed by the Grundbestandteile $D_3$, $D_2$, $D_1$, $B_1$, $B_2$, $B_3$, $B_4$. Each wall circle is again subdivided into essentially two concentric circles of grisea, in contradistinction to the more numerous concentric circles of flower structures. These griseal circles are (1) cortex ($D_s$, $B_s$) and (2) periventricular grisea D and B (pallial periventricular grisea, basal ganglia, paraterminal body), as particularly clearly displayed by Dipnoans (cf. Figs. 265, 267). As in flowers, these circles vary greatly with respect to their differentiation. Thus, in the mammalian (secondary) pallium, only the outer circle or *cortex*, $D_{3(s)}$, $D_{2(s)}$ becomes predominant, while the inner one, at most, remains recognizable as the *subependymal cell plate* (cf. vol. 3/I, section 1). *Per contra*, in the mammalian (secondary) basis, the inner circle *(corpus striatum*, etc.) predominates, the outer one being the rudimentary *basal cortex* ($B_s$).

The progressive differentiation of the Vertebrate telencephalon either in the taxonomic or in the presumptive phylogenetic series, because of the relatively simple circular arrangement of its 'initially' rather few griseal components, seems to be particularly compatible with an analysis in terms of a *prima facie* tedious and highly abstract topologic notation.

The use of comparable bauplan formulae did not, so far, seem necessary for an overall elucidation of the morphologic patterns displayed by the diencephalon or by the deuterencephalon and its different parts, but could, if required, be elaborated. Thus, the *bauplan of the diencephalon* being R (roof plate), E, Td, Tv, Hy, a topologic notation[244] for its grisea, conceived as neighborhoods, is evidently feasible. *Mutatis mutandis*, a similar symbolism may be introduced in mapping *mesencephalon*

---

[244] E, Td, Tv, Hy stand here for epithalamus, thalamus dorsalis, ventralis, and hypothalamus.

and *rhombencephalon*, their overall formulae[245] being R(o), At, As, Ai, Bi, Bv, and R, Ad, Ai, Bi, Bv, F (floor plate), respectively. Diagrams, by GERLACH (1933, 1947), indicating these topologic relationship, were shown above in Figures 158, 163, and 190B.

Although one could claim, with some justification, that the highly evolved telencephalon of mammals and man is structurally far more complicated than diencephalon, mesencephalon and rhombencephalon (including cerebellum) at any stage of vertebrate evolution, the elaboration of *detailed* diencephalic, mesencephalic and rhombencephalic pattern formulae would require somewhat more complex and awkward formulae than for telencephalic components. Such complexity is unavoidable because in diencephalon and deuterencephalon there obtains a greater morphological differentiation in depth, within the periventricular cell aggregates, of very numerous, closely adjacent and often poorly delimitable different grisea, e.g. thalamic nuclei and brain stem nuclei, with multiple neighborhood connectedness. This differentiation in depth is, moreover, combined with a morphogenesis of grisea related to an inner and an outer circle of cell layers (cf. e.g. the study on reptilian diencephalic and mesencephalic ontogenesis by SENN, 1968). In addition, cortical cell arrangements, in principle roughly similar to those of the telencephalon, occur in tectum mesencephali and cerebellum of most vertebrates.

In attempting to express the topologic connectedness of deuterencephalic and diencephalic grisea, which is of a more diversified type than that of the telencephalic ones, the resulting inordinately complicated notation might thus perhaps defeat its useful purpose.[246]

Before concluding the present section 6, on morphogenesis of the telencephalon, with a brief summary of the overall features characterizing telencephalic evolution in the vertebrate series, it seems perhaps

---

[245] The symbols have the following meaning: A: alar plate; Ad: dorsal zone of alar plate; Ai: intermediodorsal zone; As: torus semicircularis zone of mesencephalic tectum; At: dorsal tectum mesencephali; B: basal plate; Bi: intermedioventral zone; Bv: ventral zone; F: floor plate; R: roof plate; R(o): obliterated roof plate.

[246] Unfortunately, any approach leading to an advanced understanding of the highly complex aspects of neurobiology requires unremitting, highly tedious work and considerable stamina. Thus, a sufficient acquaintance with apparently unrelated subjects such, e. g., as topology and 'logistics' appears necessary. It becomes then a question of proper judgement to decide how far one should go in using and adapting intricate abstract concepts for purposes of theoretical neurology.

appropriate to add some comments on two additional topics, namely on some *general aspects of corticogenesis* in relation to 'matrix exhaustion and on the morphology of the *telencephalic commissures*.

*Corticogenesis*, namely formation of a cerebral cortex as defined above in footnote 195 (p. 476, is characterized by a separation, through a displacement conceivable as 'migration', of a cellular 'plate' from the secondary periventricular matrix. The relatively cell-poor layer between cortex and periventricular cell populations, respectively ependymal lining, affords a passage for nerve fibers. If these latter become myelinated, this layer is designated as medulla or *'subcortical medullary core' ('subkortikales Mark')*.

The distinction of primary and secondary matrix was defined in vol. 3/I, p. 31, of this series. The *primary matrix* is concerned with histogenesis, i.e. with the differentiation of specific histologic elements (e.g. neuroblasts) from 'indifferent' medullary epithelium or 'mother cells'. It can be conceived as consisting of the M, I, and S zones of S. Fujita as discussed in the preceding chapter V. The *secondary matrix* is concerned with the differentiation of particular grisea consisting of diversified elements present in that matrix. In the spinal cord, and at relatively early stages of ontogenetic development, the primary matrix corresponds to the *'ependymal matrix layer'*, and the secondary matrix represents the *'mantle layer'* of His (cf., e.g., vol. 3/I, Fig. 21, p. 23). At later stages, although a difference in density between an inner and an outer portion of the periventricular cell masses may still obtain, the activities characterizing the primary matrix have subsided.[247] At such stages, the boundary between the original primary and the secondary matrix zones may become rather indistinct. In this respect, the various regions of the neuraxis display different sorts and sequences of density patterns, precluding the formulation of an overall valid scheme of 'matrix phases'. Discounting some peripheral migrations, the noncortical periventricular grisea of the neuraxis commonly differentiate *in situ* from the secondary matrix by a process of maturation correlated with expansion, including somewhat greater spacing or dispersion of the el-

---

[247] In the course of these events, and depending on regions as well as taxonomic forms, an overlap of primary and secondary matrix activities may occur, as, e.g., suggested by the radioautographic observations of Yoshida (1967). Again, the semantic distinction between primary and secondary matrix is not meant to exclude the possibility that scattered 'immature', or even 'bipotential' elements remain included not only in secondary matrix but also in adult grisea (cf. vol. 3/I, p. 64).

ements, rather than by 'exhaustion'. A narrow, relatively undifferentiated zone, close to the ependyma, *may* remain as the *subependymal cell
plate* (cf. vol. 3/I, p. 59f.).

In addition to the regional differences in the CNS of one and the
same individual, there are substantial species, order, and class dissimi-
larities in the differentiation pattern of a given particular region.[248]

As regards the development of the telencephalic cortex cerebri in
the taxonomic Vertebrate series, a nondescript diffuse periventricular
arrangement of cells is found in *Petromyzon*, while a peculiar, rather
aberrant lamination is seen in the 'solid' telencephalon of adult *Myxinoids*. *Selachians* and *Dipnoans* display conspicuous cortical laminae externally to substantial pallial and basal periventricular cell aggregates.
The everted telencephalon of *Osteichthyes*, on the other hand, does not
seem to contain grisea which could justifiably be designated as cortical
or 'corticoid'.

*Amphibians* exhibit a particularly simple and clear-cut telencephalic
pattern with structurally rather easily distinguishable neighborhoods,
which, however, do not manifest a cortical differentiation comparable
to that found in Selachians or Dipnoans. Nevertheless, a limited degree of pallial and basal corticoid formation can be seen in Gymnophiona and perhaps (transitorily) in late Anuran larvae.[249]

---

[248] Cf. e.g. the conspicuous differences, *qua* so-called 'matrix phases' in a morphologically identical region (hypothalamus) of diverse Amphibia as shown further above in
figures 205 and 206.

[249] Thus, in his noted second *Arris and Gale Lecture*, particularly concerned with the
evolution of the cerebral cortex, ELLIOT SMITH (1910) chooses four selected types of cortical differentiation, namely in *Selachians*, in *Dipnoans*, in *Reptiles*, and in *Mammals*. Because of
their conspicuous cerebral cortex, the Dipnoans are, according to the cited author 'much
nearer the main line of amniote evolution than the amphibia and most other fishes, and so
far as the special object of these lectures is concerned, i.e., the evolution of the cortex
cerebri, the dipnoi exhibit a more highly developed – at any rate, a much more reptilian-
like – pallium than we find in any of the Amphibia. Birds again present such a highly
specialised and diversely-modified cerebrum that they have quite left the path leading to
the formation of a neopallium and evolved a mechanism peculiar to their own class, which
does much the same kind of work as the neopallium.' Despite my high esteem for ELLIOT
SMITH's considerable accomplishments, I am not inclined to accept that author's phylogenetic evaluation of Dipnoans, but believe that the 'ancestors' of Amniota were presumably represented by vertebrates more similar to extant Amphibia than to any other recent
form. Unfortunately, the relevant histologic differentiation of the telencephalon in significant fossil forms remains entirely unknown. Concerning the possibility that the phyletic
relationship of mammals to extant amphibians might possibly be less remote than the
relationship of mammals to recent Reptilia, cf. the comments on p.144 *et passim* in
volume 1 of this series.

In *Reptilians*, the differentiation of three distinctive pallial cortical laminae leads to an exhaustion of most of the periventricular matrix, although, internally to dorsal and to medial cortical plate, a thin periventricular layer may still be present, as shown in Figures 280, 288, and 291. Concerning the lateral part of the dorsal plate, and the whole extent of the lateral plate, it will be recalled that these cortices presumably receive, in the course of ontogenesis, some cellular contributions from a neighborhood $(D_1)$ mainly concerned with the development of internal, noncortical grisea. The ill-defined basal cortical plate $(B_s)$ represents a minor peripheral differentiation not significantly affecting the overall amount of the periventricular cell masses.

In the *Avian telencephalon*, the whole matrix of the $D_3$ and $D_{2a}$ neighborhoods is transformed into a more or less stratified respectively laminated cortical plate. Because, in these regions, the pallium remains very thin, its cortical grisea thereby occupy practically the whole width of the wall, their inner border remaining very close to the ependyma. Neighborhood $D_{2b}$ is not differentiated as a typical cortex but rather as diffuse cell aggregates. The corticoid plate $D_{(2+1)c}$, along much of its extent, is in close apposition to the internal griseal cell masses. The same relationship to basal nuclei is likewise displayed by the poorly developed basal cortex $B_s$.

It will be recalled that ELLIOT SMITH (1910) as quoted above in footnote 249, omitted the avian telencephalon from his elaboration of the cerebral evolution culminating in mammals, because he considered birds as having left the phylogenetic path toward formation of a neopallium, and thereby displaying a special pattern peculiar to their own class. *Mutatis mutandis*, a similar estimate could be reached with regard to the everted Osteichthyan telencephalon.

In the *Mammalian telencephalon*, which is characterized by an extensive and highly differentiated pallial cortex cerebri, including a hexalaminar isocortex or 'neocortex', most regions of pallial cortex are hologenic. This mode of development involves a peripheral migration and exhaustion of the entire periventricular matrix ('*Matrixaufbrauch*', K., 1922, 1929a), except for a thin and variable remnant representing the *subependymal cell plate* (GLOBUS and K., 1942). Only the ventrolateral fringe region of pallial cortex, comprising insular isocortex and piriform lobe allocortex, originates partly by matrix exhausion of $D_2$ and partly by accessory contributions from neighborhood $D_1$, the bulk of whose secondary matrix remains *in situ*, participating, by maturation, in the morphogenesis of the basal ganglia. The basal cortex $B_s$, com-

prising area ventralis anterior[250] and the cortical portion of the amygdalar complex, is, of course entirely merogenic, being provided by a
mere fraction of the basal secondary matrix.

Concerning the aspects of mammalian corticogenesis here under
consideration and also briefly pointed out in some of our early communications (K. and v. DOMARUS, 1920a, b; K., 1922), two relevant studies on the ontogenesis of human cerebral cortex have been undertaken
by LOO (1929) and by KAHLE (1969).

The significant events in mammalian embryonic corticogenesis
lead, first, to the formation of a relatively homogeneous cortical cell
plate, and second, to a further differentiation of this plate resulting in
stratification or lamination (Fig. 330). This sequence of events can be
conveniently described by distinguishing the following four stages.

(A) At the early stage of pallial corticogenesis (e.g. in man at about
10-mm length, respectively circa 4 weeks) only a primary matrix layer
and a narrow external marginal or zonal layer (zona marginalis of LOO,
1929) are evident.[251]

(B) Subsequently (at about 16 mm or *circa* 6 to 7 weeks) three layers can roughly be distinguished, namely matrix layer, dispersal layer
(somewhat comparable to the mantle layer of HIS in spinal cord), and
marginal or zonal layer.

(C) At the next, particularly significant stage (about 25 mm or *circa*
8 to 9 weeks) a fairly dense cortical lamina, which does not display any
clearly recognizable stratification, becomes apparent. This lamina is

---

[250] It should again be stated here that the *tuberculum olfactorium*, which, in some (macrosmatic) mammals, represents the most conspicuous part of area ventralis anterior (or
anterior basal cortex), does not comprise the entire extent of this formation pertaining
to $B_s$. The tuberculum olfactorium, commonly characterized by a pattern of cell clusters
('*islets of Calleja*') may be, both grossly and microscopically, poorly developed and not
distinctly recognizable in some forms, being replaced by a nondescript extension of
anterior perforated space. The *area ventralis anterior, in toto*, includes the *tuberculum
olfactorium*, with its surroundings, namely *transitional regions of basal cortex bordering* on
piriform lobe cortex and on bulbus (respectively tractus) olfactorius, moreover '*substantia
perforata anterior*' (which frequently overlaps with the 'tuberculum'), and the *nucleus of
Broca's diagonal band* with its undefined extension into, or fusion with, a superficial griseum
of nucleus basimedialis superior ($B_{4s}$).

[251] The *primary matrix layer*, although appearing homogeneous in routine cell preparations, may of course, be further subdivided in accordance with the histogenetic events
studied by S.FUJITA and others, as discussed in chapter V, section 1, of the preceding
volume 3/I.

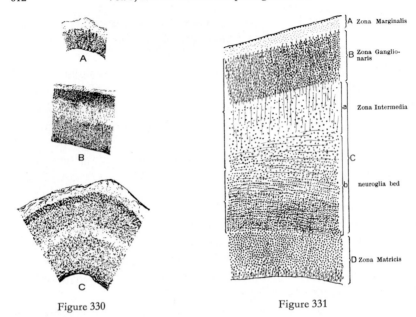

Figure 330            Figure 331

*Figure 330.* Different stages in the ontogenetic development of the mammalian isocortex (from KAPPERS, 1928). A Rabbit embryo of 1 cm. B Rabbit embryo of 2.5 cm. C Rabbit embryo of 6 cm. The inner (ventricular) surface is at bottom.

*Figure 331.* Cortical plate in the isocortical parietal region of an approximately 10-week-old (39.1 mm long) human embryo (from LOO, 1929).

formed by a crowding of cellular elements separated from the matrix by a layer of lesser cell density, and gradually increases in thickness.[252] At such stages, the following layers can be distinguished: (1). A relatively cell-poor zonal or marginal layer; (2) a dense cortical lamina ('zona ganglionaris' of Loo, 1929); (3) a relatively cell-poor subcortical layer, the anlage of the medullary core ('zona intermedia' of Loo, 1929), and (4) a periventricular matrix layer. The cell population in the

---

[252] It should be emphasized that this formation of a distinct cortical lamina displays, in man and some other mammals about which data are available, a noticeable gradient, especially in the latero-medial direction. The differentiation of the cortical lamina in the lateral part of the pallium seems to precede that taking place in the medial portion. Thus a gradual transition between the configuration of stage (C) and stage (B) can be noticed in cross-sections through the mammalian telencephalon at early periods of stage (C) corticogenesis.

zona intermedia of Loo seems to consist, to a large extent, of migrating elements contributing to the formation of the cortical plate. Within the zona intermedia, a denser inner sublayer, represents a transition to the dense matrix, and a less dense outer sublayer adjacent to the cortical plate can usually be recognized. Loo (1929) designated the outer sublayer as portion a, and the inner one as portion b of his zona intermedia (Figure 331). Toward the end of this, and at the outset of the next stage, a relatively dense row of cellular elements at the surface of the marginal or zonal layer, adjacent to the leptomeninx, may become noticeable and bears a slight resemblance to the well-known transitory superficial granular layer of the developing cerebellar cortex. The superficial cell assemblage in the marginal layer of cerebral cortex, however, is far less massive than that in the embryonic cerebellar cortex and also differs from that latter as regards origin and histogenetic significance.

(D) The definitive ontogenetic evolution of the mammalian pallial cerebral cortex begins with the 'lamination stage' of Loo (1929) characterized by the gradual differentiation of the cortical layers. In man, this developmental process becomes recognizable at approximately 12 weeks (56-mm length) of intrauterine life, and displays regional differences. It leads to the well-known subdivisions of mammalian cortex cerebri into cytoarchitecturally more or less distinctive areae.[253] In the course of these events, the periventricular matrix of the hologenic cortex disappears except for some remnants representing the *variable* subependymal cell plate.

For a phylogenetic interpretation of corticogenesis it is doubtless of interest that adult cytoarchitectural aspects of telencephalic pallium in amphibians and reptilians can be roughly compared with transitory ontogenetic stages of mammalian cortex (Figs. 332A, B). Thus, with regard to *matrix*, respectively *matrix exhaustion*, and to the formation of a *cortical plate*, stage (B) of mammalian corticogenesis corresponds approximately to the general amphibian, and mammalian stage (C) to the general reptilian configuration.

---

[253] This phase (D) of corticogenesis, namely lamination and detailed areal differentiation, particularly investigated by KAHLE (1969) in man, pertains to the final growth phase, involving 'minor' morphogenetic events in the aspect here under consideration (cf. vol. 1, p. 230 of this series). The relevant topics concerning lamination, other significant structural features, and function of mammalian cerebral cortex shall be discussed in chapter XV of volume 5.

The essential difference between the mammalian adult isocortical lamination, and the submammalian cortical respectively pallial cytoarchiteture is diagrammatically shown in Figure 333. Because of its high degree of specific differentiation combined with a hexalaminar pattern typical for, and restricted to, a pallial region of mammals, the term *neopallium*, introduced by ELLIOT SMITH, respectively the designation '*neocortex*' as a synonym for *isocortex* can evidently be justified. JAKOB and ONELLI (1911) stressed the particular evolutionary significance of mammalian cortex cerebri by distinguishing the following phylogenetic stages in corticogenesis:

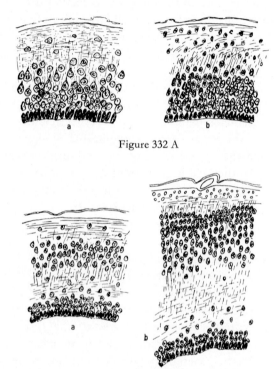

Figure 332 A

Figure 332 B

*Figure 332.* Semidiagrammatic comparison of amphibian precortex and reptilian cortex paulostratificatus with stages in ontogenetic evolution of mammalian isocortex (from K. and v. DOMARUS, 1920). A Comparison of amphibian area dorsalis pallii in adult *Salamandra maculosa* (a) with an isocortical $D_2$ pallial area in a human embryo (b) of about 7 weeks, showing periventricular layer and dispersal layer ('Schwärmschicht'). B Comparison of reptilian $D_2$ region in young adult *Lacerta agilis* (a) with an isocortical $D_2$ pallial area in a human embryo of about 4–5 months.

a) *den ependymären Praecortex der Amphibien,*

b) *den Cortex inferior monostratificatus der Reptilien und Vögel.*

c) *den Cortex superior polystratificatus der Säuger und des Menschen.*

Although the pallial cerebral cortex of birds is, at least in some regions, definitely stratified, and even the reptilian cortex pallii can be described, depending on arbitrary criteria, as manifesting at least a slight degree of stratification in some of its part, the cited evaluation by JAKOB and ONELLI appears essentially sound. One might perhaps here substitute '*paulostratificatus*' for '*monostratificatus*'.

A phylogenetic interpretation concerning the significance of the periventricular griseum in the amphibian pallium becomes, however, complicated by the fact that the pallium of taxonomically still 'lower' Anamnia such as Selachians and Dipnoans displays conspicuous cortical or corticoid plates not present in Amphibia. One would then have to assume that, if the periventricular arrangement in the amphibian pallium is 'primitive', the cortical differentiations in recent Selachians and Dipnoans are secondary. Or, assuming that a periventricular arrangement of the pallium in early Gnathostomes was primary, but evolved cortical formations at early stages of phylogenetic evolution

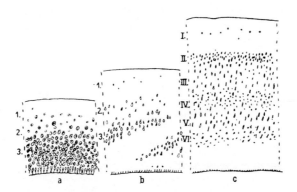

*Figure 333.* Semidiagrammatic drawing illustrating differences between mammalian isocortical lamination and cytoarchitecture of pallial regions in submammalian forms (from K., 1922). a: $D_2$ neighborhood of the urodele amphibian, *Triton;* 1: 'Zonalschicht'; 2: 'Schwärmschicht'; 3: 'Basalschicht'; b: $D_2$ neighborhood (cortex lateralis, superpositie lateralis) of the reptilian Lacerta; 1: 'Zonalschicht'; 2: 'lockere äussere Rindenplattenschicht'; 3: 'dichtere innere Rindenplattenschicht'; c: isocortex of postcentral type in the Metatherian mammal, Didelphys; I–VI standard isocortical laminae in notation adopted by BRODMANN.

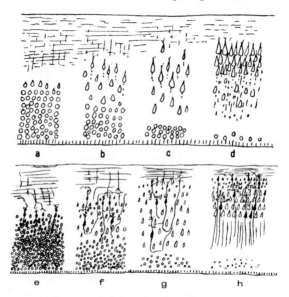

*Figure 334.* Diagram illustrating vertebrate corticogenesis and pallial matrix exhaustion in accordance with the theory of neurobiotaxis (a–d from K., 1922; e–h from K., 1924a). a and b correspond approximately to conditions in urodele amphibians, c is roughly comparable to area medialis pallii ($D_3$) of Gymnophiona and d indicates aspects of reptilian cortex dorsalis; e, f, g represent submammalian developmental stages in which axons for distant conduction remain significantly restricted to zonal layer, h shows the mammalian stage at which axons for distant conduction provide the bulk of the subcortical medullary 'center'.

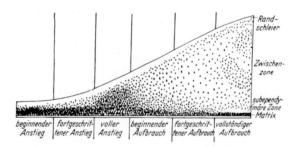

*Figure 335.* Diagram illustrating concept of 'matrix exhaustion' and 'matrix phases' according to KAHLE (from KAHLE, 1951). This author (1958) also subsumed the three first phases from left under the concept *'migration'*, namely *'beginnende'*, *'fortgeschrittene'*, and *'volle Migration'*. The three following ones, are then characterized as *'beginnende'*, *'fortgeschrittene'* and *'vollständige Exhaustion'*. KAHLE believes that *'im Höhepunkt des Anstiegs wird die Zwischenzone von indifferenten Zellen überschwemmt'*. It seems, however, rather likely that secondary matrix and *'Zwischenzone'* contain, to a significant degree, neuroblasts and spongioblasts of essentially 'determined' type, which can no longer be considered 'indifferent'.

preceding the origin of tetrapod ancestors, the perventricular griseum of recent Amphibia would have to be considered a 'regressive' feature. Since the histologic differentiation of extinct 'ancestral' forms cannot be ascertained on the basis of available data, these phylogenetic speculations remain obviously unverifiable and disputable. Nevertheless, because in a strictly periventricular configuration of the pallial grisea, the outer, marginal, ('zonal' or 'molecular') layer contains the afferent (as well as efferent) fiber systems of the pallium, I suggested an interpretation of corticogenesis and matrix exhaustion in conformity with the theory of neurobiotaxis[254] as shown in Figure 334 (K., 1922, 1924). This hypothesis was subsequently endorsed by KAPPERS (1928) who summarized my views by stating that the phylogenetic evolution of cortical gray matter from the periependymal matrix 'is a neurobiotactic result of the superficial position of the afferent cortical fibre systems, as rightly pointed out by KUHLENBECK'.

Be that as it may, and quite apart from an evaluation based on the orderliness of neurobiotaxis, the mammalian cortical plate obviously arises from the periventricular matrix in stages formally comparable with adult amphibian and reptilian conditions. This indicates a close similarity of the ontogenetic and the presumptive phylogenetic development of mammalian cerebral cortex in accordance with HAECKEL's *biogenetic rule* (K. and v. DOMARUS, 1920a, b). The concept of *matrix phases*, shown in Figure 335, has been elaborated by KAHLE (1951, 1956, 1958) as applying to the differentiation of the longitudinal zones in the human diencephalon, and is, to some extent, based on my notion of *matrix exhaustion* (K., 1922, 1929a) relating to corticogenesis. However, it can be seen from a comparison of Figures 334 and 335 that, while in mammalian corticogenesis, actual matrix exhaustion takes place by migration of elements forming a cortical plate separated from the ventricular lining by a medullary core, the diencephalic matrix phases of KAHLE do not, in any significant degree, refer to such separation of grisea from the ventricular lining. The relevant process could

---

[254] Regardless of its dubious 'causal' explanation, the theory of neurobiotaxis can be upheld as a significant morphologic concept. A critical evaluation of neurobiotaxis was given in section 7, chapter V of the preceding volume 3/I. It might here be added that, in a modified formulation, the principle of neurobiotaxis could perhaps be formulated as follows: In the course of phylogenetic evolution, certain fiber systems of particular significance appear to exert, during ontogenetic development, a field effect on some nerve cell populations.

be described as *maturation* and *expansion in depth*, but *in situ*, characterized by a variable pattern of spreading, rarefaction, as well as condensation, of an essentially periventricular cell population.[255]

It is true that, in the process of development an outer and an inner secondary matrix (area cellularis externa and interna of MIURA, 1933, cf. above p. 457) become separated, and subsequently undergo their own, distinctive further differentiation. Yet, the formation of area cellularis externa is not correlated with any appreciable degree of matrix exhaustion.

Moreover, the *matrix phases* of KAHLE, i.e. the aspects of *matrix maturation*, represent structural patterns, which differ in topologically identical regions of taxonomically disconnected vertebrate forms, and cannot be used as criteria for the establishment of morphologic homologies.[256] Nevertheless, a further systematic study of these 'matrix phases' or maturation patterns, as initiated by KAHLE, may contribute to a better understanding of the differentiation processes taking part in, and specific for, the development of the neuraxis in particular vertebrate forms, especially in mammals.

Generally speaking, it could be stated that, within deuterencephalon and spinal cord, some maturation changes in basal plate begin earlier than in the alar plate. Within the diencephalon, the thalamus ventralis and the dorsal hypothalamus display such 'maturation' before

---

[255] The transformation of a rather diffuse *secondary matrix* into groups of 'nuclei' by rarefaction and condensation of cell populations, as well as by growth of neuropil and fibers, bears some general pattern resemblance to 'crystallization' or to 'flocculation' in a colloidal system. These events occur essentially *in situ*, although some amount of displacement or shifting may take place. With respect to diencephalic ontogenesis in mammals and in birds, such processes of differentiation have been described by MIURA (1933) and myself (K., 1937). In a rather abstract formulation, these events may be conceived as morphogenetic or biologic 'field effects'.

[256] Evidently, topologically identical, i.e. morphologically homologous wall neighborhoods may display quite different structures (e.g. $D_3$, the so-called 'primordium hippocampi' of Selachians or amphibians, and the actual hippocampal formation of mammals, with its complex cytoarchitecture and synaptology). Nevertheless, differences in structure, or 'cytoarchitectural gradients', may obviously be used for the delimitation of different wall neighborhoods, and I have followed this procedure since the beginning of my studies, over 50 years ago. Thus, KAHLE (1956) justly states: '*In solchen Fällen hat auch* KUHLENBECK *die unterschiedliche Matrixstruktur zur Abgrenzung einzelner Zonen benutzt*'. KAHLE (loc. cit.) then adds the comment: '*KUHLENBECK hat diesen Gesichtspunkt jedoch nicht systematisch weiterverfolgt*'. The reason for this is quite clear: as explained above and *passim*, I did not (and do not) evaluate this viewpoint as relevant to the purpose of my investigations, which concerned the establishment of morphologic homologies.

dorsal thalamus and ventral hypothalamus. In the telencephalon, 'maturation' of the basis sets in before that of the pallium. These findings can be interpreted as an expression of 'field patterns' determined by 'organizers'. These structural patterns, however, are not identical, nor rigidly correlated, with the morphologic pattern orderliness displayed by the longitudinal zonal systems. Rather, two different sets, or types, of developmental events become superimposed upon each other.

The evolution of the *telencephalic commissures* in the taxonomic or in the presumptive phylogenetic vertebrate series is closely correlated with concomitant transformations of the *lamina terminalis*, through which the fibers of these commissures[257] take their course.

The development and significance of the lamina terminalis has been discussed in section 1 B of the present chapter. It will here be sufficient to recall that this structure, which interconnects, *ab initio*, right and left halves of the prosencephalon, becomes subdivided into a dorsal *pars telencephalica*, and a ventral *pars diencephalica*. As the commissural systems begin to take their course through the pars telencephalica, this midline bridge thickens, forming the *commissural plate*, whose ventricular prominence, known as *torus transversus*, provides a significant landmark for the definition of a morphologically appropriate *telo-diencephalic boundary*.

In all vertebrate brains, the *telencephalic commissural fibers* form two main distinctive transverse bundles crossing the midline region. In the inverted telencephalon, a *dorsal commissural bundle* predominantly carries fibers related to the *pallium*, especially but not exclusively, to its medial parts. The *ventral commissural bundle* carries mainly fibers related to the *basis*, but may also include some fibers from both lateral and medial pallial neighborhoods. Traditionally and quite justifiably, the dorsal commissure of lower vertebrates is designated as *commissura pallii*[258] although it may, in certain instances, include fibers from the basal

---

[257] In the aspect here under consideration, all fibers crossing the midline will be designated as *'commissural' sensu latiori*, regardless whether they interconnect corresponding *antimeric* neighborhoods (commissural fibers *sensu strictiori*) or whether they merely *'decussate'* in order to interconnect contralateral *nonantimeric* grisea.

[258] ELLIOT SMITH (1901, 1910) rather strongly objected to the designation of the dorsal commissure as 'commissura pallii'. Nevertheless, this term can be upheld on fairly sound morphologic grounds in accordance with a distinction of D and B neighborhoods. Because the dorsal commissure of Prototherian, Metatherian, and of submammalian vertebrates with inverted telencephalon is substantially related to $D_3$, the designation commissura hippocampi instead of commissura pallii or commissura dorsalis seems likewise quite acceptable. ELLIOT SMITH favored the term 'commissura hippocampi' for the dorsal commissure in these forms.

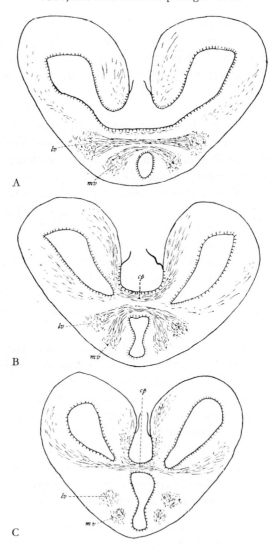

*Figure 336.* Telencephalic commissures and forebrain bundles in the forebrain of urodele amphibians, based on *Golgi pictures* of the brain in *Salamandra* (after K., 1921). A At level
of commissura anterior (not labeled) and interventricular communication respectively telencephalon medium. B and C show commissura pallii at progressively caudal levels of
caudal paired evaginated telencephalon. The commissura anterior (not labeled), interconnecting the forebrain bundles can still be seen in B below commissura pallii. The preoptic
recess (not labeled) is shown in A, B, C below the commissures of lamina terminalis. cp:
commissura pallii; lv: lateral forebrain bundle; mv: medial forebrain bundle.

neighborhood $B_4$. The ventral transverse bundle system is the well-known *commissura anterior* of standard terminology. In submammalian forms, this commissure contains crossing fibers from two longitudinal bundles connecting telencephalon and diencephalon by way of the hemispheric stalk. These bundles are: the *lateral forebrain bundle* related to the lateral hemispheric wall, and the usually somewhat more ventrally located *medial forebrain bundle*, related to the medial hemispheric wall. In mammals, the fiber masses of the internal capsule, the lamina medullaris between putamen and globus pallidus, moreover ansa lenticularis and ansa peduncularis become special differentiations of the lateral forebrain bundle, whose remaining basal components pertaining to fundus striati and tuberculum olfactorium are displaced toward the medial forebrain bundle, forming, with this latter, the relatively reduced and inconspicuous system of a *conjoint basal forebrain bundle*. Also, in mammals as well as in some submammalian forms, a fiber system related to $D_3$, and separated from the medial forebrain bundle, has become differentiated as the *fornix*.[259]

In the commissura anterior of submammalian brains, the crossing fibers of medial forebrain bundle are commonly located basally to those of lateral forebrain bundle. Figure 336 depicts telencephalic commissures and forebrain bundles in the simple configurational arrangement displayed by urodele amphibians.

In the everted telencephalon of Osteichthyes, on the other hand, the fibers of the commissura pallii run *ventrally* to those of the commissura anterior (cf. Fig. 64, I), and are frequently considered to be merely a ventral portion of anterior commissure (cf. also Fig. 62, I).

Moreover, in correlation with different modes of development of lamina terminalis, bearing upon the fate of telencephalon impar, there are additional significant differences in the location, with respect to each other, of commissura anterior and commissura dorsalis s. pallii. These diverging modes of development, discussed on p. 178 of section 1 B, have been illustrated in Figures 62, 63, and 65. Thus, in some forms, the commissura pallii has a conspicuously supraneuroporic position (e.g. Selachians, Figs. 62 II, 64 III; reptiles, Fig. 65 II), while in am-

---

[259] Additional fiber systems (e. g. such as stria terminalis), passing through neighborhoods of the hemispheric stalk, and connecting telencephalon with caudal parts of the neuraxis, are not relevant to the present topic. They shall be dealt with upon discussing the required details of fiber connections in volume 5 (chapter XII and XIII).

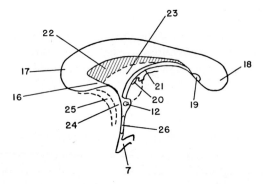

*Figure 337.* Aspect of adult human lamina terminalis and adjacent preterminal region in midsagittal view (from K., 1969). 7: optic chiasma; 12: commissura anterior; 16: rostrum corporis callosi; 17: genu; 18: splenium; 19: commissura fornicis s. hippocampi; 20: subfornical organ; 21: fornix; 22: septum gliosum; 23: septum gangliosum; 24: preterminal portion of paraterminal body (B₄, represented by gyrus subcallosus containing fibers of Broca's diagonal band); 25: precommissural hippocampus (D₃, represented by gyrus geniculatus, which is rostrally separated from parahippocampal cortex, i.e. Broca's area parolfactoria, by sulcus parolfactorius medius, and caudally from gyrus subcallosus by sulcus parolfactorius posterior, if these weak sulci are present); 26: supraoptic crest.

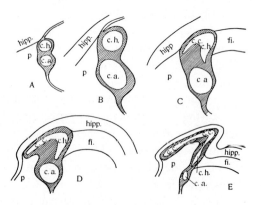

*Figure 338.* Diagrams illustrating relationships of dorsal and ventral telencephalic commissures in lamina terminalis (after ELLIOT SMITH, from JOHNSTON, 1906). A Reptile. B Monotreme (Prototherian mammal). C Marsupial (Metatherian mammal). D Bat (Eutherian, Chiroptera). E 'Higher mammal' (Eutherian, but not primate or man). The midline recess between cc and ca in D and E represents the so-called 'open cavum septi pellucidi' in mammals not possessing a true septum pellucidum characteristic for man and a few other 'higher mammals'. ca: anterior commissure; cc: corpus callosum; ch: hippocampal commissure; fi: fimbria (hippocampi); hipp: hippocampus; p: precommissural (paraterminal) body.

phibians (cf. Figs. 62 III, 64 II, 65 I) the commissura pallii, here located in the torus transversus, thereby indicates, together with commissura anterior, the basal telencephalo-diencephalic boundary. Again, an accessory 'aberrant' caudal commissura pallii posterior is present in diverse reptilian forms (cf. Fig. 63 II).

The *corpus callosum*, characteristic for Eutherian mammals, and containing *commissural fibers of isocortex* (as well as parahippocampal cortex) was defined, by ELLIOT SMITH (1910), as a neopallial commissure.[259a] It may, however, be evaluated as an expansion of the 'original' commissura dorsalis s. pallii, of which it represents the rostral part, while the caudal part becomes the *commissura hippocampi (psalterium)*. This latter, shifted ventrally to the bulging caudal splenium of the corpus callosum, is relatively small in man (cf. Fig. 337).

It should, moreover, be kept in mind that in Prototherian and Metatherian mammals the *commissura anterior* includes most if not practically all the crossing fibers of the neopallium. Even in higher mammals, such as man, the anterior commissure retains a substantial 'neocortical' component.

---

[259a] Thus, in my early investigation of the amphibian forebrain (K., 1921), I had evaluated the area lateralis pallii ($D_1$ of the subsequent notation) as being homologous to cortex lateralis of reptiles, and thereby, in accordance with the views of EDINGER, as representing the 'neopallium'. Quite logically, but, as I now believe, for topologically erroneous reasons, I therefore interpreted a part of commissura pallii related to area lateralis pallii as a primordial 'corpus callosum'. On somewhat different grounds, OSBORN (1887) had likewise considered the commissura pallii of amphibians to be a 'corpus callosum', by conclusively showing that the dorsal (or pallial) commissure in amphibians, reptiles, and birds was homologous to that in Monotremes and Marsupials. However, accepting ELLIOT SMITH's definition of corpus callosum, this structure may, with some qualifications, indeed be said to represent an exlusively Eutherian feature. These qualifications concern: (1) the probability that even in Monotremes and Marsupials the dorsal commissure could include some medial 'neocortical' or at least parahippocampal fibers, the parahippocampal cortex being considered 'neocortex' by some authors; (2) the presence of 'neocortical' fibers from the temporal region in the commissura anterior of at least numerous mammals. Again, regardless of specific fiber connections, the dorsal commissure in all inverted vertebrate hemispheres is, beyond doubt, morphologically homologous *qua* midline connecting pathway, although not 'structurally' analogous. ELLIOT SMITH (1910) indignantly dismisses the following remark by FLOWER: 'granted that only the psalterial fibers are represented in the upper commissure of the marsupial brain, why should the name "corpus callosum" be refused to it?' I opine, nevertheless, that FLOWER's remark is fully justified and that, depending on quite arbitrary semantic postulates, both FLOWER (respectively OSBORN), and ELLIOT SMITH are right. For practical purposes, however, I am inclined to accept this latter author's terminology.

The details concerning embryologic *development* of the mammalian *corpus callosum* with its neighboring so-called septal structures are difficult to ascertain and require the interpretation of serial (and in particular midsagittal) sections showing static pictures of complex morphogenetic processes occurring at certain key stages of brain ontogenesis. Thus, various conflicting views were propounded by different investigators.[260] I believe, however, that the studies of HOCHSTETTER (1919), who supplemented and amplified the work of previous authors, have substantially clarified these questions with respect to the human brain. The results of my own studies, which I therefore merely summarized in my lectures (K., 1927), are in full agreement with HOCHSTETTER's views on the origin of corpus callosum and cavum septi pellucidi in man. As regards other mammals, I concur, on the basis of my own subsequent observations, with the findings and interpretations of ELLIOT SMITH and JOHNSTON as shown in Figures 338–340. These findings rather convincingly indicate that the corpus callosum arises by a considerable thickening of the telencephalic lamina terminalis in its dorsal portion, and that the so-called *open 'cavum septi'* of lower mammals represents a recess formed in the precommissural portion of interhemispheric fissure between the external surfaces of paraterminal body. This origin of corpus callosum is likewise in accordance with the *actual findings* of GRÖNBERG (1901) shown in Figure 341, although the cited author *interpreted* the thickened commissural plate as resulting from a fusion *('concrescentia primitiva')* of the medial wall of the hemisphere.[261]

---

[260] Detailed discussions of these problems and conflicting views, together with further bibliographic references, can be found in the publications of BADINEZ and AGUIRRE (1933), GOLDSTEIN (1903), GROENBERG (1901), HOCHSTETTER (1919), KUHLENBECK (1969), RAKIC and YAKOVLEV (1968), and ZUCKERKANDL (1901, 1907, 1909).

[261] *'Die Commissuren des Hemisphärenhirns, Commissura anterior, Fornixcommissur und Corpus callosum, entstehen alle in einem verwachsenen Bezirk der medialen Hemisphärenwand. Diese verwachsene Partie, welche sowohl von der Lamina terminalis als von dem Septum pellucidum zu unterscheiden ist, habe ich mit dem Namen primitive Verwachsungsplatte (Concrescentia primitiva) bezeichnet'* (GRÖNBERG, 1901). I am rather inclined to see here merely a semantic disagreement concerning the description of the enlargement under consideration. Evidently, the *lamina terminalis*, resulting from a *fusion (concrescence)* of the *anterior neuropore*, does, in a way, originally represent a *'primitive Verwachsungsplatte'* between the two halves of the neuraxis, but subsequently enlarges by *intussusceptional and appositional growth*. I fully agree with GRÖNBERG's further remark: *'Die Fornixcommissur und das Corpus callosum zeigen ursprünglich auf Medianschnitten eine gemeinschaftliche Anlage'*.

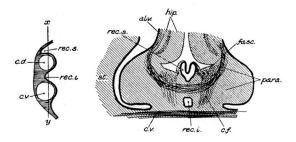

*Figure 339.* Portion of diagrammatic sections through lamina terminalis of a reptile (Hydrosaurus) illustrating recesses related to that structure (after ELLIOT SMITH, from JOHNSTON, 1906). Recesses of this type are displayed, in somewhat variable configurations by numerous and quite unrelated vertebrates. Although formed and located within the region of the original neuropore, their exact relationship to this embryonic neighborhood remains undefined. In the figure to the left, the line x–y indicates the plane of the transverse section at right. alv.: 'alveus'; cd: commissura dorsalis s. hippocampi; c.f.: 'columna fornicis'; c.v.: commissura ventralis s. anterior; fasc.: 'fasciculus marginalis'; hip: 'hippocampus' (D₃); para: paraterminal body; rec.i.: recessus inferior; rec. s.: recessus superior; st: basal ganglia.

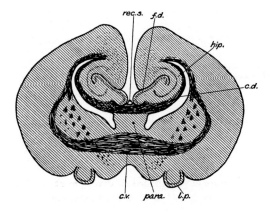

*Figure 340.* Transverse section through the telencephalic commissures in the Prototherian mammal, Ornithorhynchus (after ELLIOT SMITH, from JOHNSTON, 1906). c.d.: hippocampal commissure; c.v.: anterior commissure (including 'neopallial' fibers); f.d.: gyrus dentatus; hip.: cornu Ammonis; l.p.: lobus piriformis (medial to its prominence, area ventralis anterior s. 'tuberculum olfactorium' can be seen on both sides of the midline); para.: paraterminal body ($B_4$ or $B_{3+4}$); rec. sup.: recessus superior (cf. preceding fig. 339).

The *human corpus callosum* seems to arise, during the third month of intrauterine development, at stages of 40- to 50-mm length, as a thin fiber bundle crossing the midline in the dorsal part of lamina terminalis telencephalica. It soon becomes greatly enlarged by an expansion which does not involve any *discontinuous* secondary fusion or 'concrescence' of adjacent wall neighborhoods. Because of this continuous size

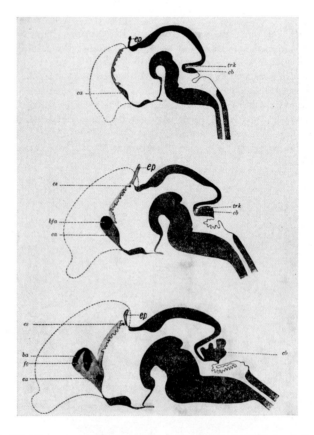

*Figure 341.* Three stages in the ontogenetic development of telencephalic commissures in the lamina terminalis of the hedgehog, *Erinaceus* (from GROENBERG, 1902). Top: early stage, only commissura anterior is indicated in torus transversus of lamina terminalis. Middle: intermediate stage, the upper part of enlarged lamina terminalis contains a bundle of corpus callosum fibers. Bottom: advanced stage, corpus callosum and fornix with its commissure assume their definitive configuration within the thickened lamina terminalis. ba, bfa: corpus callosum; ca: commissura anterior; cb: cerebellar anlage; cs: commissura habenulae; ep: epiphysis; fc: fornix and commissura fornicis; trk: decussation of trochlear nerve.

increase in the sagittal plane, the growth process of corpus callosum can be described as *intussusceptional.* The bilateral inclusion of newly arising material, particularly at the rostral and caudal extremities of the corpus callosum, may, of course, also be described as *apposition.* In this sense, *both intussusceptional and appositional growth,* or a combination of both, may be justifiably claimed. Such appositional growth of lamina terminalis, however, occurs by *continuous internal expansion* with incorporation of new mass, but not by appositional addition of material primarily separated from its original midline link (lamina terminalis) by a secondarily obliterating interface.

In the course of this growth, the human corpus callosum gradually assumes its typical configuration characterized by *rostrum, genu,* and caudal *splenium.* This latter, protruding beyond the attachment of lamina epithelialis to commissural plate, becomes located *dorsally to commissura hippocampi s. fornicis.* The main mass of the fornix system is gathered, on each side, into a conspicuous columnar bundle *(columna fornicis),* which indicates the caudo-lateral boundary of telencephalic lamina terminalis, and enters the diencephalon (hypothalamus).

In man, in some primates, and also in some other large mammals with sizable brain, the expanding midline region of the lamina terminalis, between growing mass of corpus callosum, the commissura anterior and the columns of the fornix, is subjected to extensive elongation, becoming a thin membrane-like structure, which may be partly devoid of nerve cells in its dorsal portion, and forms a narrow partition between the two lateral ventricles at the levels of lamina terminalis. This partition, resulting from the stretching of a median portion of lamina terminalis containing an intraterminal subneighborhood of $B_4$, is the *human septum pellucidum* (BNA, PNA), which commonly displays a cleft-like midline space, the *cavum septi pellucidi* (BNA, PNA) between two 'laminae' *(laminae septi pellucidi* BNA, PNA). It is well-known that, in man, both extent and width of this cleft, which may consist of several cavities,[262] and can even be completely missing, manifest consider-

---

[262] The cavities of adult human septum pellucidum may be communicating with each other or represent an interrupted sequence. A separate caudal cavity, '*Verga's ventricle*' or cavum psalterii is a not infrequent occurrence. In some cases, it is continuous with the main cavum septi pellucidi. In other instances, single or multiple complete or incomplete perforations of the septum pellucidum ('*Defektbildungen*', cf. HAHN and K., 1930) can be found. In an incomplete perforation, the cavum septi pellucidi communicates with one lateral ventricle, while a complete perforation results in a communication between the two lateral ventricles through the septum. In extreme cases, the septal wall is almost

able variations (OLIVEROS, 1965). On the other hand, such membrane-
like, thin septum pellucidum, which may contain a cavum, is not formed
within the lamina terminalis of numerous mammals and is particularly
lacking in 'lower' or small mammals with correspondingly small brain.
The caudal recess of the interhemispheric fissure extending as a pocket
toward the lamina terminalis, and mentioned above on p. 624, is morphologically and ontogenetically entirely different from a true cavum
septi pellucidi in man and some other mammals. It seems, therefore,
inadvisable to adopt the ambiguous designation 'open cavum septi' proposed by various authors.

As regards the *human cavum septi pellucidi*, the findings of HOCHSTET-
TER (1919), which I could also corroborate on the basis of my own observations (Fig. 342), rather clearly suggest that this cavity results
from a process of interstitial cleavage and tissue resorption in a
stretched or elongated portion of lamina terminalis adjacent to the
arched inferior boundary interface of corpus callosum. Among other
authors, STREETER (1912) had already noted the formation of cavum
septi pellucidi in the distended lamina terminalis, and justly remarked
that this cavity is only present in forms with large corpus callosum.[263]
A detailed fairly recent investigation on the development of the commissures in the human telencephalon and the formation of cavum septi
pellucidi, confirming HOCHSTETTER's findings, was undertaken by
BADINEZ and AGUIRRE (1953). Sagittal section depicted by J. ARIËNS
KAPPERS (1955) in a study of the human paraphysis,[264] and here re-

---

completely missing, and the two lateral ventricles communicate through a large opening
whose borders are columns of fornix, midline portion of anterior commissure, rostrum,
genu, and body of corpus callosum. This direct communication between anterior horns
of lateral ventricle is, of course, located rostrally and dorsally to columns of fornix; the
*foramina Monroi*, through which the lateral ventricles communicate with the unpaired
third ventricle, are located caudally and basally to the columnae fornicis.

[263] In a Macacus brain, I noticed a small cavum septi pellucidi between rostrum and
genu corporis callosi, but failed to find a cavum in various other specimens of this monkey.

[264] The human paraphysis seems to be a very variable and transitory rudimentary
organ, although constantly present during early ontogenetic development (J. A. KAPPERS,
1955). It probably disappears at about the 100-mm stage (i.e. roughly at 3½ months).
So-called 'paraphyseal' cystic tumors appear mostly to derive from diencephalic roof plate
structures included in choroidal folds. In a few cases, however, they may be of true
(telencephalic) paraphysial origin (KAPPERS, 1955). The (telencephalic) subfornical organ
is not related to the paraphysis, but represents an ependymal or subependymal differentiation of the dorsal part of commissural plate. Its development starts at a relatively late
stage of human ontogenesis subsequent to the 4th lunar month. It was not seen by
KAPPERS (1955) in early stages up to 145 mm (between 4 and 5 lunar months).

produced as Figure 343, likewise document the origin of corpus callosum and cavum septi pellucidi within the commissural plate of lamina terminalis.

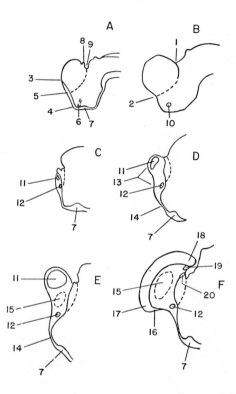

*Figure 342.* Ontogenetic development of lamina terminalis in the human brain (from K. 1969, C–I modified after Hochstetter, 1919, on the basis of original observations). A Midsagittal section of forebrain in embryo of about 12-mm length. B External view of A. C–F Midsagittal sections at about 68-, about 90-, about 100-, and about 125-mm length. The dotted line in A, extending from velum transversum to torus transversus, indicates the telo-diencephalic boundary. In C–F, this boundary has become the caudal border of foramen Monroi. 1: dorsal midline portion of sulcus hemisphaericus; 2: basal midline portion of sulcus hemisphaericus; 3: dorsal end of lamina terminalis (angulus terminalis of His); 4: basal end of lamina terminalis in preoptic recess; 5: commissural plate forming torus transversus on ventricular surface; 6: entrance into optic stalk (subsequently becoming sulcus intraencephalicus anterior); 7: anlage of optic chiasma; 8: paraphysial rudiment; 9: velum transversum; 10: cut optic stalk; 11: anlage of corpus callosum; 12: anlage of anterior commissure; 13: thickened telencephalic portion of lamina terminalis; 14: thin diencephalic (hypothalamic, preoptic) portion of lamina terminalis; 15: beginning formation of cavum septi pellucidi; 16: rostrum corporis callosi; 17: genu; 18: splenium; 19: anlage of commissura hippocampi s. fornicis; 20: anlage of subfornical organ.

The dorsal part of the adult human septum pellucidum, which may or may not contain a cavity or a set of cavities produced by tissue resorption, is essentially devoid of nerve cells, but contains glia cells and poorly differentiated neuroectodermal elements of which, at most, only a negligible amount may be neuronal. The cavum septi, if present, may become secondarily lined by fairly typical *ependymal cells* which seem to arise *in situ* (OLIVEROS, 1965). This lining is commonly incomplete rather than continuous, being interrupted by stretches devoid of a well defined and coherent cellular border of 'epithelial' or 'ependymal' type. It can be inferred from the available scattered observations that a relatively large cavum septi pellucidi is a rather constant feature in human brains at birth and during the second half of gestation. It is thus possible that this cavity becomes, to some extent, reduced in postnatal life by processes of secondary fusion within the cavitated lamina terminalis. A conspicuous *glial median raphé* extends from the dorsal, glial portion of septum pellucidum into the ventral or basal part,

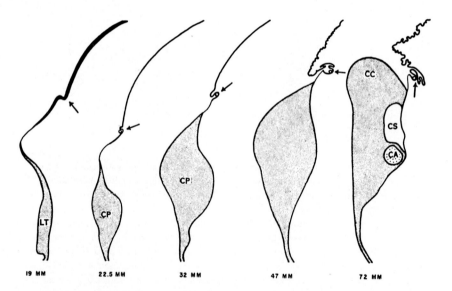

19 MM      22.5 MM      32 MM      47 MM      72 MM

*Figure 343.* Midsagittal sections illustrating developmental stages of telencephalic lamina terminalis and adjacent roof plate structures in man (from J. A. KAPPERS, 1955). The paraphysial rudiment is indicated by a little circle near the tip of the velum transversum, the latter is pointed out by arrow. CA: commissura anterior; CC: anlage of corpus callosum; CP: commissural plate; CS: cavum septi pellucidi; LT: early stage of lamina terminalis.

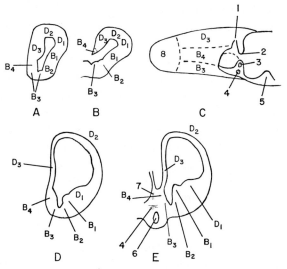

*Figure 344.* Topologic relationship of neighborhoods represented by telencephalic longitudinal zones to lamina terminalis in urodele amphibians and in man (from K., 1969). A Cross-sections through the rostral paired evaginated hemispheres of a urodele amphibian. B Cross-section, caudal to A, through lamina terminalis, showing unpaired evaginated portion of telencephalon (telencephalon medium s. impar). C Aspect of longitudinal zones and their relationship to lamina terminalis in midsagittal view of urodele amphibian forebrain. D Cross-section through hemisphere in human embryo of about 46-mm length, rostral to lamina terminalis. E cross-section caudal to D, through lamina terminalis, slightly rostrad of interventricular foramen (E and D combined after HOCHSTETTER, 1919, and personal observations). 1: paraphysis; 2: velum transversum; 3: commissura pallii s. hippocampi; 4: commissura anterior; 5: ridge of optic chiasma; 6: recessus praeopticus; 7: fibers of fornix system at level of corpus callosum anlage; 8: bulbus olfactorius.

which contains several grisea derived from caudal portions of the zone $B_4$. These grisea, again, are basally continuous with a derivative of zone $B_3$.

Accordingly, it may be stated that the adult human septum telencephali or septum pellucidum (BNA, PNA) consists of a dorsal *pars gliosa*, the *septum gliosum*, with or without cavum, and a ventral or basal *pars gangliosa*, the *septum gangliosum* (K., 1969). This latter part contains the grisea of $B_4$, which may be designated as nuclei septi *sensu strictiori*,[265] and and likewise includes, as septum *sensu latiori*, some deriva-

---

[265] A cytoarchitectural parcellation of these grisea has been proposed by ANDY and STEPHAN (1966, 1968) whose studies concern the so-called septum of mammals and particularly the phylogeny of the primate septum pellucidum.

tives of B₃. In addition, the septum gangliosum, and to some extent
also the septum gliosum, contain numerous fiber systems pertaining to
fornix *sensu latiori*, and running in dorso-basal planes rostral to the
more massive columnae fornicis. The two subdivisions of the human
septum pellucidum are shown in Figures 337. The topologic relationships of the telencephalic zonal system to the lamina terminalis in urodele amphibians and in man are indicated by the diagrammatic Figure 344. Zone B₄, i.e. the *paraterminal body sensu strictiori*, comprises a
*preterminal*, an *intraterminal*, and a *supraterminal* portion. In amphibians, the preterminal part represents the main portion, the intraterminal
part is rather negligible, and the supraterminal portion is the 'pars fimbrialis septi'. In man, the preterminal portion is relatively small, being
the variable and rather indistinct *gyrus subcallosus*.²⁶⁶ The very reduced
supraterminal portion is the *indusium verum* of Elliot Smith. The intraterminal portion is the septum pellucidum with pars gliosa and pars
gangliosa. Zone B₃ of Amphibia has, besides its large preterminal portion, a likewise substantial intraterminal one. In man, its preterminal
portion is poorly defined, being represented by anteromedial parts of
substantia perforata externally, and components of fundus striati internally. The intraterminal portion is manifested by a griseum pertaining
to nucleus accumbens.

---

²⁶⁶ In disagreement with the PNA, whose '*Area subcallosa*' represents the '*Area
parolfactoria (Brocae)*' of the BNA, I designate as '*gyrus subcallosus*', in conformity with
Elliot Smith (cf. footnote 240), the precommissural paraterminal body immediately
rostral to, and adjoining the lamina terminalis. It is separated by the variable and poorly
developed *sulcus parolfactorius posterior* from the rostrally adjacent *precommissural hippocampus*, represented by *gyrus geniculatus*. This latter, in turn, is separated by *sulcus parolfactorius medius* from the *area parolfactoria (BNA)* which is parahippocampal cortex, and,
in turn, becomes roughly delimited from the isocortex by *sulcus parolfactorius anterior*. The
ill-defined human *stria olfactoria medialis* takes its course approximately between gyrus
subcallosus and gyrus geniculatus or on the adjacent surfaces of both. In some instances,
a weak and variable accessory sulcus delimits stria olfactoria medialis either from the surface
of gyrus subcallosus or of gyrus geniculatus. This latter is continuous with the supracommissural hippocampus, which, in man, is represented by the basal fringe of parahippocampal cortex gyri cinguli and by the lateral parts of the indusium griseum (namely the
indusium spurium of Elliot Smith). The PNA apparently ignores the precommissural
hippocampus (gyrus geniculatus), and designates as '*gyrus paraterminalis*' both gyrus
subcallosus and gyrus geniculatus of the terminology which I have adopted. Considerable
discrepancy concerning terminology and interpretation obtains in the literature.

HOCHSTETTER's concepts have been recently contested by RAKIC and YAKOVLEV (1968) who concluded that the human cavum septi pellucidi arises as an open pocket of the interhemispheric fissure which becomes sealed off by the rostrum corporis callosi. In addition, these authors claimed that the development of corpus callosum 'is preceded by the infolding of the dorsal part of the lamina reuniens of HIS (1904) in the region of the prospective hippocampus into a median groove, and by fusion of its banks into a massa commissuralis which becomes the bed for corpus callosum, as was described by ZUCKERKANDL (1901)'. According to RAKIC and YAKOVLEV (1968) 'degeneration of cells in the region of fusion, with a resolution of the membrana limitans piae and degeneration of the single fold of the meninx primitiva between the juxtaposed banks' of the median fissure is supposed to occur, this degeneration being a disintegration or necrobiosis conceived as the phase of 'a normal cell cycle'. Thus, while the occurrence of tissue resorption as pointed out by HOCHSTETTER and others is admitted, the cavum septi is, however, interpreted to be a primarily open but secondarily segregated or 'closed' leptomeningeal space.

Although RAKIC and YAKOVLEV (1968), who studied an extensive human embryonic material, have presented a large number of photomicrographic illustrations in order to corroborate their interpretation, none of these figures provide any convincing evidence for their thesis. On the other hand, the illustrations provided by HOCHSTETTER (1919), supplemented by plate model reconstructions, which show, in three-dimensional aspect, the gradual thickening of the lamina terminalis, concern unambiguous key stages, and can be considered conclusive (cf. Fig. 342). This author had collected an unusually well preserved and complete material. Moreover, in so far as the photomicrographs[267] included in the paper of RAKIC and YAKOVLEV (1968) depict developmental stages of lamina terminalis, corpus callosum, and cavum septi, all of them can easily be interpreted in agreement with HOCHSTETTER's views.[268]

---

[267] A sizable proportion of the numerous photomicrographs included in the paper of RAKIC and YAKOVLEV (1968) has no bearing on the problem under consideration.

[268] There is little doubt that, as regards the ontogenetic development of certain features in the human brain, particularly of hemispheric stalk and lamina terminalis, the documentary material and the descriptions provided by HOCHSTETTER have few if any equals in the neuroanatomical literature, despite certain shortcomings displayed by this author (cf. above, footnote 117).

It might be added that, preceding the publication of HOCHSTET-TER's final studies on the development of human corpus callosum and hemispheric stalk, GOLDSTEIN (1903) had investigated these questions. This author discussed the relevant previous literature and showed, on the basis of his findings, that both corpus callosum and hemispheric stalk originate by expansive, intussusceptional growth of *ab initio* present connections between subdivisions of the brain, and not by concrescence or 'fusion' (*'Verwachsung'*) of primarily separated wall portions.[269]

Concerning the corpus callosum, GOLDSTEIN (1903) suggested '*dass der Balken innerhalb der Lamina terminalis entsteht; d. h. mit anderen Worten die Lamina terminalis als die ursprüngliche Verbindung zwischen den beiden Grosshirnhemisphären das morphologische Substrat liefert, durch welches die Commissurenfasern von einer Hemisphäre zur anderen ihren Weg nehmen*'. The cited author justly points out the characteristic relationship of ar-teria cerebri anterior (Fig. 346 A), which remains in close apposition to the expansively growing anlage of corpus callosum.

As regards the formation of cavum septi pellucidi, however, GOLD-STEIN (1903) did not assume the occurrence of a secondary cleft in the lamina terminalis, but believed that it results from inclusion of a medi-al hemispheric wall portion into the basal concavity of corpus callosum (Figs. 345, 346). I believe that, because GOLDSTEIN's material was far more incomplete and much less well preserved than that of HOCHSTET-TER, GOLDSTEIN failed to recognize, in his specimens, the poorly fixed and therefore blurred demarcation between mesodermal and neuroec-todermal tissue, thereby interpreting the locus of septum pellucidum anlage as external to the lamina terminalis (cf. Fig. 346 A, B). This in-

---

[269] This does, of course, not contradict the fact that, in some regions of the neuraxis, secondary fusions or concrescences of internal and external surfaces do indeed occur. Thus, there are obliterations of ventricular lumina, and the massa intermedia or 'commis-sura mollis' of the third ventricle may represent a massive '*Verwachsung*'. Another such instance is the unpaired 'gustatory' facial lobe in the oblongata of various Cyprinoid Teleosts. As regards fusion of external surfaces, the medial hemispheric surfaces of bulbus olfactorius and adjacent portions of lobus hemisphaericus become fused in some Anuran amphibians. GOLDSTEIN (1903) refers to the 'commissura mollis', whose occurrence might be used as an objection against his views (*'scheinbar gegen unsere Anschauungen angeführt werden könnte'*). He correctly anticipated the criticism by ZUCKERKANDL (1909), who attempted to substantiate his concept by means of this argument. It seems, however, evident that the fusion of wall surfaces in some regions and in some instances does not necessarily imply fusion in other instances and regions, particularly if such fusion can be excluded on the basis of convincing evidence.

terpretation is, of course, identical with RAKIC's and YAKOVLEV's subsequent thesis concerning origin of human cavum septi pellucidi. An
unlabelled approximately sagittal section through the embryonic human lamina terminalis in a stage closely corresponding to that depicted
by GOLDSTEIN (1903) was recently included in a paper by LOESER and
ALVORD (1968). It is here reproduced for comparison (Fig. 346B). Although the latter authors correctly explain it in their caption as 'passing through the lamina terminalis', and, moreover, describe 'gitter-cell
formation' in the wall of cavum septi pellucidi, they nevertheless express their agreement with RAKIC's and YAKOVLEV's views. Yet, an interpretation entirely in accordance with HOCHSTETTER's concepts is
evidently rather well founded on the basis of the obvious configurational relationships, as can easily be seen by the labeled sketch which I
have added to this figure.

In summarizing the overall features of vertebrate telencephalic ontogenetic and taxonomic differentiation, particularly with regard to a
phylogenetic appraisal, it becomes necessary to consider both the morphological and the functional aspects. Seen from the form analytic
viewpoint, the vertebrate neuraxis is a rostro-caudal tubular configuration, sectionalized, by identifiable and definable constant landmarks
such as folds, bulges, or evaginations, in a sequence of subdivisions. In
accordance with resonably valid criteria of classification, based on an
orderliness manifested by configurational invariants, five delimitable
regions can arbitrarily be conceived to represent the major tubular
subdivisions, namely telencephalon, diencephalon, mesencephalon,
rhombencephalon (including cerebellum), and spinal cord. These 'segments' are displayed by all craniote vertebrates in a topologically identical manner, and each 'segment' of this sequence, in any 'normal' vertebrate form, is morphologically homologous with the corresponding
'segment' in any other 'normal' vertebrate form. From a phylogenetic
viewpoint, and *qua* craniote vertebrates, none of these subdivisions is,
therefore, 'older' or 'younger' than any other.[270] In other words, no

---

[270] If, however, Cephalochorda, namely Acrania such as Amphioxus (Branchiostoma),
are evaluated as 'Vertebrata' *sensu latiori* and considered related to ancestral vertebrate
forms, a phylogenetic appraisal of neuraxis regions becomes permissible. Accordingly, a
rostral archencephalic vesicle comparable to that of Amphioxus, and homologous to the
craniote transitory prosencephalon, would be the precursor of telencephalon and diencephalon. The craniote deuterencephalon, consisting of mesencephalon and rhombencephalon can then be conceived as secondarily derived from the anterior part of an
ancestral spinal cord.

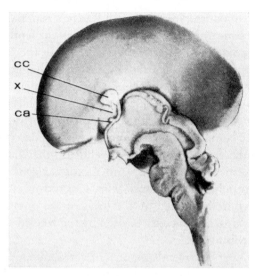

Figure 345 A

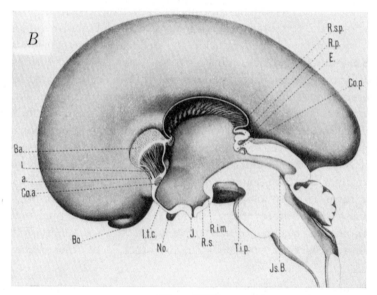

Figure 345 B

*Figure 345.* Relationship of corpus callosum and septum pellucidum in human embryos of about 14 weeks as seen by GOLDSTEIN (1903, 1904) and by HOCHSTETTER (1919). A Median view of brain in embryo of 105-mm length (reproduced, with added designations, from GOLDSTEIN, 1904). ca: anterior commissure; cc: corpus callosum; x: arrow pointing to what I believe to be an artificial defect in lamina terminalis, tearing into the for-

'new' major subdivision of the vertebrate neuraxis has been added in the course of presumptive phylogeny.

Reverting now to the *telencephalon*, which represents the most rostral major neuraxial subdivision, all craniote vertebrates may be said to display four basic telencephalic components, of which three are rele-

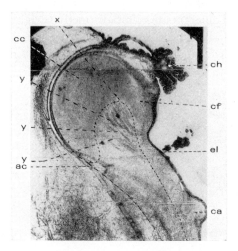

*Figure 346 A.* Composite photomicrograph of lamina terminalis with anlage of corpus callosum and cavum septi pellucidi, corresponding to stage depicted in figure 345 A, B (reproduced from GOLDSTEIN, 1903, with added outlines and altered interpretations). ac: arteria cerebri anterior; cc: anlage of corpus callosum; cf: commissura fornicis ('psalterium') and fornix fibers; ch: choroid plexus at transition of telencephalon and diencephalon; el: ependyma of telencephalic lamina terminalis; x: approximate upper (dorsal) boundary of septum anlage; y: spongy strands resulting from formation of cavum septi pellucidi (cf. fig. 345 B); dotted outline: boundary of lamina terminalis according to GOLDSTEIN's interpretation (cf. fig. 345 A); dash outline: rostral boundary of lamina terminalis interpreted in accordance with HOCHSTETTER's (1919) and my own views (cf. fig. 345 B); ca: commissura anterior.

---

mation of cavum septi (cf. fig. 346); taken at face value, GOLDSTEIN's figure would display a so-called 'open cavum septi' between adjacent external, medial B₄ surfaces of the hemispheres within the interhemispheric fissure. B Median view of brain in embryo of 102-mm length based on a careful wax plate reconstruction from well preserved material (from HOCHSTETTER, 1919). a: *'Grenzlamelle der Kommissurenplatte';* Ba: *'Balkenanlage';* Bo.: bulbus olfactorius; Co.a.: commissura anterior; Co.p.: commissura posterior; E.: epiphysial anlage; J.: 'Infundibulum'; Js.B.: *'Isthmusbucht';* l.: *'schwammiges Gewebe'* (resulting from formation of cavum septi by tissue resorption); l.t.c.: 'lamina terminalis cinerea (diencephalic portion of lamina terminalis); No.: nervus opticus; R.i.m.: 'recessus inframammillaris'; R.p.: Recessus pinealis; R.s.: 'recessus saccularis'; R.s.p.: Recessus suprapinealis.

vant neural wall portions, while one remains epithelial and can become greatly reduced or obliterated in the course of ontogenesis. These four components are: *bulbus olfactorius, dorsal zone D ('pallium'), ventral zone B ('basis'),* and *roof plate (lamina epithelialis).* In all Gnathostome vertebrates, zone D and zone B, constituting the neural wall of lobus hemisphaericus as distinguished from bulbus olfactorius, manifest a tendency toward differentiation into distinctive neighborhoods $D_1$, $D_2$, $D_3$, and $B_1$, $B_2$, $B_3$, $B_4$. These neighborhoods are sufficiently well recognizable in the telencephalon of adult Gnathostome Anamnia and of all

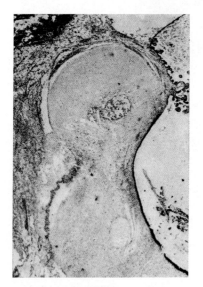 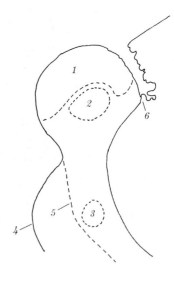

Figure 346 B                                      Figure 346 C

*Figure 346 B.* Photomicrograph of a hematoxylin-eosin stained approximately midsagittal section through the developing lamina terminalis in a human fetus of 100 mm (after LOESER and ALVORD, 1968; the rostral surface of lamina terminalis faces toward the right in the original illustration, which is here reproduced in mirror image reversion to facilitate comparison with the orientation obtaining in HOCHSTETTER's model depicted by fig. 345B).

*Figure 346 C.* Outline sketch of lamina terminalis in preceding figure, indicating my interpretation. 1: anlage of corpus callosum; 2: anlage of cavum septi pellucidi within expanded lamina terminalis; 3: commissura anterior; 4: oblique (tangential) section of medial hemispheric wall caused by slight deviation from the true midsagittal plane; 5: course of lamina terminalis in true midsagittal section (cf. fig. 345B); 6: attachment of roof plate (lamina epithelialis of choroid plexus).

Amniota at key stages of ontogenesis.[271] Corresponding neighbor-hoods in different vertebrate forms are topologically identical i.e. mor-phologically homologous. Again, in terms of topology, these neigh-borhoods, as well as their further differentiations into the subneigh-borhoods displayed by Amniota, are likewise represented by *vaguely cir-cumscribed* loci within the *undifferentiated* overall D and B zones at early stages of ontogenesis or, e.g., in the adult telencephalon of Agnatha. Despite vagueness precluding 'precise' delimitation, these loci are ap-proximately definable by their *combinatorially exact* spatial sequence.

Thus, while a certain region, such as the isocortex ($D_{2a}$ and $D_{2b}$) of mammals, is indeed, *qua* structural differentiation, characteristic for this vertebrate class, phylogenetically 'new', it does not represent a to-pologically or morphologically 'new' pallial neighborhood. Converse-ly, the postcommissural hippocampal formation ($D_3$) of mammals, al-though commonly evaluated as 'old cortex' or 'archicortex', likewise displays a structural differentiation unique for mammals[272] and is, in this respect, as '*new*' as the isocortical neocortex. Depending on mor-phological (topological) or on structural, respectively functional view-points, both cortices can be evaluated, with equal justification, as 'old' or as 'new'.

It seems evident that a clear-cut, logical and semantic distinction should be made between (1) a *morphological* (form analytic, topologi-cal), (2) a *structural*, and (3) a *functional* viewpoint or appraisal. In neu-robiology, a functional appraisal of grisea may, not uncommonly, be-come restricted to an interpretation of its significant fiber connections.

With respect to this situation, ELLIOT SMITH (1910) comments upon a 'new type of anatomist – the man who began to talk of his sub-ject as a branch of learning worthy of being pursued "for its own sake" and not merely as the necessary technical training for the surgeon and the physician. This tendency to concentrate his attention upon pure morphology and to spurn the utilitarian aspect of his subject led

---

[271] At later stages of ontogenesis, pattern distortions and blurring, caused by further differentiation and growth processes, prevent an unambiguous mapping, unless the required inter- and extrapolations are provided by comparison with the relevant earlier stages.

[272] This includes also the fully developed supracommissural hippocampal formation of e.g. 'aplacental' mammals, and of a few Eutheria, but not the supracommissural hippocampal rudiment of most Placentalia.

the anatomist to strange lengths, and he hedged himself around with a barrier shutting out communications with the physiologist and the clinician and others who might corrupt the good manners of his new path of life. He quite overlooked the fact that in thus cutting himself adrift from consideration of function and in losing intimate touch with medicine and surgery he was really emasculating his subject and losing the virility which is necessary for its proper pursuit, not "for its own sake", but for the wider advancement of science in general.

I am glad to say there are now definite signs that the era I have just referred to is passing, if it is not quite gone, and that anatomists are now beginning once more to look upon the dead material of their research as something that has lived, and to realise that if any real progress is to be made structure and function must be studied hand in hand.'

Despite my high esteem for the accomplishments of ELLIOT SMITH, I do not believe that his criticism of '*pure morphology*' is justified, and I can refer to my vindication of the rigorous formanalytic approach in section 8 (p. 284f.), chapter III, volume 1 of this series. It may here be sufficient to add that, because of the difficult and complicated problems intrinsic to configurational orderliness, and in view of this latter's basic significance for a proper understanding of organic forms, the study of 'pure morphology' represents a highly specialized branch of anatomical sciences. The pursuit of 'pure morphology', at least *as if for its own sake* becomes thereby unavoidable. The alternative whether the formanalytic investigator does or does not wish to hedge himself around with a barrier may not necessarily affect the significance of the obtained results. Although, personally, I take a rather sceptical view[273] of 'utilitarian aspects' and 'heuristic values' so dear to the prevailing

---

[273] S. R. y. CAJAL, doubtless an outstanding investigator *qua* problems of structure, and a remarkably strong personality with manifold interests, published his thoughts as an assiduous *coffee-house philosopher* in a book entitled '*Charlas de café. Pensamientos, Anécdotas y Confidencias*' (3rd ed., *Espasa-Calpe*, Buenos Aires 1944). In chapter V, p. 94 (loc. cit.) he offers the following comment:

'*Para juzgar de la mentalidad de los hombres, hablémosles de una invención científica o filosofica desprovista de aplicaciones prácticas.*

*Unos exclamaran: – ¡ Admirable!...*

*Y otros: – ¿ Para qué sirve?*

*Cultivamos la amistad de los primeros.*'

Although being somewhat at variance with many of DON SANTIAGO's views and attitudes, I am inclined to agree with the overall meaning of the quoted apophthegm.

βαναυσία of the *de facto* ruling scientific '*establishment*', I do not think that I have cut myself adrift from consideration of relevant topics pertaining to other branches of medicine and 'science'. On the contrary, despite my championing of 'pure morphology for its own sake', I consistently endeavored to counteract the effects of excessive specialization, and to present a broad synthesis of the diverse aspects of neurobiology not only related to the evolution of the brain, but to scientific thought in general.

It should be added that ELLIOT SMITH (1910) clearly realized the intrinsic complexities of the basic morphological problems, about which he made the following remarks: 'This analytical study of the morphology of the cerebral cortex cannot be wholly divested of its highly technical and intricate character, because it deals with matters difficult in themselves, but which have been rendered infinitely more so by the various interpretations put upon the facts and the confusion of nomenclature with which these facts have been enshrouded by different writers.' I believe that this alluded confusion can be avoided by following a purely form analytic approach in attempting to solve the relevant problems.

As regards a purely *functional* subdivision of the brain, Figure 347, taken from HERRICK (1926), illustrates an interesting simplified semantic procedure applied to the selachian neuraxis, and based upon the type of sensory input. Accordingly, the following designations are used: '*taste-brain*' for a medial strip of lower oblongata, '*skin-brain*' for a lateral strip of oblongata, '*ear-brain*' for a region of rhombencephalon including cerebellar auricles and a caudally adjacent strip, '*eye-brain*' for the mesencephalon,[274] respectively its tectum, and '*nose-brain*' for the

---

Again, the noted Oxford mathematician H. J. SMITH (1826–1883), particularly concerned with the theory of numbers and the study of binary and ternary quadratic forms, is said to have remarked: 'it is the peculiar beauty of this method, gentlemen, and one which endears it to the really scientific mind, that under no circumstances can it be of the smallest possible utility' (quoted by D. E. SMITH, History of Mathematics, vol. I, p. 467, Dover, New York 1951). An essentially similar quip by another leading mathematician, namely the '*formalist*' DAVID HILBERT (1862–1943), who defined mathematics as a 'meaningless formal game', was quoted in footnote 78, p. 330, vol. 2 of this series.

[274] Although the retina represents an evaginated and subsequently cup-like invaginated portion of the prosencephalon, remaining connected with the ventral diencephalic wall, it is of interest that the tectum mesencephali rather than the diencephalon becomes the significant 'eye-brain' of lower vertebrates.

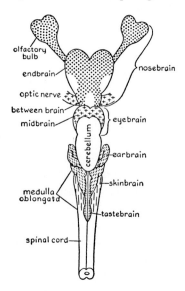

*Figure 347.* Dorsal view of the brain in the Selachian, *Squalus acanthias* (dogfish), sub-divided, on left side, according to standard anatomical nomenclature, and, on right side, in accordance with a generalized functional concept roughly corresponding to the conventional five senses, and indicated by the explained markings (from Herrick, 1926).

prosencephalon (diencephalon and telencephalon). As a rough first approximation, this terminology is doubtless justifiable.

With respect to the telencephalon, moreover, the concept 'nose-brain', namely 'rhinencephalon' assumes a special significance. There is little doubt that, in perhaps all Anamnia, the entire *telencephalon*, consisting of bulbus olfactorius and lobus hemisphaericus, *functionally* represents a *rhinencephalon* in so far as all telencephalic grisea are predominantly related to olfactory input, provided by fibers of the first order for bulbus olfactorius, and by fibers of higher order (second, third, and perhaps additional sequences) for lobus hemisphaericus.[274a]

[274a] A recent report by Ebbeson and Schroeder (1971), based on results with the *Nauta* and *Fink-Heimer degeneration methods*, claims that in the nurse shark (Ginglymostoma) the telencephalon is 'considerably less completely monopolized by the olfactory system than earlier studies by inadequate methods had seemed to indicate'. The cited authors, moreover, purport to have demonstrated 'a massive thalamo-telencephalic projection in the selachian brain'. This projection is said to be largely crossed and to include, moreover, projections to contralateral optic tectum, ipsilateral brainstem and rostral spinal cord.

In Amniota, particularly in mammals, the functional term *rhinence-phalon*, superimposed upon, used instead of, or subsuming a variety of, morphologic subdivisions,[275] has led to considerable confusion, which was pointed out and lucidly analyzed by ELLIOT SMITH (1901). The designation rhinencephalon, apparently first applied to olfactory bulb and tract by OWEN, was subsequently extended by TURNER, about 1890, to include tuberculum olfactorium, substantia perforata anterior, and piriform lobe, being meant to distinguish these configurations from the so-called 'pallium'.

The concepts of HIS, criticized by ELLIOT SMITH (1901), were incorporated into the BNA. The rhinencephalon is here subdivided into pars anterior and pars posterior. The former, rostrally bounded by sulcus parolfactorius anterior, comprises 'lobus olfactorius' with olfactory bulbus, tractus, trigonum, stria olfactoria medialis and intermedia, moreover *Broca's area parolfactoria*. Pars posterior consists of gyrus subcallosus, substantia perforata anterior, stria olfactoria olfactoria lateralis, and limen insulae. The PNA, in my opinion with full justification, entirely omit the term 'rhinencephalon'. Various authors employed this designation as including pallial regions, such as the '*lobus limbicus*' *of Broca* and the hippocampal formation (cf., e.g., VILLIGER, 1920).

From a morphological viewpoint, it seems preferable to use the word '*rhinencephalon*' as a purely functional term without definite anatomical meaning, but applicable to all those regions or grisea which, on the basis of substantial evidence, can be regarded as significantly related to olfactory input. This implies that morphologically homologous regions may be 'rhinencephalic' in some vertebrate forms, but not in others. *Mutatis mutandis*, the caudal $D_2$ region of Amphibia can

---

While, except perhaps for the claim concerning direct spinal cord connections, these statements, discounting their particular emphasis, are hardly new, considerable scepticism with regard to their accuracy seems justified. It is very easy to interpret results obtained by the cited degeneration method in accordance with highly arbitrary views on fiber connections (cf. vol. 3, part I, p. 677 of this series).

[275] Because of the limitations of language, certain terms, such as fila olfactoria, bulbus olfactorius, optic nerve, etc., even when used with a purely morphologic denotation, retain obvious and unavoidable functional connotations or implications, which are ignored in the aspect under consideration. Evidently it would be awkward or ludicrously pedantic although rigorously consistent to substitute here fila nervi primi, bulbus nervi primi, or nervus, respectively tractus retinae oculi for the cited terms.

hardly be regarded as part of the '*eye-brain*', while one of its derivatives, the area striata of mammals, quite evidently pertains to this functional brain subdivision.

Seen from a functional viewpoint, the phylogenetic evolution of the vertebrate telencephalon, manifested by a differentiation of various grisea within the D and B neighborhoods, seems doubtless related to what ELLIOT SMITH (1910) calls a 'perfecting of the smell mechanism' and its linkage with pathways coming from other sensory sources. The telencephalon is, at first, little more than a 'center' of smell input and 'a means whereby such impressions may bring their influence to bear on the nervous mechanisms which regulate movements, and so contribute their quota to the forces which control the behaviour of the animal and its reactions to the environment' (ELLIOT SMITH, 1910).

The admission,[276] through the forebrain bundles, of perhaps at first remotely indirect sensory paths from thalamic and other regions into the telencephalon is believed to have produced marked effects, namely the concomitant formation of grisea correlating this variety of input with the rhinencephalic one, and transmitting the thus processed information to the more caudal regions of the neuraxis. Hence, the telencephalon seems to have gradually evolved a mechanism 'blending' smell, taste, 'touch', and sight input, finally taking precedence, and surpassing, in significance and bulk, the somewhat similar mechanisms of tectum mesencephali with its relatively minor and distantly relayed olfactory input.

The large and complex telencephalon of birds, with a relatively much reduced olfactory system, clearly suggests 'that the greatness of the hemisphere must be due to the admission into it of the paths coming from other sense organs. Visual impulses certainly make their way into the bird's cerebrum' (ELLIOT SMITH, 1910).[277]

---

[276] One of the earliest of so-called 'phylogenetically old systems' to carry its 'impulses' into the telencephalon may be that concerned with gustatory input, particularly in the case of certain groups of fishes.

[277] By recording evoked potentials, including those from the telencephalon, after unilateral optic stimulation in the chick, we obtained definite bilateral responses from the caudal two thirds of the hemisphere. Generally speaking, responses from the surface were here of smaller amplitude than those obtained from various depths. The grisea particularly concerned with optic input seem to be various caudal portions of the epibasal complex. Except for an abstract, we had so far no opportunity to publish an elaborate analysis of our findings (K. and SZEKELY, 1963).

In a 'sketch of the origin of the cerebral hemispheres' predominantly based on 'functional' concepts, HERRICK (1921), closely following
ELLIOT SMITH, includes a concise formulation suitable for an interpretation in terms of phylogenetic evolution: 'The regional differentiation
of the anatomically distinct centers of the entire endbrain behind the
olfactory bulbs, therefore, primitively arose as a result of the invasion
of the original secondary olfactory area by diverse nonolfactory systems, and the entire history of the subsequent evolutionary differentiation of this part of the brain can be written in terms of the interaction
of these two systems of conduction fibers – those descending from the
olfactory bulb and those ascending from the between-brain. Increasingly complex correlations between the olfactory centers in front and
the nonolfactory centers of the midbrain led to the forward growth of
tracts from the lower reflex centers of touch, taste, vision, hearing,
etc., into the olfactory territory of the endbrain (and probably in still
earlier stages the ascending tracts into the between-brain were led forward by the same motive).

The details of this dramatic history cannot be recounted here. In a
general view of the process it may be said that in cyclostomes the entire
endbrain and a large part of the between-brain are dominated by the
olfactory system, the nonolfactory components entering this territory
from the midbrain being relatively small and incompletely known. As
we ascend the vertebrate scale the nonolfactory systems assume progressively greater importance. In urodeles a considerable part of the
thalamus is devoted exclusively to nonolfactory correlations, but no
part of the cerebral hemispheres is wholly free from olfactory connections. In reptiles, the ascending systems are greatly enlarged and a portion of the corpus striatum complex appears to be devoted exclusively
to them. Here there is well defined cerebral cortex, most of which is
clearly dominated by its olfactory connections (hippocampus and pyriform lobe), though in another part (the general cortex) somatic systems predominate (ELLIOT SMITH, 1910, 1919). In mammals somatic
systems with no admixture of olfactory elements come to dominate the
architecture and functions of the cerebral hemispheres, until in man,
whose olfactory organs are greatly reduced, the olfactory centers are
crowded down into relatively obscure crannies of the hemisphere by
the overgrown somatic systems' (HERRICK, 1922).

This 'crowding' into 'relatively obscure crannies' at the base of the
human hemisphere is illustrated by Figure 348, which shows my concept of what seems to represent the *'rhinencephalon'* in man, comprising

those anatomical configurations of the hemisphere that have retained a substantially or predominantly olfactory function. Recent studies on afferent connections to the reptilian telencephalon, cited above on p. 539, are in essential agreement with this appraisal of the overall functional relationships deduced from the earlier anatomical investigations, and subsequently summarized by ELLIOT SMITH and HERRICK.

As regards the conjoint functional and morphologic evolution of the mammalian cerebral hemispheres, and especially the differentiation of its pallial cortex into various distinctive types, regions, and areae, the available incomplete but at least well corroborated and basic data on fiber connections between diencephalon and telencephalon, already stressed by ELLIOT SMITH (1910), suggest some significant general conclusions.

Increasingly complex correlations between telencephalic olfactory centers and nonolfactory mesencephalic centers were presumably accompanied by a forward (rostrad) growth of optic, acoustic, tactile and other fiber tracts from more caudal centers into the olfactory territory

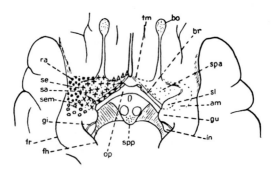

*Figure 348.* Semidiagrammatic basal view of human brain showing location of so-called 'rhinencephalon' (modified after RETZIUS and after K., 1927, from K., 1957). am: gyrus ambiens; bo: bulbus olfactorius; br: *Broca's diagonal band* (medial part of area ventralis anterior); fh: fissura hippocampi; fr: fissura rhinica; gi: *band of Giacomini* (end of gyrus dentatus); gu: gyrus uncinatus (end of cornu ammonis); in: gyrus intralimbicus (additional part of end of cornu ammonis; in, gi, gu represent the true uncus; op: optic chiasma; ra: sulcus rhinalis arcuatus; se: sulcus endorhinalis; sem: sulcus semiannularis; sl: gyrus (semi)lunaris; spa: substantia perforata anterior; spp: substantia perforata posterior; tm: tractus olfactorius medialis; black dots: cortex anterior lobi piriformis; circles: cortex posterior lobi piriformis; triangles: precommissural hippocampus; crosses: tuberculum olfactorium and diagonal band (area ventralis anterior); x-markings: nucleus amygdalae corticalis (area ventrolateralis posterior).

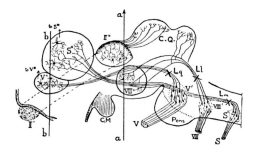

*Figure 349.* Semidiagrammatic drawing showing 'the primitive arrangement of the thalamic nuclei in the lowliest mammals and the tracts passing to them; seen from the left lateral aspect' (from ELLIOT SMITH, 1910). V: trigeminal nerve; V': its terminal nucleus in the pons; Lq: lemniscus quinti, crossing at x and proceeding to the thalamic nucleus V''; S: the posterior root of a cervical nerve; S': the nucleus cuneatus; Lm: lemniscus medialis crossing at x, and proceeding to the thalamic nucleus S''; VIII: the cochlear nerve; Ll: the lateral lemniscus, crossing at x and passing to the corpus geniculatum mediale VIII''; II': the optic tract proceeding to the corpus geniculatum laterale II''; C.M.: the corpus mammillare; aa: the plane of section shown in figure 350; bb: the plane of section shown in figure 351.

of the endbrain. This process also involved to a significant extent the diencephalon as a nodal structure mediating between telencephalon and the caudal parts of the neuraxis.

The dorsal thalamus of mammals has developed into a configuration of large grisea (nuclei), in some of which ascending pathways from the afferent cranial and spinal nerve nuclei, as well as from optic and auditory system end. From each of these thalamic grisea, namely nucleus ventralis posteromedialis and posterolateralis, corpus geniculatum laterale and mediale, a thalamo-cortical tract passes into the isocortex of the 'neopallium' (Figs. 349–353).

The isocortical neighborhoods, upon which these thalamic nuclei 'project', can therefore be considered '*sensory projection areae*', and their location displays a definite spatial sequential correspondence with the topographic relationships of their thalamic 'relay' nuclei.[278] Thus, ELLIOT SMITH (1910) believed that 'that original mapping out of the neo-

---

[278] In this simplified and generalized formulation, the reciprocal ('two-way') connections of thalamic nuclei with cerebral cortex are ignored. The more complex problems of diencephalic as well as telencephalic structure and function shall be taken up in chapters XII, XIII, and XV of volume V. A number of these topics were dealt with in the monographs on the human diencephalon and on brain and consciousness (K., 1954, 1957).

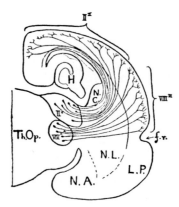

*Figure 350.* Semidiagrammatic drawing showing 'primitive arrangement of the cortical visual (II″ to II‴) and auditory (VIII″ to VIII‴) paths, seen in a hypothetical section
cut in the plane aa′ (from ELLIOT SMITH, 1910). f.r.: fissura s. sulcus rhinalis; H: hippocampus; L.P.: lobus piriformis; N.A.: amygdaloid complex; N.C.: caudate nucleus; N.L.:
'nucleus lentiformis'; Th. Op.: diencephalon (thalamus). II″: corpus geniculatum laterale;
II‴: occipital isocortex; VIII″: corpus geniculatum mediale; VIII‴: temporal isocortex;
this useful first approximation ignores the still unclarified significance of dorsal and ventral
thalamic components pertaining to the geniculate bodies; plane aa is indicated in figures 349
and 353.

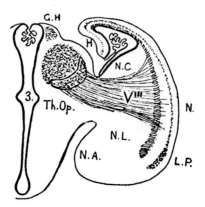

*Figure 351.* Semidiagrammatic drawing through right side of the forebrain in a fetal
platypus (Ornithorhynchus, Prototherian mammal) showing 'cortical tactile path'. The
plane of the section corresponds to bb in figures 349 and 353 (slightly modified, by dotted
outline restricted to 'sensory' thalamic grisea, from ELLIOT SMITH, 1910). V‴: the cortical
tactile path, passing from the outlined combined trigeminal and general sensory thalamic
grisea (V″ and S″ in fig. 349) to the parietal isocortex N; G.H.: ganglion habenulae; 3.:
third ventricle; other abbr. as in preceding figure.

pallium into areas, which are cultivated by the various sensory systems, is determined by this arrangement of the thalamus'.

Be that as it may, there is little doubt that the subdivision of the pallial cortex into regions and areae is closely correlated with the spatial distribution of the fiber pathways between telencephalon and dien-

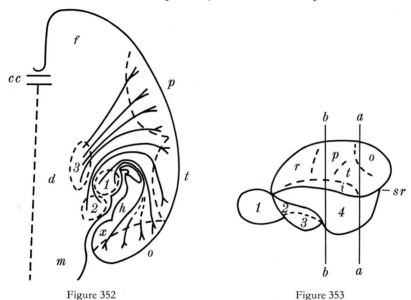

Figure 352                                    Figure 353

*Figure 352.* Diagram of relationships between isocortical regions representing 'Sinnessphären' and thalamic grisea in a 'primitive mammal' as approximately mapped upon a 'horizontal' plane, and supplementing the transverse plane mappings of figures 350 and 351 (adapted and modified after HERRICK, 1926, K., 1927, and FORTUYN and STEFENS, 1951). cc: corpus callosum; d: diencephalon (thalamus); f: regio frontalis of isocortex (presumably essentially 'motor' in 'lower' forms); h: hippocampus; m: mesencephalon; o: regio occipitalis of isocortex; p: regio parietalis; t: regio temporalis; x: parahippocampal cortex; 1: corpus geniculatum laterale; 2: corpus geniculatum mediale; 3: main 'sensory' grisea of thalamus (n. ventralis posterolateralis and n. posteromedialis complex). The ascending connections of thalamic nuclei to regio frontalis, as well as all corticofugal systems have been omitted.

*Figure 353.* Lateral aspect of hemisphere in a 'primitive' mammal, showing cortical territories (modified after ELLIOT SMITH, 1910). 1: bulbus olfactorius; 2: tractus olfactorius lateralis; 3: tuberculum olfactorium (2 and 3 correspond to regions of basal cortex); 4: lobus piriformis (its dorsal boundary is the sulcus rhinalis, its basal boundary the sulcus endorhinalis, which may frequently be 'straddled' by tractus olfactorius lateralis); aa, bb: planes indicated in figures 349, 350, 351; i: regio insularis; o: regio occipitalis (visual cortex); p: regio parietalis ('sensory cortex'); r: regio rostralis; sr: sulcus rhinalis; t: regio temporalis (auditory cortex); i, o, p, r, t are regions of isocortex.

cephalon. These channels follow more or less direct lines of connections in accordance with the principle of least constraint,[279] and are thereby loosely comparable to geodesics. The course of these pathways generally runs without crossing, as justly pointed out by Fortuyn and Stefens (1951). Accordingly, a locus on the surface of the hemisphere becomes connected, by curved lines, of 'geodesic' type, with a corresponding locus in the diencephalon, the curvatures depending on the considerable differences in shape and size between hemisphere and diencephalon.

In this manner, other, additional thalamic grisea display the following relationships to cortex telencephali. Nucleus medialis (or 'dorsomedialis') thalami is connected with the rostral ('frontal') isocortex, anterior thalamic grisea become related to parahippocampal cortex, and the hippocampal formation is linked, by means of the fornix, to the mammillary body of the hypothalamus.[280]

As regards the structural differentiation of the isocortex, some authors concerned with this topic favor a concept of exceedingly detailed cyto- and myeloarchitectural parcellation, resulting in numerous distinct areae commonly mapped with sharp linear boundaries. Others, however, while not denying significant regional characteristics of 'architecture', do not interpret the often variable and frequently rather vague transitions from one region to the other as supporting the allegedly precise overall mappings proposed by adherents of the parcellation concept. Mappings of this type, moreover, as elaborated, for brains of one and the same species by different authors, display manifest discrepancies and contradictions with regard both to boundaries and number of architectural areas, thereby clearly indicating the unreliability of this procedure.[281]

---

[279]A brief discussion of this principle was included in section 6, p. 378, vol. 3, part I of this series.

[280] Further details concerning these connections, including those of other diencephalic grisea and of particular cortical regions, shall be discussed in chapter XII, XIII, and XV of volume 5.

[281] The unreliability of detailed cyto- and myeloarchitectural parcellation seems to be a multifactorial effect whose main components can be enumerated as follows. (1) Subjective ('psychologic') factors of pattern interpretation ('*Rorschach-effect*'); (2) actual ('objective') individual structural variations; (3) distortion of 'architecture' caused by the curvature of the brain surface along gyri and sulci of gyrencephalic brains; (4) distortion of architectural appearance in microscopic preparations caused by variations in the plane

On the basis of my own observations, I became convinced that the detailed cyto- and myeloarchitectural parcellation of the mammalian cerebral cortex, and especially of the isocortex, was highly artificial and devoid of intrinsic significance. I therefore share the views of those authors (i.e. LASHLEY and CLARK, 1946, B. CAMPBELL, 1951, and others) who favor an interpretation of isocortex as a *continuum* exhibiting a rather simple but meaningful pattern of gradients of architectural structure. For lower mammals (e.g. Aplacentalia, Insectivora, rodents, Lagomorphs, etc.), cortical maps with vaguely circumscribed regions can be elaborated in accordance with this evaluation, as shown in Figure 354 and contrasted with a map based on the detailed parcellation concept.

There is, of course, no doubt that within the isocortex of 'higher' mammals frontal granular, precentral agranular, precentral gigantopyramidal, postcentral, parietal, calcarine, insular, and a few other regions have a very typical architectural structure. In the case of area striata there is even a 'hair-sharp' boundary. Likewise, in the hippocampal formation, steep structural gradients apparently also involving histochemically definable regions, have been reliably recorded and justify here some sort of more detailed areal mapping. Again, ECONOMO's distinction (1927) of five *fundamental isocortical types* seems likewise fully acceptable, expecially his concept of *coniocortex*. In view of the gradual transitions, these fundamental types might be reduced to three, namely *agranular*, *granular*, and *coniocortex*.[282]

A cautious interpretation of the isocortical cytoarchitecture in 'lower' mammals seems compatible with the distinction of *five main regions* which may be designated as *rostral*, *parietal* (dorsolateral), *occipital*,

---

of section, which may cut through the cortex with different degrees of obliqueness; (5) varying degrees of staining and differentiation in myelin-stained sections, unavoidable even in the most careful standardized procedures, and (6) differences in appearance obtaining between thick and thin cell-stained sections of one and the same area.

*Mutatis mutandis*, such factors and the thereby required restrictions apply also to the parcellation of noncortical grisea, such as diencephalic and other nuclei. Additional comments on this topic can be found in the monograph on the human diencephalon (K., 1954) and in our contribution to the treatise by HAYMAKER *et al.* on the hypothalamus (CHRIST, 1969).

[282] These terms refer to cortices with poorly developed internal granular layer IV *(agranular)*, with well developed internal granular layer *(granular)*, and with marked predominance of layer IV as well as of small 'granular' cells in other layers *(coniocortex*, i.e. 'dust-like' cortex).

*temporal,* and *insular.* Four of these regions correspond to the three 'sensory areas' described by ELLIOT SMITH (1910), who omitted an appraisal of insular cortex. Regio parietalis represents the 'general tactile' area, regio occipitalis the visual cortex, and regio temporalis the auditory cortex. As regards the regio insularis, scattered older reports, confirmed by some more recent observations (e.g. ROBERTS and AKERT, 1963), seem to suggest that this cortex receives visceral sensory and taste input from grisea of the thalamic nucleus ventralis posteromedialis com-

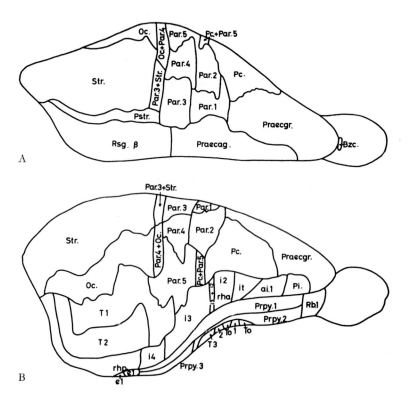

*Figure 354.* Concepts of parcellated and non parcellated cortical maps of the rabbit (from K. *et al.,* 1960). A, B: parcellated map according to M. ROSE, 1931. The detailed subdivisions can be regarded as irrelevant, their linear boundaries being, in my opinion, figments of the imagination. C, D: non parcellated map in accordance with our observations and interpretation. Ba: basal cortex; F: frontal isocortex (regio rostralis); I: insular isocortex (regio insularis); O: occipital isocortex (regio occipitalis); P: parietal isocortex (regio parietalis); Ph.a.: anterior parahippocampal cortex; Pi.a.: cortex of anterior piriform lobe; Pi.p.: cortex of posterior piriform lobe; T: temporal isocortex (regio temporalis).

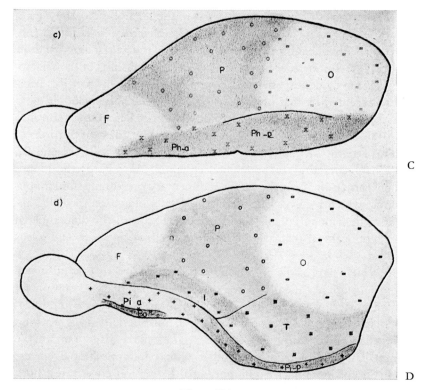

Figure 354

plex. It might thus constitute the visceral-sensory cortex. All these regions can be evaluated as sensory 'centers' in accordance with FLECH-
SIG (1896), who remarked that, in lower mammals, the *'sensory spheres'*
of the cerebral cortex are in direct continuity with each other, and may
all be considered *'projection centers'*.[282a]

---

[282a] In FLECHSIG's original interpretation, only projection 'centers', 'areas', or 'fields'
of the cerebral cortex were supposed to be connected, by afferent or efferent fiber systems,
with subcortical grisea. Association 'centers' or 'fields', on the other hand, were not assumed
to have subcortical, but only cortical connections with other homo- and contralateral
regions of the cortex. Although this concept subsequently proved to be untenable, the
distinction of projection and association centers on the basis of primary 'sensory' and
'motor' fiber systems, and of other functional considerations, discussed further below in
the text, can be said to remain valid, and represents one of several fundamental contributions to neurobiology made by FLECHSIG on the basis of his myelogenetic method (cf.
also FLECHSIG, 1927).

Since olfactory input is not relayed to the telencephalon by way of diencephalic grisea, but directly enters the olfactory bulb and is conveyed to lobus hemisphericus through the olfactory tract and its branchings, the corresponding anatomical relationships of the olfactory cortex are evidently quite different. In mammals, this cortex is laterally represented by that of *lobus piriformis*, presumably including both anterior and posterior portion. Basally, the entire basal cortex ($B_s$) may have substantial olfactory connections, shared by adjacent portions of 'fundus striati'. Medially, much of the paraterminal body ($B_3$ and $B_4$) presumably pertains to the 'rhinencephalon'. As regards the hippocampal formation, only the rather rudimentary precommissural hippocampus, which receives a component of stria olfactoria medialis, appears to be 'rhinencephalic', in contradistinction to the differentiated main hippocampal formation consisting of *cornu Ammonis* and gyrus dentatus.[283]

The main part of the hippocampal formation greatly expands within the substantial caudally paired evaginated mammalian hemisphere, located posteriorly to lamina terminalis, commissural plate and interventricular foramen. The postforaminal hemispheric region, resulting from said caudal paired evagination, represents a major part of the mammalian telencephalon. Together with parahippocampal cortex, the main and nonolfactory hippocampal formation constitutes, both morphologically and functionally, a combined region *sui generis*, whose presumed significance as a 'mechanism for emotion' was first suggested in 1937 by PAPEZ (cf. K., 1951, 1954, 1957). This region approximately corresponds to BROCA's 'limbic lobe', respectively to the 'falciform lobe' of SCHWALBE (1881).

The isocortical regio rostralis, whose posterior portion blends with the anterior one of regio parietalis, can be evaluated as a primarily 'motor cortex', giving origin to the cortico-bulbar and cortico-spinal tracts, and providing, with regio parietalis, a *sensori-motor unit*. Anterior portions of regio rostralis, under the influence of thalamic input,

---

[283] Our observations by means of virus neuronography following nasal instillation of vesicular stomatitis virus in young mice have provided rather strong evidence for this interpretation of olfactory connections in 'lower' mammals (H. K. and M. W. KIRBER, 1962). Extensive caudal parts of paraterminal body ($B_4$) or 'septum', particularly those regions related to fiber systems of the fornix, are, however, doubtless 'emancipated' from 'olfactory domination' and pertain to the system conceived as an 'emotional' or 'affectivity' mechanism by PAPEZ (1937).

may already tend toward the different functional specializations of a 'prefrontal' and a 'premotor' neighborhood in the region of the hemispheric rostral pole, but an unambiguous cytoarchitectural differentiation of these rostral subregions cannot be said to obtain in most 'lower' mammals.

The cerebral cortex of these mammals can thus be interpreted as consisting of anatomically and functionally characterized, but in part somewhat ill-defined regions, which, on the basis of subcortical (essentially thalamo-cortical) connections can, at least to some extent, be interpreted as *'projection areas'* with specialized or limited functional significance.

Yet, despite some undeniable localizatory orderliness, considerable *'equipotentiality'* of the entire cortex, especially with respect to learning, has been inferred on the basis of studies undertaken by LASHLEY in the rat. This author demonstrated that no single part of the hemisphere could be considered necessary for the acquisition of various training habits. If a rat with intact cortex is taught a visual discrimination habit and then the occipital cortex is removed, that habit becomes totally lost. Removal of no other part of the cortex appears to have this effect. However, after removal of occipital cortex the habit can be reacquired by renewed training. Although normal function of the rat's occipital cortex may be essential for some aspects of pattern vision, a rat without occipital cortex shows normal discrimination of light intensity, can avoid obstacles, and seems to recognize food by sight.

When blind rats are taught to acquire the maze habit, destruction of occipital cortex impair this habit and is followed by considerable difficulties in relearning, although the rats could not have used visual stimuli in the initial learning. The occipital (visual) cortex of the rat appears thus to be of significant import for the integration of spatial relationships, even if the animal is blind.

LASHLEY's findings seem thus to indicate that, in the cerebral cortex of the rat and presumably also of other 'lower' mammals with a similar organization of isocortex, *'equipotentiality'* may well be compatible or combined with some degree of *functional localization*. This latter is correlated with a differentiation of the isocortex into the vaguely distinguishable, five primary regions which merge into each other through gradients. The here obtaining 'equipotentiality' is not altogether identical with, but is closely related to the principles of *stability* and *ultrastability* as formulated by ASHBY. Equipotentiality in this connotation may be defined as the capacity of a given griseum or cortical region to perform a variety of different functions, similar or identical to functions performed by other regions or grisea. Equipotentiality in this sense implies unspecificity or limited specificity combined with multiplicity of functions. In addition, the topographic distribution of various performances will depend on functional demands, such that whatever region is momentarily idle will become available, either *in toto* or with a set of its elements, for the execution of a given activity. Likewise, if a certain area is destroyed, any other area will be able to function vicariously.

In higher mammals, particularly primates and man, additional isocortical areae, which may be designated as *secondary regions*, become intercalated between, or added to, the five primary regions. Equipotentiality is certainly also, to some extent, manifested by the cerebral cortex and other brain structures of these 'higher' forms, including man. But I do not believe that the human brain can be interpreted as narrowly in terms of rat brain functions as many of the views elaborated by LASHLEY[284] seem to imply. I fully agree in this respect with most of the evaluations and critical comments expressed by HERRICK (1926).

In primates and other higher mammals, the architectural characteristics of the isocortex seem to become more definite; various regions such as frontal granular, precentral agranular and gigantopyramidal cortex, koniocortex, especially area striata, and a few others are more or less readily recognizable. Concomitantly with this structural differentiation, a certain functional specialization takes place, with corresponding greater or smaller loss of equipotentiality. This latter functional characteristic, however, will still continue to obtain throughout a large extent of the cortical continuum. ASHBY (1952) has pointed out that such a type of localization is manifested by a multistable system. In a large system of this sort the total reaction will be based on step functions and activations that are both numerous and widely scattered. Removal of almost any part may cause some disturbance, but, within a certain limit, no part can be identified whose removal will be found to suppress a given function completely. This, as ASHBY remarks, agrees very closely with the actual findings of LASHLEY and others. Again, this interpretation of cerebral mechanisms is fully consistent with the concept of engrams as suggested in my lectures of 1927. The memory traces are assumed to be individually localized by means of multiple

---

[284] References to the relevant studies by LASHLEY can be found in HERRICK's publication 'Brains of Rats and Men' (1926).

traces, widely scattered over the cerebral cortex, in an apparently random dispersion. Such traces would be unified anatomically as components of a given network pattern, and functionally by the activation of such networks.

Elliot Smith (1910) justly comments on the evolution, within the cerebral hemispheres, of a mechanism for receiving and blending the various 'impressions' (i.e. input signals) pertaining 'to the sense of smell, taste, touch, sight, hearing, etc. But the cerebral hemisphere does far more than this – it provides a site far removed from the merely executive parts of the nervous system where these various sensory impressions, or the states of consciousness which they awaken, may be in some way recorded, so that later on new impressions may be influenced, compared, and tested by the results of these foregoing experiences. In other words, the cerebrum attains its dominant position in the hierarchy of the brain mainly because it provides the means for the evolution of an organ of associative memory'.[285]

With regard to the cerebral hemispheres of man, the fundamental and pioneering studies of Paul Flechsig (1896) succeeded in demonstrating, by means of the myelogenetic method, the location of the main primary sensorimotor and sensory areae, which he designated as *projection centers*. The primary sensorimotor area, related to 'touch' (in

---

[285] The special mechanisms of recording, whose nature was regarded as altogether unknown by Elliot Smith in 1910 may now be considered as at least partially intelligible with respect to its general principles. An early neurologic general theory of memory, based on concepts expressed by Semon, Pavlov, Robertson, and Herrick, and involving both structural and biochemical aspects, was elaborated in my 'Vorlesungen' (K., 1927). Very similar if not identical theories were then subsequently propounded by various other authors (e.g. Hebb, 'The organization of behavior. *Wiley, New York 1949*). Likewise, the neurological mechanisms of logical thought can now be assessed as intelligible in principle, regardless of numerous unsettled details ,on the basis of Boole's mathematical concepts elaborated about 1864, adapted to electrical circuits by Shannon in 1938, and, a few years later, to neural nets by McCulloch and Pitts, and by numerous other authors (cf. K., 1957, 1961, 1965, 1966). The nervous system, transmitting and processing signals which encode information (cf. chapter I in volume 1 of this series) operates in accordance with a mathematically definable orderliness intrinsically identical with that obtaining for artificial control or computing mechanisms. Difficulties in understanding result from the fact that the electrical and chemical transmission processes are, as it were, superimposed upon the general biologic properties of nervous tissue instead of being performed by non-living hardware. As regards *consciousness*, however, to which Elliot Smith (1910) repeatedly refers, I believe to have shown, by means of the brain paradox, that it is logically impossible to derive consciousness from activities occurring in physical space-time (K., 1959, 1961).

the wider sense) and to so-called 'voluntary' motor activities, was shown to be located on both sides of sulcus centralis. It consists of an essentially motor precentral and an essentially sensory postcentral portion (1, 2 in Fig. 355 B).

These two sensorimotor primary *subareae* include, *qua* cortical cytoarchitecture, the 'motor' area precentralis agranularis gigantopyramidalis and the 'sensory' postcentral strip of coniocortex. The *primary visual sensory area* (4) was demonstrated by FLECHSIG along and surrounding, the calcarine fissure of the occipital lobe. It corresponds to the isocortical *area striata*. The *primary auditory area* (3) was identified by FLECHSIG in the region of the transverse temporal gyri, bordering on gyrus temporalis superior, and to a large extent hidden within the fissura cerebri lateralis. This area is likewise characterized by coniocortex. The location of the primary gustatory and visceral sensory center in the insular region and that of the olfactory cortex in lobus piriformis can be deduced on the basis of additional evidence, not directly related to FLECHSIG's work, and briefly discussed further above.

FLECHSIG thus clearly located the cortical areas receiving *optic, auditory*, and *general sensory input*, and the area providing *output through the pyramidal (cortico-bulbar and cortico-spinal) tract*. These projection centers of FLECHSIG represent areae essentially corresponding to the primary areae, which, in the hemispheres of 'lower' mammals, directly adjoin each other, but became separated, in the hemispheres of 'higher' mammals, and particularly of man, by extensive cortical regions designated as *association areas* by the cited author. It is now well established that most, and taken for granted that all so-called association areas are provided with subcortical connections; nevertheless, FLECHSIG's general concept remains doubtless valid.

Cyto- and myeloarchitectural subdivisions of the cortex, if properly revised so as to avoid undue parcellation, likewise conform fairly closely with the general subdivision assumed by FLECHSIG. This latter author doubtless also greatly overemphasized the parcellation of the cortex into his myelogenetic fields. It is most unlikely that these fields can be clearly delimited and that they are as numerous as FLECHSIG indicates on his detailed brain chart. Myelinization, like other ontogenetic processes, is subject to considerable individual variations within a significant overall pattern. It is probable that many of the numerically denoted small fields of FLECHSIG's intermediate and terminal regions, like the fractional patches of highly parcellated cyto- and myeloarchitectural brain maps, have no intrinsic specialized significance. On the other hand, the particular strips of association areas closely adjacent to the sensory projection areas, and designated by FLECHSIG as 'Randzonen der Sinnessphären' are clearly distinguished by cytoarchitectural characteristics (e.g. area peristriata and parastriata). These 'Randzonen' whose peculiarities were already stressed by the cited author, seem in-

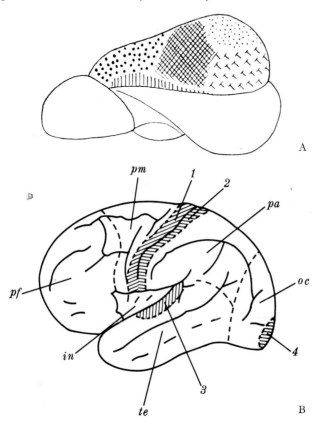

*Figure 355.* Comparison of isocortical regions in a 'primitive' mammal with those in a 'higher' mammal such as man. A: Isocortical map of the insectivore jumping shrew, *Macroscelides* (after ELLIOT SMITH, 1924, and HERRICK, 1926, from K., 1927). Coarse dots: regio rostralis; cross-hatching: regio parietalis; fine dots: regio occipitalis; vertical hatching: regio insularis; oblique Ts: regio temporalis; B Isocortical map of lateral aspect of human brain (based on a figure by FLECHSIG, 1920). The 'projection areae' (1–4), corresponding in some respects to the primaty regions in 'primitive mammals', are separated from each other by extensive 'secondary', 'association areae'. This further differentiation evidently implies that, in several respects, human 'projection areas' may significantly differ from the corresponding lower mammalian 'primary regions'; pf: prefrontal; pm: premotor.

| Primary isocortical regions of lower mammals | Isocortical regions of higher mammals |
|---|---|

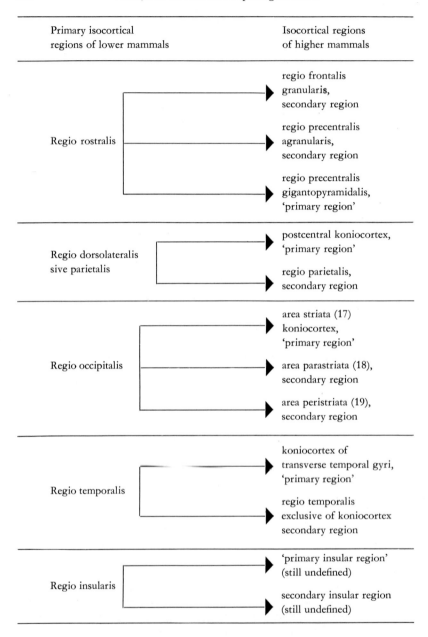

Regio rostralis
- regio frontalis granularis, secondary region
- regio precentralis agranularis, secondary region
- regio precentralis gigantopyramidalis, 'primary region'

Regio dorsolateralis sive parietalis
- postcentral koniocortex, 'primary region'
- regio parietalis, secondary region

Regio occipitalis
- area striata (17) koniocortex, 'primary region'
- area parastriata (18), secondary region
- area peristriata (19), secondary region

Regio temporalis
- koniocortex of transverse temporal gyri, 'primary region'
- regio temporalis exclusive of koniocortex secondary region

Regio insularis
- 'primary insular region' (still undefined)
- secondary insular region (still undefined)

*Figure 356.* Tabulation showing the assumed differentiation of the isocortex in the course of mammalian phylogenetic evolution (slightly modified after K., 1927). The available data concerning regio insularis are still insufficient for an unambiguous areal delimitation in accordance with the concept expressed by this tabulation.

deed to be of particular significance for the processing of the sensory input received by their adjacent projection areae.

A comparison of Figure 355 A and B indicates the presumptive phylogenetic evolution of the mammalian isocortex by a differentiation of secondary areae intercalated between, or added to the primary areae of 'lower' mammals. The tabulation of Figure 356, in overall accordance with my concepts of 1927, likewise summarizes this aspect of isocortical differentiation. One could, of course, also state that said phylogenetic evolution represents the differentiation of original primary areae *sensu strictiori* into modified primary areae *sensu latiori* and a number of secondary areae.

According to ELLIOT SMITH (1910), each (primary) sensory isocortical area is, at first, concerned with (a) the reception of a particular kind of impressions, to which its connexions and its structure fit it to respond; (b) the blending of these impressions with those simultaneously received by other neopallial areas, and (c) the excitation or inhibition of lower nervous centers, so as to bring the influence of this processing to bear upon the behavior of the animal. ELLIOT SMITH, moreover, stresses 'conscious appreciation' and 'consciousness' with regard to these cortical activities, but, by slightly rewording that author's conclusions, I have eliminated his references to consciousness, because this latter has no 'logical existence' in the here adopted behavioristic and mechanistic formulation.

ELLIOT SMITH (loc.cit.) then proceeds to remark: 'but there develops within each neopallial area a special mechanism – the nature of which is altogether unknown to us – which confers upon it the power of acting as a recording apparatus, in which impressions can in some way be stored, so that on future occasions a new stimulation can awaken memories of foregoing states of consciousness, by means of which the latest impression can be tested and compared with others which have impinged upon this neopallial mechanism on previous occasions. In other words, each neopallial area becomes not merely a receptive organ for sensory impressions, but also an organ of associative memory'.

With this double function to perform, the original isocortical areae undergo a differentiation of structure and function, insofar as one component continues to act as receiver, transmitting the processed 'impressions' to another component which, in addition to 'recording' the received input, establishes connection with different areas of the same and of the contralateral hemisphere, and also with other parts of the nervous system.

Thus each primitive isocortical area in the 'lower' mammalian brain contains both 'sensory' and 'association' elements, intermingled, according to ELLIOT SMITH, in much the same way as the sensory (fascia dentata) cells and efferent (cornu Ammonis) cells in the reptilian hippocampal formation.[286] But just as the latter, during the process of evolution of the mammal, becomes separated into two parts, of which one is essentially a receptive griseum *(gyrus dentatus)*, and the other an efferent structure *(cornu Ammonis)*, so each of the receptive isocortical areae becomes transformed into a central '*sensory*' part and a peripheral '*association*' part. The different sensory areae, which were originally co-terminous, thereby become separated the one from the other by a series of 'association' areae, i.e. by one or more developed at the periphery of each of the previously coterminous regions. It may be taken for granted that the 'superiority' of the human, over the ape's, brain as an organ of 'higher' nervous functions must be the result mainly of the higher development of the 'association' areas. Quite evidently, these views of ELLIOT SMITH (1910) are essentially in full agreement with the basic concepts elaborated by FLECHSIG for the human brain, and with my own subsequent interpretation of these concepts in terms of comparative neurology (K., 1927).

It should, however, be kept in mind that, at least in higher mammals, particularly in primates and man, the 'projection centers' comprising precentral 'motor', postcentral 'sensory', temporal auditory, and occipital visual cortices seem to display a particular topographic and functional differentiation which must also involve intricate details of synaptology. Thus, in addition to the main sensorimotor sequence displayed by precentral gigantopyramidal cortex and postcentral konio-cortex, a basally adjacent *second sensorimotor strip* has been demonstrated, and a so-called *supplementary motor area* appears to be present at the superior rostromedial border of the main precentral motor area. Again, a *secondary auditory area*, with reversed distribution of input, is adjacent to the main auditory area. Moreover, the visual 'projection

---

[286] The available data concerning connections and synaptology of the reptilian (and avian) so-called hippocampal formation ($D_3$) are not yet sufficiently complete to substantiate or to contradict ELLIOT SMITH's interpretation of 1910. This question, however, has no bearing on that author's doubtless correct appraisal of the regional differentiation in the mammalian cerebral cortex. It should be recalled that so-called 'primordium hippocampi' of Anamnia, and so-called 'hippocampus' in reptiles or birds, are indeed morphologically homologous to the mammalian hippocampal formation, but by no means necessarily analogous *qua* function and synaptologic structure.

center' (area 17 of BRODMANN) is almost completely surrounded by a strip of cortex, the area parastriata (18) which might, in some respects, be interpreted as a 'secondary visual area'. On the basis of presently available data, no answer can be given to the question whether a similar differentiation does or does not obtain with regard to the olfactory projection center.

In concluding the present section dealing with morphogenesis and presumptive phylogenetic evolution of the vertebrate telencephalon, some remarks on the prefixes *neo-*, *archi-*, and *palaeo-* are perhaps appropriate. In his brilliant treatises on the central nervous system of vertebrates, L. EDINGER (1908, 1911, 1912 and many previous other publications) proposed a distinction between *palaeencephalon* and *neencephalon*. The former comprises the basic neural structures and mechanisms present in 'essentially' similar manner from fish to man. The latter includes structures and mechanisms which reach their full development in mammals, and seem to culminate in man. Analogous considerations led ELLIOT SMITH (1901) to introduce the term *neopallium* for that part of the pallium which is the 'latest' to reach the height of its development, and which includes the isocortex. For this latter, usually together with adjacent parahippocampal cortex regions, numerous authors subsequently used the designation '*neocortex*'.

It is evident that EDINGER's neencephalon essentially or primarily consists of the neopallium but must also include a diversity of widely distributed subcortical structures directly or indirectly correlated with the neocortex and correspondingly manifesting a degree of differentiation not present in vertebrates lacking a neopallium of mammalian type. Some such subcortical 'neencephalic' components may remain intimately conjoined, 'interwoven', or 'spliced' with so-called 'palae-encephalic' ones, thus precluding a clear-cut morphologic distinction based on spatial separation respectively disjunction.

The term *archipallium*, frequently attributed to ELLIOT SMITH, but disclaimed and strongly disapproved by this author (1910), was used by EDINGER and others for the hippocampal formation. KAPPERS (1921–1922 as well as previous publications) designated a ventrolateral region, comprising, in mammals, the piriform lobe and some adjacent basal structures, as *palaeopallium*. KAPPERS furthermore distinguished, within the basal ganglia, a *neostriatum*, an *archistriatum*, and a *palaeostriatum*. EDINGER and his associates outlined a *neocerebellum* and a *palaeocerebellum*. Subsequently, the term *archicerebellum* has been added (LARSELL, 1951 and previous publications). KAPPERS *et al.* (1936) use the term *neothalamus*, HERRICK (1948) speaks of neothalamus and *palaeo-thalamus*, and GREVING (1931) distinguishes *archaeothalamus* and *neo-thalamus*.

While EDINGER's concepts of neencephalon and palaeencephalon are enlightening as well as useful for purposes of general orientation along phylogenetic lines, and ELLIOT SMITH's term 'neopallium' seemed, at the time, particularly well chosen, the indiscriminate use of the term '*phylogenetically old*', and of the prefixes '*neo-*', '*archi-*' and '*palaeo-*' has resulted in considerable confusion, engendering highly misleading notions. This became soon clear to ELLIOT SMITH, who repeatedly took issue with this ambiguous and well-nigh hopeless imbroglio.

Again, one might justifiably ask why the highly developed mammalian hippocampus, consisting of *cornu Ammonis* and fascia dentata, should not be considered a 'phylogenetically new' cortex. There is little doubt that it has reached a structural development (including synaptic and biochemical characteristics) greatly differing from that of the morphologically homologous so-called hippocampus or $D_3$ neighborhood of submammalian forms.

In the telencephalon of Selachians and Dipnoans a 'cortical' lamina, corresponding to JAKOB's and ONELLI's concept of '*cortex inferior mono-stratificatus*' is present in both pallium (D-neighborhood) and basis (B-neighborhood). A *cortex paulostratificatus* occurs in the pallium of reptiles and birds. A typical '*cortex superior polystratificatus*' of JAKOB's and ONELLI's terminology occurs only in the pallium of mammals, which comprises piriform lobe cortex, isocortex, parahippocampal cortex, and hippocampal formation. These four cortical regions, however, of which two can be regarded as 'polystratified', are also present, *qua* neighborhoods or subneighborhoods, in the pallium of all vertebrates. Seen from this viewpoint, said cortical regions are, morphologically or topologically, and thereby phylogenetically, not only quite 'old', being present in Selachians as well as even in Agnatha, but, *pari passu*, also *equally* old (cf. above, p. 639). On the other hand, each of these regions is structurally 'new' *qua* mammals, namely with respect to its mammalian type of differentiation, and regardless to differences in architectural complexity between the four pallial cortices (piriform lobe cortex, isocortex, parahippocampal cortex, main hippocampal formation).

However, if (1) the hexalaminar differentiation of the isocortex, and (2) the progressive dominance of that cortex in the taxonomic mammalian series is emphasized, then, of course, its evaluation as 'neocortex', and the designation 'neopallium' for its telencephalic wall domain, can be justifiably upheld. Whether the term 'neocortex' should or should not be extended to include the slightly less conspicuously 'polystratified' parahippocampal cortex depends on arbitrary criteria in deciding a relatively minor terminologic question.

If JAKOB's and ONELLI's *cortex inferior monostratificatus*', occurring in recent Selachians and Dipnoans,[287] is considered the phylogenetically 'oldest' type, then the 'oldest' cortex of the Mammalian brain is the basal cortex, corresponding to tuberculum olfactorium and associated structures (area ventralis anterior) and cortical amygdaloid nucleus (area ventrolateralis posterior).

Concerning the evolution of the *basal ganglia* it can be said that caudate nucleus and putamen of mammals are formed by morphologic elements already demonstrable in Anamnia. The mammalian caudate nucleus and putamen furthermore represent a *secondary fusion* of several distinctive nuclei present in Sauropsida (K., 1929 a). The amygdaloid complex of Amniota follows a different line of evolution but cannot be regarded as either phylogenetically older or more recent than caudate nucleus and putamen. The globus pallidus differentiates from the anterior interpeduncular nucleus of the hypothalamus of lower vertebrates and reaches its greatest development in higher mammals, especially in primates and man. Its relation to anterior interpeduncular nucleus is somewhat similar to the relationship between nucleus parafascicularis and centrum medianum. Thus, as regards the basal ganglia, the terms neostriatum, archistriatum and palaeostriatum are likewise devoid of any significance and, moreover, actually misleading. There is, besides, a fourth component of the basal ganglia, the claustrum, not to mention such additional morphologically semi-independent subdivisions as nucleus accumbens septi, nucleus ansae peduncularis, and perhaps others.

Granting certain premises, however, the *cerebellum* could indeed be conceived as having three main subdivisions, the neocerebellum and two older ones, the vestibular portion evolving from the auricles of lower vertebrates, and the essentially vermian part of the corpus cerebelli with its spinocerebellar and other phylogenetically old connections. There is thus justification for a convenient subdivision into *ar-*

---

[287] To a lesser degree, such 'cortex' is also displayed by some amphibians, particularly Gymnophiona, although, in general, the telencephalon of Amphibia is characterized by paraventricular or 'ependymal praecortex, (cf. above p. 615).

*chicerebellum* (auricles), *palaeocerebellum*, and *neocerebellum* as indicated by
LARSELL (1951) who fully emphasizes that there are no sharp boundaries between the stages of phylogenetic development. In the flocculo-
nodular lobe, for instance, the nodulus itself is a phylogenetically relatively new acquisition forming a functional unit with the older floccu-
lar components. An objection against the prefixes '*palaeo-*' and '*archi-*'
as judiciously employed for the cerebellum by LARSELL could yet be
raised on the ground that these terms tend to perpetuate a misleading
cliché.

With respect to the presumptive phylogenetic evolution of the *dien-*
*cephalon*, it might, *prima facie*, seem tempting to regard those thalamic
grisea with reciprocal 'neocortical' connections, such as relay nuclei
for main sensory systems, or grisea pertaining to 'neocortical' feedback
systems ('*direct cortical modulators*', K., 1954), as representing the 'neothalamus'. It would, however, be difficult to reconcile this appraisal
with the probable features of thalamic evolution as inferred from the
data provided by comparative neurology.

From a phylogenetic viewpoint one might perhaps distinguish
three fundamental stages, the *primitive stage of Anamnia* with an undifferentiated general dorsal thalamic primordium and with an additional
primordium of lateral geniculate body, the *intermediate reptilian stage*,
and the *mammalian stage*. Little would be gained by designating these
stages as archi-, palaeo-, and neothalamus, respectively. If the designation neothalamus is applied to all dorsal thalamic nuclei that are for the
first time clearly differentiated in mammals, an odd miscellany is obtained. Still, dorsal lateral geniculate body, anterior group, dorsal thalamic pretectal group, and probably nucleus medialis as well as posterior ventral complex would be excluded, these grisea being already present in reptiles. The classification of additional nuclei, such as midline
group, nucleus parataenialis and nucleus parafascicularis would be
doubtful.

If only the nuclei directly related to neocortex are meant to constitute the neothalamus, then such phylogenetically very recent nuclei as
centrum medianum and intralaminar nuclei are excluded, while the
most ancient individual nucleus of dorsal thalamus, the dorsal lateral
geniculate body, is included. The anterior group, projecting upon gyrus cinguli, is connected with a region of parahippocampal cortex,
which, however, many authors consider neocortex. The systematization of this nuclear group would be difficult on account of the transitional nature of the correlated cortex.

Turning to the allegedly 'phylogenetically old' midline nuclei we find that they cannot be recognized as such in the thalamus of Anamnia. In reptiles, depending on different species or perhaps orders, they may or may not be present. In mammals, these nuclei are conspicuous, more so in lower than in higher forms. I do not believe that this complex picture justifies the use of the terms archi- or palaeothalamus.

Moreover, many authors have the tendency to designate any center of known or suspected vegetative function as 'phylogenetically old'. There is little justification for such a sweeping generalization. It is certain that phylogenetically recent higher centers controlling vegetative functions exist, in hypothalamus as well as in cortex cerebri. In primitive vertebrates vegetative functions may be regulated by less specialized or by lower centers. The terms phylogenetically 'recent' or 'old', although doubtless at times useful for some aspects of neurobiologic theories, will here lead into semantic traps. Again, the designations 'neothalamus', 'palaeothalamus' and 'archithalamus' cannot be applied to any well defined groups of grisea, do not convey a clear meaning, and should be altogether avoided.

Concerning the limited usefulness of the phylogenetic terms *'old'* and *'new'*, this latter may evidently be employed to denote particular structural or functional characteristics evaluated as 'progressive' because displayed by higher vertebrates, especially, but not exclusively, mammals and man. In this respect, 'new' will, in general, also connote an assumed 'improvement' respectively 'refinement' of the neurobiological mechanisms. The term *'neocortex'* may thus be considered acceptable, but there seems to be no justification for any use of the prefixes *'archi-'* and *'palaeo-'*, with the barely possible exception of their application to the cerebellum, as stated above. Even here, there obtains, moreover, no phylogenetic 'age difference' between configurations characterized by the designations 'archi-' respectively 'palaeo-'.

Phylogenesis of the brain and particularly of the telencephalon implies a progressive differentiation of relatively simple topologic neighborhoods in lower or ancestral forms toward the complex configurations displayed by higher forms such as mammals and man. It would thus appear far more logical to interpret mammalian configurations in terms of their original, primordial neighborhoods than to designate, in lower forms, these primordia with names exclusively applying to highly evolved grisea of mammals, which, moreover, have in part undergone a considerable secondary spatial rearrangement.

It seems, therefore, rather inappropriate to use the term '*amygdala*' for particular grisea of lower vertebrates, since the mammalian amygdaloid complex consists of several components whose primordia are included in quite separate neighborhoods of the Anamniote telencephalon. Likewise, the terms *neostriatum*,[288] *archistriatum* and *palaeostriatum*, *cornu Ammonis*, and *gyrus dentatus* become, seen from this viewpoint, outright incongruous if applied e.g. to the hemispheres of fishes or amphibians, while the designation '*primordium hippocampi*' seems here at least permissible. Again, a 'true' *lobus piriformis*[289] and its cortex are present only in mammals. One might, at most, with some qualifications, use both terms in discussing and describing homologous Sauropsidan configurations. The designation 'piriform lobe' or 'piriform cortex' is certainly inappropriate for distinctive telencephalic grisea of Anamnia.

## 7. Meninges, Liquor, and Blood Vessels

The meninges of vertebrates represent the mesodermal membranes investing the ectodermal neuraxis, and commonly provide the adventitial tissue accompanying the penetrating blood vessels which subserve the nervous parenchyma. In addition, the meninges may be a main

---

[288] Even the term '*striatum*', characterizing a mammalian configuration, seems rather inappropriate for submammalian forms. *Mutatis mutandis*, similar objections could be raised against the use of the term 'tuberculum olfactorium'. A true 'septum', as discussed above in the text, is a configuration restricted to the hemispheres of man and of some mammals with rather large brains.

[289] *Piriform* (pear-shaped) derives from the classical Latin *pirum*, pear, respectively *pirus*, pear tree, as spelled by Latin authors such as CATO (234–149 B.C., *De agri cultura*), VARRO (116–27 B.C., *Rerum rusticarum libri tres*), and COLUMELLA (*floruit* ca.45 A.D., *De re rustica*). At a later period, the alternate spelling *pyrus*, perhaps mistakenly derived from the Greek πῦρ (fire) was introduced and especially used by medieval scribes (cf. also TRIEPEL, H.: *Die anatomischen Namen, ihre Ableitung und Aussprache*. Bergmann, München 1921). Although admitting that '*pyriformis*' represents a philological error, ELLIOT SMITH (1901) holds that it is shared by most writers of all nationalities, and so fixed by long usage, 'that it would be pedantic at this late hour to attempt to rectify the spelling'. Since some variety, including that of terminology and spelling, may be regarded as 'the spice of life', I nevertheless prefer to write '*piriform*' without prejudice to the preference of other writers. The Latin term *cortex* (bark, rind) was used with two genders (m., f.), although the masculine form predominated. I formerly employed the feminine form '*cortex olfactoria*' for the basal cortex (K., 1927) in order to add a linguistic stress to the distinction between basal and pallial cortical formations.

source for the histiocytic *Hortega cells* (microglia, mesoglia) located within that parenchyma.

Although the central nervous system of invertebrates is likewise provided with connective tissue sheaths *('perilemma')*, these membranes, as well as the thin perilemma surrounding the neuraxis in Amphioxus, are not classified as 'meninges'. It could be said that a characteristic feature of vertebrate meninges is their relationship to both central nervous system and to the skeletal structures enclosing the neuraxis.

For a brief discussion concerning classification and terminology of the different components or layers displayed by the meninges in the various vertebrate forms, it is perhaps appropriate first to consider the overall features of the ontogenetic development of this membranous investment.

Generally speaking, at fairly early embryonic stages, the neuraxis is surrounded by undifferentiated mesenchyme (cf. e.g. Figs. 50 A, 56 B–D, 308 A, B, 309 A–C). With progressive differentiation, an *outer zone* of this perineural tissue becomes the *perichondrium* respectively the *periosteum* of cranial cavity and vertebral canal, taking part, moreover, in the formation of the *inner ligaments* providing a continuity between the separate vertebral metameres. An *inner boundary zone*, in direct apposition to the neuraxis, becomes the anlage of the *leptomeninx*. Concomitantly with these events, blood vessels begin to appear within the differentiating perineural mesenchyme.

There is little doubt that in many, perhaps even most or all vertebrates, the *neural crest*, a primary ectodermal derivative, contributes to the formation of perineural mesenchyme. This, however, does in no way preclude an evaluation of that mesenchyme as a topologically, morphologically and biologically mesodermal component.[290]

---

[290] Cf. vol. 3, part I, pp. 18, 52 *et passim*, and section 1 A, p. 109 of the present volume. Even discounting the assumed contribution to histiocytic microglia, the various well corroborated instances of meningeal hematopoietic activities, occasional 'normal' ossifications within the dura mater, the occurrence of osteoblastic and lipomatous meningiomas, and other behavioral features of meningeal neoplasms provide strong arguments for a classification of meninges as typical *'mesodermal'* tissue. On the other hand it is obvious, as was repeatedly pointed out in the preceding volume, that the useful classification into ectodermal, mesodermal, and ectodermal tissues does not imply a dogmatically rigid concept of 'absolute' specificity. Again, there is little justification for the assumption, expressed by some authors, that the 'ectodermal' contribution to the meninges affects merely the leptomeninx. Finally, although the neuroectodermal elements of the disintegrating

Secondary zones of condensation and of rarefaction, which may become combined with 'separations' or 'splittings' within or between
such zones, result in the formation of more or less distinctive layers. In
this respect, various differences obtain not only with respect to taxonomic groups, but can also be displayed, in one and the same species,
between spinal and cranial meninges of at least some vertebrate categories, as well as at different intracranial locations.

An extensive investigation on the comparative anatomy of meninges in the entire vertebrate series was undertaken by STERZI (1901).
According to this author, the membranes surrounding the neuraxis in
fishes consist of a rather thin, poorly differentiated vascular *meninx
primitiva* closely investing the central nervous system, and continuous
with a similar investment of the nerve roots. The loose tissue between
meninx primitiva and periost respectively perichondrium was designated as *perimeningeal tissue*. Beginning with the phylogenetic or taxonomic stage manifested by amphibians, the meninx primitiva is said to
provide an outer, dense layer, which becomes the *dura mater*, and an inner, less dense one, the *meninx secundaria*. This latter undergoes a further differentiation into *arachnoid* and *pia mater* at the mammalian stage.
Thus, in STERZI's terminology, the *dura mater* is, *ab initio*, conceived as
a membrane *sui generis* differing from endocranial and endorhachidial
periost or perichondrium.

Other authors, however, including GEGENBAUR and VAN GELDE-
REN[291] describe the layer related to the skeletal system as pertaining to
the dura *(ectomeninx primitiva)*, separated by intermeningeal[292] tissue
from the primitive inner membrane *(endomeninx)* directly investing the
neuraxis of fishes. In higher vertebrates, the original ectomeninx could
then be regarded as becoming differentiated into (a) *endocranium* respectively *endorhachis*, and (b) *dura mater sensu proprio*. Both (a) and (b) might
also be evaluated as outer and inner layers of *pachymeninx* or dura mater

---

neural crest portions may be said to undergo a further differentiation that can be evaluated
as *'ortsgemäss'* (cf. section 1 A, p. 8), namely as subsequent mesodermal components,
some specific behavioral properties of perineural mesoderm could be related not only to
an interaction with the surrounded neuraxis, but also to the presumed assimilation of
neural crest elements.

[291] For additional details, including bibliographic references, cf. BECCARI (1943),
VAN GELDEREN (1924, 1926), KAPPERS (1925, 1947), KAPPERS *et al.* (1936), PALAY (1944).

[292] 'Intermeningeal' if conceived as located between two different meninges, namely
ectomeninx and endomeninx in this case. 'Intrameningeal' if conceived as located 'within'
the meninges considered *in toto*.

*sensu latiori*. The primitive *endomeninx* is considered to represent a single stratum of *leptomeninx* which, in higher forms, displays a further subdivision into *pia mater* and *arachnoidea*.

KAPPERS (1926, 1947, KAPPERS et al., 1936) emphasizing that the meninges in lower vertebrates are very different from those in mammals, including man, presented a critical review of the relevant problems on the basis of his own observations. As regards Cyclostomes, Selachians, and Ganoids, KAPPERS essentially confirmed the findings of STERZI, whose nomenclature he adopted. With respect to Teleosts, however, KAPPERS justly pointed out the considerable variations of meningeal differentiation displayed by the diverse forms of this multitudinous and variegated group.

KAPPERS noted that, in many small Teleosts, the meninges may show an arrangement very similar to their lamination in Selachians and Ganoids. He observed, however, a substantial differentiation of the endomeninx, associated with additional membrane formation, in other Teleosts such as, e.g., Lophius. KAPPERS also described and interpreted further details concerning the progressive differentiation of the leptomeninges in Amphibia, Sauropsida, and Mammals.

While adhering to the concept that periost respectively perichondrium should not be considered part of the dura proper, KAPPERS expressed some doubt as to the derivation of dura from meninx primitiva. It appeared to him possible 'that the real dural membrane of higher animals develops from the mesenchymatous blastema immediately adjacent to the meninx primitiva' (KAPPERS et al., 1936). Be that as it may, I prefer to consider the endocranium (respectively endorhachis) as constituting an integral part of dura mater, representing, in fact, a direct derivative of the primordial ectomeninx.

Despite several additional investigations, the comparative anatomy of the meninges with regard to the entire vertebrate phylum has remained rather neglected and incompletely elucidated. The inferences derived from various studies of ontogenetic development (FLEXNER, 1929; HARVEY and BURR, 1926; WEED, 1917) seem, at best, somewhat ambiguous and inconclusive. Investigations of the adult conditions require delicate procedures preserving exact spatial relationship between cranial or spinal cavity and neuraxis with its, in part, fragile membranes. Much of the routinely obtained material is unsatisfactory in this respect. I had no opportunity to study the relevant problems in a systematic manner and my own opinions on this topic are based on incidental observations in submammalian forms compared with the find-

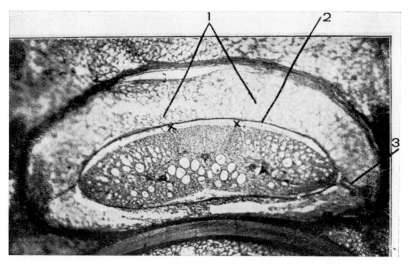

*Figure 357.* Spinal meninges in the Cyclostome Petromyzon (from KAPPERS, 1925). 1: perimeningeal tissue; 2: 'meninx primitiva'; 3: ligamentum laterale; x: shrinkage spaces between meninx and surface of spinal cord.

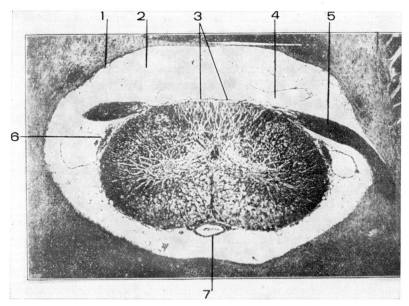

*Figure 358.* Spinal meninges in the Selachian, *Scyllium canicula* (from KAPPERS, 1925). 1: endochondrium; 2: perimeningeal tissue; 3: 'meninx primitiva'; 4: 'epidural vein'; 5: posterior root; 6: ligamentum laterale; 7: anterior spinal artery.

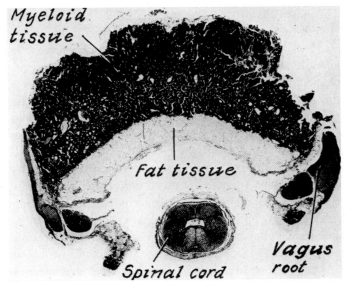

*Figure 359.* Meningeal 'myeloid organ' of the Ganoid Amia, in the region of transition between rhombencephalon and spinal cord (from E. SCHARRER, 1944).

ings recorded in the literature. With regard to mammals and man, my findings on meninges are, however, somewhat more detailed. It seems, moreover, that various discrepancies in the accounts given by different authors are essentially disagreements in the adopted semantic approach to an appropriate terminology.[293]

Figure 357 shows the meninges in the *Cyclostome* Petromyzon. In terms adapted to GEGENBAUR's original concept, this configuration may be described as follows. The intermeningeal space between *endomeninx* ('meninx primitiva') and *ectomeninx* (internal perichondrium) is wide and consists mostly of rather large rounded cells with 'mucous' or 'mucoid' appearance.[294] A recent report by ROVAINEN (1970) indi-

---

[293] E.g. concerning the question, mentioned above, whether the endocranium proper should or should not be included in the designation 'dura'. Also, to be pointed out further below, concerning a proper definition of arachnoidea, as distinguishable from pia, within the leptomeninx.

[294] HIBBARD (1963) describes this tissue layer as 'subarachnoid space', and its indistinct outer boundary zone (which actually blends with the endocranial perichondrium (his 'dura mater') as 'arachnoidea'. Needless to say, I do not believe that this terminology is morphologically appropriate.

cates that these cells contain lipid droplets and a considerable amount of glycogen. Incubated meningeal tissue is said to produce a substantial amount of glucose and to contain 'glucose-6-phosphatase'.

Bilaterally, a strip-like so-called lateral ligament is present within the intermeningeal tissue and appears to connect endomeninx with ectomeninx. The endomeninx penetrates into the brain along the rather scarce small bloodvessels.[295] The spinal cord, however, seems to be devoid of such vessels, being presumably dependent upon the vessels of the endomeninx.

Figure 358 illustrates the meninges in the *Selachian* Scyllium canicula. The relationships are here not much different from those in Petromyzon, but small blood vessels, partly accompanied by meningeal tissue, penetrate into both brain and spinal cord. An apparently interrupted or uneven ligamentum laterale is present. The intermeningeal (or 'perimeningeal') tissue seems to be essentially 'mucoid'. In some perineuraxial regions of various Selachians, the loose intermeningeal tissue may be bounded by a thin, apparently uninterrupted cellular limiting membrane. Depending on arbitrary criteria, this membrane might be interpreted as either an internal layer of dura (ectomeninx), or as a primitive limitans arachnoideae, or, perhaps preferably, as a meningeal membrane *sui generis* occurring in various Anamnia (cf. further below).

In some *Ganoids*, namely Amia and Lepidosteus, the intermeningeal tissue over the fourth ventricle displays a substantial thickening (Fig. 359), formed by hematopoietic myeloid tissue in addition to a layer of adipose tissue (E. SCHARRER, 1944).[296]

In some *Teleosts*, such as the Siluroid Corydora, I noticed a rather thin endomeninx ('meninx primitiva') investing the neuraxis, and a very loose intermeningeal tissue continuous both with endomeninx and thin endocranium respectively endorhachis (i.e. ectomeninx). The

---

[295] HIBBARD (1963), who performed perfusions of the vascular supply to brain and spinal cord of the lamprey with india ink, states that these intracerebral vessels of capillary type were never penetrated by the ink 'even in the best preparations'. The cited author did not observe 'definite capillary loops' and suggests that said intracerebral vessels might possibly 'represent parts of a lymphatic drainage system'. Tentatively, I am still inclined to regard these channels as capillary vessels pertaining to the blood vascular system.

[296] Although, according to some authors, the 'perimeningeal' tissue of lower vertebrates was interpreted as being mucous in Elasmobranchs and Ganoids, but adipose in Teleosts, KAPPERS (1925) states that he found a large amount of fat tissue in the Ganoid *Acipenser*, and mucous tissue in several Teleosts.

width of the intermeningeal tissue varied greatly at different levels of the neuraxis, and, at one and the same level, with regard to dorsal, basal, and lateral extraneural spaces of cranial cavity or vertebral canal.

In Teleosts with highly developed meninges as described by KAPPERS (1925), the endomeninx or 'leptomeninx', whose external boundary is in continuity with an internal condensation of the 'perimeningeal' tissue, appears as a wide layer with relatively loose meshes (Fig. 360). According to KAPPERS' interpretation, this condensation represents the 'true dura mater'. In Lophius, however, said dura does not seem to be separated from the endomeninx by a well defined continuous subdural space, although some fissures can be noted between the dense border limiting leptomeninx from the dural layer. In the spinal region, a rather wide layer of 'perimeningeal' tissue can be seen between 'dura proper' and endorhachis. In contradistinction to KAPPERS, I would be inclined to evaluate, with regard to the special case of Lophius and some other Teleosts, the three 'layers' endocranium or endorhachis, 'perimeningeal tissue', and 'dura proper', as together representing an ectomeninx which consists of an outer periostal or perichondral dural layer, a loose intrameningeal layer, and a

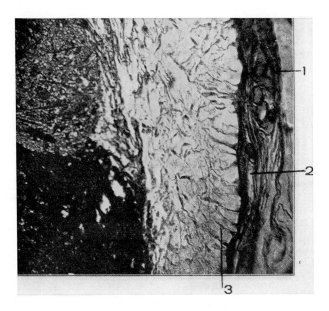

*Figure 360.* Spinal meninges in the Teleost Lophius (from KAPPERS, 1925). 1: 'dural layer'; 2: 'fissure'; 3: 'meninx secundaria'.

dense inner dural layer, this latter being incompletely separated from the outer portion of leptomeninx.

In another Teleost, namely the Gar pike *(Esox lucius)*, some regions of the brain, e.g. the tectum mesencephali, may be seen covered by a fairly wide leptomeninx, whose outer limit is provided by a continuous thin membrane comparable to the limitans arachnoideae of mammals. A definite cleft, seemingly comparable to a subdural space, may here appear between this cellular limiting membrane and an inner condensation of intrameningeal tissue, which could perhaps be evaluated as 'inner dura mater' (cf. Fig. 361).

In *Amphibians*, and depending on the investigated taxonomic forms, the meninges seem to display a development roughly comparable to that found both in lower Anamnia and in Amniota. On the whole, the meningeal membranes are more differentiated in Anurans than in Urodeles or Gymnophiona. In at least some of these latter, and in most parts of their cranial cavity, only a very thin sheet of ectomeninx is seen in close apposition with a likewise thin sheet of endomeninx investing the surface of the brain.

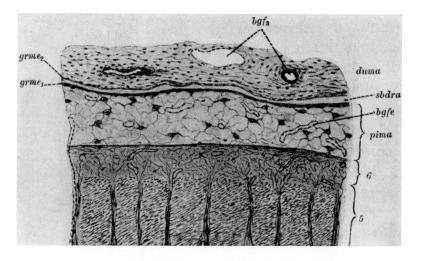

*Figure 361.* Cranial meninges of the Teleost Esox, in the region dorsal to tectum mesencephali, as seen in a cross-section (after KRAUSE, 1923). bgfe: blood vessels in leptomeninx; bgf₃: blood vessels in dura; grme₁: 'limiting membrane' of leptomeninx; grme₂: 'limiting membrane' of dura; duma: dura mater; pima: leptomeninx ('pia mater'); sbdra: subdural space; 5, 6,: layers of tectum opticum.

In various *Anurans* (Figs. 362–365), including the toad (*Bufo*) whose meninges were investigated by PALAY (1944), the dura consists of an outer layer representing endocranium respectively endorhachis. A large intrameningeal space is seen between this endocranial layer which PALAY calls '*pachymeninx*', and a membrane limiting or enclosing the connective tissue investing the neuraxis. This *limiting membrane* is designated as '*arachnoid membrane*' by PALAY (1944), and as '*dura*' by BECCARI (1943).

The intermeningeal space ('perimeningeal space' of KAPPERS *et al.*, 1936) may display thin accessory membranous lamellae and loose strands of connective tissue in addition to tubular structures which can be identified as complex extensions of the *saccus endolymphaticus*. Basally, the intermeningeal space contains the hypophysis.

Externally to the 'arachnoid' of PALAY or 'dura' of BECCARI, providing the inner boundary of intermeningeal space, irregularly distributed '*meningocytes*' may be located (Fig. 365). These elements, previous-

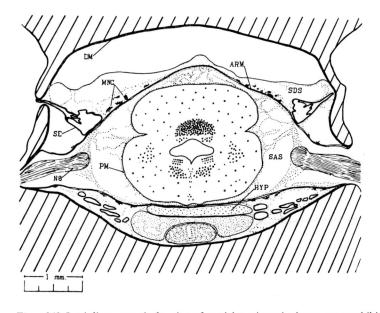

*Figure 362.* Semi-diagrammatic drawing of cranial meninges in the anuran amphibian, *Bufo*, as seen in a cross-section at level of isthmus rhombencephali (from PALAY, 1944). ARM: 'arachnoid membrane'; DM: 'dura mater'; HYP: hypophysis; MNC: 'meningocytes'; N8: nervus octavus; PM: 'pia mater'; SAS: 'subarachnoid space'; SDS: 'subdural space'; SE: saccus endolymphaticus.

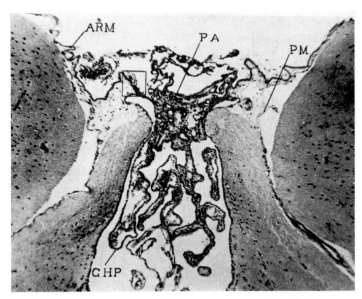

*Figure 363*. Cross-section through rostral region of the roof of the third ventricle in *Bufo*, showing relation of meninges to lamina epithelialis of roof plate (from PALAY, 1944). ARM: 'arachnoid membrane'; CHP: choroid plexus of third ventricle; PA: paraphysis; PM: 'pia mater'.

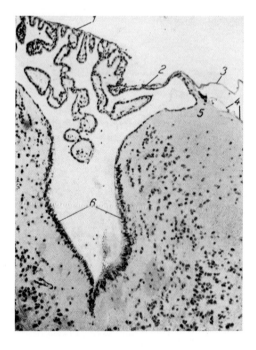

*Figure 364*. Cross-section through the roof of the rhomboid fossa in the anuran amphibian, *Rana*, showing relation of meninges to lamina epithelialis (from BECCARI, 1943). 1: region of choroid villi; 2: smooth stretch of ventricular roof; 3: 'dura'; 4: 'primitive leptomeninx'; 5: transition of roof plate to alar plate; 6: ependyma of fourth ventricle.

ly pointed out by E. SCHARRER and subsequently recorded by PALAY (1944), are characterized by acidophil cytoplasmic inclusions and seem to be macrophages of histiocytic type.

Internally to the limiting membrane a loose meshwork of connective tissue, interpreted as '*arachnoid trabeculae*', extends toward the thin membrane closely investing the neuraxis and designated as '*pia mater*'. This latter contains numerous small blood vessels and 'dips into the brain substance to form the lining of the perivascular spaces' (PALAY, 1944). As in many other vertebrates, large *melanophores* are generally

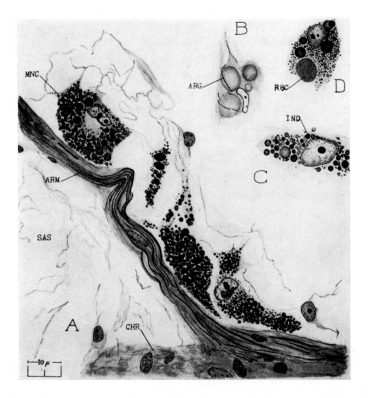

*Figure 365.* Camera lucida drawing of 'meningocytes' in the anuran amphibian, *Bufo* (from PALAY, 1944). A Enlarged drawing of square area outlined in ventricular roof of figure 363. B Meningocyte impregnated by *Foot-Bielschowsky method* for reticulum. C Meningocyte from 'arachnoid membrane over olfactory nerve' in an animal intracranially injected with india ink. D Meningocyte containing remnant of ingested erythrocyte (RBC) in an animal 'injected with autogenous blood'. ARG: argyrophil fiber; ARM: 'arachnoid membrane'; CHR: choroid roof of third ventricle; IND: ingested india ink; MNC: 'meningocyte'; SAS: 'subarachnoid space'.

scattered over the surface of the pia. The space between limiting membrane and pia is accordingly interpreted as '*subarachnoid space*'.

The relationships of limiting membrane to choroid plexuses are of interest. The membrane joins the plexus of the third ventricle (Fig. 363), 'forming a depression in which the epiphysis and the paraphysis lie' (PALAY, 1944). A similar junction of a fused membrana limitans and underlying leptomeningeal tissue with neuroectodermal choroid plexus epithelium was depicted by BECCARI[297] (1943) in the frog (Fig. 364).

The difference in opinion between the two cited authors, of whom one (BECCARI) interprets the *limiting membrane* as '*dura*', and the other (PALAY) as '*arachnoid*', is of particular significance, since it also involves a suitable interpretation of the Sauropsidan meninges.

In *reptiles*, according to KAPPERS *et al.* (1936), 'the dura is fairly well separated from the underlying leptomeninx'. The 'dura' of the cited authors represents the membrane corresponding, in Anuran amphibians, to that called 'dura' by BECCARI, but 'arachnoid membrane' by PALAY. KAPPERS *et al.* (1936) then add that their 'leptomeninx' does not display 'arachnoidal cavities'. This would agree with my own incidental observations which rarely disclosed the conspicuous 'trabeculae' noted by PALAY (1944) between his 'arachnoid membrane' and 'pia mater' in Anuran amphibians (cf. also Figs. 362, 366). I prefer to evaluate the reptilian meninges in accordance with the terminology adopted for Anuran amphibians by PALAY. In reptiles, the intermeningeal space between the ambiguous limiting membrane ('dura' respectively 'arachnoid') and the endocranium or endorhachis is relatively wide (Fig. 366). It may contain loose connective tissue in which vessels, described by KAPPERS as 'large epimeningeal veins' are located.

The meninges of *birds* are, on the whole, rather similar to those in reptiles, but show, nevertheless, a higher degree of differentiation. A detailed description of these avian membranes was given by HANSEN-PRUSS (1924), whose configurational findings essentially agree with my

---

[297] BECCARI (1943) states: '*si osserva que alla lamina corioidea epiteliale aderisce un solo strato meningeo (forse perchè qui persistono condizioni primitive) e que sui margini della membrana lo strato unico si doppia in dura madre e meninge molle. Ciò conferma, per gli Anfibi, la tesi dello* STERZI, *che le due meninge si differenzino dal primitivo strato connetivale aderente all'asse cerebrospinale.*' It might also be added that HERRICK (1935), in describing the meninges of Amblystoma, interprets the limiting membrane, i.e. the 'arachnoid' of PALAY, as representing the 'dura'.

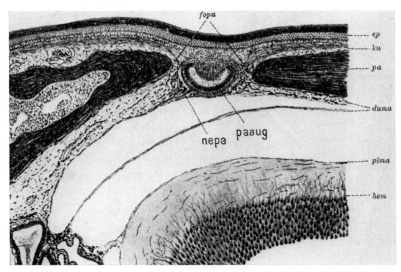

*Figure 366.* Cranial meninges of the reptilian Lacerta, as seen in a sagittal section at level of epiphysis (left bottom, unlabeled lead) and caudal telencephalon (after KRAUSE, 1921). ep: epidermis; duma: 'dura mater' (the lower lead of 'duma' points to what I would consider limitans of arachnoid); fopa: foramen parietale; hem: caudodorsal portion of telencephalic wall; ku: cutis (corium); nepa: nervus parietalis; pa: os parietale; paaug: parietal organ; pima: 'pia mater'.

own incidental observations. According to the cited author, the *dura mater* adheres closely to cranial bones and inner wall of the vertebral canal. In this latter, a separation between periosteal layer and inner dura by a continuous 'epidural' (perhaps better 'intradural') space does not seem to occur, but is merely suggested by irregularly distributed and discontinuous cleft-like spaces.

An *arachnoid membrane*, partly adhering to the inner surface of the dura, and partly separated from this latter by irregular, disconnected subdural cavities, surrounds the neuraxis. From the inner surface of the arachnoid, delicate trabeculae arise and fuse with a *pia mater* directly investing the central nervous system. In the spinal region,[298] a pair

---

[298] A peculiar characteristic of the lumbar avian spinal cord, namely the *sinus rhomboidalis*, located in the midline region dorsally to the central canal, and containing a 'glycogen body', shall be discussed in chapter VIII of volume 4. It will here be sufficient to state that the loose, jelly-like material located in this sinus, is most likely a neuroectodermal derivative originating within the dorsal portion of the spinal cord (cf. also K., 1927). The fused leptomeninges are in close apposition to the dorsal boundary surface of sinus rhomboidalis. Perhaps because of this contiguity, HANSEN-PRUSS (1924) inter-

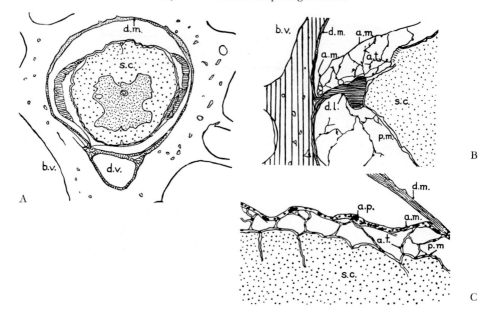

*Figure 367.* Diagrams of spinal meninges in birds, based on photomicrographs, and referring to chicks or pigeons (from HANSEN-PRUSS, 1923–1924). A Spinal cord within vertebral canal. B Region of denticulate ligament. C Further details of meninges at higher magnification. a.m.: arachnoid membrane; a.p.: 'arachnoid proliferation'; a.t.: arachnoid trabeculae; b.v.: bony vertebrae; d.l.: denticulate ligament; d.m.: dura mater; d.v.: 'dorsal blood-space'; p.m.: pia mater; s.c.: spinal cord.

of *lateral (denticulate) ligaments* extends from the dura into the inner membranes (Fig. 367 A–C). I believe that, in accordance with GEGEN-BAUR's concept, arachnoid membrane, trabeculae, and pia of HANSEN-PRUSS represent a differentiated endomeninx, comparable to the lepto-meninx of mammals, and, moreover, that the arachnoid membrane of PALAY in Anurans, and the reptilian 'dura' of KAPPERS are homologous with the above-mentioned avian arachnoid membrane. The dura, on the other hand, may be conceived as a fairly well differentiated ecto-meninx.[299]

preted the tissue in the sinus as being leptomeningeal. KAPPERS (1947) who originally shared this interpretation, subsequently acknowledged the 'glial' (neuroectodermal) nature of the cellular elements within the sinus, already regarded as '*Gliawulst*' or '*dorsale Gliawucherung*' by KÖLLIKER about 1902 (cf. ROMANOFF, 1960, pp. 285–288).

[299] A concise account of meningeal ontogenetic development in the avian embryo can be found ROMANOFF's treatise (1960, pp. 359–361).

On the whole, the arrangement of the meninges in *mammals* is better understood than that in submammalian forms. In particular the meninges of man, because of their clinical significance, have been rather closely investigated. As regards the main features of their ontogenetic development, the following overall summary, amplifying, with respect to mammals, the still more generalized statement on p. 669, appears reasonably valid. An outer condensation within the perineural mesenchyme gives rise to inner cranial and vertebral periosteum, ligaments of endorhachis, dura mater proper, and membranous arachnoid *(limitans arachnoideae)*. An inner condensation forms the pia mater. Between inner and outer condensation zones a spongy mesenchymatous layer becomes the *subarachnoid space*. A secondary separation in the outer zone, between dura and membranous arachnoid results in the formation of a *subdural space*. Within the vertebral canal an additional separation between periost respectively endorhachis and dura mater proper originates an *epidural space*.[300]

Thus, the limitans arachnoideae, although definitely constituting a component of the adult mammalian leptomeninges, appears to represent, genetically, a derivative of the outer zone, i.e. of the 'ectomeninx', rather than of the inner zone or 'endomeninx'. Hence, the so-called *arachnoid trabeculae* might perhaps better be termed '*pial trabeculae*', while the designation '*intima piae*' occasionally suggested by some authors, could be used for the innermost pial layer. In regions devoid of well defined trabeculae, the entire layer internal to the thin arachnoid limitans may then be regarded as pia.

SENSENIG (1951) in his description of the development of the spinal meninges, distinguishes, in addition to outer and inner zones of condensation, an intermediate one, from which 'arachnoid trabeculae' and inner parts ('dentate processes') of denticulate ligament arise. My own observations agree, on the whole, with SENSENIG's acount, but I prefer to evaluate this intermediate zone as a subdivision of the inner one, and the whole denticulate ligament as a joint derivative of outer and inner zones.

In the definitive (adult) configuration the meninges of man and apparently most or all mammals can be described as consisting of *three membranes, dura mater, arachnoid,* and *pia mater.* The dura or pachymen-

---

[300] Evidently, if internal vertebral periost and endorhachis are termed 'dura mater externa', and dura proper 'dura mater interna', this space could also be called 'intradural space'.

inx represents a rather thick membrane formed by a stratification of collagenous bundles whose inner layers provide the periost of cranial cavity. Some elastic connective tissue seems to be intermingled with the main collagenous substratum. The dura is supplied by its own meningeal blood vessels not shared by the neuraxis, but, in the cranial cavity, contains the venous sinuses draining the cerebral blood circulation.

Sheaths of dura mater accompany, as a perineurium, the cranial nerves where they pass through their respective foramina.[300a] Especially, but not exclusively in man and other large mammals, fibrous processes or septa extending from the dura form partially separated partitions of the cranial cavity, namely the *falx cerebri* between the cerebral hemispheres, the *tentorium cerebelli* between occipital lobes and cerebellum, the small *falx cerebelli*, and the *diaphragma sellae*, perforated by the infundibulum, and dorsal to the bulk of hypophysis.[301]

The arrangement of the dura within the vertebral canal differs from that in the cranial cavity. The outer dural membrane, namely the en-

---

[300a] This passage of cranial and spinal nerve roots through the dura, at the outer limit of the subarachnoid space, is also known as the '*subarachnoid angle*' and represents a region of transition in the course of peripheral nerves. The histological arrangement at the subarachnoid angle, where the perineurium is said to be 'open-ended', appears consistent with clinical evidence interpreted as indicating that the endoneurium of nerve trunks may become a pathway for the transmission of infections from periphery to central nervous system (McCabe and Low, 1969).

[301] The attachment of falx cerebri to endocranium contains the superior sagittal sinus, the inferior sagittal sinus being enclosed in the free border of falx. The roof of the tentorium cerebelli contains the sinus rectus. A notch in the rostro-ventral border of tentorium cerebelli forms the incisura tentorii, through which the mesencephalon extends toward the diencephalon. For clinical purposes, supratentorial and infratentorial intracranial structures can thus be distinguished.

---

*Figure 368 A.* Neonatal human spinal cord and meninges within vertebral canal (hematoxylin-eosin stain, ×17, red. $^2/_3$). 1: endorhachis (dura externa); 2: dura proper (dura interna); 3: epidural space; 4: subdural space; 5: arachnoid limitans; 6: subarachnoid space; 7: pia mater; 8: ligamentum denticulatum; 9: dorsal root; 10: ventral root (the roots are seen in their subarachnoid and in their intra- respectively extradural position); 11: 'epidural' venous plexuses; 12: spinal ganglion; x: shrinkage space between pia and surface of neuraxis.

*Figure 368 B.* Neonatal human spinal cord within dural sack (hematoxylin-eosin stain, ×35, red. $^2/_3$). t: 'arachnoid trabecula'; 1–11 as in preceding figure.

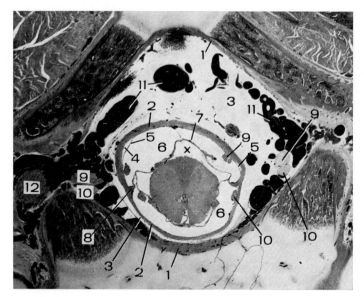

Figure 368 A

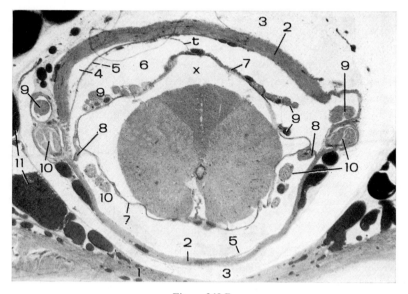

Figure 368 B

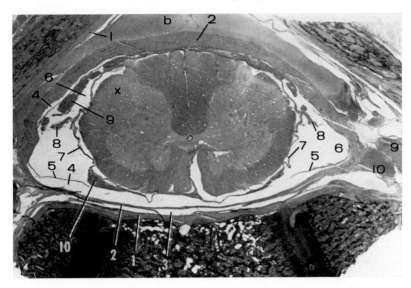

*Figure 369.* Adult human spinal cord and meninges within vertebral canal (hematoxy-lin-eosin stain, ×11, red. $^2/_3$). b: postmortal blood extravesate in 'epidural' space; other designations as in figure 368 A.

dorhachis with its periost, is separated from the inner dural membrane (or 'dura proper') by a rather wide 'epidural' (or 'intradural') space containing loose, mostly adipose connective tissue and a plexus of thin-walled veins (Figs. 368, 369). A paired lateral *denticulate ligament*, with a metameric series of serrate denticulations, connects the inner dura with the leptomeninges, reaching the pia mater. It is believed to represent a 'suspension ligament' and corresponds to the lateral liga-ments recorded by STERZI in all classes of vertebrates.[302]

In standard descriptions of the human pachymeninx, the transition of spinal dura to cranial dura in the region of foramen occipitale mag-num is generally interpreted as representing a continuity of (inner) spinal dura with inner layer of cranial dura, whose outer or periosteal layer becomes continuous, at the margin of the foramen, with the ex-ternal periost of the cranium. This view has been recently challenged

---

[302] STERZI described in vertebrates four spinal ligaments, namely the two (i.e. paired) lateral or denticulate ligaments, and two unpaired midline structures, comprising a dorsal and a ventral ligament. The midline ligaments, however, represent at best merely con-densations within the leptomeninx and are, in numerous instances, hardly distinguishable.

as inaccurate by Rogers and Payne (1961), who describe additional sublayers.[303]

Our own observations (Beasley and K., 1966), undertaken in order to obtain further evidence concerning this question, indicate that the human cranial dura is indeed stratified into a variable number of laminae. However, although more or less separable, these layers are interwoven by interlacing components. Again, several laminae (sublayers) may form two ill-defined compound layers, namely an inner and an outer one. The outer layer is, at the foramen magnum, directly continuous (a) with external cranial periost, and (b) with posterior atlanto-occipital membrane, membrana tectoria, articular capsules, ligamenta flava, posterior longitudinal ligament, and vertebral periost. With respect to the two reasonably well recognizable and definable compound 'main' layers, the conventional distinction adopted by most standard anatomical texts may be considered as representing an acceptable and useful overall description (cf. Figures 370 and 371).

The concluding remarks presented by Dr. Beasley in summarizing our joint 1966 communication at the annual meeting of the American Association of Anatomists, are perhaps relevant and may therefore be quoted in this context: 'In the vertebral canal, there is, between an outer strong fibrous layer system, which includes the internal periost of the vertebral canal, on one hand, and an inner strong fibrous layer, designated as dura mater spinalis, on the other hand, a substantial so-called epidural space containing a venous plexus and fat.

In the cranial cavity, a comparable space cannot be found, although, even if interconnected by interlacing sheets, a corresponding outer fibrous layer system and a corresponding inner one are demonstrable.

While both the so-called standard description and the formulation propounded by Rogers and Payne fail to convey a comprehensive picture of the obtaining complexity, the standard description still seems preferable, as a useful approximation, to that of Rogers and Payne.

---

[303] According to the cited authors, 'the intracranial and spinal parts of the dura mater are composed of a single continuous membrane. Both parts may readily be split into two or more layers by the surgeon in the operating room. Just below the foramen magnum the spinal theca is thick and vascular'. A detailed review of pachymeningeal sublayers in man, providing a functional framework provided by interlacing fiber systems as described by v. Lanz and others, is contained in the contribution of Schaltenbrand to vol. 4, part 2, of W. v. Möllendorff's and W. Bargmann's *Handbuch der mikroskopischen Anatomie des Menschen* (Springer, Berlin 1955).

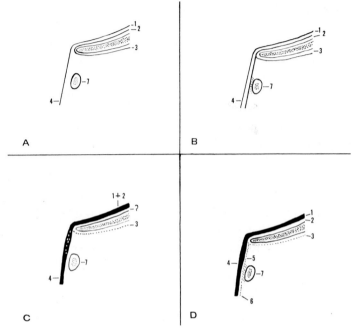

*Figure 370.* Diagrammatic representations of the arrangement of dura mater at the cranio-vertebral junction according to different concepts. A Conventional textbook represen - tation as interpreted by ROGERS and PAYNE, 1961. B Conventional textbook representation as interpeted by BEASLEY and K., 1966. C Arrangement as depicted by ROGERS and PAYNE (1961) in accordance with their observations. D Diagram of arrangement in accordance with our own observations (BEASLEY and K., 1966). 1: inner cranial dura mater; 2: outer (periosteal) cranial dura mater (endocranial periost); 3: external cranial periost; 4: (inner) spinal dura mater; 5, 6: endorhachis (outer spinal dura mater); epidural space between 4 and 5 (or 6); 7: posterior arch of atlas. In C, where we have changed the original labels, 1+2 is designated as 'cranial dura', 3 as 'periosteum', and 4 as 'spinal dural' with 'vessels'.

---

*Figure 371 A.* Sagittal section through foramen occipitale magnum of human new-born, showing relationships of dura (hematoxylin-eosin; $\times 10$; red. $^2/_3$; unpublished figure from BEASLEY and K., 1966). 1: inner cranial dura mater; 2: outer (periosteal) crani-al dura; 3: external cranial periost; 4: (inner) spinal dura; 5: endorhachis or outer spinal dura (it can be seen, at right, that it provides the membrana tectoria); 6: limitans of arach-noid; 7: pia mater; 8: epidural space; 9: subdural space; 10: subarachnoid space; a: an-terior arch of atlas; b: pars basilaris (clivus) of os occipitale; d: dens of axis; n: neuraxis at transition of oblongata to spinal cord; p: posterior arch of atlas; t: ligamentum transver-sum atlantis; x: region of separation between 4 and 5; y: region of bifurcation of 2 into 5 and 3; z: interlacing of 4 and 5.

*Figure 371 B.* Detail of preceding figure at higher magnification ($\times 40$; red. $^2/_3$). 4a, 4b: sublayers of spinal dura at transition from inner cranial dura (1); u: loose pericranial connective tissue; other designations as in preceding figure A.

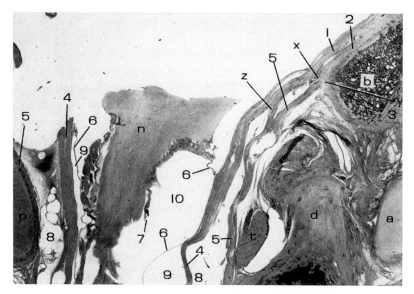

Figure 371 A

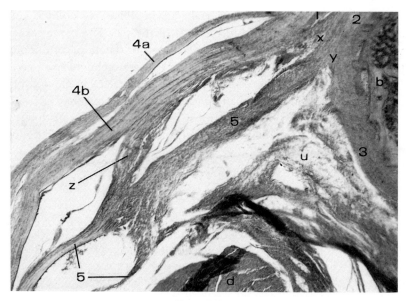

Figure 371 B

There is thus, in addition to the factual problem, involving the visualization of a complex spatial configurational relationship, the semantic problem of properly transposing such configuration, by means of a logically open transformation, into an adequate sequence of words'.

The mammalian *arachnoid* can be described as a thin, transparent membrane devoid of blood vessels, and separated from the dura by a narrow more or less 'potential' *subdural space* (cf. also further below, Figs. 372–374). Scattered, very fine strands of connective tissue may, in a few locations, display connections with the dura. In the vertebral canal, although arachnoid and dura are, on the whole, distinctively separated, stretches of fusion or adhesion between both membranes can frequently be seen. It is, therefore, convenient to speak here of a *dura-arachnoid complex*. Where dorsal and ventral nerve roots pass through the spinal dura the meninges in the angle between both roots do not form a definitive lateral boundary for the subarachnoid space which extends here peripherally as a so-called *lateral recess*. This latter could be part of a communicating pathway between subarachnoid space and peripheral nerve sheaths (cf., e.g., HIMANGO and LOW, 1971).

The *pia mater* is essentially a vascularized[304] connective tissue sheath closely investing the surface of the neuraxis, and generally attached to the inner surface of the arachnoid by a loose meshwork commonly designated as arachnoid trabeculae, which, as pointed out above, should rather be considered part of the pia. The tissue spaces within these meshes and between the elements of the more continuous pial sheath proper *(intima piae)* constitute the subarachnoid space containing the external cerebrospinal fluid. Together, arachnoid and pia represent the typical mammalian *leptomeninx* (or leptomeninges).

Since it is hardly possible, in most instances, to find a distinctive boundary between arachnoid trabeculae and pia mater, a restriction of the term arachnoid to the boundary membrane (limitans arachnoideae) between leptomeninx and subdural space seems preferable.

---

[304] The relationships of pia mater to the adventitia of the vessels penetrating or leaving the neuraxis was discussed in chapter V, section 5 (p. 299 *et passim*) of the preceding volume 3, part I. There is also some doubt concerning the question whether the pia mater proper *('intima piae')* contains a true capillary net of its own. I did, however, occasionally notice rather small vessels of capillary type in this membrane, but my incidental personal observations were not sufficiently conclusive for an opinion about that question.

While the pia mater is everywhere with its intima in direct apposition to the surface of the central nervous system, the arachnoid proper (limitans arachnoideae) spans the spaces between external prominences of the brain. Thus, in the telencephalon, the arachnoid bridges over the sulci which thereby become enlarged subarachnoid spaces. On the convexity of the convolutions, however, the arachnoid limitans is intimately fused with the pia, forming, with this latter, a relatively thin common leptomeninx of which the arachnoid merely represents the surface lining.

Particularly large subarachnoid spaces encompassed by the spreads of arachnoid limitans between externally prominent surface configurations are known as subarachnoid *cisterns*. The large *cisterna cerebello-medullaris* (cisterna magna) lies between basal surface of cerebellum and dorsal surface of oblongata. It is continuous with the basal *cisterna pontis* which encompasses the basilar artery. The *cisterna interpeduncularis* represents the subarachnoid space within the fossa between the pedes pedunculorum of mesencephalon. It is continuous dorsalward with *cisterna ambiens* (or superior) surrounding the midbrain, caudalward with cisterna pontis,[305] and rostralward with the *cisterna chiasmatis* located in the region of optic chiasma, being mostly anterior to that structure.

The cisterna ambiens continues rostrad into *cisterna corporis callosi*, on the dorsal aspect of that structure, and into the substantial subarachnoid spaces of tela chorioidea ventriculi tertii, located basally to corpus callosum, but not generally enumerated as a 'cistern', unless considered to be the ventral part of cisterna corporis callosi (or so-called '*cerebro-sagittal cistern*'). The cisterna chiasmatis is laterally connected with the paired rather large *cisterna fossae Sylvii*, and rostralward through the *cisterna laminae terminalis cinereae* with the smaller likewise paired *cisterna olfactoria* (K., 1940). The intracranial cisternal spaces, which provide channels for the external cerebrospinal fluid, communicate, at the level of foramen occipitale magnum, with the subarachnoid space surrounding the spinal cord. This latter, coextensive with the (internal) dural sack, ends, in the human adult, approximately at the level of second sacral vertebra, the (internal) dura being continued as a filum terminale durae matrix, which may fuse with the periost in the neighborhood of sacrococcygeal junction.

---

[305] This cistern, extending along the basis of oblongata toward the foramen occipitale magnum, is also occasionally designated as *cisterna basilaris*.

The *structural elements* of the *leptomeninx* are parallel or interlacing bundles of collagenous and precollagenous fibers surrounded by fine elastic networks. In addition to the fibroblasts and fibrocytes, the cell types found in the leptomeninges include histiocytes, some lymphocytes, scattered mast cells, and, mostly in the the pial layer proper, an individually variable number of pigment cells (melanocytes).

As regards the lining of the subdural and subarachnoid intrameningeal spaces, it is generally claimed that the inner surface of dura and the outer one of the arachnoid are covered by a more or less continuous layer of flat *mesothelial cells*. A similar lining is said to obtain for inner surface of arachnoid, trabeculae and the outer surface of pia.[306] Such lining would be comparable to that of the morphologically quite different true 'serous cavities' derived from the embryonic celom. Some authors, however, have contested the presence of a mesothelial covering within the meninges and describe, instead, a more or less interrupted lining by flattened 'fibroblasts'.[307]

Thus, PALAY (1944) very definitely recorded merely fibroblastic lining cells in the meninges of anuran amphibians, and, with reference to the pertinent literature, reviews the problem at issue. I consider my own incidental observations inconclusive, but believe that the available evidence for a continuous mesothelial lining of subarachnoid spaces is not entirely convincing. These compartments might rather represent wide tissue clefts. As regards the *subdural space*, however, a mesothelial lining seems not entirely improbable on the basis of recorded observations by competent investigators.[308] There appears to

---

[306] Cf. e.g. the textbook of histology by W. BLOOM and D.W. FAWCETT (p. 347, 9th ed., Saunders, Philadelphia 1968).

[307] The problem is somewhat similar to that concerning the mesothelial lining of the epineurium (cf. vol. 3, part I, p. 267 and fig. 183B, loc.cit.). It might also be recalled that in the synovial membranes of articular capsules the fibrocytes facing the articular cavity can occasionally give the appearance of a continuous 'endothelial' or 'mesothelial' surface, although it is now generally agreed that an uninterrupted cover of synovial membranes by cells of 'true' epithelial type does not obtain. Again, KAPPERS and other authors have recorded the occurrence of calcified spots ('*concretions*') within the leptomeninx, particularly in older individuals. HANSEN-PRUSS (1923–1924) described areas of calcification in the arachnoid of birds.

[308] Details concerning this and other topics relevant to the meninges (e.g. innervation and blood supply) may be found in volume IV, part 2 of v. MÖLLENDORFF's and BARGMANN's *Handbuch* cited above in Footnote 303.

be little doubt that, regardless of its actual detailed histologic structure, the limitans arachnoideae provides a continuous membrane separating subarachnoid space from subdural space, such that this latter is
not penetrated by the cerebrospinal fluid.

Various authors, since more than 100 years ago, have noted and described cell clusters, often displaying a whorl-like pattern, along the
outer and inner surface of the arachnoid limitans. These cell clusters,
whose elements assume an epithelial or epitheloid character, can undergo necrotic processes with subsequent calcification as mentioned in
footnote 307. The surface of arachnoid villi, described further below,
likewise displays *arachnoid cell clusters*. It is commonly believed that at
least some types of meningiomas may take their origin from these apparently 'normally' present cell aggregates.

The human *cranial dura mater*, besides being supplied by the arteria
meningea media and smaller anterior and posterior meningeal rami of
other arteries, is innervated by so-called recurrent (or meningeal)
branches of the trigeminal nerve ($V_{1, 2, 3}$) and of the glossopharyngeo-vagal group. Some of these fibers mediate pain sensations, especially of the more basal parts of pachymeninx, while others may include components contributed by sympathetic and parasympathetic
nervous system.

Extensive plexuses of nerve fibers within the cerebral *pia mater*, including adventitia of vascular walls, and tela chorioidea of choroid
plexuses, have been described. Some of these fibers pertain to the meningeal branches of cranial nerves, while others derive from the sympathetic periarterial carotid and vertebral plexuses. Various sorts of 'sensory' endings seem to be present in addition to ('vegetative' or 'autonomous') efferent ones. The occurrence of scattered nerve cells in the
adventitia of vessels was also reported. The adventitial nerve plexuses,
containing both non-medullated and thinly medullated fibers, appear
to accompany the blood vessels into the neuraxis; their efferent components seem to innervate the muscle cells in the vascular wall.

*Mutatis mutandis*, the *spinal meninges* display a similar pattern of innervation by recurrent meningeal rami originating from the metameric
common spinal nerve trunks. Delicate nerve plexuses comparable to
those described in the cerebral leptomeninges, probably accompany
the blood vessels into the spinal cord. As regards the peculiar vertebral
venous plexuses within the epidural space, their significance will be
pointed out further below in briefly dealing with the blood vessels of
the neuraxis.

While the *subdural space* of mammals, between inner layer of dura and limitans arachnoideae, seems to contain a very small amount of 'tissue fluid' and is commonly likened to a narrow 'serous cavity', the *subarachnoid space* of these Vertebrates includes a substantial quantity of (external) cerebrospinal fluid. It is generally assumed, in accordance with the views of KAPPERS (1947; KAPPERS *et al.*, 1936), that, in all fishes, the leptomeningeal spaces do not contain liquor cerebrospinalis externus such as fills the mammalian subarachnoid cavities. The considerable outward protrusions or bulgings of choroid plexus structures and roof plate lamina epithelialis in Petromyzonts, various Selachians, Osteichthyes, and Dipnoans seem to be correlated, in KAPPERS' opinion, with a lack of either direct ventricular communication with, or of significant liquor diffusion flow into, leptomeningeal spaces.

There is likewise considerable doubt whether, or to what extent, ventricular fluid may or may not pass into the leptomeningeal spaces of amphibians, reptiles, and birds. In these latter, HANSEN-PRUSS (1923–1924) has nevertheless claimed the presence of subarachnoid spaces with 'a true cerebrospinal fluid'. In man, however, and as could be deduced from the scattered evidence, possibly in all mammals, openings in the choroid plexus of the fourth ventricle seem to provide a clear passage between ventricular system and subarachnoid space.[309]

Because of its considerable importance for clinical neurology, the *cerebrospinal fluid* in man has been investigated by numerous authors. These observations appear to have established that the choroid plexuses of the cerebral ventricles are the main source for this fluid.[310] A 'flow' from the lateral ventricles by way of third ventricle and aque-

---

[309] These openings *(foramen Magendii* and *foramina Luschkae)* shall be dealt with in vol. 4, chapt. IX of this series. The *Foramen Magendii (Apertura mediana ventriculi quarti,* PNA) at the caudal end of the rhomboid fossa seems to be more variable than the *Foramina Luschkae* (the paired *Apertura lateralis ventriculi quarti,* PNA) of the fossa's lateral recess, and may be entirely missing in diverse Mammals.

[310] This topic and the significance of the *hemato-encephalic barriers* were dealt with in chapter V, section 5 of the preceding volume 3, part I. The possibility that paraependymal structures such as area postrema may be concerned with addition to or resorption from ventricular fluid should be kept in mind. Some absorption of fluid through the ependymal lining into neighboring veins might likewise take place. A recent report (MILHORAT, T.H.: Choroid plexus and cerebrospinal fluid production. Science *166:* 1514–1516, 1968) questions the view that the cerebrospinal fluid is primarily produced by the choroid plexus. Acute progressive hydrocephalus with essentially normal cerebrospinal fluid was observed in rhesus monkeys 'undergoing ventricular obstruction 2 to 6 months after

duct toward the fourth ventricle is generally assumed. Thence, the internal cerebrospinal fluid escapes into the cerebello-medullary and basilar (pontine) cisterns through the medial and lateral foramina of the epithelial roof of the rhomboid fossa.

The subarachnoid space, extending over the whole surface of the neuraxis, accordingly contains the external cerebrospinal fluid which presumably receives a contribution from the *Virchow-Robin spaces* of cerebral and spinal blood vessels, whose adventitial perivascular channels can be regarded as expansions of the subarachnoid compartment. Some still insufficiently elucidated connections of that fluid compartment with 'lymphatics' or tissue spaces of cranial and spinal nerves are likewise assumed.

From the cerebello-medullary and basal cisterns the external cerebrospinal fluid may diffuse in all directions, but it is assumed that a substantial amount passes over the convexity of the hemispheres, to be absorbed into the intracranial *venous sinuses*. A significant outflow toward the spinal subarachnoid spaces must also take place. Around the spinal nerve roots, connections with dural veins seem to provide a suitable passage of liquor into the blood circulation.

Along large intracranial veins and venous sinuses, especially sinus saggitalis superior, projections of the limitans arachnoideae with a core of leptomeningeal tissue protrude into the vascular lumen, forming the '*arachnoid villi*' (Figs. 372–374). These villi, which, on the basis of scattered reports, can perhaps be assumed to occur in all mammals, enlarge with increasing age, and in older human adults, represent the so-called *Pacchionian granulations*.[311] Not only increase in size, but secondary fusion of neighboring villi within the vascular lumen contribute to the

---

choroid plexectomy'. Although the cited author does not entirely deny the participation of choroid plexus in the production of cerebrospinal fluid, he assumes that most of the ventricular fluid is formed either as 'a specific secretion of the ependymal epithelium or represents a product of cerebral metabolism which enters the ventricular system across the ependymal lining, or both'. Further reports by MILHORAT *et al.* (1971) seem to indicate that, in rhesus monkeys, choroid plexectomy reduced the production of cerebrospinal fluid by an average of 33 to 40 per cent. The cited authors, moreover, claim that sodium levels in the gray matter surrounding the ventricles and in the gray matter bordering on the subarachnoid space are markedly higher than in other brain regions. Sodium exchanges in the two above-mentioned areas of the brain may be linked to cerebrospinal fluid production.

[311] The *foveolae granulares* on the inner surface of the cranial vault bones represent imprints molded by the expansion of the *Pacchionian granulations*.

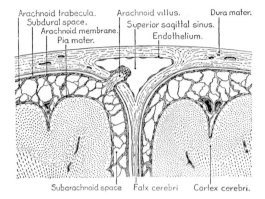

*Figure 372.* Diagrammatic representation of a coronal section through the superior sagittal sinus to illustrate an arachnoid villus and other meningeal relationships as interpreted by WEED (after WEED, 1923, and RANSON, 1947).

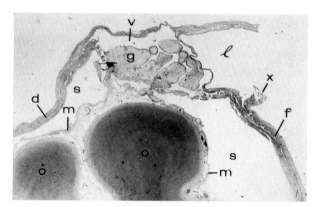

*Figure 373 A.* Section, similar to that represented in WEED's diagram, through a *Pacchionian granulation* in adult man (hematoxylin-eosin; ×6; red. ²/₃). The granulation is here not located in the lumen of the sinus, but in that of a large dural vein near its opening into the sinus. d: dura mater showing sublayers (cf. also fig. 371B); f: falx cerebri; g: *Pacchionian granulation;* l: lumen of sinus sagittalis, whose wall is torn at x; m: leptomeninx; o: gyri of cerebral hemisphere (between the two gyri, a sulcus into which the leptomeninx extends); s: subdural space; v: lumen of large dural vein; x: torn wall of sagittal sinus. The subdural space is greatly enlarged by secondary distortion caused by processing of the material.

formation of these granulations which then, accordingly, appear as a common cluster of villi with several separate stalks passing through orifices of the dura and connecting with the expanse of leptomeninx.

The surface of the villi displays *arachnoid cell clusters* and is separated from the blood stream by the endothelium of the sinuses or of the large cerebral veins, as the case may be. In preparations, occasional spaces between endothelial lining and villus surface may be seen (cf. Fig. 373B) but presumably represent shrinkage artefacts. Thus, between external cerebrospinal fluid within the leptomeningeal meshes and the venous blood stream, two membranes are interposed, namely *vascular endothelium* and *limitans arachnoideae*. Along the spinal meninges, small proliferations of arachnoidea, comparable to arachnoid villi, have been described as occurring particularly in the locations at which the segmental nerve roots pass through the dura.

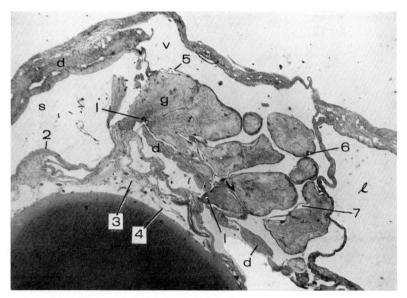

*Figure 373 B.* Detail of preceding figure at higher magnification (×12; red. ²/₃). 1: neck of villus passing through opening in dura; 2: limitans of arachnoid; 3: subarachnoid space with trabeculae; 4: pia mater; 5: endothelium of dural vein separated from arachnoid surface of granulation; 6: arachnoid cell cluster; 7: probably artificial cleft between endothelium of vein and arachnoid surface of villus (these two layers are, on the whole, closely fused, but remain distinguishable at high resolution and with proper histologic technique); other designations as in figure 373A.

The available evidence suggests that the arachnoid villi provide the main pathway for outflow of cerebrospinal fluid, which, accordingly, reaches the venous circulation. Thus, some substances injected into the subarachnoid spaces may be detected in the blood stream after less than one minute, but only after a much longer time in the lymphatics.

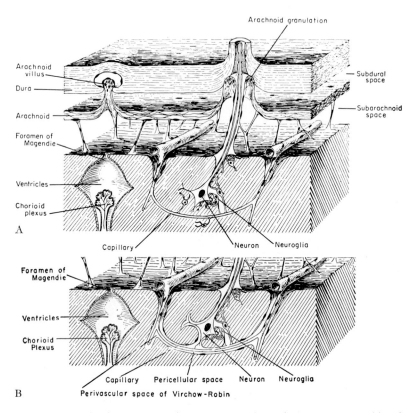

*Figure 374 A*. Diagram purporting to represent 'central nervous system with reference to the circulation of the cerebrospinal fluid' (from PEELE, 1961, modified after AYCOCK). This useful diagram is self-explanatory. The *Virchow-Robin spaces*, not labeled, are recognizable as leptomeningeal funnels at the entrance of two vessels, but do not extend along the capillaries. This concept would be in agreement with my own views expressed several years ago (K., 1951). The somewhat odd relationship of the 'neuron', as drawn in the diagram, is, however, not quite intelligible.

*Figure 374 B*. Bottom of preceding diagram as shown by PEELE in the previous 1954 edition of his treatise. At that time, the cited author had still adopted WEED's concept of pericapillary *Virchow-Robin spaces* related to the perineuronal spaces as also illustrated by figure 209 A, p. 316 in volume 3/I of the present series.

Recently, WELCH and FRIEDMAN (1960), elaborating on a previously expressed hypothesis, have claimed that arachnoid villi operate as 'valves', opening in response to a rise of pressure of the cerebrospinal fluid,[312] but the evidence presented by these authors may be considered rather unconvincing.

The total 'normal' volume of *cerebrospinal fluid* in adult man can be estimated as varying between approximately 90 to 180 cc, with 130 cc representing a fairly well substantiated average. Of this total, the internal or ventricular fluid may comprise circa 30 to 50 cc; very roughly, the ratio between internal and external cerebrospinal fluid can thus be evaluated as normally about 1:2. A recent study undertaken in children, with an average volume of cerebrospinal fluid amounting to 92 cc, disclosed the rate of fluid formation to be approximately 0.35 cc/min or 500 cc/day. Thus, as a whole, the cerebrospinal fluid seemed to be renewed about five and one-half times each day (CUTLER et al., 1968). Although the reliability of all methods used to estimate the rate of cerebrospinal fluid formation may be considered doubtful, it seems nevertheless probable that the total fluid volume is completely replaced several times a day.

The *liquor cerebrospinalis* is a clear fluid of low specific gravity[313] (1.006-7), normally containing about 1 to 6 cells (lymphocytes and monocytes) per cmm. As measured by means of lumbar puncture, its average pressure in adult man is 60–150 mm of fluid in the horizontal position, and 200–250 mm in the sitting position. The cerebrospinal fluid contains very little protein, and, in comparison with blood plasma, somewhat less glucose but a higher concentration of chlorides.

---

[312] According to WELCH and FRIEDMAN (1960) an arachnoid villus is a labyrinth of small tubes which establish open connections between the subarachnoid space and the venous blood. 'In that these tubes are opened by the pressure of the cerebrospinal fluid when this is sufficiently greater than that of the blood, and effaced when the pressure relation is reversed, the structure is, in fact, a valve'. In contradistinction to the cited authors, I am inclined to believe that the traditional concept of the arachnoid villus as a blind diverticulum into a venous channel is essentially correct. The membrane through which the cerebrospinal fluid passes into the blood seems, in fact, a double one, namely limitans arachnoideae and blood-vascular endothelium. Under special circumstances, e.g. in experiments using excessive pressure, and in some histologic preparations, parts of this membrane might evidently become ruptured.

[313] The specific gravity of whole blood measures between 1.055 to 1.066, and that of blood plasma around 1.030.

|                  | CS Fluid          | Blood plasma        |
|------------------|-------------------|---------------------|
| Proteins         | 0.01–0.045%       | 6–9%                |
| Chlorides        | 725–750 mg/100 cc | 560–620 mg/100 cc   |
| Glucose          | 50–80 mg/100 cc   | 100 mg/100 cc       |
| Specific gravity | ca. 1.006         | ca. 1.030           |

*Figure 375.* Tabulation comparing some features of human cerebrospinal fluid with those in blood plasma (compiled from various clinical treatises).

This chloride content is assumed to keep the liquor in approximate osmotic equilibrium with the blood plasma. Some of the differences between this latter and the cerebrospinal fluid are indicated by the tabulation of Figure 375. It should be added that the protein concentration shows also differences with respect to ventricular and subarachnoidal cerebrospinal fluid as obtained by ventricular, cisternal, or lumbar puncture, respectively. The commonly cited figures are 5–15 mg (per 100 ml) for ventricular fluid, 15–25 mg for cisternal fluid, and 15 to 45 mg for lumbar fluid. These differences[314] might be partly due to the fact that, in the subarachnoid space, an additional flow of another type, namely by way of the *Virchow-Robin spaces*, makes further contributions to the cerebrospinal fluid whose original ventricular source is essentially the choroid plexus.

The functional significance of cerebrospinal fluid is commonly evaluated as having a mechanical and a metabolic aspect. Thus, the central nervous system is said to be 'suspended in it as in a water bath' which protects it from jars and shocks. However, the ratio of neuraxis volume (roughly 1 200 cc) to total fluid volume (roughly 130 cc), precludes any significant 'suspension effect' in accordance with *Archimedes' law*, although some degree of 'protective effect' might be granted. In at least some vertebrate forms presumably lacking external cerebrospinal fluid (e.g. in Anamnia), 'perimeningeal' or 'intramenin-

---

[314] In this connection the difference in *protein content* between blood plasma (6–9%) and lymph (about 3.5%) might also be mentioned. As regards the proteins in the cerebrospinal fluid (0.01–0.04%), the albumin-globulin ratio (normally about 8:1) is of significance for clinical laboratory tests. Other components of the fluid, not relevant in this context, are discussed in publications on clinical neurology and clinical laboratory methods.

geal' tissue could have a comparable effect, which can also be assumed for the tissue and structures within the 'epidural' space of the mammalian vertebral canal.

As regards the *blood vessels of the vertebrate neuraxis*, the following general remarks seem perhaps appropriate. In gill-breathing Anamnia, the blood propelled by the heart flows rostrad in an unpaired median ventral aorta which gives off paired afferent aortic (branchial) arch arteries for the capillary system of the gill apparatus. The oxygenated blood leaving this capillary network is then gathered into dorsal, efferent aortic arch arteries, joining the paired dorsal aorta which becomes unpaired by median fusion at a level caudal to pharynx.[315]

From the paired dorsal aorta, a rostralward directed branch, which may be called *arteria carotis dorsalis*, supplies head structures, including brain (cf. Fig. 376). In several forms, an *arteria carotis ventralis* arises as a paired rostral branch from the system of median ventral aorta (or 'truncus arteriosus') and may anastomose with the system of a. carotis dorsalis.

The origin of a lung and the connection of that organ with the circulatory system does not introduce very significant vascular changes in Dipnoi. The lung is here supplied by an artery from the efferent portion of the sixth branchial arch. Thus, this artery carries blood which has already passed through the gill capillaries and should normally be 'oxygenated'.

---

[315] In most details, the vascular system of the Cephalochordate *Amphioxus* (Branchiostoma) can be said to resemble that of vertebrate Cyclostomes such as *Petromyzon*. In *Amphioxus*, however, a heart is not present, and the blood circulation depends upon contractility of the main vascular trunks. In Cyclostomes, a typical S-shaped heart has become differentiated, comprising three compartments, namely sinus venosus, atrium, and ventricle. In generally accepted standardized nomenclature, all vessels carrying blood from the heart are called arteries, and all vessels bringing blood to the heart are veins, regardless whether these channels convey freshly oxygenated ('arterial') or depleted, i.e. partly desoxygenated ('venous') blood. Because of the differences between branchial and pulmonary types of circulation, and the occurrence of capillary networks within channels of essentially arterial respectively venous type in addition to the general capillary network, some semantic difficulties had originally arisen in the description of the branchial circulation of Anamnia (cf. HAFFERL, 1933). The vascular system of the Acraniote Amphioxus, whose neuraxis does not seem to be penetrated by blood vessels, has been reviewed by FRANZ (1933) on the basis of his original investigations. The aortic blood of Amphioxus is apparently carried to a system of lacunae which supply the tissues, and 'true capillaries' are presumably missing.

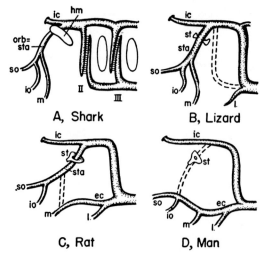

*Figure 376.* Diagrams of left side of head illustrating transformations of carotid system in representative vertebrates (from ROMER, 1950). II, III: second and third aortic arches, ec: external carotid; hm: hyomandibular; ic: internal carotid; io: infraorbital artery; l: lingual artery; m: mandibular artery; orb: orbital artery; so: supraorbital artery; st: stapes; sta: stapedial artery.

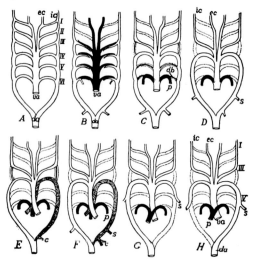

*Figure 377.* Diagrams showing transformations of aortic arches in representative vertebrates (after BOAS and KINGSLEY, from NEAL and RAND, 1936). A: 'primitive scheme'; B: Dipnoan; C: Urodele amphibian; D: Anuran amphibian (frog); E: Ophidian reptile; F: Lacertilian reptile; G: bird; H: mammal; c: celiac artery; da: dorsal aorta; db: ductus Botalli; ec, ic: external and internal carotid; p: pulmonary artery; s: subclavian; va: ventral aorta; some of the vessels carrying 'venous' blood black; some of those with 'mixed' blood shaded; disappearing arteries in dotted outlines.

In amphibians and particularly in Amniota, however, the substitution of branchial respiration by lung respiration is correlated with considerable changes in the arrangement of the blood vascular system. The rostral branchial arch arteries and the original ventral aorta become modified and reduced, such that the ventral aorta tends to remain as a channel supplying the third, fourth, and sixth arches.

The rostral part of the paired dorsal aorta and the dorsal carotid remain as *internal carotid artery*. A small ventral carotid (lingual artery) persists, and, in most mammals, expands, as *arteria carotis externa*, to take over some of the channels provided by the internal carotid in various other vertebrates. The common origin of internal and external carotid from the truncus arteriosus becomes in numerous forms, *the common carotid* (cf. Figs. 376, 377).

In submammalian vertebrates, the *brain* is supplied by the internal carotid artery which divides into rostral and caudal cerebral branches. The *spinal cord* is supplied by rami from paired segmental branches of the dorsal aorta. These vessels run laterally from the aorta and carry blood to trunk musculature, skin, axial skeleton, vertebral canal and its contents. Radicular arteries accompany the dorsal and ventral roots. In mammals as well as in a variety of other vertebrate forms, an unpaired median longitudinal anterior spinal artery, and paired longitudinal posterior spinal arteries are connected with ventral respectively dorsal radicular arteries. Transverse anastomoses may provide a so-called vasocorona. Figure 212 on p. 321 in the preceding volume 3, part I, illustrates the typical arterial vascularization of a mammalian spinal cord. At the zone of transition from brain to medulla spinalis, anastomoses between cerebral and spinal vascular supply obtain. In contradistinction to the (rather moderately) vascularized spinal cord of Myxine, that of Petromyzonts does not display intramedullary blood vessels. The brain of these latter forms, nevertheless, seems to include intracerebral vessels (cf., however, fn. 295 above).

The *human spinal cord* (Fig. 378) receives an unpaired median longitudinal anterior spinal artery, formed by union of a branch from each vertebral artery. The paired longitudinal posterior spinal arteries derive from the vertebral or from the posterior inferior cerebellar artery. Segmental radicular branches are supplied by vertebral artery, costocervical system, intercostal and lumbar arteries.

Along the vertebral canal, *except* for some cervical levels and one or two major thoracic ones, the segmental spinal arteries are commonly rather thin, and may not significantly contribute to the blood supply of spinal cord. Likewise, their end-branches, namely the radicular arteries, display considerable regional differences in size, such that some of these radicular branches cannot be definitely traced as far as the network of spinal vasocorona. Although several reports, by various authors, have been published on these details of vascularization, the available data must be evaluated with considerable caution because of

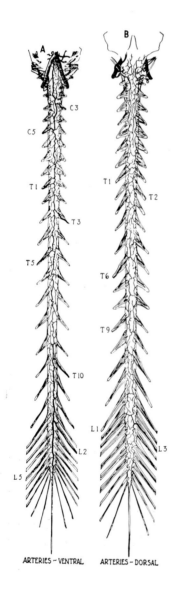

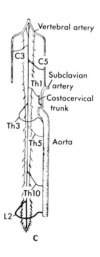

*Figure 378.* Arterial blood supply of human spinal cord (A and B after Suh and Alexander, from Steegmann, 1962; C after Lindenberg, from Haymaker, 1969).

the considerable diversity of individual variations. Moreover, the technical difficulties involved in such studies may lead to actual findings which are far less unambiguous than their interpretation or description by the observer. LAZORTHES (1961) distinguishes, with respect to spinal cord segments, three different major territories of vascularization, namely an *upper cervical* one, mainly corresponding to intumescentia cervicalis, an *intermediate thoracic* one, and an *inferior lumbo-sacral* one, mainly corresponding to intumescentia lumbalis. The intermediate or thoracic territory is said to have a relatively 'poor' arterial supply.

The direction of blood flow in the longitudinal arteries, particularly in anterior or spinal artery may not be the same throughout, being downward in some regions and upward in others, as suggested by several authors. In fact, depending upon variable functional states in combination with particular individual patterns of the network, some spinal cord segments might receive a perhaps alternating respectively reversible 'ascending' or 'descending' blood flow.

In *mammals*, the *vertebral artery*, arising from a brachiocephalic trunk or from the a. subclavia (as in man) may substantially contribute to the blood supply of the brain. This artery does not occur in submammalian forms. Although in birds a so-called vertebral artery may also send some minor branches to the brain, it is not morphologically homologous with the mammalian vertebral artery.

Because of multitudinous variations, not only between different forms but also between individuals of one and the same species, a detailed description of the arterial vascularization pattern obtaining for the vertebrate brain is exceedingly complex. Numerous problems defy here an unambiguous verbal solution and are, moreover, made more difficult by the lack of a generally acceptable standard terminology. A competent discussion of the overall questions involved can be found in the contributions by TANDLER (1933) and by HAFFERL (1933) to vol. VI of *Handbuch der vergleichenden Anatomie der Wirbeltiere* edited by BOLK, GÖPPERT *et al.* Useful short summaries based on original observations have been presented by KAPPERS (1933, 1947)[316] concerning the vascular patterns of the vertebrate neuraxis.

---

[316] Among investigations dealing with various aspects of this topic, the reports by HOFMANN (1900), STERZI (1904), DE VRIESE (1905), ROOFE (1935), VITUMS *et al.* (1965), GILLILAN (1967), and BURDA (1969) can be mentioned for further details and references.

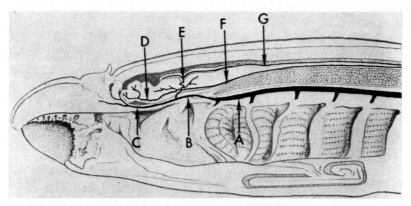

*Figure 379.* Parasagittal section through head of Ammocoetes larva of the Cyclostome, *Petromyzon marinus*, showing the pattern of arterial supply to brain (after HIBBARD, 1963). A: dorsal aorta; B: internal carotid (designated as 'common carotid' by the cited author); D: anterior cerebral branches; E: 'choroid artery'; F: caudal cerebral branch of internal carotid (designated as 'vertebral artery' by the cited author); G: dorsal branch of cerebral carotid (designated as 'dorsal spinal artery' by the cited author).

Figures 379 and 380 indicate the *arterial distribution* to the brain of representative vertebrates. Generally speaking, it can be seen that the internal carotid artery or carotis cerebralis provides rostral and caudal branches. A posterior ramus of the main caudal branch becomes unpaired, by fusion in the midline region, at the base of the rhombencephalon. This unpaired 'arteria basilaris' connects, in at least some vertebrate forms, directly with the ventral arterial supply of the spinal cord.

Although, in addition to an arteria ophthalmica, branches designated as anterior, middle, and posterior cerebral arteries can be distinguished in *Selachians*, these terms, at best, merely indicate somewhat dubious and unclarified degrees of kathomology. In particular, the posterior cerebral artery of many submammalian forms represents a channel corresponding to arteria communicans posterior rather than to arteria cerebri posterior of most mammals.[317] Anastomoses, across

---

[317] In order to perform suitable mappings for the establishment of morphological homologies, an unambiguous sequence of definable neighborhoods should be given. Branching patterns, however, displaying a variable and partly random sequences of ramifications, such as vessels, peripheral nerves, cerebral gyri and sulci, may not provide sufficiently 'stable' landmarks for a well defined type of connectedness. Such fluctuations preclude rigorous homologization, which can here merely be substituted by approximations based on arbitrary extra- and interpolations (cf. also the comments on p. 209 in chapter III, section 3, *et passim*, of volume 1 in this series).

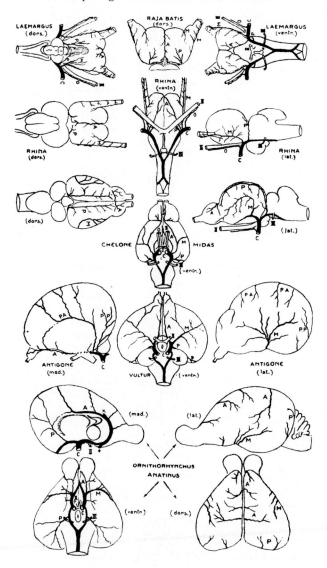

*Figure 380 A*. Arterial vascularization of brain in representative vertebrates (from KAPPERS, 1933). A: arteria cerebralis anterior; C: internal (or cerebral) carotid; M: middle cerebral artery; O: ophthalmic artery; P: posterior cerebral artery; II, III: optic and oculomotor nerves. In the interest of greater clarity, KAPPERS did not draw in these figures the small anastomoses between territories of anterior, middle, and posterior cerebral arteries.

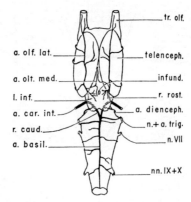

I

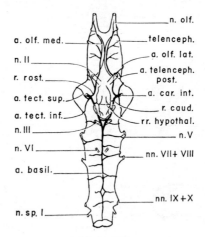

II

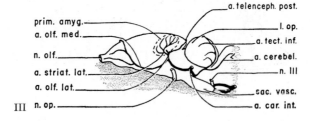

III

*Figure 380 B*. Arterial vascularization of the brain in Amphibia (after GILLILAN, 1967). I In the urodele Necturus. II and III In the anuran Rana. The designation sac. vasc. (saccus vasculosus) in III is doubtless erroneous, and prim. amyg. ('Primordium amygdalae') refers to a quite debatable concept of homologization (cf. above, section 6, p. 668). The other abbreviated designations are self-explanatory.

the midline, between basal anterior branches may provide, together with the fusion of posterior cerebral artery into the basilar one, an anastomotic circle at the base of the brain comparable to the *circulus arteriosus of Willis* characteristic for man and most mammals. However, the midline anastomosis of basal anterior branches ('a. communicans anterior') seems commonly inconspicuous and perhaps even missing in many submammalian forms. In these latter vertebrate groups, on the other hand, the arteria basilaris not infrequently becomes only partly unpair, forming, by divergence and caudal convergence leading again to midline fusion, a small caudalbasal arterial mesh or *circulus arteriosus* of variable size.

In many *Sauropsidans*, the cerebral arteries display a branching pattern in which anterior, middle and posterior cerebral arteries assume relationships roughly corresponding to those obtaining in mammals.

In at least some reptiles, and concomitantly with the expansion of a so-called hypopallial or epibasal ridge ($D_1$), a conspicuous set of pene-

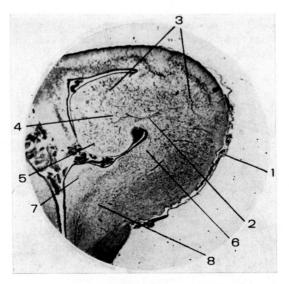

*Figure 381.* Transverse section through telencephalon of the Chelonian reptile, *Testudo graeca*, showing 'lateral striate artery' (from K., 1925). 1: arteria cerebralis antero-superior (a. cerebri media); 2: arteria epibasalis ('lateral striate artery'); 3: nucleus epibasalis; 4: nucleus (epibasalis) centralis; 5: n. epibasalis medialis (or perhaps rostral tip of n. epibasalis posterior complex); 6: n. basilateralis superior, basilateralis inferior and basimedialis inferior complex ($B_1+B_2+B_3$); 7: eminentia thalami ventralis; 8: n. entopeduncularis (hypothalami) anterior ('globus pallidus').

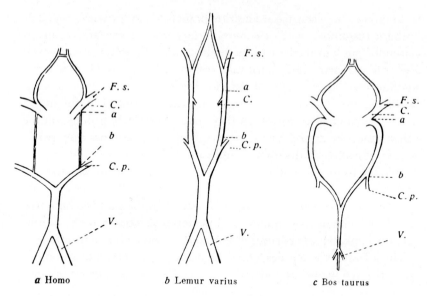

*a* Homo         *b* Lemur varius         *c* Bos taurus

*Figure 382.* Main different patterns of circulus arteriosus in mammals as conceived by
Tandler (from Hafferl, 1933). C: arteria carotis interna (rudimentary in *Lemur*); Cp:
a. cerebri posterior; Fs: a. *fossae Sylvii* (a. cerebri media); V: a. vertebralis (rudimentary in
*Bos*, but main supply for brain in Lemur); the channel between a and b in the diagrams is
a. communicans posterior.

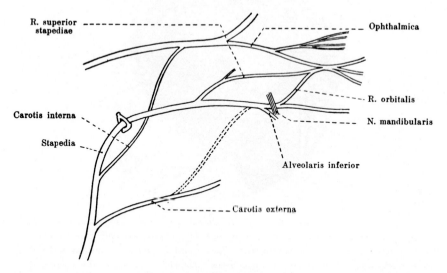

*Figure 383.* Diagram of head arteries in the insectivore mammal, *Erinaceus* (after Tandler, from Hafferl, 1933). A channel disappearing by secondary reduction is shown in
dotted outlines.

trating arteries supplying that structure originates from the arteria cerebri media (also called 'arteria cerebralis anterosuperior'). The penetrating epibasal arteries were described by several authors, including
ELLIOT SMITH (1919) and myself (K., 1925) as 'lateral striate arteries' s.
arteriae lateralis corporis striati (cf. Fig. 381). These vessels can be considered kathomologous with some of the lenticulo-striate arteries seen
in the human and mammalian brain.

As regards the distribution of cerebral arteries at the base of the
*mammalian brain*, it is possible to distinguish two main differing configurations, which, however, display a large number of intermediate
patterns. In the two extremes, the *circulus arteriosus* is provided (a) almost exclusively by the internal carotids or (b) by the vertebral arteries. In this latter case (b), the internal carotids are rudimentary, while
the former case (a) is characterized by rudimentary vertebral arteries
(cf. Fig. 382). Artiodactyla such as *Bos taurus* display pattern (a), while
pattern (b) is, e.g., displayed by some Lemuroids. In Man, an intermediate type of pattern commonly obtains.

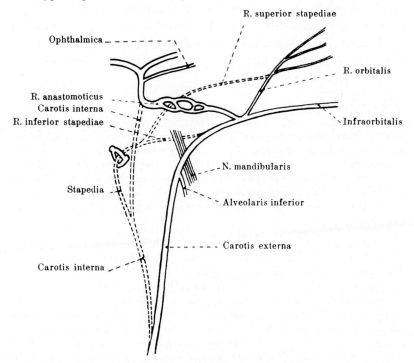

*Figure 384.* Diagram of head arteries in ungulate mammalians (after TANDLER, from
HAFFERL, 1933). Disappearing channels shown in dotted outlines.

Additional variations and complications may result from an anasto-mosis between internal carotid and arteria maxillaris interna occurring, e.g. in some ungulates (Fig. 384). This anastomosis, combined with re-duction of internal carotid, may here become a main supply channel for the basal circulus arteriosus. Again, in Ruminantia, classified as ar-tiodactyle ungulates, the internal carotid artery generally becomes re-placed by the occipital artery. This latter displays here an arrangement of pressoreceptors identical with those of the carotid sinus in man and other mammals.

The mammalian *arteria stapedia* (Fig. 383), originating from the in-ternal carotid, and passing through the stapes between crura and basis, can provide a connecting channel between internal and external catotid systems. It has a 'precursor' in submammalian forms (cf. Fig. 376). This artery, of considerable comparative and embryologic interest, is rudimentary in man, being very rarely recognizable by ordinary gross inspection.[318]

Because of its evident significance for clinical neurology, the blood supply of the human neuraxis with its manifold variations has been studied in great detail by numerous authors. Some of the relevant top-ics concern the *circle of Willis*, which displays numerous variations[318a] as well as frequent considerable asymmetries (cf. Figs. 386–387). The

---

[318] In some mammals, the arteria meningea media may originate from the arteria stapedia. Other variations of some interest are origin of a. meningea from a. ophthalmica and *vice versa*, and origin of a. occipitalis from a. carotis interna. A rare variation in man is the presence of a *arteria hypoglossi* originating from a. carotis interna and passing through *canalis hypoglossi* into the cranial cavity. This hypoglossal artery may replace a rudimentary or missing a. vertebralis. A detailed description and analysis of human variations may be found in the standard work of B. Adachi ( *Das Arteriensystem der Japaner*, 2 vols., Maruzen, Kyoto 1928).

[318a] A common unilateral or bilateral variation of the *circle of Willis* is represented by the origin of arteria cerebri posterior. The vessel designated by this name in standard human anatomical terminology usually arises as the terminal rostral bifurcation of arteria basilaris, and is connected, by a rather thin arteria communicans posterior, with the inter-nal carotid. In the alluded variation, a stout artery, providing the stem of a. cerebri posterior, arises from the internal carotid, and is joined by a thin communicating branch with arteria basilaris. Again, the channel provided by a. communicans posterior may be completely missing on at least one side. Moreover, the rami from a. carotis interna and from a. basilaris can also be of roughly identical size. These pattern variations involve a typical problem illustrating the inherent arbitrariness of semantic formulations: should the a. cerebri posterior be defined by (a) its distribution; (b) by a definite single origin, or

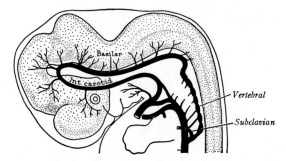

*Figures 385.* Arterial loop at the base of the embryonic human brain at about six weeks (about 12-mm body length) as seen from left side (after AREY, 1954). This instructive diagram, however, does not distinguish between unpaired median stretch of basilar artery and paired configuration of internal carotid and vertebral arteries.

vertebral arteries rather commonly manifest a conspicuous difference in size, such that either the right or the left one may be much thinner than its antimere.

During human ontogenetic development, an arterial loop is formed by the caudal, unpaired median basilar extension of internal carotid artery and its junction with the rostral extension of paired arteria vertebralis. This loop, particularly conspicuous at about the 12-mm stage (of approximately 6 weeks), represents the essential primordium of the adult circulus arteriosus (cf. Fig. 385). In very rare instances of human vascular variation, the internal carotid artery can be found missing, and may then be replaced by branches of arteria maxillaris interna entering the cranial cavity through foramen rotundum and foramen ovale.

With respect to the *human pachymeninx*, it should be recalled that the arterial blood supply of the dura mater is mainly derived from the *external carotid artery* by way of the *a. meningea media*, which, itself a branch of a. maxillaris interna, rather constantly reaches the cranial cavity through the foramen spinosum. Additional anterior and posterior meningeal rami for the dura mater arise, in a variety of patterns, from

(c) by the junction of a ramus from a. carotis interna and from a. basilaris. Evidently, (b) involves two possibilities, and (c) may not be given. Hence, in actual practice, the term 'arteria cerebri posterior' becomes ambiguous, and is flexibly adapted to the enumerated cases.

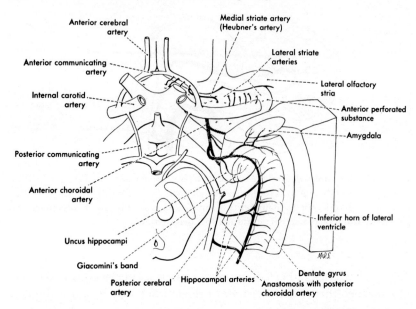

*Figure 386.* Detail of basal arterial supply of the human brain, showing a 'recurrent artery of Heubner, lateral striate arteries entering the substantia perforata anterior, and relationships of the anterior choroidal artery' (after ABBIE, from HAYMAKER, 1969).

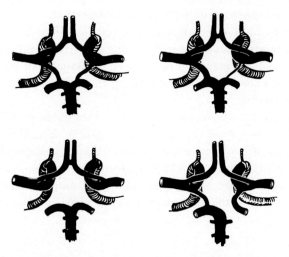

*Figure 387 A.* Common variations of the *circle of Willis* (after HAMBY, from UTTER-BACK, 1962). Upper left: so-called normal circle; upper right: unequal posterior communicating arteries; lower left: absence of posterior communicating arteries; lower right: separate circulation of the two hemispheres.

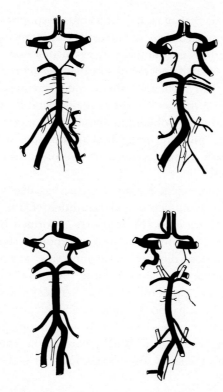

*Figure 387 B.* Additional variations of the *circle of Willis*, as recorded by HASEBE and ADACHI in adult Japanese (from ADACHI, 1928).

an ethmoidal branch of a. ophthalmica (which is a branch of *internal carotid*), from a. occipitalis, a. pharyngea ascendens, and *a. vertebralis.* The arteria meningea media is accompanied by a paired vein which joins the extracranial plexus pterygoideus, running along the arteria maxillaris interna. The other meningeal rami are likewise generally accompanied by small veins, some of which, however, may join neighboring sinuses of the dura mater. These latter, providing the channels for the venous drainage of the brain, are briefly described further below.

Other clinically significant data involve distribution of basal *lenticulo-striate arteries* arising from a. cerebri media, and the presence, course, or absence of a *recurrent striatal artery of Heubner* originating from a. cerebri anterior. In addition, the territories of regional blood supply for cerebral cortex, thalamus, hypothalamus, hypophysis, brain stem, cerebellum and spinal cord are of particular importance with respect to the diverse symptoms encountered in vascular lesions.[319] As regards these latter, COHNHEIM's concept of so-called *end-arteries* lacking collateral connections had formerly gained wide acceptance, but can no longer be upheld on the basis of the available data. It must now be assumed that the doubtless obtaining anastomotic interconnections between the terminal portions of cerebral arteries are, in the given instances, functionally insufficient for a prevention of the occurring types of vascular lesions.

A peculiar arrangement recorded in the vascular system, and already investigated by numerous older authors (cf. Fig. 388 A), is displayed by the *retia mirabilia ('Wundernetze')*, formed by networks of small vessels intercalated in the arterial or venous system, or, without being true capillaries, directly interconnecting arteries and veins. There are several types of these networks[320] which include, e.g., those of the (arterial) renal glomeruli and the (venous) portal sinusoids, as well as a variety of still poorly understood reticular, convoluted, or fascicular vascular complexes found in the extremities and, intra- or extracranially in the head region. With regard to the aspect here under consideration, only the latter sort of retia, in so far as they are of signif-

---

[319] Relevant detailed descriptions of the clinically important vascularization patterns in the various regions of the human neuraxis can be found in WEBB HAYMAKER's treatises on 'Local Diagnosis in Neurological Diseases' (1969) and on 'The Hypothalamus' (1969) as well as in various chapters of BAKER's handbook 'Clinical Neurology' (2nd ed., 1962), which also contains our own contribution on 'Disorders of the Brain Stem and its Cranial Nerves' (HAYMAKER and K., 1962) with particular consideration of that region's blood supply. A fairly recent monograph, specifically dealing with cerebral vascularization and circulation was published by LAZORTHES (1961).

[320] In addition to the retia here enumerated in the text, there are peculiar networks *(glomera)* formed by sinusoid vessels surrounded by 'epitheloid' cells and pertaining to a chemoreceptor system registering, presumably *inter alia*, the H-ion concentration of the blood stream. These glomera (at one time suspected to be 'chromaffin bodies') include *glomus caroticum* at the bifurcation of common carotid, *glomera aortica* about the arcus aortae, *glomus jugulare*, supplied by ascending pharyngeal artery, and the *tympanic glomera*. The *glomus coccygeum*, near the end of arteria sacralis media, is, despite some similarity in structure, not assumed to be a chemoreceptor organ but generally included in the group of common glomerular arteriovenous anastomoses. Depending on arbitrary semantic criteria these diverse vascular configurations may or may not be subsumed under the term *'retia mirabilia' ('Wundernetze')*.

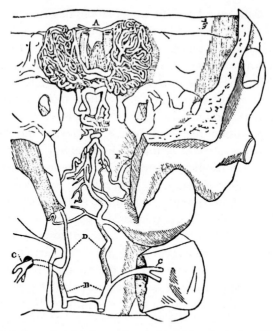

*Figure 388 A*. Rete mirabile at base of opened cranial cavity in the calf *(Bos taurus* L.,
Ruminantia, Artiodactyla, Ungulata). The brain, and the dorsal arch of the atlas have been
removed (after Rapp, from Solly, 1848). A: the cerebral carotid arising from the rete mirabile; B: vertebral arteries in vertebral canal; C: vertebral arteries passing through foramen in atlas; D: branches of vertebral artery passing rostrad to anastomose with arteria
condyloidea (E), forming thus a plexus which is connected to the rete mirabile.

icance for the blood supply of the brain, require a few further comments. Such networks have been reported in various mammals,[320a] including Edentata, Ungulata, Carnivora, Sirenia, and Cetacea.

---

[320a] A peculiar network is also present at the bifurcation of common carotid in Anuran
and Urodele (but apparently not in Gymnophione) amphibians. It is a 'cavernous' body,
improperly called '*carotid gland*' by a number of authors, which does not display glandular
structure, being rather a 'blood reservoir' (Hafferl, 1933) which receives more blood
for internal than for external carotid. This apparently contractile rete is believed to store
blood which can subsequently be 'squeezed out' at a variable rate. According to Noble
(The Biology of the Amphibia, McGraw-Hill, New York 1931, Dover, New York 1954,
pp. 191–192) it is a device regulating the flow of oxygenated blood to the head, and represented by a spongy enlargement of the third aortic arch 'which offers further resistance to
the blood and steadies the pressure by continuing to contract between beats'. Some dubious or controversial questions concerning this vascular configuration are discussed, with
numerous details, in the monograph by Adams (1958) listed in the bibliography to chapter
VII of the present treatise.

Thus, in some ungulates, TANDLER (1906) described the formation of an arterial *rete mirabile* by the development of vascular buds from the arteria carotis interna which subsequently atrophies. The network is then supplied by a branch of the external carotid system and becomes enclosed within the venous sinus cavernosus. The rete is presumed to be a secondary vascular formation, which does not result from a modification of primary embryonic capillary channels.

In the retro-orbital region of the cat, GILLILAN and MARKESBERY (1963) described a *rete mirabile* formed by a network of fine arteries intertwined in a venous plexus, and including an arteriovenous anastomosis. These authors suggest the term *rete mirabile conjugatum* for networks of this type, in contradistinction to a *rete mirabile simplex* formed by arterial vessels only. The extracranial rete mirabile surrounding the cat's arteria maxillaris interna in the fossa orbito-temporalis was also investigated by MARTINEZ (1967), who discusses various general problems pertaining to these networks. In the cat, the arteria anastomotica, formed by branches arising from the rete mirabile of a. maxillaris interna, provides the most important arterial supply of the brain.

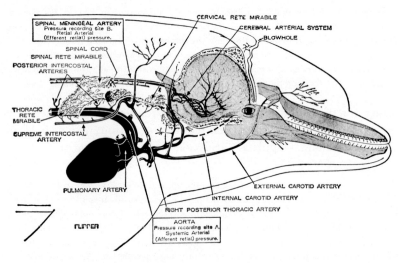

*Figure 388 B.* Diagram of *rete mirabile* complexes with principle afferent and efferent arteries in the cervico-thoracic region of the mammalian Cetacean dolphin, *Tursiops* (from NAGEL *et al.*, 1968). The afferent vessels (posterior thoracic, intercostal arteries and related branches) originate from the aorta. The efferent vessels are the spinal meningeal arteries continuing into cranial cavity and supplying the brain. In the experiments by the cited authors, catheter A was placed in the aorta via external carotid, and catheter B was introduced into spinal meningeal artery after dorsal laminectomy.

NAGEL *et al.* (1968) studied the *rete mirabile* of the bottlenose dolphin, *Tursiops*. In this Cetacean, a massive retial complex is interposed between body and cerebral circulations at cervico-thoracic levels (Fig. 388B). The network includes a cervical, a thoracic, and a spinal retial system. The spinal rete is here apparently the sole pathway for supplying blood to the brain through the so-called 'spinal meningeal vessels'. The poorly developed internal carotid artery is not functional in this respect, as could be inferred from angiography. The cited authors conclude that in the dolphin the cerebral circulation has been 'isolated' from the remainder of the arterial system by the intercalated retial system. One effect of this vascular organ is 'to markedly alter the arterial pressure profile in the thin-walled longitudinal vessels which supply the brain with blood' (Science, 161: 898–899, 1968).

Another, still poorly understood and insufficiently investigated mechanism for the regulation of arterial cerebral blood flow seems to be provided by either *valve-like* or *sphincter-like musculo-elastic structures* described within mammalian and human cerebral arteries. A radial ar-

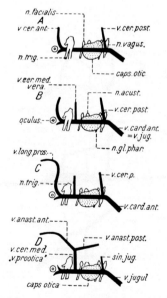

*Figure 389.* Craniocerebral veins of Anamnia as seen from left side (after VAN GELDEREN, 1933). A Adult Selachian *Acanthias*. B Adult Dipnoan *Ceratodus*. C Larval anuran *Rana*. D Later stage in *Rana*. The abbreviations are self-explanatory.

rangement of some smooth muscle fibers attached to the elastica interna in some regions is supposed to control longitudinal foldings modifying the size of the lumen. Valvular cushions and annular sphincter cushions, particularly at the site of arterial branchings presumably represent additional structures of this sort (*'dispositifs d'étranglement, de tension, de détente, de fermetures'*, LAZORTHES, 1961). Corresponding mechanisms may also obtain in submammalian vertebrates, and moreover, are not restricted to cerebral arteries. Structures of this sort (e.g. '*Drosselvorrichtungen*' etc.) have been noted by various authors in other regions of the blood circulation.

The *venous drainage* system of the brain in all craniote vertebrates (Figs. 389–391) is said to originate ontogenetically from a paired vena capitis medialis located at the base of the neuraxis and representing the cranial portion of vena cardinalis anterior. In *Cyclostomes*, a vena cerebralis posterior seems to be the only tributary, draining the brain, of vena capitis medialis. In *Selachians*, a vena cerebralis anterior, coming from the forebrain, may be added. In *Dipnoans*, *Amphibians*, and *Amniota*, a vena cerebralis media is added, while the vena cerebralis anterior tends toward partial or complete obliteration. In *Mammals*, the venous

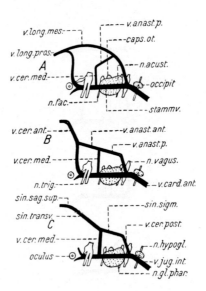

*Figure 390.* Craniocerebral veins of Amniota (after VAN GELDEREN, 1933). A Reptilian pattern. B and C Mammalian development patterns. Orientation and abbreviation as in preceding figure 389.

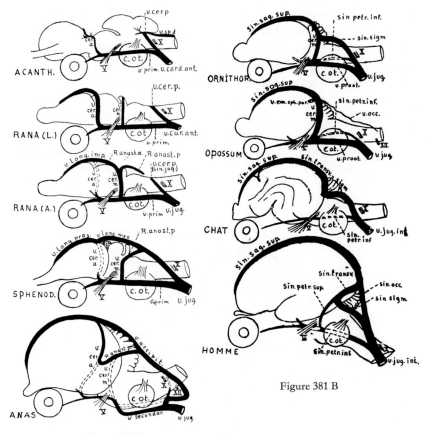

Figure 391 A

Figure 381 B

*Figure 391A.* Cerebral venous system in submammalian vertebrates, as displayed by the Selachian Acanthias, the larval (L) and adult (A) anuran amphibian *Rana*, the reptilian *Sphenodon*, and the avian *Anas* (after van GELDEREN from KAPPERS 1947). Embryonic channels which become obliterated in adults are indicated by broken lines. c.ot.: 'otic capsule'; R. anast. a.: ramus anastomoticus anterior; R. anast. p.: ramus anastomoticus posterior; v. card. ant.: vena cardinalis anterior; v. cer. a.: vena cerebralis anterior; v. cer. m.: vena cerebralis media; v. cer. post.: vena cerebralis posterior; v. jug.: vena jugularis; v. long. imp.: vena longitudinalis impar; v. long. mes.: vena longitudinal mesencephali; v. long. pros. vena longitudinalis prosencephali; v. occip.: vena occipitalis; v. prim.: vena primaria; v. secund.: vena secundaria.

*Figure 391 B.* Cerebral venous system in various mammals, as displayed by the Prototherian Ornithorhynchus, the Metatherian opossum, and the Eutherians cat and man (after van GELDEREN from KAPPERS, 1947). v. em. sph. par.: vena emissaria spheno-parietalis; v. proot.: vena prootica; other abbreviations, in the manner of those used for the preceding figure, are self-explanatory.

drainage of the brain becomes a complex system of dural venous sinuses resulting from a further evolution of the venous pattern found in submammalian forms. This tendency is already recognizable in Sauropsida, but in Anamnia, the vascular channels draining the brain remain 'ordinary' veins within the meningeal tissue.

In *man* (Fig. 392) the main sinuses of the dura mater are the following. *Sinus sagittalis superior* along the external convex border of falx cerebri, and *sinus sagittalis inferior* along internal, concave margin of falx. The *sinus rectus*, along the junction of falx cerebri and tentorium cerebelli, receives the inferior sagittal sinus, and the *vena magna Galeni*,

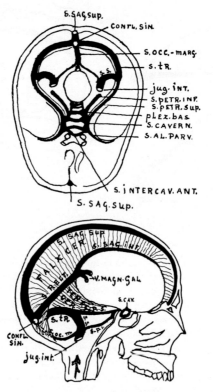

*Figure 392 A.* Arrangement of intracranial dural sinus in man (after BOEKE, from KAPPERS, 1947). S. oc. m.: sinus occipitalis ('occipito-marginalis'); S.p.i., S.p.s.: sinus petrosus inferior and superior; S.s.: sigmoid portion of sinus transversus (S.tr.); other abbreviations self-explanatory. Roughly speaking, the following four groups of human cerebral veins can be described (cf. also fig. 392B): superior, middle, and inferior cerebral veins (all 'external'), and internal cerebral veins, moreover superior and inferior cerebellar veins.

formed by the junction of internal cerebral veins; some cerebellar veins likewise join the sinus rectus.

At the occipital confluence of the sinuses, a junction of sinus rectus, superior sagittal sinus, and paired transverse sinus occurs. This latter sinus continues as *sinus sigmoideus* and joins the *bulbus (superior) venae jugularis internae*. The usually unpaired *occipital sinus* connects confluens sinuum and posterior venous plexuses of the vertebral canal.

An antero-basal group of channels includes the paired *cavernous, superior petrosal* and *inferior petrosal sinuses* as well as variable midline interconnections provided by *intercavenous* and *basilar sinuses* which form plexuses. The basilar plexus, in turn, connects, at the foramen occipitale magnum, with the anterior venous plexuses of the vertebral canal.

Figures 389 to 393 illustrate some essential features of the venous drainage pattern obtaining for the vertebrate brain. A comprehensive summary, based on his original investigations, was presented by VAN GELDEREN (1933), and a short, but still rather detailed overall descrip-

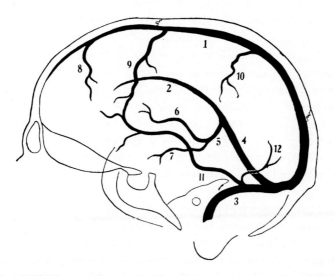

*Figure 392 B.* Diagram of human carotid angiogram at the venous phase, showing main venous outflow of brain. Note the anastomotic *veins of Trolard* (9) and *of Labbé* (11) providing a communicating system between basal and dorsal drainage (after LIST *et al.*, 1945, from PETERSON, 1962). 1: superior sagittal sinus; 2: inferior sagittal sinus; 3: transverse sinus; 4: sinus rectus; 5: *vena magna Galeni*; 6: internal cerebral vein; 7: basal *vein of Rosenthal*; 8: frontal ascending vein; 9: *vena Rolandica of Trolard* (vena anastomotica magna); 10: parietal ascending vein (8, 9, 10 also collectively: vv. cerebri superiores); 11: anastomotic *vein of Labbé* (vena anastomotica posterior); 12: descending temporo-occipital vein.

tion can be found in the posthumous treatise of Ariens Kappers (1947).

From a viewpoint of comparative anatomy related to phylogenetic speculations, the appearance of a *vena longitudinalis dorsalis impar* in the adult anuran amphibian, *Rana*, by midline fusion of a ramus pertaining to vena cerebri media after the disappearance of the original vena cerebri anterior, is interpreted as a first step toward the formation of true sinuses in Amniota (cf. Figs. 391 A and B). In all mammals, the vena cerebri anterior seems to disappear completely, while middle or posterior cerebral veins or both are involved in the drainage of the sagittal sinus. In Ornithorhynchus, both middle and posterior cerebral veins are present, in the opossum, the posterior vein is atrophic, and in the cat, the middle vein becomes reduced. In man (Figs. 391 B, 392), a condition similar to that in Ornithorhynchus obtains. The middle cerebral vein provides the sinus petrosus superior, and a basal connection between middle and posterior cerebral vein is represented by sinus petrosus inferior. The posterior connection between vena cerebri media and vena cerebri posterior forms the sinus sigmoideus.

During *human ontogeny*, the paired primary head vein, pertaining to the anterior cardinal (or precardinal) vein system, drains three pairs of tributary plexuses extending over dorsolateral regions of the brain. Concomitantly with the development of the meninges, these vascular primordia differentiate into two strata. The deeper vessels become the true cerebral veins within leptomeninx and brain substance. The channels of the superficial stratum, together with the primary head vein, are transformed into the several sinuses durae matris (cf. Fig. 393). Substantial displacements result from the considerable expansion of the human brain and particularly its cerebral hemispheres in the course of ontogenesis.

The rostral portion of primary head vein, receiving the ophthalmic vein, becomes the cavernous sinus, while an intermediate segment of head vein disappears concomitantly with expansion of inner ear. A new channel between middle and posterior plexuses arises dorsally to the internal ear anlage, and another channel subsequently originates along the course of the degenerated segment of the head vein, providing the inferior petrosal sinus.

The superior petrosal sinus, representing the stem of the middle plexus, connects cavernous sinus with the new supraotic channel. The transverse sinus, as a main line of drainage, arises from linked parts of middle and posterior plexuses in connection with supraotic channel.

The complex transformations of these venous pathways during ontogenesis, particularly investigated by STREETER (1918), despite a somewhat different terminology introduced by that author, can, nevertheless, be interpreted in essential agreement with the concepts elaborated
by van GELDEREN (1933) on the basis of a comparative anatomical approach. It is evident that, as regards ontogeny, various cenogenetic
changes arise, whose evaluation in terms of phylogeny remains rather
uncertain.

The main *extracranial outflow* in all Vertebrates remains the portion
of vena cardinalis anterior representing the *vena jugularis interna*. However, in addition to the communication between basilar plexus or occipital sinus with the system of vertebral venous plexuses, as mentioned above, connections between intracranial and extracranial veins

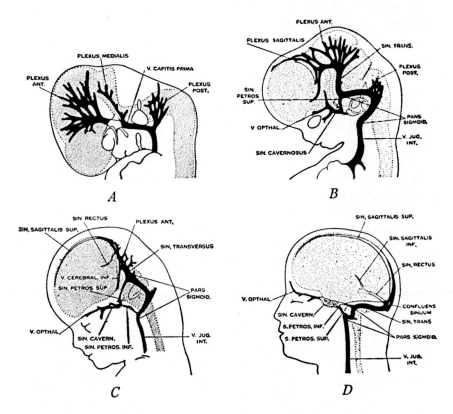

*Figure 393.* Ontogenetic transformations of primary head vein and venous plexuses
into dural sinuses (after STREETER, 1915, from AREY, 1954). A At six weeks. B At eight
weeks. C At eleven weeks. D At birth.

are provided by variable so-called *emissaries*, (e.g. emissarium temporale, occipitale, parietale), some of which also occur in Sauropsida, and by other anastomoses with the system of extracranial veins, such, e.g. as in man, between cavernous sinus and facial or pterygoid plexus veins by way of the ophthalmic veins or the foramen ovale. Again, the venous sinuses as well as the diverse extracranial venous plexuses have various anastomotic connections with the veins of the skull bones themselves (the *diploic veins* of man).

The pattern of veins directly draining the *spinal cord* is, on the whole, rather similar to that of the arteries as briefly discussed above. In man, an anterior longitudinal median trunk of variable thickness, and a nondescript posterior more or less longitudinally arranged network is present. Some authors prefer to describe six main longitudinal channels, including paired dorsolateral, paired ventrolateral, and unpaired median ventral and dorsal veins. Because of the netlike configurations and numerous variations, generalizing statements of this sort can be considered to have little practical or theoretical value.

The *longitudinal system* is said to lead upward into the intracranial sinuses durae matris around the foramen occipitale magnum, and asymmetrical *radicular veins* connect with the epidural vertebral venous plexus as well as, through the intervertebral foramen, with azygos and hemiazygos system or deep cervical veins. Whether a *lateral outflow* through intervertebral foramina or a mainly *ascending* one joining the intracranial sinuses should be considered to be normally predominating, remains a moot question.

BATSON (1940) has particularly investigated the significance of the *vertebral venous plexus* in man and Macacus monkeys. According to the cited author, the vertebral veins, with their rich, valveless ramifications and connections with cranial veins, including intracranial sinuses, pelvic and other body wall veins, as well as veins of the extremity girdles, constitute a separate, although overlapping, *spinovertebral* venous system *sui generis* (cf. Fig. 394). BATSON proposes that this vertebral vein complex should be recognized as a fourth system of veins added to the recognized pulmonary, caval, and portal venous systems. BATSON found evidence that in every act of straining, coughing or lifting with the upper extremity, the blood is not only prevented from entering the thoracico-abdominal cavity, but is actually squeezed out of the cavity, being thus forced into the vertebral system. Tumors and abscesses of the thoracico-abdominal wall, including the breast, tumors of the lung, pelvic tumors and abscesses, lesions of the shoulder girdles, and occasionally tumors and abscesses of other organs have vascular drainage connections with the vertebral vein system and may, therefore, produce metastases distributed anywhere along the system without involving the portal, the pulmonary, or the caval systems (BATSON, 1940).

In concluding the comments on the vascular pattern of the vertebrate neuraxis contained in the present section, the considerable variations displayed by the arrangement of blood vessels should again be emphasized. Although, within a given vertebrate group, some fairly constant types of vascular distribution may be encountered, significant differences between closely related species combined with remarkable

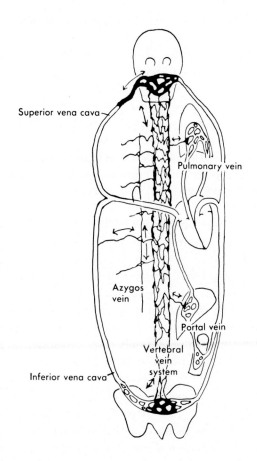

*Figure 394.* Diagram illustrating BATSON's concept of the vertebral vein system (after BATSON, 1940, and HAYMAKER, 1969).

similarities between taxonomically or phylogenetically quite unrelated taxonomic forms are frequently noted. This introduces an aspect of randomness not displayed to the same degree by the rather orderly configuration of neuraxial subdivisions and grisea. Thus, the vascular pattern does not appear particularly suitable for the elaboration of specific phylogenetic sequences, and TANDLER (1898, quoted after HAFFERL, 1933) justly remarked: '*Veränderungen des Gefässsystems, deren einzelne Stufen sich in ihrem Fortschreiten analog der Aszendenz verfolgen liessen, auf diesem Gebiete zu suchen, ist wohl vorderhand ein vollkommen vergebliches Bemühen.*'

## 8. Some Remarks on Brain Weights and Apposite Topics

If all available biological data are considered, a relationship between Vertebrate brain size and behavioral complexity, adaptability as well as learning capacity seems to obtain. Again, from a purely theoretical viewpoint, it would appear ostensible that, in order to perform functions of high intricacy, a large number of relays or similar elements (i.e. neurons) and circuits, involving synapses, must be available. This overall relationship between brain size and learning capability presumably applies to all animal organisms displaying a central nervous system, and has been well summarized in a publication by RENSCH (1956).

Numerous authors have dealt with the questions pertaining to *brain size* or *weight*, and a multitudinous literature on this topic is extant. Data, references and discussions contained in the publications of ANTHONY (1928), COBB (1965), DONALDSON (1895), KAPPERS (1929, 1947), and PORTMANN (1964) may provide useful further information for the reader interested in the ramifications and details pertaining to such quantitative studies. As regards the human central nervous system, a comprehensive compilation entitled '*Das Zentralnervensystem in Zahlen und Tabellen*' has been prepared by BLINKOV and GLEZER (1968).

From the viewpoint of the present treatise, merely some relevant overall aspects concerning this subject matter shall be pointed out. Generally speaking, it is appropriate to consider brain weight and size with respect to the following quantitative categories or features: (1) *absolute magnitude;* (2) *relation of brain to body weight*, involving also questions such as relationship of body surface to CNS mass, size of head in relation to brain, and number of neurons required to control the more or less complex non-neural body structures; (3) *relation of brain to spinal cord;* (4) *mass relations of particular brain regions or structures and,* (5) *compactness or density of the brain,* as manifested by the number of cellular and particularly of neuronal elements contained in a given volume of brain or CNS mass. It is evident that these relationships depend on the interaction, in various combinations, of a large number of different factors, several of which are insufficiently understood, and may even, so far, have remained undetected.

(1) If the crude data of *absolute magnitude* are considered, the largest brain sizes and the highest brain weights are found in *mammals.* Although, within this vertebrate group, man stands among the few animals with average adult brain weights of more than 1000 g, he is here outranked by porpoises, elephants, and whales (Fig. 395), as shown in the annexed tabulation (Fig. 396). It is obvious that these listed figures, given in round numbers, represent merely approximations indicating pertinent orders of magnitude. The variations of brain weight within one and the same species are doubtless considerable. Differences in the state of preservation, in the amount of ventricular fluid content, as well as in the procedures used by the various investigators for removal of brain, separation from spinal cord, and preparatory to actual weighing, preclude any rigorous quantitative evaluation of these data, which, nevertheless, are helpful for purposes of general orientation.[320b] Despite the intrinsic weaknesses of inferences based on the appraisal of absolute brain weights, some authors have maintained that this approach permits a few useful conclusions. However, apart from semantic and conceptual difficulties for a satisfactory definition and numerical grading of the behavioral manifestations representing 'intelligence', it is evident that this latter also depends on factors quite independent of mere brain mass. One might rightly question whether elephants or whales with a brain weight of 5000 or even 7000 g, including an extensive 'neopallium' are more 'intelligent' than man. In this respect, it seems appropriate to recall a pertinent remark expressed by YOUNG (1955), and stating that, beyond a certain limit, the working efficiency of a 'nervous centre' does not depend on its size. In agreement with YOUNG's comment I would say that the behavioral complexity and efficiency provided by the function of a vertebrate brain is not,

---

[320b] As regards preservation procedures, formalin generally involves swelling, and alcohol shrinkage of the brain. After long preservation in alcohol, the reduction in brain volume is said to approach about 50 percent.

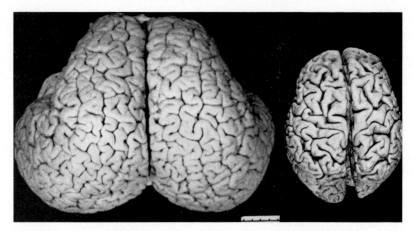

*Figure 395.* Comparison of adult brains of a fin whale *(Balaenoptera physalus* L.*)*, left, and man, right, photographed at the same magnification (after TOWER, 1954). The length of an adult human brain (here about 160 mm) amounts to between approximately 150 and 180 mm, its width to between approximately 130 and 145; the scale shown in the figure indicates intervals of about 1 cm.

within a flexible range of required minimal mass, exclusively nor necessarily correlated with its absolute size.

Similar qualifications obtain with regard the range of human adult brain volume. The minimum weight of a brain required for 'normal' behavior appears here to be of an order magnitude around 800–780 g, and a person with brain weight below this approximate minimum will almost certainly be mentally defective. In many instances a weight of 900 g seems already to involve mental deficiency. Numerous publications deal with average brain weights in different racial groups,[321] and others with the weight of the brain in persons of great intellectual accomplishments. Again, it has been claimed that the average absolute

---

[321] In all racial groups, the recorded brain weights remain within the normal overall range obtaining for present-day man in general. Moreover, differences in the numerical distribution of recorded brain weights upon the various races, and differences *qua* chance selection of the sampled individuals introduce additional statistical uncertainties frustrating a meaningful comparison. Thus, a high brain weight for Chinese, Eskimos and Palau Islanders was once claimed on the basis of findings in only 16, 7, and 4 individuals, respectively (cf. RANKE, 1923). Conclusions of this sort, which might, of course, happen to be in general accordance with subsequent more extensive series, are certainly not justified by accepted standards of statistical procedures.

|                                    | Weight, g   |
| ---------------------------------- | ----------- |
| Selachians                         |             |
| *Sphyrna zygaena* (hammerhead)     | 25          |
| *Galeus canis*                     | 12          |
| *Mustelus*                         | 4.8         |
| *Raja*                             | 3           |
| Teleosts                           |             |
| *Gadus aeglefinus*                 | 1.7         |
| *Cyprinus carpio*                  | 1–0.9       |
| Amphibians                         |             |
| *Rana catesbyana* (bullfrog)       | 0.24        |
| *Rana temporaria*                  | 0.09–0.1    |
| *Triton*                           | 0.01        |
| Reptiles                           |             |
| *Crocodilus*                       | 5–15        |
| *Chelone midas*                    | 4–7         |
| *Python molurus*                   | 1           |
| *Lacerta*                          | 0.02–0.2    |
| Birds                              |             |
| *Struthio camelus*                 | 29–35       |
| *Gallus domesticus*                | 3–4         |
| *Columba*                          | 2           |
| Mammals                            |             |
| Whales                             | 4000–7000   |
| Elephant                           | 4000–5000   |
| Porpoise (*Tursiops*, dolphin)     | 1700        |
| Man                                | 1200–1400   |
| Horse                              | 600–680     |
| Rhinoceros                         | 600         |
| Gorilla                            | 500         |
| Orang and chimpanzee               | 350–400     |
| Lion                               | 200–250     |
| Macaque                            | 90          |
| Dog                                | 40–100      |
| Cat                                | 30          |
| Hare                               | 9–10        |
| *Erinaceus* (hedgehog)             | 3–3.5       |
| Rat                                | 3           |
| Mouse                              | 0.4         |

*Figure 396.* Tabulation of approximate brain weights in some representative vertebrates, compiled from various sources (cf. K., 1927, pp. 86–88 and statements in present text).

weight of the human male exceeds that of the female by about 100 g or more. None of these data can be considered as very conclusive. It is even impossible to make a rigorously valid statement concerning the average weight of a 'typical' adult human brain, but a liberal estimate of $\pm 1400$ g for the male, and $\pm 1300$ g for the female[322] can be regarded as reasonably adequate.

Studies on brains of outstanding individuals have been interpreted to disclose some statistical correlation between high brain weight and intellectual capacity,[323] although, in given individual cases, a person with a low brain weight of about 1017 g *(Anatole France)* may be highly gifted, while another with a brain of 1800 g or more may be a dullard or congenitally mentally defective. The perhaps highest recorded human brain weight, mentioned by the psychiatrist WEYGANDT (1928), is said to have reached 2850 g, and was reported at the *post mortem* of an epileptic afflicted with idiocy.[323a] If 'normal' adult human cranial capacity of 900 to 2300 cc is interpreted in terms of brain weight, the 'normal range' of this latter would lie between about 850 to 2100 g. The above-mentioned *statistical* correlation of brain weight and 'superior intellectual ability' remains thus, if actually obtaining, at best rather inconclusive. It is, moreover, perhaps also complicated by the relationship of body weight to brain weight, discussed further below. This, for instance, might at least partially account for the statistical differences between average male and female brain weights.

Again, at least as regards mammals, big animals have, in general, a longer life-span than small ones, e.g. horses more than 40 years, dogs about 9–20 years, guinea pigs 4–7 years, rats about 3 years. Since man,

---

[322] This, e.g., was accepted by COBB (1965).

[323] The following weights are frequently quoted: *Kant* (1600 g), *Thackeray* (1658 g), *Bismarck* (1807 g), *Cuvier* (1830 g); *Daniel Webster* (1895 g), *Turgenjew* (2012 g), *Cromwell* and *Byron* likewise both over 2000 g. The figure for the two last-named has been doubted by DONALDSON (1895). In all quoted figures, the third digit must be considered approximate, and the fourth digit meaningless. This evaluation presumably also applies quite generally to all given figures purporting to state brain weights.

[323a] WEYGANDT (1928) merely cites, without further references, '*das von* WALSEM *beschriebene Gehirn eines epileptischen Idioten mit 2850 g*'. As I recall, the pathologist R. ROESSLE mentioned, in his 1920–1921 Jena lectures, a brain weight *exceeding* 2500 g, recorded at the postmortem of an institutional inmate afflicted with idiocy, as the highest known figure for man. I am uncertain whether ROESSLE's remark does or does not refer to the case cited by WEYGANDT, since I was unable to trace both statements to their actual sources.

with a heavier brain, but of much less body weight than a horse, commonly reaches an age of 70–80 years, it has been suggested, as quoted by Cobb (1965), that brain weight by itself, rather than body weight, might display significant correlation with life-span.

The brain size of *fossil animals* and precursors or ancestral forms of present-day man can be estimated on the basis of endocranial casts, either recovered as so-called 'fossil brains' or, particularly in the case of primates, artificially contrived from suitable cranial remnants. If the relationships between meningeal spaces and brain surface are sufficiently well-known, reasonably valid conclusions can be drawn. Where the meningeal spaces are not too extensive, relevant surface features of the brain may be displayed (cf. T. Edinger, 1929). For an estimate of brain weight based on endocranial capacity, it becomes necessary to consider the volume changes sustained by a dried skull, the approximate volume of the extracerebral components (meninges, blood vessels, cerebrospinal fluid), and the specific gravity of the brain. This latter, in man, seems to vary between about 1.030 and 1.047. Because of the diverse uncertainties here involved, any computation of brain weight from volume of cranial cavity remains a rough approximation.[323b]

The *phylogenetic evolution of mammals*, as inferred from paleontologic studies, is believed to have displayed four major trends in most taxonomic groups, namely, increase in body size, *allometric increase in brain size*, specialization of the teeth, and specialization of the feet. Figure 397 shows the relative sizes of the brain in several extinct Cenozoic and presently-living mammals of approximately equal size and comparable body forms. It can be seen that the size increase of the cerebral hemispheres (i.e. of the lobus hemisphaericus proper) is especially conspicuous. Details concerning the '*general law of brain growth*', formulated by Marsh about 1876, as well as the partly valid, and partly unconvincing objections to this 'law' by T. Edinger have been discussed in volume 1, p. 265f. of this series. Already in the year 1868, before Marsh elaborated his 'laws', E. Lartet had stated: '*plus les mammifères*

---

[323b] On the basis of cranial volume calculations, Weygandt (1928) has estimated the brain weight of a walrus (the Carnivore Pinnipedian *Odobenus*) at about 1040 g. Weygandt also gives an estimate of over 1400 g for the brain of the sea-cow *Rhytina stelleri* (Paenungulata, Sirenia). This animal, which lived in the Bering Sea, was completely exterminated by the year 1768. As regards the extinct ground sloth Megatherium (Edentata) living in South America until the end of the Pleistocene, the cited author recorded a cranial volume of 1200 cc, compatible with a brain weight of approximately 1000 g.

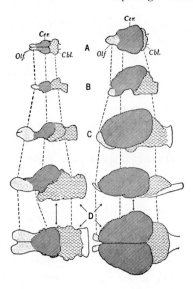

*Figure 397.* Brains of extinct Cenozoic mammals (left) and recent mammals (right), paired so that the brains of animals having equal size are side by side. Olfactory bulb stippled, cerebral hemispheres indicated by oblique lines, other brain regions marked by broken lines or unmarked. A: Arctocyon and recent Canis; B: Phenacodus and recent Sus; C: Coryphodon and recent Rhinoceros; D: Uintatherium and recent Hippopotamus (after H. F. Osborn, from Schuchert and Dunbar, Outlines of Historical Geology, New York 1941).

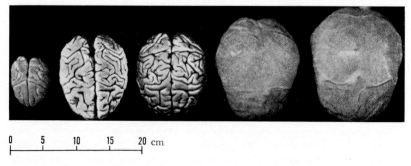

0   5   10   15   20 cm

*Figure 398.* Comparison of adult brains in (from left to right) the gibbon *Hylobates,* the gorilla, and the chimpanzee, with cranial endocasts of Pithecanthropus and the *Düsseldorf* Neandertal man (from Kappers, 1929; an approximate scale has been added).

*remontent dans l'ancienneté des temps géologiques, plus leur cerveau se réduit par rapport au volume de leur tête et aux dimensions totales de leur corps'* (quoted after JERISON, 1970).

As regards two primates presumably in some manner related to the phylogenetic evolution of man, Pithecanthropus, with a cranial capacity of approximately 1000 cc may have had a brain weighing about 970 g or less, and the brain weight of the related form Sinanthropus exceeded perhaps 1000 g in various instances. The brain of the Neandertal man was apparently as large or heavy as that of modern man, and there is no valid evidence for an increase in human brain weight since the later part of the Pleistocene, i.e. within a period probably exceeding 100 000 years. Figure 398 illustrates the relative size of brains respectively endocranial casts of several primates including Pithecanthropus and the Neandertal man.

(2) *The ratio of body weight to brain weight,* namely the simple relative brain weight, can be expressed by *Cuvier's fraction*[324] $\frac{PS}{PE}$ or by its reciprocal $\frac{E}{S}$, and seemed of obvious interest. Its computation showed that, in general, small animals within the same taxonomic group display a greater relative brain weight than larger ones. Thus, $\frac{E}{S}$ ratios of the following sort can be obtained: cat $\frac{1}{100}$, dog $\frac{1}{120}$ to $\frac{1}{300}$, lion $\frac{1}{550}$, horse $\frac{1}{600}$, hippopotamus $\frac{1}{2789}$. On the other hand, man and mouse display a roughly identical ration of about $\frac{1}{40}$, and the elephant, with $\frac{1}{560}$ stands somewhat above the horse. The highest relative weights, of approximately $\frac{1}{12}$, were found in small birds. As regards Anamnia, $\frac{1}{172}$ was recorded in the frog, $\frac{1}{248}$ in the carp, $\frac{1}{2496}$ in a large shark, and $\frac{1}{37440}$ in a large tuna fish.

It should be added that these figures for relative brain weights refer to adult individuals and represent end stages of growth processes. Thus, in newborn man, with an average body weight of about 3125 g, the brain may weigh between approximately 300 and 370g. Accordingly, its relative weight will change from around $\frac{1}{10}$ at birth to $\frac{1}{40}$ or $\frac{1}{50}$ in adult life. Finally, it has been inferred from average figures computed for different age groups that the human brain may lose, in the course of senescence, about 100 g or more of its weight at full maturity. In connection with the brain-body weight relationship it appears

---

[324] PS or S stand for body weight, PE or E for brain weight.

moreover of interest that 'human nanocephalic dwarfs with brains of less than 700 g can talk and converse fluently' (cf. COBB, 1965.)

In the second half of the 19th century, various authors, including A. BRANDT, TH. BISCHOFF, SNELL, L. MANOUVRIER, L. LAPIQUE, and particularly E. DUBOIS pointed out that additional definable factors, independent of somatic weight, must be considered in order to interpret the brain-body weight ratio. DUBOIS assumed that this ratio is significant only if used in conjunction with a so-called *cephalization factor* C which may vary from group to group. He elaborated the expression

$$E = C \cdot S^r$$

where E is the weight of the brain, S the body weight and r an empirically determined exponential constant which he assumed to be 0.56 for mammals.[325] LAPIQUE subsequently found the same value for birds, and DUBOIS again reported this constant as likewise valid for lower vertebrates. With respect to taxonomically related series, among mammals e.g. cat, cheetah, leopard, and tiger, the brain weight becomes only approximately doubled as the body weight is quadrupled. This, of course, is in agreement with a more generalized older formulation known as *Haller's rule* and dating back to about 1762 (cf. vol. 1, p. 268, footnote 60 of this series). Such behavior of brain volume in taxonomically related recent forms represents a negative allometric in-

---

[325] Originally, upon considering a relationship between body surfaces and body weight as an indicator of heat loss respectively metabolism, which might influence brain volume, previous theories had suggested $2/3$ or 0.66 as the relevant constant. In contrast to the term '*cephalization*', referring to the relative weight or volume increase of the brain, T. EDINGER (1960) has stressed '*cerebralization*' as denoting a quantitative process correlated with progressive complication involving often '*specialization*' in brain structure. A particular aspect of cerebralization, namely '*telencephalization*', was emphasized by v. ECONOMO (1928, 1929). Strictly speaking, '*encephalization*' should stand for '*cephalization*', since this latter term might be used as referring to size relationships involving the head. Evidently, a detailed study of volume or size relations between the whole head and its included brain might also disclose diverse data differing from those concerning brain-body ratios, and useful for various sorts of phylogenetic speculations. Having once removed, some 40 years ago, and with the help of several associates, the brain of a young elephant who died at the Breslau Zoological Garden, I was struck, during that difficult task, with the discrepancy between head and brain size, i.e. between '*cephalization*' and '*encephalization*'. Again, the relationship between a Selachian brain of about 3 to 25 g, and a mammalian mouse brain of about 0.4 g displays an obvious difference between the significance of mere volume and of '*cerebralization*'. Clearly, in some phylogenetic sequences or 'radiations' progressive cerebralization might have been concomitant with regressive absolute brain weight.

crease, while the increase in brain mass suggested by Figure 397, comparing some extinct with some recent forms, is of positive allometric
type.

If, in accordance with a procedure apparently first used by LA-
PIQUE, the logarithms of the brain weights are plotted on the ordinate,
and the logarithms of the body weight on the abscissa of a coordinate
system, the values fall on approximately straight lines (Fig. 399). The
angle of these so-called isoneural lines with respect to the abscissa is a
function of the relation exponent and may be used as its indicator. This

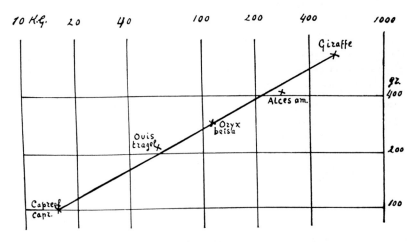

Figure 399 A

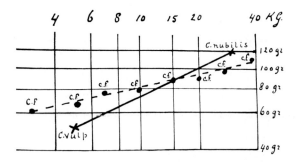

Figure 399 B

*Figure 399.* Graphic representation of cephalization coefficient (A) in interspecies relation for different ungulates and (B) in inter- as well as intraspecies relations for different
Carnivora (after LAPIQUE from KAPPERS, 1929; the intraspecial relation in B is indicated by
the broken line).

exponent (r), however, was found much smaller, namely 0.24 to 0.28 in the comparison of forms belonging to the same species, e.g. dogs of different sizes. This was explained by the assumption that among animals of the same species but differing in size there is essentially only an increase in the size of individual neuronal elements, but not of their number in the larger animals.[326] The higher exponent 0.56 calculated for *interspecies comparisons* might then be related to increase in the number of cellular elements. Again, as regards *intraspecies comparisons* the smaller coefficient (namely 0.24 instead of 0.28) has been claimed to obtain for domesticated animals, thus indicating an effect of 'domestication'.

With respect to a comparison of *fossil mammals* with recent ones, JERISON (1970) reached the conclusion 'that there has been a progressive increase in relative brain size accompanied by and correlated with increased diversity among species in relative brain size. Small-brained species have also evolved, but more large-brained species have appeared in successive epochs'.

STEPHAN (1958) has evaluated cephalization within the order Insectivora and found in Macroscelididae a brain size about 2.5 times larger than in the Tenrec of approximately equal body size. STEPHAN's series of increasing cephalizations gives the sequence Tenrec – Erinaceus – Soricidae – Potamogale – Talpa – Chlorotalpa – Macroscelididae – Tupaia.[326a] This sequence is said to be in agreement with generally accepted views concerning the relative rank assigned to these various forms on the basis of their overall biologic features.

(3) Since the *spinal cord*, because of its relatively simpler structure and function, despite significant differences in the type as well as number of ascending and descending cerebral connections, displays in con-

---

[326] The question of growth by cellular volume rather than by cell number, and related problems were particularly studied by GIUSEPPE LEVI. A reference to this topic can be found in volume 3, part I, p. 91 of this series. A few remarks on cell number are included in section 9, p. 708 f. loc. cit. Detailed data, referring to estimates by a variety of authors, can be found in the tabulations presented by BLINKOV and GLEZER (1968).

[326a] The Asiatic tree shrew, *Tupaia*, at one time classified as an insectivore, displays definite affinities with Lemuroid primates, and is now generally considered as belonging to these latter (cf. vol. 1, pp. 65 and 69 of this series). Particularly on account of its brain morphology, H. SPATZ (1965) and other evaluate Tupaiidae as representing a phylogenetically important group transitional between insectivores and primates. The term '*Subprimaten*' was also used to designate this *Grenzgruppe*.

trast with the brain a far lesser degree of evolutionary changes within the vertebrate phylum, a comparison of brain weight and weight of spinal cord can be considered relevant. VILLIGER (1920), referring to a suggestion by RANKE, states that in this respect man seems to possess the relatively heaviest brain. The following figures of spinal cord relative weight are said to obtain: man: about 2 percent, anthropoids: approximately 6 percent, macaques about 12 percent, diverse mammals: 19–47 percent. I have been unable to find specific data concerning elephant or Cetaceans. BLINKOV and GLEZER (1968) quote a few figures for submammalian forms: chicken: about 95 percent, tortoise 120 percent (i.e. the spinal cord is heavier than the brain), frog: 45 percent (this ratio being presumably related to the short but broad trunk of Anurans).

(4) Because the exponential constant 0.56 adopted by DUBOIS and others may vary with respect to different groups of animals and does not hold, as mentioned above, for intraspecies comparisons, PORTMANN (1964), who considers the use of r $=0.56$ *qua* constant incorrect, has followed another approach,[327] based on *the mass of distinctive brain regions*. These calculations, determining '*intracerebral indices*', represent, as it were, a further refinement and expansion of the rationale followed in comparing brain and spinal cord weights.

PORTMANN and his collaborator WIRZ selected brain regions, defined as '*brain stem remnants*', which they considered representative of the 'elementary vegetative functions'. In comparing the mass of different 'brain stem remnants', differences with respect to animals of identical body weight were recorded. Thus it was found 'that the stem remnant weighs twice as much in guinea pigs as it does in hamsters, twice as much in stags as in boars. The stem remnants of three birds of equal average weight, but with different degrees of cerebralization – pheasant, crow, macaw – are in the ratio 1:1.8:2.5. In other words, the development of the brain stem is a measure of the general degree of evolution' (PORTMANN, 1964).

The cited authors then attempted to determine, for all body sizes, the smallest 'stem remnant' corresponding to the 'lowest stage of evolution' of mammals and birds. Using this value as 'unit', they found,

---

[327] This approach has been extended to volumetric studies of the invertebrate central nervous system (insects, cephalopods). A brief reference to such investigations by RATZERDORFER was included in volume 2, p.194 of the present series.

e.g., that the 'stem remnant' of the tapir was 3.2, and that of the horse 6.8, while the 'stem remnant' of man appeared ten times as massive as that of a 'primitive mammal' of equal body weight.

A comparison of the mass of the 'higher brain centers' with that of the lowest 'stem remnant' in a group of mammals of equal brain weight is said to allow for the establishment of separate *indices* for other brain regions such as 'cerebrum', 'neopallium', and cerebellum. In particular, *'neopallium indices'* were computed by WIRZ in respect to mammals of different body weights and to mammals of like body weights but different levels of organization.

PORTMANN and his collaborators believe that *'intracerebral indices'* of this type provide useful information concerning heat economy and other physiological mechanism as well as diverse behavioral activities thereby indicating 'a measure of brain development'. In one series of computations, man's 'neopallial index', given as 170, compares with 49 for anthropoid apes, 38 for the baboon, 1.9 for the rat, and 0.7 for the hedgehog. Man's index is here evaluated as 'an indication of how much vaster the quantitative jump from anthropoid ape to *Homo sapiens* was than even the great jump from lemur to ape' (PORTMANN, 1964). However, because of various uncertainties involved in the principles and methods selected for the calculation of the diverse indices, one should consider the results of these studies as still somewhat inconclusive. Be that as it may, the data recorded by the cited authors are doubtless of interest and seem to supplement, in a relevant manner, those obtained by other procedures of quantitative investigations.

(5) The *compactness of the nervous parenchyma sensu latiori*, i.e. the number of cellular elements in a given volume of central nervous substance can be determined by various procedures giving useful approximate figures, none of which, however, is entirely reliable. Details concerning such computations can be found in the compilation of BLINKOV and GLEZER (1968). The packing density of nerve cells in the human cerebral cortex has been investigated by v. ECONOMO (1925, 1926) who introduced the concept of an *index* GC (*Grauzellkoeffizient*) calculated as follows

$$ GC = \frac{\text{volume of gray matter}}{\text{volume of body of nerve cells}}. $$

According to this author, the average value of GC for the human cerebral cortex is 27, indicating that for each volume of nerve cells there are 27 volumes of other substance. This latter, of course, includes the volume of the glial elements, of the intercellular spaces which may or may not contain a significant amount of 'ground substance'[328] and of the mesodermal (vascular) stroma as well as presumably of thin neuronal arborizations. The packing density of the nerve cells is thus inversely proportional to *von Economo's coefficient*, for which a partly agreeing and partly disagreeing variety of figures for different grisea has been given by a number of authors. It is obvious than an adequate determination of the volume of nerve cells and glia cells is difficult to obtain, and that it is easier to record the number of cellular elements by a count of their nuclei.

Again, the *ratio of nerve cells to glia cells* represents another relevant variable. It has been claimed that the number of glia cells in the mammalian brain is roughly ten times greater than the number of neurons. BLINKOV and GLEZER (1968) denote the *glial index* by the notation $\frac{N_g}{N_n}$, where $N_g$ is the number of neuroglia cells and $N_n$ the number of neurons in the same volume of brain. Depending on taxonomic forms and regions of the brain, values ranging between approximately 0.29 and 45 have been recorded. The high values of this index were found in grisea of brain stem, and the low values, corresponding to a high proportion of glia cells, in regions of cerebral cortex.

The *packing density of cortical neuronal elements* in 'ascending' taxonomical series of mammals seems to decrease, and the number of neuroglia cells apparently increases. In the course of ontogenetic development, a significant reduction in the packing density of cells takes place in the diverse grisea.

With regard to *brain size*, it is generally assumed that its increase becomes correlated with a decrease in the packing density of cortical neuronal elements. Such decrease in density might be associated with a greater expansion of the functional neuropil, that is, with an increase in the number of terminal arborizations and synaptic connections,[329] or merely with an increase of relatively 'inert' 'supporting' or 'filling' elements between the neuronal perikarya. These questions, which cannot

---

[328] Questions concerning 'ground substance' and intercellular spaces in the nervous parenchyma were discussed in vol. 3, part I, pp. 235f. and 489f. of this series.

[329] A relatively very large number of synaptic connections, combined with extreme compactness, and compatible with very small size, is provided by the peculiar neuropil structures in the brain of insects displaying complex behavior (cf. vol. II of this series, pp. 136, 250, 254 *et passim*).

be satisfactorily answered at the present time, have an evident bearing on the relationship of brain size, and especially of cortical, respectively 'neopallial' volume, to so-called intelligence. RIESE (1927) evaluated the structure of the dolphin's isocortex as displaying a 'primitive' type, characterized by paucity of medullated fibers, indistinct differentiation of the laminar pattern, sparsity of cells, and relatively small width.[330] These general features, also noted by some other investigators, were essentially confirmed for the cortex of the fin whale (Balaenoptera) by TOWER (1954). If the low packing density of the cortex in Cetaceans is not related to a larger amount of terminal arborizations and synaptic connections, but due to functionally rather 'inert' space-filling structural elements, then the supposition that, despite higher brain weight and more extensive 'neopallium', Cetaceans are 'less intelligent' than man would seem quite plausible. On the other hand, if the low packing were due to a more extensive 'functional neuropil', one could possibly suspect that Cetaceans might have a superior 'intelligence'. Be that as it may, a considerable degree of this behavioral ability is known to be displayed by dolphins.

With regard to some *biochemical aspects* presumably related to synaptic functions, TOWER and ELLIOTT (1952) reported that values for all components of the acetylcholine system of various mammals decreased regularly with increasing brain size. This decrease was correlated with the average total brain weight of each species and was the same function of brain weight for all types of acetylcholine activity measured. Said relationship could be characterized by the expression

$$A = K_A W^k$$

where A is the particular activity per unit weight, $K_A$ represents a constant characteristic of the particular activity measured, and W is the average particular brain weight. The exponent k is the regression coefficient for the logarithm of A versus the logarithm of W for the various species. In all cases the value for k was found to be approximately —0.3.

The possibility that decrease in activity of the acetylcholine system might be related to decrease in neuron density was tested by the cited authors, who noted that a decrease in the number of neuron perikarya

---

[330] RIESE (1927 and other publications) explicitly stated that despite what he considered a 'primitive' aspect of its structure, the dolphin's isocortex displays a typical mammalian structural pattern. An acrimonious, but in my opinion essentially unjustified polemic was directed by M. ROSE at RIESE's findings and interpretations.

per unit volume of cortex with increased brain weight gave a similar regression curve. In adaptation to the formulation given above, the expression

$$N = K_N \, W^k$$

was found to hold, N being the number of nerve cells per unit volume, with k again of approximate magnitude —0.3. These data, obtained on a variety of mammals from mouse to man, were confirmed by Tower (1954) in the whale.

On the other hand, no simple relationship between cholinesterase activity and neuron density was found. The cholinesterase activity per unit volume of cortex in a given species appeared to show no difference between the 'motor cortex' and the more compact 'striate cortex', whose density of neuron packing is about 4 times that of the former. Thus, although on the basis of interspecies comparisons, cholinesterase activity seems to be proportional to the number of neurons in a given cortical area, this proportionality is not the same for different cortical areas. Tower (1954) assumes that the size of the neurons might be a relevant parameter affecting the activity of the acetylcholine system. The cited author's calculations, based upon cholinesterase activity suggest that activity of the 'acetylcholine system' is proportional to neuron volume with respect to the following comparisons: (a) between various layers in a given cortical area (excepting layer I, i.e. the cell-poor lamina zonalis); (b) between various cortical areas in a given species, and (c) between cortices of different species.

If an overall appraisal of the quantitative data considered in the present section 8 is attempted, it could be stated that, in the course of presumptive vertebrate phylogenetic evolution, a definite general trend toward *progressive cephalization ('encephalization')* and *progressive cerebralization* (respectively *telencephalization*) is recognizable. These trends, on the whole, i.e. considering the differences obtaining between mammals and submammalian forms, have been associated with an increase in absolute brain weight. Yet, as regards recent forms, amphibians and reptiles do not display absolute brain weights reaching the maximum recorded for large Selachians. Again, progressive cephalization (i.e. increase of the relation brain size to body size), combined with progressive cerebralization, is shown to be compatible with low absolute brain weight (e.g. 0.4 g in the mouse).

While it seems likely that, for a behavioral complexity ('intelligence') as manifested by man, the necessary level of cerebralization requires a minimum range of brain volume respectively weight (very roughly $\pm 900\,\text{g}$), higher brain weights, e.g. such as $2000\,\text{g}$, considerably exceeding this range, are not necessarily correlated with a high degree of 'intelligence'. Still higher brain weights, in the 4000 to 7000 g range (Elephants and Cetaceans) are presumably not even correlated with an intelligence comparable to that of man. In the case of Elephants and Cetaceans, however, one might suspect, on the basis of the morphological and structural findings, a 'lesser', or perhaps better, *different* type of cerebralization or *telencephalization*.

A very low degree of cerebralization is found in the Chordate Amphioxus, which merely possesses an 'archencephalic' vesicle combined with some special differentiation of the rostral end of the spinal cord. As regards true vertebrates, the lowest degree of cerebralization, resulting in a fairly 'simple' type of brain, can be noted in Cyclostomes and Amphibians, particularly in some Urodeles. Whether, in this latter case, a 'regression' should or should not be assumed, remains a moot question. Some (but not all) Teleosts, on the other hand, might be said to display a degree of cerebralization far exceeding that obtaining in any known Amphibian.

Yet, in comparison with other organ systems, the central nervous system is characterized by a particularly high degree of conservatism throughout the entire vertebrate series.[331] As regards its overall bauplan, in contradistinction to its structural differentiation, and discounting the telencephalic one-many transformations $D \rightarrow D_1, D_2, D_3, B \rightarrow B_1, B_2, B_3 (B_4)$, no new Grundbestandteil becomes added in the entire taxonomic and presumably somewhat similar phylogenetic series from Cyclostomes to man. Among Invertebrates,[332] the highest degree of cerebralization seems to be manifested in some Insects and in some Cephalopods. This is combined with considerable compactness and very low weights (e.g. $\pm 1\,\text{mg}$) in the case of Insects, and with a relatively large volume in Cephalopods.

Reverting to problems concerning the presumed *human ancestry* (cf. Fig. 400), an increase in brain volume from Australopithecus to Neandertal, Cro-Magnon, and Aurignac men is evident. This occurred during the Pleistocene epoch, in the course of perhaps several hundred thousand years. The Neandertal man was characterized by a peculiar

---

[331] Cf. the references to this conservatism, stressed by WIEDERSHEIM and PLATE, in vol. 1, p. 276 and vol. 2, p. 336 of this series.

[332] Cf. vol. 2, chapt. IV (pp. 175, 194, 216 *et passim*) of this series.

development of the brow ridges, a receding forehead, massive jaws
with absence of a chin eminence, and angulated appearance of occipital
region. Le Gros Clark (1960) and others believe that these 'apparently primitive characters', nevertheless associated with a large cranial capacity and an estimated brain weight exceeding the averages given for
modern man, were the result of a 'secondary retrogression'. According
to v. Bonin (1963), the frontal lobe of the Neandertal man's brain 'is
too small'. This apparent smallness is evidently related to the receding
forehead indicated by the skull finds. In view of the great cranial capacity, the actual amount of frontal cortex may have been no less than that
in modern man despite a different shape of the frontal lobe. Reconstructions of the configuration of that lobe on the basis of endocranial

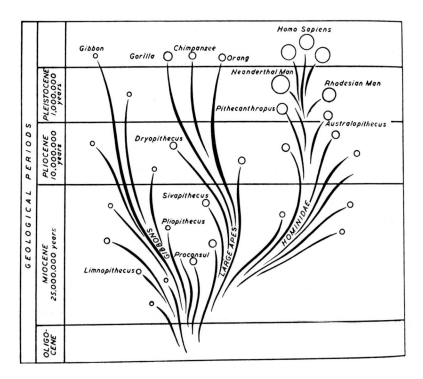

*Figure 400.* Diagram indicating hypothetic phylogenetic relationship of Hominidae
and anthropoid apes according to Le Gros Clark (from Le Gros Clark, 1960). The circles are intended to represent approximate differences in the relative size of the brain. It is
here assumed that the brain did not undergo the expansion characteristic for modern Man
until the early or middle Pleistocene, approximately 800 000 to 600 000 years ago. The author's size indication for the Neandertal man's brain is perhaps exaggerated.

casts, including attempts at identification of gyri and sulci, cannot be considered sufficiently accurate or convincing for a reasonably valid appraisal of the Neandertal man's frontal lobe.

As evidenced by finds in various regions of Europe, in North Africa, South Africa, and Palestine, the Neandertal man, of whom presumably several 'races' existed, had a wide geographical distribution, and seems to have become extinct soon after the climax of the *Würm* (i.e. the last) glacial age, perhaps about 50 000 years ago. Whether this Hominid interbred with the recent species of *Homo*, respectively whether the Neandertal man should or should not be included in that species, remains, as pointed out in volume 1 (p. 71), a difficult and perhaps undecidable question.

With regard to brain weight it could be stated that no convincingly demonstrable increase in average human capacity eventuated during a span of probably more than 100 000 years, i.e. since an undetermined period of the Pleistocene up to the present time. In addition, many of the phylogenetic inferences and speculations based upon the multitudinous quantitative studies of the brain are at best inconclusive if not outright questionable.

Concerning '*cerebralization*', it seems difficult to point out, in a rigorously satisfactory manner, indubitable *significant* structural differences between the brain of anthropoid apes, such as orang, chimpanzee or gorilla, and the human brain. Although relevant differences in 'cerebralization' may here reasonably be presumed, the clearly demonstrable differences are essentially of a quantitative nature. T. H. Huxley had claimed that, *in toto*, the structural differences which separate man from these apes are not so great as those which separate the gorilla from the 'lower' i.e. Catarrhine monkeys. E. Haeckel subsequently adopted this proposition which he called the '*pithecometra thesis*'. Clearly, said thesis is implicitly related to a justifiable taxonomic classification of *Homo sapiens* as pertaining to a recent anthropomorph superfamily comprising the anthropoids gibbon, orang, chimpanzee and gorilla (Simiidae) and the Hominidae. Seen from this aspect, man is no more than a highly developed, but by and large rather nasty, Hominid ape. However, because of the many difficulties and pitfalls involved in such comparisons, the '*pithecometra thesis*', although doubtless in some respects 'true' as a valid general estimate, defies an elaboration in detailed, rigorously definable terms. It should be recalled that the morphologic phylogenetic changes leading to the origin of man include more than brain evolution.

Important changes, which nevertheless indirectly affect the entire nervous system as the control and communication system of the body, concern the transformations of hand and foot, moreover those correlated with upright body posture and with the origin of articulated speech. Alterations pertaining to this latter faculty, and pointed out by various authors (e.g., v. BONIN, 1963) imply a caudal recession of the larynx in the microsmatic primates. In most lower mammals, particularly in macrosmatic ones, the larynx has a more rostral position, close to the choanae. Concomitantly with these laryngo-pharyngeal changes, the viscerocranium assumes a position basal to the neurocranium instead of rostral to it, leading to correlated transformations of mandible and oral cavity, allowing for *articulated speech*. These overall phylogenetic transformations[332a] also affect the course of the longitudinal axis of the central nervous system, which, during ontogenetic development, displays a variety of changing curvatures, briefly discussed in section 1 B of the present chapter VI. Bends, expressible as 'angles' between longitudinal axes of particular neuraxial subdivisions, may accordingly also be considered with regard to the phylogenetic aspect. Papers by DABELOW (1931), HOFER (1952, 1954), and SCHNEIDER (1957) contain relevant data concerning correlations of cranial configuration (e.g. '*Schädelknickung*') and brain shape (including angle between brain and spinal cord longitudinal axes), as well as further references pertaining to this general topic.

It can be taken for granted that despite considerable uncertainties about most details of man's phylogenetic history, the human brain, which engenders the human mind, is the product of a long sequence of evolutionary events. The invertebrate or prechordate episode of this process remains particularly obscure.[333] The earliest Palaeozoic vertebrate ancestors of man may have arisen during the Ordovician, about 400 million years ago.[334] Reasonably well documented inferences sug-

---

[332a] Some of these phylogenetic changes, particularly with regard to relative configurational features of neurocranium and viscerocranium, display ontogenetic aspects which could perhaps be subsumed under the concept of neoteny (cf. vol 1, p. 244 of this series).

[333] Cf. volume 2 of this series, particularly section 13, p. 286 f., and section 14, p. 300 f.

[334] A continuous course of phylogenetic evolution since pre-Cambrian time, and discounting CUVIER's theory of catastrophes, is thereby assumed. It should, however, be mentioned that, according to some recent theories concerning the evolution of the atmosphere of the earth, major and rapid fluctuations in the gradual increase in oxygen

gest a succession of Cyclostome-like (Agnathous), Gnathostome fish-like, amphibian-like, reptilian-like, and mammalian, perhaps insectivore-like stages leading to the origin of man's primate ancestors. This course of events, quite evidently, implies '*progressive cerebration*', and in particular '*telencephalization*'. It is, however, questionable whether progressive natural phenomena of this sort can indefinitely proceed. Individual growth, quite certainly, does not; population growth, for obvious reasons, must sooner or later come to an end. Comparable constraints can be assumed in the case of brain evolution, since all progressive developmental processes seem to proceed under specific limiting conditions.

Nevertheless, it might be asked whether this brain evolution, as now apparently culminating in man, has reached a maximum degree of cerebralization, or whether further evolutionary progress is likely. Although, on the basis of the available data, this question appears undecidable, some opinions concerning said topic have been expressed. *Prima facie*, three answers seem logically permissible: (1) a maximum level of cerebralization has been reached and may endure for an indefinite period of time; (2) progressive cerebration can be expected to continue at an undefined slow rate consistent with the time scale of phylogenetic evolution, and (3) the maximum now reached can be expected to change into a continuous and perhaps already incipient decline.

The first alternative, namely '*dass der heutige Zustand unseres Grosshirns ein definitiver sein könnte*', is rejected by v. ECONOMO (1928, 1929) who believes that progressive cerebration '*auch heute ihre volle Wirkung ausübt*'. The cited author assumes the possibility of a further cortical differentiation leading to the acquisition of new 'areae'. These processes, according to v. ECONOMO, need not necessarily require a significant increase of the now obtaining human brain volume, since the difference between a weight of about 1000 g in an intelligent person and 2000 g in another one seems to indicate a large amount of apparently unused '*play*' providing possibilities for progressive differentiation. In v. ECONOMO's opinion, no compelling reason justifies the assumption

---

concentration may have occurred, which could have caused the extinctions at the end of the Paleozoic and Mesozoic. Such events, to some extent, might be interpreted as justifying CUVIER's views. Nevertheless, complete extinction of Metazoan life with subsequent new lines of biologic evolution does not seem very probable (cf. vol. 1, p. 271 of this series; also: van VALEN, L.: The History and Stability of Atmospheric Oxygen. Several factors regulate its concentration, but the initial rise is unexplained. Science *171*: 439–443, 1971).

that the natural principle of 'progressive cerebration' has ceased to be effective with respect to the human brain: '*Die progressive Zerebration bedeutet aber biologisch nicht nur eine Steigerung der zu einer Zeitperiode bestehenden geistigen Fähigkeiten, sondern infolge der nachgewiesenen Möglichkeit der Entstehung neuer Hirnorgane in der Hirnrinde auch die Möglichkeit der Erwerbung ganz neuer und ungeahnter psychischer Fähigkeiten, ein Umstand, der für die künftige Entwicklung der Menschheit neue Perspektiven eröffnet*' (v. Economo, 1928).

Essentially similar views were expressed by my esteemed old friend Hugo Spatz (1888–1969), who, in some of his last studies (1955, 1962, 1964), was deeply concerned with '*Vergangenheit und Zukunft des Menschenhirns*'. Spatz assumed that evolutionary progress could be expected for particular cortical regions displaying a late development in ontogeny and phylogeny, combined with a tendency toward *promination* and '*impression*' upon the bones of the base of the cranial cavity. In Spatz' opinion, the '*basal neocortex*' of the frontal and temporal lobes represents a surface region of this type and, moreover, may be interpreted as especially significant for the higher psychic functions of man. Spatz (1964) summarized his views in the statement: '*Die Wahrscheinlichkeit spricht dafür, dass im spät entwickelten und impressionsfähigen basalen Neocortex, der in Beziehung zu höchsten menschlichen Fähigkeiten steht, der Keim zu einer weiteren Progression gelegen ist*'. Nevertheless, he cautiously added the following comment: '*es handelt sich um eine aus Tatsachen der Vergangenheit des Menschenhirns abgeleitete, hypothetische Wahrscheinlichkeitsprognose, und um nichts mehr. Biologischen Prognosen fehlt die Exaktheit, die den Prognosen in den anorganischen Naturwissenschaften unter Umständen eigen ist*'.

Being somewhat less optimistic, I have instead considered the third alternative, namely that of a *decline* which could be manifested by a *progressive schizophrenic degeneration of the human brain* (K., 1961, p. 511; 1966b, p. 353). Such degeneration[335] might possibly be based on still

---

[335] It should be understood that, from a rigorous logical and semantic viewpoint, no *axiologic evaluation* whatsoever is based on reason, but any such valuation involves affectivity (emotion), respectively teleologic concepts based upon affectivity (choice, preference, like or dislike, purpose). This was already clearly recognized by David Hume (cf. K., 1961, p. 454f.). Thus, strictly speaking, any evaluation implying a '*higher*' or '*lower*' (i. e. better or worse) status of an organism, of a racial group, of a pattern of events ('condition', 'progress', 'decline', 'degeneration', etc.) is not purely 'scientific'. Yet, for practical purposes, the application of axiologic evaluations becomes unavoidable not only in everyday common-sense verbal usage, but also in 'scientific thought'. This is particularly

poorly understood genetic trends not controllable by the corrective capabilities of the obtaining multistable organic mechanisms. Although perhaps compatible with an undefined period of survival, this deterioration[336] might finally lead to extinction of the degenerate entire human population. '*Decline*' of the envisaged type does not necessarily involve grossly as well as histologically detectable 'quantitative' reductions but could be restricted to the ultrastructural or macromolecular level.

---

the case in medicine, which requires the concepts of *pathologic states* and of '*disease*' (cf. K. 1961, p.500f.). Here, again, the definition and classification of mental diseases become especially difficult and arbitrary (cf. also K., 1961, p.511). This quandary is amusingly illustrated by a recent psychiatric treatise asking: '*is mental illness an illness?*' and claiming that mental illness is a '*myth*' (Szasz, T.S.: The myth of mental illness. Hoeber-Harper, New York 1961). Instead, the author, a competent psychiatrist, builds up a theory of personal conduct, introducing a game-playing model of human behavior. His argumentation is logically consistent and semantically fully tenable. Yet, I consider the traditional system of psychiatric nosological concepts to be a far more justifiable, useful and appropriate fiction than that represented by the system which Szasz proposes. Thus, I do not believe that the attempt undertaken by this author provides a better or more suitable substitute.

[336] Thus, contemporary pictorial and plastic art may be evaluated as displaying schizophrenic trends closely similar to those manifested by the artistic activities of the insane (cf., e.g., Prinzhorn, H., *Bildnerei der Geisteskranken*, Springer, Berlin 1922). Much the same could be said about 'modern' music (Barber, Hindemith, Khatchaturian, Piston, Prokofieff, Shostakovitch, Stravinsky, and many others whose kakophonies might be considered expressions of warped mentality). The recent publication '*The Greening of America*' by C.A.Reich (1970), glorifying a so-called '*Consciousness III*' of the new generation, represents a striking paradigm for the following apt characterization by the outstanding psychiatrist E.Kretschmer (1926): '*Im schizophrenen Denken finden wir in manchen Fällen die Abbildungsvorgänge so weit regressiv abgebaut, dass nicht nur einzelne Mechanismen, sondern weite Zusammenhänge des primitiven Weltbildes wieder vor uns lebendig werden können, wobei allerdings grosse Komplexe des fertigen Kulturdenkens dazwischen stehen bleiben und mit den primitiven Mechanismen komplizierte Interferenzen bilden. Es gibt wohl keinen der hauptsächlichen Bild- und Affektmechanismen, wie wir sie bei den Primitiven beschrieben haben, den wir nicht ausgiebig bei Schizophrenen wiederfinden könnten*'.

Prinzhorn (1922) remains very restrained and cautious in his conclusions. He rightly points out that, because schizophrenic insanes tend to express themselves in the manner of modern art, or because modern artists create works comparable to those produced by schizophrenes, it is not logically justified to deduce that said modern artists must *necessarily* be considered schizophrenic. Yet, he feels compelled to admit '*dass die Verwandtschaft zwischen dem schizophrenen Weltgefühl und dem in der letzten Kunst sich offenbarenden nur in den gleichen Worten zu schildern ist*'. In his summary, Prinzhorn states: '*Ungeübte Geisteskranke, besonders Schizophrene, schaffen nicht selten Bildwerke, die weit in den Bereich ernster Kunst ragen*

Since none of the foregoing answers to the query concerning future evolutionary events can be supported by sufficiently cogent arguments, and since I do not choose to prophesy, the question may here be left open: ''Αλλ' ἐκδιδάσκει πάνθ' ὁ γηράσκων χρόνος'. (AESCHYLUS, *Prometheus Bound*, 982)

'...*denn, was geschehn soll, ist schon vollendet*'.

(*sic fere*, HÖLDERLIN)

'ἅπανθ' ὁ μακρὸς κἀναρίθμητος χρόνος φύει τ'ἄδηλα καὶ φανέντα κρυπτεται'

(SOPHOCLES, *Ajax*, 646–647)

---

*und im einzelnen oft überraschende Ähnlichkeiten zeigen mit Bildwerken der Kinder, der Primitiven und vieler Kulturzeiten. Die engste Verwandtschaft aber besteht zu der Kunst unserer Zeit...*'.

With respect to contemporary '*Weltgefühl*' and its expression by artists such as DALI, PICASSO, and their kind, it remains true that, at earlier periods preceding the 19th century expansion of western civilization, artists like HIERONYMUS BOSCH (ca. 1460–1516), PIETER BRUEGHEL Senior (ca. 1525–1569) and Junior (1564–1638), and others, displayed what could be evaluated as schizophrenic trends. These latter were also strongly manifested by the then still dominating system of religious superstition with its obsessions, its sadomasochism glorified by DANTE (1265–1321), and its savage 'justice', as, *inter alia*, documented by the abominations of the *Malleus maleficarum* (1487) dealing with the 'crime' of alleged witchcraft.

Nevertheless, the generalized present-day tendencies toward split mentality, much more incongruous by contrast with the attained high level of factual and technological knowledge, might be interpreted, on 'circumstantial evidence', as a final bout *('Schub')* of generalized schizophrenia following a temporary remission in the period between the era of LOCKE (1632–1704) or VOLTAIRE (1694–1778) and World War I. Considering the moral and economic-ecologic bankruptcy of the 'capitalistic' and 'communistic' 'establishments', the savageries of the Fascistic and Marxistic superstitions, the mental shoddiness of the comprehensibly rebellious 'new generation', and the unmitigated barbarism of the backward or retarded nations, it seems perhaps appropriate to recall the following pertinent remark by another esteemed old friend of mine, EILHARD v. DOMARUS (1893–1958): '*Die normalen Denkprinzipien genügen, um auch Schizophren-Krankhaftes hervorzubringen. Ein Kriterium, um zu entscheiden, was "noch" als gesund, und was "schon" als schizophren zu gelten habe, liesse sich wohl kaum am Denken selbst, wennschon vielleicht an den Denkresultaten gewinnen*' (v. DOMARUS, 1925).

# 9. References to Chapter VI

ADACHI, B.: Das Arteriensystem der Japaner. 2 vols. (Maruzen, Kyoto 1928).

ADAM, K.: Der III. Ventrikel und die mikroskopische Struktur seiner Wände bei *Lampetra* (Petromyzon) *fluviatilis* L. und *Myxine glutinosa* L., nebst einigen Bemerkungen über das Infundibularorgan von *Branchiostoma* (Amphioxus) *lanceolatum* Pall; in ARIËNS KAPPERS Progr. in Neurobiology, pp.146–158 (Elsevier, Amsterdam 1956).

ADELMANN, H.B.: The embryologic treatises of *Hieronymus Fabricius ab Aquapendente* (Cornell University Press, Ithaca 1942).

ADELMANN, H.B.: *Marcello Malpighi* and the evolution of embryology (Cornell University Press, Ithaca 1966).

AIRD, R.B.: Muscular atrophies and dystrophies; in BAKER Clinical neurology, 2nd ed., vol.4, pp.1743–1810 (Hoeber-Harper, New York 1962).

ALOISI, M.: La sede ed il limite rostrale del glicogeno nel nevrasse in mammiferi durante lo sviluppo. Bol. Soc. ital. Biol. sper. *7:* 1–4 (1932).

ALPERS, B.J. and FORSTER, F.M.: Arteriovenous aneurysm of the great cerebral vein and arteries of *circle of Willis*. Arch. Neurol. Psychiat. *54:* 181–185 (1945).

AMBLER, M.; POGACAR, S. and SIDMAN, R.: *Lhermitte-Duclos disease* (granule cell hypertrophy of the cerebellum). Pathological analysis of the first familial cases. J. Neuropath. exp. Neurol. *28:* 622–647 (1969).

ANDY, O.J. and STEPHAN, H.: The septum of the cat (Thomas, Springfield 1964).

ANDY, O.J. and STEPHAN, H.: Phylogeny of the primate septum telencephali; in HASSLER and STEPHAN Evolution of the forebrain, pp.389–399 (Thieme, Stuttgart 1966).

ANDY, O.J. and STEPHAN, H.: The septum of the human brain. J. comp. Neurol. *133:* 383–409 (1968).

ANONYMOUS: Central dogma reversed. Nature, Lond. *226:* 1198–1199 (1970). Cf. also BALTIMORE, 1970, and TEMIN and MIZUTANI, 1970.

ANTHONY, R.: Leçons sur le cerveau (Doin, Paris 1928).

ANTON, G. and ZINGERLE, H.: Genaue Beschreibung eines Falles von beiderseitigem Kleinhirnmangel. Arch. Psychiat. Nervenkrankh. *54:* 8–75 (1914).

AREY, L.B.: Developmental anatomy. A textbook and laboratory manual of embryology, 6th ed. (Saunders, Philadelphia 1954).

ARISTOTLE: Generation of animals (Greek text with transl. by A.L.PECK; Loeb Classical Library, Harvard University Press 1943).

ASHBY, W.Ross: Design for a brain (Wiley, New York 1952, 1960).

ASHBY, W.Ross: An introduction to cybernetics (Wiley, New York 1956, 1957).

BADÍNEZ, O.S. y AGUIRRE, C.N.: Contribución al estudio del desarrollo de las comisuras del telencephalo en la especie humana y del significado morfologico del septum lucidum. Biológica (Fac. de Med., Univ. de Chile) *16/17:* 13–43 (1953).

BAER, K.E. VON: Über Entwicklungsgeschichte der Thiere. Beobachtung und Reflexion. 3 vols. (Bornträger, Königsberg 1828).

BAKER, R.C.: The early development of the ventral part of the neural plate of Amblystoma. J. comp. Neur. *44:* 1–27 (1927).

BALTIMORE, D.: RNA-dependent DNA polymerase in virions of RNA tumour viruses. Nature, Lond. *226:* 1209–1211 (1970).

BATSON, O.V.: The function of the vertebral veins and their role in the spread of metastases. Ann. Surg. *112:* 138–149 (1940).

BEASLEY, A.B. and KUHLENBECK, H.: Some observations on the layers of the dura mater at the cranio-vertebral transition in the human newborn (abstract). Anat. Rec. *154:* 315 (1966).

BECCARI, N.: Intorno al primo differenciamento dei nuclei motori dei nervi cranici. Mon. zool. ital. *34:* 161–166 (1923).

BECCARI, N.: Neurologia comparata anatomo-funzionale dei vertebrati, compreso l'uomo (Sanzoni, Firenze 1943).

BECKER, H.: Über Hirngefässausschaltungen. II. Intrakranielle Gefässverschlüsse; Über experimentelle Hydranencephalie (Blasenhirn). Dtsch. Z. Nervenheilk. *161:* 446–505 (1949).

BELL, E.T.: Mathematics, queen and servant of science (McGraw-Hill, New York 1951).

BERGQUIST, H.: Zur Morphologie des Zwischenhirns bei niederen Vertebraten. Acta zool. *13:* 57–303 (1932).

BERGQUIST, H.: Studies on the cerebral tube in vertebrates. The neuromeres. Acta. zool. *33:* 117–187 (1952).

BERGQUIST, H.: Die Neuromerie. Anat. Anz. *102:* 449–456 (1956).

BERGQUIST, H.: Mitotic activity during successive migrations in the diencephalon of chick embryos. Experientia *13:* 84–86 (1957).

BERGQUIST, H.: Die Entwicklung des Diencephalons im Lichte neuer Forschung. Progr. Brain Res. *5:* 223–229 (1964a).

BERGQUIST, H.: The formation of the front part of the neural tube. Experientia *20:* 92–97 (1964b).

BERGQUIST, H.: Neue Befunde über die frühe Ontogenese des ZNS; in HASSLER and STEPHAN Evolution of the forebrain, pp.175–179 (Thieme, Stuttgart 1966).

BERGQUIST, H. and KÄLLEN, B.: Studies on the topography of the migration areas in the vertebrate brain. Acta anat. *17:* 353–369 (1953a).

BERGQUIST, H. and KÄLLEN, B.: On the development of neuromeres to migration areas in the vertebrate cerebral tube. Acta anat. *18:* 65–73 (1953b).

BERGQUIST, H. and KÄLLEN, B.: Notes on the early histogenesis and morphogenesis of the central nervous system in vertebrates. J. comp. Neurol. *100:* 627–659 (1954).

BING, R. und BURCKHARDT, R.: Das Centralnervensystem von Ceratodus Forsteri. *Semon's* zoolog. Forschungsreisen, Bd.I/5. Jenaische Denkschr. Bd.4 (Fischer, Jena 1905).

BISHOP, M.: The nervous system of a two-headed pig embryo. J. comp. Neurol. *32:* 379–428 (1920–1921).

BLACK, P. and MYERS, R.E.: Visual function of the forebrain commissures in the chimpanzee. Science *146:* 799–800 (1964).

BLINKOV, S.M. und GLEZER, I.I.: Das Zentralnervensystem in Zahlen und Tabellen (Fischer, Jena 1968).

BLINKOV, S.M. and GLEZER, I.I.: The human brain in figures and tables. A quantitative handbook (Plenum Press, New York 1968).

BLOOM, W.: Electron microscopy of chromosomal changes in Ambystoma somatic cells during the mitotic cycle. Anat. Rec. *167:* 253–275 (1970).

BONE, Qu.: The central nervous system in Amphioxus. J. comp. Neur. *115:* 27–64 (1960).

BONIN, G. VON: The evolution of the human brain (University of Chicago Press, Chicago 1963).

BOULDER COMMITTEE: Embryonic vertebrate central nervous system. Revised terminology. Anat. Rec. *166:* 257–262 (1970).

BRAAK, H.: Das Ependym der Hirnventrikel von Chimaera monstrosa (mit besonderer Berücksichtigung des Organon vasculosum praeopticum). Z. Zellforsch., *60:* 582–608 (1963).

BRAIN, W.R. (Lord BRAIN): Diseases of the nervous system, 6th ed. (Oxford University Press, London 1962).

BRANDT, W.: Grundzüge einer Konstitutionsanatomie (Springer, Berlin 1931).

BRIDGMAN, P.W.: The nature of physical theory (Princeton University Press, Princeton 1936).

BRITTEN, R.J. and DAVIDSON, E.H.: Gene regulation for higher cells. A theory. Science *165:* 349–357 (1969).

BROCKHAUS, H.: Zur feineren Anatomie des Septum und des Striatum. J. Psychol. Neurol. *51:* 1–56 (1942).

BRODMANN, K.: Vergleichende Lokalisationslehre der Grosshirnrinde (Barth, Leipzig 1909).

BURCKHARDT, R.: Untersuchungen am Hirn und Geruchsorgan von Triton und Ichthyophis. Z. wiss. Zool. *52:* 369–403 (1891).

BURCKHARDT, R.: Der Bauplan des Wirbelthiergehirns. Morph. Arb. herausg. v. *Schwalbe,* Bd.4, S.131–149 (Fischer, Jena 1904).

BURDA, D.J.: Developmental aspects of intracranial arterial blood supply in the alligator brain. J. comp. Neurol. *135:* 369–380 (1969).

BURR, H.S.: Regeneration in the brain of Amblystoma. J. comp. Neurol. *26:* 203–211 (1916).

CAMPBELL, A.W.: Histological studies on the localization of cerebral function (Cambridge University Press, London 1905).

CAMPBELL, B.: Re-interpretation of the structure of the cerebral cortex (abstract). Anat. Rec. *109:* 277 (1951).

CHOVNICK, A. (ed.): Biological clocks. Cold Spring Harbor Symposia on Quantitative Biology, vol.25 (Cold Spring Harbor Laboratory, New York 1960).

CHRIST, J.F.: Derivation and boundaries of the hypothalamus, with atlas of hypothalamic grisea; in HAYMAKER, ANDERSON and NAUTA The hypothalamus, chapt.2, p.13–60 (Thomas, Springfield 1969).

CHURCHILL, J.A. and MASLAND, R.L.: Infantile cerebral palsy; in BAKER Clinical neurology, 2nd ed., vol.4, pp.1994–2019 (Hoeber-Harper, New York 1962).

CLARA, M.: Das Nervensystem des Menschen, 3.Aufl. (Barth, Leipzig 1959).

CLARK, W.E.LE GROS: The antecedents of man. An introduction to the evolution of primates (Quadrangle Books, Chicago 1960).

COBB, S.: Brain size. Arch. Neurol. *12:* 555–561 (1965).

COGHILL, G.E.: Anatomy and the problem of behaviour (Cambridge Univers. Press, Cambridge 1929).

COMBETTE, M. (1831); Quoted by RUBINSTEIN and FREEMAN (1940).

CONEL, J.L.: The development of the brain of Bdellostoma Stouti. I.External growth changes. J. comp. Neurol. *47:* 343–403 (1929).

CONEL, J.L.: The development of the brain of Bdellostoma Stouti. II.Internal growth changes. J. comp. Neurol. *52:* 365–499 (1931).

COOPER, E.R.A.: An anatomical study of hydrocephalus. Acta anat., suppl.2 (Karger, Basel 1963).

CRAIGIE, E.H.: The cell structure in the hemisphere of the humming bird. J. comp. Neurol. *56:* 135–168 (1932).

CROSBY, E.C.: The forebrain of *Alligator mississippiensis*. J. comp. Neurol. *27:* 325–402 (1917).

CROSBY, E.C.; JONGE, B.R. DE, and SCHNEIDER, R.C.: Evidence for some trends in the phylogenetic development of the vertebrate telencephalon; in HASSLER and STEPHAN Evolution of the forebrain, pp.117–135 (Thieme, Stuttgart 1966).

CROSBY, E.C. and SHOWERS, M.J.: Comparative anatomy of the preoptic and hypothalamic areas; in HAYMAKER, ANDERSON and NAUTA The hypothalamus, chapt.3, pp.61–135 (Thomas, Springfield 1969).

CURWEN, A.O.: The telencephalon of *Tupinambis nigropunctatus*. II.Amygdala. J. comp. Neurol. *71:* 613–636 (1939).

CUTLER, R.W.P.; PAGE, L.; GALICICH, J., and WATTERS, G.V.: Formation and absorption of cerebrospinal fluid. Brain *91:* 707–720 (1968).

DABELOW, A.: Über Korrelationen in der phyletischen Entwicklung der Schädelform. II. Beziehungen zwischen Gehirn und Schädelbasisform bei den Mammaliern. Morph. Jb. *67:* 84–133 (1931).

DAMMERMAN, K.W.: Der Saccus vasculosus der Fische, ein Tieforgan. Z. wiss. Zool. *96:* 654–726 (1910).

DANFORTH, C.H.: Hair and its relation to questions of homology and phylogeny. Amer. J. Anat. *36:* 47–68 (1925–1926).

DARESTE, C.: Recherches sur la production artificielles des monstruosités, 2nd ed. (Paris 1891; quoted by *Stockard*, 1920–1921).

DART, R.A.: The misuse of the term 'visceral'. J. Anat. *56:* 177–188 (1922).

DART, R.A. and SHELLSHEAR, J.L.: A new interpretation of the morphology of the nervous system (abstract). Anat. Rec. *21:* 54–55 (1921).

DART, R.A. and SHELLSHEAR, J.L.: The origin of the motor neuroblasts of the anterior cornu of the neural tube. J. Anat. *56:* 77–95 (1921–1922).

DAVIS, B.D.; DULBECCO, R.; EISEN, H.N.; GINSBERG, H.S., and WOOD, W.B.: Microbiology (Hoeber, Harper & Row, New York 1968).

DEVRIESE, B.: Sur la signification morphologique des artères cérébrales. Arch. Biol. *21:* 359–457 (1905).

DOMARUS, E. v.: Über die Beziehungen des normalen zum schizophrenen Denken. Arch. Psychiat. Nervenkr. *74:* 641–646 (1925).

DONALDSON, H.H.: Growth of the brain. Study of nervous system in relation to education (Scribner, New York 1895).

DONALDSON, H.H.: The rat. Reference tables and data for the albino rat *(Mus norvegicus albinus)* and the Norway rat *(Mus norvegicus)*. Mem. Wistar Inst. Anat. Biol., vol.6, 2nd ed. (Wistar Institute Press, Philadelphia 1924).

DORN, E.: Der Saccus vasculosus; in VON MÖLLENDORFF and BARGMANN Handb. d. mikr. Anat. des Menschen, Bd.IV, Teil 2, pp.140–185 (Springer, Berlin 1955).

DORN, E.: Über das Zwischenhirn-Hypophysensystem von *Protopterus annectens* (zugleich ein Beitrag zum Problem des Saccus vasculosus). Z. Zellforsch. *46:* 108–114 (1957).

DORN, E.: Über den Feinbau der Paraphyse von *Protopterus annectens*. Z. Zellforsch. *46:* 115–120 (1957).

DRATMAN, M.B.: Fate of thyroxin iodine in the neuraxis of *Rana pipiens* during metamorphosis. Anat. Rec. *157:* 236–237 (1967).

DRATMAN, M.B. and KUHLENBECK, H.: Interactions of thyroxin with developing skeletal tissues of the newborn rat (abstract). Anat. Rec. *163:* 180 (1969).

DRATMAN, M.B.; KUHLENBECK, H., and CRUTCHFIELD, F.: Fate of thyroxin in the neuraxis of the newborn rat (abstract). Anat. Rec. *160:* 341 (1968).

DRATMAN, M.B.; KUHLENBECK, H., and JACOB, G.: Fate of thyroxine iodine in the neuraxis of *Rana pipiens* during metamorphosis. Radioautographic studies. Confin. neurol. *31:* 226–237 (1969).

DUNN, H.L.: The growth of the central nervous system in the human fetus as expressed by graphic analysis and empirical formulae. J. comp. Neurol. *33:* 405–491 (1921).

DÜRKEN, B.: Grundriss der Entwicklungsmechanik (Bornträger, Berlin 1929).

DURSY, E.: Zur Entwicklungsgeschichte des Kopfes des Menschen und der höheren Wirbelthiere (Laupp, Tübingen 1869).

DURWARD, A.: Some observations on the development of the corpus striatum of birds, with special reference to certain stages in the common sparrow *(P. domesticus)*. J. Anat. *68:* 492–499 (1934).

EBBESON, S.O.E. and SCHROEDER, D.M.: Connections of Nurse Shark's Telencephalon. Science *173:* 254–256 (1971).

ECONOMO, C.v.: Ein Koeffizient für die Organisationshöhe der Grosshirnrinde. Klin. Wochenschr. *5:* 593–595 (1926).

ECONOMO, C.v.: Die progressive Zerebration, ein Naturprinzip. Reprint, 13 p., from Wien. med. Wochenschr. No.28 (1928).

ECONOMO, C.v.: Zellaufbau der Grosshirnrinde des Menschen (Springer, Berlin 1929).

ECONOMO, C.v. and KOSKINAS, G.N.: Die Cytoarchitektonik der Hirnrinde des erwachsenen Menschen. Textband und Atlas (Springer, Berlin 1925).

EDINGER, L.: Über die Herkunft des Hirnmantels in der Tierreihe. Berl. klin. Wschr. *43:* 1–13 (1905).

EDINGER, L.: Über die Einteilung des Cerebellums. Anat. Anz. *35:* 319–323 (1910).

EDINGER, L.: Vorlesungen über den Bau der nervösen Zentralorgane des Menschen und der Tiere. I. Das Zentralnervensystem des Menschen und der Säugetiere, 8th ed., 1911. Bd. II. Vergleichende Anatomie des Gehirns, 7th ed., 1908 (Vogel, Leipzig 1908–1911).

EDINGER, L.: Einführung in die Lehre vom Bau und den Verrichtungen des Nervensystems (Vogel, Leipzig 1912).

EDINGER, L. und FISCHER, B.: Ein Mensch ohne Grosshirn. Arch. ges. Physiol. *152:* 535–561 (1913).

EDINGER, T.: Anthropocentric misconceptions in palaeoneurology. Proc. Virchow med. Soc., New York *19:* 56–107 (1960–1962).

EIGEN, M.: Der Zeitmassstab der Natur. Jb. Max-Planck-Gesell. *1966:* 40–67 (1966).

EITSCHBERGER, E.: Entwicklung und Chemodifferenzierung des Thalamus der Ratte. Ergebn. Anat. EntwGesch. *42:* Heft 6 (1970).

ELLENBERGER, C. jr.; HANAWAY, J., and NETSKY, M.G.: Embryogenesis of the inferior olivary nucleus in the rat. A radioautographic study and a re-evaluation of the rhombic lip. J. comp. Neurol. *137:* 71–87 (1969).

ERNST, P.: Missbildungen des Nervensystems; in SCHWALBE's Handbook, vol.3, pt.2, pp.67–252 (Fischer, Jena 1909).

ESSICK, C.R.: The corpus ponto-bulbare – a hitherto undescribed nuclear mass in the human brain. Amer. J. Anat. *7:* 119–136 (1907).

ESSICK, C.R.: On the embryology of the corpus ponto-bulbare and its relation to the development of the pons. Anat. Rec. *3:* 254–257 (1909).

Essick, C.R.: The development of the nuclei pontis and the nucleus arcuatus in man. Amer. J. Anat. *13:* 25–54 (1912).

Faul, J.: The comparative ontogenetic development of the corpus striatum in Reptiles. Proc. Akad. Wetensch., Amsterdam *29:* 150–162 (1926).

Faul, J.: The ontogenetic development of the claustrum in Mammals. Proc. Akad. Wetensch., Amsterdam *29:* 642–647 (1926).

Feremutsch, K.: Die embryonale Fundamentalgliederung der Hirnrinde. Z. Anat. EntwGesch. *123:* 264–270 (1962).

Fitting, H.: Morphologie; in Strasburger Lehrbuch der Botanik, 18. Aufl., pp. 4–161 (Fischer, Jena 1931).

Flechsig, P.: Gehirn und Seele, 2. Aufl. (Veit, Leipzig 1896).

Flechsig, P.: Anatomie des menschlichen Gehirns und Rückenmarks auf myelogenetischer Grundlage (Thieme, Leipzig 1920).

Flechsig, P.: Meine myelogenetische Hirnlehre. Mit biographischer Einleitung (Springer, Berlin 1927).

Flexner, L.B.: The development of the meninges in Amphibia. A study of normal and experimental animals. Carneg. Inst. Wash., Publ. 394. Contrib. Embryol. *20:* 31–49 (1929).

Flood, P.R.: A peculiar mode of muscular innervation in Amphioxus. Light and electron microscopic studies of the so-called ventral roots. J. comp. Neurol. *126:* 181–218 (1966).

Forel, A.: Balkendefekte. In: Gesammelte Hirnanatomische Abhandlungen, pp. 225–234 (Reinhardt, München 1907).

Forel, A.: Gehirn und Seele, 13. Aufl. (Kröner, Leipzig 1922).

Fortuyn, Drooglever, A.B.: Die Ontogenie der Kerne in Zwischenhirn beim Kaninchen. Arch. Anat. Physiol., Anat. Abt. *:* 303–352 (1912).

Fortuyn, J. Drooglever: Topographical relations in the telencephalon of the sunfish *Eupomotis gibbosus*. J. comp. Neurol. *116:* 249–264 (1961).

Fortuyn, J.D. and Stefens, R.: On the anatomical relations of the intralaminar and midline cells of the thalamus. EEG Clin. Neurophysiol. *3:* 393–400 (1951).

Franz, V.: Das Gefässsystem der Acranier. In: Handb. d. vergl. Anat. d. Wirbeltiere, Bolk, L. *et al.* eds., vol. 6, p. 451–466 (Urban & Schwarzenberg, Berlin-Wien 1933).

Frederikse, A.: The lizard's brain. Acad. Proefschr. (Callenbach, Nijkerk 1931).

Friedmann, Th., and Roblin, R.: Gene therapy for human genetic disease. Science *175:* 949–955 (1972).

Frisch, D. and Farbman, A.I.: Development of order during ciliogenesis. Anat. Rec. *162:* 221–232 (1968).

Gage, S.H.: Glycogen in the nervous system of vertebrates. J. comp. Neurol. *27:* 451–465 (1917).

Gamper, E.: Bau und Leistungen eines menschlichen Mittelhirnwesens (Arhinencephalie mit Encephalocele). Zugleich ein Beitrag zur Teratologie und Fasersystematik). Z. ges. Neurol. Psychiat. *102:* 154–235; *104:* 49–120 (1926).

Gaskell, W.H.: On the structure, distribution and function of the nerves which innervate the visceral and vascular systems. J. Physiol *7:* 1–81 (1886).

Gaskell, W.H.: On the relations between the structure, function and origin of the cranial nerves, together with a theory on the origin of the nervous system of vertebrata. J. Physiol. *10:* 153–211 (1889).

GAZZANGA, M.S.: Psychologic properties of the disconnected hemispheres in man. Science *150:* 372 (1965).

GELDEREN, C. VAN: Zur vergleichenden Anatomie der Sinus durae matris. Anat. Anz. *48:* 472–480 (1924).

GELDEREN, C. VAN: Über die Entwicklung der Hirnhäute bei Teleostiern. Anat. Anz. *60:* 48–57 (1925).

GELDEREN, C. VAN: Die Morphologie der Sinus durae matris. Die vergleichende Ontogenie der Hirnhäute mit besonderer Berücksichtigung der neurokraniellen Venen. Z. Anat. EntwGesch. *74:* 432–508 (1924); *78:* 339–489 (1926).

GELDEREN, C. VAN: Venensystem; in BOLK *et al.* Handb. d. vergl. Anat. d. Wirbeltiere, vol.6, pp.685–744 (Urban & Schwarzenberg, Berlin 1933).

GERLACH, J.: Über das Gehirn von *Protopterus annectens.* Ein Beitrag zur Morphologie des Dipnoerhirnes. Anat. Anz. *75:* 310–406 (1933).

GERLACH, J.: Beiträge zur vergleichenden Morphologie des Selachierhirnes. Anat. Anz. *96:* 79–165 (1947).

GERLACH, J.: Grundriss der Neurochirurgie (Steinkopf, Darmstadt 1967).

GERLACH, J.: Individualtod – Partialtod – *Vita reducta.* Probleme der Definition und Diagnose des Todes in der Medizin von heute. Münch. med. Wochenschr. *110:* 980–983 (1968a).

GERLACH, J.: Die Definition des Todes in ihrer heutigen Problematik für Medizin und Rechtslehre. Arztrecht *6:* 83–86 (1968b).

GERLACH, J.; JENSEN, H.P.; KOOS, W. und KRAUS, H.: Pädiatrische Neurochirurgie (Thieme, Stuttgart 1967).

GIBBS, C.J.; GAJDUSEK, D.M., *et al.: Creutzfeldt-Jakob disease* (spongiform encephalopathy). Transmission to the chimpanzee. Science *161:* 388–389 (1968).

GILBERT, M.S.: The early development of the human diencephalon. J. comp. Neurol. *62:* 81–115 (1935).

GILLILAN, L.A.: A comparative study of the extrinsic and intrinsic arterial blood supply to brains of submammalian vertebrates. J. comp. Neurol. *130:* 175–196 (1967).

GILLILAN, L.A. and MARKESBERY, W.M.: Arteriovenous shunts in the blood supply to the brains of some common laboratory animals with special attention to the rete mirabile conjugatum in the cat. J. comp. Neurol. *121:* 305–311 (1963).

GILLMAN, J.; GILBERT, C.; GILLMAN, T., and SPENCE, I.: A preliminary report on hydrocephalus, spina bifida, and other congenital anomalies in the rat produced by trypan blue. Sth. afr. J. med. Sci. *15:* 125–135 (1948).

GIROUD, A.; DELMAS, A.; PROST, H. et LEFEBVRES, J.: Malformations encephaliques par carence en acide pantothénique et leur interpretation. Acta anat. *29:* 209–377 (1957).

GOLDMAN, D.E.: Potential impedance and rectification in membranes. J. gen. Physiol. *27:* 37–60 (1943–44).

GOLDSTEIN, K.: Beiträge zur Entwicklungsgeschichte des menschlichen Gehirns. I. Die erste Entwicklung der grossen Gehirncommissuren und die 'Verwachsung' von Thalamus und Striatum. Arch. Anat. EntwGesch., Anat. Abtg.: 29–60 (1903).

GOLDSTEIN, K.: Zur Frage der Existenzberechtigung der sogenannten Bogenfurchen des embryonalen menschlichen Gehirnes, nebst einigen weiteren Bemerkungen zur Entwicklung des Balkens und der Capsula interna. Anat. Anz. *24:* 579–595 (1904).

GOODWIN, B.: Physics and biology closing the gap. Sci. Res. *4* (16): 17–20 (1969). Anonymous review, signed R.B., summarizing GOODWIN's theory.

GOODWIN, B. and COHEN, M.H.: A phase-shift model for the spatial and temporal organization of developing systems. J. theoret. Biol. *25:* 49–107 (1969).

GORONOWITSCH, N.: Das Gehirn und die Cranialnerven von *Acipenser ruthenus.* Morphol. Jb. *13:* 427–574 (1888).

GORONOWITSCH, N.: Untersuchungen über die Entwicklung der sog. 'Ganglienleisten' im Kopfe der Vogelembryonen. Morphol. Jb. *20:* 187–259 (1893).

GRÄPER, L.: Die Rhombomeren und ihre Nervenbeziehungen. Arch. mikr. Anat. *83:* 371–426 (1913).

GRÄPER, L.: Die Primitiventwicklung des Hühnchens nach stereokinematographischen Untersuchungen, kontrolliert durch vitale Farbmarkierungen und verglichen mit der Entwicklung anderer Wirbeltiere. Arch. Entwmech. *116:* 382–429 (1929).

GREEN, J.D.: The comparative anatomy of the portal vascular system and of the innervation of the hypophysis; in HARRIS and DONOVAN The Pituitary gland, vol.1, pp. 127–146 (University of California Press, Berkeley 1966).

GREVING, R.: Die vegetativen Zentren im Zwischenhirn; in MÜLLER Lebensnerven und Lebenstriebe, pp.115–209 (Springer, Berlin 1931).

GRÖNBERG, G.: Die Ontogenese eines niederen Säugergehirns nach Untersuchungen an *Erinaceus europaeus.* Zool. Jb. Abt. Anat. *15:* 261–384 (1901).

GRÜNTHAL, E.: Untersuchungen zur Ontogenese und über den Bauplan des Gehirns; in FEREMUTSCH und GRÜNTHAL Beiträge zur Entw.-Gesch. u. normalen Anat. des Gehirns. Bibl. Psychiat. et Neurol., Fasc., vol.91, pp.5–32 (Karger, Basel 1952).

GURDON, J.B. and UEHLINGER, V.: 'Fertile' intestine nuclei. Nature *210:* 1240–1241 (1966).

GURWITSCH, A.: Über die nichtmateriellen Factoren embryonaler Formgestaltung. Z. Morph. Anthrop. *18* (Festschrift SCHWALBE): 111–142 (1914).

HAECKEL, E.: Natürliche Schöpfungs-Geschichte. Gemeinverständliche wissenschaftliche Vorträge über die Entwicklungslehre im Allgemeinen und diejenige von DARWIN, GOETHE und LAMARCK im Besonderen, 9th ed. (Reimer, Berlin 1898).

HAECKEL, E.: Die Lebenswunder (Kröner, Stuttgart 1906).

HAFFERL, A.: Das Arteriensystem; in BOLK *et al.* Handb. d. vergl. Anat. d. Wirbeltiere, vol.6, pp.563–684 (Urban & Schwarzenberg, Berlin 1933).

HAHN, O. und KUHLENBECK, H.: Defektbildungen des Septum pellucidum im Enzephalogramm. Fortschr. Geb. Röntgenstr. *41:* 737–742 (1930).

HALE, F.: Pigs born without eyeballs. J. Hered. *24:* 105–106 (1933).

HALLER, GRAF: Die epithelialen Gebilde am Gehirn der Wirbeltiere. Z. Anat. EntwGesch. *63:* 118–202 (1922).

HALLER, GRAF: Die Gliederung des Zwischen- und Mittelhirns der Wirbeltiere. Morph. Jb. *63:* 359–407 (1929).

HALLERVORDEN, J. und SPATZ, H.: Eigenartige Erkrankung im extrapyramidalen System mit besonderer Beteiligung des Globus pallidus und der Substantia nigra. Z. ges. Neurol. Psychiat. *79:* 253–302 (1922).

HAMBURGER, V.: A manual of experimental embryology (University of Chicago Press, Chicago 1942).

HAMBURGER, V. and HAMILTON, H.L.: A series of normal stages in the development of the chick embryo. J. Morph. *88:* 49–92 (1951).

HANDLER, P. (ed.): Biology and the future of man (Oxford University Press, New York 1970).

HANSEN, A.: Repetitorium der Botanik für Mediziner (Töpelmann, Giessen 1914).

HARVEY, S.C. and BURR, H.S.: The development of the meninges. Arch. Neurol. Psychiat. *29:* 683–690 (1926).

HAYMAKER, W. (ed.): The founders of neurology (Thomas, Springfield 1953).

HAYMAKER, W.: BING's local diagnosis in neurological diseases (Mosby, St.Louis 1956, 1969).

HAYMAKER, W. and KUHLENBECK, H.: Disorders of the brain stem and its cranial nerves; in BAKER Clinical Neurology, 2nd ed., chapt. 29, pp. 1456–1526, vol. 3 (Hoeber-Harper, New York 1962).

HAYMAKER, W.; ANDERSON, E., and NAUTA, W.J.H. (eds.): The hypothalamus (Thomas, Springfield 1969).

HEIER, P.: Fundamental principles in the structure of the brain. A study of the brain of *Petromyzon fluviatilis* (Hakan, Lund 1948; also Acta anat., suppl. VI).

HENKEL, E. and KIRSCHE, W.: Hydrocephalus internus bei *Lacerta agilis* L. Zool. Anz. *179:* 233–242 (1967).

HENRICH, G.: Untersuchungen über die Anlage des Grosshirns beim Hühnchen. Sitz. Ber. Ges. Morph. Physiol., München *12:* 96–134 (1896).

HERRICK, C.J.: The cranial and first spinal nerves of Menidia. A contribution upon the nerve components of the bony fishes. J. comp. Neurol. *9:* 153–455 (1899).

HERRICK, C.J.: The doctrine of nerve components and some of its applications. J. comp. Neurol. *13:* 301–312 (1903).

HERRICK, C.J.: The morphology of the forebrain in Amphibia and Reptilia. J. comp. Neurol. *20:* 413–547 (1910).

HERRICK, C.J.: The morphology of the cerebral hemispheres in Amphibia. Anat. Anz. *36:* 645–652 (1910).

HERRICK, C.J.: A sketch of the origin of the cerebral hemispheres. J. comp. Neurol. *32:* 429–454 (1921).

HERRICK, C.J.: What are viscera? J. Anat. *56:* 167–176 (1922).

HERRICK, C.J.: Brain of rats and men (University of Chicago Press, Chicago 1926).

HERRICK, C.J.: An introduction to neurology, 5th ed. (Saunders, Philadelphia 1931).

HERRICK, C.J.: Morphogenesis of the brain. J. Morph. *54:* 233–258 (1933).

HERRICK, C.J.: The membranous parts of the brain, meninges and their blood vessels in Amblystoma. J. comp. Neurol. *61:* 297–346 (1935).

HERRICK, C.J.: The brain of the tiger salamander, *Ambystoma tigrinum* (University of Chicago Press, Chicago 1948).

HERRICK, C.J. and OBENCHAIN, J.B.: Notes on the anatomy of a cyclostome brain. *Ichthyomyzon concolor.* J. comp. Neurol. *23:* 635–675 (1913).

HERTWIG, O.: Lehrbuch der Entwicklungsgeschichte des Menschen und der Wirbeltiere, 10. Aufl. (Fischer, Jena 1915).

HERTWIG, R.: Lehrbuch der Zoologie, 15. Aufl. (Fischer, Jena 1931).

HERZ, E. and MEYERS, R.: The extrapyramidal diseases; in BAKER Clinical neurology, 2nd. ed., vol. 3, chapt. 26, pp. 1285–1337 (Hoeber-Harper, New York 1962).

HIBBARD, E.: The vascular supply to the central nervous system of the larval lamprey. Amer. J. Anat. *112:* 93–99 (1963).

HIKIJI, K.: Zur Anatomie des Bodens der Rautengrube beim Neugeborenen. Anat. Anz. *75:* 406–442 (1933).

HIMANGO, W.A. and LOW, F.N.: The fine structure of a lateral recess of the subarachnoid space in the rat. Anat. Rec. *171:* 1–19 (1971).

HINES, M.: The development of the telencephalon in *Sphenodon punctatum*. J. comp. Neurol. *55:* 483–537 (1923).

HIS, W.: Zur Geschichte des Gehirns, sowie der centralen und peripherischen Nervenbahnen beim menschlichen Embryo. Abh. d. math.-Phys. Kl. kgl. sächs. Ges. Wiss. *14* (8): 341–392 (1899).

HIS, W.: Die Formentwicklung des menschlichen Vorderhirns vom Ende des ersten bis zum Beginn des dritten Monats. Abh. d. math.-phys. Kl. kgl. sächs. Ges. Wiss. *15* (8): 674–736 (1889).

HIS, W.: Die Entwicklung des menschlichen Rautenhirns von Ende des ersten bis zum Beginn des dritten Monats. I. Verlängertes Mark. Abh. d. math.-phys. Kl. kgl. sächs. Ges. Wiss. Bd. *17:* 3–74 (1890–1891).

HIS, W.: Zur allgemeinen Morphologie des Gehirns. Arch. Anat. EntwGesch. *1892:* 346–383 (1892).

HIS, W.: Zur Nomenclatur des Gehirns und Rückenmarkes. Arch. Anat. EntwGesch. *1892:* 425–428 (1892).

HIS, W.: Über das frontale Ende des Gehirnrohres. Arch. Anat. EntwGesch. *1893:* 157–171 (1893).

HIS, W.: Vorschläge zur Eintheilung des Gehirns. Arch. Anat. EntwGesch *1893:* 173–179 (1893).

HIS, W.: Die Entwicklung der menschlichen und thierischen Physiognomien. Arch. Anat. EntwGesch. *1893:* 384–424 (1893).

HIS, W.: Die Anatomische Nomenclatur. Nomina Anatomica. Neurologie. Suppl. Bd. Arch. Anat. EntwGesch. *1895:* 155–177 (1895).

HIS, W.: Die Entwicklung des menschlichen Gehirns während der ersten Monate (Leipzig, Hirzel 1904).

HOCHSTETTER, F.: Beiträge zur Entwicklungsgeschichte des menschlichen Gehirns, I. u. II. Teil (Deuticke, Leipzig 1919, 1923, 1929).

HODGKIN, A.L. and KATZ, B.: The effects of sodium on the electrical activity of the giant axon in the squid. J. Physiol., Lond. *168:* 37–77 (1949).

HOFER, H.: Der Gestaltwandel der Säugetiere und Vögel, mit besonderer Berücksichtigung der Knickungstypen und der Schädelbasis. Verh. anat. Ges., Jena *99:* 102–113 (1952).

HOFER, H.: Die cranio-cerebrale Topographie bei den Affen und ihre Bedeutung für die menschliche Schädelform. Homo *5:* 52–72 (1954).

HOFMANN, M.: Zur vergleichenden Anatomie der Gehirn- und Rückenmarksarterien der Vertebraten. Z. Morph. Anthrop. *2:* 246–322 (1900).

HOFMANN, M.: Zur vergleichenden Anatomie der Gehirn- und Rückenmarksvenen der Vertebraten. Z. Morph. Anthrop. *3:* 239–299 (1901).

HOLMDAHL, D.E.: Experimentelle Untersuchungen über die Lage der Grenze zwischen primärer und sekundärer Entwicklung beim Huhn. Anat. Anz. *59:* 393–396 (1925).

HOLMDAHL, D.E.: Die zweifache Bildungsweise des zentralen Nervensystems bei den Wirbeltieren. Eine formgeschichtliche und materialgeschichtliche Analyse. Arch. Entwmech. *129:* 206–254 (1933).

HOLMDAHL, D.E.: Die zweifache Morphogenese des Vertebratenorganismus. Die primäre (indirekte) und sekundäre (direkte) Körperentwicklung. Z. mikr.-anat. Forsch. *57:* 359–392 (1951).

HOLMGREN, N.: Zur Anatomie und Histologie des Vorder- und Zwischenhirns der Knochenfische. Acta zool. *1:* 137–315 (1920).

HOLMGREN, N.: Points of view concerning forebrain morphology in lower vertebrates. J. comp. Neurol. *34:* 391–459 (1922).

HOLMGREN, N.: Points of view concerning forebrain morphology in higher vertebrates. Acta zool. *6:* 413–477 (1925).

HOLMGREN, N.: On two embryos of Myxine glutinosa. Acta zool. *27:* 1–90 (1946).

HOLMGREN, N., and VAN DER HORST, C.J.: Contributions to the morphology of the brain of Ceratodus. Acta zool. *6:* 59–165 (1925).

HOLTFRETER, J.: Die totale Exogastrulation, eine Selbstablösung des Ektoderms vom Entomesoderm. Entwicklung und funktionelles Verhalten nervenloser Organe. Arch. Entwmech. *129:* 669–793 (1933).

HOLTFRETER, J. and HAMBURGER, V.: Embryogenesis progressive differentiation. Chap. I. Amphibians, pp. 230–296; in WILLIER, WEISS and HAMBURGER Analysis of development (Saunders, Philadelphia 1955).

HOPF, L.: Introduction to the differential equations of physics. Translated by W. NEF (Dover, New York 1948).

HORII, K.: Prevention of congenital defects in the offspring of alloxandiabetic mice by insulin treatment (Japanese). Folia endocrin. japon. *39:* 988–995 (1964).

HUGHES, A.F.W.: Aspects of neural ontogeny (Academic Press, New York 1968).

HUGOSSON, R.: Morphologic and experimental studies on the development and significance of the rhombencephalic longitudinal columns. Dissertation, Lund (Ohlsson, Lund 1957).

HUXLEY, J.S.: Problems of relative growth (Dial Press, New York 1932).

INNES, J.R.M. and SAUNDERS, L.Z.: Comparative neuropathology (Academic Press, New York 1962).

JACOB, F. and MONOD, J.: Genetic regulatory mechanisms in the synthesis of proteins. J. molec. Biol. *3:* 318–356 (1961a).

JACOB, F. and MONOD, J.: On the regulation of gene actions. Cold Spr. Harb. Symp. quant. Biol. *26:* 193–211 (1961b).

JACOBSHAGEN, E.: Allgemeine vergleichende Formenlehre der Tiere (Klinkhardt, Leipzig 1925).

JAKOB, A.: Über ein dreieinhalb Monate altes Kind mit totaler Erweichung beider Grosshirnhemisphären ('Kind ohne Grosshirn'). Dtsch. Z. Nervenheilk. *117–119:* 240–265 (1931).

JAKOB, C. and ONELLI, Cl.: Vom Tierhirn zum Menschenhirn. Vergleichende morphologische, histologische und biologische Studien zur Entwicklung der Grosshirnhemisphären und ihrer Rinde (Lehmann, München 1911).

JANSEN, J.: The brain of Myxine glutinosa. J. comp. Neurol. *49:* 359–507 (1930).

JANSEN, J. and BRODAL, A.: Das Kleinhirn. Erg. zu Bd. IV/1; in v. MÖLLENDORFF und BARGMANN Handb. d. mikr. Anat. d. Menschen (Springer, Berlin 1958).

JEAN PAUL (J.P.F. RICHTER): *Dr. Katzenbergers* Badereise (1822; ed. Deutsches Verlagshaus Bong, Berlin, n.d.).

JEENER, R.: Evolution des centres diencéphaliques periventriculaires des Téléostomes. Proc. Acad. Sci., Amsterdam *33:* 755–770 (1930).

JERISON, H.J.: Brain evolution. New light on old principles. Science *170:* 1224–1225 (1970).

JOFFE, J.M.: Prenatal determinants of behaviour (Pergamon Press, New York 1969).

JOHNSTON, J.B.: The brain of Petromyzon. J. comp. Neurol. *12:* 1–87 (1902).

JOHNSTON, J.B.: An attempt to define the primitive functional divisions of the central nervous system. J. comp. Neurol. *12:* 87–106 (1902).

JOHNSTON, J.B.: The nervous system of vertebrates (Blakiston, Philadelphia 1906).

JOHNSTON, J.B.: The morphology of the forebrain vesicles in vertebrates. J. comp. Neurol. *19:* 457–539 (1909).

JOHNSTON, J.B.: The cell masses in the forebrain of the turtle, *Cistudo carolina.* J. comp. Neurol. *25:* 393–468 (1915).

JONES, A.W. and LEVI-MONTALCINI, R.: Patterns of differentiation of the nerve centers and fiber tracts in the avian cerebral hemispheres. Arch. ital. Biol. *96:* 231–284 (1958).

KAHLE, W.: Studien über die Matrixphasen und die örtlichen Reifungsunterschiede im embryonalen menschlichen Hirn. Dtsch. Z. Nervenheilk. *166:* 273–302 (1951).

KAHLE, W.: Zur Entwicklung des menschlichen Zwischenhirns. Dtsch. Z. Nervenheilk. *175:* 259–318 (1956).

KAHLE, W.: Die Entwicklung der menschlichen Grosshirnhemisphäre. Neurology series, vol. 1 (Springer, Berlin 1969).

KÄLLEN, B.: Some remarks on the ontogeny of the telencephalon in some lower vertebrates. Acta anat. *11:* 537–548 (1951a).

KÄLLEN, B.: On the ontogeny of the reptilian forebrain. Nuclear structures and ventricular sulci. J. comp. Neurol. *95:* 305–347 (1951b).

KÄLLEN, B.: On the significance of the neuromeres and similar structures in vertebrate embryos. J. Embryol. exp. Morph. *57:* 111–118 (1952).

KÄLLEN, B.: On the nuclear differentiation during ontogenesis of the avian forebrain and some notes on the amniote strio-amygdaloid complex. Acta anat. *17:* 72–83 (1953).

KÄLLEN, B.: Notes on the mode of formation of brain nuclei during ontogenesis. C.R. Ass. Anatomistes 42e Réunion, pp. 1747–1756 (1955).

KÄLLEN, B.: Early morphogenesis and pattern formation in the central nervous system; in DEHAAN-URSPRUNG Organogenesis, pp. 107–128 (Holt, New York 1965).

KÄLLEN, B. and LINDSKOG, B.: Formation and disappearance of neuromery in *Mus musculus.* Acta anat. *18:* 273–282 (1953).

KALTER, H.: Teratology of the central nervous system. Induced and spontaneous malformations of laboratory, agricultural and domestic mammals (University of Chicago Press, Chicago 1968).

KAPPERS, C.U.ARIËNS: Die vergleichende Anatomie des Nervensystems der Wirbeltiere und des Menschen, 2 vols. (Bohn, Haarlem 1920, 1921).

KAPPERS, C.U.A.: The meninges in Cyclostomes, Selachians, and Teleosts, compared with those in man. Proc. roy Acad., Amsterdam *28:* 72–80 (1925).

KAPPERS, C.U.A.: Three lectures on neurobiotaxis and other subjects (Levin & Munksgaard, Copenhagen 1928).

KAPPERS, C.U.A.: The evolution of the nervous system in invertebrates, vertebrates and man (Bohn, Haarlem 1929).

KAPPERS, C.U.A.: The forebrain arteries in Plagiostomes, Reptiles, Birds, and Monotremes. Proc. roy. Acad., Amsterdam *36:* 52–62 (1933).

KAPPERS, C.U.A.: Anatomie comparée du système nerveux, particulièrement de celui des mammifères et de l'homme. Avec la collaboration de E.H.STRASBURGER (Bohn, Haarlem, & Masson, Paris 1947).

KAPPERS, C.U.ARIËNS; HUBER, G.C., and CROSBY, E.C.: The comparative anatomy of the nervous system of vertebrates, including man, 2 vols. (Macmillan, New York 1936).

KAPPERS, C.U.A. and THEUNISSEN, W.F.: Die Phylogenese des Rhinencephalons, des Corpus striatum und der Vorderhirnkommissuren. Folia neurobiol. *1:* 173–288 (1907–1908).

KARSTEN, G.: Spermatophyta oder Samenpflanzen; in STRASBURGERS Lehrbuch der Botanik, 18.Aufl., pp.439–599 (Fischer, Jena 1931).

KIESEWALTER, C.: Zur Morphologie der Ganglienkerne im Grosshirn von *Lacerta*. Jena Z. Nat. *58:* 488–532 (1922).

KIESEWALTER, C.: Basis und Pallium. Ihre mediale Grenze am Grosshirn der Amphibien und Reptilien. Jena Z. Nat. *61:* 575–406 (1925).

KIESEWALTER, C.: Zur allgemeinen und speziellen Morphogenie des Hemisphärenhirns der Tetrapoden. Jena Z. Nat. *63:* 369–454 (1928).

KIMMEL, D.L.: Differentiation of the bulbar motor nuclei and the coincident development of associated root fibers in the rabbit. J. comp. Neurol. *72:* 83–148 (1940).

KINGSBURY, B.F.: On the brain of *Necturus maculatus*. J. comp. Neurol. *5:* 139–203 (1895).

KINGSBURY, B.F.: The extent of the floor-plate of His and its significance. J. comp. Neurol. *32:* 113–135 (1920–1921).

KINGSBURY, B.F.: The fundamental plan of the vertebrate brain. J. comp. Neurol. *34:* 461–491 (1922).

KINGSBURY, B.F.: The development of the septum medullae (mammals: cat). J. comp. Neurol. *60:* 81–107 (1934).

KIRSCHBAUM, W.R.: Agenesis of the corpus callosum and associated malformations. J. Neuropath. exp. Neurol. *6:* 78–94 (1947).

KIRSCHE, W.: Einseitiger Anophthalmus mit Gehirnmissbildung bei *Lacerta agilis* (L.) Z. mikr.-anat. Forsch. *74:* 252–273 (1965).

KIRSCHE, W.: Über postembryonale Matrixzonen im Gehirn verschiedener Vertebraten und deren Beziehung zur Hirnbauplanlehre. Z. mikr.-anat. Forsch. *77:* 313–406 (1967).

KLINGMAN, W.O.: Congenital and prenatal diseases; in BAKER Clinical neurology, 1st. ed., vol. 3, chapt. 35, pp. 1695–1729 (Hoeber-Harper, New York 1955).

KNOWLTON, V.Y.: Abnormal differentiation of embryonic avian brain centers associated with unilateral anophthalmia. Acta anat. *58:* 222–251 (1964).

KRANZ, D. und RICHTER, W.: Autoradiographische Untersuchungen über die Lokalisation der Matrixzonen des Diencephalons von juvenilen und adulten *Lebistes reticulatus* (Teleostei). Z. mikr.-anat. Forsch. *82:* 42–66 (1970).

KRANZ, D. und RICHTER, W.: Autoradiographische Untersuchungen zur DNS-Synthese im Cerebellum und in der Medulla oblongata von Teleostiern verschiedenen Lebensalters. Z. mikr.-anat. Forsch. *82:* 264–292 (1970).

KRAUSE, R.: Mikroskopische Anatomie der Wirbeltiere in Einzeldarstellungen. I. Säugetiere. II. Vögel und Reptilien. III. Amphibien. IV. Teleostier, Plagiostomen, Zyklostomen und Leptokardier (De Gruyter, Berlin 1921, 1922, 1923, 1923).

KRETSCHMER, E.: Körperbau und Charakter, 1st and 19th ed. (Springer, Berlin 1922, 1948).

KRETSCHMER, E.: Medizinische Psychologie (Thieme, Leipzig 1926).

KRONSBEIN, J. and STEELE, J.H., jr.: The topological properties of structural components; in ELIAS Stereology, pp. 252–253 (Springer, New York 1967).

KRUGER, L. and BERKOWITZ, E.C.: The main afferent connections of the reptilian telencephalon as determined by degeneration and electrophysiological methods. J. comp. Neurol. *115:* 125–141 (1960).

KUHLENBECK, H.: Zur Morphologie des Urodelenvorderhirns. Jena Z. Naturwiss. *57:* 463–490 (1921).

KUHLENBECK, H.: Über den Ursprung der Grosshirnrinde. Anat. Anz. *55:* 337–365 (1922).

KUHLENBECK, H.: Über die Homologien der Zellmassen im Hemisphärenhirn der Wirbeltiere. Folia anat. japon. *2:* 325–364 (1924).

KUHLENBECK, H.: Über die morphologische Bedeutung der Arteriae corporis striati bei Reptilien. Folia anat. japon. *4:* 157–163 (1925).

KUHLENBECK, H.: Weitere Mitteilungen zur Genese der Basalganglien: Über die sogenannten Ganglienhügel. Anat. Anz. *60:* 33–40 (1925–1926).

KUHLENBECK, H.: Betrachtungen über den funktionellen Bauplan des Zentralnervensystems. Folia anat. japon. *4:* 111–135 (1926).

KUHLENBECK, H.: Vorlesungen über das Zentralnervensystem der Wirbeltiere (Fischer, Jena 1927).

KUHLENBECK, H.: Die Grundbestandteile des Endhirns im Lichte der Bauplanlehre. Anat. Anz. *67:* 1–51 (1929a).

KUHLENBECK, H.: Über die Grundbestandteile des Zwischenhirnbauplans der Anamnier. Morph. Jb. *63:* 50–95 (1929b).

KUHLENBECK, H.: Beobachtungen über das Chordagewebe bei Vogelkeimlingen. Anat. Anz. *69:* 485–520 (1930).

KUHLENBECK, H.: Bemerkungen über den Zwischenhirnbauplan bei Säugetieren, insbesondere beim Menschen. Anat. Anz. *70:* 122–142 (1930).

KUHLENBECK, H.: Über die Grundbestandteile des Zwischenhirnbauplans bei Reptilien. Morph. Jb. *66:* 244–317 (1931).

KUHLENBECK, H.: Buchbesprechung zu HESSE, R., Über die Abgrenzung des Gehirns. Anat. Anz. *75:* 124–127 (1932). *This critical review includes an epitome of basic concepts.*

KUHLENBECK, H.: Über die morphologische Stellung des Corpus geniculatum mediale. Anat. Anz. *81:* 28–37 (1935a).

KUHLENBECK, H.: Über die morphologische Bewertung der sekundären Neuromerie. Anat. Anz. *81:* 129–148 (1935b).

KUHLENBECK, H.: Über die Grundbestandteile des Zwischenhirnbauplans der Vögel. Morph. Jb. *77:* 61–109 (1936).

KUHLENBECK, H.: The ontogenetic development of the diencephalic centers in a bird's brain (chick) and comparison with the reptilian and mammalian diencephalon. J. comp. Neurol. *66:* 23–75 (1937).

KUHLENBECK, H.: The ontogenetic development and phylogenetic significance of the cortex telencephali in the chick. J. comp. Neurol. *69:* 273–301 (1938).

KUHLENBECK, H.: The development and structure of the pretectal cell masses in the chick. J. comp. Neurol. *71:* 613–636 (1939).

KUHLENBECK, H.: Cadaver encephalography with opaque contrast medium. Anat. Rec. *77:* 145–153 (1940).

KUHLENBECK, H.: Neoplastic transformation of the subependymal cell plate in the floor of the fourth ventricle (subependymal spongioblastoma). J. Neuropath. exp. Neurol. *6:* 139–151 (1947).

KUHLENBECK, H.: The derivatives of the thalamus ventralis in the human brain and their relation to the so-called subthalamus. Milit. Surg. *102:* 433–447 (1948).

KUHLENBECK, H.: The transitory superficial granular layer of the cerebellar cortex; its relationship to certain cerebellar neoplasms. J. amer. med. Women's Ass. *5:* 347–351 (1950).

KUHLENBECK, H.: *Virchow-Robin spaces, spaces of His-Held*, and their relation to the membrana limitans perivascularis (Abstract). Anat. Rec. *109:* 375 (1951).

KUHLENBECK, H.: Die Formbestandteile der Regio praetectalis des Anamnier-Gehirns und ihre Beziehungen zum Hirnbauplan. Folia anat. japon. 28 (*Nishi*-Festschrift): 23–44 (1956).

KUHLENBECK, H.: Brain and consciousness. Some prolegomena to an approach of the problem (Karger, Basel 1957).

KUHLENBECK, H.: Further remarks on brain and consciousness. The brain paradox and the meanings of consciousness. Confin. neurol. *19:* 462–486 (1959).

KUHLENBECK, H.: Mind and matter. An appraisal of their significance for neurologic theory (Karger, Basel 1961).

KUHLENBECK, H.: The concept of consciousness in neurological epistemology; in SMYTHIES Brain and mind. Modern concepts of the nature of mind, pp. 137–161 (Routledge & Kegan Paul, London; Humanities Press, New York 1965).

KUHLENBECK, H.: Weitere Bemerkungen zur Maschinentheorie des Gehirns. Confin. neurol. *27:* 295–328 (1966).

KUHLENBECK, H.: The central nervous system of vertebrates. 1. Propaedeutics to comparative neurology. (Karger, Basel 1967a).

KUHLENBECK, H.: The telencephalic zonal system of the ganoid *Amia calva*, with reference to the topologic vertebrate forebrain pattern (abstract). Anat. Rec. *157:* 368–369 (1967b).

KUHLENBECK, H.: Some comments on words, language, thought, and definition; in BUEHNE *et al*. Helen Adolf Festschrift, pp. 9–29 (Ungar, New York 1968).

KUHLENBECK, H.: Some comments on the development of the human corpus callosum and septum pellucidum. Acta anat. nippon. *44:* 245–256 (1969).

KUHLENBECK, H.: A note on the morphology of the hypophysis in the Gymnophione *Schistomepum thomense*. Okaj. Folia anat. japon. *46:* 307–319 (1970).

KUHLENBECK, H. and DOMARUS, E. v.: Zur Ontogenese des menschlichen Grosshirns. Anat. Anz. *53:* 316–320 (1920).

KUHLENBECK, H. and GLOBUS, J.H.: Arhinencephaly with extreme eversion of the end-brain. An anatomic study. Arch. Neurol. Psychiat. *36:* 58–74 (1936).

KUHLENBECK, H.; HAFKESBRING, R., and Ross, M.: Further observations on a living 'decorticate' (hydrancencephalic) child. J. amer. med. Women's Ass. *14:* 216–225 (1959).

KUHLENBECK, H. and HAYMAKER, W.: The derivatives of the hypothalamus in the human brain. Their relation to the extrapyramidal and autonomic systems. Milit. Surg. *105:* 26–52 (1949).

KUHLENBECK, H. and HAYMAKER, W.: Observations on the anatomical mechanism of hydrocephalus in tuberculous meningitis. Anat. Rec. *106:* 211–212 (1950).

KUHLENBECK, H. and WIENER KIRBER, M.: Neuroectodermal and other cells in mousebrain tissue cultures. Their morphologic relation to cell types found in human neuroectodermal neoplasms. Confin. neurol. *19:* 65–104 (1959).

KUHLENBECK, H.; MAHER, I.; Ross, M., and EASTWOOD, R.: Hydrancencephaly with univentricular telencephalic malformation. General comments, with anatomical and clinical observations on three cases. Confin. neurol *17:* 100–118 (1957).

KUHLENBECK, H.; MALEWITZ, T.D., and BEASLEY, A.B.: Further observations on the morphology of the forebrain in Gymnophiona, with reference to the topologic vertebrate forebrain pattern; in HASSLER and STEPHAN Evolution of the forebrain, pp. 9–19 (Thieme, Stuttgart 1966).

KUHLENBECK, H. and MILLER, R.N.: The pretectal region of the rabbit's brain. J. comp. Neurol. *76:* 323–365 (1942).

KUHLENBECK, H. and MILLER, R.N.: The pretectal region of the human brain. J. comp. Neurol. *91:* 369–407 (1949).

KUHLENBECK, H. and NIIMI, K.: Further observations on the morphology of the brain in the Holocephalian Elasmobranchs Chimaera and Callorhynchus. J. Hirnforsch. *11:* 265–314 (1969).

KUHLENBECK, H. and SZEKELY, E.G.: Evoked potentials from tectum mesencephali and telencephalon of the chicken after unilateral optic stimulation. Anat. Rec. *145:* 332 (1963).

KUHLENBECK, H.; SZEKELY, E.G., and SPULER, H.: Some remarks on the zonal pattern of mammalian cortex cerebri as manifested in the rabbit. Its relationship with certain electrocorticographic findings. Confin. neurol. *20:* 407–423 (1960).

KUMAMOTO-SHINTANI, Y.: The nuclei of the pretectal region in the mouse brain. J. comp. Neurol. *113:* 43–60 (1959).

KUNDRAT, H.: (1882), see KUHLENBECK, H. and GLOBUS, J.H. (1936).

KUPFFER, C. VON: Studien zur vergleichenden Entwicklungsgeschichte des Kopfes der Kranioten. 4. Zur Kopfentwicklung von Bdellostoma (Lehmann, München 1900).

KUPFFER, K. VON: Die Morphogenie des Centralnervensystems; in HERTWIG Hb. d. vergl. u. exp. Entwicklungslehre d. Wirbeltiere, 2. Bd., 3. Teil, pp. 1–272 (Fischer, Jena 1906).

LA BRUYÈRE, J. DE: Les caractères ou les mœurs de ce siècle. Suivis du discours à l'Académie et de la traduction de *Theophraste* (1688–1694; ed. Belin-Leprieur, Paris 1845).

LANDAU, E.: Anatomie des Grosshirns. Formanalytische Untersuchungen (Bircher, Bern 1923).

LANGE-COSACK, H.: Die Hydranencephalie (Blasenhirn) als Sonderform der Grosshirnlosigkeit. Arch. Psychiat. *117:* 1–51; 595–640 (1944).

LARSELL, O.: The cerebellum. A review and interpretation. Arch. Neurol. Psychiat. *38:* 580–607 (1937).

LARSELL, O.: Textbook of neuro-anatomy and the sense organs (Appleton, New York 1939).

LARSELL, O.: Anatomy of the nervous system, 2nd. ed. (Appleton, New York 1951).

LARSELL, O.: The comparative anatomy and histology of the cerebellum from myxinoids through birds. Edited by J.Jansen. (University of Minnesota Press, Minneapolis 1967).

LASHLEY, K.S. and CLARK, G.: The cytoarchitecture of the cerebral cortex of *Ateles*. A critical examination of architectonic studies. J. comp. Neurol. *85:* 223–305 (1946).

LAZORTHES, G.: Vascularisation et circulation cérébrales (Masson, Paris 1961).

LEVITAN, M. and MONTAGU, A.: Textbook of human genetics (Oxford University Press, New York 1971).

LEWIN, K.: Principles of topological psychology (McGraw-Hill, New York 1936).

LINDSAY, R.B. and MARGENAU, H.: Foundations of physics (Dover, New York 1957).

LIPPINCOTT, E.R.; STROMBERG, R.R.; GRANT, W.H., and CESSAC, G.L.: Polywater. Science *164:* 1482–1487 (1969). *Cf. also an anonymous review in Sci. Res. 4/ (9): 9 (1969), referring to studies by COPE, HAZLEWOOD et al.*

LIST, C.F.; BURGE, C.H., and HODGES, F.J.: Intracranial angiography. Radiology *45:* 1–14 (1945).

LIVY (TITUS LIVIUS): Ab urbe condita. Latin text with English translation by A. C. SCHLE-SINGER; 14 vols. Loeb Classical Library (Harvard University Press, Cambridge 1952–1959).

LOCKE, J.: An essay concerning human understanding (1689, 1690) (Oxford University Press, London 1894).

LOESER, J. D. and ALVORD, E. C.: Clinicopathological correlations in agenesis of the corpus callosum. Neurology 18: 745–756 (1968).

LOO, Y. T.: On formation of human cerebral cortex. An ontogenetic study with a discussion of the function of different cortical layers. Anat. Anz. 68: 305–324 (1929).

LUBOSCH, W.: Grundriss der wissenschaftlichen Anatomie (Thieme, Leipzig 1925).

LUBOSCH, W.: Geschichte der vergleichenden Anatomie; in BOLK, GÖPPERT et al. Handb. d. vergl. Anat. der Wirbeltiere, vol. 1, pp. 3–76 (Urban & Schwarzenberg, Berlin 1931).

MARGULIS, L.: Origin of eucaryotic cells (Yale University Press, New Haven 1970).

MARMOR, M. F. and GORMAN, A. L. F.: Membrane potential as the sum of ionic and metabolic components. Science 167: 65–67 (1970).

MARTINEZ, P. M.: Sur la morphologie du réseau admirable extracrânien. Acta anat. 67: 24–52 (1967).

MAURER, F.: Der Mensch und seine Ahnen (Ullstein, Berlin 1928).

McCABE, J. and Low, F. N.: The subarachnoid angle. An area of transition in peripheral nerve. Anat. Rec. 164: 15–34 (1969).

McKUSICK, V. A.: Mendelian inheritance in man. 3rd. ed. (Johns Hopkins Press, Baltimore 1971).

McLAUGHLIN, P. J. and DAYHOFF, M. O.: Eukaryotes versus Prokaryotes. An estimate of evolutionary distance. Science 168: 1469–1471 (1970).

METTLER, F. A.: Congenital malformation of the brain. Critical review. J. Neuropath. exp. Neurol. 6: 98–110 (1947).

MIHALKOVICS, G. V. VON: Die Entwicklungsgeschichte des Gehirns (Engelmann, Leipzig 1877).

MILHORAY, W. H.; HAMMOCK, M. K.; FENSTERMACHER, J. D., and RALL, D. P.: Cerebro-spinal fluid production by the choroid plexus and brain. Science 173: 330–332 (1971).

MILLER, R. N.: The telencephalic zonal system of the teleost Corydora paliatus. J. comp. Neurol. 72: 149–176 (1940).

MILLER, R. N.: The diencephalic cell masses of the teleost Corydora paliatus. J. comp. Neurol. 73: 345–378 (1940).

MILLOT, J. et ANTHONY, J.: Anatomie de Latimeria chalumnae. II. Système nerveux et organes des sens (Editions du Centre National de la Recherche Scientifique, Paris 1965).

MILLOT, J. et ANTHONY, J.: L'organisation générale du prosencephale de Latimeria chalumnae Smith (poisson crossopterygien coelacanthidé); in HASSLER and STEPHAN Evolution of the forebrain, pp. 50–60 (Thieme, Stuttgart 1966).

MILLOT, J.; NIEUWENHUYS, R. et ANTHONY, J.: Le diencéphale de Latimeria chalumnae Smith (poisson Coelacanthidé). C. R. Acad. Sci., Paris 248: 5051–5055 (1964).

MITCHISON, J. M.: Enzyme synthesis in synchronous cultures. Science 165: 657–663 (1965).

MIURA, R.: Über die Differenzierung der Grundbestandteile im Zwischenhirn des Kaninchens. Anat. Anz. 77: 1–65 (1933).

MONOD, J. and JACOB, F.: Teleonomic mechanisms in cellular metabolism, growth and differentiation. Cold Spr. Harb. Symp. quant. Biol. 26: 389–401 (1961).

MORIARTY, J.A. and KLINGMAN, W.O.: Congenital and prenatal diseases; in BAKER Clinical neurology, 2nd ed., vol. 4, chapt. 41, pp. 1921–1993 (Hoeber-Harper, New York 1962).

MORTON, W.R.M.: Arhinencephaly and multiple developmental anomalies occurring in a human full-term foetus. Anat. Rec. *98:* 45–58 (1947).

NAEF, A.: Allgemeine Morphologie. I. Die Gestalt als Begriff und Idee; in BOLK, GÖPPERT *et al.* Handb. d. vergl. Anat. d. Wirbeltiere, vol. 1, pp. 77–118 (Urban & Schwarzenberg, Berlin 1931).

NAUTA, W.J.H. and HAYMAKER, W.: Hypothalamic nuclei and fiber connections; in HAYMAKER, ANDERSON and NAUTA The hypothalamus, chapt. 4, p. 136–209 (Thomas, Springfield 1969).

NEAL, H.V. and RAND, H.W.: Comparative anatomy (Blakiston, Philadelphia 1936).

NEUMANN, J.v.: Probabilistic logics (California Institute of Technology, Pasadena 1952).

New York Academy of Sciences: In PETRAS and NOBACK eds., Comparative and evolutionary aspects of the vertebrate central nervous system. (Annals N.Y. Acad. Sci. 167, New York 1969).

NIEUWENHUYS, R.: The structure of the telencephalon of the Teleost *Gasterosteus aculeatus.* Proc. kon. nederl. Akad. Wet. C, *62* (4): 341–363 (1959).

NIEUWENHUYS, R.: Further studies on the general structure of the Actinopterygian forebrain. Acta morph. neerl. scand. *6:* 65–79 (1963).

NIEUWENHUYS, R.: The comparative anatomy of the actinopterygian forebrain. J. Hirnforsch. *6:* 171–192 (1963).

NIEUWENHUYS, R.: The interpretation of the cell masses in the teleostean forebrain; in HASSLER and STEPHAN Evolution of the forebrain, pp. 32–39 (Thieme, Stuttgart 1966).

NIEUWENHUYS, R. and HICKEY, M.: A survey of the forebrain in the Australian lungfish *Neoceratodus forsteri.* J. Hirnforsch. *7:* 433–452 (1965).

NOBACK, C.R.: The developmental anatomy of the human osseous skeleton during the embryonic, fetal and circumnatal periods. Anat. Rec. *68:* 91–125 (1944).

NOBACK, C.R. and ROBERTSON, G.G.: Sequences of appearance of ossification in the human skeleton furing the first five prenatal months. Amer. J. Anat. *89:* 1–28 (1957).

NOBACK, G.J.: The lineal growth of the respiratory system during fetal and neonatal life as expressed by graphic analysis and empirical formulae. Amer. J. Anat. *36:* 235–272 (1925–1926).

NORTHCUTT, R.G.: Discussion of paper on avian telencephalon by Karten. Ann. N.Y. Acad. Sci. *167:* 180–185 (1969).

OBENCHAIN, J.B.: The brains of the South American Marsupials *Caenolestes* and *Orolestes.* Field Mus. nat. Hist. Publ. 224 *14:* 175–232 (1925).

OBERSTEINER, H.: Anleitung beim Studium des Baues der nervösen Zentralorgane im gesunden und kranken Zustande, 5th ed. (Deuticke, Leipzig 1912).

OBSEQUENS, JULIUS: Ab anno urbis conditae DV prodigiorum liber. In LIVY, vol. 14, pp. 238–319 (see above: LIVY, Harvard University Press 1959).

OHNO, S.: Evolution by gene duplication (Springer, New York 1970).

O'LEAGUE, P.; DALEN, H.; RUBIN, H., and TOBIAS, C.: Electrical coupling. Low resistance junctions between mitotic and interphase fibroblasts in tissue culture. Science *170:* 464–466 (1970).

OLIVEROS, N.L.: Observations on the lining of the cavum septi pellucidi in the brain of newborn and adult man. Confin. neurol. *26:* 45–55 (1965).

OLSSON, R.: The development of *Reissner's fibre* in the brain of the Salmon. Acta zool. *37:* 235–250 (1956).

OSBORN, H.F.: The origin of the corpus callosum, a contribution upon the cerebral commissures of the Vertebrata. I, II. Morph. Jb. *12:* 223–251, 530–543 (1887).

OSBORN, H.F.: A contribution to the internal structure of the amphibian brain. J. Morph. *2:* 51–96 (1888).

OSTERTAG, B.: Grundzüge der Entwicklung und Fehlentwicklung. Die formbestimmenden Faktoren. Die Einzelformen der Verbildungen (einschliesslich Syringomyelie). Hb. d. spez. path. Anat. u. Histol., 13. Bd.; in SCHOLZ Nervensystem, Teil IV, pp. 283–601 (Springer, Berlin 1956).

PADGET, D.H.: Neuroschisis and human embryonic maldevelopment. New evidence on anencephaly, spina bifida and diverse mammalian defects. J. Neuropath. exp. Neurol. *29:* 192–216 (1970).

PALAY, S.L.: The histology of the meninges of the toad *Bufo*. Anat. Rec. *88:* 257–270 (1944).

PAPEZ, J.W.: Thalamus of turtles and thalamic evolution. J. comp. Neurol *61:* 433–475 (1935).

PAPEZ, J.W.: The embryologic development of the hypothalamic area in mammals. Proc. Ass. Res. nerv. ment. Dis. *20:* 31–51 (1940).

PATTEN, B.M.: Human embryology, 2nd ed. (Blakiston, New York 1953).

PEELE, T.L.: The neuroanatomical basis for clinical neurology, 1st and 2nd ed. (McGraw-Hill, New York 1954, 1961).

PETERSON, H.O.: Neuroroentgenography; in BAKER Clinical neurology, 3rd ed., chapt. 2, pp. 100–211, vol. 1 (Hoeber-Harper, New York 1962).

PLANCK, M.: A survey of physical theory. Translated by R.JONES and D.H.WILLIAMS (Dover, New York 1960).

PLINY (GAIUS PLINIUS SECUNDUS MAIOR): Naturalis Historia (Latin text with English transl. by H.RUCKMAN and W.H.S.JONES, 10 vols. Loeb Classical Library (Harvard University Press, Cambridge 1958–1962).

PORTMANN, A.: New paths in biology. World perspectives, vol. 30 (Harper & Row, New York 1964).

PRINZHORN, H.: Bildnerei der Geisteskranken (Springer, Berlin 1922).

RAKIC, P. and YAKOVLEV, P.I.: Development of the corpus callosum and cavum septi in man. J. comp. Neurol. *132:* 45–75 (1968).

RANKE, J.: Der Mensch, 2 vols., 3rd ed. (Bibliographisches Institut, Leipzig 1923).

RANSON, S.W.: The anatomy of the nervous system from the standpoint of development and function, 7th ed. (Saunders, Philadelphia 1943).

RASHEVSKY, N.: Mathematical biophysics. Physico-mathematical foundations of biology, 2 vols., 3rd ed. (Dover, New York 1960).

RASHEVSKY, N. (ed.): Physicomathematical aspects of biology. Introduction to the course, pp. 1–4. Proc. Int. School of Physics 'Enrico Fermi', Course 16 (Academic Press, New York 1962).

RASMUSSEN, A.T.: WILHELM HIS (1831–1904); in HAYMAKER The founders of neurology, pp. 49–52 (Thomas, Springfield 1953).

RASMUSSEN, A.T.: Anomalies of the nervous system; in BAKER Clinical neurology, 1st ed., vol. 3, chapt. 34, pp. 1656–1694 (Hoeber-Harper, New York 1955).

RAVEN, CH.P.: Zur Entwicklung der Ganglienleiste. II. Über das Differenzierungs-vermögen des Kopfganglienleistenmaterials von Urodelen. Arch. Entwmech. *129:* 179–198 (1933).

RAVEN, P.H.: A multiple origin for plastids and mitochondria. Science *169:* 641–646 (1970).

REFSUM, S.: Genetic aspects of neurology; in BAKER Clinical neurology, 1st ed., vol. 3, pp. 1927–1960; 2nd ed., vol. 4, pp. 2272–2316 (Hoeber-Harper, New York 1955, 1962).

REICH, C.A.: The greening of America. The coming of a new consciousness and the rebirth of a future (Random House, New York 1970).

REICHERT, K.B.: Der Bau des menschlichen Gehirns (Engelmann, Leipzig 1859–1861).

REMAK, R.: Untersuchungen über die Entwicklung der Wirbelthiere (Reimer, Berlin 1855).

REMANE, A.: Phylogenetische Entwicklungsregeln von Organen; in HASSLER and STEPHAN Evolution of the forebrain, pp. 1–8 (Thieme, Stuttgart 1966).

RENSCH, B.: Increase of learning capability with increase of brain-size. Amer. Naturalist *90:* 81–95 (1956).

RETZIUS, G.: Das Menschenhirn. Studien in der makroskopischen Morphologie, 2 vols. (Fischer, Jena 1896).

RETZIUS, G.: Biologische Untersuchungen, Bd. 10 (Fischer, Jena 1902).

RHINES, F.N.: Measurements of topological parameters; in ELIAS Stereology, pp. 235–249 (Springer, New York 1967).

RICHTER, E.: Die Entwicklung des Globus pallidus und des Corpus subthalamicum (Springer, Berlin 1965).

RICHTER, W. und KRANZ, D.: Die Abhängigkeit der DNA-Synthese in den Matrixzonen des Mesencephalons vom Lebensalter der Versuchstiere *( Lebistes reticulatus* – Teleo-stei). Autoradiographische Untersuchungen. Z. mikr.-anat. Forsch. *82:* 76–92 (1969).

RIEMANN, B.: Collected works. WEBER *et al.*, eds. (Dover, New York 1953).

RIESE, W.: Konvergenzerscheinungen am Gehirn, nebst Bemerkungen zu der Arbeit von ROSE: 'Der Grundplan der Cortextektonik beim Delphin' in Band 32 dieser Zeit-schrift. J. Psychol. Neurol. *33:* 84–96 (1927).

RIGGS, H.E.; McGRATH, J.J., and SCHWARTZ, H.P.: Malformation of the adult brain (albino rat) resulting from prenatal irradiation. J. Neuropath. exp. Neurol. *15:* 432–447 (1956).

ROBERTS, T.S. and AKERT, K.: Insular and opercular cortex and its thalamic projection in *Macaca mulatta.* Schweiz. Arch. Neurol. Neurochir. Psychiat. *92:* 1–43 (1963).

ROGERS, L.C. and PAYNE, E.E.: The dura mater at the cranio-vertebral junction. J. Anat. *95:* 586–588 (1961).

ROMANOFF, A.L.: The avian embryo. Structural and functional development. (Macmillan, New York 1960).

ROMER, A.S.: The vertebrate body (Saunders, Philadelphia 1950).

ROOFE, P.G.: The endocranial blood vessels of *Ambystoma tigrinum.* J. comp. Neurol. *61:* 257–293 (1935).

ROSE, M.: Über die cytoarchitektonische Gliederung des Vorderhirns der Vögel. J. Psychol. Neurol. *21:* 278–352 (1914).

ROSE, M.: Über das histogenetische Prinzip der Einteilung der Grosshirnrinde. J. Psychol. Neurol. *32:* 97–160 (1926).

ROSE, M.: Die Ontogenie der Inselrinde. Zugleich ein Beitrag zur histogenetischen Rinden-einteilung. J. Psychol. Neurol. *36:* 182–209 (1928).

Rose, M.: Entwicklungsgeschichtliche Einleitung. Ontogenie des Zentralnervensystems und des Sympathicus. Phylogenie des Zentralnervensystems. Cytoarchitektonik und Myeloarchitektonik der Grosshirnrinde; in Bumke et al. Hb. d. Neurologie, Bd. 1, pp. 1–34, 588–778 (Springer, Berlin 1935).

Röthig, P.: Beiträge zum Studium des Zentralnervensystems der Wirbeltiere. VIII. Über das Zwischenhirn der Amphibien. Arch. mikr. Anat. 98: 616–645 (1923).

Rovainen, C. M.: Glucose production by lamprey meninges. Science 167: 889–890 (1970).

Rubinstein, H. S. and Freeman, W.: Cerebellar agenesis. J. nerv. ment. Dis. 92: 489–502 (1940).

Rüdeberg, S. I.: Morphogenetic studies on the cerebellar nuclei and their homologization in different vertebrates including man. Dissertation (Ohlsonn, Lund 1961).

Saito, T.: Über das Gehirn des japanischen Flussneunauges (Entosphenus japonicus Martens). Folia anat. japon. 8: 189–263 (1930).

Saito, T.: Über die retikulären Zellen im Gehirn des japanischen Dornhaies (Acanthias mitsukurii Jordan et Fowler). Folia anat. japon. 8: 323–343 (1930).

Saller, K.: Leitfaden der Anthropologie (Springer, Berlin 1930).

Scammon, R. E.: Developmental anatomy. In: Morris' Human anatomy, 10th ed. pp. 9–52; 11th ed. pp. 11–62 (Schaeffer ed.) (Blakiston, Philadelphia 1942, 1953).

Schaltenbrand, G.: Plexus und Meningen; in v. Möllendorff und Bargmann Handb. d. mikr. Anat. des Menschen, vol. IV/2, pp. 1–139 (Springer, Berlin 1955).

Scharf, J.H.: Sensible Ganglien. Vol. IV, Part 3, Handb. d. mikr. Anat. d. Menschen, Möllendorff, W. v. and Bargmann, W., eds. (Springer, Berlin-Heidelberg 1958).

Scharrer, E.: The histology of the meningeal myeloid tissue in the Ganoids Amia and Lepisosteus. Anat. Rec. 88: 291–310 (1944).

Scherer, H.: Vergleichende Pathologie des Nervensystems der Säugetiere (Thieme, Leipzig 1944).

Schlögl, R.: Selektiver Stofftransport durch Membranen. Jb. Max-Planck-Gesellsch. 1969: 136–152 (1969).

Schneider, R.: Ein Beitrag zur Ontogenese der Basalganglien des Menschen. Anat. Nachr. 1: 115–137 (1950).

Schneider, R.: Morphologische Untersuchungen am Gehirn der Chiroptera (Mammalia). Abh. Senckenberg, naturforsch. Ges. 495: 1–92 (1957).

Schober, W.: Vergleichende Betrachtungen am Telencephalon niederer Wirbeltiere; in Hassler and Stephan Evolution of the forebrain, pp. 20–31 (Thieme, Stuttgart 1966).

Schulte, H. von W. and Tilney, F.: Development of the neuraxis in the domestic cat to the stage of twenty-one somites. Ann. New York Acad. Sci. 24: 319–346 (1915).

Schulz, E.: Zur postnatalen Biomorphose des Ependyms im Telencephalon von Lacerta agilis agilis (L.) Arch. mikr.-anat. Forsch. 81: 111–152 (1969).

Schumacher, G. A.: The demyelinating diseases; in Baker Clinical neurology, 2nd ed., vol. 3, pp. 1226–1284 (Hoeber-Harper, New York 1962).

Schwalbe, E. (ed.): Die Morphologie der Missbildungen des Menschen und der Tiere. 3. Teile, 12. Lieferungen (Fischer, Jena 1906–1927).

Schwalbe, E. and Josephy, H.: Die Cyclopie; in Schwalbe Morphologie der Missbildungen des Menschen und der Tiere, part 3, section 1 (Fischer, Jena 1913).

Schwalbe, G.: Beiträge zur Entwicklungsgeschichte des Zwischenhirns. Sitz. Ber. jen. Ges. Med. Naturwiss. 20: 2–7 (1880).

Schwalbe, G.: Lehrbuch der Neurologie (Besold, Erlangen 1881).

SELENKA, E. und GOLDSCHMIDT, R.: Zoologisches Taschenbuch für Studierende zum Gebrauch bei Vorlesungen und praktischen Übungen. 2. Wirbeltiere (Thieme, Leipzig 1923).

SENN, D. G.: Der Bau des Reptiliengehirns im Licht neuer Ergebnisse. Verh. Naturf. Ges., Basel *79*: 25–42 (1968).

SENN, D. G.: Bau und Ontogenese von Zwischen- und Mittelhirn bei *Lacerta sicula* (Rafinesque). Acta anat. *71*: suppl. 1 (1968).

SENN, D. G.: Über die Bedeutung der Hirnmorphologie für die Systematik. Verh. Naturf. Ges., Basel *80*: 49–55 (1969).

SENN, D. G.: The stratification in the reptilian central nervous system. Acta anat. *75*: 521–552 (1970).

SENSENIG, E. C.: Development of spinal meninges. Carnegie Contrib. Embryol. *34* (228): 146–157 (1961).

SHELLSHEAR, J. L.: The basal arteries of the forebrain and their functional significance. J. Anat. *55*: 27–35 (1920).

SMART, I. H. M.: The method of transformed co-ordinates applied to the deformations produced by the wall of a tubular viscus on a contained body: the avian egg as a model system. J. Anat. *104*: 507–518 (1969).

SMITH, G. ELLIOT: The morphology of the true 'limbic lobe', corpus callosum, septum pellucidum, and fornix. J. Anat. Physiol. *30*: 157–167, 185–205 (1895–1896).

SMITH, G. ELLIOT: The morphology of the indusium and striae Laneisii. Anat. Anz. *13*: 23–27 (1897).

SMITH, G. ELLIOT: The relation of the fornix to the margin of the cerebral cortex. J. Anat. Physiol. *32*: 23–58 (1898).

SMITH, G. ELLIOT: Notes upon the natural subdivision of the cerebral hemisphere. J. Anat. Physiol. *35*: 432–454 (1901).

SMITH G. ELLIOT: The cerebral cortex in Lepidosiren, with notes on the interpretation of certain features of the forebrain of other vertebrates. Anat. Anz. *33*: 513–540 (1908).

SMITH, G. ELLIOT: The term 'archipallium' – a disclaimer. Anat. Anz. *35*: 429–430 (1910).

SMITH, G. ELLIOT: Some problems relating to the evolution of the brain. (The Arris and Gale lectures.) Lancet *1910*: 1–6, 147–155, 221–227 (1910).

SMITH, G. ELLIOT: A preliminary note on the morphology of the corpus striatum and the origin of the neopallium. J. Anat. *53*: 271–291 (1918–1919).

SÖDERBERG, G.: Contributions to the forebrain morphology in Amphibians. Acta zool. *3*: 65–121 (1922).

SOLLY, S.: The human brain. Its structure, physiology and diseases. With a description of the typical forms of brain in the animal kingdom (Lea & Blanchard, Philadelphia 1848).

SPATZ, H.: Zur Anatomie der Zentren des Streifenhügels. Münch. med. Wschr. *68*: 1441–1446 (1921).

SPATZ, H.: Über die Entwicklungsgeschichte der basalen Ganglien des menschlichen Grosshirns. Erg. Bd. Anat. Anz. *60*: 54–58 (1925).

SPATZ, H.: Die Evolution des Menschenhirns und ihre Bedeutung für die Sonderstellung des Menschen. Nachr. Giessener Hochschules. *24*: 52–74 (1955).

SPATZ, H.: Der basale Neocortex und seine Bedeutung für den Menschen. Ber. physik.-med. Ges., Würzburg *71*: 7–17 (1962–1964).

SPATZ, H.: Vergangenheit und Zukunft des Menschenhirns. Jb. 1964 Akad. Wiss. Literatur, pp. 228–242 (1965).

Spatz, H.: Gehirnentwicklung (Introversion-Promination) und Endocranialausguss; in Hassler and Stephan Evolution of the forebrain, pp. 136–152 (Thieme, Stuttgart 1966).

Spemann, H.: Embryonic development and induction (Yale University Press, New Haven 1930).

Spemann, H. and Mangold, H.: Über Induktion von Embryonalanlagen durch Implantation artfremder Organisatoren. Arch. mikr. Anat. EntwMech. *100:* 599–638 (1924).

Spratt, N.T., jr., and Haas, H.: Primitive streak and germ layer formation in the chick. A reappraisal (abstract). Anat. Rec. *142:* 327 (1962).

Spratt, N.T., jr., and Haas, H.: Germ layer formation and the role of the primitive streak in the chick. I. Basic architecture and morphogenetic tissue movements. J. exp. Zool. *158:* 9–38 (1965).

Stanbury, J.B.; Wyngarden, J.B., and Frederickson, D.S.: The metabolic basis of inherited disease, 2nd ed. (McGraw-Hill, New York 1966).

Steegmann, A.T.: Vascular diseases of the spinal cord; in Baker Clinical neurology, 2nd ed., vol. 3, chapt. 35, pp. 1615–1638 (Hoeber-Harper, New York 1962).

Stein, W.D.: The movement of molecules across cell membranes. Theoretical and experimental biology, vol. 6 (Academic Press, New York 1967).

Stephan, H.: Vergleichend-anatomische Untersuchungen an Insektivorengehirnen. III. Hirn-Körper-Gewichtsbeziehungen. Morph. Jb. *99:* 853–880 (1958–1959).

Sterzi, G.: Ricerche interno all'anatomia delle meningi e considerazioni sulla filogenesi. Atti Inst. Veneto *60:* 1101–1372 (1900–1901).

Sterzi, G.: Die Blutgefässe des Rückenmarks. Untersuchungen über ihre vergleichende Anatomie und Entwicklungsgeschichte. Anat. Hefte *74:* 1–364 (1904).

Stieda, L.: Über den Bau des centralen Nervensystems des Axolotl. Z. wiss. Zool. *25:* 285–310 (1875).

Stilwell, E.F.: Observation of endomitosis in embryonic chick cells grown in vitro. Anat. Rec. *114:* 9–18 (1952).

Stingelin, W.: Vergleichend-morphologische Untersuchungen am Vorderhirn der Vögel auf cytologischer und cytoarchitektonischer Grundlage (Helbing, Basel 1958).

Stingelin, W.: Qualitative und quantitative Untersuchungen an Kerngebieten der Medulla oblongata bei Vögeln (Karger, Basel 1965).

Stockard, C.R.: Developmental rate and structural expression. An experimental study of twins, 'double monsters' and single deformities, and the interaction among embryonic organs during their origin and development. Amer. J. Anat. *28:* 115–277 (1920–1921).

Strasser, H.: Anleitung zur Gehirnpräparation (Bircher, Bern 1920).

Streeter, G.L.: The development of the nervous system; in Keibel and Mall Manual of human embryology, vol. II, chapt. XIV, pp. 1–156 (Lippincott, Philadelphia 1912).

Streeter, G.L.: The development of the venous sinuses of the dura mater in the human embryo. Amer. J. Anat. *18:* 145–178 (1915).

Streeter, G.L.: The developmental alterations in the vascular system of the brain in the human embryo. Contrib. Embryol. Carnegie. Inst., Wash. *8:* 5–38 (1918).

Streeter, G.L.: The status of metamerism in the central nervous system of chick embryos. J. comp. Neurol. *57:* 455–475 (1933).

Strong, O.S.: The cranial nerves of amphibia. A contribution to the morphology of the vertebrate nervous system. J. Morph. *10:* 101–230 (1895).

Suitsu, N.: Comparative studies on the growth of the corpus callosum. I. J. comp. Neurol. *32:* 35–60 (1920–1921).

Sumi, R.: Über die Sulci und Eminentiae des Hirnventrikels von *Diemictylus pyrrhogaster*. Folia anat. japon. *4:* 375–388 (1926).

Sumi, S. M.: Brain malformations in the trisomy 18 syndrome. Brain *93:* 821–830 (1970).

Sundberg, C.: Das Glykogen in menschlichen Embryonen von 15, 27 und 40 mm. Z. Anat. EntwGesch. *73:* 168–246 (1934).

Swank, R. L.: The relationship between the circumolivary fascicles and the ponto-bulbar body in man. J. comp. Neurol. *60:* 309–317 (1934).

Szasz, T. S.: The myth of mental illness (Hoeber-Harper, New York 1961).

Szentàgothai, J.: Growth of the nervous system. An introductory survey; in Wolsten-holme and O'Connor Growth of the nervous system, pp. 3–12 (Little Brown, Boston 1968).

Szilard, L.: Über die Entropieverminderung in einem thermodynamischen System bei Eingriffen intelligenter Wesen. Z. Physik *53:* 840–856 (1929).

Tandler, J.: Zur Entwicklungsgeschichte der arteriellen Wundernetze. Anat. Hb. *31:* 237–267 (1906).

Tandler, J.: Gefässe des Kiemenkreislaufes und ihre Umbildung; in Bolk *et al.* Hb. d. vergl. Anat. d. Wirbeltiere, vol. 6, pp. 557–562 (Urban & Schwarzenberg, Berlin 1933).

Tatum, E. L.: Some molecular aspects of congenital malformations. In: 1st Int. Conf. on Congenital Malformations, pp. 281–290 (Lippincott, Philadelphia 1961).

Tello, J. F.: Les différenciations neuronales dans l'embryon du poulet pendant les premiers jours de l'incubation. Trav. Lab. Rech. biol. Univ. Madrid *21:* 1–93 (1923).

Temin, H. M. and Mizutani, S.: RNA-dependent DNA polymerase in virions of *Rous sarcoma virus*. Nature, Lond. *226:* 1211–1213 (1970).

Theophrastus: The characters. Greek text with translation by *J. C. Edmunds*, Loeb Classical Library (Harvard University Press, Cambridge 1961).

Thompson, D'Arcy W.: On growth and form (1917) (new edition: Cambridge University Press, London 1942).

Tiedemann, F.: Anatomie und Bildungsgeschichte des Gehirns im Foetus des Menschen nebst einer vergleichenden Darstellung des Hirnbaues in den Thieren (Stein, Nürnberg 1816).

Tiedemann, H.: The molecular basis of differentiation in early development of amphibian embryos; in Moscona and Monroy Current topics in developmental biology, pp. 85–112 (Academic Press, New York 1966).

Torrey, T. W.: Morphogenesis of the vertebrates, 2nd ed. (Wiley, New York 1967).

Tower, D. B.: Structural and functional organization of mammalian cerebral cortex. The correlation of neurone density with brain size. Cortical density in the fin whale (*Balaenoptera physalus* L.) with a note on the cortical density in the Indian elephant. J. comp. Neurol. *101:* 19–51 (1954).

Tower, D. B. and Elliott, K. A. C.: Activity of the acetylcholine system in cerebral cortex of various unanaesthetized mammals. Amer. J. Physiol. *168:* 747–759 (1952).

Triepel, H.: Lehrbuch der Entwicklungsgeschichte (Thieme, Leipzig 1922).

Trinkaus, J. P.: Cells into organs. The forces that shape the embryo (Prentice-Hall, Englewood 1969).

Tyler, A.: Artificial parthenogenesis. Biol. Rev. *16:* 291–335 (1941).

UTTERBACK, R.A.: Extracerebral vascular disease; in BAKER Clinical neurology, 2nd ed., chapt. 11, pp. 616–641, vol. 2 (Hoeber-Harper, New York 1962).

VAAGE, S.: The segmentation of the primitive neural tube in chick embryos *(Gallus domesticus)*. A morphological, histochemical and autoradiographical investigation. Ergebn. Anat. EntwGesch. *41* (3): 1–88 (1969).

VEIT, O.: Alte Probleme und neuere Arbeiten auf dem Gebiete der Primitiventwicklung der Fische. Ergebn. Anat. EntwGesch. *24:* 414–490 (1923).

VILLIGER, E.: Gehirn und Rückenmark, 7th ed. (Engelmann, Leipzig 1920).

VITUMS, A.; MIKAMI, S., and FARNER, D.S.: Arterial blood supply to the brain of the white-crowned sparrow *Zonotrichia leukiphrys gambelii.* Anat. Anz. *116:* 309–326 (1965).

VOGT, C. and VOGT, O.: Allgemeinere Ergebnisse unserer Hirnforschung. J. Psychol. Neurol. *25:* 279–462 (1919).

VOGT, W.: Gestaltungsanalyse am Amphibienkeim mit örtlicher Vitalfärbung. Arch. Entwmech. *120:* 385–706 (1929).

VRAA-JENSEN, G.: On the correlation between the function and structure of nerve cells. (Quoted after HUGOSSON, 1957.) Acta psychiat. neurol. scand. suppl. 109 (1956).

VRIES, E. DE: Bemerkungen zur Ontogenie und vergleichenden Anatomie des Claustrums. Folia neurobiol. *4:* 481–513 (1910).

WADDINGTON, C.H.: Principles of development and differentiation (Macmillan, New York 1966).

WARKANY, J. and NELSON, R.C.: Appearance of skeletal abnormalities in the offspring of rats reared on a deficient diet. Science *92:* 383–384 (1940).

WARNER, F.J.: The development of the diencephalon in the American water snake *(Natrix sipedon)*. Trans. roy. Soc. Canada *36:* 53–70 (1942).

WARNER, F.J.: The development of the diencephalon in *Trichosurus vulpecula*. Okaj. Folia anat. japon. *46:* 265–295 (1969).

WARNER, F.J.: The brain stem in a case of dicephaly in a new born human foetus. Okajimas Folia anat. japon. *49:* 271–350 (1972).

WEED, L.H.: The development of the cerebrospinal spaces in pig and man. Carnegie Inst. Wash. Publ. 225. Contrib. Embryol. *5:* 1–115 (1917).

WEISS, P. (ed.): Genetic neurology. Problems of the development, growth, and regeneration of the nervous system and of its function (University of Chicago Press, Chicago 1950).

WEISS, P.: Nervous system (neurogenesis). VII. Special vertebrate neurogenesis; in WILLIER, WEISS and HAMBURGER Analysis of development, chapt. I, pp. 346–401 (Saunders, Philadelphia 1955).

WEISSENBERG, R.: Grundzüge der Entwicklungsgeschichte des Menschen in vergleichender Darstellung, 12th ed. (Thieme, Leipzig 1931).

WELCH, K. and FRIEDMAN, V.: The cerebrospinal fluid valves. Brain *83:* 454–469 (1960).

WELSCH, U.: Die Feinstruktur der *Josephschen Zellen* im Gehirn von Amphioxus. Z. Zellforsch. *86:* 252–261 (1968).

WENKEBACH, K.F.: Der Gastrulationsprozess bei *Lacerta agilis.* Anat. Anz. *6:* 57–61, 72–77 (1891).

WESTON, J.A.: A radioautographic analysis of the migration and localization of trunk neural cells in the chick. Develop. Biol. *6:* 279–310 (1963).

WESTON, J.A.: Cell marking; in WILT and WESSELLS Methods of developmental biology, pp. 723–736 (Crowell, New York 1967).

WETZEL, R.: Untersuchungen am Hühnchen. Die Entwicklung des Keims während der ersten beiden Bruttage. Arch. Entwmech. *119:* 188–321 (1929).

WEYGANDT, W.: Tierhirngrösse (Reprint: Forsch. u. Fortschr., Nachrichtenblatt d. Deutschen Wissensch. u. Technik, 4.Jahrg. Nr.34, Dez. 1928).

WHITTAKER, R.H.: New concepts of kingdoms of organisms. Science *163:* 150–160 (1969).

WIEDERSHEIM, R.: Der Bau des Menschen als Zeugniss für seine Vergangenheit, 2.Aufl. (Mohr, Freiburg 1893).

WIEDERSHEIM, R.: Vergleichende Anatomie der Wirbeltiere, 7.Aufl. (Fischer, Jena 1909).

WILLIAMS, R.T.; SCHUMACHER, H.; FABRO, S., and SMITH, R.L.: The chemistry and metabolism of thalidomide; in ROBSON *et al.* Embryopathic activity of drugs, pp. 167–182 (Little Brown, Boston 1965).

WILLIS, R.A.: The borderland of embryology and pathology (Butterworth, London 1958).

WILSON, J.G. and GAVAN, J.A.: Congenital malformations in nonhuman primates. Spontaneous and experimentally induced. Anat. Rec. *158:* 99–110 (1967).

WINFREE, A.T.: Spiral waves of chemical activity. Science *175:* 634–636 (1972).

WINDLE, W.F.: Neurofibrillar development in the central nervous system of cat embryos between 8 and 12 mm long. J. comp. Neurol. *58:* 643–723 (1933).

WINGSTRAND, K.G.: Comparative anatomy and evolution of the hypophysis; in HARRIS and DONOVAN The pituitary gland, vol. 1, pp. 58–126 (University of California Press, Berkeley 1966).

WOLPERT, L.: Positional information and the spatial pattern of cellular differentiation. J. theoret. Biol. *25:* 1–47 (1969).

WOLSTENHOLME, G.E.W. and O'CONNOR, M. (eds.): Growth of the nervous system. Ciba Symp. (Little Brown, Boston 1968).

YAKOVLEV, P.I. and WADSWORTH, R.C.: Schizencephalies. A study of the congenital clefts in the cerebral mantle. J. Neuropath. exp. Neurol. *5:* 116–130, 169–206 (1946).

YOSHIDA, Y.: A radioautographic study of histogenesis during the development of the telencephalon, thalamus, optic tectum and cerebellum from the neural tube in the chick (Japanese with English summary). Acta anat. nippon. *42:* 311–329 (1967).

YOUNG, J.Z.: The life of vertebrates (Clarendon Press, Oxford 1950, 1955).

ZIEHEN, T.: Die Morphogenie des Zentralnervensystems der Säugetiere; in HERTWIG Hb. d. vergl. u. exp. Entwicklungslehre der Wirbeltiere, Bd. 2 (Fischer, Jena 1906).

ZUCKERKANDL, E.: Zur Entwicklung des Balkens und des Gewölbes. S. Ber. Akad. Wiss. Wien math.-nat. Kl. *110* (3): 233–307 (1901).

ZUCKERKANDL, E.: Zur Anatomie und Entwicklungsgeschichte des Indusium griseum corporis callosi. Arb. neurol. Inst. Univ. Wien *15:* 17–51 (1907).

ZUCKERKANDL, E.: Zur Entwicklung des Balkens. Arb. neurol. Inst. Univ. Wien *17:* 373–409 (1909).

ZWILLING, E.: Teratogenesis; in WILLIER, WEISS and HAMBURGER Analysis of development, section XIV, pp. 699–719 (Saunders, Philadelphia 1955).

# VII. The Vertebrate Peripheral Nervous System and its Morphological Relationship to the Central Neuraxis

## 1. Remarks on the Peripheral Nerve Endings

From a viewpoint of control and communication engineering, the peripheral nerve endings can be regarded as *input* or *output transducer structures*. Input is provided by the activities of *receptors*, and output is manifested by the behavior of *effectors*. In a different terminology, receptors are represented by 'sense' organs and 'sensory' nerve endings, and effectors by muscles and glands.[1]

The classification of receptors,[2] depending upon the adopted criteria, can be formulated in a number of different ways, some of which involve considerable semantic ambiguities and highly arbitrary interpretations. In a relatively simple and consistent interpretation, based upon the nature of the relevant stimuli, mechanical receptors, chemo-receptors, and receptors to radiation[3] can be distinguished. Depending upon the relationship of the stimuli to configurational aspects of the organism (e.g. external surface, internal surface, depth of tissues related to some motor activities) *exteroceptive*, *interoceptive* and *proprioceptive* receptors can be defined in accordance with SHERRINGTON's views. A slight-

---

[1] Very special cases of *effectors*, apparently restricted, among vertebrates, to some fishes, are *electrical* and *bioluminescent organs*, which shall briefly be discussed in section 5 of this chapter.

[2] Cf. the introductory remarks on p. 4 *et passim* in volume 1, of this series. As regards the term 'receptor', it should be understood that it can here be applied to the input terminals of true neuronal elements (e.g. free nerve endings), to those of neuronal elements *sensu latiori* (neuroepithelial olfactory cells, rods and cones of retina, cells of saccus vasculosus), and to non-neuronal elements ('taste cells', 'hair cells' of otic organ, etc., to some extent perhaps also to the muscle fibers of muscle spindles).

[3] In this connection, *thermoreceptors* are usually enumerated as a special category. *Heat*, nevertheless, represents a *radiation*, although its energy may also be transmitted by *conduction* and *convection*. The problems concerning thermoreceptors are briefly discussed further below in dealing with the significance of the diverse types of sensory endings. Again, as a special case, *electroreceptors* seem to be present in electric fishes (cf. below p. 796 and p. 884 f. of this chapter).

ly similar, but somewhat more ambiguous classification, discussed in section 4 of the preceding chapter, recognizes *somatic* and *visceral* input respectively output.

SHERRINGTON (1906, 1947), moreover, stressed the concept of *distance receptors*, adapted to the registration of 'objects' at a distance, and including the optic, the auditory, and the olfactory sense, this latter being 'taste at a distance'. Although distance receptors may indeed in some respect be contrasted with touch or contact receptors registering 'objects' directly abutting on the body, SHERRINGTON did not overlook the obvious fact that all receptors are stimulated by 'material' events 'in direct contact with them'. Like all semantic formulations, this concept of 'distance receptors', doubtless valid to a certain extent, involves intrinsic weaknesses. Quite evidently, the ectodermal taste buds of some fishes can be considered distance receptors. Also, the lateralis system of Ichthyopsida, discussed further below, would have a rather ambiguous status in respect to that classification. Much the same could be said about cutaneous thermo-receptors.

From a generalized anatomical viewpoint, receptors may be classified as (a) *cutaneous or surface and membrane receptor structures* in the *wider sense*, thereby including receptors within *mucous membranes and glands*, (b) *muscular, tendinous and articular membrane* receptors, and as two rather complex sensory organs provided by (c) the *otic apparatus*, and (d) *the eye*.

The traditional enumeration of the *five senses* of touch, taste, smell, hearing, and sight, refers, of course, to the experienced modalities of human sensory consciousness, that is, to a set of P-events. Despite its oversimplification, this classification remains of considerable relevance with respect not only to commonsense but also to the epistemic problems thereby involved. It is understood that '*touch*' must here be taken in a wider connotation, that is, as subsuming an undefined variety of '*body senses*' exclusive of the four others (i.e. of sight, hearing, smell, and taste). Further distinctions on the basis of actual (conscious, including sphairal) perceptual experience have added heat, cold, pain, sexual voluptuousness (lust, '*Wollustgefühle*'), muscular and visceral perceptions, gravity and acceleration-deceleration perceptions respectively equilibrium, nausea, thirst, hunger, and fatigue, as 'minor' senses to the traditional 'major' ones. With additional refinement in analysis of this type, a much larger number of 'special senses', perhaps exceeding 30, could be distinguished, including, *inter alia*, the so-called '*Gemeingefühle*' of classical German authors.

It is generally conceded that receptors are particularly 'adapted' to 'adequate' or 'specific' stimuli, for which a relatively low threshold obtains. However, regardless of this adaptation, inadequate or unspecific stimuli above a certain strength can elicit a response producing the transmission of a nervous impulse. JOHANNES MÜLLER (1840) formulated, about 1838, the principle of so-called *'specific energies'* displayed by sensory nerves. According to this concept, any effective stimulation of a given type of receptor by a specific or an unspecific stimulus elicits always a *'sensation'* (e.g. P-event) of the same sort. Thus, electrical or mechanical stimulation of the optic nerve, e.g. by a blow on the eye or by cutting the optic nerve, produces a flash of light,[4] or stimulation of a cutaneous 'cold point' by heat may evoke a sensation of cold.

Although the term *'energy'* as here used by MÜLLER must now be considered a misnomer, his concept seems indeed based upon valid observations. However, in agreement with the subsequently recorded data, the particular specificity postulated by this author can be attributed to the cortical (or cortico-thalamic) N-events correlated with the relevant sensations (P-events) and occurring in particular regions (areae) of the cerebral cortex, such e.g. as the visual or auditory cortical fields.[5]

In other words, and in a simplified formulation, it can be assumed that the kind of experienced (conscious) sensation produced by a sense organ is not dependent on the type of receptor but on this latter's central connection with a particular griseum in the brain. Only certain neural events of a postulated but at least at present not definable specificity, characteristic for, and restricted to a region comprising that griseum, seem to be 'correlated' with (conscious) sensory perceptions of a given type. As regards (mental, psychic) sensory percepts, and very generally speaking, the following sorts of *differences* are experienced: (1) *qualitative*, (2) *quantitative*, (3) *spatial*, (4) *temporal*. Since the result-

---

[4] Also, in cases involving loss of both eyebulbs, irritation of the optic nerve stumps during the cicatrization process is known to be accompanied by optic sensations. Such patients, despite both 'empty' orbits, may then 'see' lights, sparks, fiery circles and dancing shapes. Again, in a more pointed fashion, it has been stated that, if the peripheral part of the cochlear nerve could be connected with the central portion of the optic nerve, and vice-versa, we would then hear the lightning and have a visual percept of the thunder.

[5] Cf. also p. 590 in vol. 3, part I, of this series, and p. 210 f. of the monograph *'Brain and Consciousness'* (K., 1957).

ing sensation or percept is a strictly private phenomenon whose comparable occurrence in other living beings is most clearly indicated by verbal communication, restricted to man, similar sensations or percepts in animals can merely be inferred on the basis of observed relevant nonverbal behavior. Evidently, because the central connections in lower vertebrates are significantly different from those in man and closely related mammals, inferences concerning presumed sensations and their specific nature in lower animal forms become increasingly uncertain. It is, moreover, understood that 'nerve terminations on effectors' refer to junctions concerned with peripheral neural output (efferent) activity, but not to afferent endings related to effector structures.

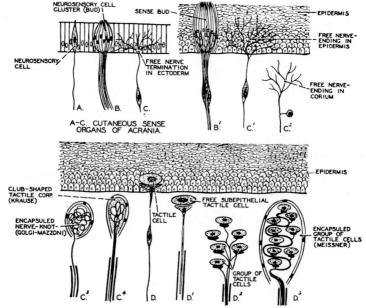

*Figure 401.* Diagrams depicting various types of cutaneous receptors in Chordates (after PLATE, from NEAL and RAND, 1936). A, B, C show receptors in the skin of Acrania (Amphioxus). B[1] to D[3] show receptors in Craniota; some aspects of C[1], C[2], D, and the sequence D–D[3] purport to represent stages in the hypothetical evolution of spinal ganglion cells and of encapsulated nerve terminations.

In accordance with the purpose of the present treatise, only a few topics pertaining to peripheral nerve endings shall be considered, with particular reference (1) to *cutaneous receptors* in the wider sense; (2) to *receptors of the muscular system*, and (3) to the *nerve terminations on effectors*. It is intended to bring appropriate comments on vascular presso- and chemoreceptors, on the otic apparatus, the eye, the olfactory organ, and on intracerebral receptors[6] in the special part of this treatise, dealing with the relevant particular brain regions (vol. 4, brain stem, cerebellum; vol. 5, diencephalon and telencephalon).

(1) *Cutaneous, mucosal and visceral receptors.* Cutaneous receptors in the Cephalochordate *Amphioxus* include neurosensory cells, either single or in clusters, within the ectodermal epithelium, as well as free nerve terminations branching in that skin layer (Figure 401) and perhaps other regions. The cutaneous receptors of Craniote vertebrates are provided by free endings in epidermis and corium, by nerve terminations on epithelial or subepithelial non-neural 'tactile' or 'gustatory' (chemoreceptor) cells, and by a variety of encapsulated nerve endings (Figure 401). On the other hand, true *peripheral* neurosensory (neuro-epithelial) cells, comparable to those found in the skin of Amphioxus, are only represented by the chemoreceptors in the olfactory mucosa of Craniote vertebrates. Concerning the peripheral sensory innervation of *man*, a detailed study of endings by means of a silver impregnation method has been published as a monograph by SETO (1963).

With regard to some histologic details, as seen in light microscopy, peripheral input nerve terminations are generally provided by branchings of an 'afferent' or 'sensory' axon. These ultimate ramifications, even if their immediate stem fiber is medullated, are commonly non-medullated.

Afferent *free endings* in skin, cornea and some mucous membranes form plexuses in epithelium and subjacent connective tissue (Fig. 401, $C_1$, $C_2$). Endings of this type are also present in the dental pulp. As regards mammals, intricate arrangements of free terminations can be seen surrounding the hair follicles. Palisade- and ring-like configura-

---

[6] The still poorly understood *intracerebral receptors* include those of saccus vasculosus, moreover presumed osmoreceptors in hypothalamus (cf. K., 1954, p.205) and perhaps brain stem, suspected chemoreceptors in area postrema, and, possibly within the reticular formation, additional elements acting as chemoreceptors for the blood stream ($CO_2$ respectively $H_2 CO_3$ concentrations, etc.). The presence of hypothalamic thermoreceptors, suggested by some authors, seems likewise not improbable.

tions are particularly conspicuous in the region of the hair bulb (Fig. 402). The vibrissae *('Spürhaare')* of some mammals display not only a complex innervation, but are also provided with an elaborate blood supply of their follicles, including relatively large venous sinuses. Such vibrissae have been designated as *'Sinushaare'*.

A type of receptor structure *transitional* between *free* and *encapsulated endings* is represented by *'tactile discs'* (tactile cells or tactile menisci), which are formed by disc-like terminal swellings of a nerve fiber in close apposition to an epithelial cell (Fig. 403). Somewhat similar nerve endings, 'sandwiched' between two cells of epitheloid type, which are surrounded by a more or less distinct connective tissue capsule are known as *corpuscles of Merkel* or *Grandry*, and could already be

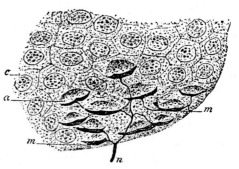

Figure 403

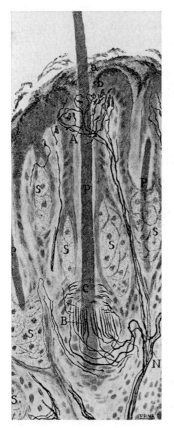

Figure 402

*Figure 402.* Nerve terminations about a hair follicle in a young rat as shown by silver nitrate impregnation (after TELLO, from CAJAL-TELLO, 1933). A: 'foraminal termination'; B: palisade endings; C: annular endings; N: small cutaneous nerve fiber bundle; P: hair root; S: sebaceous gland; a: epidermal stratum corneum; b: stratum germinativum Malpighii.

*Figure 403.* Tactile cells in cutaneous epithelium of the groin in a guinea pig as shown by gold-chloride technique (after RANVIER, from LARSELL, 1939). a: tactile cell; c: epithelial cell; m: tactile meniscus formed by nerve termination; n: nerve fiber.

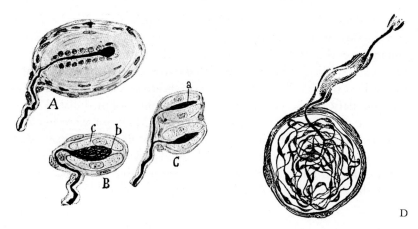

*Figure 404 A–C.* Receptors in the borders of the tongue and beak of the duck *( Anas)*, as displayed by silver nitrate impregnation (from CAJAL-TELLO, 1933). A: *corpuscle of Herbst;* B, C: *corpuscles of Merkel* (or *Grandry-Merkel*).

*Figure 404 D. End-bulb of Krause*, from margin of mammalian conjunctiva bulbi, as seen by the methylene blue technique (after DOGIEL, from LARSELL, 1939).

classified as encapsulated nerve endings. Among these latter, a variety of different forms, some of which are not easily distinguishable from each other,[7] have been described (Fig. 404). The globular *end-bulbs of Krause* contain glomerular loops of nerve terminals and are located in the cutaneous connective tissue; somewhat similar but smaller '*genital*

---

[7] Sceptical recent authors have claimed that many of the elaborate distinctions between types of encapsulated nerve endings might be based on artefacts produced by histological processing (cf., e. g., YOUNG, 1957; WEDDELL and SINCLAIR, 1953). Again, RUFFINI (1905) stressed the occurrence of numerous intermediate forms, precluding a rigid classification. The elongated encapsulated nerve endings represented by bulbous corpuscles (corpuscula terminali bulboidea), and found in the dermis of the Pig's snout as well as in several other locations, have been investigated by FITZGERALD (1962) with regard to their structure, distribution and life history in that Mammal. The cited author, who considers said corpuscles to be 'mechano-recorders', reports that their development begins shortly before birth. New corpuscles are then formed throughout the first year in snout and buccal membranes. Later the development of new corpuscles appears to proceed at a slower rate in the snout and to cease entirely in the cheek. Degeneration of bulbous corpuscles was observed in all the postnatal material. Local populations appear to depend on a 'balance struck between new formation and loss of end-organs'. Thus, as in the somewhat different case of 'taste buds', briefly discussed further below, a 'turnover' of these corpuscles is suggested. In the case of Meissner's and Pacinian corpuscles, there is apparently no evidence for a comparable 'turnover'.

*corpuscles*' are found in glans penis and in clitoris. Cylindrical *end-bulbs of Krause* are present in connective tissue (fasciae) of muscles, tendons, and in the oral mucosa. In the connective tissue of various viscera, TI-MOFEEW described oblong encapsulated corpuscles, with double inner-vation by a thicker fiber, whose myelin sheath extends about as far as the boundary of the corpuscle, and by a thin, nonmedullated fiber. Globular or elongated corpuscles of GOLGI-MAZZONI were likewise described in tendons, periost, and articular capsules. The elongated or roughly fusiform *corpuscles of Ruffini*, located in deeper layers of corium at transition to tela subcutanea, display connective tissue bundles in which the nerve endings ramify (Fig. 405).

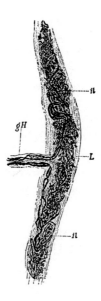

*Figure 405. Ruffini's end-organ*, as seen by that author's gold-chloride method (after RUFFINI, from LARSELL, 1939). gH: myelin sheath; il: terminal arborizations of the axon; L: connective tissue capsule.

---

*Figure 406 A. Meissner's corpuscle* in dermal papilla of a human finger as shown by the methylene blue technique (redrawn after DOGIEL, from MAXIMOW, 1930). a: thick myeli-nated fiber; b: thin myelinated or non myelinated fiber; ep: epithelium.

*Figure 406 B, C. Meissner's corpuscles* from dermal papillae of human fingertip. B Silver proteinate impregnation, showing nerve endings. C Routine hematoxylin-eosin stain (B × 600, C × 370, red. $^2/_3$).

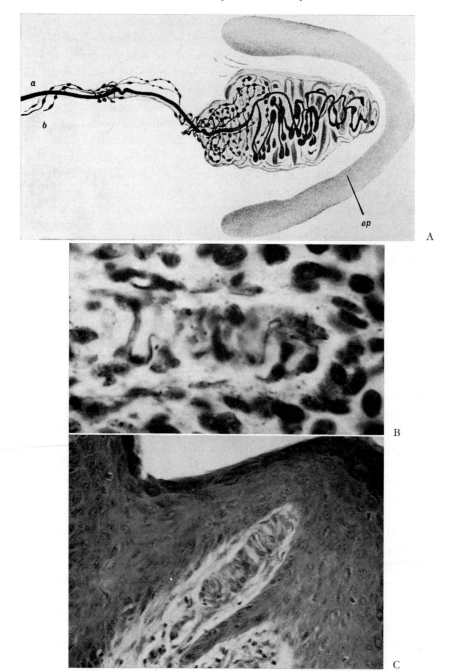

Figure 406

Rather characteristic encapsulated afferent nerve endings are the *tactile corpuscles of Meissner* and the large *lamellar Pacinian corpuscles*. The former are found in the corium papillae of the skin, especially in regions particularly sensitive to touch. These elongated or cylindrical structures, enclosed by a connective tissue capsule, contain a more or less columnar arrangement of transversely oriented flattened, plate-like 'epitheloid' cells between which the terminal nerve arborizations end (Fig. 406). A *corpuscle of Meissner* can be supplied by more than one medullated fiber. These relatively large myelinated axons envelop the lower portion of the corpuscle with spirals. The myelin sheath ends externally to the corpuscle, and the *sheath of Schwann* is said to join the capsule. The naked axons penetrate into the corpuscle. In addition, nonmedullated fibers, or fine medullated fibers losing their myelin sheath at a distance from the corpuscle, enter this latter, and form an accessory network of varicose fibers which apparently remains independent of the meshes pertaining to the above-mentioned main fibers. This 'additive' or 'auxiliary' arrangement is also known as the *apparatus of Timofeew*.

In contradistinction to *Meissner's bodies*, the fairly large[8] lamellated *corpuscle of Pacini* (or *Vater-Pacini*) are located in deeper structures, namely in the tela subcutanea of the skin, in the mesenteries, in the interstitial tissue of the pancreas, and in or near articular membranes. These corpuscles consist of a rather narrow, cylindrical inner bulb, approached by a thick medullated fiber which loses its myelin at the locus of entrance of the axon into the bulb (Fig. 407). Within this latter, it may display a reticular arrangement of neurofibrillar bundles. The main mass of the corpuscle is provided by the capsule, formed by concentrically arranged plates of lamellated connective tissue with flattened fibroblasts of 'endothelial' type and very fine interlaced fibrils. Elastic networks have been described in the external lamellae, which may also be supplied by small blood vessels entering the corpuscle in company of the nerve. In at least some *Pacinian corpuscles*, an *apparatus of Timofeew* has been observed; it is formed by thin nonmedullated fibers penetrating the inner bulb together with the main fiber, but ending in an independent terminal network.

---

[8] *Meissner's corpuscles* of man generally vary in length between 35 to 120 μ, and between 25–60 μ in width. Human *Pacinian corpuscles* may reach a size of 1 to 4 mm in length and up to 2 mm in width. A skilled dissector in the gross anatomy laboratory can obtain a preparation of volar digital nerves connected with a series of these corpuscles (cf., e.g., figure 943 in vol. 3 of the SPALTEHOLZ atlas, Hirzel, Leipzig 1933).

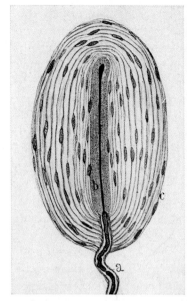

Figure 407 A

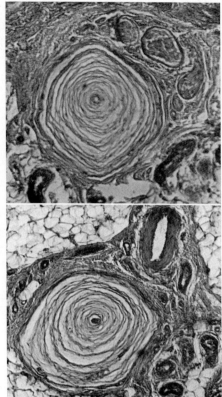

Figure 407 B

*Figure 407 A.* Pacinian corpuscle in man (from CAJAL-TELLO, 1933). a: '*sheath of Henle*' of medullated nerve fiber; b: 'central granular material'; c: concentric lamellae of capsule.

*Figure 407 B.* Cross-sections through *Pacinian corpuscles* in tela subcutanea of human fingertip as seen in routine hematoxylin-eosin preparations ($\times 95$, red. $^2/_3$). Inner bulb, thin bundles of cutaneous nerve fibres, and the lumina of capillaries and other small vessels can easily be recognized.

Another type of afferent peripheral nerve endings is displayed by the structures known as '*taste buds*', which evidently represent chemo-receptors.[9] These structures are flask-like or ovoid clusters of epithelial cells imbedded in the ectoderm of the skin or in the entoderm of a mucous membrane (Fig. 408). In many vertebrates, including man, the location of taste buds is restricted to the oral cavity and its neighborhoods. These structures occur particularly in some regions of the tongue, on the epiglottis, in the pharynx and in the palate. The parent tissue is here to a large extent, but not exclusively, *entodermal*. In some Teleost fishes, however, notably in Siluroids and Cyprinoids, the taste buds are greatly increased in number, and not only distributed within the oral cavity and gut entrance, but upon the entire *ectodermal* body surface,[10] from head to tail fin, being especially numerous on the barblets around the mouth (cf., e.g., HERRICK, 1924). On the other hand, it was shown that the skin of Selachians, which lacks taste buds or other distinctive sensory structures identifiable as specific chemoreceptors, is everywhere quite sensitive to a number of chemical substances (SHELDON, 1909). Quite evidently, this response must here be mediated by some of the other available afferent nerve endings.[11]

---

[9] The qualities of the sensory modality of taste, i.e. of experienced taste sensations (P-events) in man, have been reasonably well established as subsuming four different qualities, namely salty, sour, bitter and sweet. Inferences based upon behavioral manifestations seem to indicate that a comparable discrimination of chemical stimuli by the receptor structures obtains in other vertebrates, including fishes. Experimental data obtained in man have here disclosed a topographic distribution of the receptors mediating qualitatively specific taste sensations. Receptors sensitive to sweet material are said to be mainly located at the tongue's tip, and those responding to bitter substances at the base. Receptors to acid and salty substances overlap along the tongue's margin, with an additional overlap of responses to sweet and salty materials at the tip. 'Alkaline' and 'metallic' tastes may represent compound sensations. Taste and smell perceptions likewise become compounded in manifold patterns. Some doubts, moreover, remain as to numerous details of the taste sense, including its possible mediation by additional unspecified nerve endings in other regions of oral cavity (e.g. palate) and parts of the pharynx.

[10] The innervation of these structures is mediated by branches of a cranial nerve (VII, n. facialis) and shall be dealt with in section 3 of this chapter.

[11] Free neuromasts of the pit organs, pertaining to the lateralis system, discussed further below, have been suspected to serve as external taste buds, but the evidence remains inconclusive. Again, free nerve endings, some of which, in Amphibians, seem to act as chemoreceptors (cf. NOBLE, 1931, 1954) might also have a similar functional significance in Selachians.

*Taste buds* (Fig. 408), which open on the surface with a small pore, include two types of cells, namely sensory cells with a short, hair-like process projecting into the pore, and sustentacular cells. Both cell sorts display transitional forms. Afferent nerve endings are in close apposition to the taste cells, and arborizations of additional, *'perigemmal fibers'* surround the outer surface of the taste bud. Since, in contradistinction to the chemoreceptor olfactory cells, the sensory cells of taste buds are not provided with a neurite, but seem to transmit their excitation to the free nerve endings, these elements could be regarded as not representing true 'neuroepithelial cells'. Taste buds degenerate after their nerve is cut, and are not replaced prior to the regeneration of that nerve. The close functional relationship between nerve endings and sensory cells, which can be inferred on the basis of the structural arrangement, is thereby substantiated.

Recent studies have confirmed previous views suggesting that *'taste cells'* have a short life duration (allegedly as short as 'several days'), thereby representing typical 'labile elements' (cf. vol. 3, part I, p. 705f) which are continually replaced, apparently by the undifferentiated 'basal cells' of taste buds. It is believed that as a new taste cell becomes innervated by a terminal nerve fiber, which may just have lost its previous receptor cell contact, it could receive, from the nerve cell, 'information' determining the adaptation of the taste cell to a particular sort of chemical stimulus (cf., e.g., HANDLER, 1970).

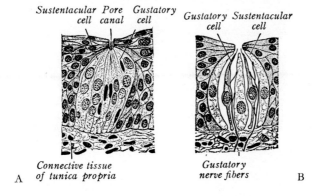

A Sustentacular Pore Gustatory cell canal cell — Connective tissue of tunica propria

B Gustatory Sustentacular cell cell — Gustatory nerve fibers

*Figure 408.* Taste buds from tongue of human neonatus (after AREY, 1954). A As drawn from actual section. B Interpretative diagram, omitting, however, the 'perigemmal' nerve arborizations in the epithelium surrounding the taste bud.

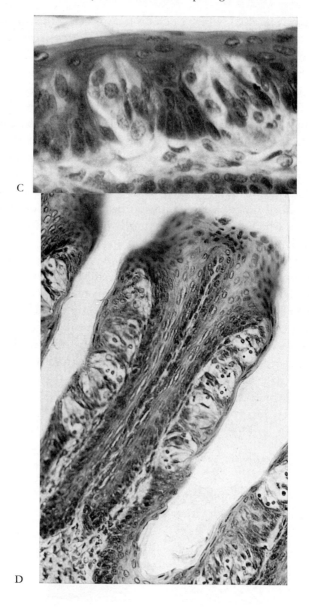

*Figure 408 C, D.* Taste buds from human (C) and from rabbit (D) papilla foliata (hematoxylin-eosin, C × 390, D × 150, red. ²/₃).

A peculiar system of cutaneous receptors is provided by the *lateral line organs* present in all groups of aquatic Anamnia, namely in fishes and Amphibia, and is thus restricted to Ichthyopsidan vertebrates (Fig. 409). These receptor structures, also designated as *neuromasts*, consist of isolated patches or rows of clustered epithelial sensory cells along certain lines on head and trunk, innervated by input (afferent) endings of cranial nerves (VII, IX, X).

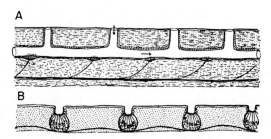

*Figure 409.* Diagrammatic representation of lateral organs in Gnathostome fishes. A Longitudinal section through lateral line canal showing sensory areas and their nerve. Arrows indicate direction of water current. B Isolated groups of neuromasts located in the skin (after Dean, from Romer 1950).

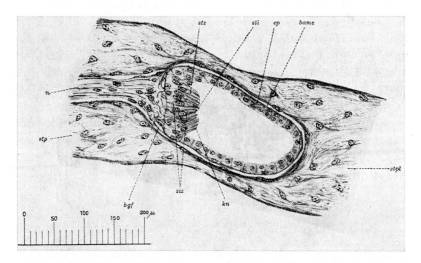

*Figure 409 C.* Cross-section through a lateral line canal in the tail of the Plagiostome *Torpedo* (from Krause, 1923). bame: basal membrane; bgf: blood vessel; ep: epithelium; kn: neuromast; n: nerve fiber bundle; siz: sensory cells; sti: sensory cell process; stpl: 'supporting plate'; stz: supporting cell.

A *neuromast* displays an evident, but essentially superficial resemblance to taste buds, and is a roughly pear-shaped aggregation of columnar sensory cells whose central ends, as a rule, do not reach the inner
limit of the epithelium, and whose outer extremities are commonly
provided with fine hairs. These latter can be imbedded in a gelatinous
or mucous cupula. The sensory cells of neuromasts in ordinary lateral
line organs usually possess a number of stereocilia and one kinocilium,
while those of so-called specialized lateral organs may lack cilia. Deformation of the sensory hairs toward the kinocilium is said to cause
depolarization of the sensory cell, directly or indirectly resulting in
the conduction of an impulse in the nerve fibers connected with the
organ.

*Neuromasts* are distributed as isolated *pit organs*, as linear rows within a slit-like groove, and as structures within a closed canal, opening at
intervals with pores, which can pierce the scales. Along the trunk, dorsal, lateral, and ventral rows of lateral line organs may occur, and be
enclosed in grooves or canals. Similar linear arrangements can form a
complex pattern on the head, including a supraorbital, an infraorbital
and a hyomandibular line. A main trunk line or canal may be continuous with the cranial system.

In *Cyclostomes* such as lampreys, the lateral line organs, distributed
in partly somewhat irregular rows, are very simple, each neuromast
being located in the center of a rather shallow open pit which is not
sunk into a canal. In *Selachians*, free neuromasts, neuromasts in slit-like
grooves, and canal neuromasts can be found. Single pit organs commonly lie between the bases of modified scales and may be partly covered by the exposed scale crowns. Other organs of the lateralis system,
located on the head, are highly modified as *ampullae of Lorenzini*, representing deep canals filled with mucus. The slightly vesicular blind end
contains the receptor elements[12] differentiated as a group of so-called
'terminal cells'. In addition, but apparently only on the head (rostrum)
of *Torpedo* there occur the entirely closed *vesicles of Savi* filled with fluid

---

[12] According to some observations (SAND, 1938) heat increases and cold decreases the
rate of their 'spontaneous' discharge. Since the nerve impulses originating in the *ampullae
of Lorenzini* are altered by very slight changes of temperature, these organs might act as
thermoreceptors (unless the thermal effect is coincidental and unrelated to an unknown
normal functioning of the ampullae under conditions of rather constant temperature).
MURRAY (1960, 1962), moreover, has shown that the *ampullae of Lorenzini* in Elasmobranchs
are sensitive to stimuli of different sort, including mechanical and *electrical* stimuli.

and containing a patch of sensory cells whose hairs are connected with a gelatinous 'cupula terminalis'. The detailed functional significance of these receptors, located in the subepidermal connective tissue and first described by SAVI, more than hundred years ago as '*appareil folliculaire nerveux*' (cf. LEYDIG, 1857), still remains unknown.

The *lateral line system* displays, on the whole, its highest and most typical development in *Osteichthyes* (Teleosts and Ganoids). This system may here be mostly contained in canals, entirely closed, except for their series of pores, and located within or below the body scales or within the substance of the head plates. In some Osteichthyans, however, the lateralis system is superficially situated. *Dipnoi* likewise possess a lateral line system consisting of pit organs and canals.

Among *Amphibia*, lateral line organs are present in all thoroughly aquatic larval and adult Urodeles, moreover in aquatic larvae of Anurans, being usually absent in the adult forms.[13] However, some aquatic types, such as Pipidae and Bombina, retain a lateralis system, which, on the other hand, is lacking in various terrestrial Urodeles. Gymnophiona, whose *larvae* are predominantly aquatic, include also some *adult* aquatic forms. A lateral line system with neuromasts is present and seems to function in the free-swimming larvae, but becomes lost at an early period. Nevertheless, in the larger river forms, it may be retained through much of the free-swimming larval life (TAYLOR, 1968).

Generally speaking, the lateral line organs of Amphibia are less specialized than those of most fishes and lie usually entirely within the epidermis as shallow pits. Only in some forms the neuromasts have partly sunk into the corium and are found at the bottom of 'pores'. In certain salamander larvae, the pits are located in the center of unpigmented, light round spots. Three trunk rows may be related to the lateral line nerves of which there is a postauditory component from the ninth and tenth cranial nerve, while the variable rostral rows are innervated by the preauditory component provided by the facial nerve. According to GOODRICH (1958), the distribution of the neuromasts in recent Amphibians agrees in general with that of 'primitive fishes' and is remarkably like that of Dipnoi.

---

[13] Lateral line organs are readily lost when amphibia become terrestrial. Thus, in the common newt *(Triton)*, these larval sense organs partially atrophy and are covered by adjacent epidermis during an intermediate terrestrial stage, but reappear again on the surface as fully functional structures when the newt takes up an aquatic habitat in adult life (NOBLE, 1931, 1954). Nevertheless, lateral line organs are not present in all aquatic Amphibia.

As regards the *functional significance* of the lateral line system, it seems well established, on the basis of inferences from structure and connections, as well as on experimental evidence, that these sense organs are essentially mechanoreceptors responding to water pressure against the side of the body. This involves, as it were, a sense of 'water touch', including the registration of vibrations at relatively low frequency. The lateral line system is thus rather obviously a neurosensory mechanism related to swimming activities. Its central connections, moreover, evidence a close functional affinity between lateral line and vestibularis systems. The function of the vestibular apparatus is commonly classified as proprioceptive, this latter term subsuming neural input providing information concerning movements and position of the body.[14] Yet, although the lateralis system can, accordingly, be considered *proprioceptive*, its anatomical arrangement and its functional 'adaptation' to external stimuli would also justify a classification as exteroceptive.

YOUNG (1950) quotes experimental results obtained by SAND (1937), indicating that lateral line organs may constantly discharge impulses, even when not under the influence of any external stimulation. A tailward flow of water is said to check, and a headward flow to accelerate this spontaneous discharge of impulses.[15] With regard to *electric fishes*, whose relevant organs shall briefly be considered in section 5 of this chapter, it seems possible that some sensory cells of the lateralis system have a dual function as mechano- and *electroreceptors* (cf. also above, footnote 12). In fact, certain so-called tuberous and ampullary lateral line organs of electric fishes may probably be pure *electroreceptors* (cf. SUGA, 1967). It has been justly and repeatedly remarked that, in animals significantly differing from man, and particularly in lower, in-

---

[14] Cf. the definition in *Dorland's Medical Dictionary*, edited by competent authors. Again, proprioception (proprioceptive input) as originally defined by SHERRINGTON (1906, 1947) is related to 'changes going on in the organism itself, particularly in its muscles and their accessory organs (tendons, joints, blood vessels, etc.)'. Strictly speaking, this would include some aspects of interoception, or of 'visceral afferent' activities. As now generally conceived by many authors, proprioception subsumes (a) input from muscle-spindles; (b) so-called deep sensibility, and (c) vestibular input.

[15] Concerning the dubious significance of *Lorenzini's ampullae* cf. footnote 12. Whether some neuromasts are or are not involved in other sorts of sensory registration still remains a moot question. The *vesicles of Savi*, whose relationship to the lateralis system remains questionable, and whose significance is not understood, were mentioned above in the text.

cluding invertebrate forms, there may be a variety of receptor structures which give responses of a type unfamiliar to us, and difficult for
us to understand. In this connection, one could also cite the *pit organ*
(BULLOCK and COWLES, 1952) of certain Ophidian *reptiles*, namely pit
vipers, of which the rattlesnake (*Crotalus*) is an example. Between eye
and nose on either side of the head is a pit filled with a peculiar highly
vascular tissue containing numerous nerve endings. Experimental evidence has shown this organ to be a heat-sensitive thermoreceptor by
means of which the snake can register the movement of a moderately
warm body past its head at a distance of several feet. Since rattlesnakes appear to be somewhat poorly endowed with 'normal' sense
organs, such a thermoreceptor is of evidently substantial use for the
capture of warm-blooded rodents.

Before dealing with the receptors of the muscular system, some additional remarks on the cutaneous nerve endings are perhaps appropriate, since their structure was briefly described above without particular
reference to their specific functional significance *qua* elicited sensations.
A meaningful discussion of this topic is evidently restricted to man,
whose experienced sensations can be regarded as generally known.

Reverting to the overall arrangement, it may be taken for granted
that single afferent nerve fibers contained in a peripheral nerve divide
into preterminal branches which link up with a group of receptors related to, or even provided by, the terminal arborizations respectively
networks of these branches. Such nerve fiber may therefore be conceived as related to a *receptive field*. Before and shortly after the turn of
the century, fundamental experimental studies by GOLDSCHEIDER
(1898), v. FREY (1910) and others demonstrated that four sorts of *cutaneous sensations*, namely *touch* (pressure), *cold*, *warmth*, and *pain* can only
be elicited by the stimulation of more or less densely distributed discontinuous, discrete *spots* (*Druckpunkte*, *Kaltpunkte*, *Warmpunkte*,
*Schmerzpunkte*). The number of *tactile spots* on the human body surface
has been estimated at about 640000, that of *cold spots* at about 250000
or more, and that of *warmth spots* at about more than 30000. *Pain spots*,
although lacking in some regions, e.g. an area within the cheek mucosa, seem to the particular numerous and may exceed two million. Other estimates, depending on the regions concerned, have given average
figures such as 3 warm spots, 15 cold spots, 25 tactile spots, and 100 to
200 pain spots per cm² for some locations.

In attempting to correlate these findings with the available histologic data, it has been widely assumed that *touch* sensations are mediated through the *free nerve endings* on *hairs*, through the *corpuscles of Meissner*, and through the *Pacinian corpuscles*, which, however, may particularly be concerned with pressure sensations. *Cold* sensations are believed to be mediated by the *end bulbs of Krause*, and sensations of *warmth* by the *corpuscles of Ruffini*. Yet, it appears that, in the external ear of man,[16] the only nerve endings present in the skin are of the 'free' type (including endings around hair bulbs). Nevertheless, as shown by tests, the external ear is no less sensitive to touch, cold, warmth, and pain than the skin of the forearm (SINCLAIR et al., 1952). Moreover, the nature of the actual stimulus registered by the *thermoreceptors* is improperly understood. It is presumably not the electromagnetic radiation itself, as, e.g., within a certain range, directly registered by retinal rods and cones, but rather an ambient molecular energy change (thermal movement, conduction) related to a *thermal diffusion coefficient* of the skin. Again, it should be kept in mind that the responses of thermoreceptors not only elicit temperature sensations, but also participate, at the level of unconscious N-events, in the thermoregulatory mechanisms of the body.

*Pain* sensations are believed to be mediated by *free endings* responding to various stimuli either damaging the tissues or incompatible with their function within a 'normal' range. Such stimuli may be mechanical, chemical, thermal, or electrical, and the receptors registering these 'disruptive' events have been designated by SHERRINGTON (1906, 1947) as *noci-ceptive* or *nocireceptors*. Free endings which could mediate pain are not only found in skin, tympanic membrane, and cornea, but also in serous membranes, viscera, blood vessels (arteries), periost, articular membranes, and muscles.

*Pain sensu strictiori* is, of course, a mental (or conscious) state, namely a P-event, which might be classified as a consciousness modality closely related to both affectivity (emotion) and sensory percepts.[17]

---

[16] Likewise, in the cornea of man, only free nerve endings are found. Although the cornea is preponderantly pain-sensitive (cf. further below), it seems well established that touch and cold sensations can also here be elicited by appropriate stimulation. Concerning the distribution of cold and warm spots, respectively the *sensitivity to temperature*, it is claimed that from the conjunctiva bulbi only cold, and from the glans penis only warmth sensations can be elicited (cf., e.g., SCHNEIDER, 1962).

[17] Further comments on pain and affectivity from the viewpoint of neurologic epistemology can be found in the author's publications '*Brain and Consciousness*' (1957) and '*Mind and Matter*' (1961). In these publications, and in the preceding monograph on the

With respect to affectivity, pain is characterized by qualities of various intensities which can be subsumed under the general terms avoidance, annoyance, unpleasantness, displeasure, dislike and distress. With respect to sensory percepts, there are painful body sensations of various sorts (mechanical and thermal), painful intensity of light, painful loudness,[18] intolerable odors, and excessively strong taste sensations.

In comparison with consciousness modalities such as sound, one might next ask whether a nonperceived, i.e. not-experienced pain has 'objective' or 'physical' existence. Quite evidently, and depending on arbitrary semantic use, sound can be described in terms (a) of physical vibrations (R-events); (b) of physical (biochemical, bioelectrical) neural signals encoding sound (N-events), and (c) of mentally experienced sound (P-events). With regard to *pain*, however, comparable particular physical events (R-events) cannot be properly defined, except (a) in the very abstract manner indicated above ('damaging stimuli', 'disruptive events'). Nevertheless, a certain group or pattern of neural signals, resulting from 'disruptive' stimuli, could, if sufficiently well identified, be defined as pertaining to, or encoding pain. These N-events, being likewise physical, would then represent another conceivable physical aspect of pain (b).The loosely used term *'pain'* refers thus to a complex set of multifactorial phenomena, including (a) extraneural physical (or chemical) events; (b) physical neural events in various patterns as well as at different levels, and (c) correlated mental events.[18a]

---

human diencephalon (1954), it was emphasized that every mental percept (including pain) is the 'parallel' correlate of a specific and complex pattern of neural events ('impulses', 'discharges', or signals) encoding the perception. The multifactorial aspect of said neural events, and the *'filtering'* activities of various grisea, previously suggested by LHERMITTE, LE GROS CLARK, and other authors' were likewise stressed.

[18] The pain produced by dazzling light has been attributed to spasms of the pupillary constrictor, and painful noise likewise to spasms of the small muscles connected with the auditory ossicles. These explanations may be considered highly unconvincing or, at best, as involving only a minor component of visual respectively auditory pain.

[18a] In recent time, particularly in connection with the increasingly sophisticated techniques of neurosurgery as well as with progressively neoprimitive trends in psychiatry and psychology, a mushrooming growth of so-called symposia on pain by diverse mutual admiration clubs of the 'establishment' has become noticeable. Despite the proliferation of new data, few of the elaborations presented by the participating authors attain the far higher level of many older publications on that topic, such, e.g. as that by LIVINGSTON (1943). In one of the typical recent symposia (edited by CRUE, 1970), a psychiatrist, quite gratuitously, and in very poor taste, illustrates (on p. 176) a semantic problem by the use

Since 'excessive' intensity of various sensory modalities respectively qualities is experienced as 'painful', some authors do not classify
pain as a modality, but merely as a 'too much'. I do not share this view
and prefer to consider the P-event of pain as a specific modality, which,
however, blends with, or is superimposed upon, other modalities (cf.
K., 1957, pp. 178–179). The available data, moreover, suggest that
even at the level of peripheral N-events, bodily or touch 'pain' represents a 'specific' sort of event, such that, e.g., the 'burning' 'pain' of
heat or extreme cold is not transduced by hot or cold spots, but by the
endings located in 'pain spots'.

As regards *body pain* which shall exclusively be considered in this
context, the experienced sensations can be described in numerous
ways. A widely accepted overall classification distinguishes *sharp* and
*dull pains* ('*stechend-heller*' *und* '*dumpfer*' *Schmerz*). With regard to localization, '*deep pain*' of viscera and arteries, pain of muscles, bone, periosteum, joints and tendons, toothache, and pain of cutaneous regions can
be distinguished. Again, there are well localized and *indistinctly or poorly
localized pain sensations.*[19]

*Body pain*, particularly of the cutaneous type *sensu strictiori*, can be
elicited by pricking, by plucking of hairs, by pinching or cutting, by
blows, bruising and mutilating, by heat or by cold, by electrical stimulation, and by action of strong chemicals. The *pain of inflamnation* may
result in part from the tension of tissues caused by hyperemia and ede-
ma, in part by vascular pain, and in part by the chemical effects of
substances 'liberated' or 'produced' in the tissues as a result of 'damaging events'. *Deep pain*, of visceral type, appears especially caused by the

---

of a vulgar simile including an ugly English four-letter word, which, in my opinion, has
a much more repulsive sound than the corresponding German 8-letter word or more
elegant French equivalent (*merde, le mot de Cambronne*), but which, nevertheless, can be
symbolically taken, to represent, in a *semi Freudian* way, a self-appraisal of the gentleman's
own symposium contribution. The alluded author's avowed purpose was 'to jolt'.
Having been, during a transitory stage of my World War I 1914–1918 military career, a
Royal Prussian Field Artillery master sarge and drill instructor, I failed to become shocked
or 'jolted' by this familiar barrack-yard 'household word' but was merely reminded of
the maxim in the *Ecclesiastes* (3, 1–22) stating that there is a proper time (and place) for
everything.

[19] The phenomenon of *referred pain*, namely the localization of a pain in a region quite
distant from the actual locus of the receptor nerve endings stimulated by a pathologic
process, shall be discussed in section 6 of this chapter (the Vegetative Nervous System)
and again in volume 4 (Spinal Cord and Deuterencephalon).

distention of hollow organs, by the stretching of such structures, as well as by chemical effects. Also, excessive dilatation, contraction, or blocking of arteries seems to elicit pain, which may likewise be triggered in some instances of tissue ischemia. Pinching and squeezing not only of skin, but of striated muscles can likewise be registered as painful. Moreover, the muscular apparatus is, in this respect, particularly susceptible to painful stimulation by chemical agents, which might also be involved in muscular pain during activity under ischemic conditions.

As regards the *'adequate' stimulus* directly acting on the pain receptors, some authors (e.g. SCHNEIDER, 1962) assume that it is of chemical nature, the relevant substances being released or produced by tissue damage, as mentioned above in the special case of inflammation. It remains, nevertheless, an open question whether direct mechanical stimulation (including stretching) might not also be commonly effective in a number of instances.

The experienced sensations of *tickling* and *itching*, particularly the latter, have been interpreted as a variety of cutaneous pain, mediated by the stimulation of pain receptors. Intense itching is known to be caused by the effect of histamine-like substances. In the case of tickling, a combined stimulation of light touch and of pain receptors has been suggested.

A further and still insufficiently clarified problem concerning pain sensations is presented by the fact that a single painful stimulus, applied to the skin, can elicit two sensations separated by a short time interval. Accordingly, a *'first'* or *'fast pain'* and a *'second'* or *'slow pain'* may be distinguished, the first being short and sharp, the second 'more prolonged and severe'. An additional, still later 'third type' of pain, persisting and of 'dull' nature is stressed by some authors. 'Fast pain' impulses are believed to be transmitted by medullated nerve fibers of the A-group, and 'slow pain' impulses by thin non-medullated fibers of the C-group.[20]

Elaborating on the well-known multifactorial aspects of pain, and the 'filtering' as well as 'modulating' activities of grisea suggested by a number of previous authors (cf. above, footnote 17), MELZACK and WALL (1968) have formulated what they call a *'code selection'* or so-called *'gate control theory'* of pain. The cited authors assume that, in the

---

[20] Cf. Volume 3, part I, chapter V, p.579 of this series.

spinal cord, the substantia gelatinosa[20a] functions as a gate control system modulating transmission of impulses from peripheral fibers to the
dorsal horn cells whose axons provide the long ascending tracts. The
afferent patterns in the dorsal column system are also presumed to act,
in part at least, as 'a central control trigger which activates selective
brain processes that influence the modulating properties of the gate
control system'. Finally, the long tract cells are generally said to 'activate neural mechanisms which comprise the action systems responsible
for response and perception'. This, of course, was already known since
the early studies of L. EDINGER before the turn of the 19th century.
MELZACK's and WALL's theory 'proposes that pain phenomena are determined by interactions among these three systems'. It goes without
saying that, despite its particular emphasis on so-called 'gate control',
the theory of the cited authors is based on familiar concepts which
have repeatedly been stated in essentially similar general terms.

In summarizing the available data on cutaneous and related internal
nerve endings of essentially exteroceptive and interoceptive type, it
could be said that, at least in man, some *free endings* are most likely *pain
receptors*, while others seem to be receptors for *touch* and even *thermoreceptors*.[21]

According to WEDDEL (1945), the network of nerve endings sub-
serving the sensation of pain in normal skin is disposed in such a manner that any one small cutaneous area is supplied by several overlapping terminals derived from different axons. Subsequent studies by
this author and his associates indicated that, when, under certain pathological conditions, an abnormal reaction to painful stimuli was obtained in some areas, the terminals were invariably isolated from each
other, in contradistinction to normal skin, in which a characteristic in-
terweaving of adjacent terminal fibers obtains.

It was therefore suggested that alterations in the peripheral pattern
of cutaneous innervation might be responsible for alterations in the

---

[20a] The close relationship of the nonmedullated dorsal root fibers and of the short
relay fibers in *Lissauer's tract* to the *substantia gelatinosa Rolandi* of the mammalian and human
spinal cord, as well as the significance of these structures for the transmission of pain
signals was particularly stressed by KARPLUS and KREIDEL, by RANSON, and by others
since about 1914.

[21] E.g. touch receptors in human auricle and cornea, thermoreceptors in auricle, and
thermoreceptors for cold in cornea, cf. above p. 798 and footnote 16. Again, it is assumed
that free endings may commonly form an *'overlapping mosaic'*.

quality of the sensation perceived following the application of a painful stimulus. Thus, WEDDELL *et al.* (1948) explained thereby the occurrence of '*protopathic*' pain [21a] in many clinical conditions affecting
the peripheral nerves.

As regards the pain sensibility of deep 'somatic' structures such as
muscles, fasciae and periost, FEINDEL *et al.* (1948) found that the thin
nerve terminations in these tissues differ both *qua* general pattern and
density from the arrangement in the skin. It appears thus not improbable that the difference in quality observed by LEWIS (1942) between
skin pain and deep somatic pain could be due to differences between
the pattern of innervation in skin and deeper tissue.[21b] Much the same
might be assumed with regard to some aspects of deep visceral pain
not referred to the surface (cf. above, footnote 19) but vaguely localized within the body's depth.

Reverting to the *free nerve endings of fishes*, it seems well established
that positive reactions to food substances has been obtained in some
Teleosts by stimulation of skin areas lacking taste buds (E. SCHARRER
*et al.*, 1947). It appears here likely that free endings, at least in these
vertebrate forms, can act as chemoreceptors. Again, some authors
have considered, with respect to fishes, 'general chemical sense' as distinct from the sense of 'taste', the former being mediated by free endings, particularly spinal nerves, and the latter by taste buds. Chemical
'sensitivity' is supposed to evoke 'negative' or 'defensive' reactions,
while 'positive' reactions are exclusively mediated by the 'taste' sense.
The findings by SCHARRER *et al.* (1947) contradict such distinction
based on behavioral reactions.

*Tactile discs*, and, among encapsulated nerve endings, *corpuscles of
Merkel, of Meissner* and *of Pacini* can be considered *touch receptors*.[22] The
significance of the *apparatus of Timofeew* associated with these two sorts

---

[21a] Protopathic and epicritic sensibility, as defined by HEAD (1920), was discussed in
the preceding volume 3, part I, p.695f.

[21b] LEWIS (1942 and other publications) also postulated a particular and peculiar type
of innervation by so-called '*nocifensors*' related to the clinical phenomenon of hyperalgesia.
LIVINGSTONE (1943) and others, however, believe that the anatomic characteristics of the
subepithelial nerve net as described by WOOLLARD (1936) and WEDDELL (1941) and presumed
to subserve pain sensations, would fulfill all the requirements demanded by LEWIS to
explain the activities he has ascribed to the nocifensors.

[22] *Meissner's corpuscles* perhaps for relatively light touch, pressure or stroking, and
*Pacinian corpuscles* for stronger pressure and 'traction', involving 'deformation' of the skin.
One might also consider the question whether the experienced sensation of mechanical
vibration (disregarding sound) is elicited by a particular set of receptor structures.

of corpuscles remains obscure. One could here suspect superimposed pain receptors, but some observers claim that stimulation of *Meissner's corpuscles* does never elicit pain. The *end-bulbs of Krause* can be interpreted as *cold receptors*, and, on somewhat less satisfactory evidence, the *corpuscles of Ruffini* might be considered warmth receptors. It should be added that the pain caused by excessive heat or cold cannot be elicited through the stimulation of warmth or cold spots, but only through stimulation of pain spots. Accordingly, such pain is not registered by *Ruffini's* or *Krause's corpuscles*, but presumably by nociceptor free endings.

Other peripheral nerve endings whose functional significance seems reasonably well established are those in chemoreceptor taste buds and in the essentially (but perhaps not exclusively) mechanoreceptor *neuromasts* discussed further above. Yet, discounting the just enumerated partially understood special cases, the classification of cutaneous respectively body sensations and the detailed identification of the receptors involved remain in an unsatisfactory state (cf. YOUNG, 1957). It should be recalled that in addition to touch, warmth, cold, and pain, and apart from proprioception, numerous other aspects of exteroceptive and interoceptive body sensations obtain, as referred to on p. 780. If the differences between the various sorts of described encapsulated endings are significant, a satisfactory functional classification for a number of these structures cannot be given.

RUFFINI (1905), also cited by Granit (1955), remarked that, '*de 1891 à aujourd'hui, par l'application des méthodes plus électives au chlorure d'or et au bleu de méthylène, le nombre des formes connues s'est énormément accru, si bien que nous ne croyons pas exagérer en disant que la peau, tant en surface qu'en épaisseur, est littéralement remplie d'expansions nerveuses. Et tandis qu'avant cette époque les fonctions étaient beaucoup plus nombreuses que les formes, aujourd'hui, par contre, cettes-si ont pris leur revanche sur les premières*'. Thus, GRANIT (1955) quotes attempts by BAZETT *et al.* (1932) to determine as completely as possible the number of sensory experiences obtainable from a piece of skin (prepuce) that afterward was sacrificed for histological study. They could not record more than *four types of sensation*, namely heat, cold, touch, and pain, but the piece was found to contain *seven distinct types of sensory endings*. One might, however, doubt whether the preputial skin can be regarded as a significantly representative cutaneous region. Yet, reverting to the overall problem under discussion, one can accept as still valid YOUNG's conclusion of 1957, namely that no general solution of these difficulties is possible at present.

Although a *relative specificity*, i.e. adaptation of peripheral receptor structures to certain types and ranges of stimuli is known to obtain, this specificity is not entirely exclusive or absolute. Experienced sensations, moreover, depend on the activities of central mechanisms[23] and on the relevant transmission channels within the neuraxis. Generally speaking, it seems that the lateral spino-thalamic tract transmits already partly processed signals related to pain and temperature registration,[24] while ventral spino-thalamic tract, and the components of bulbo-thalamic lemniscus medialis, appear to be essentially channels for signals registering touch, spatial configuration and detailed (e.g. finer discriminative) qualities of somesthesis *sensu strictiori*. Accessory channels, including relay systems are presumed to be present.[25]

(2) *Receptors of the muscular system*. In HERRICK's (1931) interpretation of SHERRINGTON's terminology, the proprioceptors are a system of sense organs 'found in the muscles, tendons, joints, etc., to regulate the movements called forth by the stimulation of the exteroceptors'. All reactions concerned with motor coordination, with muscular tonus, with maintenance of posture or attitude of the body, and with equilibrium implicate the proprioceptive system. Quite obviously, proprioceptive activities involve, to a very substantial extent, typical feedback processes (cf. vol. 1, p. 8f of this series).

Again, others have defined the proprioceptors as those end organs which are stimulated by movements of the body itself, and comprise receptors in the muscular system, the labyrinth, as well, to some ex-

---

[23] These central mechanisms, including the presumed action of the *cortical modulating circuits* mediating the *affective tone* (prefrontal circuit, parahippocampal and hippocampal *circuit of Papez*) have been dealt with in the author's monographs on the *Human Diencephalon* (1954) and on *Brain and Consciousness* (1957). On p. 183f. of this latter publication, the rare condition characterized by *congenital insensitivity to pain* was also dealt with. This anomaly may be due to functional, biochemical alterations of cortico-thalamic processes without gross or microscopic structural changes, but perhaps with some still undefined ultrastructural modifications. A more recent case was reported by BAXTER *et al.* in Brain, *83:* 381, 1960.

[24] No significant data seem to be available concerning the question whether this tract is composed of distinctive respectively separate fibers for the transmission of the encoded pain, warmth, and cold signals.

[25] The relevant fiber tracts and their connections within the central neuraxis shall be discussed in volumes 4 and 5 of this treatise.

tent, in the viscera. Still others[26] distinguish general proprioceptors concerned with sensations of joint movement and position (kinesthesis), and special proprioceptors represented by the vestibular apparatus. Muscle spindles and *Golgi tendon organs*, although they can evidently be considered 'general proprioceptors', are said to make no contribution to the 'kinesthetic sense' (as experienced by man).

Moreover, should afferent endings in the heart, in the musculature of bloodvessels, and in smooth musculature in general, be classified as proprioceptive or as interoceptive respectively as viscero-receptive structures? Afferent endings of various sorts have been described in epicardium, endocardium and myocardium, and likewise within bloodvessels. Terminations of that type seem also to be present within the smooth musculature of the gastrointestinal, respiratory, and urogenital tracts. Relevant data on that whole group of afferent endings, however, are rather scarce, scattered, and still quite ambiguous. Further references to this topic can be found in the treatises by KUNTZ (1953) and PICK (1970).

In the terminology which I am inclined to adopt, the proprioceptive system may be tentatively defined to subsume the muscular and tendon receptors, those in or about articular capsules, and the vestibular receptors. Postponing, as mentioned above, further comments on the vestibular system, and omitting the previously considered receptor endings, including *Pacinian corpuscles*, of fasciae and articular membranes, only muscle spindles and *tendon organs of Golgi* need here to be briefly dealt with.

*Muscle spindles* represent a particular set of afferent nerve endings which are commonly spirally wound around specialized muscle fibers, several of which form a fusiform bundle surrounded by a sheath. Spindles of this sort are found only in amphibians and Amniota, and thus seem not to be present in fishes. These structures, moreover, are re-

---

[26] Because of the semantic traps related to the intrinsic limitations of language, not only the distinction, by definition, between somatic and visceral organ systems, but also that between proprioceptive, exteroceptive and interoceptive functions can be, and has been, variously expressed by different authors, depending on the arbitrarily selected criteria. Thus, e.g., an evidently exteroceptive system such as that of the lateral line organs, could be considered to overlap with, or to be included among the proprioceptive mechanisms, particularly in view of the close relationship between lateral line nerves and vestibularis system (cf. above, p. 796). Finally, v. BONIN (1963, see ref. to chapter VI) seems to consider *proprioception sensu strictiori* only as a function of muscle spindles, in contradistinction to '*deep sensibility*' mediated by tendon receptors, endings in articular capsules, etc.

stricted to striated musculature (of so-called 'skeletal type'), being apparently absent in cardiac and smooth muscles.

Typical muscle spindles of mammals and man are each composed of about two to six or more thin intrafusal muscle fibers enclosed within their common connective tissue sheath (Figs. 410 A, B). Each fiber consists of two contractile striated 'polar' portions separated by an 'equatorial' non-contractile and non-striated nuclear bag containing

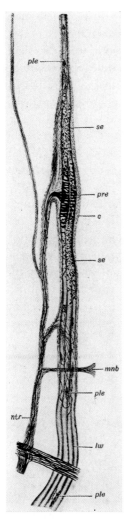

Figure 410 A

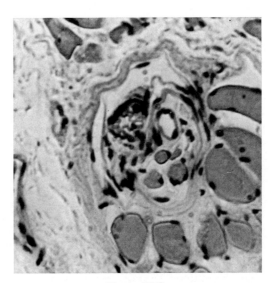

Figure 410 B

*Figure 410 A.* Muscle spindle of a cat as displayed by *Ruffini's gold chloride technique* (redrawn after RUFFINI, from MAXIMOW, 1930). c: capsule; lw: axial muscle fibers of the spindle; mnb: motor nerve fibers; ntr: nerve fiber bundle; ple: motor end plates; pre: spiral ('annulospiral') endings on nuclear bag; se: branched ('flowerspray') endings on myotube.

*Figure 410 B.* Cross-section through muscle spindle in human omohyoid muscle, as seen in a routine hematoxylin-eosin preparation ($\times 225$; red· $^7/_{10}$).

several nuclei. Each of the two contractile regions receive a motor innervation by thinly medullated efferent fibers (so-called γ-fibers) terminating with an end plate. The afferent or receptor fibers are provided by large, heavily medullated A-fibers whose terminal arborizations form a helicoid primary (or annulo-spiral) ending on the nuclear bag, and two secondary (or flower spray) endings on an intermediate region (myotube) transitional between nuclear bag and contractile 'polar' regions. The primary receptor fiber is thus wound around a region that can be passively stretched by the pull of the two ends.

Depending on the muscles concerned (e.g. extrinsic ocular muscles), and on the different taxonomic groups of terrestrial Vertebrates, various simpler but essentially similar types of muscle receptors have been described. Thus, in *amphibians*, the muscle spindles apparently do not receive their motor innervation by specific γ-fibers, but rather by collateral terminal branches of the 'ordinary' motoneurons. The development of two distinctive and separate systems, namely a set of efferent neurons innervating the spindle muscle fibers in addition to the motoneurons innervating the 'ordinary' muscle fibers seems to be a subsequent evolutionary refinement reaching a high degree of efficacy in mammals and man.

The *Golgi tendon organs* are specialized receptor structures, commonly located at the junction of a muscle fiber with the tendon, and consisting of palisade-like terminal branches (Fig. 411). Their innervation is mediated by heavily medullated fibers of the A-type. It is noteworthy that the muscle spindle receptors are arranged *in parallel* with the contractile elements, while those of *Golgi tendon organs* (also occasionally called *neurotendinal spindles*) are connected *in series* with the active structures. The significant response of a muscle spindle appears to be a

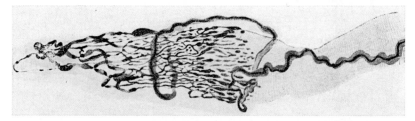

*Figure 411.* Golgi tendon organ located at junction of muscle fiber and tendon, showing palisade-like terminal branchings, as visualized by the methylene blue technique (redrawn after Dogiel, from Maximow, 1930).

signal discharge when the muscle is stretched or the spindle itself contracts. The discharge would cease upon contraction of the muscle itself but is then maintained if the small muscle fibers of the spindle are activated. The large afferent fibers from the muscle spindle receptors (particularly the annulo-spiral endings) enter the dorsal horns of the spinal cord and their branches directly connect with large ventral horn motoneurons, thus mediating a monosynaptic reflex. Ascending branches have ultimate connections with the cerebellum, either through spinocerebellar tracts or through posterior funiculi and the links of their terminal nuclei with the cerebellum.

The *tendon organs of Golgi* appear to become stimulated when the muscle is passively stretched as well as when it contracts. The response of the *muscle spindles*, however, seems to be *adjustable*, by means of their motor $\gamma$-fibers, to a large variety of changing conditions. These fibers can control the 'bias' of the receptor in such a way that it fires impulses at the same rate at any length while the muscle contracts. A complex pattern of feedback activities is thereby made possible, and, in this respect, the *modus operandi* of muscle spindles has been likened to that of the pupil, whose variable aperture controls the illumination of the retina.

If a muscle is stretched, the resulting afferent impulses from the spindle receptors, by means of the above-mentioned monosynaptic reflex, elicit a discharge of the motoneurons resulting in a contraction of the stretched muscle ('monosynaptic stretch reflex). This counteraction to stretch appears significant for the maintenance of a stable posture. The discharge of spindle effectors seems also to inhibit, presumably by way of an internuncial connection, the motoneurons of antagonistic muscles. Further details concerning the functional significance of the proprioceptive muscle receptors can be found in the publications of LEKSELL (1945) GRANIT (1955, 1970) and YOUNG (1957).

In addition to their segmental and intersegmental distribution, the messages from muscle spindle receptors may also transmit information, by way of some of the ascending channels, to grisea within the brain, as mentioned above. These messages, however, do not seem to represent a direct or significant input for the human or mammalian cerebral cortex, while deep sensibility (kinesthesis), as mediated by lemniscus medialis or spinothalamic systems involves substantial cortical connections. With regard to possible proprioceptive functions of afferent endings in cardiac and smooth muscle, relevant detailed data do not seem to be available. Nevertheless, LARSELL (1921), NETTLESHIP

(1936) and others have described structures interpreted as receptor endings in the musculature of viscera.

Although it is intended to postpone a discussion of vascular chemo- and pressoreceptor mechanisms for inclusion into the special part of this treatise, concerned with the pertinent central grisea, it seems perhaps appropriate, before dealing with the nerve terminations on effectors, to mention a few preliminary general data pertaining to the relevant vascular receptor nerve endings.

The *carotid* and *aortic bodies*, particularly developed in mammals,[26a] and also known as *glomus caroticum* respectively *glomus aorticum*, represent small configurations involving arteriovenous anastomoses *(retia mirabilia)* by sinusoid vessels lined with epitheloid cells intimately adjacent to the endothelial lining. The carotid body, innervated by n. glossopharyngeus (r. intercaroticus) is usually located within the bifurcation of common carotid, and the generally paired aortic bodies, innervated by a vagus branch (commonly from the n. recurrens), are closely adjacent to the concavity of arcus aortae. The receptor nerve terminals end, with a variety of ramification types, on the epitheloid cells or between these latter and the endothelium. It is not unlikely that the epitheloid cells, which may perhaps derive from (mesodermal) modified smooth muscle cells, represent the receptors sensu proprio. These chemoreceptors seem to be stimulated by lowering of the oxygen level of the blood as well as, to an apparently lesser degree, by an increase in $CO_2$-tension, respectively H-ion concentration. The nerve terminals on the presumed receptor cells can be classified as *'free endings'*.

---

[26a] These bodies were formerly classified as chromaffine paraganglia, but that interpretation could not be upheld. The either neuroectodermal ('parasympathetic') or mesodermal vascular origin of its epitheloid cells is not yet clarified; on the basis of my own incidental observations I am inclined to favor the probability of a mesodermal, vascular provenance. A *carotid body* seems to be present in at least some birds and reptiles, but not in Anamnia. It does not correspond to the so-called *'carotid gland'* or *'carotid labyrinth'*, which is a peculiar type of *rete mirabile* found in Anuran and Urodele *amphibians* (cf. footnote 320a, p. 717, in section 7 of the preceding chapter VI).

The *glomus coccygeum* of man and mammals, although histologically somewhat similar to carotid or aortic glomera, appears to be an arteriovenous anastomosis *(rete)* not related to chemoreceptor functions. I was unable to find conclusive or serviceable data on the presence, in Submammalian forms, of vascular pressoreceptor structures comparable to the Mammalian carotid sinus. A detailed monograph on vertebrate carotid body and sinus has been published by ADAMS (1958).

The *carotid sinus* is a more or less fusiform enlargement of the internal carotid's initial portion, and is characterized by a structure of elastic arterial type. Its afferent nerves (in man) are branches of n. glossopharyngeus (r. intercaroticus) and of n. vagus. The initial trunk and the arch of the aorta, of typical elastic arterial type, receive afferent nerve terminal of n. vagus (respectively its so-called n. depressor branch, which is commonly a quite distinctive ramus of n. vagus in various mammals). In some mammals, particularly in those whose internal carotid is poorly developed or vestigial, an initial segment of the *occipital artery* is believed to be a significant vascular pressoreceptor region. The pressoreceptors, presumably represented by various sorts of *'free endings'* in adventitia or between the elastic lamellae of media, are stimulated by a stretching force, as exerted by a rise in arterial pressure.

These vascular chemoreceptors and pressoreceptors provide impulses for central feedback mechanisms affecting respiratory rate and blood pressure.[26b] The cell bodies of their afferent nerve fibers are presumably located in the glossopharygeal and vagal ganglia, possibly in their more distal components (ganglion petrosum respectively ggl. nodosum). Spontaneous activity displayed by both the chemoreceptors and the pressoreceptors here under consideration has been recorded (cf. Fig. 334, p. 592 of the preceding vol. 3, part I, showing the rhythmic discharges obtained from the rabbits carotid sinus). The inferred central mechanisms related to the activities of these receptors shall be discussed in chapter IX of volume 4.

(3) *Nerve terminations on effectors.* Disregarding output of electricity and phenomena of bioluminescence, both to be briefly considered in section 5 of this chapter, the vertebrate effectors, concerned, either 'spontaneously' or under control of neural signals, with motion of masses, comprise muscles and glands. In contradistinction, electrical organs and luminescent organs represent effectors essentially emitting 'energy'.

---

[26b] Generally speaking, stimulation of the pressoreceptors slows the heart rate and reduces blood pressure. As an additional effect, respiration is likewise somewhat inhibited. Stimulation of the chemoreceptors mainly increases rate and depth of breathing but may also cause a rise in blood pressure. Relevant data on these vascular receptors can be found in the publications by SUNDER-PLASSMANN (1930), HEYMANS *et al.* (1933, 1953), CASTRO (1928, 1951), and ADAMS (1958).

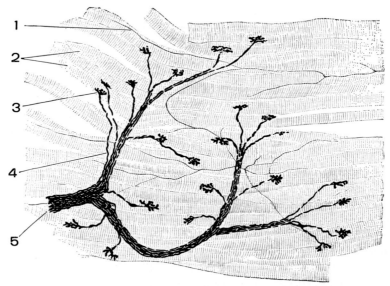

*Figure 412.* Motor end-plates in the rabbit's intercostal musculature, as visualized by the gold-chloride technique (from Stöhr-Möllendorff, 1933). 1: 'sensory nerve fiber'; 2: muscle fibers; 3: motor end-plate; 4: medullated nerve fibers; 5: nerve fiber bundle.

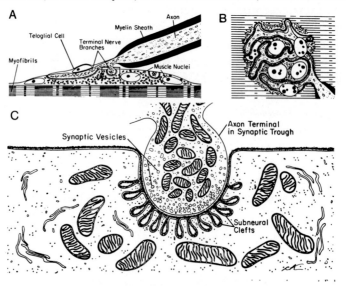

*Figure 413.* Diagrammatic interpretation of motor end plate on the basis of electron microscopy. A Total view of end-plate in longitudinal section, combining data obtained by high-resolution light microscopy and by electron microscopy. B As seen in surface view. C Area indicated by the rectangle in A, as seen in electron micrographs (modified after Couteaux, from Bloom and Fawcett, 1968).

Generally speaking, the striated musculature (also including what is commonly called, in man, the 'voluntary musculature') is innervated by nerve endings designated as '*motor end plates*' or *myoneural junctions*, which represent synapses between the efferent nerve and the effectors (Fig. 412, 413). Medullated nerve fibers of A-type undergo repeated divisions before terminating, and a single axon of a motoneuron may thus innervate, in numerous instances, significantly more than 100 individual muscle fibers. The myelin sheath ends in rather close vicinity of the end plate. It was formerly believed, on the basis of the observations by means of light microscopy, that the terminal nerve arborization were 'hypolemmal', namely located internally to the sarcolemm within an expanded region of sarcoplasm. However, some authors, using refined techniques already expressed doubts concerning this intrasarcoplasmic termination before the data provided by electron microscopy conclusively demonstrated that the nerve endings remain *epilemmal*.

The *motor end plate* is a slightly elevated plaque on the muscle fiber, containing the club-like terminal arborizations and an accumulaton of nuclei. Distally to the end of the myelin sheath, the *Schwann cell cytoplasm*, enclosing the axon as it approaches the myoneural junction, pulls away from the axon terminal and, over the junction area, merely covers the outer surface of the axonal branches. This lid-like cover contains the so-called '*teloglial nuclei*'. The axonal terminations are separated by a synaptic cleft from the sarcolemma lining of the muscle fiber and local accumulation of sarcoplasm including the '*sole nuclei*' of the muscle fiber. The expanded axon terminals contains synaptic vesicles and mitochondria. The primary synaptic cleft (or '*synaptic trough*') expands into narrow secondary synaptic clefts. Within the synaptic clefts, axolemma and sarcolemma remain, moreover, separated by a thin glycoprotein boundary layer apparently similar to that investing the rest of the surface of the muscle fiber.

In the years between 1911 and about 1940, BOEKE and a number of other authors claimed, on the basis of experimental as well as histological data, a double efferent innervation of the striated musculature by medullated somatic and nonmedullated vegetative sympathetic or parasynpathetic fibers. Thus, some efferent, autonomic, unmyelinated so-called '*accessory nerve fibers*' had been described as joining the motor end plate. The vegetative nerve terminations on striated muscle fibers were commonly interpreted as mediating the 'plastic tonus' of the musculature. Rather elaborate theories on this topic, e.g. by KEN

KURÉ and his school, and by LANGELAAN, have been propounded, including triple, namely parasympathetic, sympathetic, and somatic efferents, but both the histologic and the experimental data could not be corroborated by a number of independent observers. These perhaps somewhat unconvincing concepts of 'double' or 'triple' innervation are therefore no longer generally accepted. A discussion of the problem can be found in chapter XVII of KUNTZ' treatise on the autonomic nervous system (1953, 4th ed.) and in REGELSBERGER's contribution to MÜLLER's 'Lebensnerven und Lebenstriebe' (1931, 3rd ed., pp. 793–804).

In contradistinction to this innervation of striated musculature in Craniote Vertebrata by motor end plates representing the terminations of efferent peripheral nerve fibers, a peculiar mode of muscular innervation is described for the Cephalochordate Amphioxus (FLOOD, 1966). The trunk muscles of this Chordate form an unbroken series of segments (myotomes) extending throughout the entire length of the animal and are innervated by fibers which, as seen in light microscopy, appear to be identical with ventral root fibers in Craniote vertebrates. However, because of considerable difficulties in tracing these fibers to their assumed cells of origin, SCHNEIDER (1879) and ROHDE (1892) had already suggested that the ventral root fibers of Amphioxus were actually processes of the muscle fibers reaching the surface of the spinal cord and obtaining their innervation at this location. Recent studies by means of electron microscopy confirmed this interpretation. FLOOD (1966) found that the ventral root fibers of Amphioxus appear to be structures pertaining to muscle fibers of the myotome, and extending toward the spinal cord. On the surface of this latter, the *fibrous muscular projections* end in conical expansions, separated from an extensive layer of axon endings by an extracellular cleft containing a basement membrane.

It is well-known (cf., e.g., MAURER, 1915) that this peculiar mode of innervation, whereby 'the muscle goes to the nerve' occurs, among invertebrates, in Nematoda (roundworms). Long processes of the muscle cells reaching the main nerve cords have been described by many observers (SCHNEIDER, 1866, and others). It is possible that a few instances of such innervation migth also occur in some particular structures of certain Echinoderms (FLOOD, 1966), but, as far as is generally known, this sort of neuromuscular relation seems to be essentially restricted to Amphioxus and Nematodes. However, *qua* peculiar modes of muscular innervation, and *mutatis mutandis*, one is here reminded of the unconvincing and apparently unsubstantiated claim by DART and SHELLSHEAR (1921, cf. footnote 156, p. 391 of preceding chapter VI, and references to that chapter) who interpreted spinal motoneurons as cells of the mesodermal primitive somites secondarily incorporated into the neural tube.

As regards *efferent endings on cardiac*[27] *and smooth*[28] *musculature*, specialized structures comparable to motor end plates have not been recorded. Nevertheless, nonmedullated axons presumably representing postganglionic fibers of the vegetative nervous system appear to make synaptic connections on the surface of such muscle cells, controlling respectively mediating various sorts of activities such as vasomotor, pilomotor, peristaltic, etc.

Electron microscopic observations show that, in the myocardium, vesiculated processes of nonmedullated nerve fibers make contact with the surfaces of muscle cells (THAEMERT, 1969). In smooth musculature, WATANABE (1969) noted that as the unmyelinated nerve bundles pass into the narrow clefts between the muscle cells, the *Schwann cell* covering disappears. The naked axons may run along grooves on the surface of the muscle cells, and form elongated, slightly bulbous nerve endings in close apposition to the innervated cell (Fig. 414). Many synaptic vesicles of different sorts are packed within the nerve endings. These latter could be classified into two main types. One contains *small synaptic vesicles without core* and occasional larger ones with a central core of moderate density. The other type includes small vesicles with core of high density, mixed with larger cored and small noncored vesicles. WATANABE (1969) believes that the small cored vesicles contain noradrenaline and its precursors. Accordingly the second type of ending is interpreted as being adrenergic, while the first type might be choliner-

---

[27] It should be recalled that the initiation of the heart beat does not seem to be dependent upon the nervous system. Nevertheless, parasympathetic autonomous (vagus) and sympathetic autonomous innervation exert a negative respectively positive bathmotropic, inotropic, dromotropic and chronotropic influence upon cardiac muscular activity.

[28] With regard to smooth muscle innervation, there is some uncertainty concerning the question whether, in all instances, each single cell receives its own motor nerve ending, or whether excitation transmission, resulting in contraction, might in part be mediated by coupling of several muscular elements. Be that as it may, the observations of WATANABE (1969) discussed further below, disclosing a very profuse innervation of the smooth musculature in the guinea pig's vas deferens, are compatible with the assumption of individual innervation for each element in this particular instance.

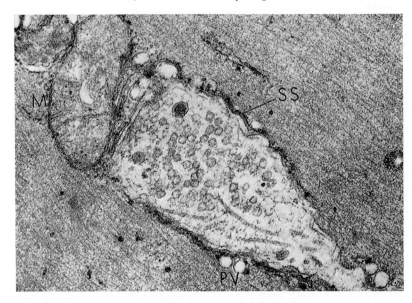

*Figure 414*. Efferent nerve terminal on smooth muscle cell of vas deferens in the guinea pig as seen in electron photomicrograph at a magnification of about ×48 000 (from WATANABE, 1969). M: mitochondria; PV: pinocytic vesicles; SS: elongated subsynaptic sac just beneath plasma membrane of muscle cell; the mitochondria and pinocytic vesicles are found in the vicinity of this sac.

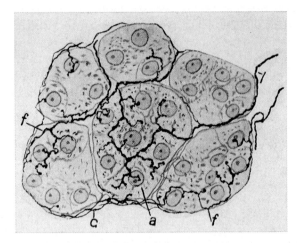

*Figure 415*. Efferent nerve fiber endings on glandular epithelial cells in the pancreas of the adult rat, as seen by means of the *Golgi technique* (after CASTRO, from CAJAL-TELLO, 1933). a: glandular fundus with periacinous terminations; c: fibers of the periacinous plexus; f: interepithelial terminations; l: nerve fiber proximal to ultimate branchings.

gic. The cited author noted, in some instances, the innervation of a single smooth muscle cell by both types of nerve endings.

In addition to the 'motor' endings on muscular effectors, there are 'autonomous' postganglionic *efferent nerve endings on the secretory effectors*, i.e. on the essentially epithelial gland cells. It will be recalled that nerve endings in epithelial layers of skin and mucous membranes are generally of afferent type, while those in epithelial glands may be afferent as well as efferent. It is, as a rule, not possible to distinguish, with a sufficient degree of accuracy, these two types of endings by means of the techniques used for standard light microscopy. However, further clarification can be expected from the current investigations with electron microscopy, which discloses relevant ultrastructural details of the synaptic junctions.

Glandular nerve endings are frequently provided by more or less dense plexuses of nonmedullated fibers within the connective tissue membrane or stroma. Rami separate from these bundles, and their terminations end freely in close contact with the basal surfaces of the glandular elements, often also between these latter, within the narrow intercellular space (Fig. 415). Some of the slight swellings of these terminal arborizations can be interpreted as neuroglandular synapses. The chromaffine cells of adrenal medulla and comparable paraganglia seem to be innervated by preganglionic sympathetic fibers, and are generally considered to represent modified 'postganglionic' neuronal elements. This interpretation agrees with the well substantiated neuro-ectodermal ontogenetic derivation of these cells (cf. chapter V, section 1, pp. 53–55, and Fig. 44, p. 58 in volume 3, part I of this series).

## 2. Spinal Nerves; Morphologic Problems of Neuromuscular Relationship

The vertebrate central neuraxis, whose activities control the reactions upon events occurring in the environment and interior of the body, is connected with surface and extraneural internal parts of the organism through the peripheral nerves. These latter include *afferent* or *efferent fibers* or *both*.

From a morphologic viewpoint, the peripheral nerves of Craniote vertebrates may be subsumed under *three main groups*, corresponding to the 'natural' subdivision of the neural tube into *spinal cord*, *deuterencephalon* and *archencephalon*. The first group comprises the *segmental spinal nerves,* the second is represented by the cranial or *cerebral nerves of the deu-terencephalon.* This latter can be regarded as a modified and specialized rostral portion of the spinal cord, partly or completely enclosed within the cranium (cf. sections 2 and 5 of chapter VI). Accordingly, these cranial nerves may be interpreted as modified respectively specialized spinal nerves. The third group subsumes the *nerves of the archencephalon,* and is related to the special senses of olfaction and vision. The fila olfactoria, which form the olfactory nerve, are the only peripheral nerves of Craniote vertebrates provided by neurites of primary sensory cells that can be evaluated as neurons. The nervus terminalis, absent or rudimentary in various forms, is closely related to the olfactory apparatus. The so-called optic nerve, namely the cranial respectively cerebral

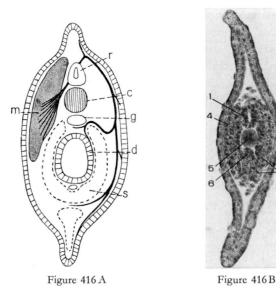

Figure 416 A                    Figure 416 B

*Figure 416 A.* Diagrammatic section through the body of Amphioxus caudal to gills, but rostral to anal opening (modified after HATSCHEK, from K., 1927). c: notochord; d: gut; g: vessel; m: myotomic segment; r: spinal cord; arrangement of dorsal and ventral roots, not labeled, is self-explanatory.

*Figure 416 B.* Cross-section through tail region in a 6-mm frog larva, showing neural tube, notochord, tail-gut manifestation, and somite-like mesodermal configuration, for comparison with Fig. 416 A. The characteristic vertebrate bauplan displayed by this section is topologically homeomorph with that obtaining in Amphioxus (hematoxylin-eosin, $\times 120$, red. $^1/_2$). 1: neuraxis; 4: epimere (somite); 5: notochord; 6: gut (tail gut); 8: hypomere (somatopleure and splanchnopleure). Cf. figures 416 A and C.

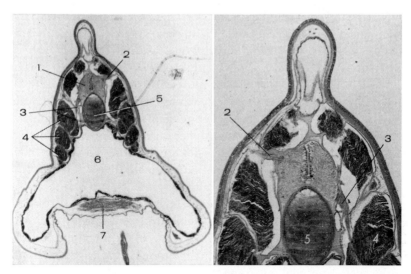

*Figure 416C.* Cross-section through the anterior region of the body in Amphioxus, showing dorsal and ventral roots (hematoxylin-eosin, I×120, II×360, red. $^1/_2$). 1: spinal cord; 2: dorsal root; 3: ventral root; 4: myotomes; 5: notochord; 6: oral cavity near transition to pharynx; 7: 'visceral' musculature.

nerve II is not a genuine peripheral nerve, since the retina constitutes an evaginated or protruded portion of the archencephalic tube. Only the spinal nerves and those of the deuterencephalon shall be considered in this chapter.[29]

Depending upon arbitrary postulates of definition, a *fourth group* of peripheral nerves can be recognized as pertaining to the *vegetative or autonomic nervous system*, briefly discussed further below in section 6. The peripheral components of this system, however, are represented by partly overlapping constituents of spinal and of deuterencephalic cranial nerves.

As regards the *spinal nerves*, a very simple respectively 'primitive, arrangement is found in the Cephalochordate Acranian *Amphioxus* whose overall bauplan, regardless of the dubious phylogenetic relationships, displays a remarkable isomorphism with respect to Craniote vertebrates (cf. Fig. 416). Two segmentally arranged series of fiber bundles, which do not join, thus remaining completely separate from each other, here connect the spinal cord with the peripheral structures.

[29] Pertinent comments on olfactory, terminal, parietal, and optic nerve are to be included in the relevant chapters (XII, XIII) of volume 5.

The *ventral roots* link the ventral portion of the neural tube with the myotomic muscle segments and give the appearance of essentially motor nerves innervating the 'somatic' musculature. However, as discussed above in section 1, these ventral roots do not represent nerves and have been identified as fibrous muscular processes (FLOOD, 1966), apparently effecting junctions of synaptic type at the surface of the spinal cord. These ventral roots, despite their striking overall similarity to those of Craniote vertebrates (if seen within the average resolving ranges of light microscopy) are thus not homologous but merely *analogous* to the corresponding structures in Craniota.[30]

According to AYERS (1921; cf. also KAPPERS *et al.*, 1936) these so-called ventral roots display four main branches, namely an anterior one connected with dorsal 'fork' of the preceding (rostral) myotome, two branches linked with dorsal and ventral 'fork' of the myotome pertaining to the 'root' segment, and one branch running to the ventral 'fork' of the adjacent caudal myotome. A ventral 'root' would thus be connected with a succession of three myotomes. The majority of the 'ventral root' fibers, however, are related to the (middle) myotome which can be considered as belonging to the segment of the 'root'. The third of these branches was described as containing fibers supplying 'visceral organs' respectively part of the vascular system. Other authors, including myself (K., 1927), assumed that some afferent, proprioceptive fibers might also be contained in the ventral roots. In view of the discontinuity between 'ventral roots' and neuraxis, as recorded by FLOOD (1966) in the electron microscopic picture, the presence of some actual nerve fibers in these 'roots' appears now quite doubtful, but cannot be entirely excluded.

The *dorsal roots* link the dorsal portion of the spinal cord with the integument, the nonmyotomal muscles of the ventral part of the body, with the gut, and perhaps additional structures. Accordingly, these dorsal roots[31] include afferent and efferent nerve fibers. The cell bodies

[30] Assuming, of course, that the in my opinion rather convincing findings and conclusions of FLOOD (1966) are correct. YOUNG (1955) regarded, like myself (1927), the ventral roots of *Amphioxus* as identical with those of vertebrates, i.e. carrying motor fibers, which 'end on the muscle fibers with motor end plates similar to those of vertebrates' (YOUNG, 1955, p.37). My own *Amphioxus* material, prepared for overall morphologic purposes, was not suitable for the study of finer structural relationships even at high light microscopic resolution.

[31] BONE (1960) lists 'somatic-sensory', visceral-sensory', and 'viscero-motor' fibers. PETERS (1963) identified the cellular elements scattered within the dorsal roots as sheath cells on the basis of electron microscopic findings.

of these latter, and likewise those of perhaps most afferent fibers are located within the neuraxis, but some afferent fibers originate from in-
tradermal nerve cells situated beneath the epidermal epithelium, and
perhaps, as claimed by DOGIEL (1903) but contested by others (e.g.
FRANZ, 1923), from epidermal, neuroepithelial primary sensory cells.
Although, on somewhat inconclusive evidence, nondescript cells scattered within the first and second dorsal spinal roots have been interpreted as rudimentary spinal ganglia by some authors, it seems most
likely that the dorsal roots of Amphioxus do not display a spinal
ganglion comparable to that of Craniote vertebrates (BONE, 1960).
Again, KAPPERS (1947) considers the first and second spinal dorsal
roots of Amphioxus to be homologous with n. ophthalmicus and n.
maxillo-mandibularis, respectively, in Craniota. Stressing that such
correspondence, at best, should be evaluated as kathomology, I am
inclined to agree with the cited author. These two dorsal roots, moreover, are located rostrally to the first ventral one.

The connection of the independent dorsal and ventral roots of Amphioxus with the spinal cord is not located at the same transverse plane
level, but follows an alternating order. The ventral roots essentially correspond to the adjacent myotomic segment, while the dorsal roots
link with the neuraxis at the level of the intermuscular septa between
adjacent myotomes. In addition to this alternating arrangement, a further asymmetry obtains, right and left halves of the body being somewhat obliquely displaced against each other, as if by a shearing distor-

*Figure 417.* Diagram showing arrangement of ventral and dorsal spinal nerve roots in
Petromyzon. The dorsal root displays a spinal ganglion (after JOHNSTON and KAPPERS,
from K., 1927).

tion, such that in addition to their alternation, dorsal and ventral roots are not located at one and the same transverse level with respect to their antimeres.

Among the true vertebrates, the spinal nerves of the *Cyclostome Petromyzon* manifest an arrangement which includes some features rather similar to those obtaining in Amphioxus.[31a] Dorsal and ventral spinal nerve roots remain entirely independent from each other and enter, respectively leave, the spinal cord at alternating transverse levels, but in a more or less symmetric manner (cf. Fig. 417). The *ventral roots* are essentially *efferent*,[32] supplying the myotomic musculature, but seem also to contain smaller, presumably afferent fibers reaching the septa between the muscle blocks, and perhaps concerned with 'proprioceptive' function.

The *dorsal roots* consist of *afferent fibers* from the skin and possibly other, e.g. visceral, organs. In contradistinction to Amphioxus, however, whose sensory nerve cells are mostly intramedullary, the bipolar cell bodies of the afferent nerve fibers are here gathered in a typical *spinal ganglion* forming a nodular or spindle-shaped swelling of the dorsal root. Nevertheless, some primary afferent neuronal elements are also present in the spinal cord of Cyclostomes. In addition to the afferent fibers, it is believed that, as in Amphioxus, *visceral efferent fibers* pass through the dorsal roots (cf., e.g., KAPPERS *et al.*, 1936).

*Selachians* display a configuration of the spinal nerve segments essentially corresponding to that found in all Gnathostome vertebrates (Fig. 418). Dorsal and ventral roots are linked with the spinal cord at approximately one and the same transverse level in a fairly symmetrical fashion. Both roots join distally to the spinal ganglion, thereby forming a *segmental common spinal nerve* which divides into 'mixed' branches supplying different body regions, and comprising a ramus dorsalis, a

---

[31a] In Myxinoids, however, a variable degree of union between dorsal and ventral roots has been described (cf. KAPPERS *et al.*, 1936). The ventral roots, moreover, display a number of peculiarities, such, e.g., that, at certain levels, one dorsal and two ventral roots can be found on each side.

[32] In contradistinction to the peculiar mesodermal respectively 'muscular' ventral roots of *Amphioxus* discussed further above, those of Cyclostomes and all other Craniote vertebrates are, of course, neurites of intraspinal motoneurons, i.e. true nerve fibers. It might here be added that, generally speaking, the nerve fibers in the nervous system of Cyclostomes and of *Amphioxus* do not display myelin sheaths.

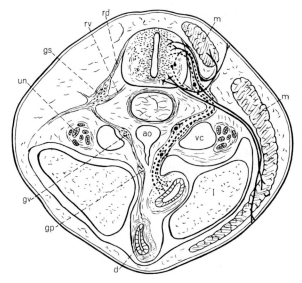

*Figure 418 A.* Diagrammatic section through the body of a Selachian embryo at advanced developmental stage, showing the configuration of a spinal nerve segment (modified after Froriep and Edinger, from K., 1927). Broken lines indicate visceral (afferent and efferent) fibers. ao: aorta; d: gut; gp and gv: sympathetic ganglia (paravertebral ganglia; a true differentiation into vertebral and collateral ganglia comparable to the arrangement in mammals does not obtain in Selachians); gs: spinal ganglion; l: liver; m: musculature; rd: radix dorsalis; rv: radix ventralis; un: mesonephros; vc: cardinal vein. The ramus recurrens and the postganglionic sympathetic fibers (so-called gray ramus communicans) joining dorsal and ventral common spinal nerve rami are not shown.

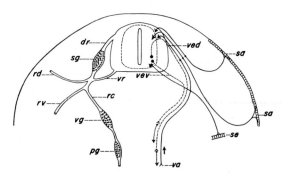

*Figure 418 B.* Simplified diagram representing features of preceding figure. dr: dorsal root; pg: prevertebral sympathetic ganglion (the distinction between vertebral and prevertebral sympathetic ganglia is poorly manifested and barely recognizable in Selachians); rc: ramus communicans; rd: ramus dorsalis; rv: ramus ventralis; sa: somatic afferent fibers from skin; se: somatic efferent fibers to musculature; va: visceral afferent fiber from gut; ved: visceral efferent fiber in dorsal root; vev: visceral efferent fiber in ventral root; vg: vertebral ganglion of sympathetic; vr: ventral root.

ramus ventralis, and a visceral branch, designated as ramus communicans, connecting with the sympathetic trunk.[33] In addition, a ramus recurrens s. meningeus is directed toward the vertebral canal with its enclosed meninges and blood vessels.

As regards the muscles of the trunk, the dorsal musculature is innervated by the rami dorsales, and the ventral musculature, which includes that of paired appendages (pectoral and pelvic fins, extremities
of Tetrapods), by the rami ventrales. The ventral rami innervating
these vertebrate appendages commonly interweave by means of anas-
tomosing branches, forming plexuses, such as plexus cervicobran-
chialis and plexus lumbosacralis of man.

Since, in the aspect here under consideration, the overall arrangement of a spinal nerve segment can be regarded as significantly identical in all Gnathostome vertebrates, further comments on this topic
may be restricted to a few remarks on segmental innervation by the
*spinal nerves of man* (Fig. 419 A.)

With respect to the striated musculature it should be recalled that
most individual muscles derive, directly or indirectly[34], by one-
many as well as many-one transformations, from *several* respectively
*different* myotomes and are correspondingly *innervated by several spinal
nerve segments*. Again, one and the same myotome or genetically equivalent segmental blastema can participate in the development of different
muscles which then receive nerve fibers from a common segment
(Fig. 419 B).

---

[33] The innervation of the '*viscera*' involves the so-called *vegetative nervous system*,
briefly discussed further below in section 6 of this chapter. Although, strictly speaking
and by definition, said system represents only the efferent component, the relevant
questions of afferent 'visceral' innervation shall likewise be dealt with in that section.

[34] In the embryos of lower vertebrates, even including at least some Amniote reptiles,
the myotomes, as they grow ventrally into the body wall and reach the level of the lateral
folds from which the appendages develop, give off lateral buds into the appendicular
folds. Within the anlage of the appendage these buds subdivide into dorsal and ventral
portions, from which the levator (extensor) and depressor (flexor) muscles of the appendage develop. In mammals and man, on the other hand, many muscles, including those of
the extremities, do not arise from myotomes, but by cell migrations from mesodermal
blastemas. These modes of development, however, also occur in lower forms, such as
Amphibia, and seem to overlap. Similarity of innervation, which will be discussed further
below, supports the concept of homology for appendicular muscles throughout the vertebrate series (cf. also RAND and NEAL, 1936).

As regards cutaneous innervation, each spinal nerve segment supplies a zone of integument designated as *dermatome*, which displays overlap (Fig. 419 C). A significant difference obtains here between trunk and extremities. In the trunk, each dermatome corresponds to the course of its peripheral nerve, which is a posterior or anterior ra-

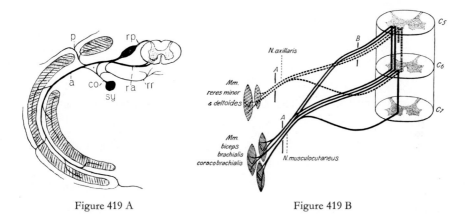

Figure 419 A                              Figure 419 B

*Figure 419 A.* Diagram of a typical spinal nerve segment (thoracic segment) in man (after VILLIGER, from K., 1927). a: ramus anterior innervating ventral (intercostal) musculature; co: ramus communicans; p: ramus posterior innervating dorsal musculature; ra: radix anterior; rp: radix posterior with spinal ganglion; rr: ramus recurrens; sy: sympathetic trunk ganglion.

*Figure 419 B.* Diagram illustrating human segmental (radicular) and peripheral motor innervation of muscles (from VILLIGER, 1933). A: peripheral nerve; B: segmental radicular bundle of motor fibers; $C_5$ to $C_7$: cervical spinal segments participating in formation of brachial plexus.

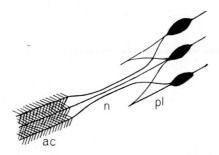

*Figure 419 C.* Diagram illustrating human segmental radicular and peripheral innervation (after VILLIGER, from K., 1927). ac: three overlapping dermatomes; n: peripheral cutaneous nerve; pl: plexus, of which the dorsal root components, including their spinal ganglia, are shown.

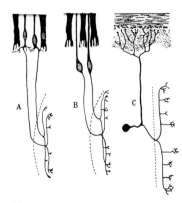

*Figure 420.* Diagrammatic illustration indicating the phylogenetic origin of (spinal or cranial) dorsal root ganglia according to the interpretation by RETZIUS, CAJAL, and others (modified after RETZIUS by CAJAL, from SCHARF, 1958). A: neurosensory cells (primary sensory cells) of the earthworm *Lumbricus;* B: bipolar neurosensory cells of the Oligochaete Annelid worm, *Nereis;* C: typical afferent neuron of vertebrate spinal (or cranial) ganglion.

mus of the segmental spinal nerve (e.g. n. intercostalis). In the extremities, however, since, because of plexus formation, segmental fibers are distributed upon several different peripheral nerves, which, in turn carry fibers from different segments, the dermatomes do not correspond to the areas supplied by individual nerves.

Thus, with respect to motor and sensory innervation by peripheral nerves originating from plexuses, damage to a root or a spinal cord segment will cause disturbances in regions or structures innervated by several nerves, and damage to a peripheral nerve involves disturbances in regions or structures which may be innervated by several segments.[35]

Before concluding the present section with a short discussion on the significance of neuromuscular relationship in respect to problems of morphologic homology, a brief comment on two additional topics concerning the vertebrate spinal nerve segment is perhaps appropriate, namely on the *phylogenetic origin of the spinal ganglia* and on the so-called *law of Bell-Magendie.*

---

[35] '*Der Ausfall einer Wurzel erzeugt Störungen in Gebieten, die mehreren Nerven angehören, und der eines sensiblen Nerven trifft Felder, die von mehreren Wurzeln her versorgt sein mögen*' (EDINGER, 1912). The details of these relationships, of special importance for clinical neurology, are concisely dealt with in the publications by VILLIGER (1919, 1933) and by HAYMAKER-BING (1956, 1969).

Assuming that the receptor or sensory elements in the skin of invertebrate precursors were essentially neuroepithelial cells located in the epidermis, one could draw the conclusion that bipolar and subsequently pseudounipolar afferent nerve cells of vertebrates originated by a displacement or migration of these elements toward the neuraxis as indicated by Figure 420. The displaced elements, aggregating in the dorsal roots near its entrance into the central nervous system, thus formed the spinal or dorsal root ganglia. A number of authors, such as CAJAL, HANSTRÖM, v. LENHOSSEK, RETZIUS, and others have adopted this view (cf. EDINGER, 1912; SCHARF, 1958, with detailed bibliographic references). HANSTRÖM's concept is illustrated by Figure 195, p. 281 in volume 2 of this series.

On the other hand, assuming that most afferent neurons of the early Chordate forms which evolved into Craniote vertebrates were intramedullary, as in Amphioxus, one would then be led to infer an origin of dorsal root ganglia by peripheral migration of these cells (cf. Fig. 426). The occurrence of dorsal primary afferent cells within the spinal cord of some adult lower forms, the transitory so-called *Rohon-Beard first order afferent cells* in the spinal cord at larval respectively embryonic stages of various Anamnia and Amniota, and the peculiar primary afferent elements of the mesencephalic trigeminal root located within the brain stem of all vertebrates can be interpreted as providing substantial evidence for this view.[36] Such migration, moreover, would be in accordance with the general principle of neurobiotaxis, dealt with in section 7, chapter V, of the preceding volume.

Nevertheless, the concept of the opposite, central migration, which is also supported by the relationship of the placodes to various cranial nerves, cannot be easily discounted, particularly since the principle of

---

[36] Further data concerning the intramedullary dorsal primary sensory neurons, including the transitory *Rohon-Beard cells* shall be included in chapter VIII of volume 4. The primary afferent nerve cells of radix mesencephalica trigemini will be discussed in chapter XI *et passim* of that volume. This general topic, with relevant references, is also dealt with in sections C and D of SCHARF's (1958) *Handbuch* contribution. It may here be added that one of the earliest identifications of dorsal primary sensory cells was made in *Petromyzon* by SIGMUND FREUD (1877), who, subsequently turning to other fields of neurobiology, later developed his theory of psychoanalysis.

Still earlier, however, is the identification of these cells by REISSNER (1860), likewise in Petromyzon. Thus, some authors, e. g. BURCKHARDT (Arch. mikr. Anat. 34, 1889) have referred to some large dorsal cells in the spinal cord of Anamnia as '*Reissner-cells*'.

neurobiotaxis, despite its presumptive morphogenetic significance, should not be considered as having absolute validity, but rather as a frequently manifested general tendency allowing for numerous exceptions. It is not unlikely that *both* central and peripheral migration of elements might have taken place in the phylogenetic evolution of dorsal root ganglia. In connection with the discussion of cranial nerve morphology in the following section 3 of this chapter, further comments on this topic will be included. Whether intraspinal or intramedullary afferent perikarya of the first order, respectively of *Rohon-Beard type* should or should not be considered homologous to spinal ganglion cells depends on arbitrary definitions of homology. The question whether genetically different 'generations' of cells are hereby involved can here be discounted as topologically irrelevant.

Reverting to the functional significance of peripheral nerves, the distinction between *afferent (input)* and *efferent (output)* fibers is self-evident and can, to a substantial extent, be recognized on the basis of merely anatomical (structural) features displayed by their connections. Additional functional concepts, such as somatic and visceral, or exteroceptive, interoceptive, and proprioceptive, were dealt with in section 4 of the preceding chapter *et passim*.

In accordance with the widely used classification of nerve fibers under the categories somatic and visceral with their further subsets, one can interpret the *dorsal spinal nerve roots* in Amphioxus, fishes, amphibians and perhaps some Amniota as somatic afferent, visceral afferent, and visceral efferent. Discounting the peculiar *ventral roots* of Amphioxus, those of Petromyzon (Cyclostomes) are somatic efferent and probably also somatic afferent (proprioceptive). In the ascending taxonomic as well as presumably phylogenetic series, and beginning with the Selachians, the spinal ventral roots[37] show the tendency to become essentially or exclusively output channels, composed of somatic efferent

---

[37] AICHEL (1895) demonstrated afferent fibers, which he considered to be of visceral nature, in the ventral root of Teleosts. Proprioceptive fibers (and even cells) have been suspected to occur in spinal ventral roots of mammals as well as in cranial nerve roots comparable to spinal ventral roots (cf. further below in section 3). A summary on the problem concerning the possible presence of afferent visceral or proprioceptive type in the ventral spinal root can be found in SCHARF's (1958) *Handbuch* contribution. This problem is complicated by the possibility that, at least in Mammals, recurrent afferent fibers from posterior roots may, in various and segmentally overlapping patterns, enter the ventral roots.

and visceral efferent (preganglionic) fibers. The dorsal spinal roots,[38] on the other hand, tend to mediate essentially afferent (input) functions. Although not only in fishes, but also at least in certain Amniota, visceral efferent fibers appear to pass through dorsal spinal roots, their number seems to become reduced in comparison to those included in the ventral roots. With regard to mammals, the overall but presumably not absolute validity of *Bell-Magendie's law* can be accepted. This principle is commonly formulated as stating that the dorsal spinal roots are afferent ('sensory') and the ventral roots efferent ('motor').

The morphogenetic events leading to an arrangement for which *Bell-Magendie's rule* approximately holds can be interpreted, in accordance with KAPPERS, as an 'improvement' or simplification of conduction pathway patterns *('Leitungsverbesserung in der Phylogenese')*, related to the general orderliness of neurobiotaxis.

The concept of distinct afferent and efferent nerves can be traced back to GALEN (130–ca. 200 A.D.), who performed experiments on pigs and came close to a recognition of the significant difference between ventral and dorsal roots. Many centuries later, BELL (1811) stated that the ventral spinal roots were concerned with motor function, and subsequently referred to the dorsal roots as being sensory. MAGENDIE (1822) conclusively demonstrated the motor function of ventral, and the sensory function of dorsal roots, but did not claim an exclusive functional significance for either root.[39] The term *Bell's law* or, more justifiably, *Magendie-Bell's law*, was introduced by later authors and commonly interpreted in an extreme and exclusive sense which cannot be traced to the original propounders of that rule.

---

[38] In addition to the scattered but rather convincing evidence for the presence of efferent fibers in the dorsal roots of Gnathostome Anamnia, morphologic (structural) data for the presence of such fibers in the dorsal roots of some Amniota (birds) were obtained by CAJAL (1890), by van GEHUCHTEN (1893), and v. LENHOSSEK (1890). Cf. also SCHARF (1958).

Additional topics concerning the spinal nerve roots, namely (a) the numerical ratio of dorsal root fibers to spinal ganglion cells, and (b) the numerical ration of dorsal root fibers to ventral root fibers, shall briefly be discussed in chapter VIII (volume 4), dealing with the spinal cord.

[39] In contradistinction to MAGENDIE's results, BELL's conclusions were based on rather unsatisfactory evidence and also left open the question of exclusive motor or sensory root function. A rather acrimonious controversy as regards priority was initiated by BELL, but is generally considered unjustified (cf., e.g., C.C. METTLER's History of Medicine, pp. 152–153, 1947).

Generally speaking, and from a phylogenetic viewpoint, the *ventral roots* may be considered as the *primary myotomic nerves* and the *dorsal roots* as originally the *nerves for skin, viscera, and nonsomitic respectively nonmyotomic musculature*. The overall arrangement of peripheral afferent 'somatic' and 'visceral, and of efferent 'somatic' spinal nerves is relatively simple. In all vertebrates from Cyclostomes to man, the cell body of the first order afferent neuron is commonly located in the spinal ganglion, one of its cell processes reaching the periphery and the other one joining the neuraxis.[39a] The cell body of the somatic efferent neuron is located in the ventral horn of the spinal cord, its neurite representing the final motor end path leading without interruption to the effector. The arrangement of the spinal root (general) visceral efferent pathway is more complicated, consisting of a preganglionic and a postganglionic sequence. These connections form part of the vegetative nervous system which includes significant cranial nerve components, and shall be discussed in the 6th section of the present chapter.

*Neuromuscular relationship* with regard to striated musculature may be considered from a structural and functional, as well as from a morphological, formanalytic viewpoint. The structural aspect, particularly concerning the nerve endings, was dealt with in the preceding section 1 of this chapter.[40] The morphologic aspect, which involves difficult problems of homology, requires some additional comments. It is well-known that the comparative anatomy of musculature discloses a considerable amount of variability manifested by this organ system, and conspicuously related to so-called 'functional' or 'adaptive' factors. Again, a given muscle in one animal form may be represented, *qua* identifiable and comparable morphologic unit by two or more distinct entities in another form and *vice versa*, thus implying, as the case may be, either splitting or fusion in the course of presumed phylogenetic development. Although, on the basis of the topologic neighborhood connectedness displayed by muscular entities and by proper interpreta-

---

[39a] Exceptions are here e.g. the *Rohon-Beard cells* mentioned above. Again, as regards the *Magendie-Bell rule*, the possible presence of some efferent fibers in dorsal roots, and of some afferent (proprioceptive) fibers in ventral roots must be kept in mind. These relationships were likewise pointed out in the preceding discussion and shall again be taken up, with respect to the cranial nerves, in sections 3 and 4 of this chapter.

[40] This particular topic was also discussed by HINES (1927), whose review, summarizing the status of the relevant problems at that time, now more than 40 years ago, is of some historical interest.

tion of the obtaining one-one, one-many, or many-one transformations, relevant ortho- or kathomologies can be established as a general
rule, such interpretations remain ambiguous in some instances.

While ontogenetic origin may supply relevant clues for the recognition of homologies, so-called cenogenetic changes in the mode of
ontogenesis[41] can blur, obscure, or even modify the topologic relationships at early developmental (embryonic) stages. Because of the morphologically and phylogenetically perhaps significant primary correlation between nerves and myotomes, the *innervation of muscles* can evidently be considered an important criterion in determining the homology of individual muscles. There is little doubt that, in many instances,
this innervation tends to remain constant throughout a series of secondary configurational changes.

The *primary connection of muscle and nerve*, constituting, as it were, a
single basic unit, was already postulated by CARL GEGENBAUR (1874)
whose disciple MAX FÜRBRINGER (1888) elaborated on this concept,
formulating a fundamental ontogenetic and phylogenetic relationship
between striated muscles and particular nerves,[42] such that homology
of a given muscle *must* include homology of its nerve supply. CUNNING-
HAM (1882, 1890) suggested a constant relationship between given motoneurons and given muscles, including, however, the possibility that
the axons, in reaching the muscle, might vary in their pathways, thus
producing variations in the gross pattern of peripheral innervation.

FÜRBRINGER was well acquainted with the variability in muscular
innervation, which he interpreted as due to abnormal anastomoses of
nerves and to the formation of 'vicarious' or 'aberrant' muscles. According to this concept, a muscle can disappear in phylogeny, being replaced by another with different innervation. Since a muscle could thus
be substituted for another, but no change in the innervation of a given
muscle might occur, a muscle with identical topologic relationships in
two species, but with different innervation, must then be evaluated as
morphologically analogous, but not homologous.[43]

---

[41] Cf. above p. 824, footnote 34.

[42] FÜRBRINGER considered HENSEN's concept of primary protoplasmic continuity between nerve cells and end organs as particularly important for his theory of neuromuscular
relationship. This latter, however, does not necessarily depend on, or require the validity of
HENSEN's interpretation, which has been generally rejected on rather convincing evidence.

[43] In other words, and using JACOBSHAGEN's terminology (cf. vol. 1, chapt. III of this
series), the significant bauplan relationship of a muscle is here postulated to include its
innervation.

In an extensive survey of this topic, STRAUS (1945) stressed the fact that the pattern of muscular nerve supply is *inconstant*. He therefore concluded that the concept of unalterably fixed nerve-muscle specificity should not be upheld. Nevertheless, STRAUS admitted that a number of relevant questions remain poorly understood. Moreover, he did not deny the evident tendency toward the maintenance of a particular type of innervation. Hence, the cited author concluded that, although not of decisive significance, the pattern of innervation is of great help in homologizing the muscles of vertebrates belonging to a given group (e.g. class), but because of its *flexibility* or 'looseness', much less relevant in homologizing muscles of vertebrates pertaining to different classes.

My esteemed old friend SEIHO NISHI (1961, 1963, 1968), who, like myself, is a representative of the *Gegenbaur-Fürbringer*, or *Jena-Heidelberg School of Comparative Anatomy*,[44] and, moreover, a personal disciple of FÜRBRINGER, has nevertheless fully upheld the concept of nerve-muscle relationship elaborated by his teacher.

NISHI's formulation of '*typologic*' comparative anatomy, although not explicitly defined in terms identical with *Jacobshagen*'s '*Bauplanlehre*', uses the same basic formanalytic approach, which, of course, is implicitly the procedure elaborated by GEGENBAUR on an not formally recognized, but intuitively topologic basis. However, while JACOBSHAGEN and myself completely eliminate phylogenetic considerations from the actual establishment of homologies, NISHI attempts to preserve, to some extent, a phylogenetic interpretation within the procedural framework of his '*typology*'.[45]

In retaining innervation respectively '*neurotopy*' as the only reliable criterion for the morphologic evaluation of a muscle, NISHI is fully aware of the problems posed by differences respectively shifts of inner-

---

[44] NISHI's fundamental studies on the comparative anatomy of vertebrate trunk and head muscles are summarized in his contribution to volume V of BOLK *et al.*: Handbuch der vergleichenden Anatomie der Wirbeltiere (1938).

[45] Cf. vol. 1, p. 186 of this series. My own procedure, on the other hand, which differs from JACOBSHAGEN's concept merely by the explicit elementary *topologic* formulation, attempts, on said basis, to establish homologies as if phylogeny had no existence. The thereby obtained results (*Baupläne*) can then, with full justification, be used for the elaboration of phylogenetic inference. Very minor differences between JACOBSHAGEN's rigorous definitions of the various sorts of homology and my own interpretations, as indicated in the cited chapter III, p. 200 *et passim* of this series, can here be discounted.

vation and even by replacement of muscles (e.g. in metamorphosis of larval Amphibia). Nishi, moreover, recognizes that, as regards the topic under consideration, an essentially semantic problem is involved which might be arbitrarily solved by considering as homologous only those morphologically corresponding muscles which display identical innervation, and by designating such corresponding muscles with different innervation as *homomorphous*.[46] In a personal communication (1970) elaborating on his viewpoint, Nishi stated: '*So bin ich jetzt der Meinung: 2 Organe werden miteinander homolog genannt, wenn sie je zu den benachbarten Organen in gleicher Lagebeziehung stehen (homotop). Wenn es sich aber um die Muskelindividuen handelt, ist die Frage nicht so einfach. Man darf sie stets mit den sie versorgenden Nerven in Zusammenhang betrachten, weil ein Muskel ohne Nerv morphologisch sowie physiologisch undenkbar ist, und der Muskel sozusagen das Endorgan des betreffenden Nerven darstellt. Wenn also 2 miteinander homolog erscheinende Muskeln von nichthomologen Nerven versorgt sind, in solchem Fall allein werden sie miteinander homomorph genannt. Kurz, die Homologisierung der Muskeln ist nur durch die sie versorgenden Nerven statthaft*'.

It is evident that, if identical innervation is postulated as essential for homologization, this would represent a special case of orthohomology in Jacobshagen's sense. If, on the other hand, homotopic and therefore, *qua* muscular system *per se*, homologous muscles display differences in innervation, such case of homomorphy could be regarded as a special case of Jacobshagen's kathomology.

In contradistinction to Nishi, our common former pupil and close friend whose untimely death we deeply deplored, the late Tsunetarô Fujita (1962), following Straus (1945), criticized Fürbringer's theory of nerve-muscle specificity, stressing that muscles and other structures can receive their innervation from such sources 'as are topographically proximate to them'. Fujita regarded innervation as a phenomenon of secondary or incidental nature. 'Its pattern may therefore fluctuate in individuals and species to give rise to variations and anomalies and can also be changed by experiments' (Fujita, 1962).

The cited author illustrates the homology concept in the light of the variability of segmentation by taking the diaphragm as an example (Fig. 421). With respect to diaphragm types I, II, and III, it is obvious

---

[46] '*Es handelt sich also am Ende um einen Wortstreit, ob man den Homologie-Begriff bis auf das Gebiet der Homomorphie erweitern will oder nicht*' (Nishi, 1968).

that, if homology strictly depends on identical innervation, no homology relation between the three types obtains. Yet, on account of their *homotopy*, the different types of diaphragm may reasonably be considered homologous. FUJITA concludes therefore that the segmental relationship should not be 'considered too seriously'. It is here evident

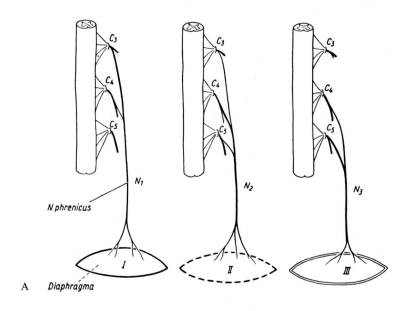

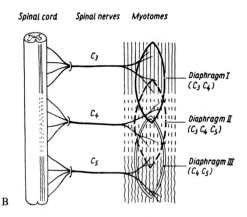

*Figure 421.* Diagrams showing three types of mammalian diaphragm innervated by different cervical spinal nerve segments (from FUJITA, 1962). A Adult condition. B Assumed ontogenetic relationships between nerve segments and myotomes.

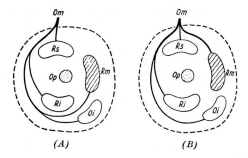

*Figure 422.* Different distribution of oculomotor nerve to extrinsic eye muscles in Ganoids (A) and in Selachians (B), as seen in diagrammatic transverse sections through orbit (from Fujita, 1962). Oi: m. obliquus inferior; Om: oculomotor nerve; Op: optic nerve; Ri: m. rectus inferior; Rm: m. rectus medialis; Rs: m. rectus superior.

that, in Jacobshagen's terminology, which apparently had here escaped Fujita's attention, the three types of diaphragm represent a clear-cut case of *allomeric kathomology*.[47]

Concerning the extrinsic eye muscles, whose peculiar morphologic status shall be pointed out further below in section 3, Fujita referred to Nishi's (1922, 1938) investigations disclosing two different types of innervation characteristic for Selachians and for Ganoids (Fig. 422). In the former type the m. rectus medialis is reached by an upper branch of n. oculomotorius, and in the latter type, also displayed by mammals, a lower branch of n. oculomotorius supplies said muscle. If the difference *qua* innervation is considered fundamental, then a difficulty in homologizing the homotopic mm. recti mediales of Selachians and Ganoids arises.

The solution of this question, in my opinion, merely depends on the adopted logical respectively semantic postulates. If the muscles are considered *per se*, regardless of innervation, then the mm. recti mediales of Selachians and Ganoids are obviously orthohomologous. If Nishi's postulates are accepted, then these muscles should, accordingly, be regarded *homomorph*. This, especially since the relevant innervating branch is still a ramus of n. oculomotorius, can here still be subsumed, as a *kathomology*, under the homology concept. Such particular form of kathomology, implying here the innervation of basically homologous,

---

[47] Cf. vol. 1, chapter III, pp. 200–201 of this series.

i.e. topologically homeomorph structures by a topologically nonhomeomorph nerve, may be designated as an *allotypic kathomology*.[48]

It is finally of interest that Fujita (1965), in his posthumous paper
on the human and mammalian mm. levatores costarum, identified
these muscles *exclusively on the basis of their innervation* by the dorsal rami
of the segmental spinal nerves, as pertaining to the dorsal, and not to
the ventral trunk musculature,[49] under which they had been traditionally subsumed.

Moreover, in his relevant paper on neuro-muscular relationship,
Fujita's (1962) conclusion were propounded not as an outright rejec-
tion, but as a 'revision' and 'enlargement' of Fürbringer's nerve-muscle theory. The main points of Fujita's formulation can here be summarized as follows: (1) The nerve-muscle connection is not inherently
fixed, but is, to some extent 'free'. (2) During ontogeny a nerve takes a
course presenting minimum resistance to its outgrowth and 'penetration'. (3) Under normal conditions, the neuromuscular connection results in accordance with 'formative and developmental' factors which
remain insufficiently elucidated. (4) Much the same could be said about
the innervation of smooth muscles, skin, bones, etc., regardless whether the nerve concerned is efferent or afferent. The homology problem
in reference to nerve supply was apparently never considered with regard to these structures, perhaps because (discounting the distinctly
patterned arrangement of bones), many of these configurations lack
well definable '*morphologic boundaries*'.[50] (5) For a homologization of

---

[48] On p. 200, chapter III, vol. 1 of this series, four types of *kathomology* were listed and
defined, namely (a) augmentative; (b) defective; (c) allomeric, and (d) mixed forms of a–c.
Accordingly, a fifth type, designated as (e) *allotypic kathomology*, may be added. This can
be defined as the relationship between two formelements whose topologic connectedness
are to a relevant extent homeomorph, but include some additional nonhomeomorph
connections.

[49] The dorsal or *epaxial* trunk musculature, innervated by branches from the *dorsa
rami* of segmental common spinal nerves, includes the m. erector spinae (PNA) consisting
of iliocostalis, longissimus, transverso-spinalis and others. The ventral or *hypaxial* trunk
musculature, innervated by the *ventral rami* (e. g. intercostal nerves) respectively by plexus
branches pertaining to the ventral rami system, includes the mm. intercostales externi and
interni as well as the anterolateral abdominal muscles, and the m. quadratus lumborum.

[50] Fujita (1960) speaks here of 'clean' morphological boundaries. It should, however,
be emphasized that, in agreement with topological concepts involving neighborhoods,
well-definable or linear boundaries are not a necessary requirement for the homologization
of configurations consisting of such neighborhoods representing open sets (e. g. the D and
B *Grundbestandteile* of the telencephalic longitudinal pattern).

striated muscles, the segmental relation of the nerve supply 'should not be taken into account too rigorously'.[51] Segmentation, in FUJITA's view, 'can be just an incidental marking of the body'. (6) The topographic relation of a muscle to a nerve does not possess such a decisive significance as believed by previous comparative anatomists. It can vary to a considerable extent without changing very much the morphological homology status of the muscle.

My own view on this question, as already indicated above, tends toward a compromise between NISHI's and FUJITA's formulations of the nerve-muscle relationship problem. Evidently, if NISHI's *homomorphy* is, in accordance with JACOBSHAGEN's concepts, equated with *allomeric* or *allotypic kathomology*, both formulations, apparently incompatible in an extreme semantic interpretation, are not mutually exclusive.

### 3. Cranial or Cerebral Nerves and the Head Problem

A morphologic evaluation of the *cranial nerves* involves questions pertaining to the so-called *head problem*, which concerns ontogenetic and presumptive phylogenetic origin of the vertebrate head characterized by a brain case or cranium. Although the rostral end of the Acraniote Cephalochordate Amphioxus may be said to represent a head *sensu latiori*, the notochord extends here beyond the neuraxis as far as the rostrum. Again, among the Craniote vertebrates, the Cyclostomes lack jaws, being thus classified as Agnatha. All other vertebrates (Gnathostomata) are provided with jaws.

As regards the skull or *cranium*, three components can be distinguished. The *neurocranium* (1) or *chondrocranium* is an originally cartilaginous brain case related to or combined with three capsules[52] developing around (a) the olfactory sacs; (b) the eyes, and (c) the otic vesicles. Some overall aspects of the chondrocranium are illustrated by Figure 423. During ontogenetic development in various groups of Anamnia and in all Amniota, the chondrocranium ossifies as cartilage bone.

---

[51] This, of course, although loosely stated, is in full agreement with JACOBSHAGEN's more rigorous formulation of allomeric kathomology.

[52] As shown in figure 423, paired parachordal and prechordal components are believed to represent the fundamental elements of the chondrocranium in addition to the three 'sensory capsules'.

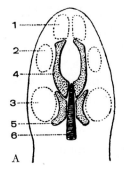

*Figure 423.* Diagram of the development of the chondrocranium (A from SELENKA-GOLDSCHMIDT, 1923, B and C from HYMAN, 1942). A Diagram showing rostral trabeculae cranii and caudal parachordalia in relation to the more anterior or lateral, essentially 'vesicular' or pocket-like main sensory structures (olfactory pit or diverticulum, eye, otic vesicle). 1: olfactory pit; 2: bulbus oculi; 3: otic vesicle; 4: trabeculae cranii; 5: parachordale; 6: notochord. B, C Two (early and later) developmental stages.

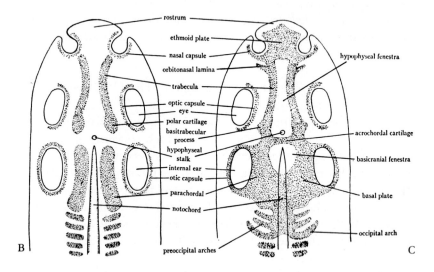

It is likely that, at some stages of phylogenetic evolution, a few segmented elements, presumably modified vertebrae, have been added to the chondrocranium in the region behind the otic capsules.[53] The *dermatocranium* (2) can be interpreted as a complex set of dermal bones which becomes connected, respectively may become integrated, with the chondrocranium. The *splanchnocranium* (3) consists of the endoskeletal branchial arches which support the Anamniote gills and also dis-

---

[53] A short comment on the reduction in the number of (essentially dermal) cranial bones displayed by the ascending taxonomic and presumably phylogenetic scale from primitive palaeozoic fishes to mammals and man, as well as on the addition of vertebral elements to the cranium can be found in vol.1, pp. 275–276 of this series.

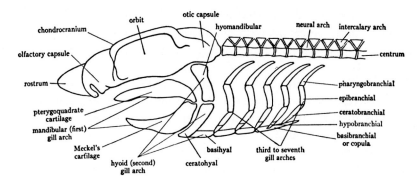

*Figure 424.* Diagram of chondrocranium, vertebral column, and branchial arches of a pentanch Elasmobranch in lateral view (based on diagrams by GEGENBAUR and adaptations by VIALLETON, from HYMAN, 1942).

play connections with the chondrocranium, undergoing considerable transformations in Amniota. The relationship of branchial arches, chondrocranium, and vertebral column, as manifested in Selachians, is shown by the diagram of Figure 424.

The earliest vertebrates provided with jaws are recorded from the Silurian. It is believed that the jawless ancestors of Gnathostomes possessed a nearly terminal mouth, either at the rostral end of the head or on its ventral surface. By a transformation of first and second branchial arches the grasping jaws with their modified branchial musculature may have evolved. The relevant bracing attachment of the jaw apparatus to the rostral end of the axial skeleton (i.e. to the cranium) was obtained in three different ways, designated as (a) *autostylic;* (b) *hyostylic,* and (c) *amphistylic.* In type (a) the upper jaw (pertaining to first branchial arch) is firmly fused with the brain case, while in type (b) the hyomandibular portion of the second arch (including a ligament) provides this support. The amphistylic (c) type can be described as a combination of (a) and (b). Within the recent class Chondrichthyes, all three distinctive types of jaw arrangement occur, being displayed by different forms.[54]

[54] Type (a) is displayed by Holocephalians, and type (b) by most other Elasmobranchs except a few forms such as Notidanoid sharks (e.g. *Hexanchus*) which are characterized by the amphistylic condition. Cf. also our paper (K. and NIIMI, 1969) on the Holocephalian brain.

The vertebrate *body* is characterized by a *metameric* or segmental arrangement involving a longitudinal sequence of configurations which, to various degrees, manifest promorphologic homology.[55] Such metamerism is particularly evident in the mesomery exhibited by the sequence of the somites appearing during embryonic development. A likewise conspicuous metamerism, representing peripheral neuromery, obtains in the segmental arrangement of spinal roots, respectively spinal nerve segments.

The ontogenetic and assumed phylogenetic development of the vertebrate *head* includes the evolution of particular *segmentally arranged configurations* pertaining to the rostral or cephalic gut, *videlicet* the branchial arches, respectively the gill clefts, manifesting *branchiomery*. This latter is correlated with an obvious metamerism of the deuterencephalic branchial nerves. The non-branchial nerve roots of this brain portion likewise display a sequence which implies a segmental orderliness. A still poorly understood central neuromery, involving segmentation of the cerebral neuraxis, was discussed in section 3 of the preceding chapter VI.

The *head* of adult higher vertebrates, which, except for the sequence of deuterencephalic nerves, does not show any obvious signs of segmentation, is generally rather well demarcated from neck, respectively trunk. In lower vertebrates, the distinction between head and trunk becomes less distinct; in Cyclostomes 'it is difficult to say where one begins and the other ends; and finally, in Amphioxus, the body is clearly segmented from end to end' (GOODRICH, 1930, 1958).

Since, nevertheless, even in higher vertebrates, ontogenetic development discloses not only *branchiomery* but also an at least partly corresponding peripheral *neuromery*, it seems evident that metameric components are involved in the process of cephalization. This was already pointed out, in the second half of the 19th century, by C. GEGENBAUR and T. H. HUXLEY.

Moreover, the investigations of BALFOUR (1881) had shown, since about 1876, that *somite-like head cavities*, whose walls seem to originate muscles, are formed in the cephalic region of Elasmobranch fishes. Numerous subsequent authors reported similar findings not only in Anamnia but also in Amniota. Although their results and conclusions were not wholly consistent and led to various disagreements, there re-

---

[55] Defined in vol. 1, pp. 219–220.

mains little doubt about the actual occurrence of rostral, *preotic somite-like configurations* at certain stages of vertebrate head ontogenesis.

NEAL and RAND (1936), in a concise review of the questions at issue, stressed a threefold aspect presented by the head problem: 'First, is the head, like the trunk, metameric? Second, if metameric, are segments of head and trunk fundamentally similar? Third, how many segments are contained in the head?'

With respect to these three questions, the first one can obviously be answered in the affirmative, since somitic, branchial, peripheral and central neural metamery are convincingly displayed. The cited authors, moreover, have omitted a further question which, in my opinion, is morphologically relevant, namely, *third*, is there a direct *correspondence between somitic metamery, branchiomery, and neuromery* of the head? NEAL's and RAND's (1936) third question becomes thereby the *fourth*. My own tentative answers to said questions 2, 3 and 4 will be given further below.

To their cited queries, NEAL and RAND (loc.cit.) have added some pertinent comments, emphasizing that the complexity of the head problem and the diversity of answers have combined to make it one of the persistent riddles of vertebrate morphology. 'The problem is not identical with that of the ancestry of vertebrates, although its elucidation may throw light upon the latter. Even if it were demonstrated that metamerism characterizes both head and trunk of vertebrates, this would by no means prove that vertebrates are derived from metameric ancestors. For metamerism, like any other animal characteristic, may be a convergent trait acquired *de novo* within the Chordate group' (NEAL and RAND, 1936).

Because of the repeatedly stressed '*conservatism*' of the nervous system,[56] the arrangement displayed by the cranial nerves seems particularly relevant for an appraisal of the head problem. Reverting to the configuration obtaining in Amphioxus, it seems reasonable to subsume *archencephalic* and *deuterencephalic* cranial nerves under two separate and quite distinct morphologic groups.[57] The archencephalic vesi-

---

[56] Cf. vol. 2, p. 336 of this series.

[57] Cf. above, p. 817 in the introductory remarks to the preceding section 2 of this chapter. With regard to their recognition and enumeration, ERASISTRATUS (about 290 B.C.), who joined the *Alexandrian school*, was probably the first to trace cranial nerves to their cerebral 'origin'. GALEN (130–200 A.D.) distinguished seven pairs, namely optic (1),

cle of Amphioxus displays a paired nerve whose connection with the neuraxis is located somewhat more ventrally than that of the spinal dorsal roots. It carries fibers related to the receptor structures of oral hood and its tentacles, including, moreover, fibers from a peculiar unpaired epidermal pit lined by a distinctive ciliated epithelium. Said structure, known as '*Kölliker's pit*', is also designated as '*Riechgrube*' and considered to be a chemoreceptor. The archencephalic nerve of Amphioxus could be interpreted as homologous to the nervus terminalis of craniote vertebrates. In these latter, the archencephalon, becoming differentiated into telencephalon and diencephalon, displays, in addition to a variable or rudimentary nervus terminalis, the paired olfactory bulbs with their fila olfactoria, the paired optic 'nerves', and insufficiently understood traces of a parietal nerve. None of these archencephalic components can properly be conceived as comparable true 'segmental' configurations nor as being promorphologically ('serially') homologous with deuterencephalic nerve roots.

Corresponding to three groups of head structures, namely to *eye musculature*, *branchial arches* and their muscles, and to the *hypobranchial musculature*, the nerves of the deuterencephalon can be described as comprising three distinct and well-defined natural groups, namely,

(1) *Eye muscle nerves*,
(2) *Branchial nerves*, and,
(3) *Nerves of hypobranchial musculature*.

---

oculomotor (2), trigeminus (as 3rd and 4th pair), facial-auditory (5), vagus group (6), and hypoglossus (7). VESALIUS (1514–1565) followed GALEN's enumeration and interpretation. The olfactory tract with its bulb was described as a process extending from the cerebrum and not enumerated as a cerebral nerve. THOMAS WILLIS (1621–1675), in his treatise '*Cerebri anatome*' (1664) was one of the first to include the olfactory nerve in the nine pairs of his enumeration, which also referred to trochlear and abducens, the sequence being now (1) olfactory; (2) optic; (3) oculomotor; (4) trochlear; (5) trigeminus; (6) abducens; (7) facial and acustic; (8) vagus group, and (9) hypoglossus with spinal accessory. It is uncertain to which extent WILLIS, who also apparently coined the term '*neurology*', appropriated previous findings by others as purporting to be his own. There is little doubt that discovery of n. accessorius and circulus arteriosus cannot be attributed to WILLIS (cf. also METTLER, 1947). The conventional enumeration of the cerebral or cranial nerves as twelve pairs (I–XII) from nervi olfactorii (fila olfactoria) to n. hypoglossus as standardized by BNA and PNA goes back to the description by S. T. SOEMMERING (1755–1830) in his work '*De basi encephali et originibus nervorum cranio egrendentium*' (*Goettingae*, 1778). It is evident that the early investigators attempting to dissect brain and cranial nerves encountered severe difficulties because of the lack of proper fixatives. J. C. REIL (1759–1813) was perhaps one of the first to use preserving and hardening solutions, while, according to EDINGER (1912), formalin was not introduced into neuroanatomical technique until after 1893.

All these nerves may easily be compared with those of the spinal cord and evaluated as being promorphologically (serially or metamerically) homologous to these latter, although displaying a combination of 'primitive' and of self-evident 'progressive' ('adapted' or 'specialized') features. The *'primitive'* characteristics can be enumerated as follows:

(a) dorsal and ventral roots remain separate (unconnected) and independent of each other;

(b) this independence, moreover, is so pronounced that it cannot be recognized with any degree of certainty whether such ventral roots as are present correspond to any particular dorsal root;

(c) the dorsal roots include a substantial amount of efferent fibers.

(1) The *eye muscle nerves* are represented by *nervus oculomotorius* (III), *n. trochlearis* (IV), and *n. abducens* (VI). The oculomotor emerges basally from the midbrain, and the abducens nerve, likewise basally, from the rostral portion of rhombencephalon. Both on account of their ventrally located roots, and because these nerves supply striated musculature whose origin is related to somites, the classification of oculomotorius and abducens as radices ventrales appears fully justified. These two nerves seem to include proprioceptive fibers,[57a] which, as mentioned above in section 2, may also occur in some spinal ventral roots. In addition, the oculomotor carries preganglionic efferent fibers of the vegetative nervous system (cf. further below, section 6 of this chapter).

The morphologic status of *nervus trochlearis*, located in the mesencephalo-rhombencephalic isthmus region, appears less evident. It is the only cranial nerve emerging from the dorsal surface of the brain,[58] and after undergoing a decussation which appears to be complete in most, if not all vertebrates. Thus, some authors, including GEGENBAUR and MAURER, considered the trochlear nerve to represent a dorsal root split off from the trigeminus, the musculus obliquus superior being interpreted as being of branchial origin. Nevertheless, the superior oblique

---

[57a] WINCKLER (1956) has recorded evidence indicating that, at least in several mammals, the extrinsic eye muscles receive proprioceptive fibers from the trigeminal nerve. According to this author, the facial musculature likewise receives proprioceptive fiber from the fifth nerve.

[58] Dorsal even in respect to the branchial nerves which represent dorsal roots and are connected with the neuraxis in a dorsolateral location.

muscle rather seems to originate from the second head somite (cf. e.g. NEAL and RAND, 1936). The nucleus of the trochlear nerve, moreover, is located in line (i.e. serially) with those of oculomotorius and abducens, at least in most vertebrates.[59] Although, because of its peculiar arrangement, the morphologic significance of this unique nerve remains still doubtful (cf. FÜRBRINGER, 1902), it may be classified as an atypical ventral root which possibly includes, as presumed concerning some other such radices, proprioceptive components. YOUNG (1955), who interprets the trochlear nerve as 'the ventral root of the second segment', believes that its dorsal emergence is related to the dorsal position of m.obliquus superior, the nerve being 'modified so as to reach its muscle by running partly within the tissues of the brain'. Having formerly tended to favor GEGENBAUR's and MAURER's views (K., 1927), I would now prefer the classification adopted by YOUNG (1955) and others, while remaining somewhat sceptical about the various proffered explanations purporting to account for peculiarities of said nerve.

(2) The *branchial nerves* supply the branchial arches with afferent and efferent fibers. Although essentially metameric (branchiomeric), some of these nerves have, in addition, an extensive polymeric distribution. All branchial nerves can justifiably be regarded to represent *dorsal roots* which, nevertheless, differ in some respects from the spinal ones, being, therefore, promorphologically kathomologous rather than orthohomologous with these latter.

In mammals and essentially also in Sauropsida, the afferent neurons of branchial nerves derive mainly from the neural crest, fairly evident exceptions being the cells of nucleus radicis mesencephalicae trigemini and (at least in birds) those of the apparently placodal cochlear respectively vestibular ganglion. In Anamnia, however, the cells of the branchial nerve ganglia seem to arise from three different substantial sources, namely (a) neural crest, (b) dorso-lateral placodes, and (c) epibranchial placodes (cf. Fig. 425).

The term, '*placode*', refers to thickenings of the embryonic ectoderm. In the wider sense, the neural plate is, itself, a placode. *Sensu stric-*

---

[59] Concerning the somewhat difficult problem of properly identifying the nucleus nervi trochlearis in Cyclostomes, cf. the investigation of my collaborator SAITO (1930) and the monograph by HEIER (1948). This latter author also brings a general review, with pertinent references, of the '*trochlearis-problem*'. A similar problem obtains here with regard to the abducens.

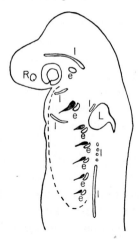

*Figure 425.* Diagram showing the series of epibranchial and of dorsolateral (or lateral) placodes in the head region of a Selachian embryo (modified after FRORIEP and EDINGER, from K., 1927). e: epibranchial placodes; l: dorsolateral placodes; L: otic vesicle (developed from otic placode); R: olfactory placode.

*tiori*, however, this designation is restricted to thickenings lateral to neural tube region with its neural crest, and becoming separated from the epidermis. Such placodes, in addition to those mentioned above under (b) and (c) are: the olfactory placode, the (non-neural) lens placode, and the otic placode forming otic vesicle, respectively ectodermal labyrinth.

(a) The *neural crest*, presumably the only primordium for the dorsal ganglia of spinal nerves, develops in the cerebral region much as in the trunk, becoming broken up into segmental portions,[60] some of its cells, which do not provide neuroectodermal elements, being then assimilated by the mesenchyme.[61]

---

[60] In certain forms, transitory strands of cells have been observed to occur in the course of this process. Their interpretation as rudiments of vanished ancestral nerves by PLATT (1891) and TRETJAKOFF (1909) seems to me highly unconvincing. The latter author suspected a *nervus mesencephalicus* in *Petromyzon* (Ammocoetes), and the former a *nervus thalamicus* in embryonic Selachians (Acanthias).

[61] Cf. vol. III, part I, pp. 18, 52, 299, and p. 109 of the present volume. GOODRICH (1958, p. 764, footnote 1) expresses the opinion that the 'doctrine of the formation of special "mesectoderm" in the head is, however, almost certainly founded on misinterpretations and erroneous observations on unsuitable material'. There is, nevertheless, little doubt that, on substantial evidence, neural crest material has been shown to take part in the formation of mesenchyme, apparently becoming assimilated by '*ortsgemässe*'

(b) The *dorso-lateral placodes* are serial ectodermal thickenings in line with the otic vesicle (otic placode), which can be regarded as the largest and most constant formelement of this group. The dorso-lateral placodes *sensu strictiori* are best developed in Anamnia, and provide the afferent neurons of the lateralis system. This latter, because of its close relationship to the acoustic (vestibular) apparatus is also frequently designated as acustico-lateralis (or vestibulo-lateralis) system. Nevertheless, rudimentary dorso-lateral placodes are also displayed during the ontogenetic development of various Amniota. Thus, one of the preotic placodes in this series, the so-called 'ophthalmic placode', seems here to participate in the formation of the nervus ophthalmicus portion of the ganglion nervi trigemini.

(c) The *epibranchial placodes* are represented by thickening of the ectoderm at the dorsal edge of the corresponding branchial pouches or gill slits. Each placode contributes to the ganglia of branchial nerves, particularly of facial, glossopharyngeus, and vagus.

Each of these apparent sources of branchial nerve afferent neurons have been associated with particular functional components (e.g. STRONG, 1895; LANDACRE, 1910, 1912, 1933; cf. also chapt. VI, section 4). Thus, the neural crest is presumed to provide the general somatic (cutaneous) and visceral components, the dorso-lateral placodes supplying the special somatic (acoustico-lateral) components, and the epibranchial placodes the special visceral afferent ones. Although a number of findings by various authors tend to support this interpretation, several complicating factors seem to obtain, apart from intrinsic difficulties concerning the observation of ambiguous and transitory embryonic configurations. These uncertainties have led to many contradictory descriptions and interpretations. Moreover, Anamnia and Amniota display considerable differences in the relevant developmental processes (cf. ADELMANN, 1925; STONE, 1922, 1924; TELLO, 1923). Thus, the problems of branchial nerve ontogenesis and phylogenesis cannot be regarded as sufficiently clarified.[62]

---

induction. The question whether such secondary mesenchyme (mesoderm) should or should not be particularly designated as 'mesectoderm' can be dismissed as an arbitrary and meaningless semantic argumentation.

[62] The *epibranchial placodes* were apparently first pointed out by BEARD and by FRORIEP about 1887. Detailed studies on dorso-lateral and epibranchial placodes were undertaken by v. KUPFFER (1906). Further discussions of the problems at issue can be found in the publications by NEAL and RAND (1936), and by GOODRICH (1958, with selected bibliography).

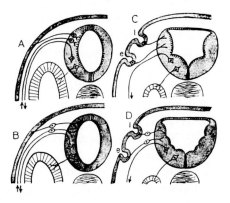

*Figure 426.* Diagram purporting to illustrate peripheral migration of afferent elements in the formation of spinal ganglia, and centralward migration of such elements in the formation of cranial (branchial) nerve ganglia (from K., 1927). A and B: spinal tube; C and D: deuterencephalon (rhombencephalon); e: epibranchial placode; l: dorsolateral placode. The relation of efferent elements to the longitudinal zones is also suggested in this diagram.

Following this short discussion of lateral and epibranchial placodes, an additional reference to the problem concerning *phylogenetic origin of dorsal ganglia*, dealt with in section 2 (p. 827) is perhaps appropriate. It will be recalled that both peripheral and central displacement or migration were considered. With regard to the placodal derivatives of branchial nerve ganglia an interpretation assuming central displacement of originally primary (neuroepithelial) sensory cells, as illustrated by Figure 420, does not seem improbable, while most or at least some of the neural crest derivatives could be considered as resulting from peripheral displacement of neural tube elements during phylogeny (cf. Fig. 426).

With respect to the *efferent* components of branchial nerves it appears significant that none of the effectors innervated by these fibers are derived from somites respectively myotomes, whose innervation is typically provided by ventral roots. In vertebrate embryonic development, the organized mesoderm, which includes the ontogenetic substratum of muscle tissue,[63] manifests a differentiation into three main

---

[63] Discounting rare exceptions, such as, e.g., the ectodermal iris musculature of the eye (cf. also vol. 3, part I, p. 705).

regions: (1) the dorsal *epimere* located laterally to the neuraxis; (2) the ventral *hypomere* or lateral plate, an extensive region on each side of the archenteron, and (3) the *nephrotome ('mesomere')* arising ventrolaterally to the somites between these latter and the hypomere.[64]

Depending on taxonomic forms and ontogenetic stages, epimere and hypomere may display a common cavity, but soon become separated from each other (cf. Fig. 209 A, p. 298 of vol. 2, and Fig. 11, p. 13 of vol. 3, part I). The epimere undergoes segmentation forming the metameric series of somites, which, in the body region caudally to the otic capsule, generally further differentiate into *myotome*, *dermatome* and *sclerotome*. The hypomere does not display segmentation. Its cavity provides the coelom; its walls form the parietal somatopleure and the visceral splanchnopleure, adjacent portions of right and left hypomere becoming, as the case may be, dorsal or ventral mesenteries.

The *gill slits* of Anamnia develop by the meeting of corresponding evaginations of entodermal pharyngeal wall and invaginations of ectodermal surface, followed by disappearance of the thin intervening ecto-entodermal double membrane. Identical entodermal pharyngeal and ectodermal pouches develop in Amniota, but their bottom may or may not become pierced to provide actual transitory slits. The number of gill slits, respectively pouches, seems larger in 'lower' than in 'higher' vertebrates. As regards Amphioxus, up to 180 pairs have been recorded in adults, but this number can hardly be regarded as 'primitive'. No more than 8 can be found in recent fishes, the first pair becoming either specialized as the *spiraculum* of various Selachians, or becoming reduced.[65] In Amniota, including man, about five branchial arches are

---

[64] The nephrotome, from which tubules develop, becomes, in the course of ontogeny, respectively phylogeny, the common primordium of pronephros, mesonephros, and to some extent (ureteric bud) also of metanephros.

[65] The *spiraculum*, open in most Selachians and some Ganoids, becomes closed in Holocephalians, in most if not all Teleosts, in Dipnoans, and all Tetrapoda. In various amphibians (e.g. some Anurans), this closure of the spiraculum is provided by a tympanic membrane homologous with that of Amniota. The first pharyngeal pouch likewise forms a middle ear cavity and an *Eustachian tube* in diverse amphibians.

Again, *gill slits* are not *gills* (branchia). These latter are folded respiratory lamellae or filaments attached to the gill slit (branchial slit) septa. External and internal gills are formed, depending on taxonomic forms. The gills of Cyclostomes are particularly deeply located and have been suspected, on still uncertain evidence, to be covered by entoderm instead of by ectoderm as in Gnathostome Anamnia. Thus, a distinction between 'endobranchiate' Cyclostomes, and 'ectobranchiate' Gnathostome Anamnia was suggested (cf. GOODRICH, 1958).

commonly displayed during ontogeny, corresponding to four ectodermal visceral pouches, the most caudal ones being poorly developed. The first ectodermal groove, corresponding to the spiraculum of Anamnia, represents the external auditory meatus, the correlated pharyngeal pouch becoming middle ear cavity and Eustachian tube, and the above-mentioned intervening dissepiment, the tympanic membrane.[66]

The metameric development of branchial arches, respectively slits, interferes with the unsegmented lateral plate or hypomere of the head region, which becomes broken up. Remnants of its coelom can be seen at some stages, at least in Cyclostomes (Fig. 427) but obliterate or never become manifested in most other forms. The lateral plate mesoderm enclosed within the branchial arches *sensu latiori*[67] provides the branchial musculature.

As regards the *sequence of branchial nerves* in the vertebrate series, two subsets can be distinguished, namely (A) the rostral trigeminus group and (B) the caudal vagus group. Except for the trigeminus, each branchial nerve is related to a particular gill slit or corresponding branchial pouch, with a *pretrematic* branch rostral to the slit and a *posttrematic* branch caudal to it.[67a] Figures 428 to 433 illustrate the overall configuration of branchial nerves including their relation to brain and to other cerebral nerves.

A. *Trigeminus group.* The *first branchial nerve* is the *fifth cranial nerve or n. trigeminus*, which pertains to mesencephalon and rhombencephalon, its peripheral root being connected with rostral rhombencephalon respectively mammalian pons. Its dorsal root ganglion is the *Gasserian* or semilunar ganglion of human anatomical nomenclature. Some of the

---

[66] The complex phylogenetic evolution of the auditory ossicles characteristic for mammals is still poorly understood. However, morphologic evidence clearly indicates that the *stapes* (columella of some submammalian forms) derives from the *hyoid* (or second branchial) arch, while the *incus* represents the *quadrate*, and the *malleus* the *articular* of the *maxillo-mandibular* (or first branchial) arch.

[67] Branchial arches or 'bars' *sensu latiori* represent the configurations between gill *slits* respectively between the pharyngeal or the ectodermal *pouches*. Branchial arches *sensu strictiori* are the skeletal bars supporting, or within, the arches *sensu latiori* (cf. fig. 424). There are, in addition, the highly relevant aortic (or vascular) branchial arches briefly discussed in section 7 of the preceding chapter VI (cf. also figs. 376 and 377).

[67a] In Cyclostomes such as *Petromyzon*, however, the pretrematic rami are lacking or poorly developed (cf. also CORDS, 1929, and figs. 428, 439, 445 A of the present chapter).

primary afferent cells of this nerve, representing the nucleus of radix
mesencephalica trigemini, have an intracerebral location and can be regarded as corresponding to the dorsal sensory cells discussed above in
section 2 (p. 827), respectively to 'permanent' *Rohon-Beard cells*.

The trigeminus represents the nerve of the *maxillo-mandibular arch*,
which is the branchial bar rostral to the spiraculum. The primordial
maxilla (palatoquadrate) seems to represent the dorsal, epibranchial
part, and the mandible the ventral, so-called ceratobranchial one, of a
rostrally concave, jointed branchial bar. Since the unpaired and median

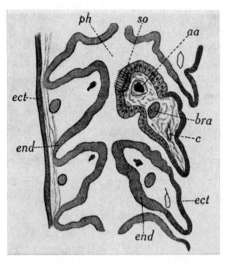

*Figure 427.* Semidiagrammatic, slightly oblique horizontal section through pharynx
and gill slits of Petromyzon at the larval Ammocoetes stage, passing through openings on
right, and dorsal to them on left (from GOODRICH, 1958). aa: arterial arch; bra: skeletal
branchial arch; c: coelomic rudiment of hypomere; ect: ectoderm; end: entoderm; ph: lumen of pharynx; so: sensory patch.

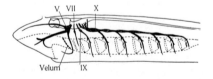

*Figure 428.* Simple diagram of branchial nerves in a Cyclostome (from JOHNSTON,
1906). The so-called 'velum' forms a pair of muscular flaps involved in directing the water
current toward the gill slits.

oral opening cannot be likened to a paired gill slit, there are evidently no pretrematic and post-trematic rami. On the other hand, the rostral branch of the trigeminus or nervus ophthalmicus profundus, with its ganglion, seems to derive from separate *'premandibular'* dorsal root ($V_1$) secondarily fused with the *maxillo-mandibular* dorsal root ($V_{2+3}$) whose main branches are the ramus maxillaris and the ramus mandibularis. This latter includes the motor fibers for the primordial branchial jaw musculature and its derivatives. The ramus ophthalmicus superficialis commonly present in Anamnia (cf. Fig. 430) is a more dorsal branch of the maxillo-mandibular portion, closely associated with a rostral branch of the facial nerve, and with the n. ophthalmicus profundus. In Amniota, it seems to be integrated with the ramus frontalis derived from the profundus.

The *second branchial nerve* is the *seventh cranial nerve or n. facialis*, related to the first branchial slit respectively pouch, respresented by the spiraculum of Selachians. Because of its posttrematic branch it is also referred to as the nerve of the hyoid or second branchial arch. As a dorsal root, the facial displays a ganglion usually termed *ganglion geniculi* with reference to its designation in human anatomic nomenclature. In various Anamnia, the facial root and its ganglion are closely connected

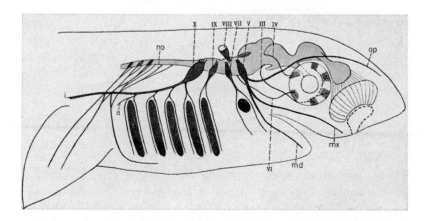

*Figure 429.* Semidiagrammatic sketch showing the arrangements of cranial nerves in a pentanch Selachian. The nervus terminalis and the stato-acustico-lateralis system have been omitted (from K., 1927). Roman numerals indicate cranial nerves according to their conventional sequential order. a: nervus accessorius vagi (XI); i: ramus intestinalis vagi; md: ramus mandibularis trigemini; mx: ramus maxillaris trigemini; no: nervi (spino-) occipitales; op: nervus ophthalmicus profundus trigemini ($V_1$).

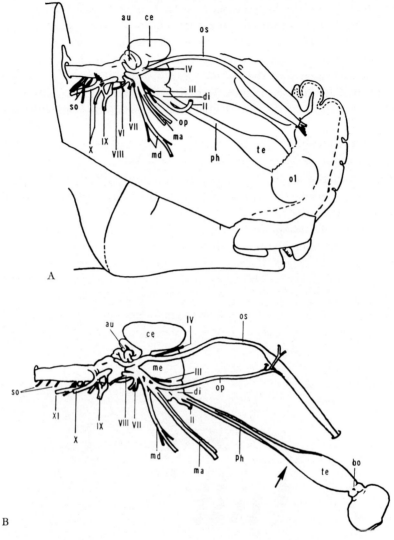

*Figure 430 A, B.* Semidiagrammatic lateral view sketches of brain and cranial nerves of adult Elasmobranch Holocephalians (from K. and Niimi, 1969) A *Callorhynchus antarcticus*, dissection *in situ* within cranium and head, based on dissection and sketch by Prof. Tsunetarô Fujita (Tokyo) done at the Woman's Medical College of Pennsylvania. B Removed brain of Chimaera Colliei. Roman numerals indicate cranial nerves, XI being the ramus accessorius vagi; au: cerebellar auricle; ce: cerebellum; di: diencephalon; ma: maxillary branch of V; md: mandibular branches of V; me: mesencephalon; ol: olfactory organ; op: n. ophthalmicus profundus; os: n. ophthalmicus superficialis; ph: preoptic portion of hypothalamus; so: spino-occipital nerve roots; te: telencephalon. Arrow indicates approximate location of external telo-diencephalic boundary.

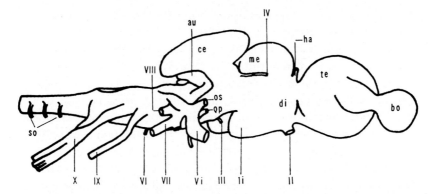

*Figure 430C*. Semidiagrammatic sketch of the brain of *Squalus acanthias* in lateral view (from K. and NIIMI, 1969). bo: olfactory bulb; ha: ganglion habenulae; li: lobi inferiores s. laterales hypothalami; Vi: mandibular branches of trigeminus. Other abbr. as in preceding figures A and B.

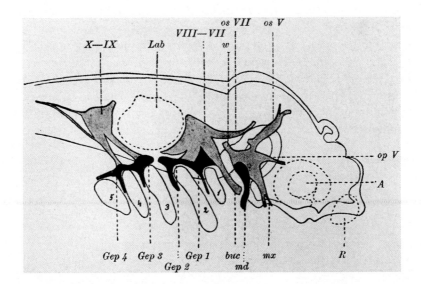

*Figure 430 D*. Reconstruction of branchial nerve anlagen in a 5-day-old embryo of the Ganoid *Acipenser* (from NEUMAYER, 1906). A: eye; Gep 1–4: epibranchial placode ganglia; Lab: otic vesicle; R: olfactory pit; buc: n. buccalis; md: n. mandibularis; mx: n. maxillaris; opV: n. ophthalmicus profundus; osV, osVII: nn. ophthalmici superficiales of trigeminus and facial; w: radix of trigeminal nerve; V, VII–VIII, IX–X: cranial nerves, respectively their ganglia.

with trigeminal root and ganglion, forming what appears grossly as a fused, common complex (cf. Fig. 430 A–C). In mammals, the root of the facial is located at the cerebello-pontine angle. The branchial musculature of the hyoid arch, which the seventh nerve innervates, is represented in man by facial musculature and platysma. Again, in Teleosts with a cutaneous gustatory system distributed over the skin of the whole body, the facial nerve includes an extensive dorsolateral posterior branch from which the taste buds of that system receive their sensory innervation (cf. also section 4 and fig. 444 of this chapter).

From an ontogenetic and phylogenetic viewpoint the *eighth cranial nerve* or *nervus acusticus* can be regarded as an offshoot or specialized development of the facial or second branchial nerve (cf. Fig. 430 D), becoming, as it were, secondarily separated from this latter. In Anamnia, the eighth nerve pertains to the vestibulo-lateralis system. The lateralis portion of this system has an extensive polymeric distribution upon the territory of other branchial nerves, contributing component fibers to these latter. In Amniota, the lateralis system has disappeared, and the eighth nerve becomes the vestibulo-cochlearis, related to equilibrium (vestibular) and hearing (cochlea). Although it appears certain that among Anamnia, such, e.g. as fishes, numerous forms distinctly respond to sound, i.e. have a 'sense of hearing', a cochlea is not present in lower vertebrates. However, an outgrowth of the vestibular sacculus, a so-called lagena, can be found in many Anamnia. In amphibians, another outgrowth of the sacculus, the basilar papilla, is present, which, together with lagena, can be regarded as a primordium of the Amniote cochlea.

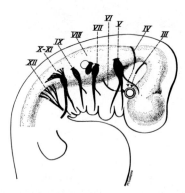

*Figure 431.* Simplified semidiagrammatic sketch of cranial nerves in a human embryo of approximately 8- to 10-mm length (about five weeks) for comparison with the preceding figures 424 and 428–430.

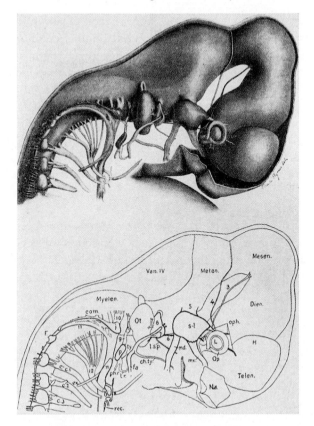

*Figure 432.* Reconstruction of brain and cranial nerves in a 12-mm pig embryo (after MINOT and LEWIS, from JOHNSTON, 1906). 3–12: cranial nerves in standard enumeration; c. 1–3: first three cervical spinal nerve segments; ch. ty.: chorda tympani; com.: fused segmental caudal vagus roots (so-called 'ganglionic commissure'); Dien.: diencephalon; ex.: spinal accessory branch to trapezius; F.: *Froriep's ganglion;* fa: facial branch of seventh nerve; H: pallial portion of cerebral hemisphere; j: ganglion jugulare of vagus; L: lens of eye bulb; l.r.: lingual branch of glossopharyngeus; l.s.p.: n. petrosus superficialis major of facial nerve; md.: mandibular branch of trigeminus; Mesen.: mesencephalon; Meten.: metencephalon; mx.: maxillary branch of trigeminus; Myelen: myelencephalon; n.: ganglion nodosum of vagus; Na.: nasal sac; Op.: optic cup; oph.: ophthalmic branch of trigeminus; Ot: otic vesicle (labyrinth anlage); p: ganglion petrosum of glossopharyngeus; ph.r.: pharyngeal rami of glossopharyngeus; rec.: recurrent nerve of vagus; s: ganglion superius of glossopharyngeus; s–l: semilunar ganglion of trigeminus; Telen.: telencephalon; ty.: nervus tympanicus of glossopharyngeus; Ven. IV: fourth ventricle.

B. *Vagus group*. In Cyclostomes, the location of this group remains essentially or mainly external to the cartilaginous cranium ('palaeocranium') which ends about caudally to seventh and eighth (facial and acoustic) nerves. In Selachians, the vagus group becomes included into the skull ('neocranium'). The *third branchial nerve* is the *ninth cranial nerve* or *n. glossopharyngeus*. Disregarding the spiraculum, it may be designated as the nerve of the first true gill cleft in fishes, its post-trematic ramus being the nerve of the third branchial arch. In many forms the glossopharyngeal ganglion is closely connected with that of the vagus. In man and many other mammals the ganglion of the ninth cranial nerve becomes subdivided into a ganglion superius and a slightly more peripheral ganglion petrosum.

The *fourth branchial nerve* is the *tenth cranial nerve* or *n. vagus*, with several rootlets, and corresponding, as a compound of several dorsal roots, to an undefined number of gill slits or equivalent pouches caudal to the third branchial arch (cf. Figs. 428 and 429). In Man and other

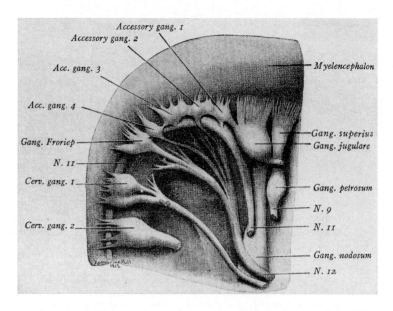

*Figure 433*. Dissection of head of a 15-mm pig embryo seen from right side to show *Froriep's ganglion* and accessory ganglia with peripheral roots passing to hypoglossal nerve (after PRENTISS and AREY, from GOODRICH, 1958). The interrelation between caudal accessory and hypoglossus may be interpreted as the result of the complex morphogenetic events involved in the phylogenetic and ontogenetic development of nerve roots at the spino-rhombencephalic border region.

Mammals, its ganglion, like that of glossopharyngeus, is subdivided into upper ganglion jugulare and lower ganglion nodosum. The vagus is a large nerve with a caudal distribution far beyond the branchial region. Its *ramus intestinalis*[67b] supplies the gut and represents an important component of the vegetative nervous system to be dealt with below in section 6 of this chapter. In Anamnia, the *ramus lateralis vagi* extends far caudalward along the lateral line system and carries fibers related to the lateralis component of the eighth cranial nerve, as mentioned above.

The *eleventh cranial nerve* or *n. accessorius* is represented in fishes by a branch of the vagus which supplies a musculature ('primordial trapezius and sternocleidomastoid') believed to be of branchial arch origin and connecting the caudal end of these arches with the pectoral fin girdle (shoulder girdle, scapula of higher vertebrates). Concomitantly with the further differentiation of this musculature in the taxonomic respectively phylogenetic series, efferent components of adjacent dorsal spinal roots appear to have been added to the accessory vagus branch. There are numerous controversial questions concerning the evolution of the n. accessorius (HALLER V. HALLERSTEIN, 1934; STRAUS AND HOWELL, 1936; BECCARI, 1943; and others), but it appears justified to assume that this nerve is a phylogenetic derivative of the vagus with added modified dorsal spinal nerve roots providing efferent fibers.

In man and other mammals this added spinal component of the accessorius, namely the *n. accessorius spinalis*, is believed to participate in the innervation of the trapezius and sternocleidomastoid muscles which have their phylogenetic origin in the primitive branchiogenic shoulder musculature apparently combined with additional components from myotomic muscle primordia. In conformity with such an assumed dual derivation these muscles are innervated by the spinal accessory, interpreted as representing a series of modified dorsal roots, and also by the ventral roots respectively ventral rami of the spinal nerve segments contributing to the cervical plexus, i.e. by typical spinal nerves independent of the accessory. Peripheral anastomoses between spinal innervation and accessory innervation are, however, commonly present. The modified spinal accessory root fibers emerge in several segmental bundles from the lateral aspect of the cervical

---

[67b] This ramus seems to be very poorly developed in Petromyzont Cyclostomes (cf. also CORDS, 1929).

spinal cord, dorsal to the denticulate ligament (thus manifesting their
dorsal root derivation), but somewhat more ventral than the regular
cervical dorsal roots. There may occur rudimentary or transitory embryonic spinal accessory root dorsal ganglia, comparable to *Froriep's
ganglion* of the hypoglossus, which is discussed below. This *spinal accessory nerve*, running rostrad, enters the cranial cavity through the foramen occipitale magnum, joining the *cerebral n. accessorius*, and together
with this latter, the vagus, and the glossopharyngeus, again leaving the
cranial cavity through the anterior portion of foramen jugulare. The
*cerebral n. accessorius* or *accessorius vagi* (radices craniales, PNA) provides
the ramus internus (BNA, PNA) of human anatomy, participating in
the innervation of larynx through n.laryngeus inferior. The *spinal accessory* (radices spinales, PNA), through ramus externus (BNA, PNA),
participates in the innervation of the mm.trapezius and sternocleidomastoideus. According to STRAUS and HOWELL (1936), the spinal accessory, although 'originally' a mixed nerve, has exhibited a strong
trend toward loss of its sensory elements through their migration into
the dorsal roots of adjacent cervical nerves.

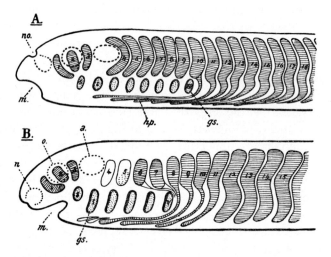

*Figure 434.* Diagrams of rostral body region of Cyclostomes (A) and Anamniote Gnathostomes (B) purporting to show position and development of somites 1–18 according to
GOODRICH's interpretation (from GOODRICH, 1958). Somites (epimeres) providing muscular tissue ('myomeres') respectively, in the case of postotic somites, typical 'myotomes' are
shaded. a.: otic vesicle; gs.: gill-slits (I–VII, respectively I–V): hp.: hypobranchial ('hypoglossal') musculature derived from postotic myotomes; m.: mouth; no.: nasal opening;
s: spiraculum.

(3) *Nerves of hypobranchial musculature.* In the postotic region, the branchial arch development seems to interfere with the differentiation of the somites and their myotomes, but several of the most rostral ones 'adapt' to this constraint by sending myotomic buds ventrally to the branchial arches and anteriorly as far as the mouth, forming the *hypobranchial musculature* (figs. 434, 436 A). The functionally important 'piston-like' tongue of Cyclostomes is not a 'true' tongue morphologically comparable (homologous) with that of Gnathostomes, and, among these latter, Fishes lack a muscular tongue, which, in the taxonomic series, is first displayed by Amphibians. Nevertheless, the tongue musculature of Anamniote and Amniote Tetrapods, as well as some correlated neck muscles (sternohyoid, omohyoid and others) represent differentiations respectively derivatives of the *hypobranchial muscular primordium.* In fishes, the hypobranchial musculature is accordingly innervated by the ventral roots, respectively anterior rami, of the corresponding rostral spinal nerve segments. A propensity of these segments to lose their dorsal roots and spinal ganglia is already evident in fishes,

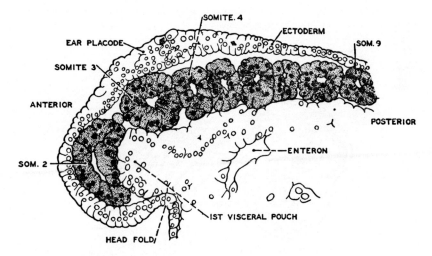

*Figure 435.* Parasagittal section through head region of eight-day Petromyzon embryo showing the mesodermal somites (epimeres) 2 to 9 in KOLTZOFF's enumeration for comparison with figure 434 (A). The first somite, identified in an adjacent section, is not included in the figure. According to the cited author, all head somites in Petromyzon form permanent musculature, the first three providing eye muscles, while the fourth is the first myomere of the (epimeric) lateral trunk musculature (after KOLTZOFF, from NEAL and RAND, 1936).

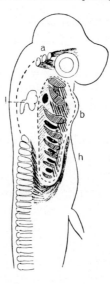

*Figure 436 A.* Simplified semidiagrammatic sketch showing genetic and configurational relationship of extrinsic eye musculature, branchial musculature, and hypobranchial musculature in the embryo of a pentanch Selachian (slightly modified after FRORIEP and EDINGER, from K., 1927). a: eye musculature; b: branchial musculature; h: hypobranchial musculature; l: otic vesicle (laryrinth anlage).

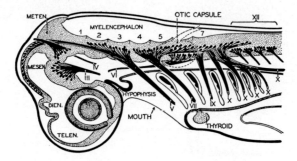

*Figure 436 B.* Semidiagrammatic parasagittal section through the head of a 13-mm Selachian embryo *(Squalus acanthias)*, based on serial sections processed with a modification of the *Cajal silver impregnation* technique (from NEAL and RAND, 1936). The relationship of several motor nuclei to the rhombomeres is shown. The evidence, as interpreted by NEAL and RAND, does not indicate any primary metameric correspondence between rhombomeres, nerve roots, myotomes, and visceral arches.

and some of the most rostral single ventral roots tend toward inclusion into the cranium. The group of 'specialized' rostral spinal nerves located between branchial nerves and 'typical' spinal nerves shows numerous variations in the diverse groups of Gnathostome Anamnia. Insofar as some of these nerves emerge from the cranial cavity, the assimilation of one or more vertebrae during the phylogenetic evolution of the skull may be assumed to have occurred concomitantly with a rostral migration of those spinal nerves which become transformed into the n. hypoglossus.

The evolutionary history of these so-called *spino-occipital nerves* (Fürbringer, 1897) shows complex and not entirely elucidated features. In amphibians, the tongue musculature (of hypobranchial origin), is innervated, by way of the ventral roots, from the rostral (commonly the first two) spinal nerve segments. In Amniota, the nerve innervating the tongue and related hypobranchial muscles is provided by several spinal nerve segments which have lost their dorsal roots[68] while their ventral roots have become included into caudal medulla oblongata respectively into cranium, thus forming a true *twelfth cranial nerve* or *n. hypoglossus*. Rudimentary dorsal root fibers with a small spinal ganglion related to the hypoglossus were described by Froriep (1882) and others in mammalian embryos (cf. Figs. 432, 433). These rudimentary or transitory embryonic structures can be interpreted as vestigial elements of the ancestral complete spinal nerve segments from which the hypoglossus presumably derived.

Following the preceding comments on the cranial nerves, and taking into account their morphologic arrangement, the discussion of the *head problem* may be resumed by some concluding remarks. Reverting to the four questions outlined above on p. 841, it seems evident that the vertebrate head displays *segmentation* manifested by *myomery*, *branchiomery*, and *neuromery*.

As regards the other questions it could be stated that, generally speaking, two conflicting opinions have been expressed. According to one view, based on older concepts of Gegenbaur and interpretations by Fürbringer and by van Wijhe, as also subsequently summarized by Ziegler (1923), the vertebrate head is strictly metameric such that

---

[68] This tendency toward loss of dorsal roots is also manifested in mammals by the 'true' first spinal nerve segment caudal to n. hypoglossus. Thus, in rats, the dorsal root of this segment was found missing in about 97% of examined specimens (Yagasaki, 1924).

preotic and postotic segments can be compared with those of the trunk. Moreover, branchiomery, myomery and neuromery are believed to correspond with each other. According to the other view, based, *inter alia*, on some of FRORIEP's studies, and epitomized by VEIT (1924), the vertebrate head, particularly with respect to its preotic portion representing a region *sui generis*, is, *ab initio*, not segmented in a manner identical with trunk metamerism.

Although, in the head region, formation of somites occurs indeed as far as the rostral end of notochord,[69] the pharyngeal and corresponding branchial pouches take their origin in an unsegmented neighborhood rostral to the somites. This region expands concomitantly with the development of the branchial apparatus in a dorsal and caudal direction, displaying its own metamerism, and disrupting the preotic series of epimeric somites, as well as the correlated unsegmented hypomeric components. These mesodermal configurations disintegrate into a diffuse, 'unspecific' mesenchyme. The hypomeric component then provides the branchial musculature (cf. Fig. 436). Thus, the branchiomery of the foregut results in an independent segmentation, unrelated to the mesomery of the somites (cf. also AHLBORN, 1884).

The available evidence seems to indicate that the number of transitory or rudimentary *head somites* recognizable during ontogenetic stages of Anamnia varies between about 9 to 11 in different taxonomic forms. Apparently, at least three of these somites[70] are definitely preotic (Figs. 434, 435), without, however, subsequent differentiation into typical myotomes, sclerotomes and dermatomes, but, nervertheless, giving origin to the eye musculature. The mm.rectus superior, inferior, medialis, and obliquus inferior, supplied by n.oculomotorius, seem to be derivatives of the first somite. Musculus obliquus superior, innervated by n.trochlearis, appears to originate from the second somite, and m.rectus lateralis, supplied by n.abducens, seems to derive from the third somite. Configurations corresponding to these three somites of Anamnia can also be seen, as so-called head cavities, in em-

---

[69] In Amphioxus, however, the notochord extends rostrally not only beyond the neuraxis, but also beyond the first myotome. Cf., however, footnote 71.

[70] There is some evidence that, in addition to the preotic somites 1–3 a still more rostral somite-like configuration or so-called 'anterior cavity' (PLATT, 1891) may be displayed by some Anamnia (certain Selachians). Its significance remains obscure as well as controversial.

bryos of Amniota, being rather well substantiated anlagen of eye muscles in the sequence just mentioned for Anamnia.

It seems reasonable to infer that the somites in the otic region have been eliminated by the morphogenetic events related to the branchiomeric expansion (cf. Fig. 436). There is general agreement that the *otic capsule* represents, in this respect, an important morphologic or morphogenetic landmark, and that there are, at least, three somites (epimeres) rostral to it, and an undefined *postotic* number of '*head somites*'. These latter, in addition to the hypobranchial musculature, seem to provide, at least in some vertebrate forms, poorly defined epibranchial musculature of 'somitic' derivation and apparently innervated by way of 'spino-occipital nerves'.

My own opinion, based on incidental, but rather extensive firsthand acquaintance with the embryonic configurations at issue acquired during my investigations of neuraxial ontogenesis, essentially coincides with the second interpretation, more or less as expressed by VEIT (1924). I would thus answer the second question (cf. p. 841) as follows: although the *vertebrate head* is originally *metameric*, head segments and trunk segments are fundamentally dissimilar insofar as *branchiomery* and *somitic myomery* represent *two different sorts of segmentation*, substantially independent of each other.

As regards the third question (introduced in addition to the three formulated by NEAL and RAND, 1936), I am inclined to assume, because of the behavior of branchial arches, the independent ventral and dorsal manifestations of peripheral neuromery, the *ventral root neuromery* directly corresponding to the *somitic myomery*, and the *dorsal root neuromery* directly corresponding[71] to the *branchiomery*. Thus, although somitic and branchial metamery can be regarded as essentially unrelat-

---

[71] The segmental alternation of dorsal and ventral roots in *Amphioxus* may be interpreted as a feature not comparable with conditions in Craniote vertebrates. It is evident (cf., e.g., the peculiar muscular ventral 'roots' of the neuraxis), that, although recent *Amphioxus* doubtless displays a number of significant 'ancestral' vertebrate configurational features, it cannot be taken, because of many substantial divergences from typical vertebrate configurational arrangements, as representing a suitable link in a direct phylogenetic sequence.

Again, as regards correspondence of dorsal peripheral neuromery with branchiomery, it seems obvious that processes of 'fusion', blending', 'amalgamation' or 'absorption' (i.e. many-one transformations) seem to have commonly occurred in vertebrate phylogeny, as e.g. manifested by the relationship of vagus nerve to several branchiomeric structures.

ed to each other, both display their own type of corresponding peripheral neuromery. The numerous and highly ambiguous details of central neuromery were discussed in section 3 (pp. 304–350) of the preceding chapter VI. With respect to the *head problem*, it could be said that the so-called archencephalic neuromeres differ from the deuterencephalic ones, and represent a phenomenon *sui generis*. Stretching the point, one might, in a somewhat unconvincing interpretation, relate telencephalon to an 'olfactory' head segment, and parencephalon to an 'optic' one, this 'segmentation' being essentially 'autonomous'. Similarly, and likewise in a not entirely convincing way, deuterencephalic central neuromery could be related to the peripheral neuromery displayed by the sequence of cranial nerves III–X, as suggested by some authors and indicated by the tabulation of Figure 113 in section 3, chapter VI.

The fourth question (NEAL's and RAND's third), namely, how many segments are contained in the head, appears to me entirely meaningless for the following two reasons. (1) Evidently, on the basis of the available data, it is quite impossible to reconstruct the morphologic details, including the number of myotomes and branchiomeres, of an 'ancestral' form from which, assuming a monophyletic sequence, the vertebrate phylum might have originated. (2) It is likewise obvious that segmentation, involving e.g. the number of vertebrae, of branchia,[72] and of myotomes, depends upon 'fluctuating', 'secondary' factors or parameters. It will here be sufficient to mention differences such as pentanch, hexanch, and heptanch Elasmobranchs within one and the same overall taxonomic group, or the variable number of vertebrae displayed by taxonomically related forms. In some instances, a 'reduction', and in others, a 'multiplication' of diverse metameric structures seems compatible with 'progressive' phylogenetic evolution.

---

[72] Generally speaking, a tendency toward reduction of the number of branchial arches seems to be manifested in phylogenetic evolution, even discounting the very large number displayed by *Amphioxus*, whose phylogenetic status is rather ambiguous or dubious. Thus, in craniote vertebrates, Agnatha display about 8 branchial arches. The same number is found in Gnathostome heptanch Selachians, believed to be 'primitive'. It becomes reduced to 7 in hexanchs, and to 6 in pentanchs. In mammals, whose most caudal branchial arches are poorly developed and merely suggested, depending on taxonomic forms, respectively ontogenetic stage, or criteria of identification, about 4–5 remain recognizable.

## 4. The Secondary Deuterencephalic Longitudinal Zones
and the Concept of Functional Nerve Components

Section 4 (p. 350 f.) of the preceding chapter VI dealt with the transformations of the primary longitudinal zones of the deuterencephalon into a definitive secondary system, which is particularly conspicuous in various Anamnia. Two different interpretations respectively descriptions of the said longitudinal system were considered, namely (1) on the basis of the *doctrine of functional nerve components*, and (2) on a purely morphologic *(form analytic)* basis.

As seen from this latter viewpoint, the *basal plate* displays a ventral zone (BV) and an intermedioventral zone (BI), which, in Amniota, is further subdivided into internal (BIi) and external (BIe) subzones. The *alar plate* shows an intermediodorsal zone (DI) and a dorsal zone (DLS) comprising two subzones, namely, the pars of area superior (DS lm), which includes lateral and medial neighborhoods, and the pars or area lateralis seu inferolateralis (DLi).

With respect to functional peripheral nerve components, the motoneurons for striated musculature, directly or indirectly (phylogenetically) derived from somites, are located in BV. The motoneurons inervating branchial or branchiogenic musculature and the preganglionic elements of the vegetative nervous system are found in BI of Anamnia, respectively BIe (branchiomotor elements) and BIi (preganglionics) of Amniota.

The *longitudinal bundles of primary afferent fibers* (root fibers), respectively their endings in the primary grisea related to innervation of viscera including their mucosa, of taste buds, and apparently also of blood vessels, are located in DI. The corresponding primary fibers respectively grisea related to general cutaneous innervation assume a location in DLi. Vestibular, lateralis, and cochlear root fibers respectively all or most of their primary terminal grisea generally aggregate in DS lm.

If the terminology of functional nerve components is adopted, then, at least to some extent, a designation of the above-mentioned configurational arrangements in terms of this doctrine appears justified. Accordingly, one might distinguish the following zones or subzones: a *somatic motor column* (BV), a *visceral motor or efferent column* (BI), which can be further subdivided into *special visceral motor or branchiomotor* (BIe) and *general visceral motor* (BIi) components, moreover a *visceral sensory column* (DI), and a *somatic sensory column* (DLS). This latter, again, comprises a *general somatic sensory component* (DLi) and a *special somatic sensory* one (DS lm). Figure 437 shows, in JOHNSTON's conception (1906), somatic motor, visceral motor, visceral sensory, and somatic sensory columns in the rhombencephalon of the Ganoid Acipenser. The functional relevance of these columns is furthermore evidenced by the findings of C. L. and C. J. HERRICK illustrated by Figure 438. The considerable development of the 'visceral sensory' gustatory system in Cyprinoid Teleosts is here correlated with an expansion of the visceral

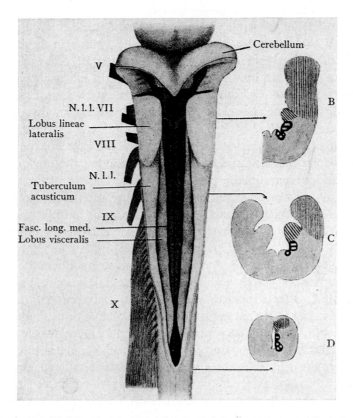

*Figure 437.* Rhombencephalon of the Ganoid Acipenser (sturgeon), displaying the secondary deuterencephalic longitudinal zones related to the so-called functional nerve components (from JOHNSTON, 1906). At left rhombencephalon in dorsal view, with choroid plexus removed. B, C, D: cross-sections at levels indicated by reference lines; dark area with white dots: 'somatic motor column'; dark area with white dashes: 'visceral motor column'; area with oblique hatching: 'visceral sensory column'; area with vertical hatching: 'somatic sensory column'; N.l.l.: nervus lineae lateralis; Roman figures: cranial nerves.

sensory column forming the conspicuous so-called vagal lobe. The columnar arrangement in the spinal cord is rather similar to that in deuterencephalon, but in a somewhat simplified manner, as illustrated by Figure 155 of chapter VI, and does not require further comments in this context.

The preceding section 3 dealt with the nerves of the deuterencephalon and their metameric arrangement in particular reference to the head problem. In the present section 4, the cranial nerves shall briefly be considered from the viewpoint of *functional components*. As already pointed out in section 4 of chapter VI, a restriction of any detailed application of the doctrine of functional components with their related longitudinal columns to the *nerves of deuterencephalon* seems indicated, particularly also since the correlated griseal columns are confined to deuterencephalon (and spinal cord). Thus, nn. olfactorii (fila olfactoria), n. terminalis, n. opticus, both *qua* special structures and *qua* archencephalic (prosencephalic) components, shall here be excluded.

Concerning the deuterencephalic peripheral nerves, their components could evidently be described in terms of their neuraxial longitudinal zones (BV, BI, DI, DLS, etc.). I believe, nevertheless, despite my inclination toward pure morphology, and notwithstanding the semantic weaknesses displayed by the doctrine of *functional nerve components*, that this latter is practically more suitable for a relevant description of

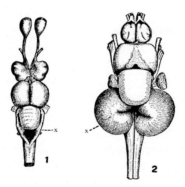

*Figure 438.* Dorsal view of brains in two taxonomically closely related Cyprinoid Teleosts, displaying conspicuous differences in the development of their primary 'taste centers' contained in the 'visceral sensory column' of medulla oblongata (after C. L. HERRICK, from C. J. HERRICK, 1931). 1: the 'moon eye' *Hyodon tergisus;* 2: the carp-like *Carpiodes tumidus;* x: corresponding regions of oblongata.

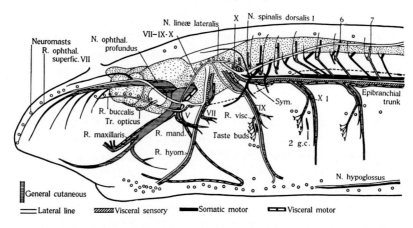

*Figure 439.* Reconstruction of the cranial nerves in the Cyclostome Petromyzon, showing the distribution of 'functional nerve components' (from JOHNSTON, 1906). 2 g.c.: second gill cleft; Sym.: ramus intestinalis nervi vagi; the most rostral dorsal spinal nerve roots 1–7 are shown.

deuterencephalic cranial nerves and it shall therefore be adopted in the present context. In connection with the following generalized summary it should also be emphasized that, depending on the taxonomic vertebrate forms, considerable differences are displayed concerning relative amount, presence or absence, and distribution of fibers pertaining to the diverse functional components of cranial nerves.

As regards the eye muscle nerves, the *oculomotor* (III) includes three components, namely somatic motor, general visceral motor (preganglionic), and (probably) special somatic sensory (proprioceptive) fibers. The *trochlear nerve* (IV), and the *nervus abducens ( VI )* include somatic motor and (probably) proprioceptive fibers.

*The spino-occipital nerves* respectively the comparable[73] *n. hypoglossus* (XII) of Amniota have somatic motor and possibly, or in certain instances, some special somatic sensory (proprioceptive) components.

Concerning the *branchial nerves* it must be kept in mind that, because of the often very substantial lateral line structure of Anamnia, and the system of cutaneous taste buds in some fishes, considerable differences between Amniota and Anamnia obtain. In these latter, all branchial

---

[73] Depending on arbitrary criteria or postulates, the n. hypoglossus of Amniota may be considered either orthohomologous or kathomologous with the spino-occipital nerves of Anamnia.

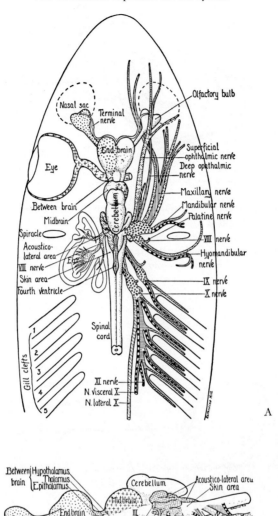

*Figure 440*. Diagrams of brain and cranial nerves in the Pentanch Selachian, *Squalus acanthias;* showing 'functional nerve components' (from HERRICK, 1931). Olfactory region indicated by coarse dots, optic region by crosses, acoustico-lateral components by broken oblique lines, visceral sensory components by horizontal lines, general somatic sensory (cutaneous) components by vertical lines, visceral efferent components by black and white rectangles. A: dorsal view, with brain *in situ;* b: lateral view of removed brain with its nerve roots.

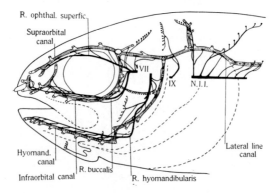

*Figure 441.* Diagram of the lateral line canals and pit organs together with their nerve supply in the Ganoid, *Amia calva* (after E. PHELPS ALLIS, from JOHNSTON, 1906). The canals are shaded with cross lines and the canal organs are shown as black discs in the course of the canals. The pit organs are indicated by black dots. Only the peripheral nerve trunks are shown, the ganglia and roots being omitted. N.l.l.: ramus lateralis vagi.

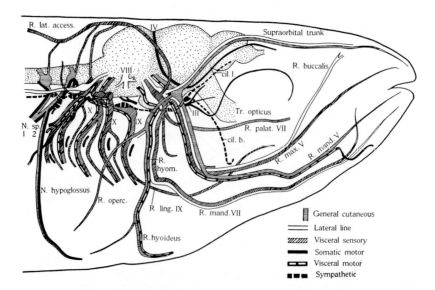

*Figure 442.* Reconstruction of the main rami of the cranial nerves in the Teleost Menidia, showing the distribution of functional nerve components (after C. J. HERRICK, from JOHNSTON, 1906).

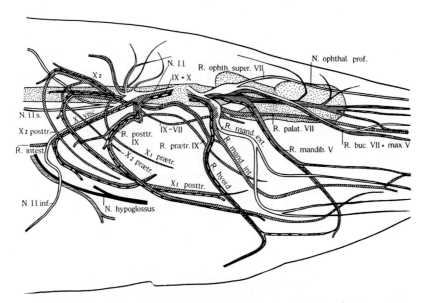

*Figure 443*. Reconstruction of the cranial nerve components in the urodele amphibian Amblystoma (after COGHILL, from JOHNSTON, 1906). Notation as in figures 439 and 442. N.l.l.: nervus lineae lateralis; posttr.: posttrematic; praetr.: pretrematic; other abbr. self-evident.

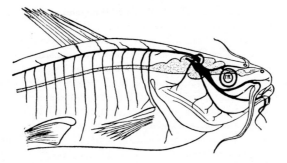

*Figure 444*. Cutaneous gustatory branches arising from the ganglion nervi facialis ('geniculate ggl.') of the Siluroid Teleost *Ameiurus* (catfish). Gustatory nerves drawn in black, stippled outline of neuraxis projected upon surface. The gustatory rami of VII supplying mucosa of oral cavity are omitted. Taste buds are found in all regions of ecto-dermal skin to which the indicated branches are distributed (from HERRICK, 1931).

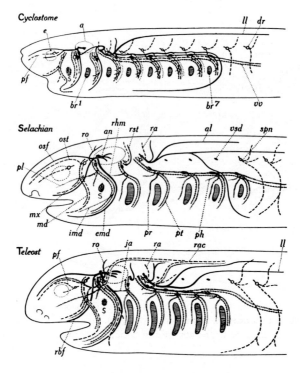

Figure 445 A

*Figure 445 A, B.* Diagrams of functional component of branchial nerves and of rostral spinal dorsal roots in the vertebrate series according to GOODRICH's interpretation (from GOODRICH, 1958). The ventral root nerves to eye muscles and hypobranchial musculature are omitted. Black lines: lateral line component; broken lines: general cutaneous (somatic afferent); cross-hatched lines: visceral afferent; beaded lines: visceral efferent; a: auditory capsule; al: supplementary dorsal lateral line; an: auditory nerve; br [1-7]: branchial slits; cht: chorda Tympani; clbr: closed branchial slit; dr: dorsal root nerve of trunk; e: eye; emd: external mandibular; fa: facial and auditory rootlets; gl: glossopharyngeal; imd: internal mandibular; ja: *Jacobson's anastomosis;* ll: main lateral line nerve; md: mandibular; mx: maxillary; osf: superior ophthalmic branch of facial; ost: superior ophthalmic branch of trigeminus; pf: profundus branch of trigeminus; ph: pharyngeal; pl: palatime; pr: pretrematic; pt: posttrematic; ra: dorsal ramus; rac: recurrent accessory ramus of facial and of vagus; rbf: buccal; rfr: ramus frontalis; rhm: hyomandibular; rn: orbitonasal; ro: ramus oticus; rst. r.: ramus supratemporalis; s: spiracular slit; spn: dorsal root of spinal nerve; tg: trigeminal; tp: tympanic membrane; vsd: vestigial dorsal root; vg: vagus; vv: visceral ramus of vagus.

nerves may include 'special somatic sensory' lateralis fibers, either directly with their roots, or, particularly in the case of r. ophthalmicus superficialis trigemini, by peripheral anastomosis close to their origin (cf. Figures 439–443, 445). In the case of the cutaneous gustatory system the 'special visceral sensory fibers' seem to have their cell bodies in the ganglion nervi facialis (ggl. geniculi), and join *rostrally* the trigeminal domain, being distributed *caudalward* upon the body through a long dorsolateral posterior branch of n. facialis (cf. Fig. 444).

The *first branchial nerve* or *n. trigeminus* (V), whose suspected derivation from two originally separate dorsal roots ($V_1$ and $V_{2,\,3}$) was mentioned above in section 3 (p. 851), contains *special visceral motor* (branchiomotor) fibers, supplying, in mammals, the muscles of mastication, mylohyoid muscle, anterior belly of m. digastricus, and the mm. tensor tympani and tensor veli palatini. All these motor fibers are distributed through the mandibular branch. *General somatic sensory* fibers originating in the trigeminal ganglion supply the skin of the head respectively

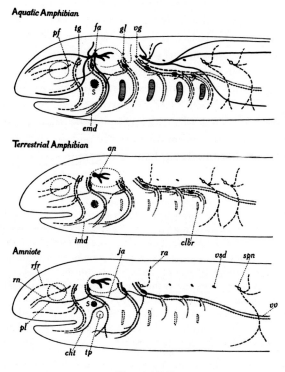

Figure 445 B

face including orbit, the teeth, and the ectodermal mucosa of oral and nasal cavities. *Special somatic afferent* (proprioceptive) fibers supply particularly the muscles of mastication but perhaps also various other head muscles.[74] Their cells of origin form, as discussed in the preceding section, the intracerebral nucleus of radix mesencephalicae trigemini.

The *second branchial nerve* or *n. facialis* (VII) contains *special visceral motor* fibers for the musculature derived from the second branchial, i.e. the hyoid arch, namely, for facial musculature, platysma, and m. stapedius.[75] Preganglionic fibers for lacrimal and some salivary glands represent a *general visceral 'motor'* or better, efferent, component.

*General somatic sensory* facial nerve fibers supply, in man and other mammals, the skin in back of the external ear. The presence of *'special somatic'* (proprioceptive) fibers for the facial musculature has been suspected. *General visceral sensory* fibers for the mucous membrane of the palate, and *special visceral sensory* (gustatory) fibers for the taste buds are other afferent components of this nerve. In human anatomy, the term n. intermedius or n. glossopalatinus is sometimes applied to the afferent and the general visceral (preganglionic) component, which frequently represents a grossly recognizable strand of fibers separated from the main 'branchiomotor' portion of n. facialis. The sensory ganglion, associated with the intermedius portion, is known as ganglion geniculi and contains the cell bodies of somatic sensory, general visceral sensory as well as special visceral sensory fibers. As regards the cell bodies of the suspected proprioceptive fibers, their location is unknown, but these elements might possibly pertain to the general system of nucleus radicis mesencephalicae trigemini.

---

[74] It will be noted that, in contradiction to HERRICK (1931) and others, who classify such *proprioceptive* innervation as *'general somatic'* afferent, I prefer to subsume all proprioceptive innervation under the classification *'special somatic afferent'*. Nevertheless, in either case the qualifier *'somatic'* for proprioceptive innervation of branchiogenic musculature, required by the *Procrustean* terminology introduced by the doctrine of 'functional nerve components' is obviously awkward. It implies that (either general or special) 'somatic' afferents supply branchiogenic muscles innervated by special visceral motor nerves. Yet, the classification of these proprioceptive fibers as 'special visceral afferents' would be still more awkward, equating such fibers with the altogether functionally different special visceral afferent taste fibers. Cf. also the comments on this topic in section 4 of chapter VI.

[75] The phylogenetic origin of the auditory ossicles and their musculature (m. tensor tympany, m. stapedius) was briefly pointed out above in section 3, p. 849, footnote 66.

The *acoustic nerve* (VIII) with its *vestibular* and *lateralis* subdivisions in Anamnia, respectively its *vestibular* and *cochlear*[76] subdivisions in Amniota represents a nerve with almost exclusively *special somatic sensory* components. It can be interpreted, from the viewpoint on the head problem expressed in the foregoing section 3, as a secondarily separated specialized subdivision of the second branchial nerve (n. facialis, VII). The dorsal ganglion of nervus octavus becomes differentiated into vestibular, lateralis, and cochlear (*ggl. spirale Corti*) portions, as the case may be with regard to the diverse taxonomic groups. Even in all Amniota, including mammals and man, the neuronal elements in the dorsal ganglia of this highly specialized cranial nerve retain a 'primitive' bipolar shape. Efferent components of nervus acusticus seem to be fibers of still poorly identified respectively understood intracerebral origin, apparently ending on, and 'modulating', receptor structures (cf., e.g., HELD, 1893; RASMUSSEN and WINDLE, 1961). A classification of these efferent fibers in accordance with the phraseology used by the doctrine of functional nerve components involves semantic difficulties. The qualification 'special somatic efferent', which various authors have applied, without convincing reasons, to the innervation of eye muscles and hypobranchial (tongue) musculature, might be more properly reserved for fibers of this type.[77]

The functional components of the *third branchial nerve* or *n. glossopharyngeus* (IX) are *special visceral motor* fibers for striated branchiogenic musculature of the pharynx, and *general visceral efferent* (preganglionic) fibers for salivary glands (parotid of man and other mammals). *General somatic sensory* fibers innervate, in man, skin areas of the external ear. The presence of *'special somatic afferent'* (proprioceptive) fibers for the glossopharyngeal musculature can be suspected, but definite evidence of their inclusion in this nerve is not available. *General visceral afferent* fibers supply the mucous membrane of pharynx and part of the tongue. *Special visceral afferent* fibers innervate taste buds in posterior regions of tongue.

---

[76] The rudiment of a cochlear apparatus is, of course, already displayed by amphibians in general, and a cochlearis subdivision can be easily recognized in various Anuran amphibians.

[77] Comparable efferent (centrifugal) fibers reaching the retina are present in the optic nerve and were described about 1894 by CAJAL (1911, 1955), DOGIEL (1895), AREY (1916), and others. It is not impossible that mechanisms of this sort, as e.g. also involved in the calibration of muscle spindles, are rather widespread with respect to other sorts of receptor structures, but few useful or convincing data have, so far, come to my attention.

The *fourth branchial nerve* or *n. vagus* (X) includes the following components. *General somatic afferent* fibers, reported in man for the skin of the external ear and the external auditory meatus.[78] *Special* (according to HERRICK and others, 'general') *somatic afferent*, proprioceptive for the striated musculature innervated by the vagus are also believed to be present in this nerve. In Anamnia, of course, the extensive *ramus lateralis nervi vagi* represents a 'proprioceptive' component with whose classification by HERRICK (1931) as '*special somatic afferent*' in the here tentatively adopted phraseology I would agree. *General visceral afferent* fibers are distributed to the mucous membrane of the pharynx and larynx, and furthermore innervate the trachea and lung, the greater part of the digestive tract, and the heart.[79] *Special visceral afferent* fibers of the vagus innervate, e.g. in man, taste buds of the epiglottic region.

Turning to the efferent components, *special visceral efferent* fibers innervate the corresponding branchiogenic musculature, which, e.g. in man, includes the muscles of the larynx and those of pharynx not pertaining to the glossopharyngeus. *General visceral efferent* (preganglionic) fibers of the vagus supply, like the corresponding afferent ones mentioned above, the respiratory organs, the greater part of the digestive tract, and the heart. A long *ramus visceralis vagi* innervating the gastro-intestinal tract and the heart is already present in Anamnia.

The *nervus accessorius* (XI) of human anatomical nomenclature arises with *radices craniales* and *radices spinales* (PNA). The former provide the *ramus internus* which joins the vagus and essentially contains the *special visceral efferent* fibers for the laryngeal muscles innervated by the *nervus recurrens vagi*. These radices craniales can be considered a subdivision of the vagus proper *(n. accessorius vagi)*. The radices spinales provide the *ramus externus accessorii* (BNA, PNA) essentially containing *special visceral efferent* fibers for the (directly or indirectly) branchiogenic components of mm. trapezius and sternocleidomastoideus. Whether '*special*

---

[78] It is doubtful, however, in which proportion such fibers, also contained in other branchial nerves, are distributed upon nn. VII, IX, and X.

[79] The *pressoreceptor* and *chemoreceptor* fibers of vagus and glossopharyngeus which innervate carotid sinus, aortic arch, carotid glomus and aortic glomus, are difficult to classify according to the scheme postulated by the 'doctrine of functional nerve components' but might be included with the general visceral afferent fibers.

---

*Figure 446.* Simplified synoptic tabulation of vertebrate cranial nerves as particularly applying to man (adapted and modified from K., 1927). The open space at right is meant to indicate the open boundary neighborhood transitional between brain stem and spinal cord.

Trigeminus group

Vagus group

| | Dorsal roots | Ventral roots |
|---|---|---|
| n. oculomotorius (N. III) | | somatic motor; general visceral efferent; special somatic afferent (proprioceptive) |
| n. trochlearis (N. IV) | | somatic motor; special somatic afferent (proprioceptive) |
| n. trigeminus (N. V) | special visceral motor; general somatic afferent; special somatic afferent (proprioceptive) | |
| n. acusticus (N. VIII) | special somatic afferent | |
| n. facialis (N. VII) | special visceral motor; general visceral efferent; general somatic afferent; special and general visceral afferent | |
| n. abducens (N. VI) | | somatic motor; special somatic afferent (proprioceptive) |
| n. glossopharyngeus (N. IX) | special visceral motor; general visceral efferent; general somatic afferent; special and general visceral afferent | |
| n. vago-accessorius (N. X et XI) | special visceral motor; general visceral efferent; general somatic afferent; special and general visceral afferent — only in X (?) | |
| n. hypoglossus (N. XII) | | somatic motor; special somatic afferent? (proprioceptive?) |

*somatic afferent'* proprioceptive fibers are also included in ramus externus remains a moot question. The ramus externus accessorii of man and other mammals can be considered kathomologous with the ramus accessorius vagi of fishes (cf. Fig. 429). The above-mentioned n. accessorius vagi of man, on the other hand, which innervates laryngeal musculature, does not correspond to the accessorius vagi of fishes which supplies musculature related to the pectoral fin girdle (shoulder girdle) as discussed in section 3 with reference to the head problem.[80]

The simplified diagrams of Figure 445 display, in Goodrich's (1958) interpretation, the functional components of branchial nerves in various Anamnia and in a generalized Amniote. These sketches, summarizing, as it were, significant features of the doctrine of functional nerve components, should be compared with the more elaborate illustrations of this topic by Figures 439–443. The synoptic tabulation of Figure 446 summarizes morphologic and functional features of vertebrate cranial nerves as particularly applying to human anatomy.

## 5. Electrical Nerves and Organs; Bioluminescence

Specialized organs, innervated by cranial or spinal nerves, and generating external electric fields substantially stronger than those produced by the 'normal' electrical activities of nerves, muscles, and glands, are present in a number of different fishes. The electrical nature of the discharge producing the well-known peculiar shock experienced upon contact with some of these aquatic organisms was apparently first clearly recognized about 1751 by the French botanist Michel Adanson, who observed the electric catfishes of the Senegal river. He compared their discharge with that of the *Leyden flask* which had been just recently contrived.[80a]

---

[80] Because of the confusing complexities inherent in both head problem and the widely accepted doctrine of functional nerve components, a separate discussion of cranial nerves from both viewpoints seemed mandatory, despite a few thereby unavoidable partial repetitions.

[80a] About 1745, apparently simultaneously and independently by E. G. von Kleist *(Kamin, Prussia)* and P. van Musschenbroek (Leyden, Holland). Further data on the history of the subsequent theories and studies on electric fishes, of considerable import for the development of present-day electrophysiologic concepts, can be found in the publications by Biedermann (1895) and Fessard (1958). A concise summary was also included in older editions (e.g. 1929) of Landois' and Rosemann's Textbook of Physiology. Strangely enough, except for a perfunctory reference to *Electrophorus*, I could find no discussion concerning electric fishes in the recent (28th) edition of 1962.

As regards their taxonomic distribution, electric fishes are found among two families of *Elasmobranchs*, namely *Torpedinidae* and *Rajidae*, as well as among five families of *Teleosts*, including *Siluroids*, *Gymnotid Apodes*, *Mormyridae*, *Gymnarchids*,[81] and *Astroscopidae* (Acanthopterygians related to Uranoscopus). The head shield of some *Silurian Cephalaspids*, classified as Agnatha, displays fossil evidence of a peculiar area which is suspected, by some authors, to have been either an electric or a 'sensory' organ (cf. YOUNG, 1955). It is estimated that at least approximately 250 (FESSARD, 1958) to 500 (GRUNDFEST, 1960a) species of fish are provided with electric organs, but, as far as I could ascertain at the time of this writing, only about a score or so have been investigated in a detailed manner. As regards habitat, only the Elasmobranch and Astroscopidean electric fishes are marine, while all others, so far as hitherto known, are fresh-water dwellers.

With respect to the histogenetic origin of the electric organs, almost all those, whose provenance has been definitely established, are *derivatives of striated musculature*, either branchial or myotomic. It seems possible, however, that one 'aberrant' type of Gymnotid fish (Stenarchus) has electric organs derived from modifications of nerve fibers (FESSARD, 1958, quoting a personal communication by COUCEIRO). There is, moreover, some doubt concerning the origin of the electric organ in the Siluroid Malapterurus.[82] The structural elements of electric organs are provided by commonly multinucleated (syncytial) cellular units designated as *electroplaques*, which, upon being 'triggered' by efferent nerve endings, respond with an electric discharge. Although, depending on the diverse taxonomic forms, the electroplaques display numerous variations *qua* structural details, they generally can be described as thin, 'wafer-like' configurations whose two main surfaces differ in their anatomical relationships, only one of the surfaces being innervated.[83] The efferent nerves provide cholinergic synaptic endings comparable to neuromuscular junctions. These synapses are either

---

[81] Some authors (e.g. BALLOWITZ, 1938) subsume *Gymnarchus niloticus* under the group *Mormyrids* ('*Nilhechte*').

[82] FRITSCH (1887) first believed that it was derived from the cutaneous muscle system but subsequently regarded it as originating from cutaneous gland cells. BALLOWITZ (1899) assumed that it might be formed by 'electroblasts' (i.e. transformed striated myoblasts) which secondarily migrated into the skin.

[83] In some aberrant *Gymnotidae* ('knife fishes'), however, an exception to this general rule seems to occur.

formed by dense arborizations of branching terminals, or by connections with stalk-like extensions of the electroplaques, occurring in several unrelated forms such as African Siluroids and Mormyrids and at least in one of the South American Gymnotids.

The *modus operandi* of electroplaques was elucidated, with respect to its essential features, by BERNSTEIN (1912). According to this author the membranes at the two opposed main surfaces of the plaque display, at rest, equal interior negative potentials (cf. Figure 447). Upon stimulation by the nerve ending, the corresponding membrane would become depolarized, the resting potential of the opposite membrane remaining unaffected. This potential gradient must evidently result in an electric 'current' flowing at a right angle to the main surfaces, the outside of the innervated one being, as it were, negative to that of the non-innervated surface (*Pacini's rule*, dating back to 1852). The electromotive force *(emf)* generated by each plaque corresponds approximately to that of the resting potential whose order of magnitude is roughly 0.1 volt (more accurately about 70–90 mV, cf. vol. 3, part I, p. 565f.). In contemporary phraseology, the response of an electroplaque is a postsynaptic potential (PSP) generated by an electrically inexcitable membrane. This also agrees with older as well as recent observations indicating that the electric organ of *Torpedo* is, itself, unresponsive to electric stimuli.

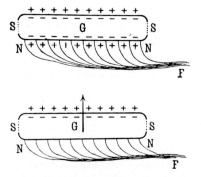

*Figure 447.* BERNSTEIN's concept of electroplaque discharge (from GRUNDFEST, 1967). Upper diagram represents condition at rest, lower diagram indicates depolarization of innervated surface upon a nerve impulse. F: nerve fiber; G: electroplaque; N: synaptic nerve terminals; S: side wall, as contrasted to main surfaces of electroplaque.

The subsequent detailed studies with present-day sophisticated gadgetry disclosed that BERNSTEIN's original hypothesis very closely corresponds to the behavior of electroplaques in *Torpedinidae* and *Astroscopidae*, and rather closely but with some modifications, to plaque behavior in the Gymnotid *Electrophorus*. In this latter, only the innervated surface is electrogenically reactive, but it responds to depolarization by an applied stimulus as well as to a 'neural volley', being thus electrically excitable (KEYNES and MARTINS-FERREIRA, 1953). Moreover, during its response, the inside potential changes from a resting negative value to a positivity as large as 60–75 mV. The response amplitude across the active membrane may therefore attain a magnitude of about 150 mV contrasting with an ordinary maximal PSP of approximately 70 mV. Such temporary polarity reversal instead of simple depolarization results thus in a so-called '*overshoot*'.[84]

Further investigations of different electric fishes revealed additional varieties in the *modus operandi* of electroplaques, representing, as it were, variants of the basic '*Bernstein-mechanism*'. The general concept of opposed membranes generating different potential can still be regarded as correct, but both faces of the plaques may become active in a diversity of ways. In some instances, *Pacini's rule*, stating that, during the discharge, the negative pole corresponds to the side on which the plaque is innervated, does not hold *qua* net discharge. Nevertheless, even here some of the complex events partly correspond to said rule. Details concerning the electrical phenomena[84a] involved in the diverse modes of operations of electroplaques are summarized in the publications by FESSARD (1958), BENNETT (1961), and GRUNDFEST (1960a, b, 1967). Figure 448 illustrates outlines of the electroplaques in seven groups of electric fishes.

Proceeding from *electroplaques*, representing structural and functional units, to *electric organs*, the design of these latter can be described

---

[84] Cf. vol.3, part I, p.566, where, for the magnitude of the 'overshoot', an estimate of 30 to 60 mV was mentioned. As regards postsynaptic potentials (PSP) etc., cf. also loc.cit., p.618f.

[84a] In the aspect under consideration, two different modes of bioelectrical manifestation are significant. (1) Graded '*membrane potentials*' and (2) *all-or-none* '*spike potentials*'. The first are local, but affect the neighborhood by producing an electric field with 'lines of force' respectively of 'current flow'. The second are characterized by an 'explosive', propagated electrochemical disturbance resulting in the transmission of the 'neurokym' or 'nerve signal'.

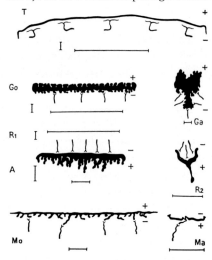

*Figure 448.* Outlines of electroplaques in different electric fishes, showing innervation and polarity during discharge (from FESSARD, 1958, and GRUNDFEST, 1967). A: Astroscopidae; Ga: Gymnarchus; Go: Gymnotidae; Ma: Malapterurus; Mo: Mormyrus; R: Rajidae (R$_1$: disk-type; R$_2$: cup-type); T: Torpedinidae; Horizontal lines of scales: approx. 1 mm; vertical lines: approx. 0.1 mm. Note innervation of plaque on dorsal surface in Astroscopus, on ventral surface in *Torpedo*, on rostral surface in Rajidae, and on caudal surface in the other fishes.

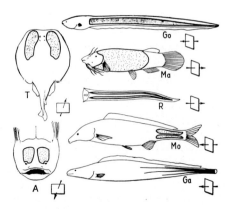

*Figure 449.* Electric organs in representatives of seven families of electric fishes. A, Ga, Go, etc. as in preceding figure. In Rajidae (R) only the tail of the animal is shown. Arrows indicate direction of current flow (arrowhead directed toward positive side, as in fig. 447). (From FESSARD, 1958, and GRUNDFEST, 1967.)

as closely corresponding to that of the original electric piles devised by
VOLTA (1745–1827) and his friend GALVANI (1737–1798). In other
words, the plaques are stacked in columns ('*Voltaische Säulen*') which
may comprise up to several thousand units, whereby a corresponding
summation of the individual plate PSP is obtained. Moreover, the
plaques are arranged in *serial* and in *parallel* array, the former providing
the total electromotive power *(voltage)*, and the latter the total current
intensity *(amperage)*. In freshwater electric fishes, the arrangement is
predominantly in series, thus yielding a high voltage, while in the marine species the arrangement is prevailingly in parallel. This divergence
can evidently be interpreted as an 'adaptation' to the different conductivity of fresh and sea water. Concerning the electric organs taken *in
toto*, their location and overall outline in the seven main types of electric fishes is illustrated by Figure 449.

As regards output of electricity, *strongly* and *weakly electric fishes* are
known to occur. The *strong* ones include particularly *Torpedo, Electrophorus (Gymnotus) electricus*, and *Malapterurus*. It is generally assumed
that these fishes use the electric discharges for defense or capture of
prey, or both. Among *weak* electric fishes, the *Rajids*, various species of
*Mormyrus, Gymnarchus*, and a number of *Gymnotids* might be mentioned. Before the functional significance of weakly electric organs was
understood, some of these latter fishes were commonly classified as
'*pseudo-electric*'. Since about 1950, however, it became evident that, in at
least several, and possibly all of these forms, the electric pulses emitted
by the weak organs provide a *guidance system* for navigation (e.g. especially in murky waters) or for the detection of predators respectively
prey, or both sorts of behavioral activities (cf. COATES, 1950; LISS-
MANN, 1958, 1963). Guidance systems of this overall type, in principle
corresponding to sonar or radar devices, and used by invertebrates as
well as vertebrates, were briefly discussed in chapter IV (volume 2,
p. 185 f.) of the present series.

The general principle of *electrical guidance systems*, illustrated by Figure 450, seems to be based on the generation of low voltage pulses in a
more or less continuous sequence, and is, e.g., comparable to the ultra-
sonic cries emitted by bats during their flight. The electrical field produced by the fish in its environment becomes distorted by objects
within that field, such that the 'lines of force' respectively of 'current
flow' diverge from a poor conductor or converge toward a good one.
Without disturbance by objects in the surrounding water, the undis-
torted 'lines' essentially display the pattern of a dipole field. The fish

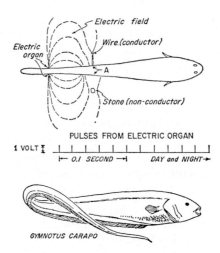

*Figure 450.* Diagram illustrating electric field produced by the electric organ of *Gymnotus carapo* and its distortion by conductors or non-conductors in the environment of the fihs (from GALAMBOS, 1962).

apparently responds to the distribution of potential over the surface of its body.[85]

The identification of the *electroreceptors* required for the registration of the relevant potential changes has presented considerable difficulties which cannot be considered properly solved at this time. Nevertheless, it seems likely that specialized, adapted, or modified *lateral line organs* act as detector devices (cf. also p. 794, and 796 respectively foot-notes 12 and 15 in section 1 of this chapter). In *Mormyrids* and *Gymnarchus*, cutaneous pores lead through canals filled with a jelly-like sub-stances to 'glandular sense organs' or mormyromasts innervated by lat-eral line nerves. Sense organs very similar to those of the just-men-tioned African fishes have been found in at least some South American *Gymnotids* (cf. Fig. 451). *Savi's vesicles* in *Torpedo*, and the *ampullae of Lorenzini* in *Torpedo* and *Raja* display conspicuous similarities to mormy-

---

[85] The sensitivity of the presumed receptor structures to field changes seems extremely high, implying a very low threshold of about 0.003 microvolts per mm (GALAMBOS, 1962).

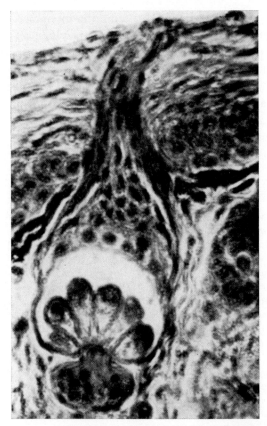

*Figure 451.* Section through skin and electric sense organ of a Gymnotid. A tube containing jelly-like substance widens at the base into a gland-like vesicular structure holding a cluster of specialized cells (from LISSMANN, 1963).

romasts. However, some Selachians possessing *ampullae of Lorenzini* are not known to be equipped with specific electric organs.[86]

Following these general comments, a few remarks on some particular forms of the diverse electric fishes might be appropriate. In this respect, the marine *Torpedinidae*, occurring in European, including Mediterranean waters, as well as in those of Asia and America are perhaps

[86] The only Teleosts known to possess typical *ampullae of Lorenzini* are Siluroids, namely the marine species *Plotosus anguillaris* (FRIEDRICH-FRESKA, 1930, cf. LISSMANN, 1958). Only the freshwater Siluroid *Malapterurus*, however, is known to have an electric organ.

the best known, and were commented upon by numerous authors of classical antiquity.[87] Post-Renaissance authors who dealt with anatomy or with the peculiar potency of *Torpedo* are, among others, Nicolaus Steno (Niels Stensen, 1638–1686), Francesco Redi (1626–ca. 1697), Stefano Lorenzini (*floruit* ca. 1678), and René-Antoine Ferchault de Réaumur (1683–1757). John Walsh demonstrated in 1772 at *La Rochelle*, by means of experiments involving transmission through appropriate conductors, the electric nature of the discharge, thus confirming views of Adanson concerning the activity of Malapterurus.

---

[87] Aristotle (384–322 B.C.) refers to *Torpedo*, using the Greek term *Narce (Narke)* in his biologic writings, and Pliny (23–79 A.D.) likewise brings a number of comments on this fish, called *Torpedo* by the Romans, in his *Naturalis historia* (Books IX and XXXII). Among Graeco-Roman (Hellenistic) writers, Oppian (*floruit* ca. 180 A.D.) in his poem '*Halieutica*', and Aelian (170– ca. 230 A.D.) in his work '*On Animals*' include various reports on *Narce*. Athenaeus (*floruit* about 230 A.D.), in his verbose '*Deipnosophistae*', repeatedly expatiates on the *cramp fish* which was also considered quite edible. In some of his remarks, however, the designation *Narce* may perhaps also subsume the electric catfish *Malapterusus* of the Nile. The Greek terms νάρκη and ναρκάω, from which *narcosis* is derived, and the Latin words *torpedo, torpeo, torpor*, all refer to cramp, numbness, stunning, or stupor.

The late Roman writer Claudian (*floruit* ca. 400 A.D.), a remarkable stylist of Silver Latinity, composed a short poem '*De torpedine*':

'*Quis non indomitam dirae torpedinis artem*
*audiit et merito signatas nomine vires?*
    *Illa quidem mollis segnique obnixa natatu*
*reptat et attritis vix languida serpit harenis.*
*sed latus armavit gelido natura veneno,*
*et frigus, quo cuncta rigent animata, medullis*
*miscuit et proprias hiemes per viscera duxit.*

. . . . . . . . . . . . . . . . . . . . . . . . . . . . . . . . .

*sed propius nigrae iungit se callida saetae*
*et meminit captiva sui longeque per undas*
*pigra venenatis effundit flamina venis.*
*per saetam vis alta meat fluctusque relinquit*
*absentem victura virum: metuendus ab imis*
*emicat horror aquis et pendula fila secutus*
*transit harundineos arcano frigore nodos*
*victricemque ligat concreta sanguine dextram.*
*damnosum piscator onus praedamque rebellem*
*iactat et amissa redit exarmatus avena.*'

Finally, it might be mentioned that E. du Bois-Reymond (1818–1896) a prominent '*Founding Father*' of electrophysiology, began his scientific career at the University of Berlin in 1843 with a doctor's dissertation entitled '*Quae apud veteres de piscis electricis extant argumenta.*'

In the species of *Torpedinidae*, the electric organs are bilateral, roughly bean-shaped thick discs located in the branchial region of the flattened head-body configuration (cf. Figs. 449, 452). The polarity obtains between ventral and dorsal surface, the latter being positive during the discharge. Each organ consists of many prismatic columns, numbering approximately 400–500 in *Torpedo ocellata*, and up to about 1000 in *Torpedo marmorata*. Each prism is formed by stacked electroplaques (cf. Fig. 453), whose number, depending on the species, varies between 500 or less and about 1000. In one American species *(Narcine brasiliensis)*, a mediocaudally adjacent smaller *accessory electrical organ*, essentially similar despite a number of secondary differences, has been described (MATHEWSON *et al.*, 1958; cf. Fig. 452B). The 'electrical nerves' innervating the overall organ, which is derived from branchial musculature, are rami of facial (VII), glossopharyngeal (IX), and vagus (X). There are generally four electric nerves or rami, one each from VII and IX, and two from X (STUART and KAMP, 1934). The electromotor fibers within these branches are neurites of large multipolar cells pertaining to the intermedioventral (so-called visceral-efferent) deuterencephalic longitudinal column, which is greatly enlarged and forms the characteristic '*lobus electricus*' in the rhombencephalon of *Torpedo* (cf. Fig. 454). Each lobe may perhaps contain up to 60000 or more large cells. Discharges of about 30 to 200 volts and high amperage (up to ca. 50 amp) have been recorded. The electrical activity of *Torpidinidae* is said to tire somewhat rapidly, and the responses evoked in their electroplaques by a single nerve impulse seem to be maximal, 'or nearly so' (GRUNDFEST, 1967).

*Rajidae* (rays), pertaining to the *Batoid* group of *Elasmobranchs*, comprise only species with *weak electric organs* located in the tail (cf. Fig. 449), and derived from somitic, axial musculature, being innervated, through the radices anteriores, by corresponding segmental spinal nerves. The Rajid electroplaques exhibit some degree of diversity and include structural elements of either disc or cup type. As a vestige of their muscular origin, Rajid electroplaques commonly retain a striation. Although it can be suspected that, in these fishes, the weak electric discharges, with an amplitude not exceeding a few volts, is related to a navigational guidance system such as discussed above, no evidence concerning a behavioral activity of this sort is available. Yet, skates or rays are known to possess *ampullae of Lorenzini* responding to delicate electrical stimulation. LISSMANN (1963) offers the following comment: 'Unfortunately, either skates are rather uncooperative ani-

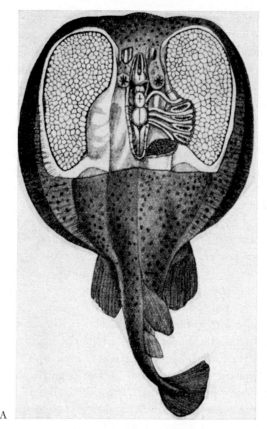

A

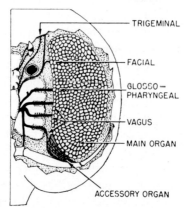

B

*Figure 452.* Electric organs of Torpedinidae and their innervation as seen in dorsal view. A *Torpedo marmorata* (after GARTEN, from BALLOWITZ, 1938). B *Marcine brasiliensis*, with accessory electric organ (from GRUNDFEST, 1967).

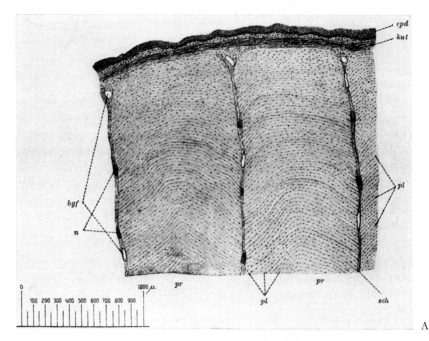

A

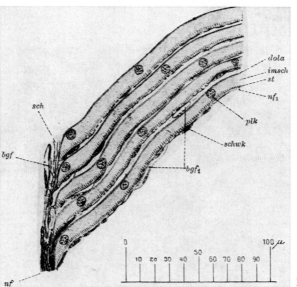

B

*Figure 453.* The stacked electroplaques of *Torpedo ocellata* (from KRAUSE, 1923). A transverse (dorso-ventral) section through two columns or prisms of the organ. B A few electroplaques at higher manification. bgf: blood vessel; dola: dorsal surface of plaque; epd: epidermis; imsch: cytoplasm of plaque; kut: corium; n, nf: nerve fibers; pl: electroplaques; pr: columns or prisms; sch: connective tissue septa; schwk: *nuclei of Schwann;* st: so-called palisade border at ventral surface of plaque.

mals or we have not mastered the trick of training them; we have been unable to repeat with them the experiments in discrimination in which Gymnarchus performs so well'.

The *Teleostean Astroscopus*, like the preceding Selachians, another but quite rare marine fish, found off the Atlantic coast from the southern United States to Brazil, assumes an intermediary status between strongly electric and weakly electric forms. It was generally known to

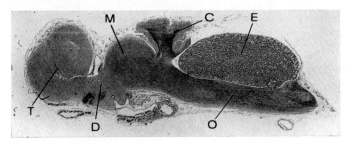

*Figure 454 A.* Parasagittal section through the brain of *Torpedo ocellata*, showing lobus electricus (from the author's sample material, hematoxylin-eosin stain, ×6.5, red. $^3/_5$). C: cerebellum; D: diencephalon; E: lobus electricus; M: mesencephalon; O: medulla oblongata; T: telencephalon.

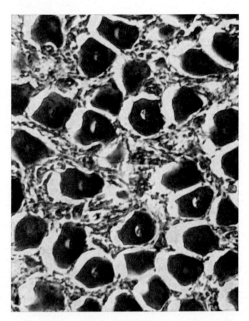

*Figure 454 B.* Large efferent cells of lobus electricus in *Torpedo* (from the author's sample collection, hematoxylin-eosin, ×180, red. $^2/_3$).

fishermen that upon touching their head, a feeble electric shock could be experienced. Subsequent histologic studies disclosed a paired electric organ derived from eye musculature, and filling out most of the orbits, which, however, still contain the eyebulbs and a few reduced true muscular remnants. Each organ, innervated by the hypertrophied oculomotor nerve (III), consists of about 150 to 200 electroplaque layers, whose arrangement is parallel to the cranial surface. The joint nucleus of the oculomotor and trochlear nerves is greatly enlarged, most of its cells resembling those of lobus electricus in *Torpedo* (KAPPERS, 1947). The trochlear nerve, much smaller than the oculomotor, is nevertheless well developed and might perhaps contribute to the innervation of the electric organ.

Among the *freshwater electric fishes*, the *Gymnotid Electrophorus (Gymnotus) electricus* or 'electric eel' *(Temblador)* of northern South America is both the most powerful and the best known. Its discharge, although not generally considered lethal for large animals, can stun man or horse. Smaller animals may be instantaneously killed.

The electric organs which are positive toward the head (Figs. 449, 455) take up a large part of the body, and are derived from somitic, myotomic musculature, being innervated, through the radices anteriores, by segmental spinal ('intercostal') nerves. There are three pairs of organs, namely the principal organ producing the bulk of the electricity, the smaller fusiform ventral *organ of Hunter*, and the likewise smaller dorsal and caudal *organ of Sachs*. All three organs display a columnar arrangement. About 6000 to 10000 electroplaques may be contained in one column, and there are approximately altogether 70 columns in the organs on each side of the body. The electric eel can generate a discharge of about 1 ampere driven by up to 700 volts. In comparison with the discharge of *Torpedo*, the potential is thus much higher, but the intensity much lower. Also, *Electrophorus* can sustain its discharge activity for a much longer period than *Torpedo*, without exhibiting fatigue. When first disturbed, the electric eel may discharge up to 300 times per second. If the activity continues for some time, the discharge rate subsequently decreases to about 50 per second but it is reported that the fish can discharge at this rate for many hours. Each discharge is said to last about 2 msec.

In addition to *Electrophorus*, there are several *Gymnotids ('Knife fishes')* with electric organs providing relatively *weak* discharges. Although the body type of these fishes is, generally speaking, not very variable, their electroplaques are conspicuously diverse, and display

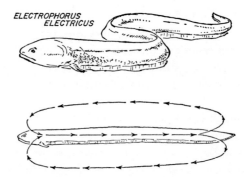

*Figure 455 A.* General aspect of the Gymnotid *Electrophorus electricus.* Some adult specimens can reach a length of about 2 m. The lower sketch shows direction of current flow in electric organ and surrounding water (from GALAMBOS, 1962). It will be noted the arrowheads indicate direction of current in the conventional manner from positive to negative pole, in contradistinction to the arrows in figures 447 and 449.

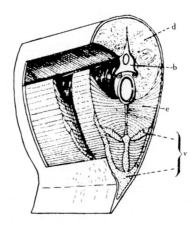

*Figure 455 B.* Diagram illustrating the general arrangement of the electric organ in the Gymnotid, *Electrophorus electricus* (after WIEDERSHEIM, from SELENKA-GOLDSCHMIDT, 1923). b: body of vertebra; d: dorsal musculature; e: main electric organ (the identically marked, but unlabeled lower electric organ is known as *Hunter's organ*); v: ventral musculature.

structural details with little intergrading between them. FESSARD (1958) comments in this respect: '*la diversité structurale de ces formations ne peut manquer d'être un sujet d'étonnement*'.

Thus, the organ of *Gymnotus carapo*, discussed above with regard to '*electrical guidance systems*', differs greatly from that of *Electrophorus*, and is more closely related to the electric organ of the entirely unrelated African *Gymnarchus*, to be briefly dealt with further below. Other 'weakly electric' *Gymnotidae* are *Stenoarchus albifrons* (whose electroplaques may be modified nerve fibers), *Steatogenes elegans*, moreover species of *Eigenmannia*, *Hypopomus*, and *Sternopygus*. Detailed electrophysiological data on these two latter forms have been presented by BENNETT (1961).

The African *Siluroid Malapterurus (Malopterurus) electricus* or '*electric catfish*', depicted on old Egyptian tomb sculptures, doubtless intrigued, already several thousand years ago, the inhabitants of the Nile valley. This fish displays a cutaneous electric organ, whose derivation still remains dubious (cf. above, footnote 82). Beginning at the caudal boundary neighborhood of the head it encases most of the body (cf. Figs. 449, 456), being subdivided into a rostro-caudal series of compartments by connective tissue septa. The electroplaques (cf.

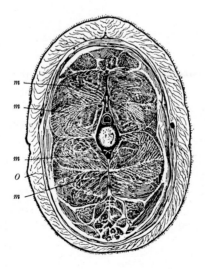

*Figure 456 A*. Cross-section through the body of Malapterurus (after FRITSCH from BIEDERMANN, 1895). m: musculature; O: cutaneous electric organ.

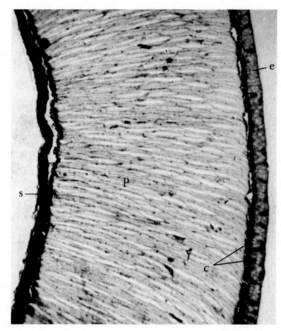

*Figure 456 B*. Low-power view of section through the electric organ of Malapterurus. The plane of the section is parallel to the longitudinal axis of the body and perpendicular to the surface of the integument (from the author's sample material, *Mallory stain*, ×32, red. ²/₃). e: ectoderm; c: chromatophores; p: electroplaques; s: subcutaneous stratum, separating organ from musculature.

Fig. 456 B, C) are structurally arranged in parallel order (which, of course, means electrotechnically *in series*), being somewhat imbricated, and generally perpendicular to the surface of the integument. Between the plaques, a variable amount of delicate connective tissue can be found. Considerable regional differences in the regularity of the arrangement obtain. Particularly in the anterior portions of the organs, the plaques may be 'shoved together' without any regularity, so that a definite orientation is not recognizable.[88] The total number of electroplaques on each side has been estimated at somewhat less than 2 million (FESSARD, 1958). The discharges, of relatively low amperage,

---

[88] '*Dans la région antérieure, les plaques prennent d'autres directions, elle finissent même par s'entasser parallelement à la surface du corps*' (FESSARD, 1958).

may attain an electromotive force of about 300 volts. Although evidently used in the catching of prey, the discharges may also be related to a not yet properly clarified guidance system.

The innervation of the electric organ in *Malapterurus* is peculiar; it receives, on each side, efferent endings from the axon of a single, large neuron, described by FRITSCH (1887), and located in the ventral horn of the spinal cord at a level just caudal to the brain, namely, according

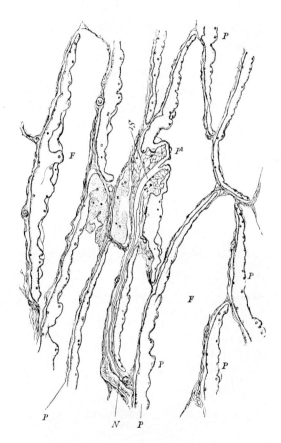

*Figure 456 C*. Electroplaques of Malapterurus at higher magnification (after FRITSCH, from BIEDERMANN, 1895). The plane of section is perpendicular to the main surfaces of the electroplaques. Caudal surface left, rostral surface right. The organ displays interstitial compartments containing loose connective tissue and nerve endings. F: compartment; N: terminal nerve branches; P: electroplaques; P¹: region of plaque at origin of stalk; St: stalk.

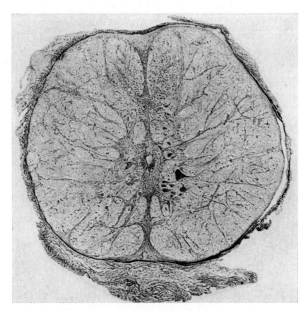

*Figure 457 A.* Cross-section through the third segment of spinal cord in *Malapterurus*. The large cell on the right is the root cell of the electric nerve of that side. Below the cornu anterius, Mauthner's fiber is recognizable (from STUART and KAMP, 1934).

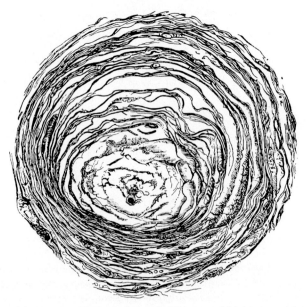

*Figure 457 B.* Cross-section through the electric nerve of *Malapterurus* (after FRITSCH, from BIEDERMANN, 1895). The axon with its myelin sheath is surrounded by a dense arrangement of concentric connective tissue lamellae.

to Stuart and Kemp (1934), in the third spinal segment (Figure 457 A). Although there remains some doubt whether the neurite of that '*giant cell*' emerges by way of the dorsal or of the ventral root, Ballowitz (1938) and Fessard (1958) assume that the neurite joins the radix ventralis.[89] This neurite, at any rate, leaves the vertebral canal through the first intervertebral foramen. The electrical nerve, thus consisting of a single axon, nevertheless has a diameter of about 1 mm or more because its myelin sheath is surrounded by thick additional concentric connective tissue investments (Fig. 457 B). Accompanied by an artery and a vein, the nervus electricus runs caudalward, roughly parallel to the lateral line, toward the posterior end of the electric organ. Along its course, numerous collaterals are given off in close proximity to the organ. The branchings of these collaterals provide the nerve endings on the electroplaques. Each of these latter has, on its caudal side, a single stalk-like process whose tip receives the synaptic nerve terminal which triggers the discharge. The myelin sheath ends close to the terminal knobs.

Both rostral and caudal electroplaque faces in Malapterurus and in the Mormyrids are electrogenic. In Malapterurus, the spike of the rostral (uninnervated) surface is slightly larger, and lasts longer than the response of the caudal surface from which the stalk emerges. The potential of the organ discharge thus breaks *Pacini's rule* since the caudal

---

[89] Stuart and Kemp (1934) remark with respect to the not entirely settled question of dorsal or ventral root exit, that the neurite's dorsal emergence would correspond to the relationship in *Torpedo*, where the electric organ is supplied by branchial nerves (which represent modified dorsal roots). The cited authors then add: 'The innervation may probably give an indication about the origin of the organ. For if it is innervated by visceral nerves, this points to its having originated from smooth musculature, eventually glands. If on the other hand the nerve emerges with the ventral root, which is most likely considering the origin of the cell, this would point rather to its originating from somatic elements which have secondarily acquired a position in the skin'. To this, one might reply (1) that, in the Gnathostome spinal cord, efferent ' visceral innervation' may occur through either dorsal or ventral roots, and (2) that branchial muscles are the only ('special') visceral effectors directly innervated, through deuterencephalic dorsal roots, without an intercalated postganglionic fiber originating in a peripheral ganglion. All other ('general') visceral effectors are either supplied by a postganglionic fiber or, themselves, represent a 'postganglionic element' (e.g. chromaffin cells). The electroplaques of *Malapterusus* seem, very definitely, to be directly innervated by the central neuron. Thus, although innervation *via* dorsal root could be interpreted to suggest a 'visceral' origin, ventral root innervation would still be inconclusive in this respect, and the question of the origin of the organ, remaining open, must be solved by decisive ontogenetic evidence.

surface is then positive to the rostral one.[90] The events contributing to the total discharge elicit a brief diphasic potential (GRUNDFEST, 1960b).

The electrical organs of the *African Mormyrids* (Fig. 449) consist, on each side, of two columns, a dorsal and a ventral one. These organs are derived from a portion of the tail musculature, and the number of electroplaques in a column is said to vary between approximately 120 and 200. The amplitude of the Mormyrid's weak electric discharge may attain a maximum of about 12 volts. The innervation is provided, by way of the radices anteriores, through ventral and dorsal rami of segmental spinal nerves. In the closely related *Gymnarchids*, which some authors subsume, as a family, together with the family Mormyridae under the order *Mormyriformes*, the weak electric organs likewise derive from part of the tail musculature, but consist of four pairs, two dorsal and two ventral ones.

*Mormyrids* are characterized not only by a particularly well developed lateral line system but also by a greatly 'hypertrophied' *cerebellum*.[91] Especially noteworthy is here the valvula cerebelli, which displays numerous transverse ridges and a number of peculiar cellular structures. To some extent, but apparently less so than in Mormyrids, the Gymnarchid lobus lineae lateralis and cerebellum likewise manifest a high degree of development.

The use of the weak discharges produced by *Mormyrids* and *Gymnarchids* (moreover by the entirely unrelated weakly electric *Gymnotids* including, e.g. *Gymnotus carapo* ) as a *guidance system* for navigation and prey location in connection with receptors presumably pertaining to the lateral line, has been particularly investigated by LISSMANN (1958, 1963). These studies, undertaken in the natural habitat of these fishes and in the laboratory, disclosed two *discharge types* in both the quite unrelated African and South American forms.

The *first type* of electric discharge consists of very regular sequences of continuously emitted, monophasic pulses, varying from species to

---

[90] Cf. above on p. 880 the general comments concerning the *modus operandi* of electroplaques.

[91] LISSMANN (1958) and other authors, e.g. FRANZ (1921) consider, on reasonable grounds, the peculiar '*mormyro- cerebellum*' as a significant 'association center' or 'integrating center'. Detailed, but *qua* specific significance rather inconclusive data on the *Mormyrid* '*gigantocerebellum*' with reference to the 'electrosensory system' as presented by several authors can be found in the symposium volume on cerebellar neurobiology edited by LLINÁS (1969). The Mormyrid cerebellum will be dealt with in volume 4.

species in frequency, and with narrower limits from individual to individual, within a range of 60 to 400 per sec. The duration of a single discharge measures about 2–10 msec. The *second type* of discharge is less regular in frequency, the pulse duration being much shorter (e.g. 0.2 msec), and the pulse shape more complex.

Discharges of the first type were recorded by LISSMANN (1958) in the *African Gymnarchus* and in the *American Gymnotids Hypopomus* and *Eigenmannia*. Those of the second type are reported by the cited author in all examined *Mormyridae* and in the *Gymnotids, Gymnotus carapo* and *Staetogenes elegans*. The basic discharge rate of a resting Mormyrid was found to be somewhat variable between 1 and 6 pulses per second, and not strictly rhythmical. Stimuli exciting the Mormyrids caused an increase of frequency up to a maximum of about 130 pulses per second, while inhibiting stimuli caused the discharges to cease for prolonged periods.

Again, the electric discharge of the South American gymnotid *Sternopygus macrurus* was recently shown to display sex differences (HOPKINS, 1972). The electric activity of this fish is distinctly different from the discharges of ten taxonomically related gymnotids likewise found in Guyana. According to the cited author, 'Sternopygus is the first known example of a fish with sexually different electric discharges. Males and females differ in the steady-state frequency of their discharges, and males produce variations in their discharge during courtship. Playback experiments demonstrate that species and sex difference in electric discharges have communicative significance'.

In all electric fishes, the discharge of the specific organs is triggered by 'special visceral' efferent respectively 'somatic efferent' neurons of the first order, representing the actual *'discharge centers' (centres moteurs de la décharge* or CMD of FESSARD, 1958). The activity of these centers,[92] however, seems to be controlled by additional grisea of a higher order, apparently located in the reticular formation of the deuterencephalon[93] and designated as *'noyau de commande de la décharge'* (NCD, FES-

---

[92] The total number of efferent elements in the CMD grisea varies considerably, between only two (i.e. one on each side) in *Malapterurus*, perhaps 400 in *Mormyrus*, possibly about 120,000 or more in *Torpedo*, and maybe some two or three hundered thousand in the Gymnotid *Electrophorus*.

[93] It is intended to include further remarks on structure and location of these centers in the relevant chapters of volume IV of this series.

SARD, 1958). This griseum is believed to exert a synchronizing, frequency-modulating, and integrative function. Yet, quite evidently, and particularly with respect to the electroreceptor apparatus of major importance for the guidance or navigational systems, the NCD must, in turn, be controlled by an additional circuitry related to input processing grisea whose details and modus operandi still remain poorly understood.

Another unsettled question, related to the behavior of strongly electric fishes, concerns their apparent *'immunity' to their own discharges*. The problem, together with several inconclusive explanations, was critically examined by BIEDERMANN (1895). With respect to its contemporary status, FESSARD (1958) remarks that this doubtless obtaining 'immunity' is relative and reports observations of muscular twitchings as well of contractions in large electric fishes concomitant with discharge activity. The cited author believes that two factors must be considered. (1) An indeed much higher threshold of electrical excitability for organic structures of electric fishes in comparison with those of other animals. (2) The particular spatial arrangement of the electrically excitable structures in relation to the 'lines of force' respectively of 'current flow' pertaining to the electric field resulting from the discharge. Such arrangement might minimize the stimulating effect of the field.

Seen from the viewpoint of *phylogenetic theories*, the differentiation of electric organs and electroreceptor mechanisms in fishes raises a diversity of difficult problems concerning origin and evolution of these systems. LISSMANN (1958) believes that the first step in this process was the possession of receptors sensitive to electric stimuli. Such sensibility might have been an 'incidental' property of lateral line structures responding to the weak muscular action potentials. A recent report by BARHAM et al. (1969) seems to lend support to this suggestion. The cited authors recorded pulses in the 0.01 to 40 microvolt range, probably generated by muscle action potentials, from several fishes and one amphibian in aquarium tests. The pulses were remotely received through dipole antennae, and seemed to be merely by-products of the animals' normal functions.

Although inconclusive, these observations could be interpreted as suggesting evolutionary intermediates between 'nonelectric' and 'electric' fishes. A variety of genetic factors may have resulted in a mutual 'adaptation' of electric receptor and effector structures, thereby, 'via weak electric organs used for orientation' (LISSMANN, 1963), the strong electric organs suitable for defense and aggression evolved. The apparent lack of electric organs in aquatic Amphibia might be related to the greater sluggishness and to the by and large terrestrial habits of these animals. Moreover, the relative rarity of true (weak or strong) electric fishes would indicate that only a rather exceptional combination of factors led to the evolution of electric organs. These evolutionary events, moreover, displayed conspicuous trends of 'convergence' in a number of quite unrelated forms.

Yet, even if this suggested process of evolution should correspond to the course of the actual phylogenetic events, no significant explanation of the evolutionary mechanism would thereby be given. The invocation of suitable mutations and of natural selection does not seem to be particularly helpful. Initially, the obtaining 'adaptive value' could hardly have been significant but was presumably neutral. Again, with increasingly detailed elucidation of the macromolecular interactions providing the genetic mechanisms, the doubtless useful, if vague, standard concept of 'mutation' has shifted to the status of an almost obsolescent oversimplification subsuming a variety of complex and ill-definable events.

In addition to effectors discharging electricity, there are *effectors emitting light*. Among *vertebrates*, such phenomena of *bioluminescence* have been observed in a variety of *fishes*. The hitherto recorded luminous species pertain to the *Gnathostome* classes *Elasmobranchii* and *Osteichthyes*. Within this latter, only *Teleosts* seem to be represented. All these fishes are marine, the great majority being oceanic, and particularly, deep-sea forms.[93a] 'Lower' organisms displaying bioluminescence are diverse Bacteria, Fungi, Protozoa, Coelenterates, Annelids, Arthropods (e.g. the well-known fireflies), and Chordates (Hemichorda, Urochorda) such as, e.g., Pyrosoma. Although emission of light can be considered a very special case of effector activity, in this respect comparable to the substantial electric discharges, it is, by and large, a rather widespread one, possibly more so than marked production of electricity, and occurs in perhaps more than several scores of orders distributed upon diverse phyla.

---

[93a] There are, however, some well-known luminescent shallow-water fishes, such e.g. as the Teleosts *Photoblepharon* and *Anomalops*, which are indigenous to the Banda Sea of the Indonesian Archipelago, and are briefly referred to further below (cf. also HANEDA and TSUJI, 1971).

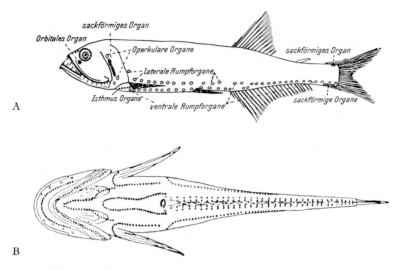

*Figure 458.* Arrangement of the luminescent organs in two Teleosts. A Lateral view of the Sternoptychid, *Gonostoma denudatum* (after Brauer, 1906, from Penners, 1931). B Ventral view of *Porichthys notatus* (modified after Greene, from Strum, 1969).

In some chemical reactions involving oxydation of organic compounds, energy is emitted in form of light, namely as '*chemiluminescence*' without significant heat production *('cold light')*. Such reactions occurring in living organisms are designated as *bioluminescence* or, less accurately, as '*phosphorescence*',[94] and, roughly speaking, imply a direct conversion of chemical energy into radiation, 'without passing through heat' (Bayliss, 1924). In many cases, the cessation of light emission in the absence of oxygen and its reappearance upon subsequent oxygen admission has been demonstrated. At least two different substances seem to be involved, an enzyme *luciferase* and an oxidisable compound, *luciferin*.[95] The process can take place within the cell itself, but in some cases it might occur after the secretion is extruded. In numerous in-

---

[94] The term *phosphorescence*, in the widely accepted or 'true' sense, implies previous exposure to light, and is rather akin to '*fluorescence*'. This latter designation subsumes the property of emitting light during exposure to it, such that the emitted wavelength differs (being generally longer) from that of the absorbed radiation.

[95] A pertinent summary of the biochemical aspects relevant to bioluminescence was presented by Harvey (1953).

stances, the light is limited to the middle region of the visible spectrum, with maximum in the green. However, there are manifestations of blue, yellow, purple, orange and even red bioluminescence.

As displayed by *vertebrates*, i.e. by *fishes*, luminescent organs are apparently always provided by modified *mucous glands* of the skin, generally located within the subepithelial connective tissue (corium). In the *Selachian*, *Spinax niger*, however, the phosphorescent organs remain in the epithelial layer and thus represent intraepidermal glands. The luminescent organs vary greatly as regards shape, complexity of differentiation, and number. In some forms, merely a single one is present, and in others a pair, but in most instances the organs are distributed in diverse linear arrangements (Fig. 458) along the body, upon the head, and as far as the tip of barbels or of fins.

Typically differentiated luminescent organs are provided by acinous or cup-like extensions of epidermis into corium. In some instances there remains a glandular duct reaching the surface (Fig. 459). Within the acini of certain fishes, photogenic cells and supporting cells have been described. A prominent basal membrane may separate the acinar elements from the surrounding connective tissue (STRUM, 1969). Externally to the gland cells, but on the internal aspect of the organ, a perhaps mesodermal pigmented cup is usually present. Aggregations of epithelial or of connective tissue cells form lens-like as well as reflec-

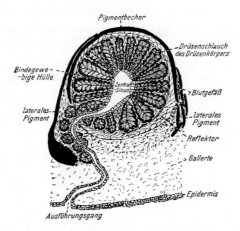

*Figure 459*. Section through a large photophore of the Sternoptychid, *Gonostoma elongatum* (after BRAUER, 1908, from PENNERS, 1931).

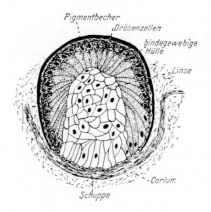

*Figure 460.* Section through a photophore of the Stomiatid, *Chauliochus sloanei* (after
BRAUER, 1908, from PENNERS, 1931).

tor-like structures. The reflectors may be composed primarily of
strongly birefringent cells containing guanine crystals. The overall configuration of such organs is thus comparable to the build of a search-
light or motor vehicle headlight (cf. Fig. 460). Further data on the luminescent organs are contained in the publications by BRAUER (1906,
1908), BÜTSCHLI (1921), PENNERS (1931), HARVEY (1952), and NICOLS
(1967). Short comments on this topic are included in the treatises of
YOUNG (1955) and ANDREW (1959).

The relevant constituents of the light-emitting organs *( photophores )*
are the *photogenic 'glandular' epithelial elements* which either *produce* the
effective bioluminescent substances (e.g. the enzyme luciferase) or *contain symbiotic luminescent bacteria*. One could, accordingly, distinguish *actively* or *passively* photogenic organs.[96] The photophores, and especially
the more complex ones, are richly supplied by blood vessels.

According to HARVEY (1952), '*bacterial light organs* are usually found
in surface or medium depth fish and are relatively simple, either single
or paired. Their light shines continually but may be decreased or com-

---

[96] HARVEY (1952) comments on various sources of error in some reports on luminescent fishes. Thus, dead marine forms may serve as a 'culture medium' for saprophytic
luminous bacteria, and swimming fish can stimulate the luminescence of surrounding
marine dinoflagellates, thereby appearing 'in a fiery glow'. Again, the brightly shining
tapeta lucida of eyes and other reflecting surface may give an erroneous impression of true
luminescence.

pletely excluded by secondary control, as by chromatophores, or movable screens, or by shifting the position of the organ'.

In true deep-sea fishes, on the other hand, the *photophores* are numerous and highly complex in structure, with lenses and reflectors. The light appears only on stimulation, although the response may not be as sudden as in the case of many invertebrates (HARVEY, 1952).

A few years ago, NICOL (1967) pointed out that no comprehensive and critical account of luminescence in fishes existed, and added that perhaps the time is not yet opportune for composing one. This comment can be considered still valid. Nevertheless, there is little doubt concerning the control of the luminescent organs, respectively of light emission, by the *nervous system*, and the cited author reviews, in this respect, a number of pertinent data together with the presentation of relevant suggestions.

As regards the serial photophores of *Teleosts*, their innervation has been well established in several instances. Thus, nerve fibers supplying the photophores pierce the reflector and ramify among the cells of the photogenic tissue.[97] In the head, the light organs are innervated by branches of the facial nerve (mandibular, buccal, and hyoid rami), some of whose bundles join the r. maxillaris of the trigeminus. In the trunk, the photophores are supplied by branches of the spinal nerves.

In a recent investigation of the luminescent organs in *Porichthys notatus*, the terminal arborizations of nerves were noted in pockets of connective tissue which penetrate among the photogenic, supportive, and lens cells. None of these endings could be seen to pass through the basal lamina and to make direct intercellular contact with either a supportive or a photogenic cell (STRUM, 1969). The author concluded that the nerve endings, which contained dense-cored vesicles displayed in electron photomicrographs, remained outside of the basal membrane.

NICOL (1967, and previous publication) had suggested that the serial photophores of *Porichthys* and other *Teleosts* are innervated by the sympathetic system through adrenergic postganglionic sympathetic fibers distributed by way of spinal or cranial nerves as shown in Figure 461. This mode of innervation is likewise assumed for the chromatophores. It has been repeatedly substantiated that the injection of ad-

---

[97] NICOL (1967) believes that although actual nervous terminations on photocytes have not been recorded, these cells might nevertheless be the goal of these fibers. According to STRUM (1969), however, and at least in *Porichthys*, the terminations remain outside the basal membrane (cf. further below).

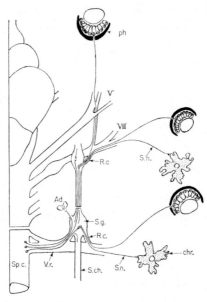

*Figure 461.* Diagrammatic representation of hypothetical photophore innervation in a luminescent Teleost (after NICOL, 1967). Ad.: suprarenal gland cell; chr.: chromatophore; ph: photophore; R.c.: ramus communicans; S. ch.: sympathetic (postganglionic) fiber; Sp.c.: spinal cord; V.r.: spinal ventral root; V, VII: cranial nerves.

renaline into various luminescent fishes causes the photophores to lighten. Thus, regardless of the moot question whether the photogenic cells are or are not in direct synaptic contact with the nerve endings, a nervous control in the manner suggested by NICOL does not appear improbable. As regards other kinds of nonserial photophores, the cited author believes that somewhat different, but not yet elucidated types of innervation might obtain.

With respect to the control of *bacterial light organs* whose luminescence is constant, several *screening mechanisms* have been described. Thus, in *Anomalops*, the organ, provided with a hinge-like edge, can be rotated into a pigmental pocket, whereby the emitting surface becomes hidden. In *Photoblepharon*, a fold of black tissue on the ventral edge of the organ socket can be drawn over the luminescent surface like an eyelid, thus 'turning off' the light. In other instances, the controling devices involve expansion and contraction of chromatophores, or muscular action which opens and closes fenestrae.

Concerning the significance of luminescent organs for the behavioral activities of fishes, a number of more or less plausible and to some extent not mutually exclusive theories have been suggested. These include (1) illumination for vision at night or in dark waters; (2) courtship as well as mutual recognition in shoaling, (3) defense by 'frightening' or 'distracting' predators, and (4) capture of prey by attracting other animals; thus, e.g. in the deep-sea 'angler' *Ceratius*, the luminous tip of the dorsal fin is interpreted to serve as a lure.[98] No convincing interpretation can be given for some small photophores which are so placed as to shine into the eyes of their bearer.

The *phylogenetic evolution* of luminescent organs poses a set of particularly perplexing problems. More than a hundred years ago, DARWIN, in his '*Origin of Species*' (1859), included both luminous[99] and electric organs in the category of 'special difficulties of the theory of Natural Selection', and refrained from offering any explanation. The contemporary status of this subject matter does not appear very different and the relevant factors have perhaps remained more obscure than those pertaining to electric organs. Since photophores occur in taxonomically widely separate groups and in a great variety of different degrees of differentiation, quite independent phylogenetic origins and lines of evolution can be assumed. Some of these latter, again, display rather conspicuous examples of '*parallel*' or '*convergent*' evolution. As regards extinct fishes, a few Oligocene species, displaying structures interpreted as fossilized photophores, have been described (HARVEY, 1952, quoting ARAMBOURG, 1920, and PANCA, 1929, 1931).

---

[98] In the *Selachian Spinax niger*, the ventral photophores seem to throw out parallel beams of light such that the luminescence will only shine upon objects immediately beneath the ventral surface of the fish, resulting, as it were, in a sudden flash of light upon a prey and causing it to hesitate in that location for just one short moment in which *Spinax* 'can make a successful snatch' (quoted from HICKLING by HARVEY, 1952). Again in the Teleost *Leiognathus equulus* ('pony fish') of Indo-Pacific waters, light (of symbiotic bacterial origin), controlled by a 'shutter-mechanism', is reflected within the swim bladder and emitted from a broad area of the ventral body surface. It appears that this luminescent system functions during the daytime, matching the background light from the surface, and thereby obscuring the silhouette of the animal (HASTINGS, 1971).

[99] DARWIN, however, referred to 'the luminous organs which occur in a few insects, belonging to widely different families, and which are situated in different parts of the body'. Although DARWIN concisely discussed electric fishes, he did not mention the luminescent ones, but his general remarks can be equally well applied to these latter.

## 6. The Vegetative Nervous System

The *vegetative* or *autonomic nervous system* can be defined as that structural and functional subdivision of central and peripheral nervous system which is concerned with *output* to *smooth musculature, cardiac musculature* and *glands*. It thereby exerts a control on the behavior of gastro-intestinal tract with its glands, respiratory system, urogenital system, skin, and cardiovascular system, which latter, in turn, involves all other systems of the organism. Hence the autonomic nervous system, whose peripheral components are widely distributed throughout the body, subserves fundamental activities conventionally classified as '*vegetative*'.[100] Moreover, as experienced by man, these activities are, to a large extent, not consciously or voluntarily performed. Thus, the term '*involuntary nervous system*' is occasionally used, although numerous activities of 'somatic' or 'animal' nervous system are likewise 'involuntary' or 'unconscious'. Since many 'vegetative' functions of the body's organ systems can be performed independently of central nervous output, and such functions might still be controlled by the peripheral components of the vegetative nervous system, the term '*autonomic*' refers to a presumed independence of peripheral vegetative ganglia with respect to the central neuraxis. Such independence, however, if at all obtaining to any degree, is, at best, merely relative. Moreover, the central components of the vegetative nervous system are commonly also designated as 'autonomic' grisea or centers, although representing intrinsic structural and functional components of brain and spinal cord.

The French anatomist JACQUES BÉNIGNE WINSLOW (1669–1760), a native of Odense, Denmark, was perhaps the first to introduce the term '*sympathetic*' (1732). He designated the vertebral chain of ganglia as '*grand sympathique*', the vagus as '*moyen sympathique*', the facial nerve as '*petit sympathique*', and regarded the large peripheral ganglia as '*cerebra secundaria subordinata sive parva*'.[101] MÜLLER (1931) justly points out that the term 'sympathetic' can be considered well chosen, because

---

[100] MÜLLER *et al.* (1931) use the German term '*Lebensnerven*' or '*Lebensnervensystem*', '*da durch dieses Nervensystem der richtige Ablauf der "vegetativen", d.h. zur Unterhaltung des Lebens notwendigen Leistungen der Organe gewährleistet werden*'.

[101] The not particularly suitable concept of such ganglia (e.g. plexus coeliacus s. solaris, '*Sonnengeflecht*') as secondary brains was eagerly and conveniently adopted by *aficionados* of the occult for the elaboration of various fanciful theories of clairvoyance, somnambulism, and spiritism.

through mediation of this system the viscera are made to participate (συμπαθεῖν) in the emotional experiences *(Freude und Leid)* of mental events. This, of course, essentially agrees with the well-known *James-Lange theory of emotions*, which, however, may be interpreted as only relatively valid, namely as referring to a 'reinforcing' feedback mechanism providing emotional 'resonance' or 'reverberations' (cf. K., 1957, p. 177f).

Following LANGLEY (1921), the designation *autonomic (vegetative)* nervous system refers exclusively to a class of *efferent elements* corresponding to the general visceral efferent fibers in the classification based upon GASKELL's views. Some authors, however, also speak of afferent sympathetic or autonomic fibers. These latter, of course, essentially correspond to the group of general afferent visceral fibers, which do not, in their overall arrangement, significantly differ from the somatic afferent ones. Moreover, these latter, carrying input from the skin and other 'somatic' structures, likewise contribute, by way of their central connections, to the activities of the autonomic system. Since autonomic outflow, on the other hand, significantly differs from somatic and 'special visceral' outflow, there are substantial reasons for retaining LANGLEY's definition, restricting the term *'autonomic'* to the *output* components as enumerated above. One might, accordingly, speak of a loosely defined *'visceral nervous system'* of which the more rigorously definable *autonomic* (or *vegetative*) system represents the *efferent subdivision*.

The essential feature of this efferent innervation can be defined as the *interruption* of the pathway from central nervous system to effector by a synapse in a peripheral ganglion or plexus. In LANGLEY's terminology,[102] this pathway consists, therefore, of a sequence of two neurons respectively of two fibers, designated as *preganglionic* and as *postganglionic*.

---

[102] LANGLEY (1893, 1921) started about 1889 his highly important work concerning the effect of nicotine on the transmission of impulses through autonomic ganglia. The results of this author's investigations not only clarified the structural arrangement of the vegetative nervous system but also provided fundamental data for the subsequently established basic functional concepts of synaptic activity and transmitter substances. Additional significant contributions to the understanding of the autonomic nervous system and its central outflow are those by GASKELL (1885, 1916) with whom LANGLEY, at one time, was associated in Cambridge. LANGLEY (1898) moreover, performed highly interesting experiments demonstrating nerve regeneration into functionally different pathways by connecting the cut central vagus stump with the distal stump of the sympathetic trunk and observing a subsequent vagus effect on organs innervated by the sympathetic.

The preganglionic fibers leaving the neuraxis do not come from every rostro-caudal subdivision or segment of the neuraxis, but only from three definite regions. Their cells of origin can be regarded as pertaining to the intermedio-ventral deuterencephalic and spinal longitudinal zone. In mammals and man, where these regions of outflow were elucidated in considerable detail on the basis of experimental respectively clinical evidence, the following arrangement has been ascertained.

(1) *The cranial autonomic nervous system* provides preganglionic fibers from cell groups, respectively grisea, associated with the nuclei of oculomotor (III), facial (VII), glossopharyngeal (IX) and vagus (X) nerves. These fibers may peripherally join either the branches of some other cranial nerves, or the visceral peripheral plexuses, and become intermingled with components pertaining to the next subdivision.

(2) *The thoracolumbar autonomic nervous system* provides preganglionic fibers originating from neurons in the intermedio-ventral zone of a spinal cord region extending from the seventh, eighth or first thoracic to about the fourth lumbar segment (C VII, C VIII or Th I to L IV). These fibers generally pass through the ventral roots[103] into the common spinal nerve and then reach, by way of the ramus communicans, the sympathetic trunk.

(3) *The sacral autonomic nervous system* provides preganglionic fibers originating from neurons in the intermedio-ventral zone of the spinal cord in a region extending from about second to fourth sacral segment (S II–IV). These fibers likewise generally pass through ventral roots, the common spinal nerves, and, in most instances, then pass through the sympathetic trunk.

The cranial and sacral autonomic components are also designated as (cranial respectively sacral) *parasympathetic* nervous systems, the thoracolumbar components then representing the *sympathetic* system in the narrower sense. The general arrangement in these subdivisions is illustrated by the diagrams of Figures 462–463.

As regards the *peripheral ganglia* or *plexuses*, in which the preganglionic fibers terminate with synaptic endings on dendrites or cell bodies of the postganglionic neurons, three different groups can be distin-

---

[103] Cf., however, the probable presence of some visceral efferent (preganglionic) fibers in the dorsal roots of Anamnia as well as at least of some Amniota, discussed in section 3 (footnote 38) *et passim*. The preganglionic fibers in branchial nerves are, of course, morphologically speaking, dorsal root fibers.

guished, namely (1) the *vertebral ganglia* of the sympathetic trunk or chain; (2) *prevertebral ('collateral') ganglia* or *plexuses*, and (3) *terminal* respectively *intramural ganglia or plexuses*. The ontogenetic origin of the neuroectodermal cells within the peripheral ganglia of the vegetative nervous system from neural tube or neural crest was briefly discussed in chapter V, section 1 of the preceding volume of this series. With respect to sympathetic as well as sacral parasympathetic, and cranial parasympathetic components, it can be assumed that these peripheral elements reach their final location by migration along the outgrowing nerve bundles. It should here be added that the presence of postganglionic elements in the cerebrospinal ganglia of at least some Vertebrates, including Mammals, and the connection of certain such cells with preganglionic fibers taking their exit through the corresponding posterior roots have been assumed by various authors. Although this possibility cannot be entirely ruled out, the data leading to said supposition may be considered insufficiently convincing. A discussion of pertinent questions, with relevant references and illustra-

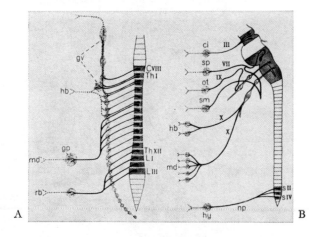

*Figure 462 A, B* Diagrams showing general arrangement of human autonomic or vegetative nervous system (slightly modified after VILLIGER, from K., 1927). A: sympathetic nervous system; B: parasympathetic nervous system; ci: ganglion ciliare; gv: sympathetic trunk with vertebral ganglia; gp: prevertebral ganglia (pl. coeliacus, pl. mesentericus sup.); hb: postganglionic fibers to heart and bronchi; hy: pelvic terminal plexuses (pl. hypogastricus); md: postganglionic fibers to gastro-intestinal tract (in A from prevertebral, in B from terminal or intramural ganglia); np: nervus pelvicus; ot: ganglion oticum; rb: postganglionic fibers to lower intestinal tract (from pl. mesentericus inf.); sm: ganglion submaxillare; sp: ganglion sphenopalatinum.

tions can be found in SCHARF's *Handbuch* contribution (1958). Some comments on this topic shall also be included in section 1, chapter VIII of volume 4.

(1) The *sympathetic trunk* consists of a paired more or less segmental series of ganglia located close to the ventrolateral surface of the vertebral column, linked together by a longitudinal strand, and connected

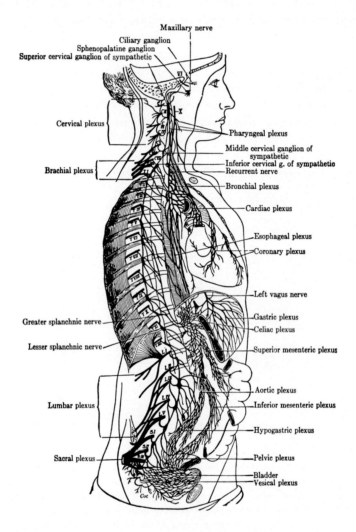

*Figure 462 C.* Simplified drawing of the peripheral distribution and plexuses of the vegetative nervous system as seen in an 'idealized' dissection of the human body (after SCHWALBE, from HERRICK, 1931).

with each spinal nerve segment by a communicating ramus. It extends, in man, from the base of the skull to the coccyx, where the two trunks commonly converge to form an unpaired coccygeal ganglion. As a rule, the paired vertebral ganglia include a superior, a middle, and an inferior cervical ganglion, eleven thoracic, four lumbar, and four sacral ganglia. Inferior cervical and first thoracic ganglia may be fused to a common 'stellate ganglion', the longitudinal connective between stellate and middle cervical ganglion being commonly double or split and forming a loop through which the subclavian artery takes its course

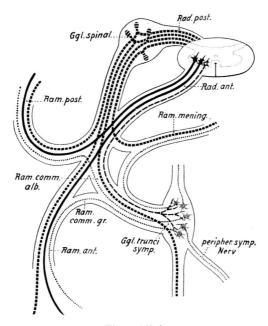

Figure 463 A

*Figure 463.* Three diagrams showing the general arrangement of a spinal nerve segment and its relation to the sympathetic nervous system (A after VILLIGER, 1933; B after RANSON, 1947; C after HERRICK, 1931). In A, the postganglionic fibers in ramus posterior and ramus meningeus are indicated. Heavy lines: somatic efferent fibers; heavy dots: somatic and visceral afferent fibers; dashes: preganglionic fibers; small dots: postganglionic fibers. In B both vertebral and prevertebral ganglia are indicated in simplified fashion as a single combined unit. In C the indicated afferent fibers are only the visceral ones, drawn in dot-and-dash lines; continuous lines are preganglionic fibers, and broken lines represent postganglionic fibers; int. lat.: intermedio-lateral column; p.m.: pilomotor fibers; s.g.: fibers to sweat glands, v.m.: vasomotor fibers. In all three diagrams the separation between gray and white communicating rami has been exaggerated for greater clarity.

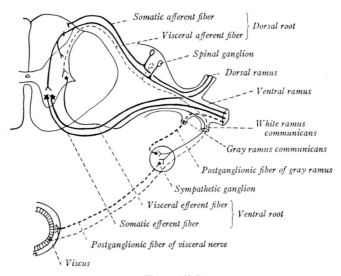

Figure 463 B

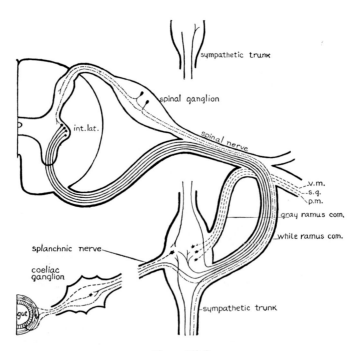

Figure 463 C

*(ansa subclavia Vieusseni)*. At or near the cranial end of the sympathetic trunk, branches of this chain, carrying postganglionic fibers, extend along the major arteries, forming the nervus respectively plexus caroticus internus and externus, as well as the nervus jugularis. Various anastomoses between n. glossopharyngeus and vagus and sympathetic trunk occur in this region. Plexus cavernosus, plexus ophthalmicus, and n. petrosus profundus represent extensions of the internal carotid plexus.

(2) The *prevertebral ganglia* are more or less extensive nondescript plexuses including flat or plate-like ganglionic swellings containing the postganglionic neurons, and are closely related to the stem of large arteries, particularly at their origin from the aorta. These ganglia, also known as collateral ganglia, comprise plexus coeliacus (semilunar ganglion, plexus solaris), superior and inferior mesenteric plexus, plexus hypogastricus, plexus renalis and suprarenalis. Fibers from the said plexuses usually join the vascular adventitia, follow the course of the main arteries and their branches, and reach the corresponding organs. Scattered postganglionic sympathetic nerve cells occur in these plexus extensions. Branches arising from thoracic ganglia of the sympathetic trunk, passing through the dorsal origin of the diaphragm, form the splanchnic nerves and join the celiac plexus (n. splanchnicus major generally from thor. ggl. 5 or 6–9, splanchnicus minor from thor. ggl. 10–11, also to plexus renalis). Plexus coeliacus, renalis, and mesentericus superior receive vagus branches, while plexus hypogastricus and particularly its pelvic portion ('pelvic plexus') receive sacral parasympathetic fibers. These latter, originating grom S 2–4 and passing through or close to the sacral sympathetic trunk, form the so-called pelvic nerve or nerves ('presacral nerves', 'nn. erigentes').

(3) The *terminal* or *intramural ganglia* or plexuses are located close to, or within, the substance or walls of the various organs. In the head region, such terminal ganglia include the ciliary ganglion within the orbit, related to oculomotor nerve, the sphenopalatine ganglion, related to facial nerve, the otic ganglion, related to glossopharyngeal nerve, and the submaxillary ganglion, again related to facial nerve. These ganglia, although not functionally pertaining to the trigeminus, display close topographic relations or anastomoses with branches of this latter nerve and are also traversed by postganglionic sympathetic fibers.

Within the thorax, the cardiac plexus, displaying several subdivisions, is located within the arch of the aorta and extends along the coronary arteries. It receives postganglionic fibers from cervical ganglia including stellate ggl. of sympathetic trunk as well as preganglionic fibers of vagus branches. Other plexuses of the thoracic cavity are plexus oesophageus and plexus bronchio-pulmonalis, in which sympathetic and cranial parasympathetic (vagus) components overlap.

Within the abdominal cavity, terminal ganglia are found in close proximity to the various organs, e.g. in hilus and adventitia, or in the interior of the organs. Thus, the hilus portion of plexus renalis represents a terminal as well as prevertebral plexus. Other such terminal plexuses are plexus vesicalis or vesico-prostaticus, plexus deferentialis, and plexus utero-ovaricus or uterinus (*Frankenhäuser's plexus* within the duplicature of ligamentum latum).

Truly intramural plexuses[104] are plexus myentericus (*Auerbach's pl.*) and plexus submucosus (*Meissner's pl.*) in the wall of the gut, from esophagus to rectum. The former plexus (Fig. 464) is located between outer longitudinal and inner circular musculature, the latter is found in the tela submucosa, closely adjacent to the muscularis mucosae. The nerve cells of these plexuses are presumably all postganglionic neurons of the parasympathetic (of n.vagus, respectively of sacral parasympathetic), but, in addition, postganglionic sympathicus fibers and probably visceral afferent fibers take their course through these neural networks.

With regard to course and synaptic connections, the moderately or thinly medullated *preganglionic* fibers of the *sympathetic* (or *thoraco-lumbar autonomic*) system reach the *sympathetic trunk* through the so-called white rami communicantes connected with inferior cervical ganglion, respectively stellate ganglion, the thoracic and the four lumbar ganglia. Some of these fibers effect synaptic connections within their segmental vertebral ganglia, while others proceed craniad or caudad for synaptic connections in middle and upper cervical, or in the sacral ganglia. Still others proceed through the vertebral ganglia, their terminal

---

[104] Comparable '*intramural*' or '*interstitial*' networks of nerve cells, similar to the ectodermal and entodermal networks of *Coelenterates* (cf. fig. 16, p. 26, of vol. 2 in this series), were, at one time, presumed to be present within the skin (subepidermal) and within the striated musculature, as so-called dermal and muscular intramural systems ('*Wandnervensystem*'). This concept, proposed by a number of older authors, and which, during my early studies, I was inclined to accept (cf. K., 1927, p. 107) has not been supported by sufficiently reliable evidence, and may be discounted. It seems likely that, as in the case of CAJAL's *interstitial cells*, briefly discussed further below, diverse, particularly mesodermal reticular elements were mistaken for nervous structures.

synaptic arborizations ending on postganglionic neurons in the prevertebral (or collateral) ganglia. Some preganglionic sympathetic fibers
pass apparently without interruption as far as the *adrenal medulla* or
comparable *paraganglia*, ending on the chromaffine cells. These latter,
of neuroectodermal origin (cf. chapter V, p. 39 of the preceding volume), can therefore be evaluated as representing postganglionic neural
(neurosecretory) elements.

Generally speaking, one preganglionic fiber seems to have synaptic
connections with numerous postganglionic neurons, and may reach
several ganglia. In some instances a ratio of one preganglionic fiber to
about 30 or more postganglionic nerve cells has been inferred on substantial evidence and there seems little doubt that the postganglionic
neurons, which innervate the effectors, outnumber the preganglionic
ones by a wide margin. Moreover, one postganglionic cell may receive
terminals from serveral preganglionic ones.

It can be assumed that, perhaps with few exceptions the *postganglionic* fibers are nonmedullated. Those originating in the vertebral
ganglia either join, by way of the so-called grey communicating rami,

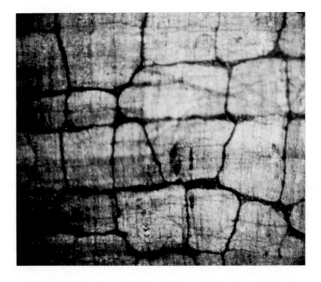

*Figure 464.* Intramural vegetative plexus as exemplified by the plexus myentericus
(*Auerbach's plexus*) in the small intestine of the rat. Surface view of total mount from an
area of muscularis in a routine preparation for the author's histologic course at the Woman's
Medical College of Pennsylvania (Goldchloride-formic acid technique; ×60, red. $^2/_3$).

the peripheral spinal nerves, or the peripheral perivascular plexuses, particularly the above-mentioned cranial ones. Still others, as rami originating from various vertebral ganglia, form several sympathetic cardiac nerves related to plexus cardiacus, or various shorter branches to bronchio-pulmonar plexus and esophagus. The postganglionic fibers joining the spinal nerve segments form the so-called grey communicating rami. Thus, while white communicating rami, containing preganglionic fibers, are restricted to the segments from C VIII to L IV, respectively to the lower cervical and the thoracic and lumbar, sympathetic vertebral ganglia, all segments and all vertebral ganglia display grey communicating rami.

It should, however, be stressed that, at the levels provided with 'white' communicating rami, these latter may not run separately from the 'grey' ones, but form a joint mixed ramus. Moreover, the significant difference between medullated white, and nonmedullated grey fibers is often rather indistinct upon gross inspection by the unaided eye. Finally, many communicating rami also carry *visceral afferent* fibers, representing peripheral neurites of spinal ganglion cells. Such fibers may be either nonmedullated or (more or less thinly) medullated ones.

The thoraco-lumbar autonomic preganglionic fibers passing through or close to the ganglia of the sympathetic trunk without there effecting synaptic connections reach the *prevertebral ganglia* such as plexus coeliacus, mesentericus, and hypogastricus, where they terminate on postganglionic neurons whose neurites reach the effectors. Except for the special case of adrenal medulla respectively true chromaffine paraganglia, no sympathetic preganglionic fibers seem to reach the innervated organs, nor are cell bodies of sympathetic postganglionic neurons believed to be present in most intramural or terminal plexuses (such, e.g., as plexus cardiacus). It seems possible, however, that some postganglionic perikarya of the sympathetic are scattered toward the periphery in some of the larger perivascular plexuses.

As regards the *parasympathetic* system, the *preganglionic* fibers seem, as a general rule, to run as far as the innervated organs and there to undergo their synaptic connections with the postganglionic elements either in the *terminal* or in the strictly *intramural* ganglia. Thus, in contradistinction to the sympathetic system, whose postganglionic fibers may have a considerable length, the parasympathetic ones are commonly quite short. As regards the ratio of preganglionic to postganglionic neurons, it is believed that one preganglionic vagus neurite can effect, through its branchings, synaptic contacts with up to 8 000 postganglionic neurons in plexus entericus *(Auerbach's plexus)*, while, for some other organs, a one-to-one relationship has been claimed.

As regards the *histologic structure* of vertebral, collateral, and terminal ganglia of the vegetative nervous system, the nerve cells are, as a rule, multipolar[105] with a variable number of cell processes (cf. Figs. 465–467). One of these latter represents the postganglionic axon. Numerous different types with respect to size of perikaryon, and number as well as pattern of cell processes have been described. In most ganglia, the nerve cell body is enclosed in a capsule presumably formed by peripheral neuroectodermal lemnoblasts (cf. chapter V, section 1, in vol. 3, part I). The larger dendrites may extend outside this capsule. Terminal ramifications of preganglionic fibers, displaying diverse types, including glomerular ones, end, with presumably synaptic structures, on the extra- and intracapsular portions of the postganglionic neurons. It should be added that the data obtained by the techniques of light microscopy remain insufficient for a satisfactory clarification of finer structural details, and that it is particularly difficult to obtain reasonably convincing pictures of the relevant synaptic connections.

As mentioned in the discussion of the neuron theory in chapter V, section 7, p. 527f., of the preceding volume, STÖHR JR. (1957) and a number of other authors assume that the entire peripheral autonomic system consists essentially of a syncytial network whose endings at the effector level displays the structure described as *'terminal reticulum'*. HERZOG (1966) as well as many others, however, have not accepted the hypotheses concerning the significance of that syncytial reticulum and interpret the structure of the peripheral autonomic system in accordance with the neuron theory. The data provided by electron microscopy, although including a number of somewhat ambiguous ultrastructural patterns, likewise are, on the whole, far more consistent with an interpretation assuming true synaptic connections in terms of the neuron concept. Personally, I am inclined to regard the terminal reticulum as an artefact related to the technical procedures, and especially to the poorly understood factors involved in silver impregnations.[106]

---

[105] Some of the older authors have also described unipolar and bipolar elements on the basis of well substantiated observations.

[106] In addition to the metallic impregnation procedures, the so-called vital, but actually supravital *methylene blue staining method*, introduced by EHRLICH about 1886, essentially displays, if properly used, the nervous structures (nerve cells, axons, terminal ramifications),

The so-called *interstitial cells* described by CAJAL (1911, 1955) and other authors have raised an additional problem. These elements were regarded as small nerve cells of primitive character within connective tissue of intestinal mucosa, including that of the villi (cf. Figs. 468 A, B). CAJAL assumed that some of these elements were interconnected by true anastomoses of their cell processes. The significance of a nerve net formed by the 'interstitial cells', if actually existing, could not be properly assessed in terms of the views based on LANGLEY's concepts and postulating preganglionic and postganglionic synaptic transmission. However, another group of investigators regarded the 'interstitial cells' as lemnoblasts of the terminal nerve plexus. A third group of investigators interpreted the 'interstitial cells' as connective tissue cells displaying special characteristics and being well stainable by means of the supravital methylene blue technique. I share this latter view and believe that the so-called 'interstitial cells' are most likely mesodermal elements pertaining to the 'reticular' or 'argyrophil' connective tissue.

Further details concerning the various conflicting concepts concerning the structure of the peripheral autonomic nervous system can be found in the publications by BOTAR (1966), HERZOG (1954, 1960, 1966), J. KAPPERS (1964), PICK (1970), REISER (1959), and STÖHR JR. (1957). For the brief account of this system, as given in the present section 6, I have tentatively but rather confidently adopted the widely accepted and reasonably well substantiated views based on neuronal synaptic transmission.

The theory of *neurohumoral transmission*[107] assumes that, with respect to the autonomic nervous system, all terminals of *preganglionic* fibers are *cholinergic*. As regards *postganglionic* fibers, all those of the *parasympathetic* systems are likewise considered *cholinergic*, while, in the *sympathetic (thoraco-lumbar)* system, the *postganglionic* fibers to most organs

---

while non-nervous element are, as a rule, only very faintly colored. Yet certain connective tissue structures are, not infrequently, also quite distinctly stained. The staining effect depends on external oxygenation of the methylene blue which is initially reduced by the tissue to a colorless '*leukobase*'. Although, at times, spectacular pictures may be obtained by means of this technique, this latter is, like the *Golgi methods*, subject to many vagaries and rather capricious.

[107] Significance and mechanisms of transmitter substances were dealt with in chapter V, p. 629 of the preceding volume in this series and require no further comments in the context. It may, however, be recalled that effectors innervated by adrenergic sympathetic postganglionic fibers may possess, at their synaptic input loci, either alpha-receptors eliciting excitation, or beta-receptors mediating inhibition.

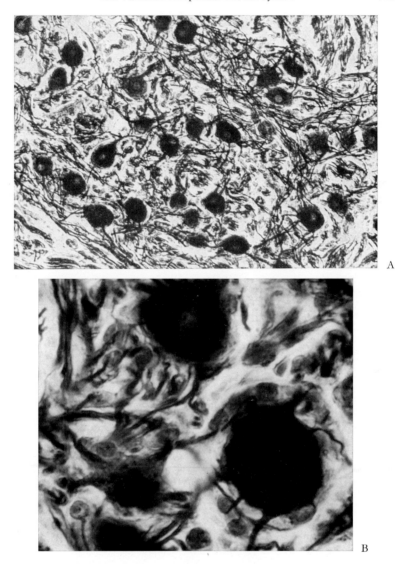

A

B

*Figure 465.* Multipolar postsynaptic sympathetic nerve cells in human coeliac ganglion as seen in a preparation from the author's collection *(Cajal's silver impregnation;* A×150, red. ²/₃; B×560, red. ²/₃). Preganglionic fibers form plexuses around the postganglionic nerve cell bodies. Some of the nerve fiber endings seen at the higher magnification may represent synapses, but the interpretation of such silver impregnation pictures remains uncertain.

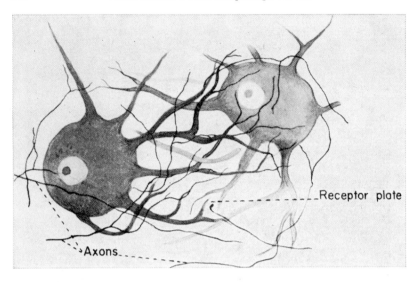

*Figure 466.* Drawing interpreting synapses between pre- and postganglionic elements within a 'dendritic glomerulus' of a human sympathetic ganglion (from KUNTZ, 1953). This drawing should be compared with the preceding photomicrograph.

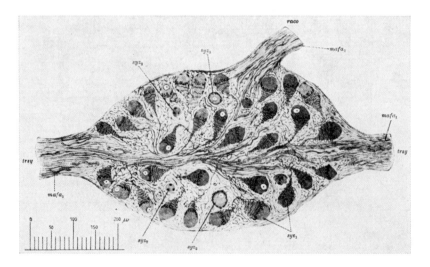

*Figure 467.* Small sympathetic vertebral ganglion from pars abdominalis of the frog's sympathetic trunk as displayed by the supravital methylene blue technique. The peculiar spiral pattern of terminal preganglionic fibers around portions of the essentially unipolar postganglionic elements is shown (from KRAUSE, 1923). mafa: medullated preganglionic fibers; raco: ramus communicans; syz: various postganglionic nerve cells; trsy: sympathetic trunk (stumps of the longitudinal connectives).

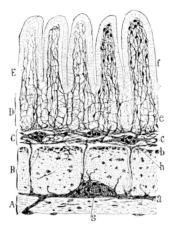

*Figure 468 A.* Semidiagrammatic drawing, by CAJAL, of the intramural plexuses in the small intestine of the guinea pig based on *Golgi impregnations* (from CAJAL, 1911, 1955). A: muscularis externa (longitudinalis); B: muscularis interna (circularis); C: tela submucosa; D: *crypts* (glands) *of Lieberkühn;* E: intestinal villi; a: plexus myentericus; b: deep components of plexus myentericus; c: components of plexus submucosus *( Meissner's plexus ) ;* e: extensions of plexus submucosus (probably including reticular fibers of mesodermal propria mucosae); f: 'intravillous plexus' (probably mesodermal reticular elements of propria mucosae); g: ganglion of *Auerbach's plexus;* h: ganglion of *Meissner's plexus.*

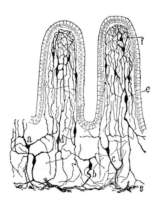

*Figure 468 B.* So-called interstitial and intravillous 'nerve cells' forming a periglandular and intravillous plexus in the small intestine of the guinea pig, as shown by the *Golgi method* (from CAJAL, 1911, 1955). It is not unlikely that these elements represent reticular connective tissue cells rather than nerve cells as presumed by CAJAL. It should be recalled that the *Golgi method* impregnates both nervous and connective tissue structures. The supravital methylene blue technique likewise displays this tendency. a, b, c, d: triangular and fusiform cells in subvillous layer of propria mucosae; e, f: fusiform intravillous cells; g: fibers of *Meissner's plexus submucosus.*

are presumed to be *adrenergic*. However, the sympathetic postganglionic endings on sweat glands[108] and in *some* instances of vascular innervation, being *cholinergic*, represent here an exception. Numerous experimental and clinical data have confirmed these concepts of neurohumoral transmission, which, in turn, provide strong circumstantial evidence for the neuronal structure of the autonomic nervous system, and against the various syncytial or reticularistic hypotheses.

Many organs of the body are known to have a *double innervation*, being controlled by parasympathetic and sympathetic nervous systems. In these instances, a mutual *antagonism* between both systems is usually evident. Thus, with respect to eye and orbit, the sympathetic, through postganglionic fibers originating in the superior cervical ganglion, dilates the pupil, widens the palpebral fissure, and may cause exophthalmos. Cranial parasympathetic outflow from the oculomotor nerve, on the other hand, constricts the pupil.[109]

Concerning the vegetative innervation of the cardiovascular system, the parasympathetic cranial autonomic vagus nerve retards the heart by exerting a 'retarding' or 'inhibiting' bathmotropic, inotropic, dromotropic and chronotropic influence, while the thoracolumbar autonomic (sympathetic), through the 'stimulating' cardiac nerves (nn. accelerantes) increases the rate of heart beat by an opposite effect.

Cranial parasympathetic outflow seems to cause vasodilation for vessels of salivary glands, and probably also of cerebral arteries, but constriction of cardiac, coronary vessels. The sacral parasympathetic, through the pelvic nerves (nn. erigentes), is assumed to dilate vessels of the external genitals and thus to mediate the phenomenon of erection. The sympathetic, on the other hand, constricts most blood ves-

---

[108] In man and at least in some other mammals. There are, however, still some poorly understood problems concerning 'adrenergic sweating', respectively the innervation of sweat glands. In the sheep, sweating is said to be stimulated by epinephrine. The distribution of sweat glands of either eccrine or apocrine type varies greatly among the different mammalian groups, and some forms display little or practically no perspiration. The full development of eccrine sudoriferous glands has been interpreted as a particular feature of primate evolution. Sweat glands are missing in Cetaceans and Sirenia. Rare cases of lacking sweat glands (anidrosis) combined with additional cutaneous anomalies occur in man as a genetic defect.

[109] As regards the intrinsic eye musculature, each muscle seems to be innervated by a single system, namely m. dilatator pupillae by the sympathetic, and m. sphincter pupillae as well as m. ciliaris (for accommodation) by postganglionic parasympathetic oculomotor fibers, originating in ciliary ganglion.

sels, especially in skin and abdominal viscera, but dilates the coronary arteries. It also seems to constrict the spleen, and, by its widespread vasomotor action, causes a rise of blood pressure combined with increased blood flow through many organs. By discharging adrenaline from the suprarenal medulla these general sympathetic activities are presumably reinforced. Moreover, the blood sugar level is thereby raised by liberation of glucose from the liver.

In the respiratory tract, the sympathetic apparently dilates the bronchioles, which seem to be constricted by parasympathetic vagus action. In the gastrointestinal tract, the parasympathetic subdivisions stimulate peristalsis and inhibit contraction of the main sphincters, while the sympathetic inhibits peristalsis and promotes sphincter activity.

In the urinary tract, the sympathetic contracts the detrusor of the bladder, and inhibits its sphincter, the sacral parasympathetic having the opposite effect. Thus, generally speaking in respect to both gastrointestinal and urinary tract, the craniosacral *parasympathetic* system can be regarded as '*emptying*', and the *sympathetic* as '*retaining*'.

The details of structural connections and functional relationships of parasympathetic and sympathetic systems with respect to male and female genital organs still remain poorly elucidated. The action of sympathetic innervation on the uterus seems to differ in the pregnant and the nonpregnant state.

The cranial parasympathetic promotes submaxillary and sublingual salivary, as well as lacrimal secretion through the facial nerve, and parotid salivary secretion through n. glossopharyngeus. A sympathetic innervation of lacrimal gland has not been conclusively demonstrated, but the blood vessels of salivary glands receive postganglionic sympathetic fibers, and some of these latter may also have an effect on the secretory elements. The bronchial, gastric, at least some intestinal glands, and the pancreas appear to be mainly to be controlled by the parasympathetic system which seems to promote secretion, including production of insulin, which lowers the blood sugar level.

'Vegetative' effector structures presumed to lack double autonomic innervation are the *pilomotor muscles* of the skin supplied by postganglionic sympathetic fibers, and, as mentioned above, the adrenal medulla, whose secretory elements receive preganglionic thoraco-lumbar fibers. Sweat gland, at least to some extent, may likewise only be under postganglionic sympathetic control. Nothing certain is known about the innervation of the sebaceous glands which are closely related to hair follicles and to the pilomotor muscles controlled by the sympathetic.

CANNON (1915, 1939) has particularly emphasized the antagonism between sympathetic and parasympathetic systems which are both concerned with 'adjustments of the internal environment of the body'. The cited author stressed that the changes produced by sympathetic stimulation are an appropriate preparation for violent activity, involving 'expenditure of energy'. The parasympathetic action, on the other hand, is considered as related to conservative and restorative processes, including anabolic, excretory and reproductive activities. Accordingly, the sympathetic is described as an activator *'for fight or flight'*, and the parasympathetic a promotor of *'repose and repair'*. BRAIN (1969) justly comments on this formulation by stating that it is 'a suggestive generalization though in some respects it oversimplifies the facts'.

*Reflex arcs* of the autonomic system evidently involve afferent components, which, by definition, are not included in that exclusively efferent 'system'. Generally speaking, the afferent arcs of autonomic reflexes are provided by either somatic or visceral afferent fibers of spinal or cranial ganglion nerve cells. Such reflexes, whose *efferent* arc is represented by the sequence of preganglionic and postganglionic fibers, are mediated, in various degrees of complexity, through the central neuraxis and its grisea.

The *afferents* are medullated or nonmedullated spinal nerve fibers reaching the viscera through communicating rami and sympathetic nerves such as splanchnics and cardiacs, or, e.g. included in the pelvic nerves. Others supply the blood vessels, the skin, or various somatic structures directly through the spinal nerves. Still other afferent fibers are carried by the branchial cranial nerves, particularly vagus and glossopharyngeus, their cell bodies being located in the corresponding ganglia, such e.g. as ganglion nodosum vagi. In this respect, significant difference can be said to obtain between 'somatic' and 'visceral' afferent fibers. No afferent fibers, reaching the neuraxis, are known to arise in the ganglia of the sympathetic trunk or in the more peripheral autonomic ganglia.

Nevertheless, it seems that some degree of independent or local *'reflex activity'* can be carried out through the peripheral autonomic plexuses, particularly the intramural ones, after transection of the main connections with the neuraxis. LANGLEY postulated that afferent and efferent limbs of autonomic reflexes could be provided by the branchings of a single nerve fiber, thus not involving two neurons, and designated such impulse transmissions as *'axon reflexes'* or *'pseudoreflexes'*. The cited author obtained reflex activity by stimulating preganglionic fibers separated from their central origin, as well as reflex action presumably mediated by postganglionic neurons alone. It appeared therefore that stimulation of an axon or axon collateral might originate impulses transmitted to widely distributed branches of that axon in different regions, thereby producing reactions of reflex type. Preganglionic axon reflexes were essentially observed under rather artificial experimental conditions, while reflexes of apparently postganglionic type also seem to occur under conditions comparable to the 'normal' state.

If 'axon reflexes' are indeed instances of normal behavior, this would imply that 'neurites' or 'neurite-like' processes ('apotiles') of the corresponding neurons are provided with both input and output loci. The postulated occurrence of axon reflexes plays also a role in the explanations of some phenomena of vasodilation discussed further below in the concluding part of the present section 6.

On the other hand, it is not impossible that in the peripheral ganglia manifesting behavior of reflex type, e.g. in coeliac and mesenteric plexuses as well as in the myenteric plexuses, local afferent respectively internuncial elements, mediating these activities, are present. Evidence for the presence of such nerve cells has been reported by and is discussed in the treatise by KUNTZ (1953) on the basis of his own findings.

Reverting from the better known autonomic nervous system of mammals or man to that of 'lower' vertebrates, it should be sufficient, in the aspect here under consideration, to supplement the foregoing discussion with the following generalized summary.[110]

In *Cyclostomes*, autonomic fibers seem to be present in facial, glossopharyngeal, and vagus nerves as well as (essentially dorsal) spinal nerve roots. Postganglionic neurons are believed to be present in the peripheral nerve ramifications or plexuses, but a true sympathetic trunk has not been recorded. In *Myxine*, the vagus may be the main or only cranial autonomic component, and is said to extend as far as about

---

[110] As regards invertebrates, the so-called 'stomatogastric' and similar peripheral nerves, which include components comparable to the 'autonomic system' of vertebrates, were briefly dealt with on pp. 76, 98, 123, 145 and 240–241 of volume 2 in this series. Concerning the poorly understood visceral nerves of *Amphioxus*, some comments were included above in section 2 of this chapter. A fairly detailed summary reviewing the peripheral autonomic nervous system in the Vertebrate series, supplemented by references to the literature, is included in the treatise of PICK (1970).

the anal region, while, in *Petromyzon*, this nerve has a lesser caudal extension, and spinal nerve segments may supply the autonomic innervation of that region. KAPPERS (1947) and others assume that a clear-cut differentiation of terminal cholinergic and adrenergic action with regard to cranial and spinal autonomic systems does not obtain in Cyclostomes.[111] This view is supported by the results of some inconclusive reports on vagus action upon the heart of Petromyzon.

On the other hand, FÄNGE *et al.* (1963), who particularly investigated the conditions obtaining in Myxine, reached the conclusion that the 'autonomic' nervous system consists here of adrenergic and cholinergic components. 'The cholinergic system innervating visceral smooth muscle is localized within the nervus vagus and corresponds to the 'parasympathetic', vagal, system of higher vertebrates. The adrenergic system consists of spinal sympathetic fibres, some elements of the vagus nerve trunk and chromaffine cell system. The function of the adrenergic system may be vasomotor, although it should be pointed out that cranial nerves other than the glossopharyngeal-vagus nerve may contain autonomic fibres' (FÄNGE *et al.*, 1963). The cited authors, moreover, evaluate the autonomic system of Myxinoids as similar to that of Lampreys, and as being, in some respects, intermediate 'between that present in Amphioxus and in Gnathostomes'. As regards some other aspects of these very poorly elucidated relationships, FÄNGE *et al.* (1963) speculate that the nervus vagus probably contains a number of sensory fibers which 'serve as afferent pathways for reflexes controlling the gill- and cardiac constrictor muscles, while the intestinal mucosa of the anal region seems to have a sensory intervation of spinal origin. This may serve as afferent pathways for reflexes controlling the sphincter cloacal muscles'.

In those *Selachians* whose autonomic nervous system has been investigated in sufficient detail, the oculomotor nerve contains autonomic fibers for the dilator of the iris,[112] and the vagus such fibers for heart and gut. The sympathetic system consists of an irregular series of approximately segmental ganglia containing postganglionic cell bodies,

---

[111] '*Le contraste para- et orthosympatique en sense fonctionel n'est pas toujours tel que l'on suppose d'appres les experiences physiologiques et pharmacologique effectuées chez les Mammifères, Ces deux systèmes autonomes peuvent fonctionner d'une manière cholénergique, ou bien encore*' ... '*de facon adrénergétique*' (KAPPERS, 1947).

[112] According to YOUNG (1955), the sphincter pupillae is not controlled by any nervous mechanism, but 'works as an independent effector', 'stimulated to contract by light'.

the preganglionic fibers emerging essentially through the ventral roots. Although a few irregular connections between segmental paravertebral ganglia exist, a true sympathetic trunk is not present. Moreover, the sympathetic (spinal thoraco-lumbar autonomic) system does not extend into the head (YOUNG, 1933). According to PICK (1970) it is likely that a spinal autonomic outflow through the dorsal roots innervates the blood vessels of the body wall by way of branches of spinal nerves whose fibers do not traverse the paravertebral ganglia.

The autonomic nervous system of *Teleosts*, insofar as it has been studied within this very large and diversified group, differs from that of Selachians by the presence of a true sympathetic trunk with segmental ganglia. The sympathetic, moreover, extends into the head (cf. Fig. 469), forming ganglia closely adjacent to vagus, glossopharyngeus, facial, and trigeminus roots. Each sympathetic ganglion of the trunk region is connected by a white and grey communicating ramus with its spinal nerve segment. Most preganglionic fibers leave the neuraxis through the ventral roots, but the exit of some such fibers through the dorsal root does not seem unlikely. Some of the grey postganglionic fibers in the ramus communicans are believed to innervate respectively contract the cutaneous melanophores. Cranial autonomic fibers are contained in n. oculomotorius and in n. vagus, which supplies the heart

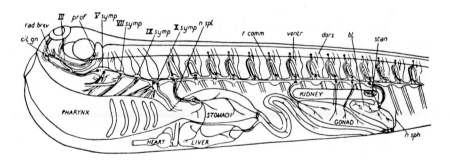

*Figure 469.* Diagram of peripheral autonomic nervous system in the Teleost *Uranoscopus* (after YOUNG, 1955). bl: mesonephric bladder; cil. gn.: ciliary ganglion; dors.: dorsal root; n. sph.: nerve to anal sphincter; n. spl.: splanchnic nerve; prof.: n. ophthalmicus profundus; rad. brev.: short root of ciliary ganglion; r. comm.: ramus communicans; stan.: adrenal cortical tissue *('corpuscle of Stannius')*; ventr.: ventral root; III: n. oculomotorius; V–X symp.: sympathetic ganglia associated with cranial nerves (the connection of such ganglia with cranial nerves is also shown above in figure 461 illustrating the innervation of Teleostean luminescent organs).

and the gut as far as the stomach, the caudal part of gut and the caudally located viscera being innervated by spinal autonomic fibers which YOUNG (1933, 1955), who investigated the Teleostean autonomic system, particularly in Uranoscopus, considers to be 'sympathetic'.

As regards the intrinsic eye musculature, the Teleostean m.dilator pupillae, as in Selachians, is innervated by the cranial autonomic n. oculomotorius, whose postganglionic fibers, at least in Uranoscopus, originate in a ciliary ganglion comparable to that of mammals, but which may also contain sympathetic postganglionic elements. The m. sphincter pupillae is innervated by the sympathetic. It will be noted that this innervation of dilator and sphincter pupillae by parasympathetic is just opposite to the type of innervation obtaining in Mammals.

'Electrical stimulation of the vagus nerve produces movements of the stomach but not of the intestine; the latter, however, shows movements when the splanchnic nerve is stimulated. Throughout the viscera acetylcholine causes initiation of rhythmic contractions and these are inhibited by adrenaline.' 'This is equally true of the stomach, with motor fibers from the vagus, intestine, with sympathetic motor fibres, and of the muscles of the bladder, which contract or stimulation of the hinder sympathetic ganglia. The effect of these nerves to the bladder is prevented by atropine but not by ergotoxine, though the latter is the drug which in mammals often inhibits sympathetic motor fibres. In these fishes, therefore, it is not possible to divide up the autonomic nervous system into sympathetic and parasympathetic divisions by either anatomical, physiological, or pharmacological criteria. Presumably the two 'antagonistic' systems found in mammals are a late development, allowing for a delicate balancing of activities for the maintenance of homeostasis' (YOUNG, 1955).

In *Dipnoans*, the cranial autonomic outflow seems to be limited to the vagus nerve. A sympathetic trunk with rami communicantes and other branches is present, but further useful specific data are still lacking.[113] With regard to *Amphibia*, the vegetative nervous system of *Anurans*, as represented by the frog, has been studied in some detail, while few pertinent data concerning Urodeles and Gymnophiona are available. In the frog, the arrangement of the vegetative nervous system is rather similar to that in the higher forms (Amniota). A cranial parasympathetic subdivision is provided by components of the oculomotor, facial, glossopharyngeal, and vagus nerves. The sympathetic

---

[113] The available data are reviewed and discussed by PICK (1970).

trunk with paravertebral ganglia, communicating rami, cardiac and splanchnic nerves, is well differentiated (cf. Fig. 470). Collateral ganglia of the sympathetic, however, are not developed. The preganglionic fibers seem to emerge from the neuraxis by way of both dorsal and ventral roots. Moreover, some postganglionic cells appear to be included within the spinal ganglia. The synapses between pre- and postganglionic neurons are commonly provided by spiraling preganglionic

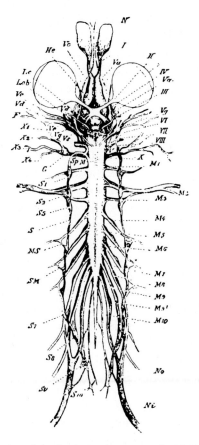

*Figure 470.* Dissection of the frog's nervous system showing the paired sympathetic trunk (after ECKER, from SELENKA-GOLDSCHMIDT 1923). I–X: cranial nerves; Va–e: trigeminal branches; Vg: trigeminal ganglion; Vs: connection of sympathetic with trigeminal ganglion; $X_{1-4}$: vagus branches; F: n. facialis; G: vagus ganglion; He: telencephalon; Lc: optic tract; Lob: mesencephalon; M: spinal cord; $M_{1-10}$: spinal nerves; MS: ramus communicans; N: nasal sack; Ni: n. ischiadicus; No: n. obturatorius; S, $S_{1-10}$: sympathetic trunk and ganglion; SM: ramus communicans, Sp: loop of sympathetic trunk; o: bulbus oculi.

terminals surrounding the axon and the axonal pole of the postganglionic cells. These latter give the appearance of unipolar elements (cf.
Fig. 467). Similar 'spiraling' preganglionic terminals have also been
described in some other vertebrate forms such as reptiles. A spinal (sacral) parasympathetic outflow is assumed to be mediated by the ventral
roots of 9th and 10th segment, passing to the intramural rectovesical
plexus, and seems comparable to the system of pelvic nerves in mammals.[114]

The autonomic nervous system of *Sauropsida* includes cranial parasympathetic, sacral parasympathetic[115] and sympathetic subdivisions
essentially similar to those of mammals. As regards the available details
and a number of differences which can be considered minor in the aspect here under consideration, reference may again be made to the
treatises by KAPPERS (1947) and PICK (1970). It will here be sufficient
to point out that Birds seem to be the only submammalian forms
whose sympathetic system displays, in addition to the vertebral (or
paravertebral) trunk ganglia, typical collateral (or prevertebral) ganglia. Moreover, the cervical sympathetic in Sauropsida commonly displays a superficial and a deep portion. This latter, in birds, may run
through the transverse foramina of the cervical vertebrae.

The complex ontogenetic events leading to the formation of the
*avian sympathetic trunk* have been investigated especially in the chick
(TELLO, 1925, 1945). Two successive waves of cellular migration[116]
can be recognized, which provide so-called *primary* and *secondary sympathetic trunks* (Fig. 471). The primary trunk develops between about 68
and 84 h of incubation, being well established at approximately 4 to 5
days. During the fourth day of incubation a further migration of the
sympathoblasts toward the periphery takes place and as the primary
trunk begins to disappear, a new migration of cells, at about the sixth
day of incubation, provides a second series of segmental ganglion.

---

[114] Numerous details together with a discussion of the many still unclarified problems
are contained in the treatise by PICK (1970).

[115] According to KAPPERS (1947) who refers in particular to a study by STIEMENS.
According to PICK (1970), however, the evidence concerning a possible sacral parasympathetic outflow bypassing the sympathetic system in Sauropsidans remains insufficient.

[116] The histogenetic aspect of this type of peripheral migration was discussed in section
1, chapter V, of the preceding volume 3, part I, and a further short reference was included
in section 1 A, p. 109, chapter VI of the present volume. Again, in the cited chapter V,
section 1, the still unsettled topic concerning a possible 'nonectodermal' origin of some
peripheral nervous tissue elements was pointed out (p. 18 loc. cit.).

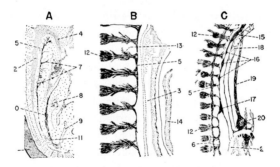

*Figure 471.* Development of primary and secondary sympathetic trunks in the chick embryo as seen in longitudinal sections (modified after TELLO, from ROMANOFF, 1960). A Primary trunk and *ganglion of Remak* at 4 days of incubation. B Primary trunk and its relation to spinal ganglia at 5 days. C Primary and secondary trunks and vagus nerve at 6 days. 1: spinal cord; 2: dorsal aorta; 3: carotid artery; 4: omphalomesenteric artery; 5: primary sympathetic trunk; 6: secondary sympathetic trunk; 7: suprarenal branches; 8: large intestine; 9: cloaca; 10: ganglion of Remak; 11: pelvic plexus; 12: spinal ganglion; 13: ventral root; 14: esophagus; 15: superior cervical ganglion; 16: middle cervical ganglion; 17: inferior cervical ganglion; 18: carotid plexus; 19: n. vagus; 20: thoracic plexus of vagus.

Whether the definitive vertebral sympathetic ganglia derive from both primary and secondary trunk, or exclusively from this latter, still remains a moot question. Details of this topic are reviewed and discussed by ROMANOFF (1960). Another peculiarity of the avian autonomic nervous system is the generally unpaired *nerve of Remak (n.intestinalis)* with its ganglion, located in mesorectum, mesocolon and mesentery. The paired roots of this nerve arise from the lower sympathetic trunk, but the resulting n. intestinalis has been interpreted as pertaining to the spinal ('sacral') parasympathetic component.

Evaluating the available data of comparative anatomy, one could state that the presumptive *phylogenetic evolution* of the vertebrate vegetative nervous system is characterized by two main trends, namely (1) the progressive differentiation of a cranial parasympathetic division, of which the vagus nerve represents one of the most conspicuous components. This nerve is, moreover, in several respects analogous to the stomatogastric nerve found in many invertebrates (cf. above, footnote 110). The other trend (2) is manifested by the development, within the spinal vegetative division, of a sympathetic trunk with paravertebral and further collateral ganglia, as well as by a differentiation of this spinal outflow into a main sympathetic and a caudal (sacral) parasympathetic component.

From a still more generalized viewpoint, the *intramural nerve plexuses*, particularly the enteric ones, may be said to display analogies with the primitive nerve nets of Coelenterates. In the vertebrates, however, such nerve plexuses are essentially formed by postganglionic autonomic elements. Yet, as BOTAR (1964, 1966) attempted, it is possible to outline an overall concept of the phylogenesis of the vegetative nervous system by elaborating on progressive patterns of relationship between (a) vegetative components of central neuraxis; (b) enteric nervous network ('*enterales Nervenzellsystem*'), and (c) perivascular networks ('*vaskuläres Nervenzellsystem*').

In concluding the present section on the vegetative nervous system brief comments on two still poorly elucidated problems are perhaps appropriate, namely concerning the question of *spinal posterior root vasodilators* and the phenomena of '*referred pain*'.

Parasympathetic and sympathetic *vasodilator activities* that can easily be explained, on substantial evidence, in accordance with the generally accepted notions of preganglionic and postganglionic innervation, are well-known. Such vasodilation, mediated by cranial nerves, cardiac and other sympathetic nerves, and sacral parasympathetic nerves were mentioned further above. STRICKER (1876), however, had reported vasodilation in peripheral blood vessels of the dog upon stimulation of the peripheral stump in transected dorsal roots. Subsequent studies by various authors including BAYLISS (1909, 1923) confirmed these observations and suggested, moreover, that such vasodilatory effect was mediated by 'sensory' spinal ganglion cells.[117] LANGLEY and others therefore postulated an antidromic action exerted by the processes of these neurons, combined with a mechanism of axon reflex type as illus-

---

[117] Sectioning the root between spinal ganglion and spinal cord, and allowing time for degeneration of possibly present efferent posterior root fibers with cell body in the spinal cord did not abolish the vasodilator response upon stimulation. Again, stimulation of a posterior root transected distal to the ganglion at first still elicited the vasodilator response, but failed to do so after a period sufficient for neurite degeneration. These observations conclusively demonstrated that the vasodilator effect was mediated by neuron cell bodies within the spinal ganglion and could not be attributed to neurons either in spinal cord or more peripheral vegetative ganglia. Nicotine did not abolish vasodilator response upon stimulation on the central side of the ganglion, and synaptic transmission within the ganglion could therefore be excluded on rather convincing grounds.

trated in Figure 472. Additional experiments demonstrating superficial vasodilation as a result of cutaneous stimuli provided data which can be interpreted in conformity with this view. The reactions of the vessels to antidromic stimulation apparently differ from 'ordinary' autonomic vasodilation by longer latent period and prolonged aftereffect. Such difference could be explained by the assumption of a different 'transmitter-substance' which might be histamine (DALE, 1929). This substance is also believed to be liberated by local tissue injury and to act as a persistent stimulus upon sensory endings. According to others, the substance in question might be related to adenosine triphosphate. It was, moreover, also suggested that the fibers displaying such axon reflexes belong to the *'pain conducting' system*. Be that as it may, the postulate of axon reflexes involved in vasodilation of this type would require the presence of input and of output loci at terminal branches of the peripheral spinal ganglion cell (or trigeminal ganglion cell) neurite.

In the anuran amphibian bullfrog, on the other hand, KOT-SUKA and NAITO (1961) have made experimental observations suggesting a conventional sympathetic vasodilator action on peripheral and intestinal blood vessels by way of spinal dorsal roots. The preganglionic fibers are presumed to end on postganglionic neurons included in the spinal ganglia. The postganglionic fibers are said to be cholinergic and, at least as far as the intestinal blood vessels are concerned, to pass by way of the communicating rami through the sympathetic trunk. This posterior root system is considered antagonistic with respect to the ventral root sympathetic vasoconstrictor system, whose postganglionic endings are adrenergic. LANGLEY's concept of anti-

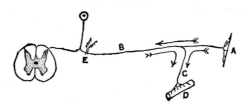

*Figure 472.* Diagram illustrating concepts of antidromic vasodilation and of axon reflex (after BAINBRIDGE and MENZIES, from BAYLISS 1924). A: afferent ending in skin; B: peripheral neurite of spinal ganglion cell; C: vascular branch of peripheral neurite; D: efferent ending in vascular wall; E: locus of artificial stimulation producing vasodilation (this could, of course, also be produced by stimulation of the central neurite).

dromic action is rejected by Kotsuka and Naito (1962). Yet, even if the conclusions of the cited authors should be confirmed, the presence of their postulated vasodilator system in Anurans, or, for that matter, in some other forms, would not necessarily invalidate the concept of anti-dromic action suggested for mammals and man.

Although thoracic and abdominal viscera, innervated by branches of the vagus, and by nerves originating from the sympathetic trunk, or by the pelvic nerves, are seemingly 'insensitive' to a number of 'pain-ful' stimuli, there is little doubt that *pain sensations* can and do arise from effects upon afferent nerve endings within these viscera.[118] In fact, pain generally represents a predominant manifestation of visceral disease. Clinicians assume, on convincing grounds, that, *inter alia*, a potent cause of such pain is an increase in the tension of the viscus. *Prima facie*, it would seem possible that afferent vagus fibers as well as visceral-afferent spinal ganglion cells, whose peripheral fibers accom-pany sympathetic or sacral parasympathetic nerves, mediate visceral pain sensations. Many clinical and experimental observations, on the other hand, suggest that, with some exceptions such, e.g., as laryngeal innervation, most visceral afferent vagus impulses do not elicit pain nor, for that matter, any conscious sensation. It is thus believed that the visceral afferent fibers mediating pain are generally represented by those pertaining to spinal ganglion cells. Yet, some clinical observa-tions, quoted by Kuntz (1953), indicate that pain from the lung can be mediated through the vagus nerve.

Visceral pain, as a rule, is a diffuse and poorly localized sensation, frequently associated with pain or tenderness in other regions of the body, and especially in those which are innervated from the same seg-ments of the nervous system as the affected viscus. The term *'referred pain'* is used to characterize this phenomenon, for which several differ-ent explanations have been offered by a number of authors.

A plausible theory of referred pain assumes that visceral afferent fi-bers mediating pain sensations excite the central conducting systems of

---

[118] Some clinicians (e.g. Mackenzie, 1893; Lennander, 1906) have denied the occurrence of 'visceral pain' and claimed that pain associated with visceral diseases was due to concomitant irritation of parietal peritoneum, pleura, or other structures, inner-vated by 'somatic afferent' spinal nerves. This explanation is, at present, no longer generally accepted. Likewise, the existence of afferent fibers, mediating pain sensations from blood vessels, repeatedly denied by various authors, appears, nevertheless, well substantiated, particularly as regards arteries. Although not all blood vessels pertain to the 'viscera', vascular pain is generally subsumed under 'visceral pain'.

the neuraxis concerned with somatic pain sensations, thereby producing the experience of peripheral, somatic, e.g. cutaneous pain. In other words, such visceral afferents do not seem to possess a central ascending channel of their own and transmit, through collaterals within the neuraxis, their impulses to tract cells of, e.g., the lateral spino-thalamic tract mediating somatic pain sensations.[119] In some instances, these tract cells may not become directly activated, but the visceral afferent impulses lower their threshold, so that additional slight somatic-afferent input from the periphery will cause a discharge. The referred pain is then not constant, but experienced as a corresponding *hypersensitive (hyperalgetic)* cutaneous region or so-called '*zone of Head*'. This generalized concept (MacKenzie, 1909; Szemzo, 1927) has been expressed in a variety of slight modifications subsumed under terms such as 'convergence facilitation' or 'convergence projection'. Depending on the 'intensity' of the visceral discharge, anesthetization of the pathway from the somatic periphery would or would not abolish the referred pain, in full conformity with recorded clinical observations.

Sinclair *et al.* (1948), however, do not believe that the theories implicating the spinal cord account in a satisfactory manner for the just-mentioned clinical findings. The cited authors postulate, as the essential factor in the production of referred pain, a branching of relevant peripheral afferent axons, so that the same distal stem fiber supplies both somatic and visceral structures. This, of course, is merely an expansion of the axon reflex concept as applied to spinal (or some cranial) ganglion cells and illustrated above in Figure 472. Sinclair *et al.* (1948), who have supplemented that assumption by including visceral branches in diagrams of this type, claim that their own hypothesis removes the weaknesses displayed by the theories assuming a 'convergence' within the spinal cord grisea.

---

[119] The experienced sensation is, of course, a mental event (P-event), of which, in turn, peripheral localization ('projection') represents here a characteristic aspect. The P-event can be interpreted as the ('parallel') correlate of a specific central (cortico-thalamic) N-event. This latter is triggered by a particular input discharge. Regardless of the actual cause of this input discharge, the resulting P-event will be characterized by the same localisatory aspect. In this regard, 'referred pain' pertains to the same category as the phenomenon of phantom limb pain, of peripheral pain in spinal root disturbances, or of pain 'irradiating' into forearm and hand upon a blow against the ulnar nerve in the groove between epicondylus medialis and trochlea humeri. Again, this whole group of phenomena is related to Müller's concept of so-called 'specific nerve energies' discussed above on p. 781, section 1 of this chapter.

A somewhat different explanation is based on the assumption of *vis-cero-cutaneous reflex phenomena* (POLLOCK and DAVIS, 1935). Afferent 'no-ciceptive' impulses from the viscera may set up, in the spinal cord, a preganglionic discharge reaching, by way of the sympathetic trunk and its postganglionic communicating rami, the peripheral nerve endings. Unspecified effects, including liberation of 'metabolites' or 'transmitter substances' resulting from this reflex activity are presumed to produce the cutaneous pain.

In addition to 'viscero-cutaneous' reflexes, there are doubtless also 'viscero-muscular' ones, in which true visceral afferent impulses[120] are transmitted to somatic motoneurons, thereby causing muscle spasms or rigidity associated with visceral disease. Finally, it does not seem entirely impossible that, as presumed by some authors, unspecified cerebral mechanisms are involved in some of the aspects of referred pain. A brief reference to this question, quoting some previous suggestions of this type, is included in the critical review by SINCLAIR *et al.* (1948). Since it seems reasonable to infer that the numerous and diversified manifestations of referred pain or of hyperalgesic *zones of Head* may depend on 'multifactorial' events, the various proffered explanations are not necessarily mutually exclusive.

As regards typical *localizations of referred pain*, the following instances may be mentioned. In diseases of the coronary arteries (e.g. angina pectoris), the pain is frequently experienced in the area of the left arm innervated by the ulnar nerve, and occasionally in a shoulder region approximately over the scapula, or in similar locations on the right side, or, again, the pain can be bilaterally felt. These regions correspond roughly to dermatomes innervated by third to eighth cervical and first to third thoracic segments, that is, to levels at which visceral afferent fibers from the heart join the spinal cord together with somatic afferent ones of peripheral nerves. In some instances, cardiac pain may even be referred to areas of trigeminal nerve distribution and involve forehead, eye region, and cheek. Pain of this type could be mediated by afferent vagus fibers with 'convergence' connections in nucleus radicis descendentis trigemini similar to those assumed for spinal cord theories of referred pain.

---

[120] 'True visceral afferent impulses' subsume here those originating in structures other than parietal serosa (parietal peritoneum, pleura, pericard) which, although obviously related to the viscera, is considered, by some authors, to have a 'somatic 'sensory innervation.

In liver affections, the pain can be experienced in superficial dorsal areas on the right side extending from the shoulder to the lower thorax, (spinal segments Th. 1–10). Pathologic gastric conditions may be accompanied by pain on both sides of the dorsal midline in dermatomes Th. 7–9, while referred menstrual or uterine pain is frequently similarly localized in the dermatome region Th. 10 to L. 3. Still other areas of reference are epigastric region in pulmonary affections, and right hypogastric region in gall bladder conditions, which may cause attacks of pain wrongly suggestive of appendicitis.

Although referred pain is frequently of visceral origin it should be kept in mind that '*reference*' or '*projection*' of pain to regions distant from the affected structure is not restricted to pathologic conditions of viscera. Thus, a toothache is not infrequently localized in an unaffected tooth even pertaining to another (lower or upper) jaw of the same side, in the ear, or in the mastoid region. Also, deep 'somatic' lesions (in muscles, tendons, or periosteum) can cause '*irradiating*' vaguely localized pain in extensive and fairly distant regions corresponding to the segmental innervation of the affected structures. *Mutatis mutandis*, the various explanations offered for visceral referred pain can be assumed to hold for these phenomena.

Again, as pointed out by KUNTZ (1953), the differentiation of true referred pain from diffuse local visceral pain is frequently difficult, since this latter, although poorly 'localized' can be experienced in the region occupied by the visceral organs in question. Thus, 'local' cardiac pain may be felt behind the sternum or in the left submammary region. Pain due to gastric ulcers can be experienced in the midepigastric region. Other instances of vague but still fairly accurate visceral pain localization occur in cases of duodenal ulcer, renal colic, ureteric calculus, appendicitis, and gall bladder disease. The complexities of interrelated factors involved in the experience of pain sensations still preclude any relevant explanation why in some cases approximately accurate localization, and in other cases 'reference' to distant regions obtain.

The autonomic nervous system, moreover, may provide an alternative path mediating painful sensations from areas deprived of their somatic sensory nerves or of their ascending spinal channel, e.g. (lateral) spino-thalamic tract after bilateral tractotomy or spinal cord transection). It seems possible that afferent fibers supplying blood vessels, and, as the case may be, running craniad for some distance in the sympathetic trunk, could be concerned with such modes of impulse transmission.

# 7. References to Chapter VII

ADAMS, W.E.: The comparative morphology of the carotid body and carotid sinus (Thomas, Springfield 1958).

ADELMANN, H.B.: The development of the neural folds and cranial ganglia of the rat. J. comp. Neurol. *39:* 19–171 (1925).

ADELMANN, H.B.: The development of the premandibular head cavities and the relations of the anterior end of the notochord in the chick and robin. J. Morph. *42:* 371–427 (1926).

AHLBORN, F.: Über die Segmentation des Wirbeltierkörpers. Z. wiss. Zool. *40:* 309–330 (1884).

AICHEL, O.: Zur Kenntnis des embryonalen Rückenmarks der Teleostier. Sitz. Ber. Ges. Morph. Physiol., München *11:* 25–40 (1895).

ANDREW, W.: Textbook of comparative histology (Oxford University Press, New York 1959).

AREY, L.B.: The functions of the efferent fibers of the optic nerve of fishes. J. comp. Neurol. *26:* 213–245 (1916).

AREY, L.B.: Developmental anatomy, 6th ed. (Saunders, Philadelphia 1954).

AYERS, H.: Ventral spinal nerves in Amphioxus. J. comp. Neurol. *33:* 155–162 (1933).

BALFOUR, F.M.: Comparative embryology, 2 vols. (Macmillan, London 1881).

BALLOWITZ, E.: Das elektrische Organ des afrikanischen Zitterwelses *(Malapterurus electricus Lacépède)*. (Fischer, Jena 1899).

BALLOWITZ, E.: Elektrische Organe; in BOLK, GÖPPERT et al. Hb. d. vergl. Anat. d. Wirbeltiere, pp. 657–682 (Urban & Schwarzenberg, Berlin 1938).

BARHAM, E.G.; HUCKABAY, W.B.; GOWDY, R. and BURNS, B.: Microvolt electric signals from fishes and the environment. Science *164:* 965–968 (1969).

BAYLISS, W.M.: On the origin from the spinal cord of the vasodilator fibres of the hindlimb, and on the nature of these fibers. J. Physiol., Lond. *26:* 173–207 (1901).

BAYLISS, W.M.: The vasomotor system (Longmans Green, London 1923).

BAYLISS, W.M.: Principles of general physiology (Longmans Green, London 1924).

BAZETT, H.C,; McGLONE, B.; WILLIAMS, R.G., and LUFKIN, H.M.: Sensation. I.Depth, distribution and probable identification in the prepuce of sensory end-organs concerned in sensations of temperature and touch; thermometric conductivity. Arch. Neurol. Psychiat. *27:* 489–517 (1932).

BEARD, J.: The transient ganglion cells and their nerves in *Raja batis.* Anat. Anz. *7:* 191–206 (1892).

BEARD, J.: The history of a transient nervous apparatus in some Ichthyopsida. Zool. Jb. Anat. ontog. Abt. *9:* 319–426 (1896).

BECCARI, N.: Neurologia comparata anatomo-funcionale dei vertebrati, compreso l'uomo (Sanzoni, Firenze 1943).

BELL, C.: Idea of a new anatomy of the brain submitted for the observations of his friends (Strahan & Preston, London 1811).

BENNETT, M.V.L.: Modes of operation of electric organs. Ann. N. Y. Acad. Sci. *94* (2): 458–509 (1961).

BERNSTEIN, J.: Electrobiologie (Vieweg, Braunschweig 1912).

BIEDERMANN, W.: Elektrophysiologie, vol. 2 (Fischer, Jena 1895).

BLOOM, W. and FAWCETT, D.W.: A textbook of histology, 9th ed. (Saunders, Philadelphia 1968).

BOEKE, J.: Die doppelte (motorische und sympathische) Innervation der quergestreiften Muskelfasern. Anat. Anz. *44:* 343–356 (1913).

BONE, Q.: The central nervous system in Amphioxus, J. comp. Neurol. *115:* 27–64 (1960).

BOTAR, J.: Entwurf der Phylogenese des vegetativen Nervensystems. Anat. Anz. *115:* 156–160 (1964).

BOTAR, J.: Phylogenetische Evolution des vegetativen Nervensystems. Anat. Anz. *118:* 488–494 (1966).

BOTAR, J.: The autonomic nervous system. An introduction to its physiological and pathological histology (Akadémiae Kladó, Budapest 1966).

BRAIN, W.R.: Diseases of the nervous system, 7th ed. Revised by the late Lord BRAIN and J.N.WALTON (Oxford University Press, London 1969).

BRAUER, A.: Die Tiefseefische; in CHUN Wiss. Erg. d. Deutschen Tiefsee-Exp. Valdivia, Bd. 15, 2 vols. (Fischer, Jena 1906, 1908).

BUDDENBROCK, W. VON: The senses (University of Michigan Press, Ann Arbor 1958).

BULLOCK, T.H. and COWLES, R.B.: Physiology of an infrared receptor. The facial pit of pit vipers. Science *115:* 541–543 (1952).

BÜTSCHLI, O.: Vorlesungen über vergleichende Anatomie, 3rd ed. (Springer, Berlin 1921).

CAJAL, S.R. y: A quelle époque apparaissent les expansions des cellules nerveuses de la moelle épinière du poulet? Anat. Anz. *5:* 609–613, 631–539 (1890).

CAJAL, S.R. y: Histology. Revised by J.F.TELLO-MUÑOZ and translated by M.FERNAN-NUÑEZ (Wood, Baltimore 1933).

CAJAL, S.R. y: Histologie du système nerveux de l'homme et des vertébrés, 2 vols. (Maloine, Paris 1909, 1911; Instituto Ramon y Cajal, Madrid 1952, 1955).

CANNON, W.B.: Bodily changes in pain, hunger, fear and rage (Appleton, New York 1915, 1929).

CANNON, W.B.: The wisdom of the body (Norton, New York 1932, 1939).

CASTRO, F., DE: Sur la structure et l'innervation du sinus carotidien de l'homme et des mammifères. Nouveaux faits sur l'innervation et la fonction du glomus caroticum. Etudes anatomiques et physiologiques. Trab. Lab. Invest. biol. Univ. Madrid *25:* 331–380 (1928).

CASTRO, F., DE: Sur la structure de la synapse dans les chemorecepteurs. Leur mécanismes d'excitation et rôle dans la circulation sanguine locale. Acta physiol. scand. *22:* 14–43 (1951).

COATES, C.W.: Electric fishes. Electr. Eng., N.Y. *69:* 47–50 (1950).

CORDS, E.: Die Kopfnerven der Petromyzonten (Untersuchungen an *Petromyzon marinus*) Z. Anat. EntwGesch. *89:* 201–249 (1929).

CRUE, B.L., ed.: Pain and suffering. Selected aspects (Thomas, Springfield 1970).

CUNNINGHAM, D.J.: The relation of nerve-supply to muscle homology. J.Anat. Physiol. *16:* 1–9 (1882).

CUNNINGHAM, D.J.: The value of nerve-supply in the determination of muscular homologies and anomalies. J. Anat. Physiol. *25:* 31–40 (1890).

DALE, H.H.: Some chemical factors in the control of the circulation *(Croonian* Lectures). Lancet *1929:* 1179–1183, 1233–1237, 1285–1290 (1929).

DART, R.: Double innervation of striped muscles. A review of its implications. J. nerv. ment. Dis. *60:* 553–567 (1924).

DITTLER, R.: Allgemeine Sinnesphysiologie; in LANDOIS-ROSEMANN Lehrbuch der Physiologie des Menschen, 28th ed., vol. 2, pp. 786–797 (Urban & Schwarzenberg, München 1962).

DOGIEL, A.S.: Die Retina der Vögel. Arch. mikr. Anat. *44:* 622–648 (1895).

DOGIEL, A.S.: Das periphere Nervensystem des Amphioxus *(Branchiostoma lanceolatum)*. Anat. Hefte I, 147–213 (1903).

DOGIEL, A.S.: Über die Nervenendigungen in den *Grandry'schen* und *Herbtschen* Körperchen im Zusammenhang mit der Frage der Neuroentheorie. Anat. Anz. *25:* 558–574 (1904).

DOGIEL, A.S.: Der fibrilläre Bau der Nervenendapparate in der Haut des Menschen und der Säugetiere und die Neuronentheorie. Anat. Anz. *27:* 97–118 (1905).

DU BOIS, REYMOND, E.: Quae apud veteres de piscibus electricis extant argumenta. Diss. inaug. (University of Berlin, Berlin 1843).

EDINGER, L.: Einführung in die Lehre vom Bau und den Verrichtungen des Nervensystems (Vogel, Leipzig 1912).

FÄNGE, R., JOHNELS, A.G. and ENGER, P.S.: The autonomic nervous system. Chapter II, p.124–136 in: The Biology of Myxine, BRODAL, A. and FÄNGE, R., eds. (Universitetsforlaget, Oslo 1963).

FEINDEL, W.H.; WEDDELL, G. and SINCLAIR, D.C.: Pain sensibility in deep somatic structures. J. Neurol. Neurosurg. Psychiat. *11:* 113–117 (1948).

FESSARD, A.: Les organes électriques; in GRASSÉ Traité de zoologie, vol. 13/ II: pp. 1143–1238 (Masson, Paris 1958).

FITZGERALD, M.J.T.: On the structure and life history of bulbous corpuscles (corpuscula nervorum terminalia bulboidea). J. Anat. *96:* 189–208 (1962).

FLOOD, P.R.: A peculiar mode of muscular innervation in Amphioxus. Light and electron microscopic studies of the so-called ventral roots. J. comp. Neurol. *126:* 181–218 (1966).

FRANZ, V.: Zur mikroskopischen Anatomie der Mormyriden. Zool. Jb. Abt.2 *42:* 91–148 (1921).

FRANZ, V.: Haut, Sinnesorgane und Nervensystem der Akranier. Jena. Z. Naturwiss. *59:* 401–526 (1923).

FREUD, S.: Über den Ursprung der hinteren Nervenwurzeln im Rückenmark von Ammocoetes (Petromyzon Planeri). Sitz. Ber. Kaiserl. Akad. Wiss., mathem.-naturw. Cl. *75:* 15–27 (1877).

FREUD, S.: Über Spinalganglien und Rückenmark des Petromyzon. Sitz. Ber. Kaiserl. Akad. Wiss. Wien, mathem.-naturw. Cl. *78:* 81–167 (1879).

FREY, M. VON: Physiologie der Sinnesorgane der menschlichen Haut. Ergebn. Physiol. *9:* 351–368 (1910).

FRITSCH, G.: Die elektrischen Fische. I. *Malapterurus electricus* (Veit, Leipzig 1887).

FRITSCH, G.: Die elektrischen Fische. II. Die Torpidineen (Veit, Leipzig 1890).

FRORIEP, A.: Über ein Ganglion des Hypoglossus und Wirbelanlagen in der Occipitalregion. Beitrag zur Entwicklungsgeschichte des Säugetierkopfes. Arch. Anat. Physiol., Anat. Abt. *1882:* 279–303 (1882).

FRORIEP, A.: Zur Entwicklungsgeschichte des Wirbeltierkopfes Verh. anat. Ges. Ergebn. H., Anat. Anz. *21:* 34–36 (1902).

FUJITA, T.: Problems on the neuro-muscular relationship. Morph. *102:* 312–326 (1962).

FUJITA, T.: Über die Innervation der Mm. levatores costarum nebst morphologischen Bemerkungen dieser Muskeln. Anat. Anz. *116:* 327–339 (1965).

Fürbringer, M.: Untersuchungen zur Morphologie und Systematik der Vögel, zugleich ein Beitrag zur Anatomie der Stütz- und Bewegungsorgane, 2 vols. (Fischer, Jena 1888).

Fürbringer, M.: Über die spino-occipitalen Nerven der Selachier und Holocephalen und ihre vergleichende Morphologie. Festschr. *Gegenbaur*, vol. 3, pp. 351–788 (Engelmann, Leipzig 1897).

Fürbringer, M.: Morphologische Streitfragen. 1. N. trochlearis. Morph. Jb. *30:* 85–274 (1902).

Galambos, R.: Nerves and muscles (Doubleday, New York 1962).

Gaskell, W.H.: On the structure, distribution and function of the nerves which innervate the visceral and vascular systems. J. Physiol., Lond. *7:* 1–80 (1885).

Gaskell, W.H.: The involuntary nervous system (Longmans Green, London 1916).

Gegenbaur, C.: Grundriss der vergleichenden Anatomie (Engelmann, Leipzig 1874).

Gegenbaur, C.: Die Metamerie des Kopfes und die Wirbeltheorie des Kopfskelettes. Morph. Jb. *13:* 1–114 (1887–1888).

Gegenbaur, C.: Vergleichende Anatomie der Wirbeltiere mit Berücksichtigung der Wirbellosen, 2 vols. (Engelmann, Leipzig 1898–1901).

van Gehuchten, A.: Les éléments moteurs des racines postérieures. Anat. Anz. *8:* 215–223 (1893).

Gerlach, J.: Über das Gehirn von *Protopterus annectens*. Ein Beitrag zur Morphologie des Dipnoerhirnes. Anat. Anz. *75:* 310–406 (1933).

Gerlach, J.: Beiträge zur vergleichenden Morphologie des Selachierhirnes. Anat. Anz. *96:* 79–165 (1947).

Goldscheider, A.: Gesammelte Abhandlungen (Barth, Leipzig 1898).

Goldscheider, A.: In Bethe *et al.* Hb. norm. u. pathol. Physiol., vol. II/1: 181–202 (Springer, Berlin 1926).

Goodrich, E.S.: Studies on the structure and development of vertebrates, 2 vols. (Constable, London 1930; Dover, New York 1958).

Granit, R.: Receptors and sensory perception (Yale University Press, New Haven 1955).

Granit, R.: The basis of motor control. Integrating the activity of muscles, alpha and gamma motoneurons and their leading control systems (Academic Press, New York 1970).

Grundfest, H.: Electric fishes. Sci. Amer. *203* (4): 115–124 (1960a).

Grundfest, H.: Electric organs. In: McGraw-Hill Encyclopedia of Science and Technology, vol. 4, pp. 427–432 (McGraw-Hill, New York 1960b).

Grundfest, H.: Comparative physiology of electric organs of Elasmobranch fishes; in Gilbert *et al.*, Sharks, skates, and rays, pp. 399–432 (Johns Hopkins Press, Baltimore 1967).

Haller v. Hallerstein, V.: Kranialnerven; in Bolk *et al.*, Hb. d. vergl. Anat. d. Wirbelt., vol. 2, pp. 541–684 (Urban & Schwarzenberg, Berlin 1934).

Handler, P.(ed.): Biology and the future of man (Oxford University Press, NewYork 1970).

Haneda, Y. and Tsuji, F.I.: Light production in the luminous fishes Photoblepharon and Anomalops from the Banda Islands. Science *173:* 143–145 (1971).

Harvey, E.N.: Bioluminescence (Academic Press, New York 1952).

Harvey, E.N.: Bioluminescence. Evolution and comparative biochemistry. Fed. Proc. *12:* 597–611 (1953).

Hastings, J.W.: Light to hide by. Ventral luminescence to camouflage the silhouette. Science *173:* 1016–1017 (1971).

HAYMAKER, W.: *Bing's* Local diagnosis in neurological diseases, 14th and 16th eds. (Mosby, Saint Louis 1956, 1969)

HEAD, H.: Studies in neurology, 2 vols. (Oxford University Press, 1920).

HELD, H.: Die centrale Gehörleitung. Arch. Anat. Physiol., Anat. Abtg. *1893:* 201–248 (1893).

HERRICK, C.J.: The cranial and first spinal nerves of Menidia. A contribution upon the nerve components of the bony fishes. J. comp. Neurol. *9:* 153–455 (1899).

HERRICK, C.J.: Neurological foundations of animal behavior (Holt, New York 1924).

HERRICK, C.J.: An introduction to neurology, 5th ed. (Saunders, Philadelphia 1931).

HERZOG, E.: Über die periphere Glia in den sympathischen Ganglien. Z. Zellforsch. *40:* 199–206 (1954).

HERZOG, E.: Über die Morphologie des peripheren vegetativen Nervensystems. Dtsch. med. Wschr. *85:* (45): 1965–1981 (1960).

HERZOG, E.: Die orthologische und pathologische Morphologie der neurovegetativen Funktionen; in BUCHER *et al.* Hb. allg. Pathol., vol. I/2, pp. 285–343 (Springer, Berlin 1966).

HEYMANS, C.; BOUCKAERT, J.J. et REGNIERS, P.: Le sinus carotidien et la zone homologue cardio-aortique (Doin, Paris 1933).

HEYMANS, C.; DELAUNOIS, A.L.; MARTINI, L., and JANSSEN, P.: The effect of certain autonomic drugs on the chemoreceptors of the carotid body and the baroreceptors of the carotid sinus. Arch. int. Pharmacodyn. *96:* 209–219 (1953).

HINES, M.: Nerve and muscle. Quart. Rev. Biol. *2:* 149–180 (1927).

HOPKINS, C.D.: Sex differences in electric signaling in an electric fish. Science *176:* 1035–1037 (1972).

HYMAN, L.H.: Comparative vertebrate anatomy, 2nd ed. (University of Chicago Press, Chicago 1942).

JOHNSTON, J.B.: The Nervous System of Vertebrates (Blakiston, Philadelphia 1906).

KAMP, C.J. and STUART, C.: The electric organ and its centres in *Gymnotus electricus.* Proc. kon. nederl. Akad. Wet. *37:* 245–252 (1934).

KAPPERS, C.U.A.: Anatomie comparée du système nerveux, particulièrement de celui des mammifères et de l'homme. Avec la collaboration de E.W.STRASBURGER (Bohn, Haarlem & Masson, Paris 1947).

KAPPERS, C.U.A.; HUBER, G.C., and CROSBY, E.C.: The comparative anatomy of the nervous system of vertebrates, including man (Macmillan, New York 1936).

KAPPERS, J.ARIËNS: A survey of different opinions relating to the structure of the peripheral autonomic nervous system. Acta neuroveg. *26:* 145–171 (1964).

KEYNES, R.D. and MARTINS FERREIRA, H.: Membrane potentials in the electroplates of the electric eel. J. Physiol., Lond. *119:* 315 351 (1953).

KOLTZOFF, N.K.: Metamerie des Kopfes von *Petromyzon planeri.* Anat. Anz. *16:* 510–523 (1899).

KOLTZOFF, N.K.: Entwicklungsgeschichte des Kopfes von *Petromyzon planeri.* Bull. Soc. imp. Nat. Moscow, N.S. *15:* 259–589 (1901).

KOTSUKA, K. and NAITO, H.: On the vasodilator action of efferent 'sympathicus via posterior root'...'sympathetic double innervation' of frog's blood vessel (abstract). Angiology *12:* 329 (1961).

KOTSUKA, K. and NAITO, H.: On the vasodilator action of efferent 'sympathicus via posterior roots'. An efferent sympathetic double innervation of the bull frog's blood vessel. Acta neuroveg. *23:* 454–478 (1962).

KRAUSE, R.: Mikroskopische Anatomie der Wirbeltiere in Einzeldarstellungen. IV. Teleostier, Plagiostomen, Zyklostomen und Leptokardier (De Gruyter, Berlin 1923).

KUHLENBECK, H.: Betrachtungen über den funktionellen Bauplan des Zentralnervensystems. Folia anat. japon. 4: 111–135 (1926).

KUHLENBECK, H.: Vorlesungen über das Zentralnervensystem der Wirbeltiere (Fischer, Jena 1927).

KUHLENBECK, H.: The human diencephalon. A summary of development structure, function and pathology (Karger, Basel 1954).

KUHLENBECK, H.: Brain and consciousness. Some prolegomena to an approach of the problem (Karger, Basel 1957).

KUHLENBECK, H.: Mind and matter. An appraisal of their significance for neurologic theory (Karger, Basel 1961).

KUHLENBECK, H. and NIIMI, K.: Further observations of the morphology of the brain in the Holocephalian Elasmobranchs Chimaera and Callorhynchus. J. Hirnforsch. 11: 265–314 (1969).

KUNTZ, A.: The autonomic nervous system, 4th ed. (Lea & Febiger, Philadelphia 1953).

KUPFFER, C. VON: Die Morphogenie des Centralnervensystems; in HERTWIGS Hb. d. vergl. u. exper. Entwicklungslehre d. Wirbeltiere, Bd. 2, 3. Teil (Fischer, Jena 1906).

LANDACRE, F.L.: The origin of the cranial ganglia in Ameiurus. J. comp. Neurol. 20: 309–411 (1910).

LANDACRE, F.L.: The fate of the neural crest in the head of the Urodeles. J. comp. Neurol. 33: 1–43 (1921).

LANDACRE, F.L.: The epibranchial placode of the seventh cranial nerve in *Amblystoma jeffersionianum*. J. comp. Neurol. 58: 289–311 (1933).

LANGLEY, J.N.: Preliminary account of the arrangement of the sympathetic nervous system based chiefly on observations upon pilo-motor nerves. Proc. roy. Soc. Lond. 52: 547–556 (1893).

LANGLEY, J.N.: On the union of cranial autonomic (visceral) fibers with the nerve cells in the superior cervical ganglion. J. Physiol., Lond. 23: 240–270 (1898).

LANGLEY, J.N.: On axon reflexes in the preganglionic fibers of the sympathetic nervous system. J. Physiol., Lond. 25: 364–398 (1900).

LANGLEY, J.N.: The autonomic nervous system, vol. I (Heffer, Cambridge 1921).

LARSELL, O.: Nerve terminations in the lung of the rabbit. J. comp. Neurol. 33: 105–131 (1921).

LARSELL, O.: Textbook of neuro-anatomy and the sense organs (Appleton, New York 1939).

LEKSELL, L.: The action potential and excitatory effect of the small ventral root fibres to the skeletal muscle. Acta physiol. scand. 10: suppl. 31, pp. 1–84 (1945).

LENHOSSÉK, M. VON: Über Nervenfasern in den hinteren Wurzeln, welche aus dem Vorderhorn entspringen. Anat. Anz. 5: 360–363 (1890).

LENNANDER, K.G.: Leibschmerzen, ein Versuch, einige von ihnen zu erklären. Mitt. Grenzgeb. Med. Chir. 16: 24–46 (1906).

LEWIS, T.: Pain (Macmillan, New York 1942).

LEYDIG, F.: Lehrbuch der Histologie des Menschen und der Thiere (Meidinger, Frankfurt a. M. 1857).

LISSMANN, H.W.: On the function and evolution of electric organs in fish. J. exp. Biol. 35: 156–191 (1958).

LISSMANN. H.W.: Electric location by fishes. Sci. Amer. 208: (3) 50–59 (1963).

LIVINGSTON, W.K.: Pain mechanisms (Macmillan, New York 1943).

LLINÀS, R.R., ed.: Neurobiology of cerebellar evolution and development (American Medical Association, Chicago 1969).

LUBOSCH, W.: Vergleichend-anatomische Untersuchung über den Ursprung und die Phylogenese des N. accessorius Willisii. Arch. mikr. Anat. *54:* 514–602 (1899).

MACKENZIE, J.: Some points bearing on the association of sensory disorders and visceral disease. Brain *16:* 321–354 (1893).

MACKENZIE, J.: Symptoms and their interpretation (Shaw, London 1909).

MAGENDIE, F.: Expériences sur les fonctions des racines des nerfs rachiciens. J. Physiol. Exp. *2:* 276–279 (1822).

MAGENDIE, F.: Expériences sur les fonctions des racines des nerfs qui naissent de la moelle épinière. J. Physiol. Exp. *2:* 366–371 (1822).

MATHEWSON, R.F.; MAURO, A.; AMATNIK, E. and GRUNDFEST, H.: Morphology of main and accessory electric organs of *Narcine brasiliensis* (Olfers) and some correlations with their electrophysiological properties. Biol. Bull. *15:* 115–135 (1958).

MAURER, F.: Grundzüge der vergleichenden Gewebelehre (Reinicke, Leipzig 1915).

MAXIMOW, A.A.: A text-book of histology. Completed and edited by W.BLOOM (Saunders, Philadelphia 1930).

MELZACK, R. and WALL, P.D.: Gate control theory of pain; SOULAIRAC *et al.* Pain. Proc. Int. Symp. on Pain, Paris 1967, pp. 11–31 (Academic Press, New York 1968).

METTLER, C.C.: In METTLER History of medicine (Blakiston, Philadelphia 1947).

MÜLLER, J.: Handbuch der Physiologie des Menschen, 2 vols. (Holscher, Koblenz 1834–1840).

MÜLLER, L.R.: Lebensnerven und Lebenstriebe (Springer, Berlin 1931).

MURRAY, R.W.: The response of the *ampullae of Lorenzini* of Elasmobranchs to mechanical stimulation. J. exp. Biol. *37:* 417–424 (1960).

MURRAY, R.W.: The response of the *ampullae of Lorenzini* of Elasmobranchs to electrical stimulation. J. exp. Biol. *39:* 119–128 (1962).

NEAL, H.V.: The history of the eye muscles. J. Morph. *30:* 433–453 (1918).

NEAL, H.V.: Neuromeres and metameres. J. Morph. *31:* 293–315 (1919).

NEAL, H.V. and RAND, H.W.: Comparative anatomy (Blackiston, Philadelphia 1936).

NETTLESHIP, W.A.: Experimental studies on the afferent innervation of the cat's heart. J. comp. Neurol. *64:* 115–131 (1936).

NEUMAYER, L.: Histogenese und Morphogenese des peripheren Nervensystems; in HERTWIGS Hb. d. vergl. u. exper. Entwicklungslehre d. Wirbeltiere, Bd. 2, 3.Teil (Fischer, Jena 1906).

NICOL, J.A.C.: The luminescence of fishes; in MARSHALL Symp. Zool. Soc., Lond., No.19, pp. 27–55 (Academic Press, London 1967).

NISHI, S.: Beiträge zur vergleichenden Anatomie der Augenmuskulatur. Arb. anat. Inst. kaiser. japan. Univ. Sendai *7:* 65–82 (1922).

NISHI, S.: Muskeln des Rumpfes. Muskeln des Kopfes: parietale Muskulatur; in BOLK *et al.* Hb. d. vergl. Anat. d. Wirbeltiere, vol. 5, pp. 351–466 (Urban & Schwarzenberg, Berlin 1938).

NISHI, S.: Einige vergleichend-myologische Notizen. Zweiter Nachtrag zu meiner Abhandlung: Muskeln des Rumpfes usw., 1938. Gunma J. med. Sci. *10:* 75–82 (1961).

NISHI, S.: Systematisierung der spinalen Stammuskeln des Menschen im Licht der typologischen Anatomie. Dritter Nachtrag zu meiner Abhandlung: Muskeln des Rumpfes usw., 1938. Gunma J. med. Sci. *12:* 1–5 (1963).

NISHI, S.: Einige Zusätze zu meinem myologischen Beitrag in: BOLKS Handbuch der vergleichenden Anatomie. Acta anat. nippon. *43:* 387–393 (1968).

NOBLE, G.K.: The biology of the Amphibia (McGraw-Hill, New York 1931; Dover, New York 1954).

PACINI, F.: Sulla struttura intima dell organo elettrico del Gymnoto e di altri pesci elettrici (Cecchi, Firenze 1852).

PENNERS, A.: Die Leuchtorgane; in BOLK et al. Handb. d. vergl. Anat., vol. 1, pp. 693–702 (Urban & Schwarzenberg, Berlin 1931).

PETERS, A.: The structure of the dorsal root nerves in Amphioxus. An electron microscope study. J. comp. Neur. *121:* 287–304 (1963).

PICK, J.: The autonomic nervous system. Morphological, comparative, clinical and surgical aspects (Lippincott, Philadelphia 1970).

PLATT, J.B.: A contribution to the morphology of the vertebrate head, based on a study of *Acanthias vulgaris*. J. Morph. *5:* 79–112 (1891).

PLATT, J.B.: Further contribution to the morphology of the vertebrate head. Anat. Anz. *6:* 251–265 (1891).

POLLOCK, L.F. and DAVIS, L.: Visceral and referred pain. Arch. Neurol. Psychiat. *34:* 1041–1054 (1935).

RASMUSSEN, G. and WINDLE, W.F. (eds.): Neural mechanisms of the auditory and vestibular systems (Thomas, Springfield 1961).

REISER, K.A.: Die Nervenzelle; in v. MÖLLENDORFF and BARGMANN Handb. d. mikr. d. Menschen, Bd. IV, 4. Teil, Erg. z. Bd. IV/1, pp. 185–514 (Springer, Berlin 1959).

REISSNER, E.: Beiträge zur Kenntnis vom Bau des Rückenmarkes von Petromyzon fluviatilis. Arch. Anat. Physiol. wiss. Med. *1860:* 545–588 (1860).

RHODE, E.: Muskel und Nerv. II. Mermis und Amphioxus. Zool. Beitr. *3:* 161–182 (1892).

ROHON, J.V.: Zur Histiogenese des Rückenmarks der Forelle. Sitz. Ber. bayr. Acad. Naturwiss., math.-physik. Kl. *14:* 39–56 (1884).

ROMANOFF, A.L.: The avian embryo. Structural and functional development (Macmillan, New York 1960).

ROMER, A.S.: The vertebrate body (Saunders, Philadelphia 1950).

RUFFINI, A.: Les dispositifs anatomiques de la sensibilité cutanée. Sur les expansions nerveuses de la peau chez l'homme et quelques autres mammifères. Rev. gén. histol. *1:* 421–540 (1905).

SAITO, T.: Über das Gehirn des japanischen Flussneunauges *(Entosphenus japonicus Martens)*. Fol. anat. japon *8:* 189–263 (1930).

SAND, A.: The mechanism of the lateral sense organs of fishes. Proc. roy. Soc. Lond. B *123:* 472–495 (1937).

SAND, A.: The function of the *ampullae of Lorenzini*, with some observations on the effect of temperature on sensory rhythms. Proc. roy. Soc. Lond. B *125:* 524–553 (1938).

SCHARF, J.H.: Sensible Ganglien; in v. MÖLLENDORFF and BARGMANN Handb. d. mikr. Anat. d. Menschen, vol. IV, part 3 (Springer, Berlin 1958).

SCHARRER, E.; SMITH, S.W. and PALAY, S.L.: Chemical sense and taste in the fishes, Prionotus and Trichogaster. J. comp. Neurol. *86:* 183–198 (1947).

SCHNEIDER, A.: Monographie der Nematoden (Reimer, Berlin 1866).

SCHNEIDER, A.: Beiträge zur vergleichenden Anatomie und Entwicklungsgeschichte der Wirbelthiere (Reimer, Berlin 1879).

SCHNEIDER, D.: Haut- und Enterorezeptoren; in LANDOIS-ROSEMANN Lehrbuch der Physiologie des Menschen, 28th ed., vol. 2, pp. 798–825 (Urban & Schwarzenberg, München 1962).

SETO, H.: Studies on the sensory innervation (human sensibility), 2nd ed. (Igaku Shoin, Tokyo 1963).

SHELDON, R.E.: The reactions of the dogfish to chemical stimuli. J. comp. Neurol. *19:* 273–311 (1909).

SHERRINGTON, C.: The integrative action of the nervous system (Yale University Press, New Haven 1906, 1947).

SINCLAIR, D.C.; WEDDELL, G. and FEINDEL, W.H.: Referred pain and associated phenomena. Brain *71:* 184–211 (1948).

STÖHR, P., jr.: Mikroskopische Anatomie des vegetativen Nervensystems; in v. MÖLLENDORFF and BARGMANN Handb. d. mikr. Anat. d. Menschen, Bd. IV, 5. Teil, Erg. z. Bd. IV/1 (Springer, Berlin 1957).

STÖHR, P. and MÖLLENDORFF, W. VON: Lehrbuch der Histologie, 23rd, ed. (Fischer, Jena 1933).

STONE, L.S.: Experiments on the development of the cranial ganglia and the lateral line sense organs in *Amblystoma punctatum.* J. exp. Zool. *35:* 421–496 (1922).

STONE, L. S.: Experiments on the transplantation of the cranial ganglia in the amphibian embryo. J. comp. Neurol. *38:* 73–105 (1924–1925).

STRAUS, W.L., jr.: The concept of nerve-muscle specificity. Biol. Rev., Cambridge *27:* 75–91 (1946).

STRAUS, W.L., jr. and HOWELL, A.B.: The spinal accessory nerve and its musculature. Quart. Rev. Biol. *11:* 387–405 (1936).

STRICKER, S.: Untersuchungen über die Gefässnerven-Wurzeln des Ischiadicus. Sitz. Ber. Akad. Wiss. Wien, math.-naturwiss. Kl., Abt. 3 *74:* 173–185 (1876).

STRONG, O.S.: The cranial nerves of Amphibia. A contribution to the morphology of the vertebrate nervous system. J. Morph. *10:* 101–230 (1895).

STRUM, J.M.: Fine structure of the dermal luminescent organs, photophores, in the fish, *Porichthys notatus.* Anat. Rec. *164:* 433–462 (1969).

STRUM, J.M.: Photophores of *Porichthys notatus.* Ultrastructure of innervation. Anat. Rec. *164:* 463–478 (1969).

STUART, C. and KAMP, C.J.: The electric organ and its innervation in *Malapterurus electricus.* Proc. Proc. kon. nederl. Akad. Wet. *37:* 106–113 (1934).

STUART, C. and KAMP, C.J.: The electric organ and its centres in *Torpedo marmorata.* Proc. kon. nederl. Akad. Wet. *37:* 342–347 (1934).

SUGA, N.: Coding in tuberous and ampullary organs of a Gymnotid electric fish. J. comp. Neurol. *131:* 437–452 (1967).

SUGA, N.: Electrosensitivity of canal and free neuromast organs in a Gymnotid electric fish. J. comp. Neurol. *131:* 453–458 (1967).

SUNDER-PLASSMANN, P.: Untersuchungen über den Bulbus carotidis bei Mensch und Tier im Hinblick auf die 'Sinusreflexe' nach H.E.HERING; ein Vergleich mit anderen Gefässstrecken; die Histopathologie des Bulbus carotidis; das Glomus caroticum. Z. ges. Anat. EntwGesch. *93:* 567–622 (1930).

SZEMZO, G.: Der Schmerz als führendes Symptom. Theoretische Überlegungen und praktische Beobachtungen zum Schmerzproblem. Z. klin. Med. *106:* 365–405 (1927).

TAYLOR, E.H.: The Caecilians of the world. A taxonomic review (University of Kansas Press, Lawrence 1968).

TELLO, J.F.: Les différenciations neuronales dans l'embryon du poulet, pendant les premiers jours de l'incubation. Trav. Rech. Biol., Madrid *21:* 1–93 (1923).

TELLO, J.F.: Algunas observaciones más sobre las primeras fases del desarrollo del simpatico en el pollo. Trab. Inst. Cajal *37:* 103–149 (1945).

THAEMERT, J.C.: Fine structure of neuromuscular relationships in mouse heart. Anat. Rec. *163:* 575–585 (1969).

TRETJAKOFF, D.: Nervus mesencephalicus bei Ammocoetes. Anat. Anz. *34:* 151–157 (1909).

VEIT, O.: Beiträge zur Kenntnis des Kopfes der Wirbeltiere. II. Frühstadien der Entwicklung des Kopfes von Lepidosteus osseus und ihre prinzipielle Bedeutung für die Kephalogenese der Wirbeltiere. Morph. Jb. *53:* 319–390 (1924).

VEIT, O.: Entwicklungsgeschichte und vergleichende Anatomie, erörtert an dem Problem des Wirbeltierkopfes. Anat. Anz. *58:* 374–393 (1924).

VERMEULEN, H.A.: Die Accessoriusfrage. Feestbundel *Winkler*, psychiat. en neurol. Bladen, Amsterdam: 729–742 (1918).

VILLIGER, E.: Die periphere Innervation, 3rd ed.; 6th ed. revised by E. LUDWIG (Engelmann, Leipzig 1919, 1933).

WATANABE, H.: Electron microscopic observations on the innervation of smooth muscle in the guinea pig vas deferens. Japanese, with English summary. Acta anat. nippon. *44:* 189–202 (1969).

WEDDELL, G.: Clinical significance of pattern of cutaneous innervation. Proc. roy. Soc. Med. *34:* 776–778 (1941).

WEDDELL, G.: The anatomy of cutaneous sensibility. Brit. med. Bull. *3:* 167–172 (1945).

WEDDELL, G.; SINCLAIR, D.C., and FEINDEL, W.H.: An anatomical basis for alterations in quality of pain sensibility. J. Neurophysiol. *11:* 99–109 (1948).

van WIJHE, J.W.: Über die Mesodermsegmente und die Entwicklung der Nerven des Selachierkopfes. Verh. kon. nederl. Akad. Wet. *22* (E): 1–50 (1883).

van WIJHE, J.W.: Über die Kopfsegmente und die Phylogenese des Geruchsorgans der Wirbelthiere. Zool. Anz. *9:* 678–682 (1886).

WINCKLER, G.: L'innervation proprioceptive des muscles extrinsèques du globe oculaire chez le Bouquetin et le Chevreuil. Folia anat. japon. *28* (Nishi-Festschrift:) 341–351 (1956).

WINSLOW, J.B.: Exposition anatomique de la structure du corps humain (Desprez, Paris 1732).

WOOLLARD, H.H.: Anatomy of peripheral sensation. Brit. med. J. *2:* 861–862 (1936).

YAGASAKI, K.: Über den ersten Zervikalnerv (abstract). Ber. 32. Vers. japan. anat. Ges. Folia anat. japon. *2:* 378–379 (1924).

YOUNG, J.Z.: On the autonomic nervous system of the teleostean fish, *Uranoscopus scaber*. Quart. J. micr. Sci. *74:* 491–525 (1931).

YOUNG, J.Z.: The autonomic nervous system of Selachians. Quart. J. micr. Sci. *75:* 571–634 (1933).

YOUNG, J.Z.: The life of vertebrates (Clarendon Press, Oxford 1950, 1955).

YOUNG, J.Z.: The life of mammals (Oxford University Press, Oxford 1957).

ZIEGLER, H.E.: Der jetzige Stand des Kopfproblems. Anat. Anz. *57:* 62–72 (1923).

## Addendum to Chapter VII, Section 6

Concerning the vegetative nervous system, as discussed in Section 6, a reference to recent studies on the nerve growth factor (NGF) and its antiserum should have been included. The investigations of Professor LEVI-MONTALCINI (briefly mentioned in vol. 3/I of this series) have shown that the nerve growth factor is a protein which can be obtained from mouse salivary glands and other sources. It displays a remarkable biologic activity as a potent and specific stimulant of the growth of postganglionic sympathetic and of peripheral afferent neurons such as spinal ganglion cells. *Per contra*, an antiserum obtained from NGF and administered to young animals causes a permanent loss of adrenergic neurons in (peripheral) sympathetic ganglia, an effect designated as 'immunosympathectomy'. Details about these topics can be found in the following two recent publications:

STEINER, G., and SCHÖNBAUM, E., eds.: Immunosympathectomy. A meeting (Elsevier, New York 1969).

ZAIMIS, E., and KNIGHT, J., eds.: Nerve growth factor and its antiserum. A symposium (Athlone, London 1972).

## Corrigenda to Volume 3/I, Chapter V

P. 105, in footnote 35, line 2 from bottom, instead of: no cytosine, read: no uracil.

P. 230, in footnote 92, line 7 from bottom, instead of: 'causual', read: 'causal'.

P. 300, in footnote 122, line 5 from bottom, instead of: 'intravascular', read: 'intervascular'.

P. 380, on line 3 from top, the expression

$$\frac{12.24 \times 10^{-8}}{\sqrt{10^5}} \quad 0.0384 \,\text{Å}. \qquad \text{should correctly read:} \qquad \frac{12.24 \times 10^{-8}}{\sqrt{10^5}} = 0.0384 \,\text{Å}.$$

P. 404, Figure 245 B, which was correctly oriented in the final page proof, subsequently became accidentally reversed.

P. 409, in footnote 182, line 7 from bottom, instead of: that this technique, read: than that this technique.

P. 514, on the two bottom lines of footnote 222 instead of: pro- dounded by HIS, read: pro- pounded by HIS.

P. 519, on line 7 from bottom, instead of: 'cell chain hypothesis' (3), read: 'cell chain hypothesis' (2).

P. 715 on line 13 from top, instead of: worker bees, read: queen bees.

Cf. also the emendations referring to the preceding volumes and included on p. 14 (fn. 12), p. 60 (fn. 31, 32), p. 62, and p. 63 (fn. 34, 35, 36) of the present volume 3/II.